STUDENT'S SOLUTIONS MANUAL

DAVID ATWOOD
Rochester Community and Technical College

ALGEBRA AND TRIGONOMETRY
WITH MODELING AND VISUALIZATION
FIFTH EDITION

PRECALCULUS
WITH MODELING AND VISUALIZATION
FIFTH EDITION

Gary Rockswold
Minnesota State University, Mankato

PEARSON

Boston   Columbus   Indianapolis   New York   San Francisco   Upper Saddle River
Amsterdam   Cape Town   Dubai   London   Madrid   Milan   Munich   Paris   Montreal   Toronto
Delhi   Mexico City   São Paulo   Sydney   Hong Kong   Seoul   Singapore   Taipei   Tokyo

The author and publisher of this book have used their best efforts in preparing this book. These efforts include the development, research, and testing of the theories and programs to determine their effectiveness. The author and publisher make no warranty of any kind, expressed or implied, with regard to these programs or the documentation contained in this book. The author and publisher shall not be liable in any event for incidental or consequential damages in connection with, or arising out of, the furnishing, performance, or use of these programs.

www.pearsonhighered.com

PEARSON

Preface

This solutions manual is written as an aid for students studying the text *College Algebra with Modeling and Visualization 5/e* by Gary Rockswold. It contains answers and solutions to the odd-numbered exercises.

This manual is consistent with the text and gives solutions that are accessible to college algebra students. The solutions are written in a clear format with hundreds of graphs and tables to help with the learning process. Solutions frequently include not only symbolic solutions but also graphical and numerical solutions to help students understand the problem completely. It is recommended that you make a genuine attempt to solve a problem before looking in this solutions manual for help. Regular class attendance is also recommended. Following these suggestions will promote insight and understanding of college algebra.

It is the author's hope that you will find this manual helpful when studying the text *College Algebra with Modeling and Visualization 5/e*. Please feel free to send comments to the email address below. Your opinion is important. Best wishes for an enjoyable and successful college algebra course.

David Atwood

david.atwood@roch.edu

Table of Contents

Chapter 1: Introduction to Functions and Graphs

1.1: Numbers, Data, and Problem Solving

1. $\frac{21}{24}$ is a real and rational number.

3. 7.5 is a real and rational number.

5. $90\sqrt{2}$ is a real number.

7. Natural number: $\sqrt{9} = 3$; integers: -3, $\sqrt{9}$; rational numbers: $-3, \frac{2}{9}, \sqrt{9}, 1.\overline{3}$; irrational numbers: $\pi, -\sqrt{2}$

9. Natural number: None; integer: $-\sqrt{4} = -2$; rational numbers: $\frac{1}{3}, 5.1 \times 10^{-6}, -2.33, 0.\overline{7}, -\sqrt{4}$; irrational number: $\sqrt{13}$

11. Shoe sizes are normally measured to within half sizes. Rational numbers are most appropriate.

13. Speed limit is measured using natural numbers.

15. Temperature is typically measured to the nearest degree in a weather forecast. Since temperature can include negative numbers, the integers would be most appropriate.

17. $|5 - 8 \cdot 7| = |5 - 56| = |-51| = 51$

19. $-6^2 - 3(2 - 4)^4 = -6^2 - 3(-2)^4 = -36 - 3(16) = -36 - 48 = -84$

21. $\sqrt{9 - 5} - \frac{8 - 4}{4 - 2} = \sqrt{4} - \frac{4}{2} = 2 - 2 = 0$

23. $\sqrt{13^2 - 12^2} = \sqrt{169 - 144} = \sqrt{25} = 5$

25. $\frac{4 + 9}{2 + 3} - \frac{-3^2 \cdot 3}{5} = \frac{13}{5} - \frac{-27}{5} = \frac{40}{5} = 8$

27. $-5^2 - 20 \div 4 - 2 = -25 - 5 - 2 = -32$

29. 4×10^1

31. 3.65×10^{-3}

33. $2450 = 2.45 \times 10^3$

35. $0.56 = 5.6 \times 10^{-1}$

37. $-0.0087 = -8.7 \times 10^{-3}$

39. $206.8 = 2.068 \times 10^2$

41. $10^{-6} = 0.000001$

43. $2 \times 10^8 = 200,000,000$

45. $1.567 \times 10^2 = 156.7$

47. $5 \times 10^5 = 500,000$

49. $0.045 \times 10^5 = 4500$

51. $67 \times 10^3 = 67,000$

53. $(4 \times 10^3)(2 \times 10^5) = 4 \cdot 2 \times 10^{3+5} = 8 \times 10^8$; 800,000,000

55. $(5 \times 10^2)(7 \times 10^{-4}) = 5 \cdot 7 \times 10^{2-4} = 35 \times 10^{-2} = 3.5 \times 10^{-1}$; 0.35

57. $\dfrac{6.3 \times 10^{-2}}{3 \times 10^{1}} = \dfrac{6.3}{3} \times 10^{-2-1} = 2.1 \times 10^{-3}$; 0.0021

59. $\dfrac{4 \times 10^{-3}}{8 \times 10^{-1}} = \dfrac{4}{8} \times 10^{-3-(-1)} = 0.5 \times 10^{-2} = 5 \times 10^{-3}$; 0.005

61. $\dfrac{8.947 \times 10^{7}}{0.00095}(4.5 \times 10^{8}) \approx 42381 \times 10^{15} = 4.2381 \times 10^{19} \approx 4.24 \times 10^{19}$

63. $\left(\dfrac{101 + 23}{0.42}\right)^2 + \sqrt{3.4 \times 10^{-2}} \approx 87166 + 0.2 \approx 87166.2 \approx 8.72 \times 10^{4}$

65. $(8.5 \times 10^{-5})(-9.5 \times 10^{7})^2 = (8.5 \times 10^{-5})(9.025 \times 10^{15}) \approx 76.7 \times 10^{10} = 7.67 \times 10^{11}$

67. $\sqrt[m]{192} \approx 5.769$

69. $|\pi - 3.2| \approx 0.058$

71. $\dfrac{0.3 + 1.5}{5.5 - 1.2} \approx 0.419$

73. $\dfrac{1.5^3}{\sqrt{2} + \pi - 5} \approx \dfrac{3.375}{-2.732} \approx -1.235$

75. $15 + \dfrac{4 + \sqrt{3}}{7} \approx 15.819$

77. $0.14 = 1.4 \times 10^{-1}$ watt

79. The distance Mars travels around the sun is $2\pi r = 2\pi(141{,}000{,}000) \approx 885{,}929{,}128$ miles.

The number of hours in 1.88 years is $365 \times 1.88 \times 24 \approx 16{,}469$ hours. So Mars' speed is

$\dfrac{885{,}929{,}128}{16{,}469} \approx 53{,}794$ miles per hour.

81. (a) First we will write both numbers in scientific notation. 208 million $= 2.08 \times 10^{8}$, $3{,}000{,}000 = 3 \times 10^{6}$.

Then, divide the numbers to find the percentage. $\dfrac{3 \times 10^{6}}{2.08 \times 10^{8}} \approx 0.144 = 1.44\%$

(b) First we will write both numbers in scientific notation. 300 million $= 3 \times 10^{8}$, $11{,}700{,}000 = 1.17 \times 10^{7}$.

Then, divide the numbers to find the percentage. $\dfrac{1.17 \times 10^{7}}{3 \times 10^{8}} = 0.39 = 3.9\%$

83. (a) It would take $\dfrac{5.54 \times 10^{12}}{100} \approx 5.54 \times 10^{10}$ or 55.4 billion \$100-dollar bills to equal the federal debt. The

height of the stacked bills would be $\dfrac{5.54 \times 10^{10}}{250} \approx 2.216 \times 10^{8}$ inches or $\dfrac{2.216 \times 10^{8}}{12} \approx 18{,}466{,}667$ feet.

(b) There are 5280 feet in one mile, so the stacked bills would span $\dfrac{18{,}466{,}667}{5280} \approx 3497$ miles. It would reach

farther than the distance between Los Angles and New York.

85. (a) $V = \pi r^2 h \Rightarrow V = \pi(1.3)^2(4.4) \Rightarrow V = 7.436\pi \approx 23.4 \text{ in}^3$

(b) $1 \text{ in}^3 = 0.55$ fluid ounce $\Rightarrow 23.4 \cdot 0.55 = 12.87$ fluid ounces; Yes, it can hold 12 fluid ounces.

1.2: Visualizing and Graphing Data

1. (a)

    (b) Maximum: 6; minimum: -2

    (c) $\dfrac{3 + (-2) + 5 + 0 + 6 + (-1)}{6} = \dfrac{11}{6} = 1.8\overline{3}$

3. (a)

    (b) Maximum: 30; minimum: -20

    (c) $\dfrac{-10 + 20 + 30 + (-20) + 0 + 10}{6} = 5$

5.

-30	-30	-10	5	15	25	45	55	61

(a) The maximum is 61 and the minimum is –30.

(b) The mean is $\dfrac{-30 - 30 - 10 + 5 + 15 + 25 + 45 + 55 + 61}{9} \approx 15.1$ and the median is 15.

7. $\sqrt{15} \approx 3.87, \; 2^{2.3} \approx 4.92, \; \sqrt[m]{69} \approx 4.102, \; \pi^2 \approx 9.87, \; 2^{\pi} \approx 8.82, \; 4.1$

$\sqrt{15}$	4.1	$\sqrt[3]{69}$	$2^{2.3}$	2^{π}	π^2

(a) The maximum is π^2 and the minimum is $\sqrt{15}$.

(b) The mean is $\dfrac{\sqrt{15} + 4.1 + \sqrt[m]{69} + 2^{2.3} + 2^{\pi} + \pi^2}{6} \approx 5.95$ and the median is $\dfrac{\sqrt[m]{69} + 2^{2.3}}{2} \approx 4.51$.

9. (a)

    (b) Mean $= \dfrac{31.7 + 22.3 + 12.3 + 26.8 + 24.9 + 23.0}{6} = 23.5$; Median $= \dfrac{23.0 + 24.9}{2} = 23.95$. The average area of the six largest freshwater lakes is 23,500 square miles. Half of the lakes have areas larger than 23,950 square miles and half have less. The largest difference in area between any two lakes is 19,400 square miles.

    (c) The freshwater lake with the largest area is Lake Superior.

11. *Answers may vary.* 16, 18, 26; No

13. (a) $S = \{(-1, 5), (2, 2), (3, -1), (5, -4), (9, -5)\}$

     (b) $D = \{-1, 2, 3, 5, 9\}$

     $R = \{-5, -4, -1, 2, 5\}$

15. (a) $S = \{(1, 5), (4, 5), (5, 6), (4, 6), (1, 5)\}$

     (b) $D = \{1, 4, 5\}$

     $R = \{5, 6\}$

17. (a) The domain is $D = \{1, 3, -5, 8, 0\}$ and the range is $R = \{1, 0, -5, -2, 3\}$.

     (b) The minimum x-value is –5, and the maximum x-value is 8. The minimum y-value is –5, and the maximum y-value is 3.

(c) The axes must include at least $-5 \le x \le 8$ and $-5 \le y \le 3$. Extend each axis slightly beyond these intervals and let each tick mark represent 2 units. Plot the points $(1, 1)$, $(3, 0)$, $(-5, -5)$, $(8, -2)$, and $(0, 3)$.

(d) See Figure 17.

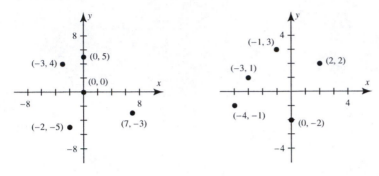

Figure 17   Figure 19

19. (a) The domain is $D = \{2, -3, -4, -1, 0\}$ and the range is $R = \{2, 1, -1, 3, -2\}$.

    (b) The minimum x-value is -4, and the maximum x-value is 2. The minimum y-value is -2, and the maximum y-value is 3.

    (c) The axes must include at least $-4 \le x \le 2$ and $-2 \le y \le 3$. Extend each axis slightly beyond these intervals and let each tick mark represent 1 unit. Plot the points $(2, 2)$, $(-3, 1)$, $(-4, -1)$, $(-1, 3)$, and $(0, -2)$.

    (d) See Figure 19.

21. (a) The domain is $D = \{10, -35, 0, 75, -25\}$ and the range is $R = \{50, 45, -55, 25, -25\}$

    (b) The minimum x-value is -35, and the maximum x-value is 75. The minimum y-value is -55, and the maximum y-value is 50.

    (c) The axes must include at least $-35 \le x \le 75$ and $-55 \le y \le 50$, and let each tick mark represent 25 units. It would be appropriate to extend the axes slightly beyond these intervals. Plot the points $(10, 50)$, $(-35, 45)$, $(0, -55)$, $(75, 25)$, and $(-25, -25)$.

    (d) See Figure 21.

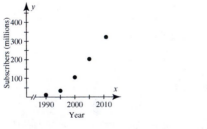

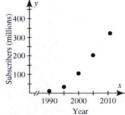

Figure 21   Figure 23a

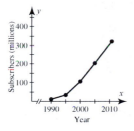

Figure 23b

23. See Figures 23a and 23b

25. $d = \sqrt{(5-2)^2 + (2-(-2))^2} = \sqrt{3^2 + 4^2} = \sqrt{25} = 5$

27. $d = \sqrt{(9-7)^2 + (1-(-4))^2} = \sqrt{2^2 + 5^2} = \sqrt{29} \approx 5.39$

29. $d = \sqrt{(-2.1-3.6)^2 + (8.7-5.7)^2} = \sqrt{(-5.7)^2 + 3^2} = \sqrt{41.49} \approx 6.44$

31. $d = \sqrt{(-3-(-3))^2 + (10-2)^2} = \sqrt{0^2 + 8^2} = \sqrt{64} = 8$

33. $d = \sqrt{\left(\frac{3}{4} - \frac{1}{2}\right)^2 + \left(\frac{1}{2} - \left(-\frac{1}{2}\right)\right)^2} = \sqrt{\left(\frac{1}{4}\right)^2 + 1^2} = \sqrt{\frac{1}{16} + 1} = \sqrt{\frac{17}{16}} = \frac{\sqrt{17}}{4} \approx 1.03$

35. $d = \sqrt{\left(-\frac{1}{10} - \frac{2}{5}\right)^2 + \left(\frac{4}{5} - \frac{3}{10}\right)^2} = \sqrt{\left(-\frac{1}{2}\right)^2 + \left(\frac{1}{2}\right)^2} = \sqrt{\frac{1}{4} + \frac{1}{4}} = \sqrt{\frac{1}{2}} = \frac{\sqrt{2}}{2} = 0.71$

37. $d = \sqrt{(-30-20)^2 + (-90-30)^2} = \sqrt{(-50)^2 + (-120)^2} = \sqrt{2500 + 14{,}400} = \sqrt{16{,}900} = 130$

39. $d = \sqrt{(0-a)^2 + (-b-0)^2} = \sqrt{(-a)^2 + (-b)^2} = \sqrt{a^2 + b^2}$

41. $d = \sqrt{(3-0)^2 + (4-0)^2} = \sqrt{3^2 + 4^2} = \sqrt{9+16} = \sqrt{25} = 5$

　　$d = \sqrt{(7-3)^2 + (1-4)^2} = \sqrt{4^2 + (-3)^2} = \sqrt{16+9} = \sqrt{25} = 5$

　　The side between (0,0) and (3,4) and the side between (3,4) and (7,1) have equal length, so the triangle is isosceles.

43. (a) See Figure 43.

　　(b) $d = \sqrt{(0-(-40))^2 + (50-0)^2} = \sqrt{40^2 + 50^2} = \sqrt{1600 + 2500} = \sqrt{4100} \approx 64.0$ miles.

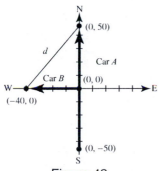

Figure 43

45. Use the midpoint formula $M = \left(\dfrac{24 + 6}{2}, \dfrac{44 + 10}{2} \right) = (15, 27)$. According to the midpoint estimate the number of Nintendo Wii units sold after 15 months was 27 million units.

47. Assuming the distance of 0 meters requires 0 seconds to run, the midpoint between the data points $(0, 0)$ and $(200, 20)$ is equal to $M = \left(\dfrac{0 + 200}{2}, \dfrac{0 + 20}{2} \right) = (100, 10)$. According to the midpoint estimate the time required to run 100 meters is 10 seconds or half the time required to run 200 meters.

49. $M = \left(\dfrac{1 + 5}{2}, \dfrac{2 + (-3)}{2} \right) = (3, -0.5)$

51. $M = \left(\dfrac{-30 + 50}{2}, \dfrac{50 + (-30)}{2} \right) = (10, 10)$

53. $M = \left(\dfrac{1.5 + (-5.7)}{2}, \dfrac{2.9 + (-3.6)}{2} \right) = (-2.1, -0.35)$

55. $M = \left(\dfrac{\sqrt{2} + \sqrt{2}}{2}, \dfrac{\sqrt{5} + (-\sqrt{5})}{2} \right) = (\sqrt{2}, 0)$

57. $M = \left(\dfrac{a + (-a)}{2}, \dfrac{b + 3b}{2} \right) = (0, 2b)$

59. $x^2 + y^2 = 25 \Rightarrow (x - 0)^2 + (y - 0)^2 = 5^2 \Rightarrow$ Center: $(0, 0)$; Radius: 5

61. $x^2 + y^2 = 7 \Rightarrow (x - 0)^2 + (y - 0)^2 = (\sqrt{7}\,)^2 \Rightarrow$ Center: $(0, 0)$; Radius: $\sqrt{7}$

63. $(x - 2)^2 + (y + 3)^2 = 9 \Rightarrow (x - 2)^2 + (y - (-3))^2 = 3^2 \Rightarrow$ Center: $(2, -3)$; Radius: 3

65. $x^2 + (y + 1)^2 = 100 \Rightarrow (x - 0)^2 + (y - (-1))^2 = 10^2 \Rightarrow$ Center: $(0, -1)$; Radius: 10

67. Since the center is $(1, -2)$ and the radius is 1, the equation is $(x - 1)^2 + (y + 2)^2 = 1$.

69. Since the center is $(-2, 1)$ and the radius is 2, the equation is $(x + 2)^2 + (y - 1)^2 = 4$.

71. $(x - 3)^2 + (y - (-5))^2 = 8^2 \Rightarrow (x - 3)^2 + (y + 5)^2 = 64$

73. $(x - 3)^2 + (y - 0)^2 = 7^2 \Rightarrow (x - 3)^2 + y^2 = 49$

75. First find the radius using the distance formula: $r = \sqrt{(4 - 3)^2 + (2 - (-5))^2} = \sqrt{50}$.

$(x - 3)^2 + (y - (-5))^2 = (\sqrt{50}\,)^2 \Rightarrow (x - 3)^2 + (y + 5)^2 = 50$

77. First find the center using the midpoint formula: $C = \left(\dfrac{-5 + 1}{2}, \dfrac{-7 + 1}{2} \right) = (-2, -3)$

Then find the radius using the distance formula: $r = \sqrt{(-2 - 1)^2 + (-3 - 1)^2} = \sqrt{25} = 5$.

$(x - (-2))^2 + (y - (-3))^2 = 5^2 \Rightarrow (x + 2)^2 + (y + 3)^2 = 25$

79. Find the center using the midpoint formula: $C = \left(\dfrac{5 + 2}{2}, \dfrac{5 + 1}{2} \right) = \left(\dfrac{7}{2}, 3 \right)$. Then find the radius using the

distance formula: $r = \sqrt{\left(\dfrac{7}{2} - 2 \right)^2 + (3 - 1)^2} = \sqrt{\dfrac{25}{4}} = \dfrac{5}{2}$.

$\left(x - \dfrac{7}{2} \right)^2 + (y - 3)^2 = \left(\dfrac{5}{2} \right)^2 \Rightarrow \left(x - \dfrac{7}{2} \right)^2 + (y - 3)^2 = \dfrac{25}{4}$

81. See Figure 81.

83. See Figure 83.

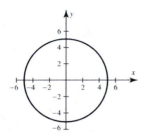

Figure 81

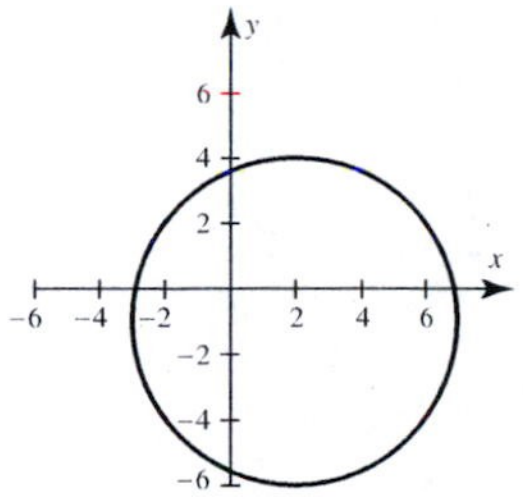

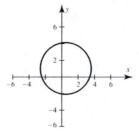

Figure 83

85. See Figure 85.

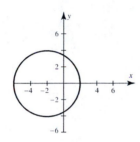

Figure 85

87. See Figure 87.

89. See Figure 89.

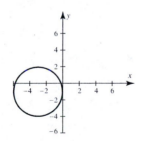

Figure 87

Figure 89

91. x-axis: 10 tick marks; y-axis: 10 tick marks. See Figure 91.

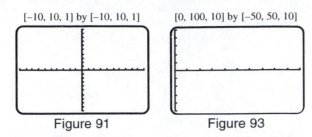

Figure 91   Figure 93

93. *x*-axis: 10 tick marks; *y*-axis: 5 tick marks. See Figure 93.

95. *x*-axis: 16 tick marks; *y*-axis: 5 tick marks. See Figure 95.

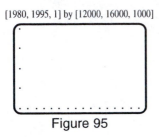

Figure 95

97. Graph b

99. Graph a

101. Plot the points $(1, 3)$, $(-2, 2)$, $(-4, 1)$, $(-2, -4)$ and $(0, 2)$ in $[-5, 5, 1]$ by $[-5, 5, 1]$. See Figure 101.

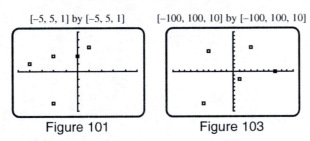

Figure 101   Figure 103

103. Plot the points $(10, -20)$, $(-40, 50)$, $(30, 60)$, $(-50, -80)$, and $(70, 0)$

in $[-100, 100, 10]$ by $[-100, 100, 10]$.   See Figure 103.

105. (a) The maximum number of Netflix subscriptions is 19.4 million, and the minimum number is 6.1 million.

(b) *x*-min: 2006; *x*-max: 2010; *y*-min: 6; *y*-max: 19.5. [2005, 2001, 1] by [5, 20, 5]. Answers may vary.

(c) See Figure 105c

(d) See Figure 105d

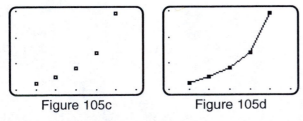

Figure 105c   Figure 105d

107. (a) The maximum Myspace U.S. advertising revenue is $590 million, and the minimum is $180 million.

(b) *x*-min: 2006; *x*-max: 2011; *y*-min: 180; *y*-max: 590. [2005, 2012, 1] by [150, 600, 50]. Answers may vary.

(c) See Figure 107c

(d) See Figure 107d

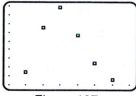

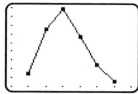

Figure 107c   Figure 107d

Checking Basic Concepts for Sections 1.1 and 1.2

1. (a) $\sqrt{4.2(23.1 + 0.5^3)} \approx 9.88$

    (b) $\dfrac{23 + 44}{85.1 - 32.9} \approx 1.28$

3. (a) $348{,}500{,}000 = 3.485 \times 10^8$

    (b) $-1237.4 = -1.2374 \times 10^3$

    (c) $0.00198 = 1.98 \times 10^{-3}$

5. $M = \left(\dfrac{-2 + 4}{2}, \dfrac{3 + 2}{2} \right) = \left(1, \dfrac{5}{2} \right)$

7. Mean $= \dfrac{13{,}215 + 12{,}881 + 13{,}002 + 3953}{4} = 10{,}762.75$; Median $= \dfrac{12{,}881 + 13{,}002}{2} = 12{,}941.5$. The

    mean average depth of the four oceans is 10,762.75 feet. Half of the oceans have average depths of

    more than 12,941.5 feet and half have average depths of less than 12.941.5 feet. The largest difference in aver-

    age depths between any two oceans is 9262 feet.

1.3: Functions and Their Representations

1. If $f(-2) = 3$, then the point $(-2, 3)$ is on the graph of f.

3. If $(7, 8)$ is on the graph of f, then $f(7) = 8$.

5. See Figure 5.

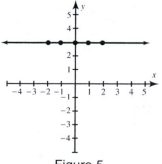

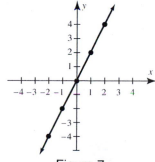

Figure 5   Figure 7

7. See Figure 7.

9. See Figure 9.

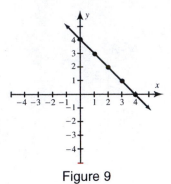

Figure 9

11. See Figure 11.

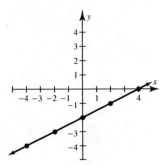

Figure 11

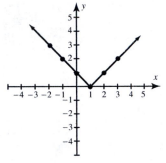

Figure 13

13. See Figure 13.

15. See Figure 15.

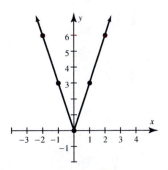

Figure 15

17. See Figure 17.

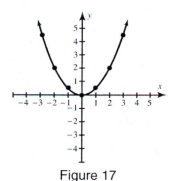

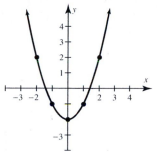

Figure 17   Figure 19

19. See Figure 19.

21. (a) $g = \{(-1, 0), (2, -2), (5, 7)\}$

   (b) $D = \{-1, 2, 5\}, R = \{-2, 0, 7\}$

23. (a) $g = \{(1, 8), (2, 8), (3, 8)\}$

   (b) $D = \{1, 2, 3\}, R = \{8\}$

25. (a) $g = \{(-1, 2), (0, 4), (1, -3), (2, 2)\}$

   (b) $D = \{-1, 0, 1, 2\}; \ R = \{-3, 2, 4\}$

27. (a) $f(x) = x^3 \Rightarrow f(-2) = (-2)^3 = -8$ and $f(5) = 5^3 = 125$.

   (b) All real numbers.

29. (a) $f(x) = \sqrt{x} \Rightarrow f(-1) = \sqrt{-1}$ which is not a real number, and $f(a + 1) = \sqrt{a + 1}$.

   (b) All non-negative real numbers.

31. (a) $f(x) = 6 - 3x \Rightarrow f(-1) = 6 - 3(-1) = 6 + 3 = 9$ and

   $f(a + 1) = 6 - 3(a + 1) = 6 - 3a - 3 = 3 - 3a$

   (b) All real numbers.

33. (a) $f(x) = \dfrac{3x - 5}{x + 5} \Rightarrow f(-1) = \dfrac{3(-1) - 5}{-1 + 5} = -\dfrac{8}{4} = -2$ and $f(a) = \dfrac{3a - 5}{a + 5}$.

   (b) All real numbers not equal to –5 $(x \neq -5)$.

35. (a) $f(x) = \dfrac{1}{x^2} \Rightarrow f(4) = \dfrac{1}{4^2} = \dfrac{1}{16}$ and $f(-7) = \dfrac{1}{(-7)^2} = \dfrac{1}{49}$.

   (b) All real numbers not equal to 0 $(x \neq 0)$.

37. (a) Domain and Range = All real numbers.

   (b) $g(x) = 2x - 1 \Rightarrow g(-1) = 2(-1) - 1 = -2 - 1 = -3$ and $g(2) = 2(2) = 2(2) - 1 = 4 - 1 = 3$

   (c) $g(-1) = 3$ and $g(2) = 3$

39. (a) $D = $ All real numbers; $R = \{y \mid y \leq 2\}$

   (b) $g(x) = 2 - x^2 \Rightarrow g(-1) = 2 - (-1)^2 = 2 - 1 = 1$ and $g(2) = 2 - (2)^2 = 2 - 4 = -2$

   (c) $g(-1) = 1$ and $g(2) = -2$

41. (a) $D = \{x \mid -2 \leq x \leq 2\}; R = \{-3 \leq y \leq 1\}$

   (b) $g(x) = x^2 - 3 \Rightarrow g(-1) = (-1)^2 - 3 = 1 - 3 = -2$ and $g(2) = (2)^2 - 3 = 4 - 3 = 1$

   (c) $g(-1) = -2$ and $g(2) = 1$

43. $D = \{x \mid -3 \leq x \leq 3\}$

    $R = \{y \mid 0 \leq y \leq 3\}$

    $f(0) = 3$.

45. $D = \{\text{all real numbers}\}$

    $R = \{y \mid y \leq 2\}$

    $f(0) = 2$

47. $D = \{x \mid x \geq -1\}$

    $R = \{y \mid y \leq 2\}$

    $f(0) = 0$

49. (a) $f(2) = 7$

    (b) $f = \{(1, 7), (2, 7), (3, 8)\}$

    (c) $D = \{1, 2, 3\}; \; R = \{7, 8\}$

51. Graph $f(x) = 0.25x^2$ in $[-4.7, 4.7, 1]$ by $[-3.1, 3.1, 1]$ by letting $Y_1 = 0.25X\wedge2$. See Figure 51.

    (a) From the graph, it appears that $f(2) = 1$.

    (b) $f(2) = 0.25(2)^2 = 0.25(4) = 1$

    (c) See Figure 51c.

[−4.7, 4.7, 1] by [−3.1, 3.1, 1]

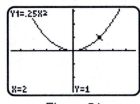

Figure 51 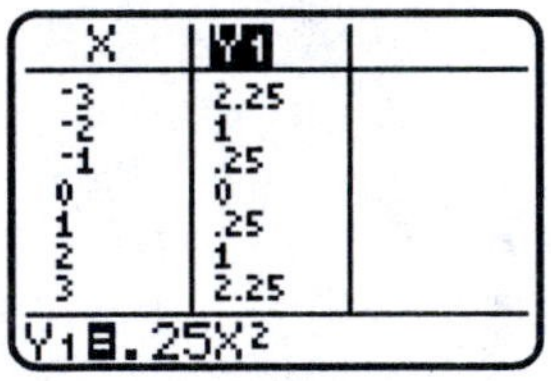   Figure 51c

53. Graph $f(x) = \sqrt{x + 2}$ in $[-4.7, 4.7, 1]$ by $[-3.1, 3.1, 1]$ by letting $Y_1 = \sqrt{(X + 2)}$. See Figure 53.

    (a) From the graph, it appears that $f(2) = 2$.

    (b) $f(2) = \sqrt{2 + 2} = \sqrt{4} = 2$

    (c) See Figure 53c.

[−4.7, 4.7, 1] by [−3.1, 3.1, 1]

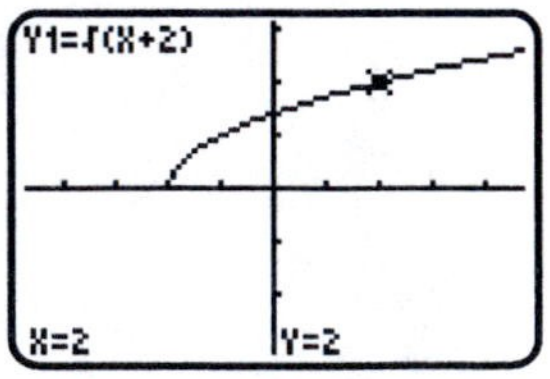

Figure 53 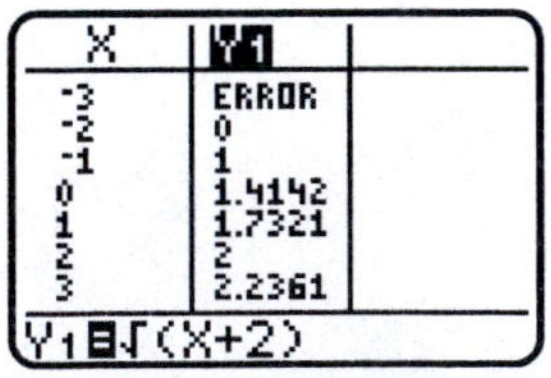   Figure 53c

55. Verbal: Square the input x.

Graphical: Graph $Y_1 = X^2$. See Figure 55.

Numerical:

x	-2	-1	0	1	2
y	4	1	0	1	4

$f(2) = 4$

[−10, 10, 1] by [−10, 10, 1]   [−6, 6, 1] by [−4, 4, 1]

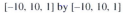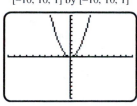

Figure 55   Figure 57

57. Verbal: Multiply the input x by 2, add 1, and then take the absolute value.

Graphical: Graph $Y_1 = \text{abs}(2X + 1)$. See Figure 57.

Numerical:

x	-2	-1	0	1	2
y	3	1	1	3	5

$f(2) = 5$

59. Verbal: Subtract the input x from 5.

Graphical: $Y_1 = 5 - X$. See Figure 59.

Numerical:

x	-2	-1	0	1	2
y	7	6	5	4	3

$f(2) = 3$

[−10, 10, 1] by [−10, 10, 1]   [−6, 6, 1] by [−4, 4, 1]

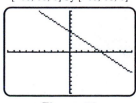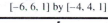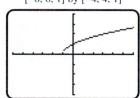

Figure 59   Figure 61

61. Verbal: Add 1 to the input x and then take the square root of the result.

Graphical: Graph $Y_1 = \sqrt{(X + 1)}$ See Figure 61.

Numerical:

x	-2	-1	0	1	2
y	—	0	1	$\sqrt{2}$	$\sqrt{3}$

$f(2) = \sqrt{3}$

63. It costs about $0.50 per mile.

Symbolic:$f(x) = 0.50x$.

Graphical: See Figure 63a.

Numerical: See Figure 63b.

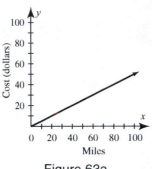

Figure 63a

Miles	1	2	3	4	5	6
Cost	0.50	1.00	1.50	2.00	2.50	3.00

Figure 63b

65. This is a graph of a function because every vertical line intersects the graph at most once. Both the domain and the range are all real numbers.

67. This is not a graph of a function because some vertical lines can intersect the graph twice. Because a vertical line can intersect the graph twice, two functions are necessary to create this graph.

69. This is a graph of a function because every vertical line intersects the graph at most once.
The domain is $\{x \mid -4 \le x \le 4\}$. The range is $\{y \mid 0 \le y \le 4\}$.

71. a.) Yes

b.) Each real number has exactly one real cube root.

73. a.)No.

b.) More than one student can have score x.

75. Yes, because the IDs are unique.

77. No. The ordered pairs (1, 2) and (1, 3) belong to the set S. The domain element 1 has more than one range element associated with it.

79. Yes. Each element in its domain is associated with exactly one range element.

81. No. The ordered pairs (1, 10.5) and (1, −0.5) belong to the set S. The domain element 1 has more than one range element associated with it.

83. No, for example, the ordered pairs (1, −1) and (1, 1) belong to the relation. The domain element 1 has more than one range element associated with it.

85. Yes. Each element in the domain of f is associated with exactly one range element.

87. No, for example, the ordered pairs$(0, \sqrt{70})$ and$(0, -\sqrt{70})$ belong to the relation. The domain element 0 has more than one range element associated with it.

89. Yes. Each element in the domain of f is associated with exactly one range element.

91. $g(x) = 12x \Rightarrow g(10) = 12(10) = 120$; there are 120 inches in 10 feet.

93. $g(x) = 0.25x \Rightarrow g(10) = 0.25(10) = 2.50$; there are 2.5 dollars in 10 quarters.

95. $g(x) = 60 \cdot 60 \cdot 24 \cdot x \Rightarrow g(x) = 86{,}400x \Rightarrow g(10) = 86{,}400(10) = 864{,}000;$ there are

   864,000 seconds in 10 days.

97. (a) $\{(R, 37), (N, 30), (S, 17)\}$

   (b) $D = \{N, R, S\};\ R = \{17, 30, 37\}$

99. $f(x) = 40x \Rightarrow f(5) = 40(5) = 200;$ About 200 million tons of electronic waste will be piled up after 5 years.

101. $N(x) = 2200x;\ N(3) = 2200(3) = 6600;$ in 3 years the average person uses 6600 napkins.

103. Verbal: Multiply the input x by -5.8 to obtain the change in temperature.

   Symbolic: $f(x) = -5.8x$.

   Graphical: $Y_1 = -5.8X$. See Figure 103a.

   Numerical: Table $Y_1 = -5.8X$. See Figure 103b.

   $[0, 3, 1]$ by $[-20, 20, 5]$

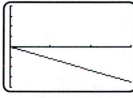

   Figure 103a   Figure 103b

1.4: Types of Functions and Their Rates of Change

1. $f(x) = 5 - 2x \Rightarrow f(x) = -2x + 5;\ m = -2, b = 5$

3. $f(x) = -8x \Rightarrow f(x) = -8x + 0;\ m = -8, b = 0$

5. $m = \dfrac{5 - 6}{2 - 4} = \dfrac{-1}{-2} = \dfrac{1}{2} = 0.5$

7. $m = \dfrac{-2 - 4}{5 - (-1)} = \dfrac{-6}{6} = -1$

9. $m = \dfrac{-8 - (-8)}{7 - 12} = \dfrac{0}{-5} = 0$

11. $m = \dfrac{0.4 - (-0.1)}{-0.3 - 0.2} = \dfrac{0.5}{-0.5} = -1$

13. $m = \dfrac{7.6 - 9.2}{-0.3 - (-0.5)} = \dfrac{-1.6}{0.2} = -8$

15. $m = \dfrac{8 - 6}{-5 - (-5)} = \dfrac{2}{0} =$ undefined

17. $m = \dfrac{\frac{7}{10} - (-\frac{3}{5})}{-\frac{5}{6} - \frac{1}{3}} = \dfrac{\frac{13}{10}}{-\frac{7}{6}} = \dfrac{13}{10} \cdot \left(-\dfrac{6}{7}\right) = -\dfrac{39}{35} \approx -1.143$

19. Slope $= 2;$ the graph rises 2 units for every unit increase in x.

21. Slope $= -\dfrac{3}{4}$; the graph falls $\dfrac{3}{4}$ unit for every unit increase in x, or equivalently, the graph falls 3 units for

   every 4-unit increase in x.

23. Slope $= -1$; the graph falls 1 unit for every unit increase in x.

25. (a) Buying no carpet should and does cost $0.

    (b) Slope $= \dfrac{100}{5} = 20$

    (c) The carpet costs $20 per square yard.

27. (a) $D(x) = 150 - 20x \Rightarrow D(5) = 150 - 20(5) = 50.$ After 5 hours the train is 50 miles from the station.

    (b) Slope equals -20. The train is traveling toward the station at 20 mph.

29. (a) $D(2) = 75(2) = 150 \, \text{miles}$

    (b) Slope $= 75$; the car is traveling away from the rest stop at 75 miles per hour.

31. $f(x) = -2x + 5$ is a linear function, but not a constant function, with a slope of $m = -2$. See Figure 31.

[–10, 10, 1] by [–10, 10, 1]   [–10, 10, 1] by [–10, 10, 1]

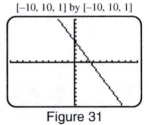

Figure 31   Figure 33

33. $f(x) = 1$ is a constant (and linear) function. See Figure 33.

35. From its graph, we see that $f(x) = |x + 1|$ represents a nonlinear function. See Figure 35.

[–10, 10, 1] by [–10, 10, 1]   [–10, 10, 1] by [–10, 10, 1]

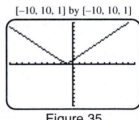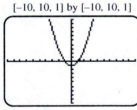

Figure 35   Figure 37

37. From its graph, we see that $f(x) = x^2 - 1$ represents a nonlinear function. See Figure 37.

39. Between each pair of points, the y-values increase 4 units for each unit increase in x. Therefore, the data is linear. The slope of the line passing through the data points is 4.

41. The y-values do not increase by a constant amount for each 2-unit increase in x. The data is nonlinear.

43. (a) Slope $= \dfrac{\text{rise}}{\text{run}} = \dfrac{2}{1} = 2$; y-intercept: -1; x-intercept: 0.5

    (b) $f(x) = ax + b \Rightarrow f(x) = 2x - 1$

    (c) 0.5

45. (a) Slope $= \dfrac{\text{rise}}{\text{run}} = \dfrac{-1}{3} = -\dfrac{1}{3}$; y-intercept: 2; x-intercept: 6

    (b) $f(x) = ax + b \Rightarrow f(x) = -\dfrac{1}{3}x + 2$

    (c) 6

47. $f(x) = ax + b \Rightarrow f(x) = -\dfrac{3}{4}x + \dfrac{1}{3}$

49. $f(x) = ax + b \Rightarrow f(x) = 15x + 0,$ or $f(x) = 15x$

51. $[5, \infty)$

53. $[4, 19)$

55. $[-1, \infty)$

57. $(-\infty, 1) \cup [3, \infty)$

59. $(-3, 5]$

61. $(-\infty, -2)$

63. $(-\infty, -2) \cup [1, \infty)$

65. f is decreasing on $(-\infty, \infty)$. In set-builder notation, the interval is $\{x \mid -\infty < x < \infty\}$.

67. f is increasing on $(2, \infty)$ and decreasing on $(-\infty, 2)$. In set-builder notation, these intervals are $\{x \mid x \geq 2\}$ and $\{x \mid x \leq 2\}$ respectively.

69. f is increasing on $(-\infty, -2), (1, \infty)$ and decreasing on $(-2, 1)$. In set-builder notation, these intervals are $\{x \mid x \leq 2 \text{ or } x \geq 1\}$ and $\{x \mid -2 \leq x \leq 1\}$ respectively.

71. f is increasing on $(-8, 0), (8, \infty)$ and decreasing on $(-\infty, -8), (0, 8)$. In set-builder notation, these intervals are $\{x \mid -8 \leq x \leq 0 \text{ or } x \geq 8\}$ and $\{x \mid x \leq -8 \text{ or } 0 \leq x \leq 8\}$ respectively.

73. The graph of this equation is linear with a slope of $2 \Rightarrow$ it is increasing: $(-\infty, \infty)$ and decreasing: never. In set builder notation, the interval is $\{x \mid -\infty < x < \infty\}$.

75. The graph of this equation is a parabola with a vertex $(0, -2) \Rightarrow$ it is increasing: $(0, \infty)$ and in set builder notation, these intervals are $\{x \mid x \geq 0\}$ and $\{x \mid x \leq 0\}$ respectively.

77. The graph of this equation is a parabola with a vertex $(1, 1)$, because of the negative coefficient before x^2 it opens downward $\Rightarrow$ it is increasing: $(-\infty, 1)$ and decreasing: $(1, \infty)$. In set builder notation, these intervals are $\{x \mid x \leq 1\}$ and $\{x \mid x \geq 1\}$ respectively.

79. The square root equation graph has a starting point $(1, 0)$, all x-values less than 1 are undefined $\Rightarrow$ it is increasing: $(1, \infty)$ and decreasing: never. In set builder notation, the interval is $\{x \mid x \geq 1\}$.

81. The absolute value graph has a vertex $(-3, 0) \Rightarrow$ it is increasing: $(-3, \infty)$ and decreasing: $(-\infty, -3)$. In set builder notation, these intervals are $\{x \mid x \geq -3\}$ and $\{x \mid x \leq -3\}$ respectively.

83. The basic x^3 function is always increasing $\Rightarrow$ it is increaing: $(-\infty, \infty)$ and decreasing: never. In set builder notation, the interval is $\{x \mid -\infty < x < \infty\}$.

85. The graph of this cubic equation has turning points $\left(-2, \dfrac{16}{3}\right)$ and $\left(2, \dfrac{-16}{3}\right) \Rightarrow$ it is increasing: $(-\infty, -2)$ or $(2, \infty)$; and decreasing: $[-2, 2]$. In set builder notation, these intervals are $\{x \mid x \leq -2 \text{ or } x \geq 2\}$ and $\{x \mid -2 \leq x \leq 2\}$ respectively.

87. The graph of this x^4 graph has a negative lead coefficient therefore is reflected through the x-axis has turning points $\left(-1, \dfrac{5}{12}\right)$, $(0, 0)$ and $\left(2, \dfrac{8}{3}\right)$ $\Rightarrow$ it is increasing: $(-\infty, -1)$ or $(0, 2)$; and decreasing: In set builder notation, these intervals are $\{x \mid x \leq -1 \text{ or } 0 \leq x \leq 2\}$ and $\{x \mid -1 \leq x \leq 0 \text{ or } x \geq 2\}$ respectively.

89. According to the graph the water levels are increasing on the time intervals $(0, 2.4)$, $(8.7, 14.7)$ and $(21, 27)$.

    In set builder notation, these intervals are $\{x \mid 0 \leq x \leq 2.4 \text{ or } 8.7 \leq x \leq 14.7 \text{ or } 21 \leq x \leq 27\}$.

91. The average rate of change from –3 to –1 is $\dfrac{f(-1) - f(-3)}{-1 - (-3)} = \dfrac{3.7 - 1.3}{2} = \dfrac{2.4}{2} = 1.2$.

    The average rate of change from 1 to 3 is $\dfrac{f(3) - f(1)}{3 - 1} = \dfrac{1.3 - 3.7}{2} = -\dfrac{2.4}{2} = -1.2$.

93. (a) $f(x) = x^2 \Rightarrow f(1) = 1^2 = 1$ and $f(2) = 2^2 = 4 \Rightarrow (1, 1), (2, 4)$; using the slope formula for rate of change we get $\dfrac{4 - 1}{2 - 1} = \dfrac{3}{1} = 3$.

    (b) See Figure 93.

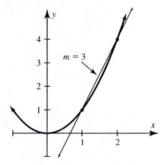
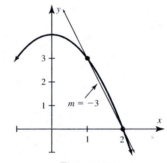

Figure 93   Figure 94

95. If $f(x) = 7x - 2$ then $\dfrac{f(4) - f(1)}{4 - 1} = 7$. The slope of the graph is 7.

97. If $f(x) = \sqrt{2x - 1}$ then $\dfrac{f(3) - f(1)}{3 - 1} \approx 0.62$. The slope of the line passing through the points is approximately 0.62.

99. (a) The average rate of change from 1900 to 1940 is calculated as $\dfrac{182 - 3}{1940 - 1900} = \dfrac{179}{40} = 4.475$; from 1940 to 1980: $\dfrac{632 - 182}{1980 - 1940} = \dfrac{450}{40} = 11.25$; from 1980 to 2010: $\dfrac{315 - 632}{2010 - 1980} = \dfrac{-317}{30} \approx -10.6$.

    (b) The average rates of change in cigarette consumption in the time periods 1900 to 1940, 1940 to 1980, and 1980 to 2010 were 4.475, 11.25, and –10.06, respectively.

101. See Figure 101.

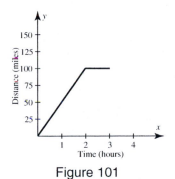

Figure 101

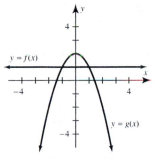

Figure 103

103. See Figure 103. *Answers may vary.*

Extended and Discovery Exercises for Section 1.4

1. (a) $C(r) = 2\pi r \Rightarrow C(r+1) = 2\pi(r+1) = 2\pi r + 2\pi$; for every 1 inch increase in the radius, the

    circumference increases by 2π inches. So the circumference increases at a constant rate of 2π inches per second.

    (b) No, because the area function, $A(r) = \pi r^2$, depends on the radius squared. The area function is not linear

    and thus does not increase at a constant rate.

Checking Basic Concepts for Sections 1.3 and 1.4

1. Symbolic: $f(x) = 5280x$

    Numerical: Use a table f starting at $x = 1$, incrementing by 1. See Figure 1a.

    Graphical: Graph $Y_1 = 5280X$ as shown in Figure 1b.

x	1	2	3	4	5
$f(x)$	5280	10,560	15,840	21,120	26,400

Figure 1a

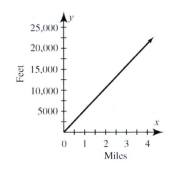

Figure 1b

3. The slope is calculated as follows: $m = \dfrac{-5-4}{4-(-2)} = -\dfrac{9}{6} = -\dfrac{3}{2}$

    If the graph of the linear function $f(x) = ax + b$, passes through the points $(-2, 4)$ and $(4, -5)$, its slope must

    be equal to $-\dfrac{3}{2}$; therefore, $a = -\dfrac{3}{2}$.

5.   (a) $(-\infty, 5]$

(b) $[1, 6)$

7. $\dfrac{f(-1) - f(-3)}{-1 - (-3)} = \dfrac{4 - 18}{2} = -7$

Chapter 1 Review Exercises

1. -2 is an integer, rational number, and real number. $\dfrac{1}{2}$ is both a rational and a real number. 0 is an integer,

rational number, and real number. 1.23 is both a rational and a real number. $\sqrt{7}$ is a real number. $\sqrt{16} = 4$ is

a natural number, integer, rational number, and real number.

3. $1,891,000 = 1.891 \times 10^6$

5. $1.52 \times 10^4 = 15,200$

7.   (a) $\sqrt[m]{1.2} + \pi^3 \approx 32.07$

(b) $\dfrac{3.2 + 5.7}{7.9 - 4.5} \approx 2.62$

(c) $\sqrt{5^2 + 2.1} \approx 5.21$

(d) $1.2(6.3)^2 + \dfrac{3.2}{\pi - 1} \approx 49.12$

9   $.4 - 3^2 \cdot 5 = 4 - 9 \cdot 5 = 4 - 45 = -41$

11.

-23	-5	8	19	24

(a) Maximum $= 24$; Minimum $= -23$

(b) Mean $= \dfrac{-23 + (-5) + 8 + 19 + 24}{5} = 4.6$; Median $= 8$

13. (a) $S = \{(-15, -3), (-10, -1), (0, 1), (5, 3), (20, 5)\}$

(b) $D = \{-15, -10, 0, 5, 20\}$ and $R = \{-3, -1, 1, 3, 5\}$

15. The relation $\{(10, 13), (-12, 40), (-30, -23), (25, -22), (10, 20)\}$ is plotted in Figure 15. It is not a function

since both $(10, 13)$ and $(10, 20)$ are contained in the set. Notice that these points are lined up vertically.

[−50, 50, 10] by [−50, 50, 10]   ]

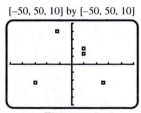

Figure 15

17. $d = \sqrt{(2 - (-4))^2 + (-3 - 5)^2} = \sqrt{6^2 + (-8)^2} = \sqrt{36 + 64} = \sqrt{100} = 10$

19. $M = \left(\dfrac{24 + (-20)}{2}, \dfrac{-16 + 13}{2}\right) = \left(\dfrac{4}{2}, \dfrac{-3}{2}\right) = \left(2, \dfrac{-3}{2}\right)$

21. Use the distance formula to find the side lengths:

$d_1 = \sqrt{(1 - (-3))^2 + (2 - 5)^2} = \sqrt{4^2 + (-3)^2} = \sqrt{16 + 9} = \sqrt{25} = 5$

$d_2 = \sqrt{((-3) - 0)^2 + (5 - 9)^2} = \sqrt{(-3)^2 + (-4)^2} = \sqrt{9 + 16} = \sqrt{25} = 5$

$d_3 = \sqrt{(1 - 0)^2 + (2 - 9)^2} = \sqrt{1^2 + (-7)^2} = \sqrt{1 + 49} = \sqrt{50} = 7.07$

Two sides are 5 units long and the other side is 7.07 units long, therefore, the triangle is isosceles.

23. First calculate the midpoint of the diameter: $x = \dfrac{6 + (-2)}{2} = \dfrac{4}{2} = 2$ and $y = \dfrac{6 + 4}{2} = \dfrac{10}{2} = 5 \Rightarrow$ the center

is (2, 5). Then calculate the length of the radius using points (2, 5) and (6, 6):

$d = \sqrt{(6 - 2)^2 + (6 - 5)^2} = \sqrt{4^2 + 1^2} = \sqrt{17}.$ The equation is $(x - 2)^2 + (y - 5)^2 = 17.$

25. See Figure 25.

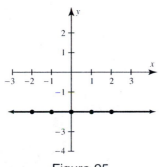

Figure 25

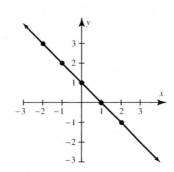

Figure 27

27. See Figure 27.

29. See Figure 29.

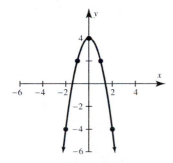

Figure 29

31. See Figure 31.

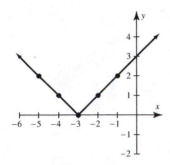

Figure 31

33. Symbolic: $f(x) = 16x$.

Numerical: Table f starting at $x = 0$, incrementing by 25. See Figure 33a.

Graphical: Graph $Y_1 = 16X$ in [0, 100, 10] by [0, 1800, 300]. See Figure 33b.

[0, 100, 10] by [0, 1800, 300]

x	0	25	50	75	100
$f(x)$	0	400	800	1200	1600

Figure 33a   Figure 33b

35. (a) $f(x) = 5 \Rightarrow f(-3) = 5$ and $f(1.5) = 5$

(b) All real numbers

37. (a) $f(x) = x^2 - 3 \Rightarrow f(-10) = (-10)^2 - 3 = 97$ and $f(a + 2) = (a + 2)^2 - 3 =$

$a^2 + 4a + 4 - 3 = a^2 + 4a + 1$

(b) All real numbers

39. (a) $f(-3) = \dfrac{1}{-3 - 4} = -\dfrac{1}{7}; f(a + 1) = \dfrac{1}{a + 1 - 4} = \dfrac{1}{a - 3}$

(b) $D = \{x \mid x \neq 4\}$

41. No, for example, an input $x = 6$ produces outputs of $y = \pm 1$.

43. (a) Using the points (0, 6) and (2, 2), $m = \dfrac{2 - 6}{2 - 0} = \dfrac{-4}{2} = -2$; y-intercept: 6; x-intercept: 3.

(b) $f(x) = -2x + 6$

(c) The zeros of f are the same as the x-intercepts. That is $x = 3$.

45. Since any vertical line intersects the graph of f at most once, it is a function.

47. Yes, it is a function. Each input produces a single output.

49. $a = 0$, so the slope $= 0$.

51. $m = \dfrac{4 - 7}{3 - (-1)} = -\dfrac{3}{4}$

53. $m = \dfrac{4 - 4}{-2 - 8} = \dfrac{0}{-10} = 0$

55. $f(x) = 8 - 3x$ represents a linear function.

57. $f(x) = |x + 2|$ represents a nonlinear function.

59. See Figure 59.

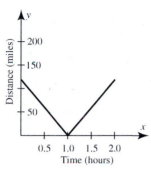

Figure 59

61. Yes, $m = \dfrac{50 - 26}{-2 - 4} = \dfrac{24}{-6} = -4$. The best model is linear, but not constant, since the y-values decrease 8 units

for every 2-unit increase in x.

63. $f(x + h) = 5(x + h) + 1 = 5x + 5h + 1$

$$\frac{f(x + h) - f(x)}{h} = \frac{5x + 5h + 1 - (5x + 1)}{h} = \frac{5h}{h} = 5$$

65. $\dfrac{2.28 \times 10^8}{3 \times 10^5} = 760$ seconds $= 12\dfrac{2}{3}$ minutes

67. (a) Sketching a diagram of the pool and sidewalk (Not Shown) gives the following dimensions:

$l = 62$ ft. and $w = 37$ ft. Thus, $P = 2(62) + 2(37) = 198$ ft.

(b) The area of the sidewalk would consist of the area of four 6×6 squares, two 50×6 rectangles, and two

25×6 rectangles. $A = 4(6 \cdot 6) + 2(50 \cdot 6) + 2(25 \cdot 6) = 4(36) + 2(300) + 2(150) = 1044$ ft^2

69. (a) Plot the points $(0, 100)$, $(1, 10)$, $(2, 6)$, $(3, 3)$, and $(4, 2)$ and make a line graph. See Figure 69. The survival

rates decrease rapidly at first. This means that a large number of eggs never develop into mature adults.

(b) Since any vertical line could intersect the graph at most once, this graph could represent a function.

(c) From 0 to 1, $\dfrac{10 - 100}{1 - 0} = -90$; from 1 to 2, $\dfrac{6 - 10}{2 - 1} = -4$; from 2 to 3, $\dfrac{3 - 6}{3 - 2} = -3$;

from 3 to 4, $\dfrac{2 - 3}{4 - 3} = -1$;   during the first year, the population of sparrows decreased, on average, by 90

birds. The other average rates of change can be interpreted similarly.

[−1, 5, 1] by [0, 110, 10]   [1, 5, 1] by [40, 70, 5]   [1, 5, 1] by [50, 64, 5]

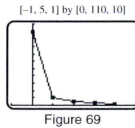
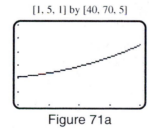
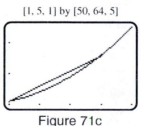

Figure 69   Figure 71a   Figure 71c

71. (a) See Figure 71a. f is nonlinear.

(b) $f(x) = 0.5x^2 + 50 \Rightarrow \dfrac{f(4) - f(1)}{4 - 1} = \dfrac{58 - 50.5}{3} = 2.5$

(c) The average rate of change in outside temperature from 1 P.M. to 4 P.M. was 2.5° F per hour. The slope of the line segment from (1, 50.5) to (4, 58) is 2.5. See Figure 71c. The temperature increased, on average, by 2.5° F per hour

Extended and Discovery Exercises for Chapter 1

1.   For (–3, 1.8) and (–1.5, 0.45): $d = \sqrt{(-1.5 + 3)^2 + (0.45 - 1.8)^2} \approx 2.018$

     For (–1.5, 0.45) and (0, 0): $d = \sqrt{(1.5)^2 + (-0.45)^2} \approx 1.566$

     For (0, 0) and (1.5, 0.45): $d = \sqrt{(1.5)^2 + (0.45)^2} \approx 1.566$

     For (1.5, 0.45) and (3, 1.8): $d = \sqrt{(3 - 1.5)^2 + (1.8 - 0.45)^2} \approx 2.018$

     Curve length $\approx 2(2.018) + 2(1.566) \approx 7.17$ km.

3.   For (–1, –1) and (0, 0): $d = \sqrt{(0 + 1)^2 + (0 + 1)^2} = \sqrt{2}$

     For (0, 0) and (1, 1): $d = \sqrt{(1 - 0)^2 + (1 - 0)^2} = \sqrt{2}$

     For (1, 1) and $(2, \sqrt[m]{2})$: $d = \sqrt{(2 - 1)^2 + \left(\sqrt[m]{2} - 1\right)^2} \approx 1.033$

     Curve length $\approx 2\sqrt{2} + 1.033 \approx 3.862$. See Figure 3.

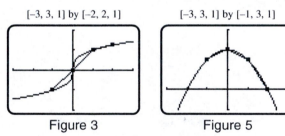

[–3, 3, 1] by [–2, 2, 1]   [–3, 3, 1] by [–1, 3, 1]

Figure 3   Figure 5

5.   For (–1, 1.5) and (0, 2): $d = \sqrt{(0 + 1)^2 + (2 - 1.5)^2} \approx 1.118$

     For (0, 2) and (1, 1.5): $d = \sqrt{(1 - 0)^2 + (1.5 - 2)^2} \approx 1.118$

     For (1, 1.5) and (2, 0): $d = \sqrt{(2 - 1)^2 + (0 - 1.5)^2} \approx 1.803$

     Curve length $\approx 2(1.118) + 1.803 \approx 4.039$. See Figure 5.

7.   The graph of $y = \sqrt{x}$ is nearly linear from (1, 1) to (4, 2). It would be reasonable to estimate the distance along the curve by finding the linear distance from (1, 1) to (4, 2).

     $d = \sqrt{(4 - 1)^2 + (2 - 1)^2} = \sqrt{(3)^2 + (1)^2} = \sqrt{10} \approx 3.162$

9.   (a) Determine the number of square miles of the Earth's surface that are covered by the oceans. Then divide the total volume of water from the ice caps by the surface area of the oceans to get the rise in sea level.

     (b) The surface area of a sphere is $4\pi r^2$, so the surface area of Earth is $4\pi(3960)^2 \approx 197,061,000$ sq. mi.;

         Part of surface covered by the oceans is $(0.71)(197,061,000) \approx 139,913,000$ sq. mi.;

         rise in sea leve $= \dfrac{\text{volume of water}}{\text{surface area of the oceans}} = \dfrac{680,000 \text{ mi.}^3}{139,913,000 \text{ mi.}^2} \approx 0.00486$ mi. ≈ 25.7 ftl

     (c) Since the average elevations of Boston, New Orleans, and San Diego are all less than 25 feet, these cities would be under water without some type of dike system.

     (d) Rise in sea level $= \dfrac{6,300,000 \text{ mi.}^3}{139,913,000 \text{ mi.}^2} \approx 0.04503$ mi. ≈ 238 ft.

Chapter 2: Linear Functions and Equations

2.1: Equations of Lines

1. Find slope: $m = \dfrac{-2 - 2}{3 - 1} = \dfrac{-4}{2} = -2$. Using $(x_1, y_1) = (1, 2)$ and point-slope form $y = m(x - x_1) + y_1$,

   we get $y = -2(x - 1) + 2$. See Figure 1.

3. Find slope: $m = \dfrac{2 - (-1)}{1 - (-3)} = \dfrac{3}{4}$. Using $(x_1, y_1) = (-3, -1)$ and point-slope form $y = m(x - x_1) + y_1$,

   we get $y = \dfrac{3}{4}(x + 3) - 1$. See Figure 3.

5. The point-slope form is given by $y = m(x - x_1) + y_1$. Thus, $m = -2.4$ and $(x_1, y_1) = (4, 5) \Rightarrow$

   $y = -2.4(x - 4) + 5 \Rightarrow y = -2.4x + 9.6 + 5 \Rightarrow y = -2.4x + 14.6$ and $f(x) = -2.4x - 14.6$.

7. First find the slope between the points $(1, -2)$ and $(-9, 3)$: $m = \dfrac{3 - (-2)}{-9 - 1} = -\dfrac{1}{2}$.

   $y = -\dfrac{1}{2}(x - 1) - 2 \Rightarrow y = -\dfrac{1}{2}x + \dfrac{1}{2} - 2 \Rightarrow y = -\dfrac{1}{2}x - \dfrac{3}{2}$ and $f(x) = -\dfrac{1}{2}x - \dfrac{3}{2}$

9. $(4, 0), (0, -3)$; $m = \dfrac{-3 - 0}{0 - 4} = \dfrac{3}{4}$. Thus, $y = \dfrac{3}{4}(x - 4) + 0$ or $y = \dfrac{3}{4}x - 3$ and $f(x) = \dfrac{3}{4}x - 3$.

11. Using the points $(0, -1)$ and $(3, 1)$, we get $m = \dfrac{1 - (-1)}{3 - 0} = \dfrac{2}{3}$ and $b = -1$; $y = mx + b \Rightarrow y = \dfrac{2}{3}x - 1$.

13. Using the points $(-2, 1.8)$ and $(1, 0)$, we get $m = \dfrac{0 - 1.8}{1 - (-2)} = \dfrac{-1.8}{3} = -\dfrac{18}{30} = -\dfrac{3}{5}$; to find b, we use $(1, 0)$

   in $y = mx + b$ and solve for b: $0 = -\dfrac{3}{5}(1) + b \Rightarrow b = \dfrac{3}{5}$; $y = -\dfrac{3}{5}x + \dfrac{3}{5}$.

15. c

17. b

19. e

21. $m = \dfrac{2 - (-4)}{1 - (-1)} = 3$; $y = 3(x + 1) - 4 = 3x + 3 - 4 = 3x - 1$

23. $m = \dfrac{-3 - 5}{1 - 4} = \dfrac{8}{3}$; $y = \dfrac{8}{3}(x - 4) + 5 = \dfrac{8}{3}x - \dfrac{32}{3} + 5 = \dfrac{8}{3}x - \dfrac{17}{3}$

25. $b = 5$ and $m = -7.8 \Rightarrow y = -7.8x + 5$.

27. The line passes through the points $(0, 45)$ and $(90, 0)$.

   $m = \dfrac{0 - 45}{90 - 0} = -\dfrac{1}{2}$; $b = 45$ and $m = -\dfrac{1}{2} \Rightarrow y = -\dfrac{1}{2}x + 45$

29. $m = -3$ and $b = 5 \Rightarrow y = -3x + 5$

31. $m = \dfrac{0 - (-6)}{4 - 0} = \dfrac{6}{4} = \dfrac{3}{2}$ and $b = -6$; $y = mx + b \Rightarrow y = \dfrac{3}{2}x - 6$

33. $m = \dfrac{\frac{2}{3} - \frac{3}{4}}{\frac{1}{5} - \frac{1}{2}} = \dfrac{-\frac{1}{12}}{-\frac{3}{10}} = \dfrac{5}{18}$; using the point-slope form with $m = \dfrac{5}{18}$ and $\left(\dfrac{1}{2}, \dfrac{3}{4}\right)$, we get

$y = \dfrac{5}{18}\left(x - \dfrac{1}{2}\right) + \dfrac{3}{4} \Rightarrow y = \dfrac{5}{18}x - \dfrac{5}{36} + \dfrac{3}{4} \Rightarrow y = \dfrac{5}{18}x + \dfrac{11}{18}$.

35. The line has a slope of 4 and passes through the point $(-4, -7)$; $y = 4(x + 4) - 7 \Rightarrow y = 4x + 9$.

37. The slope of the perpendicular line is equal to $\dfrac{3}{2}$ and the line passes through the point $(1980, 10)$;

$y = \dfrac{3}{2}(x - 1980) + 10 \Rightarrow y = \dfrac{3}{2}x - 2960$

39. $y = \dfrac{2}{3}x + 3 \Rightarrow m = \dfrac{2}{3}$; the parallel line has slope $\dfrac{2}{3}$; since it passes through $(0, -2.1)$,

the y-intercept $= -2.1$; $y = mx + b \Rightarrow y = \dfrac{2}{3}x - 2.1$.

41. $y = -2x \Rightarrow m = -2$; the perpendicular line has slope $\dfrac{1}{2}$; since it passes through $(-2, 5)$, the equation is

$y = \dfrac{1}{2}(x + 2) + 5 = \dfrac{1}{2}x + 1 + 5 = \dfrac{1}{2}x + 6$.

43. $y = -x + 4 \Rightarrow m = -1$; the perpendicular line has slope 1; since it passes through $(15, -5)$, the equation is

$y = 1(x - 15) - 5 = x - 15 - 5 = x - 20$.

45. $m = \dfrac{1 - 3}{-3 - 1} = \dfrac{-2}{-4} = \dfrac{1}{2}$; a line parallel to this line also has slope $m = \dfrac{1}{2}$. Using

$(x_1, y_1) = (5, 7), m = \dfrac{1}{2}$, and point-slope form $y = m(x - x_1) + y_1$, we get $y = \dfrac{1}{2}(x - 5) + 7 \Rightarrow$

$y = \dfrac{1}{2}x + \dfrac{9}{2}$.

47. $m = \dfrac{\frac{2}{3} - \frac{1}{2}}{-3 - (-5)} = \dfrac{\frac{1}{6}}{2} = \dfrac{1}{12}$; a line perpendicular to this line has slope $m = -\dfrac{12}{1} = -12$.

Using $(x_1, y_1) = (-2, 4), m = -12$, and point-slope form $y = m(x - x_1) + y_1$, we get

$y = -12(x + 2) + 4 \Rightarrow y = -12x - 24 + 4 \Rightarrow y = -12x - 20$.

49. $x = -5$. It is not possible to write as a linear function since a vertical line does not represent a function.

51. $y = 6$ and $f(x) = 6$.

53. Since the line $y = 15$ is horizontal, the perpendicular line through $(4, -9)$ is vertical and has equation $x = 4$.

It is not possible to write as a linear function since a vertical line does not represent a function.

55. The line through $(19, 5.5)$ and parallel to $x = 4.5$ is also vertical and has equation $x = 19$. It is not possible to

write as a linear function since a vertical line does not represent a function.

57. Let $4x - 5y = 20$.

x-intercept: Substitute $y = 0$ and solve for x. $4x - 5(0) = 20 \Rightarrow 4x = 20 \Rightarrow x = 5$; x-intercept: 5

y-intercept: Substitute $x = 0$ and solve for y. $4(0) - 5y = 20 \Rightarrow -5y = 20 \Rightarrow y = -4$; y-intercept: -4

See Figure 57.

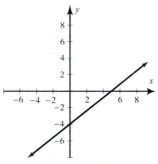

Figure 57

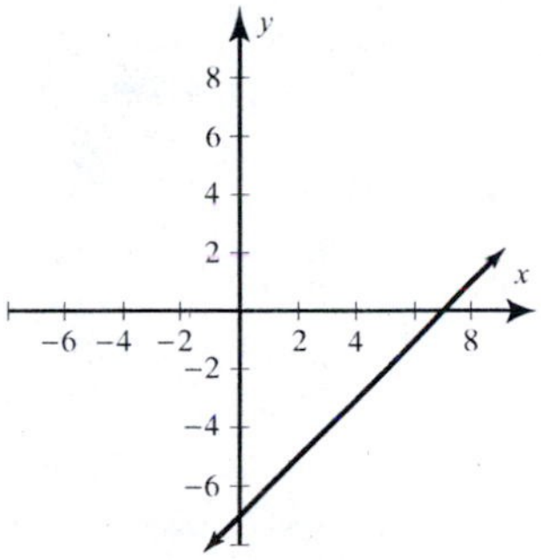

Figure 59

59. Let $x - y = 7$.

   x-intercept: Substitute $y = 0$ and solve for *x*. $x - 0 = 7 \Rightarrow x = 7$; *x*-intercept: 7

   y-intercept: Substitute $x = 0$ and solve for *y*. $0 - y = 7 \Rightarrow -y = 7 \Rightarrow y = -7$; *y*-intercept: -7

   See Figure 59.

61. Let $6x - 7y = -42$.

   x-intercept: Substitute $y = 0$ and solve for *x*. $6x - 7(0) = -42 \Rightarrow 6x = -42 \Rightarrow x = -7$; *x*-intercept: -7

   y-intercept: Substitute $x = 0$ and solve for *y*. $6(0) - 7y = -42 \Rightarrow -7y = -42 \Rightarrow y = 6$; *y*-intercept: 6

   See Figure 61.

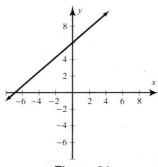

Figure 61

63. Let $y - 3x = 7$.

   x-intercept: Substitute $y = 0$ and solve for *x*. $0 - 3x = 7 \Rightarrow -3x = 7 \Rightarrow x = -\dfrac{7}{3}$; *x*-intercept: $-\dfrac{7}{3}$

   y-intercept: Substitute $x = 0$ and solve for *y*. $y - 3(0) = 7 \Rightarrow y - 0 = 7 \Rightarrow y = 7$; *y*-intercept: 7

   See Figure 63.

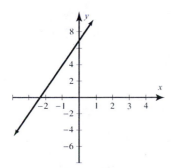

Figure 63

65. Let $0.2x + 0.4y = 0.8$.

x-intercept: Substitute $y = 0$ and solve for x. $0.2x + 0.4(0) = 0.8 \Rightarrow 0.2x = 0.8 \Rightarrow x = 4$; x-intercept: 4

y-intercept: Substitute $x = 0$ and solve for y. $0.2(0) + 0.4y = 0.8 \Rightarrow 0.4y = 0.8 \Rightarrow y = 2$; y-intercept: 2

See Figure 65.

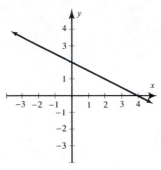

Figure 65

67. Let $y = 8x - 5$.

x-intercept: Substitute $y = 0$ and solve for x. $0 = 8x - 5 \Rightarrow 5 = 8x \Rightarrow x = \dfrac{5}{8}$; x-intercept: $\dfrac{5}{8}$

y-intercept: Substitute $x = 0$ and solve for y. $y = 8(0) - 5 \Rightarrow y = -5$; y-intercept: -5

See Figure 67.

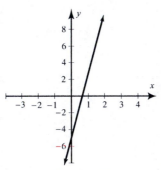

Figure 67

69. Let $\dfrac{x}{5} + \dfrac{y}{7} = 1$.

x-intercept: Substitute $y = 0$ and solve for x. $\dfrac{x}{5} + \dfrac{0}{7} = 1 \Rightarrow \dfrac{x}{5} = 1 \Rightarrow x = 5$; x-intercept: 5

y-intercept: Substitute $x = 0$ and solve for y. $\dfrac{0}{5} + \dfrac{y}{7} = 1 \Rightarrow \dfrac{y}{7} = 1 \Rightarrow y = 7$; y-intercept: 7

a and b represent the x- and y-intercepts, respectively.

71. Let $\dfrac{2x}{3} + \dfrac{4y}{5} = 1$.

x-intercept: Substitute $y = 0$ and solve for x. $\dfrac{2x}{3} + \dfrac{4(0)}{5} = 1 \Rightarrow \dfrac{2x}{3} = 1 \Rightarrow x = \dfrac{3}{2}$; x-intercept: $\dfrac{3}{2}$

y-intercept: Substitute $x = 0$ and solve for y. $\dfrac{2(0)}{3} + \dfrac{4y}{5} = 1 \Rightarrow \dfrac{4y}{5} = 1 \Rightarrow y = \dfrac{5}{4}$; y-intercept: $\dfrac{5}{4}$

a and b represent the x- and y-intercepts, respectively.

73. $\dfrac{x}{a} + \dfrac{y}{b} = 1$; x-intercept: $5 \Rightarrow a = 5$, y-intercept: $9 \Rightarrow b = 9$; $\dfrac{x}{5} + \dfrac{y}{9} = 1$

75. (a) Since the point $(0, -3.2)$ is on the graph, the y-intercept is -3.2. The data is exactly linear, so one can use

any two points to determine the slope. Using the points $(0, -3.2)$ and $(1, -1.7)$, $m = \dfrac{-1.7 - (-3.2)}{1 - 0} = 1.5$.

The slope-intercept form of the line is $y = 1.5x - 3.2$.

(b) When $x = -2.7$, $y = 1.5(-2.7) - 3.2 = -7.25$. This calculation involves interpolation.

When $x = 6.3$, $y = 1.5(6.3) - 3.2 = 6.25$. This calculation involves extrapolation.

77. (a) Since the data is exactly linear, one can use any two points to determine the slope. Using the points

$(5, 94.7)$ and $(23, 56.9)$, $m = \dfrac{56.9 - 94.7}{23 - 5} = -2.1$. The point-slope form of the line is

$y = -2.1(x - 5) + 94.7$ and the slope-intercept form of the line is $y = -2.1x + 105.2$.

(b) When $x = -2.7$, $y = -2.1(-2.7) + 105.2 = 110.87$. This calculation involves extrapolation.

When $x = 6.3$, $y = -2.1(6.3) + 105.2 = 91.97$. This calculation involves interpolation.

79. (a) Using the points $(2008, 3)$ and $(2011, 24)$, $m = \dfrac{24 - 3}{2011 - 2008} = \dfrac{21}{3} = 7$. The point slope form of the line

is $f(x) = 7x - 14053$. The function approximately models the given data.

(b) $f(2007) = 7(2007 - 2008) + 3 = -7 + 3 = -4 \Rightarrow -4\%$

(c) The calculation involved extrapolation. The result was a negative so it is not possible.

Numbers were decreasing but increased after 911.

81. (a) Find the slope: $m = \dfrac{37{,}000 - 25{,}000}{2010 - 2003} = \dfrac{12{,}000}{7}$. Using the first point $(2003, 25000)$ for (x_1, y_1) and

$m = \dfrac{12{,}000}{7}$, we get $y = \dfrac{12{,}000}{7}(x - 2003) + 25{,}000$. The cost of attending a private college or

university is increasing by $\dfrac{12{,}000}{7} \approx \$1714$ per year on average.

(b) $y = \dfrac{12{,}000}{7}(2007 - 2003) + 25{,}000 \Rightarrow y = \dfrac{12{,}000}{7}(4) + 25{,}000 \Rightarrow y \approx 6857 + 25{,}000 \Rightarrow$

$y \approx \$31{,}857$

83. (a) Water is leaving the tank because the amount of water in the tank is decreasing. After 3 minutes there are

approximately 70 gallons of water in the tank.

(b) The x-intercept is 10. This means that after 10 minutes the tank is empty. The y-intercept is 100. This

means that initially there are 100 gallons of water in the tank.

(c) To determine the equation of the line, we can use 2 points. The points $(0, 100)$ and $(10, 0)$ lie on the line.

The slope of this line is $m = \dfrac{0 - 100}{10 - 0} = -10$. This slope means the water is being drained at a rate of 10

gallons per minute. Since the y-intercept is 100, the slope-intercept form of this line is given by

$y = -10x + 100$.

(d) From the graph, when $y = 50$ the x-value appears to be 5. Symbolically, when $y = 50$ then

$-10x + 100 = 50 \Rightarrow -10x = -50 \Rightarrow x = 5$. The x-coordinate is 5.

85. (a) First calculate the slope: $m = \dfrac{15 - 9}{1999 - 2013} = \dfrac{6}{-14} = -\dfrac{3}{7}$, using the first point we have

$$y = -\frac{3}{7}(x - 1999) + 15.$$

(b) The sales decreased, on average, by $\dfrac{3}{7} \approx \$0.43$ billion per year.

(c) $f(2008) = -\dfrac{3}{7}(2008 - 1999) + 15 \approx 11.1 \Rightarrow \$11.1$ billion. The estimate is about $0.7 billion higher than

the true value of $10.4 billion. The calculation involves interpolation.

87. (a) See Figure 87a.

(b) Using the points (2006, 160) and (2010, 425), $m = \dfrac{425 - 160}{2010 - 2006} = \dfrac{265}{4} = 66.25$. The point slope form of

the line is $f(x) = 66.25(x - 2006) + 160.$

(c) See Figure 87c.

(d) Bankruptcies increased, on average, by 66,250 per year.

(e) $f(2014) = 66.25(2014 - 2006) + 160 = 690$; The calculation involves extrapolation.

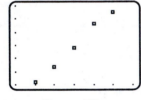

Figure 87a   Figure 87c

89. (a) Using the points (1970, 2000) and (2010, 1590), $m = \dfrac{1590 - 2000}{2010 - 1970} = \dfrac{-410}{40} = -10.25$. Since x

represents the number of years after 1970, we have a y-intercept of 2000, and the function is

$$f(x) = -10.25x + 2000.$$

(b) Hours worked decreased, on average, by 10.25 hours per year.

(c) Since 2014 is 44 years after 1970, we will let $x = 44$ and $f(44) = -10.25(44) + 2000 = 1549$. The result

is about 1549 hours.

91. (a) Graph $Y_1 = X/1024 + 1$ in $[0, 3, 1]$ by $[-2, 2, 1]$ as in Figure 93. The line appears to be horizontal in

this viewing rectangle, however, we know that the graph of the line is not horizontal because its slope

is $\dfrac{1}{1024} \neq 0$.

(b) The resolution of most graphing calculator screens is not good enough to show the slight increase in the

y-values. Since the x-axis is 3 units long, this increase in y-values amounts to only $\dfrac{1}{1024} \times 3 \approx 0.003$

units, which does not show up on the screen.

$[0, 3, 1]$ by $[-2, 2, 1]$

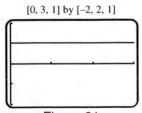

Figure 91

93. (a) From Figure 93a, one can see that the lines do not appear to be perpendicular.

    (b) The lines are graphed in the specified viewing rectangles and shown in Figures 93b-d, respectively. In the windows $[-15, 15, 1]$ by $[-10, 10, 1]$ and $[-3, 3, 1]$ by $[-2, 2, 1]$ the lines appear to be perpendicular.

    (c) The lines appear perpendicular when the distance shown along the x-axis is approximately 1.5 times the distance along the y-axis. For example, in window $[-12, 12, 1]$ by $[-8, 8, 1]$, the lines will appear perpendicular. The distance along the x-axis is 24 while the distance along the y-axis is 16. Notice that $1.5 \times 16 = 24$. This is called a "square window" and can be set automatically on some graphing calculators.

| $[-10, 10, 1]$ by $[-10, 10, 1]$ | $[-15, 15, 1]$ by $[-10, 10, 1]$ | $[-10, 10, 1]$ by $[-3, 3, 1]$ | $[-3, 3, 1]$ by $[-2, 2, 1]$ |

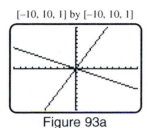
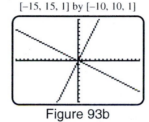
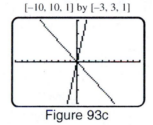
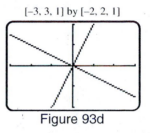

Figure 93a   Figure 93b   Figure 93c   Figure 93d

95. (i)   The slope of the line connecting $(0, 0)$ and $(2, 2)$ is 1. Let $y_1 = x$.

    (ii)   A second line passing through $(0, 0)$ has a slope of -1. Let $y_2 = -x$.

    (iii) A third line passing through $(1, 3)$ has a slope of 1. Let $y_3 = (x - 1) + 3 = x + 2$.

    (iv)   A fourth line passing through $(2, 2)$ has a slope of -1. Let $y_4 = -(x - 2) + 2 = -x + 4$.

97. (i)   The slope of the line connecting $(-4, 0)$ and $(0, 4)$ is 1. Let $y_1 = x + 4$.

    (ii)   A second line passing through $(4, 0)$ and $(0, -4)$ has a slope of 1. Let $y_2 = x - 4$.

    (iii) A third line passing through $(0, -4)$ and $(-4, 0)$ has a slope of -1. Let $y_3 = -x - 4$.

    (iv)   A fourth line passing through $(0, 4)$ and $(4, 0)$ is -1. Let $y_4 = -x + 4$.

99. Enter the x-values into the list L_1 and the y-values into the list L_2. Use the statistical feature of your graphing calculator to find the correlation coefficient r and the regression equation. $r \neq -0.993$; $y \neq -0.789x + 0.526$ See Figure 99.

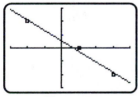

Figure 99

101.(a) Enter the x-values into the list L_1 and the y-values into the list L_2 in the statistical feature of your graphing calculator; the scatterplot of the data indicates that the correlation coefficient will be positive (and very close to 1).

    (b) $y = ax + b$, where $a \approx 3.25$ and $b \approx -2.45$; $r \approx 0.9994$

    (c) $y \approx 3.25(2.4) - 2.45 = 5.35$

103.(a) Enter the x-values into the list L_1 and the y-values into the list L_2 in the statistical feature of your graphing

calculator; the scatterplot of the data indicates that the correlation coefficient will be negative (and very

close to -1).

(b) $y = ax + b$, where $a \approx -3.8857$ and $b \approx 9.3254$; $r \approx -0.9996$

(c) $y \approx -3.8857(2.4) + 9.3254 = -0.00028$. *Due to rounding answers may very slightly.*

105.(a) The data points (50, 990), (650, 9300), (950, 15000) and (1700, 25000) are plotted in Figure 105. The data

appears to have a linear relationship.

(b) Use the linear regression feature on your graphing calculator to find the values of a and b in the equation

$y = ax + b$. In this instance, $a \approx 14.680$ and $b \approx 277.82$.

(c) We must find the x-value when $y = 37,000$. This can be done by solving the equation

$37,000 = 14.680x + 277.82 \Rightarrow 14.680x = 36,722.18 \Rightarrow x \approx 2500$ light years away. One could also

solve the equation graphically to obtain the same approximation.

[−100, 1800, 100] by [−1000, 28000, 1000]

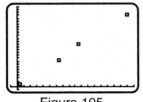

Figure 105

107.(a) Positive. See Figure 107a.

(b) Enter the x-values into the list L_1 and the y-values into the list L_2 in the statistical feature of your graphing

calculator. $f(x) \approx 0.085x + 2.08$

(c) See Figure 107c. The slope indicates the number of miles traveled by passengers per year.

(d) Since 2015 is 45 years after 1970, we let $x = 45$ and $f(45) = 0.085(45) + 2.08 = 5.905$.

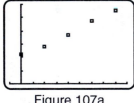

Figure 107a

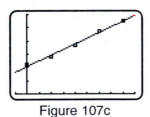

Figure 107c

2.2: Linear Equations

1. $ax + b = 0 \Rightarrow ax = -b \Rightarrow x = \dfrac{-b}{a}$. This shows that the equation $ax + b = 0$ has only one solution.

3. $4 - (5 - 4x) = 4 - 5 + 4x = -1 + 4x = 4x - 1$

5. The zero of f and the x-intercept of the graph of f are equal. The zero of f and the x-intercept of the graph of f

are both found by finding the value of x when $y = 0$.

7. $3x - 1.5 = 7 \Rightarrow 3x - 1.5 - 7 = 0 \Rightarrow 3x - 8.5 = 0$; the equation is linear.

9. $2\sqrt{x} + 2 = 1$; since the equation cannot be written in the form $ax + b = 0$, it is nonlinear.

11. $7x - 5 = 3(x - 8) \Rightarrow 7x - 5 = 3x - 24 \Rightarrow 4x + 19 = 0$; the equation is linear.

13. $2x - 8 = 0 \Rightarrow 2x = 8 \Rightarrow x = 4$ Check: $2(4) - 8 = 0 \Rightarrow 8 - 8 = 0 \Rightarrow 0 = 0$

15. $-5x + 3 = 23 \Rightarrow -5x = 20 \Rightarrow x = -4$ Check: $-5(-4) + 3 = 23 \Rightarrow 20 + 3 = 23 \Rightarrow 23 = 23$

17. $4(z - 8) = z \Rightarrow 4z - 32 = z \Rightarrow 3z = 32 \Rightarrow z = \frac{32}{3}$ Check: $4\left(\frac{32}{3} - 8\right) = \frac{32}{3} \Rightarrow 4\left(\frac{8}{3}\right) = \frac{32}{3} \Rightarrow$

$\frac{32}{32} = \frac{32}{32}$

19. $-5(3 - 4t) = 65 \Rightarrow -15 + 20t = 65 \Rightarrow 20t = 80 \Rightarrow t = 4$ Check: $-5[3 - 4(4)] = 65 \Rightarrow$

$-5(3 - 16) = 65 \Rightarrow -5(-13) = 65 \Rightarrow 65 = 65$

21. $k + 8 = 5k - 4 \Rightarrow -4k = -12 \Rightarrow k = 3$ Check: $3 + 8 = 5(3) - 4 \Rightarrow 11 = 15 - 4 \Rightarrow 11 = 11$

23. $2(1 - 3x) + 1 = 3x \Rightarrow 2 - 6x + 1 = 3x \Rightarrow -6x + 3 = 3x \Rightarrow -9x = -3 \Rightarrow x = \frac{1}{3}$

Check: $2\left[1 - 3\left(\frac{1}{3}\right)\right] + 1 = 3\left(\frac{1}{3}\right) \Rightarrow 2(1 - 1) + 1 = 1 \Rightarrow 0 + 1 = 1 \Rightarrow 1 = 1$

25. $-5(3 - 2x) - (1 - x) = 4(x - 3) \Rightarrow -15 + 10x - 1 + x = 4x - 12 \Rightarrow 11x - 16 = 4x - 12 \Rightarrow$

$7x = 4 \Rightarrow x = \frac{4}{7}$ Check: $-5\left[3 - 2\left(\frac{4}{7}\right)\right] - \left(1 - \frac{4}{7}\right) = 4\left(\frac{4}{7} - 3\right) \Rightarrow$

$-5\left(\frac{13}{7}\right) - \frac{3}{7} = 4\left(-\frac{17}{7}\right) \Rightarrow -\frac{65}{7} - \frac{3}{7} = -\frac{68}{7} \Rightarrow -\frac{68}{7} = -\frac{68}{7}$

27. $-4(5x - 1) = 8 - (x + 2) \Rightarrow -20x + 4 = 8 - x - 2 \Rightarrow -20x + 4 = 6 - x \Rightarrow -19x = 2 \Rightarrow$

$x = -\frac{2}{19}$ Check: $-4\left[5\left(-\frac{2}{19}\right) - 1\right] = 8 - \left(-\frac{2}{19} + 2\right) \Rightarrow -4\left(-\frac{10}{19} - 1\right) = 8 + \frac{2}{19} - 2 \Rightarrow$

$\frac{40}{19} + 4 = 6 + \frac{2}{19} \Rightarrow \frac{116}{19} = \frac{116}{19}$

29. $\frac{2}{7}n + 2 = \frac{4}{7} \Rightarrow \frac{2}{7}n = -\frac{10}{7} \Rightarrow n = -5$ Check: $\frac{2}{7}(-5) + 2 = \frac{4}{7} \Rightarrow \frac{-10}{7} + \frac{14}{7} = \frac{4}{7} \Rightarrow \frac{4}{7} = \frac{4}{7}$

31. $\frac{1}{2}(d - 3) - \frac{2}{3}(2d - 5) = \frac{5}{12} \Rightarrow \frac{1}{2}d - \frac{3}{2} - \frac{4}{3}d + \frac{10}{3} = \frac{5}{12} \Rightarrow -\frac{5}{6}d + \frac{11}{6} = \frac{5}{12} \Rightarrow$

$-\frac{5}{6}d = -\frac{17}{12} \Rightarrow d = \frac{17}{10}$

Check: $\frac{1}{2}\left(\frac{17}{10} - 3\right) - \frac{2}{3}\left[2\left(\frac{17}{10}\right) - 5\right] = \frac{5}{12} \Rightarrow \frac{1}{2}\left(-\frac{13}{10}\right) - \frac{2}{3}\left(\frac{34}{10} - 5\right) = \frac{5}{12} \Rightarrow$

$\frac{1}{2}\left(-\frac{13}{10}\right) - \frac{2}{3}\left(-\frac{16}{10}\right) = \frac{5}{12} \Rightarrow -\frac{13}{20} + \frac{32}{30} = \frac{5}{12} \Rightarrow \frac{25}{60} = \frac{5}{12} \Rightarrow \frac{5}{12} = \frac{5}{12}$

33. $\dfrac{x-5}{3} + \dfrac{3-2x}{2} = \dfrac{5}{4} - 2(1-x) \Rightarrow 4(x-5) + 6(3-2x) = 15 - 24(1-x) \Rightarrow$

$4x - 20 + 18 - 12x = 15 - 24 + 24x \Rightarrow -8x - 2 = -9 + 24x \Rightarrow 7 = 32x \Rightarrow x = \dfrac{7}{32}$

Check: $\dfrac{\frac{7}{32} - 5}{3} + \dfrac{3 - 2\left(\frac{7}{32}\right)}{2} = \dfrac{5}{4} - 2\left(1 - \dfrac{7}{32}\right) \Rightarrow 4\left(\dfrac{7}{32} - 5\right) + 6\left(3 - 2\left(\dfrac{7}{32}\right)\right) \Rightarrow$

$-\dfrac{153}{8} + \dfrac{123}{8} = 15 - \dfrac{75}{4} \Rightarrow -\dfrac{15}{4} = -\dfrac{15}{4}$

35. $0.1z - 0.05 = -0.07z \Rightarrow 0.17z = 0.05 \Rightarrow z = \dfrac{0.05}{0.17} \Rightarrow z = \dfrac{5}{17}$ Check:

$0.1\left(\dfrac{5}{17}\right) - 0.05 = -0.07\left(\dfrac{5}{17}\right) \Rightarrow \dfrac{5}{170} - \dfrac{5}{100} = -\dfrac{35}{1700} \Rightarrow \dfrac{50}{1700} - \dfrac{85}{1700} = -\dfrac{35}{1700} \Rightarrow$

$-\dfrac{35}{1700} = -\dfrac{35}{1700}$

37. $0.15t + 0.85(100 - t) = 0.45(100) \Rightarrow 0.15t + 85 - 0.85t = 45 \Rightarrow -0.7t = -40 \Rightarrow t = \dfrac{40}{0.7} \Rightarrow$

$t = \dfrac{400}{7}$ Check: $0.15\left(\dfrac{400}{7}\right) + 0.85\left(100 - \dfrac{400}{7}\right) = 0.45(100) \Rightarrow \dfrac{6000}{700} + 85 - \dfrac{34{,}000}{700} = 45 \Rightarrow$

$\dfrac{60}{7} + 85 - \dfrac{340}{7} = 45 \Rightarrow 85 - \dfrac{280}{7} = 45 \Rightarrow 85 - 40 = 45 \Rightarrow 45 = 45$

39. (a) $5x - 1 = 5x + 4 \Rightarrow -1 = 4 \Rightarrow$ there is no solution.

   (b) Since no x-value satisfies the equation, it is contradiction.

41. (a) $3(x-1) = 5 \Rightarrow 3x - 3 = 5 \Rightarrow 3x = 8 \Rightarrow x = \dfrac{8}{3}$

   (b) Since one x-value is a solution and other x-values are not, the equation is conditional.

43. (a) $0.5(x-2) + 5 = 0.5x + 4 \Rightarrow 0.5x - 1 + 5 = 0.5x + 4 \Rightarrow 0.5x + 4 = 0.5x + 4 \Rightarrow$ every x-value
   satisfies this equation.

   (b) Since every x-value satisfies the equation, it is an identity.

45. (a) $\dfrac{t+1}{2} = \dfrac{3t-2}{6} \Rightarrow 6\left(\dfrac{t+1}{2} = \dfrac{3t-2}{6}\right) \Rightarrow 3t + 3 = 3t - 2 \Rightarrow 3 = -2 \Rightarrow$ there is no solution.

   (b) Since no x-value satisfies the equation, it is contradiction.

47. (a) $\dfrac{1-2x}{4} = \dfrac{3x-1.5}{-6} \Rightarrow -6(1-2x) = 4(3x-1.5) \Rightarrow -6 + 12x = 12x - 6 \Rightarrow 0 = 0 \Rightarrow$ every
   x-value satisfies this equation.

   (b) Since every x-value satisfies the equation, it is an identity.

49. In the graph, the lines intersect at $(3, -1)$. The solution is the x-value, 3.

51. (a) From the graph, when $f(x)$ or $y = -1, x = 4$; the solution is the x-value, 4.

   (b) From the graph, when $f(x)$ or $y = 0, x = 2$; the solution is the x-value, 2.

   (c) From the graph, when $f(x)$ or $y = 2, x = -2$; the solution is the x-value, -2.

53. Graph $Y_1 = X + 4$ and $Y_2 = 1 - 2X$. See Figure 53. The lines intersect at $x = -1$.

   $x + 4 = 1 - 2x \Rightarrow 3 = -3x \Rightarrow x = -1$

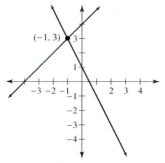

Figure 53

55. Graph $Y_1 = -X + 4$ and $Y_2 = 3X$. See Figure 55. The lines intersect at $x = 1$.

$-x + 4 = 3x \Rightarrow 4 = 4x \Rightarrow x = 1$

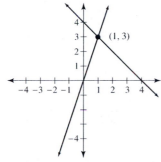   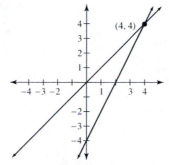

Figure 55   Figure 57

57. Graph $Y_1 = 2(X - 1) - 2$ and $Y_2 = X$. See Figure 57. The lines intersect at $x = 4$.

$2(x - 1) - 2 = x \Rightarrow 2x - 2 - 2 = x \Rightarrow 2x - 4 = x \Rightarrow -4 = -x \Rightarrow x = 4$

59. Graph $Y_1 = 5X - 1.5$ and $Y_2 = 5$. Their graphs intersect at $(1.3, 5)$. The solution is 1.3. See Figure 59.

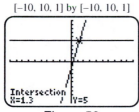

Figure 59

61. Graph $Y_1 = 3X - 1.7$ and $Y_2 = 1 - X$. Their graphs intersect at $(0.675, 0.325)$. The solution is 0.675. See Figure 61.

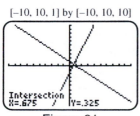

Figure 61

63. Graph $Y_1 = 3.1(X - 5)$ and $Y_2 = X/5 - 5$. Their graphs intersect near $(3.621, -4.276)$. The solution is approximately 3.621. See Figure 63.

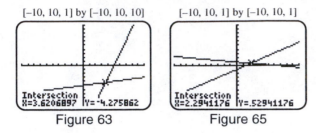

[−10, 10, 1] by [−10, 10, 10]   [−10, 10, 1] by [−10, 10, 1]

Figure 63   Figure 65

65. Graph $Y_1 = (6 - X)/7$ and $Y_2 = (2X - 3)/3$. Their graphs intersect near $(2.294, 0.529)$. The solution is approximately 2.294. See Figure 65.

67. One way to solve this equation is to table $Y_1 = 2X - 7$ and determine the *x*-value where $Y_1 = -1$. See Figure 67. This occurs when $x = 3$, so the solution is 3.

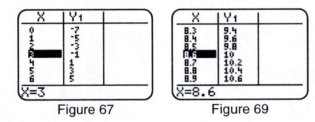

Figure 67   Figure 69

69. Table $Y_1 = 2X - 7.2$ and determine the *x*-value where $Y_1 = 10$. See Figure 69. This occurs when $x = 8.6$, so the solution is 8.6.

71. Table $Y_1 = \sqrt{(2)}(4X - 1) + \pi X$ and determine the *x*-value where $Y_1 = 0$. See Figure 71. This occurs when $x \neq 0.2$,   so the solution is 0.2.

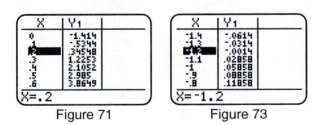

Figure 71   Figure 73

73. Table $Y_1 = 0.5 - 0.1(\sqrt{(2)} - 3X)$ and determine the *x*-value where $Y_1 = 0$. See Figure 73. This occurs when $x \neq -1.2$,   so the solution is −1.2.

75. (a) $5 - (x + 1) = 3 \Rightarrow 5 - x - 1 = 3 \Rightarrow 4 - x = 3 \Rightarrow x = 1$

    (b) Using the intersection of graphs method, graph $Y_1 = 5 - (X + 1)$ and $Y_2 = 3$. Their point of intersection is shown in Figure 75b as $(1, 3)$. The solution is the *x*-value, 1.

    (c) Table $Y_1 = 5 - (X + 1)$ and $Y_2 = 3$. Figure 75c shows a table where $Y_1 = Y_2$ at $x = 1$.

[−10, 10, 1] by [−10, 10, 1]

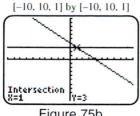

    Figure 75b   Figure 75c

77. (a) $x - 3 = 2x + 1 \Rightarrow -x = 4 \Rightarrow x = -4$

(b) Using the intersection-of-graphs method, graph $Y_1 = X - 3$ and $Y_2 = 2X + 1$. Their point of intersection is shown in Figure 77b as $(-4, -7)$. The solution is the x-value, -4.

(c) Table $Y_1 = X - 3$ and $Y_2 = 2X + 1$ starting at $x = -7$, incrementing by 1. Figure 77c shows a table where $Y_1 = Y_2$ at $x = -4$.

[−12, 8, 1] by [−12, 8, 1]

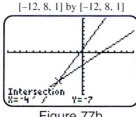 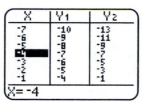

    Figure 77b   Figure 77c

79. (a) $\sqrt{3}(2 - \pi x) + x = 0 \Rightarrow 2\sqrt{3} - \sqrt{3}\pi x + x = 0 \Rightarrow -\sqrt{3}\pi x + x = -2\sqrt{3} \Rightarrow$

$x(-\sqrt{3}\pi + 1) = -2\sqrt{3} \Rightarrow x = \dfrac{-2\sqrt{3}}{(-\sqrt{3}\pi + 1)} \approx 0.8.$

(b) Using the intersection of graphs method, graph $Y_1 = \sqrt{(3)}(2 - \pi X) + X$ and $Y_2 = 0$. Their point of intersection is shown in Figure 79b as approximately $(0.8, 0)$. The solution is the x-value, 0.8.

(c) Table $Y_1 = \sqrt{(3)}(2 - \pi X) + X$ and $Y_2 = 0$. Figure 79c shows a table where $Y_1 = Y_2$ at $x \neq 0.8$.

[−10, 10, 1] by [−10, 10, 1]

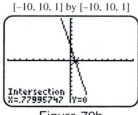 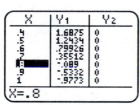

    Figure 79b   Figure 79c

81. (a) $6x - 8 = -7x + 18 \Rightarrow 13x = 26 \Rightarrow x = 2$

(b) Using the intersection-of-graphs method, graph $Y_1 = 6X - 8$ and $Y_2 = -7X + 18$. Their point of intersection is shown in Figure 81b as $(2, 4)$. The solution is the x-value, 2.

(c) Table $Y_1 = 6X - 8$ and $Y_2 = -7X + 18$ starting at $x = 0$, incrementing by 1. Figure 81c shows a table where $Y_1 = Y_2$ at $x = 2$.

[–10, 10, 1] by [–10, 10, 1]

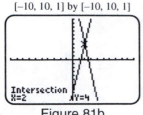

Figure 81b   Figure 81c

83. $A = LW \Rightarrow W = \dfrac{A}{L}$

85. $P = 2L + 2W \Rightarrow P - 2W = 2L \Rightarrow L = \dfrac{P - 2W}{2} \Rightarrow L = \dfrac{1}{2}P - W$

87. $3x + 2y = 8 \Rightarrow 2y = 8 - 3x \Rightarrow y = 4 - \dfrac{3}{2}x$

89. $y = 3(x - 2) + x \Rightarrow y = 3x - 6 + x \Rightarrow y = 4x - 6 \Rightarrow 4x = y + 6 \Rightarrow x = \dfrac{1}{4}y + \dfrac{3}{2}$

91. (a) $2x + y = 8 \Rightarrow y = 8 - 2x \Rightarrow y = -2x + 8$

   (b) $f(x) = -2x + 8$

93. (a) $2x - 4y = -1 \Rightarrow -4y = -1 - 2x \Rightarrow y = \dfrac{1}{4} + \dfrac{1}{2}x$

   (b) $f(x) = \dfrac{1}{4} + \dfrac{1}{2}x$

95. (a) $-9x + 8y = 9 \Rightarrow 8y = 9 + 9x \Rightarrow y = \dfrac{9}{8} + \dfrac{9}{8}x$

   (b) $f(x) = \dfrac{9}{8} + \dfrac{9}{8}x$

97. $f(x) = 19,000$ and $f(x) = 1000(x - 1980) + 10,000 \Rightarrow 1000(x - 1980) + 10,000 = 19,000 \Rightarrow$

   $1000(x - 1980) = 9000 \Rightarrow x - 1980 = 9 \Rightarrow x = 1989$. The per capita income was $19,000 in 1989.

99. Using the intersection of graphs method, graph $Y_1 = 51.6(X - 1985) + 9.1$ and

   $Y_2 = -31.9(X - 1985) + 167.7$. Their approximate point of intersection is shown in Figure 99 as (1987, 107).

   In approximately 1987 the sales of LP records and compact discs were equal.

101.(a) The percentage against were decreasing at 1.6% per year; the percentage for were increasing at 1.5% per year.

   (b) $-1.6x + 3265 = 1.5x - 2968 \Rightarrow 3265 = 3.1x - 2968 \Rightarrow 6233 = 3.1x \Rightarrow x \approx 2011$

103. To calculate the sale price subtract 25% of the regular price from the regular price.

   $f(x) = x - 0.25x \Rightarrow f(x) = 0.75x$.   An item which normally costs $56.24 will be on sale for

   $f(56.24) = 0.75(56.24) = \$42.18$.

105.(a) The number of skin cancer cases is given by $0.048x$.

   (b) $76,000 = 0.048x \Rightarrow \dfrac{76,000}{0.048} = x \Rightarrow x \approx 1,583,000$

107. (a) It would take a little less time than the faster gardener, who can rake the lawn alone in 3 hours. It would

       take both gardeners about 2 hours working together. *Answers may vary.*

(b) Let $x =$ time to rake the lawn working together. In 1 hour thge first gardener can rake $\frac{1}{3}$ of the lawn, whereas the second gardener can rake $\frac{1}{5}$ of the lawn; in x hours both gardeners working together can rake $\frac{x}{3} + \frac{x}{5}$ of the lawn; $\frac{x}{3} + \frac{x}{5} = 1 \Rightarrow 5x + 3x = 15 \Rightarrow 8x = 15 \Rightarrow x = \frac{15}{8} = 1.875$ hours.

109. Let $t =$ time spent traveling at 55 mph and $6 - t =$ time spent traveling at 70 mph. Using $d = rt$, we get $d = 55t + 70(6 - t) \Rightarrow 372 = 55t + 420 - 70t \Rightarrow -48 = -15t \Rightarrow t = 3.2$ and $6 - t = 2.8$; the car traveled 3.2 hours at 55 mph and 2.8 hours at 70 mph.

111. Let $t =$ time traveled by car at 55 mph and $t + \frac{1}{2} =$ time traveled by runner at 10 mph; since $d = rt$ and the distance is the same for both runner and driver, we get $55t = 10\left(t + \frac{1}{2}\right) \Rightarrow 55t = 10t + 5 \Rightarrow 45t = 5 \Rightarrow$ $t = \frac{1}{9}$; it takes the driver $\frac{1}{9}$ hour or $6\frac{2}{3}$ minutes to catch the runner.

113. This problem can be solved using similar triangles or a proportion. Let x be the height of the tree; then, $\frac{5}{4} = \frac{x}{33} \Rightarrow x = \frac{5 \times 33}{4} = 41.25$. The height of the tree is 41.25 feet.

115. Use similar triangles to find the radius of the cone when the water is 7 feet deep: $\frac{r}{3.5} = \frac{7}{11} \Rightarrow r \approx \frac{49}{22}$ ft. Use $V = \frac{1}{3}\pi r^2 h$ to find the volume of the water in the cone at $h = 7$ ft.: $V = \frac{1}{3}\pi\left(\frac{49}{22}\right)^2 (7) \approx 36.4$ ft^3.

117. Let $x =$ amount of pure water to be added and $x + 5 =$ final amount of the 15% solution. Since pure water is 0% sulfuric acid, we get $0\%x + 40\%(5) = 15\%(x + 5) \Rightarrow 0.40(5) = 0.15(x + 5) \Rightarrow$ $40(5) = 15(x + 5) \Rightarrow 200 = 15x + 75 \Rightarrow 15x = 125 \Rightarrow x = \frac{125}{15} \approx 8.333$; about 8.33 liters of pure water should be added.

119. $P = 2w + 2l \Rightarrow 180 = 2w + 2(w + 18) \Rightarrow 180 = 4w + 36 \Rightarrow 4w = 144 \Rightarrow w = 36$ and $w + 18 = 54$; the window is 36 inches by 54 inches.

121.(a) The linear function S must fit the coordinates (2011, 192) and (2014, 249).
$m = \frac{249 - 192}{2014 - 2011} = \frac{57}{3} = 19; S = 19(x - 2014) + 249 \Rightarrow S = 19x - 38,017$

(b) The slope shows that the sales increased, on average, by $19 billion per year.

(c) $230 = 19x - 38,017 \Rightarrow 38,247 = 19x \Rightarrow x = 2013$

123. $C = \frac{5}{9}(F - 32)$ and $F = C \Rightarrow F = \frac{5}{9}(F - 32) \Rightarrow F = \frac{5}{9}F - \frac{160}{9} \Rightarrow \frac{4}{9}F = -\frac{160}{9} \Rightarrow F = -40$;
$-40°$ F is equivalent to $-40°$ C.

125. (a) It is reasonable to expect that f is linear because if the number of gallons of gas doubles so should the amount of oil. Five gallons of gasoline requires five times the oil that one gallon of gasoline would. The increase in oil is always equal to 0.16 pint for each additional gallon of gasoline. Oil is mixed at a constant rate, so a linear function describes this amount.

(b) $f(3) = 0.16(3) = 0.48$; 0.48 pint of oil should be added to 3 gallons of gasoline to get the correct mixture.

(c) $0.16x = 2 \Rightarrow x = 12.5$; 12.5 gallons of gasoline should be mixed with 2 pints of oil.

127. Linear regression gives the model:

$y = 0.36x - 0.21.$ $y = 2.99 \Rightarrow 2.99 = 0.36x - 0.21 \Rightarrow 3.2 = 0.36x \Rightarrow x \approx 8.89$

Extended and Discovery for Section 2.2

1. Let x = number of fish in the sample and y = number of tagged fish. Then $y = kx$, where k represents the proportion of fish tagged. From the data point (94, 13), get $13 = k(94) \Rightarrow k \approx 0.138298$. Letting the sample represent the entire number of fish, we get $y = 0.138298x \Rightarrow 85 = 0.138298x \Rightarrow x \approx 615$.

Checking Basic Concepts for Sections 2.1 and 2.2

1. $f(x) = 4 - 2x$.   See Figure 1. Slope: –2; y-intercept: 4; x-intercept: 2

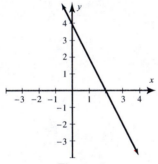

Figure 1

3. $5x + 2 = 2x - 3 \Rightarrow 3x + 2 = -3 \Rightarrow 3x = -5 \Rightarrow x = -\dfrac{5}{3}$

Check: $5\left(-\dfrac{5}{3}\right) + 2 = 2\left(-\dfrac{5}{3}\right) - 3 \Rightarrow -\dfrac{25}{3} + 2 = -\dfrac{10}{3} - 3 \Rightarrow -\dfrac{19}{3} = \dfrac{19}{3}$

5. Let $-3x + 2y = -18$.

x-intercept: Substitute $y = 0$ and solve for x. $-3x + 2(0) = -18 \Rightarrow -3x = -18 \Rightarrow x = 6$; x-intercept: 6

y-intercept: Substitute $x = 0$ and solve for y. $-3(0) + 2y = -18 \Rightarrow 2y = -18 \Rightarrow y = -9$; x-intercept: -9

Extended and Discovery Exercises for Section 2.3

1. (a) Yes; since multiplication distributes over addition, doubling the lengths gives double the sum of the lengths.

   (b) No; If the length and width are doubled, the product of the length and width is multiplied by 4.

3. (a) $(100 \text{ ft}^2)(140 \, \mu g/\text{ft}^2) = 14{,}000 \, \mu g$; $f(x) = 14{,}000x$.

   (b) $(800 \text{ ft}^3)(33 \mu g/\text{ft}^3) = 26{,}400 \, \mu g$; $f(x) = 14{,}000x \Rightarrow 26{,}400 = 14{,}000x \Rightarrow x \approx 1.9$; it takes about 1.9 hours for the concentrations to reach $33 \, \mu g/\text{ft}^3$.

2.3: Linear Inequalities

1. $(-\infty, 2)$

3. $[-1, \infty)$

5. $[1, 8)$

7. $(-\infty, 1]$

9. $2x + 6 \geq 10 \Rightarrow 2x \geq 4 \Rightarrow x \geq 2$; $[2, \infty)$; set-builder notation the interval is $\{x \mid x \geq 2\}$.

11. $-2(x - 10) + 1 > 0 \Rightarrow -2x + 21 > 0 \Rightarrow -2x > -21 \Rightarrow x < 10.5$; $(-\infty, 10.5)$; set-builder notation the interval is $\{x \mid x < 10.5\}$.

13. $\dfrac{t + 2}{3} \geq 5 \Rightarrow t + 2 \geq 15 \Rightarrow t \geq 13$; $[13, \infty)$ set-builder notation the interval is $\{t \mid t \geq 13\}$.

15. $4x - 1 < \dfrac{3 - x}{-3} = 7 \Rightarrow -12x + 3 > 3 - x \Rightarrow -11x > 0 \Rightarrow x < 0$; $(-\infty, 0)$; set-builder notation the interval is $\{x \mid x < 0\}$.

17. $-3(z - 4) \geq 2(1 - 2z) \Rightarrow -3z + 12 \geq 2 - 4z \Rightarrow z \geq -10$; $[-10, \infty)$; set-builder notation the interval is $\{z \mid z \geq -10\}$.

19. $\dfrac{1 - x}{4} < \dfrac{2x - 2}{3} \Rightarrow 3(1 - x) < 4(2x - 2) \Rightarrow 3 - 3x < 8x - 8 \Rightarrow -11x < -11 \Rightarrow x > 1$; $(1, \infty)$; set-builder notation the interval is $\{x \mid x > 1\}$.

21. $2x - 3 > \dfrac{1}{2}(x + 1) \Rightarrow 2x - 3 > \dfrac{1}{2}x + \dfrac{1}{2} \Rightarrow \dfrac{3}{2}x > \dfrac{7}{2} \Rightarrow x > \dfrac{7}{3}$; $\left(\dfrac{7}{3}, \infty\right)$; set-builder notation the interval is $\left\{x \mid x > \dfrac{7}{3}\right\}$.

23. $5 < 4t - 1 \leq 11 \Rightarrow 6 < 4t \leq 12 \Rightarrow \dfrac{3}{2} < t \leq 3$; $\left(\dfrac{3}{2}, 3\right]$ set-builder notation the interval is $\left\{t \mid \dfrac{3}{2} < t \leq 3\right\}$.

25. $3 \leq 4 - x \leq 20 \Rightarrow -1 \leq -x \leq 16 \Rightarrow 1 \geq x \geq -16$; $[-16, 1]$; set-builder notation the interval is $\{x \mid -16 \leq x \leq 1\}$.

27. $-7 \le \dfrac{1-4x}{7} < 12 \Rightarrow -49 \le 1-4x < 84 \Rightarrow -50 \le -4x < 83 \Rightarrow 12.5 \ge x > -20.75$;

    $(-20.75, 12.5]$; set-builder notation the interval is $\{x \mid -20.75 < x \le 12.5\}$.

29. $5 > 2(x+4)-5 > -5 \Rightarrow 5 > 2x+8-5 > -5 \Rightarrow 2 > 2x > -8 \Rightarrow 1 > x > -4$; $(-4, 1)$;

    set-builder notation the interval is $\{x \mid -4 < x < 1\}$.

31. $3 \le \dfrac{1}{2}x + \dfrac{3}{4} \le 6 \Rightarrow 12 \le 2x+3 \le 24 \Rightarrow 9 \le 2x \le 21 \Rightarrow \dfrac{9}{2} \le x \le \dfrac{21}{2}$; $\left[\dfrac{9}{2}, \dfrac{21}{2}\right]$; set-builder

    notation the interval is $\left\{x \mid \dfrac{9}{2} \le x \le \dfrac{21}{2}\right\}$.

33. $5x - 2(x+3) \ge 4 - 3x \Rightarrow 5x - 2x - 6 \ge 4 - 3x \Rightarrow 6x \ge 10 \Rightarrow x \ge \dfrac{5}{3}$; $\left[\dfrac{5}{3}, \infty\right)$; set-builder

    notation the interval is $\left\{x \mid x \ge \dfrac{5}{3}\right\}$.

35. $\dfrac{1}{2} \le \dfrac{1-2t}{3} < \dfrac{2}{3} \Rightarrow \dfrac{3}{2} \le 1 - 2t < 2 \Rightarrow \dfrac{1}{2} \le -2t < 1 \Rightarrow -\dfrac{1}{4} \ge t > -\dfrac{1}{2} \Rightarrow -\dfrac{1}{2} < t \le -\dfrac{1}{4}$;

    $\left(-\dfrac{1}{2}, -\dfrac{1}{4}\right]$; set-builder notation the interval is $\left\{t \mid -\dfrac{1}{2} < t \le -\dfrac{1}{4}\right\}$.

37. $\dfrac{1}{2}z + \dfrac{2}{3}(3-z) - \dfrac{5}{4}z \ge \dfrac{3}{4}(z-2) + z \Rightarrow \dfrac{1}{2}z + 2 - \dfrac{2}{3}z - \dfrac{5}{4}z \ge \dfrac{3}{4}z - \dfrac{3}{2} + z \Rightarrow$

    $-\dfrac{17}{12}z + 2 \ge \dfrac{7}{4}z - \dfrac{3}{2} \Rightarrow -\dfrac{38}{12}z \ge -\dfrac{7}{2} \Rightarrow z \le \dfrac{21}{19}$; $\left(-\infty, \dfrac{21}{19}\right]$; set-builder notation the interval is

    $\left\{z \mid z \le \dfrac{21}{19}\right\}$.

39. Graph $y_1 = x + 2$ and $y_2 = 2x$. See Figure 39. $y_1 \ge y_2$ when the graph of y_1 is above the graph of y_2, which

    is left of the intersection point $(2, 4)$ and includes point $(2, 4) \Rightarrow \{x \mid x \le 2\}$.

$[-10, 10, 1]$ by $[-10, 10, 1]$ $[-10, 10, 1]$ by $[-10, 10, 1]$

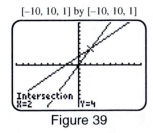   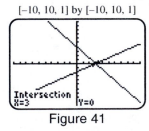

Figure 39   Figure 41

41. Graph $y_1 = \dfrac{2}{3}x - 2$ and $y_2 = -\dfrac{4}{3}x + 4$. See Figure 41. $y_1 > y_2$ when the graph of y_1 is above the graph of y_2,

    which is right of the intersection point $(3, 0)$ and does not include point $(3, 0) \Rightarrow \{x \mid x > 3\}$.

43. Graph $y_1 = -1$, $y_2 = 2x - 1$, and $y_3 = 3$. See Figure 43. $y_1 \le y_2 \le y_3$ when the graph of y_2 is in between the

    graphs of y_1 and y_3, which is in between the intersection points $(0, -1)$ and $(2, 3)$ and it does include each

    point $\Rightarrow \{x \mid 0 \le x \le 2\}$.

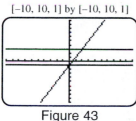

[-10, 10, 1] by [-10, 10, 1]

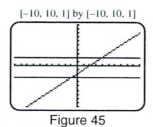

[-10, 10, 1] by [-10, 10, 1]

Figure 43   Figure 45

45. Graph $y_1 = -3$, $y_2 = x - 2$, and $y_3 = 2$. See Figure 45. $y_1 < y_2 \le y_3$ when the graph of y_2 is in between the

graphs of y_1 and y_3, which is in between the intersection points $(-1, -3)$ and $(4, 2)$ and it does not include the

point $(-1, -3)$ but does include $(4, 2) \Rightarrow \{x \mid -1 < x \le 4\}$.

47. (a) $y = \dfrac{3}{2}x - 3$, then $ax + b = 0$ gives us $\dfrac{3}{2}x - 3 = 0 \Rightarrow \dfrac{3}{2}x = 3 \Rightarrow x = 2$

(b) $ax + b < 0$ gives us $\dfrac{3}{2}x - 3 < 0 \Rightarrow x < 2 \Rightarrow (-\infty, 2)$ or in set builder notation, $\{x \mid x < 2\}$.

(c) $ax + b \ge 0$ gives us $\dfrac{3}{2}x - 3 \ge 0 \Rightarrow x \ge 2 \Rightarrow [2, \infty)$ or in set builder notation, $\{x \mid x \ge 2\}$.

49. (a) $y = -x - 2$, then $ax + b = 0$ gives us $-x - 2 = 0 \Rightarrow -x = 2 \Rightarrow x = -2$

(b) $ax + b < 0$ gives us $-x - 2 < 0 \Rightarrow x > -2 \Rightarrow (-2, \infty)$ or in set builder notation, $\{x \mid x > -2\}$.

(c) $ax + b \ge 0$ gives us $-x - 2 \ge 0 \Rightarrow x \le -2 \Rightarrow (-\infty, -2]$ or in set builder notation, $\{x \mid x \le -2\}$.

51. $x - 3 \le \dfrac{1}{2}x - 2 \Rightarrow x - 3 - \dfrac{1}{2}x + 2 \le 0 \Rightarrow \dfrac{1}{2}x - 1 \le 0$. Figure 51 shows the graph of $y_1 = \dfrac{1}{2}x - 1$.

The solution set for $y_1 \le 0$ occurs when the graph is on or below the x-axis, or when $x \le 2$. The solution set

is $(-\infty, 2]$. Solving symbolically, $x - 3 \le \dfrac{1}{2}x - 2 \Rightarrow \dfrac{1}{2}x \le 1 \Rightarrow x \le 2 \Rightarrow$

$(-\infty, 2]$. In set-builder notation the interval is $\{x \mid x \le 2\}$.

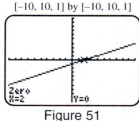

[-10, 10, 1] by [-10, 10, 1]

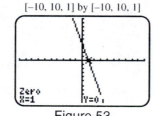

[-10, 10, 1] by [-10, 10, 1]

Figure 51   Figure 53

53. $2 - x < 3x - 2 \Rightarrow 2 - x - 2x + 2 < 0 \Rightarrow -4x + 4 < 0$.   Figure 53 shows the graph of $y_1 = -4x + 4$.

The solution set for $y_1 < 0$ occurs when the graph is below the x-axis, or when $x > 1$.   The solution set is

$(1, \infty)$. Solving symbolically, $2 - x < 3x - 2 \Rightarrow -4x < -4 \Rightarrow x > 1 \Rightarrow (1, \infty)$. In set-builder notation

the interval is $\{x \mid x > 1\}$.

55. Graph $Y_1 = 5X - 4$ and $Y_2 = 10$ The graphs intersect at the point $(2.8, 10)$. The graph of Y_1 is above

the graph of Y_2 for x-values to the right of this intersection point or where $x > 2.8$, $\{x \mid x > 2.8\}$.

See Figure 55.

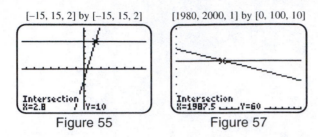

$[-15, 15, 2]$ by $[-15, 15, 2]$ $[1980, 2000, 1]$ by $[0, 100, 10]$

Figure 55   Figure 57

57. Graph $Y_1 = -2(X - 1990) + 55$ and $Y_2 = 60$ The graphs intersect at the point $(1987.5, 60)$. The graph

of Y_1 is above the graph of Y_2 for x-values to the left of this intersection point, so $y_1 \geq y_2$ when $x \leq 1987.5$,

$\{x | x \leq 1987.5\}$. See Figure 57.

59. Graph $Y_1 = \sqrt{(5)}(X - 1.2) - \sqrt{(3)}X$ and $Y_2 = 5(X + 1.1)$ The graphs intersect near the point

$(-1.820, -3.601)$. The graph of Y_1 is below the graph of Y_2 for x-values to the right of this intersection

point or when $x > k$, where $k \approx -1.82$, $\{x | x > -1.82\}$. See Figure 59.

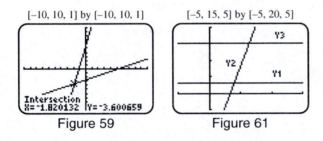

$[-10, 10, 1]$ by $[-10, 10, 1]$ $[-5, 15, 5]$ by $[-5, 20, 5]$

Figure 59   Figure 61

61. Graph $Y_1 = 3$, $Y_2 = 5X - 17$ and $Y_3 = 15$ as shown in Figure 61. The graphs intersect at the points $(4, 3)$

and $(6.4, 15)$. The solutions to $Y_1 \leq Y_2 < Y_3$ are the x-values between 4 and 6.4, including $4 \Rightarrow [4, 6.4)$. In

set-builder notation the interval is $\{x | 4 \leq x < 6.4\}$.

63. Graph $Y_1 = 1.5$, $Y_2 = 9.1 - 0.5X$ and $Y_3 = 6.8$ as shown in Figure 63. The graphs intersect at the points

$(4.6, 6.8)$ and $(15.2, 1.5)$. The solutions to $Y_1 \leq Y_2 \leq Y_3$ are the x-values between 4.6 and 15.2 (inclusive) or

$4.6 \leq x \leq 15.2 \Rightarrow [4.6, 15.2]$. In set builder notation the interval is $\{x | 4.6 \leq x \leq 15.2\}$.

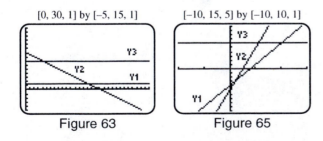

$[0, 30, 1]$ by $[-5, 15, 1]$ $[-10, 15, 5]$ by $[-10, 10, 1]$

Figure 63   Figure 65

65. Graph $Y_1 = X - 4$, $Y_2 = 2X - 5$ and $Y_3 = 6$ as shown in Figure 65. The graph of y_2 intersects the graphs of

y_1 and y_3 at $(1, -3)$ and $(5.5, 6)$. The solutions to $Y_1 < Y_2 < Y_3$ are the x-values between 1 and 5.5 or

$1 < x < 5.5 \Rightarrow (1, 5.5)$. In set-builder notation the interval is $\{x | 1 < x < 5.5\}$.

67. (a) The graphs intersect at the point (8, 7). Therefore, $g(x) = f(x)$ is satisfied when $x = 8$. The solution is 8.

    (b) $g(x) > f(x)$ whenever the y-values on the graph of g are above the y-values on the graph of f. This occurs to the left of the point of intersection. Therefore the x-values that satisfy this inequality are $x < 8$. In set-builder notation the interval is $\{x \mid x<8\}$.

69. From the table,

    $Y_1 = 0$ when $x = 4$. $Y_1 > 0$ when $x < 4 \Rightarrow \{x \mid x < 4\}$; $Y_1 \leq 0$ when $x \geq 4 \Rightarrow \{x \mid x \geq 4\}$.

71. Let $Y_1 = -4X - 6$. From the table shown in Figure 71, $Y_1 = 0$ when $x = -1.5$ or $-\dfrac{3}{2}$. $Y_1 > 0$ when $x < -\dfrac{3}{2}$. In set-builder notation the interval is $\left\{x \mid x<-\dfrac{3}{2}\right\}$.

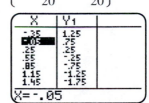

Figure 71   Figure 73

73. Let $Y_1 = 3X - 2$. From the table shown in Figure 73, Y_1 is between 10 and 4 (inclusive) for x-values between 1 and 4 (inclusive) $\Rightarrow [1, 4]$. In set-builder notation the interval is $\{x \mid 1 \leq x \leq 4\}$.

75. Let $Y_1 = (2 - 5X)/3$. From the table shown in Figure 75, Y_1 is between -0.75 and 0.75 for x-values between -0.05 and 0.85 and $Y_1 = 0.75$ when $x = -0.05 \Rightarrow \left[-\dfrac{1}{20}, \dfrac{17}{20}\right)$. In set-builder notation the interval is $\left\{x \mid -\dfrac{1}{20} \leq x<\dfrac{17}{20}\right\}$.

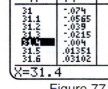

Figure 75   Figure 77

77. Let $Y_1 = (\sqrt{11} - \pi)X - 5.5$. From the table shown in Figure 77, $Y_1 \approx 0$ when $x = 31.4$. $Y_1 \leq 0$ when $x \leq 31.4$; $(-\infty, 31.4]$. In set-builder notation the interval is $\{x \mid x \leq 31.4\}$.

79. Symbolically: $2x - 8 > 5 \Rightarrow 2x > 13 \Rightarrow x > \dfrac{13}{2}$. The solution set is $\left(\dfrac{13}{2}, \infty\right)$. In set-builder notation the interval is $\left\{x \mid x>\dfrac{13}{2}\right\}$.

81. Graphically: Let $Y_1 = \pi X - 5.12$ and $Y_2 = \sqrt{2}X - 5.7(X - 1.1)$ Graph Y_1 and Y_2 as shown in Figure 81. The graphs intersect near $(1.534, -0.302)$. The graph of Y_1 is below Y_2 for $x<1.534$, so $Y_1 \leq Y_2$ when $x \leq 1.534$ The solution set is $(-\infty, 1.534]$.

[−10, 10, 1] by [−10, 10, 1]   [0, 3, 1] by [0, 70, 10]

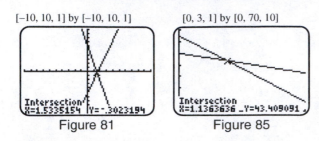

Figure 81   Figure 85

83. (a) Car A is traveling faster since it passes Car B. Its graph has the greater slope.

(b) The cars are the same distance from St. Louis when their graphs intersect. This point of intersection occurs at (2.5, 225). The cars are both 225 miles from St. Louis after 2.5 hours.

(c) Car B is ahead of Car A when $0 \le x < 2.5$.

85. (a) Since the air temperature cools at $19°F$ for 1 mile increase in altitude we have a slope of -19. The ground level air temperature is $65°F$, and we will have the function $T(x) = 65 - 19x$.

(b) Since the dew point decreases by $5.8°F$ for 1 mile increase in altitude we have a slope of -5.8. The ground level dew point is $50°F$, and we will have the function $D(x) = 50 - 5.8x$.

(c) $65 - 19x > 50 - 5.8x \Rightarrow 15 > 13.2x \Rightarrow x < 1.14$. The result is below approximately 1.14 miles.

(d) See Figure 85d.

87. (a) Revenue increased, on average, by $0.86 billion per year.

(b) $0.86x + 1.2 > 3 \Rightarrow 0.86x > 1.8 \Rightarrow x > 2.09$, the revenue is expected to be more than $3 billion from 2012 to 2015.

89. (a) The linear function U must fit the following points (0, 100) and (4, 1000). Using
$m = \dfrac{1000 - 100}{4 - 0} = \dfrac{900}{4} = 225$ and the point (0, 100) we have $U(x) = 225x + 100$.

(b) $225x + 100 \ge 550 \Rightarrow 225x \ge 450 \Rightarrow x \ge 2$, since x represents the number of years after 2008, the result is 2010 or later.

91. (a) The linear function V must fit the following points (2006, 33) and (2011, 71). Using
$m = \dfrac{71 - 33}{2011 - 2006} = \dfrac{38}{5} = 7.6$ and the point (2011, 71) we have $71 = 7.6(2011) + b \Rightarrow b = -15{,}212.6$
and $V(x) = 7.6x - 15{,}212.6$.

(b) $40 \le 7.6x - 15{,}212.6 \le 55 \Rightarrow 15{,}252.6 \le 7.6x \le 15{,}267.6 \Rightarrow 2007 \le x \le 2009$

93. The graph of linear function will intersect the points (90, 6.5) and (129, 5.5).
$m = \dfrac{6.5 - 5.5}{90 - 129} = -\dfrac{1}{39} \Rightarrow f(x) = -\dfrac{1}{39}(x - 129) + 5.5 \Rightarrow 5.75 < -\dfrac{1}{39}(x - 129) + 5.5 < 6 \Rightarrow$

$0.25 < -\dfrac{1}{39}(x - 129) < 0.5 \Rightarrow -9.75 > x - 129 > 79.5 \Rightarrow 119.25 > x > 109.5$

95. $r = \dfrac{C}{2\pi}$ and $1.99 \le r \le 2.01 \Rightarrow 1.99 \le \dfrac{C}{2\pi} \le 2.01 \Rightarrow 3.98\pi \le C \le 4.02\pi$

97. (a) $m = \dfrac{4.5 - (-1.5)}{2 - 0} = \dfrac{6}{2} = 3$ and y-intercept $= -1.5$; $f(x) = 3x - 1.5$ models the data.

    (b) $f(x) > 2.25 \Rightarrow 3x - 1.5 > 2.25 \Rightarrow 3x > 3.75 \Rightarrow x > 1.25$

99. (a) Using the linear regression function on the calculator the function P is found to be

       $P(x) = 0.658x - 1290.76$.

    (b) Let Y_1 be $0.658X - 1290.76$, $Y_2 = 18$, and $Y_3 = 28.5$. The points of intersection are near (1989, 18) and

       (2005, 28.5).

    (c) The answer was a result of interpolation.

Extended and Discovery Exercises for Section 2.3

1. $a < b \Rightarrow 2a < a + b < 2b \Rightarrow a < \dfrac{a + b}{2} < b$

2.4: More Modeling with Functions

1. (a) $f(x) = \dfrac{x}{16}$

    (b) $f(x) = 10x$

    (c) $f(x) = 0.06x + 6.50$

    (d) $f(x) = 500$

3. Using the points (0, 3) and (4, 1) we have $m = \dfrac{1 - 3}{4 - 0} = \dfrac{-2}{4} = -\dfrac{1}{2}$ and a y-intercept of 3.

    Therefore, the equation of the line is slope intercept form is $y = -\dfrac{1}{2}x + 3$.

5. Using the points (3, 11) and (1, 7) we have $m = \dfrac{11 - 7}{3 - 1} = \dfrac{4}{2} = 2$. Substituting the point (1, 7) into the

    equation $y = 2x + b \Rightarrow 7 = 2(2) + b \Rightarrow b = 5$. Therefore, the equation of the line is slope intercept form is

    $y = 2x + 5$.

7. The height of the Empire State Building is constant; the graph that has no rate of change is d.

9. As time increases the distance to the finish line decreases; the graph that shows this decline in distance as time

    increases is c.

11   $B(t) = 1.2t + 27$; t represents years after 2010; $D = \{t \mid 0 \le t \le 4\}$

13. $T(t) = 0.5t + 0.4$; t represents months after January; $D = \{t \mid 0 \le t \le 11\}$

15. $P(t) = 21.5 + 0.6t$; t represents years after 1900; $D = \{t \mid 0 \le t \le 111\}$

17. (a) $W(t) = -10t + 300$

(b) $W(7) = -10(7) + 300 = 230$ gallons

(c) See Figure 17. x-intercept: 30, after 30 minutes the tank is empty; y-intercept: 300, the tank initially
contains 300 gallons of water.

(d) $D = \{t \mid 0 \le t \le 30\}$

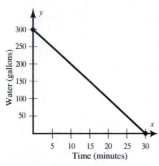

Figure 17

19. (a) $f(x) = 0.04x + 1.2$

(b) Since 2007 corresponds to $x = 0$, 2014 corresponds to $x = 7$; $f(7) = 0.04(7) + 1.2 = 1.48$, which means
that about 1,480,000 may be infected by 2014.

21. (a) $f(x) = 0.25x + 0.5$

(b) $f(2.5) = 0.25(2.5) + 0.5 = 1.125$ inches

23. (a) $(5, 84), (10, 169) \Rightarrow$ slope $= \dfrac{169 - 84}{10 - 5} = 17$

$(10, 169), (15, 255) \Rightarrow$ slope $= \dfrac{255 - 169}{15 - 10} = 17.2$

$(15, 255), (20, 338) \Rightarrow$ slope $= \dfrac{338 - 255}{20 - 15} = 16.6$

(b) $f(x) = 17x$

(c) See Figure 23. The slope indicates that the number of miles traveled per gallon is 17.

(d) $f(30) = 17(30) = 510$ miles. This indicates that the vehicle traveled 510 miles on 30 gallons of gasoline.

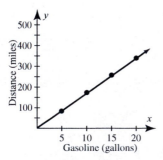

Figure 23

25. (a) The maximum speed limit is 55 mph and the minimum is 30 mph.

    (b) The speed limit is 55 for $0 \le x < 4, 8 \le x < 12$, and $16 \le x < 20$. This is $4 + 4 + 4 = 12$ miles.

    (c) $f(4) = 40, f(12) = 30,$ and $f(18) = 55$.

    (d) The graph is discontinuous when $x = 4, 6, 8, 12,$ and 16. The speed limit changes at each discontinuity.

27. (a) $P(1.5) = 1.10$; it costs \$1.10 to mail 1.5 ounces. $P(3) = 1.30$; it costs \$1.30 to mail 3 oz.

    (b) See Figure 27. $D = \left\{ x | 0 < x \le 5 \right\}$

    (c) $x = 1, 2, 3, 4$

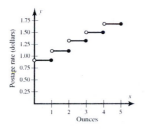

Figure 27

29. (a) $f(1.5) = 30; f(4) = 10$

    (b) $m_1 = 20$ indicates that the car is moving away from home at 20 mph; $m_2 = -30$ indicates that the car is moving toward home at 30 mph; $m_3 = 0$ indicates that the car is not moving; $m_4 = -10$ indicates that the car is moving toward home at 10 mph.

    (c) The driver starts at home and drives away from home at 20 mph for 2 hours. The driver then travels toward home at 30 mph for 1 hour. Then the car does not move for 1 hour. Finally, the driver returns home in 1 hour at 10 mph.

    (d) Increasing: $0 \le x \le 2$; Decreasing: $2 \le x \le 3$ or $4 \le x \le 5$; Constant: $3 \le x \le 4$

31. (a) $D = \{ x | -5 \le x \le 5 \}$

    (b) $f(-2) = 2, f(0) = 0 + 3 = 3, f(3) = 3 + 3 = 6$

    (c) See Figure 31.

    (d) f is continuous.

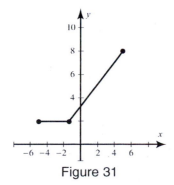

Figure 31

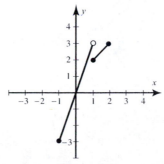

Figure 33

33. (a) $D = \{x\,|-1 \le x \le 2\}$

    (b) $f(-2)$ is undefined, $f(0) = 3(0) = 0, f(3)$ is undefined

    (c) See Figure 33.

    (d) f is not continuous.

35. (a) $D = \{x\,|-3 \le x \le 3\}$

    (b) $f(-2) = -2, f(0) = 1, f(3) = 2 - 3 = -1$

    (c) See Figure 35.

    (d) f is not continuous.

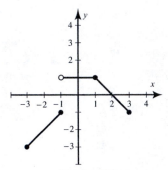

Figure 35

37. $f(-4) = -\dfrac{1}{2}(-4) + 1 = 3, (-4, 3);\ f(-2) = -\dfrac{1}{2}(-2) + 1 = 2, (-2, 2);$ graph a segment from

    $(-4, 3)$ to $(-2, 2)$, use a closed dot for each point. See Figure 37.

    $f(-2) = 1 - 2(-2) = 5, (-2, 5);\ f(1) = 1 - 2(1) = -1, (1, -1);$ graph a segment from

    $(-2, 5)$ to $(1, -1)$, use an open dot at $(-2, 5)$ and a closed dot for $(1, -1)$. See Figure 37.

    $f(1) = \dfrac{2}{3}(1) + \dfrac{4}{3} = 2, (1, 2);\ f(4) = \dfrac{2}{3}(4) + \dfrac{4}{3} = 4, (4, 4);$ graph a segment from $(1, 2)$ to $(4, 4)$, use an

    open dot at $(1, 2)$ and a closed dot for $(4, 4)$. See Figure 37.

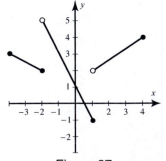

Figure 37

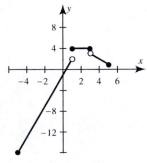

Figure 39

39. (a) $f(-3) = 3(-3) - 1 = -10, f(1) = 4, f(2) = 4,$ and $f(5) = 6 - 5 = 1$

    (b) The function f is constant with a value of 4 on the interval $[1, 3]$.

    (c) See Figure 39. f is not continuous.

41. We must determine two equations for the lines that represent the two line segments. Given the slope and y-intercept we can write the slope intercept form of the line. The first line segment has a slope of $\frac{3}{4}$ and y-intercept of 3, so $f(x) = \frac{3}{4}x + 3$. The second line segment has a slope of $-\frac{2}{3}$ and a y-intercept of 2, so $f(x) = -\frac{2}{3}x + 2$. The piecewise function is $f(x) = \begin{cases} \frac{3}{4}x + 3 & -4 \le x < 0 \\ -\frac{2}{3}x + 2 & 0 \le x \le 3 \end{cases}$

43. (a) See Figure 43.

    (b) $R(2009) = 700$, there were 700 people for each housing start. A small number will indicate a strong housing market.

    (c) R is continuous on its domain.

    (d) From the graph we can see that R is increasing on the interval $(2005, 2011)$, decreasing on the interval $(2000, 2005)$, and is never constant.

    (e) The piecewise function is $R(x) = \begin{cases} -9(x - 2000) + 225 & 2000 \le x \le 2005 \\ 130(x - 2005) + 180 & 2005 < x \le 2009 \\ 13.5(x - 2009) + 700 & 2009 < x \le 2011 \end{cases}$

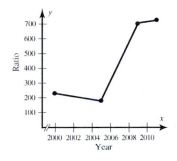

Figure 43

45. (a) Graph $Y_1 = \text{int}(2X - 1)$ as shown in Figure 45.

    (b) $f(-3.1) = [2(-3.1) - 1] = [-7.2] = -8$ and $f(1.7) = [2(1.7) - 1] = [2.4] = 2$

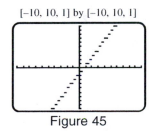
$[-10, 10, 1]$ by $[-10, 10, 1]$
Figure 45

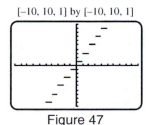
$[-10, 10, 1]$ by $[-10, 10, 1]$
Figure 47

47. (a) Graph $Y_1 = 2(\text{int}(X)) + 1$ as shown in Figure 47.

    (b) $f(-3.1) = 2[-3.1] + 1 = 2(-4) + 1 = -7$ and $f(1.7) = 2[1.7] + 1 = 2(1) + 1 = 3$

49. (a) $f(x) = 0.8\left[\!\left[\dfrac{x}{2}\right]\!\right]$ for $6 \leq x \leq 18$

   (b) Graph $Y_1 = 0.8(\text{int}(X/2))$ as shown in Figure 49.

   (c) $f(8.5) = 0.8\left[\!\left[\dfrac{8.5}{2}\right]\!\right] = 0.8[\![4.25]\!] = 0.8(4) = \$3.20; f(15.2) = 0.8\left[\!\left[\dfrac{15.2}{2}\right]\!\right] = 0.8[\![7.6]\!] = 0.8(7) = \$5.60$

   [6, 18, 1] by [0, 8, 1]   [–2, 3, 1] by [–2, 7, 1]

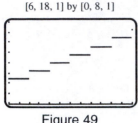
Figure 49

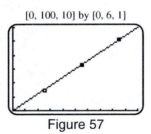
Figure 51

51.   Since y is directly proportional to x, the variation equation $y = kx$ must hold. To find the value of k, use the value $y = 7$ when $x = 14$. Solve the equation $7 = k(14) \Rightarrow k = \dfrac{1}{2}$. Then $y = \dfrac{1}{2}(5) = \dfrac{5}{2} = 2.5$.

53.   Since y is directly proportional to x, the variation equation $y = kx$ must hold. To find the value of k, use the value $y = \dfrac{3}{2}$ when $x = \dfrac{2}{3}$. Solve the equation $\dfrac{3}{2} = k\left(\dfrac{2}{3}\right) \Rightarrow k = \dfrac{9}{4}$. Then $y = \dfrac{9}{4}\left(\dfrac{1}{2}\right) = \dfrac{9}{8}$.

55.   Since y is directly proportional to x, the variation equation $y = kx$ must hold. To find the value of k use the value $y = 7.5$ when $x = 3$ from the table. Solve the equation $7.5 = k(3) \Rightarrow k = 2.5$. The variation equation is $y = 2.5x$ and hence $y = 2.5(8) = 20$ when $x = 8$. A graph of $Y_1 = 2.5X$ together with the data points is shown in Figure 55.

   [0, 10, 1] by [0, 24, 2]   [0, 100, 10] by [0, 6, 1]

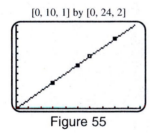
Figure 55

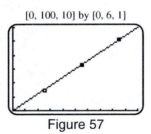
Figure 57

57.   Since y is directly proportional to x, the variation equation $y = kx$ must hold. To find the value of k use the value $y = 1.50$ when $x = 25$ from the table. Solve the equation $1.50 = k(25) \Rightarrow k = 0.06$. The variation equation is $y = 0.06x$ and hence when $y = 5.10$, $x = \dfrac{5.1}{0.06} = 85$. A graph of $Y_1 = 0.06X$ together with the data points is shown in Figure 53.

59.   Let y represent the cost of tuition and x represent the number of credits taken. Since the cost of tuition is directly proportional to the number of credits taken, the variation equation $y = kx$ must hold. If cost $y = \$720.50$ when the number of credits $x = 11$, we find the constant of proportionality k by solving $720.50 = k(11) \Rightarrow k = 65.50$. The variation equation is $y = 65.50x$. Therefore, the cost of taking 16 credits is $y = 65.50(16) = \$1048$.

61. (a) Since the points $(0, 0)$ and $(300, 3)$ lie on the graph of $y = kx$, the slope of the graph is $\dfrac{3 - 0}{300 - 0} = 0.01$

    and $y = 0.01x$, so $k = 0.01$.

    (b) $y = 0.01(110) = 1.1$ millimeters.

63. (a) Using $F = kx \Rightarrow 15 = k(8) \Rightarrow k = \dfrac{15}{8}$

    (b) The variation equation is $y = \dfrac{15}{8}x$; $25 = \dfrac{15}{8}(x) \Rightarrow x = 13\dfrac{1}{3}$ inches.

65. (a) For $(150, 26)$, $\dfrac{F}{x} = \dfrac{26}{150} \approx 0.173$; for $(180, 31)$, $\dfrac{F}{x} = \dfrac{31}{180} \approx 0.172$; for $(210, 36)$, $\dfrac{F}{x} = \dfrac{36}{210} \approx 0.171$;

    for $(320, 54)$, $\dfrac{F}{x} = \dfrac{54}{320} \approx 0.169$; the ratios give the force needed to push a 1 lb box.

    (b) From the table it appears that approximately 0.17 lb of force is needed to push a 1 lb cargo box $\Rightarrow$

    $k = 0.17$.

    (c) See Figure 65.

    (d) $F \approx 0.17(275) \Rightarrow F = 46.75$ lbs of force.

[125, 350, 25] by [0, 75, 5]

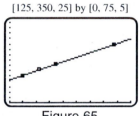

Figure 65

67. (a) Linear regression gives the model:

    $S(x) \approx 3.974x - 14{,}479$. **Answers may vary.**

    (b) $S(x) = 6 \Rightarrow 6 = 3.974x - 14.479 \Rightarrow 20.479 = 3.974x \Rightarrow x \approx 5.15$; The circumference of a

    finger with ring size 6 is approximately 5.15 cm.

69. (a) Linear regression gives the model: $f(x) \approx 0.12331x - 244.75$ (Answers may vary)

    (b) $f(2009) = 0.12331(2009) - 244.75 \Rightarrow f(2009) \approx 3$. The estimate was found using interpolation.

    (c) $4 - 0.12331x - 244.75 \Rightarrow 248.75 = 0.12331x \Rightarrow x \approx 2017$

Extended and Discovery Exercises for Section 2.4

1. Let x = number of fish in the sample and y = number of tagged fish. Then $y = kx$, where k represents the

   proportion of fish tagged. From the data point $(94, 13)$, get $13 = k(94) \Rightarrow k \approx 0.138298$. Letting the sam-

   ple represent the entire number of fish, we get $y = 0.138298x \Rightarrow 85 = 0.138298x \Rightarrow x \approx 615$.

3. *Answers may vary.*

[1.580, 1.584, 0.001] by [−6.252, −6.248, 0.001]

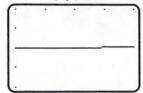

Figure 5

5. (a) Graph $Y_1 = X\char`^4 - 5X\char`^2$. If one repeatedly zooms in on any portion of the graph, it begins to look like a straight line. See Figure 5 for an example.

    (b) A linear approximation will be a good approximation over a small interval.

Checking Basic Concepts for Sections 2.3 and 2.4

1. $2(x - 4) > 1 - x \Rightarrow 2x - 8 > 1 - x \Rightarrow 3x > 9 \Rightarrow x > 3$; $\{x \mid x > 3\}$

3. (a) $-3(2 - x) - \dfrac{1}{2}x - \dfrac{3}{2} = 0$ when $x = 3$; symbolically, $-3(2 - x) - \dfrac{1}{2}x - \dfrac{3}{2} = 0 \Rightarrow$

    $-6 + 3x - \dfrac{1}{2}x - \dfrac{3}{2} = 0 \Rightarrow \dfrac{5}{2}x - \dfrac{15}{2} = 0 \Rightarrow \dfrac{5}{2}x = \dfrac{15}{2} \Rightarrow x = 3$

    (b) $-3(2 - x) - \dfrac{1}{2}x - \dfrac{3}{2} > 0$ when $x > 3$; symbolically, $-3(2 - x) - \dfrac{1}{2}x - \dfrac{3}{2} > 0 \Rightarrow$

    $-6 + 3x - \dfrac{1}{2}x - \dfrac{3}{2} > 0 \Rightarrow \dfrac{5}{2}x - \dfrac{15}{2} > 0 \Rightarrow \dfrac{5}{2}x > \dfrac{15}{2} \Rightarrow x > 3 \Rightarrow (3, \infty)$. In set-builder notation

    the interval is $\{x \mid x > 3\}$.

    (c) $-3(2 - x) - \dfrac{1}{2}x - \dfrac{3}{2} \leq 0$ when $x \leq 3$; symbolically, $-3(2 - x) - \dfrac{1}{2}x - \dfrac{3}{2} \leq 0 \Rightarrow$

    $-6 + 3x - \dfrac{1}{2}x - \dfrac{3}{2} \leq 0 \Rightarrow \dfrac{5}{2}x - \dfrac{15}{2} \leq 0 \Rightarrow \dfrac{5}{2}x \leq \dfrac{15}{2} \Rightarrow x \leq 3 \Rightarrow (-\infty, 3]$. In set-builder

    notation the interval is $\{x \mid x \leq 3\}$.

    (b) $f(39) = 27(39) = 1053$; the number of people 15 to 24 years old who die from heart disease is 1053.

5. Since the car is initially 50 miles south of home and driving south at 60 mph, the y-intercept is 50 and

    $m = 60$; $f(t) = 60t + 50$, where t is in hours.

2.5: Absolute Value Equations and Inequalities

1. $|x| = 3 \Rightarrow x = 3$ or $x = -3$

3. $|x| > 3 \Rightarrow x > 3$ or $x < -3$; $(-\infty, -3) \cup (3, \infty)$

5. The graph of $y = |ax + b|$ is V-shaped with the vertex on the x-axis.

7. $\sqrt{36a^2} = |6a|$ since 36 and a^2 are always positive values.

9. (a) $x + 1 = 0 \Rightarrow x = -1 \Rightarrow$ the vertex is $(-1, 0)$. Find any other point, $x = 0 \Rightarrow (0, 1)$; graph the

    absolute value graph with the vertex $(-1, 0)$, point $(0, 1)$ and its reflection through $x = -1$.

    See Figure 9.

    (b) $y = |x + 1|$ is increasing on $x > -1$ or $(-1, \infty)$ and decreasing on $x < -1$ or $(-\infty, -1)$.

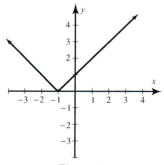

Figure 9

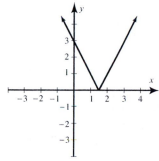

Figure 11

11. (a) $2x - 3 = 0 \Rightarrow 2x = 3 \Rightarrow x = \dfrac{3}{2} \Rightarrow$ the vertex is $\left(\dfrac{3}{2}, 0\right)$. Find another point such as $(0, 3)$; graph the

    absolute value function. See Figure 11.

    (b) $y = |2x - 3|$ is increasing on $x > \dfrac{3}{2}$ or $\left(\dfrac{3}{2}, \infty\right)$ and decreasing on $x < \dfrac{3}{2}$ or $\left(-\infty, \dfrac{3}{2}\right)$.

13. (a) The graph of $y_1 = 2x$ is shown in Figure 13a.

    (b) The graph of $y = |2x|$ is similar to the graph of $y = 2x$ except that it is reflected across the x-axis

    whenever $2x < 0$. The graph of $y_1 = |2x|$ is shown in Figure 13b.

    (c) The x-intercept occurs when $2x = 0$ or when $x = 0$. The x-intercept is 0.

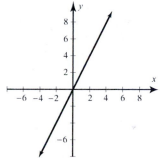

Figure 13a

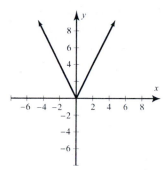

Figure 13b

15. (a) The graph of $y_1 = 3x - 3$ is shown in Figure 15a.

    (b) The graph of $y = |3x - 3|$ is similar to the graph of $y = 3x - 3$ except that it is reflected across the x-axis

    whenever $3x - 3 < 0$ or $x < 1$. The graph of $y_1 = |3x - 3|$ is shown in Figure 15b.

    (c) The x-intercept occurs when $3x - 3 = 0$ or when $x = 1$. The x-intercept is located at 1.

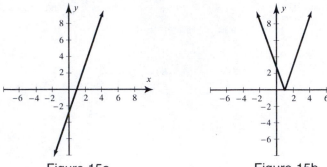

Figure 15a   Figure 15b

17. (a) The graph of $y_1 = 6 - 2x$ is shown in Figure 17a.

    (b) The graph of $y = |6 - 2x|$ is similar to the graph of $y = 6 - 2x$ except that it is reflected across the x-axis whenever $6 - 2x < 0$ or $x > 3$. The graph of $y_1 = |6 - 2x|$ is shown in Figure 17b.

    (c) The x-intercept occurs when $6 - 2x = 0$ or when $x = 3$. The x-intercept is located at 3.

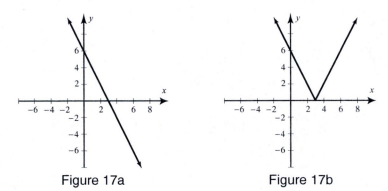

Figure 17a   Figure 17b

19. $|-2x| = 4 \Rightarrow -2x = 4$ or $-2x = -4 \Rightarrow 2x = -4$ or $2x = 4 \Rightarrow x = -2$ or 2

21. $|5x - 7| = 2 \Rightarrow 5x - 7 = -2$ or $5x - 7 = 2$. If $5x - 7 = -2$, then $5x = 5 \Rightarrow x = 1$;
If $5x - 7 = 2$, then $5x = 9 \Rightarrow x = \dfrac{9}{5}$; $1, \dfrac{9}{5}$.

23. $|3 - 4x| = 5 \Rightarrow 3 - 4x = -5$ or $3 - 4x = 5$. If $3 - 4x = -5$, then $-4x = -8 \Rightarrow x = 2$;
If $3 - 4x = 5$ then $-4x = 2 \Rightarrow x = -\dfrac{1}{2}$; $-\dfrac{1}{2}, 2$.

25. $|-6x - 2| = 0 \Rightarrow -6x - 2 = 0 \Rightarrow -6x = 2 \Rightarrow x = -\dfrac{1}{3}$.

27. $|7 - 16x| = 0 \Rightarrow 7 - 16x = 0 \Rightarrow 7 = 16x \Rightarrow x = \dfrac{7}{16}$.

29. $|17x - 6| = -3$ has no solutions since the absolute value of any quantity is always greater than or equal to 0.

31. $|1.2x - 1.7| - 5 = -1 \Rightarrow |1.2x - 1.7| = 4$, then $1.2x - 1.7 = -4$ or $1.2x - 1.7 = 4$.
If $1.2x - 1.7 = -4$ then, $1.2x = -2.3 \Rightarrow x = -\dfrac{2.3}{1.2} \Rightarrow x = -\dfrac{23}{12}$; if $1.2x - 1.7 = 4$ then

$1.2x = 5.7 \Rightarrow x = \dfrac{5.7}{1.2} \Rightarrow x = \dfrac{19}{4}$; $\Rightarrow -\dfrac{23}{12}, \dfrac{19}{4}$.

33. $|4x - 5| + 3 = 2 \Rightarrow |4x - 5| = -1$ has no solution since the absolute value of any quantity is always greater than or equal to 0.

35. $|2x - 9| = |8 - 3x| \Rightarrow 2x - 9 = 8 - 3x$ or $2x - 9 = -(8 - 3x) \Rightarrow$

    $2x + 3x = 8 + 9$ or $2x - 3x = -8 + 9 \Rightarrow 5x = 17$ or $-x = 1 \Rightarrow x = \dfrac{17}{5}, -1$

37. $\left|\dfrac{3}{4}x - \dfrac{1}{4}\right| = \left|\dfrac{3}{4} - \dfrac{1}{4}x\right| \Rightarrow \dfrac{3}{4}x - \dfrac{1}{4} = \dfrac{3}{4} - \dfrac{1}{4}x$ or $\dfrac{3}{4}x - \dfrac{1}{4} = -\left(\dfrac{3}{4} - \dfrac{1}{4}x\right) \Rightarrow$

    $\dfrac{3}{4}x + \dfrac{1}{4}x = \dfrac{3}{4} + \dfrac{1}{4}$ or $\dfrac{3}{4}x - \dfrac{1}{4}x = -\dfrac{3}{4} + \dfrac{1}{4} \Rightarrow x = 1$ or $\dfrac{1}{2}x = -\dfrac{1}{2} \Rightarrow x = -1, 1$

39. $\left|15x - 5\right| = \left|35 - 5x\right| \Rightarrow 15x - 5 = 35 - 5x$ or $15x - 5 = -(35 - 5x) \Rightarrow$

    $15x - 5 = 35 - 5x \Rightarrow 20x = 40 \Rightarrow x = 2$ or $15x - 5 = -(35 - 5x) \Rightarrow 15x - 5 = -35 + 5x \Rightarrow$

    $10x = -30 \Rightarrow x = -3 \Rightarrow x = -3$ or 2

41. (a) $f(x) = g(x)$ when $x = -1$ or 7 .

    (b) $f(x) < g(x)$ between these x-values or when $-1 < x < 7$; $(-1, 7)$

    (c) $f(x) > g(x)$ outside of these x-values or when $x < -1$ or $x > 7$; $(-\infty, -1) \cup (7, \infty)$

43. (a) $|2x - 3| = 1 \Rightarrow 2x - 3 = 1$ or $2x - 3 = -1$. If $2x - 3 = 1$, then $2x = 4 \Rightarrow x = 2$; If

    $2x - 3 = -1$ then $2x = 2 \Rightarrow x = 1$; $x = 1$ or $x = 2$

    (b) $|2x - 3| < 1 \Rightarrow -1 < 2x - 3 < 1 \Rightarrow 2 < 2x < 4 \Rightarrow 1 < x < 2$; $(1, 2)$

    (c) $|2x - 3| > 1 \Rightarrow 2x - 3 > 1$ or $2x - 3 < -1$.   If $2x - 3 > 1$, then $2x > 4 \Rightarrow x > 2$.

    If $2x - 3 < -1$, then $2x < 2 \Rightarrow x < 1$. $x < 1$ or $x > 2$ or $(-\infty, 1) \cup (2, \infty)$

45. (a) Graph $Y_1 = \text{abs}(2X - 5)$ and $Y_2 = 10$ See Figures 45a and 45b. The solutions are -2.5 and 7.5.

    (b) Table $Y_1 = \text{abs}(2X - 5)$ starting at -5, incrementing by 2.5. See Figure 45c. The solutions are -2.5 and 7.5.

    (c) $|2x - 5| = 10 \Rightarrow 2x - 5 = -10$ or $2x - 5 = 10 \Rightarrow x = -\dfrac{5}{2}$ or $\dfrac{15}{2}$

    From each method, the solution to $|2x - 5| < 10$ lies between -2.5 and 7.5, exclusively: $-2.5 < x < 7.5$

    or $\left(-\dfrac{5}{2}, \dfrac{15}{2}\right)$.

[−10, 10, 1] by [−5, 15, 1]   [−10, 10, 1] by [−5, 15, 1]

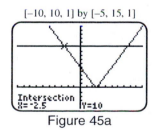
Figure 45a

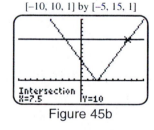
Figure 45b

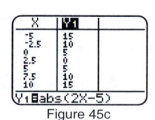
Figure 45c

47. (a) Graph $Y_1 = \text{abs}(5 - 3X)$ and $Y_2 = 2$ See Figures 47a and 47b. The solutions are 1 and $\frac{7}{3}$.

(b) Table $Y_1 = \text{abs}(5 - 3X)$ starting at $-\frac{1}{3}$, incrementing by $\frac{2}{3}$. See Figure 47c. The solutions are 1 and $\frac{7}{3}$.

(c) $|5 - 3x| = 2 \Rightarrow 5 - 3x = -2$ or $5 - 3x = 2 \Rightarrow x = \frac{7}{3}$ or 1

From each method, the solution to $|5 - 3x| > 2$ lies outside of 1 and $\frac{7}{3}$, exclusively: $x < 1$ or $x > \frac{7}{3}$ or

$(-\infty, 1) \cup \left(\frac{7}{3}, \infty\right)$.

[-1, 4, 1] by [-1, 3, 1]   [-1, 4, 1] by [-1, 3, 1]

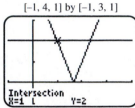
Figure 47a

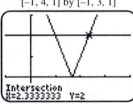
Figure 47b

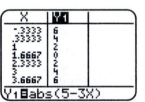
Figure 47c

49. $|2.1x - 0.7| = 2.4 \Rightarrow 2.1x - 0.7 = -2.4$ or $2.1x - 0.7 = 2.4 \Rightarrow x = -\frac{17}{21}$ or $\frac{31}{21}$

The solution to $|2.1x - 0.7| \geq 2.4$ lies outside of $-\frac{17}{21}$ and $\frac{31}{21}$, inclusively: $x \leq -\frac{17}{21}$ or $x \geq \frac{31}{21}$ or

$\left(-\infty, -\frac{17}{21}\right] \cup \left[\frac{31}{21}, \infty\right)$.

51. $|3x| + 5 = 6 \Rightarrow |3x| = 1 \Rightarrow 3x = -1$ or $3x = 1 \Rightarrow x = -\frac{1}{3}$ or $\frac{1}{3}$.

The solution to $|3x| + 5 > 6$ lies outside of $-\frac{1}{3}$ and $\frac{1}{3}$, exclusively: $x < -\frac{1}{3}, x > \frac{1}{3}$,

$\left(-\infty, -\frac{1}{3}\right) \cup \left(\frac{1}{3}, \infty\right)$.

53. $\left|\frac{2}{3}x - \frac{1}{2}\right| = -\frac{1}{4}$ has no solutions since the absolute value of any quantity is always greater than or equal to 0.

There are no solutions to $\left|\frac{2}{3}x - \frac{1}{2}\right| \leq -\frac{1}{4}$.

55. The solutions to $|3x - 1| < 8$ satisfy $s_1 < x < s_2$ where s_1 and s_2, are the solutions to $|3x - 1| = 8$.

$|3x - 1| = 8$ is equivalent to $3x - 1 = -8 \Rightarrow x = -\frac{7}{3}$ or $3x - 1 = 8 \Rightarrow x = 3$.

The interval is $\left(-\frac{7}{3}, 3\right)$.

57. The solutions to $|7 - 4x| \leq 11$ satisfy $s_1 \leq x \leq s_2$, where s_1 and s_2 are the solutions to $|7 - 4x| = 11$.

$|7 - 4x| = 11$ is equivalent to $7 - 4x = -11 \Rightarrow x = \frac{9}{2}$ or $7 - 4x = 11 \Rightarrow x = -1$.

The interval is $\left[-1, \frac{9}{2}\right]$.

59. The solutions to $|0.5x - 0.75| < 2$ satisfy $s_1 < x < s_2$, where s_1 and s_2 are the solutions to $|0.5x - 0.75| = 2$.

    $|0.5x - 0.75| = 2$ is equivalent to $0.5x - 0.75 = -2 \Rightarrow x = -\dfrac{5}{2}$ or $0.5x - 0.75 = 2 \Rightarrow x = \dfrac{11}{2}$.

    The interval is $\left(-\dfrac{5}{2}, \dfrac{11}{2}\right)$.

61. The solutions to $|2x - 3| > 1$ satisfy $x < s_1$ or $x > s_2$, where s_1 and s_2 are the solutions to $|2x - 3| = 1$.

    $|2x - 3| = 1$ is equivalent to $2x - 3 = -1 \Rightarrow x = 1$ or $2x - 3 = 1 \Rightarrow x = 2$.

    The solution set is $(-\infty, 1) \cup (2, \infty)$.

63. The solutions to $|-3x + 8| \geq 3$ satisfy $x \leq s_1$ or $x \geq s_2$, where s_1 and s_2 are the solutions to $|-3x + 8| = 3$.

    $|-3x + 8| = 3$ is equivalent to $-3x + 8 = -3 \Rightarrow x = \dfrac{11}{3}$ or $-3x + 8 = 3 \Rightarrow x = \dfrac{5}{3}$

    The solution set is $\left(-\infty, \dfrac{5}{3}\right] \cup \left[\dfrac{11}{3}, \infty\right)$.

65. The solutions to $|0.25x - 1| > 3$ satisfy $x < s_1$ or $x > s_2$, where s_1 and s_2 are the solutions to $|0.25x - 1| = 3$.

    $|0.25x - 1| = 3$ is equivalent to $0.25x - 1 = -3 \Rightarrow x = -8$ or $0.25x - 1 = 3 \Rightarrow x = 16$.

    The solution set is $(-\infty, -8) \cup (16, \infty)$.

67. $|-6| = 6$

69. Since the inputs of absolute values can be positive or negative, the domain of $|f(x)|$ is also $[-2, 4]$.

71. Since all solutions or the range of absolute values must be non-negative, all negative solutions will change to

    positive solutions; therefore, if the range of $f(x)$ is $(-\infty, 0]$, the range of $|f(x)|$ is $[0, \infty)$.

73. $|S - 57.5| = 17.5 \Rightarrow S - 57.5 = 17.5$ or $S - 57.5 = -17.5$ If $S - 57.5 = 17.5$, then $S = 75$. If

    $S - 57.5 = -17.5$, then $S = 40$. Therefore, the maximum speed limit is 75 mph and the minimum speed limit

    is 40 mph.

75. (a) $0 \leq 80 - 19x \leq 32 \Rightarrow -80 \leq -19x \leq -48 \Rightarrow \dfrac{80}{19} \geq x \geq \dfrac{48}{19} \Rightarrow \dfrac{48}{19} \leq x \leq \dfrac{80}{19}$. The air temperature

    is between $0°\text{F}$ and $32°\text{F}$ when the altitudes are between $\dfrac{48}{19}$ and $\dfrac{80}{19}$ miles inclusively.

    (b) The air temperature is between $0°\text{F}$ and $32°\text{F}$ inclusively when the altitude is within $\dfrac{16}{19}$ mile of $\dfrac{64}{19}$ miles.

    $\left|x - \dfrac{64}{19}\right| \leq \dfrac{16}{19}$.

77. (a) $|T - 43| = 24 \Rightarrow T - 43 = -24$ or $T - 43 = 24 \Rightarrow T = 19$ or 67 . The average monthly temperature

    range is $19°\text{F} \leq T \leq 67°\text{F}$.

    (b) The monthly average temperatures in Marquette vary between a low of $19°\text{F}$ and a high of $67°\text{F}$.

    The monthly averages are always within $24°$ of $43°\text{F}$.

79. (a) $|T - 50| = 22 \Rightarrow T - 50 = -22$ or $T - 50 = 22 \Rightarrow T = 28$ or 72. The average monthly temperature

    range is $28°F \le T \le 72°F$.

    (b) The monthly average temperatures in Boston vary between a low of 28°F and a high of 72°F.

    The monthly averages are always within 22° of 50°F.

81. (a) $|T - 61.5| = 12.5 \Rightarrow T - 61.5 = -12.5$ or $T - 61.5 = 12.5 \Rightarrow T = 49$ or 74 The average monthly

    temperature range is $49°F \le T \le 74°F$.

    (b) The monthly average temperatures in Buenos Aires vary between a low of 49°F and a high of 74°F.

    The monthly averages are always within 12.5° of 61.5°F.

83. (a) $|T - 10.5| < 0.05$

    (b) $|T - 10.5| < 0.05 \Rightarrow -0.05 < T - 10.5 < 0.05 \Rightarrow 10.45 < T < 10.55$, the actual thickness must be

    greater than 10.45 mm and less than 10.55 mm.

85. (a) $|D - 2.118| \le 0.07$

    (b) $|D - 2.118| \le 0.07 \Rightarrow -0.07 \le D - 2.118 \le 0.07 \Rightarrow 2.111 \le T \le 2.125$, the diameter must be greater

    than or equal to 2.111 inches and less than or equal to 2.125 inches.

87. $\left|\dfrac{Q - A}{A}\right| \le 0.02 \Rightarrow \left|\dfrac{Q - 35}{35}\right| \le 0.02$, so $-0.02 \le \dfrac{Q - 35}{35} \le 0.02 \Rightarrow -0.7 \le Q - 35 \le 0.7 \Rightarrow$

    $34.3 \le Q \le 35.7$

Extended and Discovery Exercises for Section 2.5

1. The distance between points x and c on a number line can be shown by $|x - c|$. This distance is given to be

   less than some positive value δ. Then $|x - c| < \delta$.

Checking Basic Concepts for Section 2.5

1. $\sqrt{4x^2} = |2x|$

3. (a) Graphically: Graph $Y_1 = abs(2X - 1)$ and $Y_2 = 5$. Their graphs intersect at the points $(-2, 5)$ and $(3,5)$.

   The solutions are $-2, 3$. See Figures 3a & 3b.

   Numerically: Table $Y_1 = abs(2X - 1)$ starting x at -2 and incrementing by 1. The solutions are $-2, 3$.

   See Figure 3c.

Symbolically: $|2x - 1| = 5 \Rightarrow 2x - 1 = 5$ or $2x - 1 = -5 \Rightarrow 2x = 6$ or $2x = -4 \Rightarrow x = 3, -2$.

The solutions are $-2, 3$.

(b) The solutions to $|2x - 1| \leq 5$ lie between $x = -2$ and $x = 3$, inclusively. Thus, $-2 \leq x \leq 3$ or $[-2, 3]$.

The solutions to $|2x - 1| > 5$ lie left of $x = -2$ or right of $x = 3$. Thus, $x < -2$ or $x > 3$ or

$(-\infty, -2) \cup (3, \infty)$.

[-10, 10, 1] by [-10, 10, 1]   [-10, 10, 1] by [-10, 10, 1]

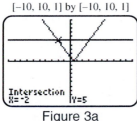   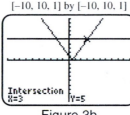   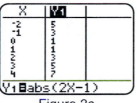

Figure 3a   Figure 3b   Figure 3c

5. $|x + 1| = |2x| \Rightarrow x + 1 = 2x$ or $x + 1 = -2x$. If $x + 1 = 2x$, then $x = 1$;

$x + 1 = -2x \Rightarrow 1 = -3x \Rightarrow x = -\dfrac{1}{3}; \ -\dfrac{1}{3}, 1$

Chapter 2 Review Exercises

1. The slope of the line is $m = \dfrac{5 - 4}{2 - (-3)} = \dfrac{1}{5}$, and then we substitute the slope and the point $(-3, 4)$ into the

equation $y = m(x - x_1) + y_1$ to find $y = \dfrac{1}{5}\left(x - (-3)\right) + 4 \Rightarrow y = \dfrac{1}{5}(x + 3) + 4$.

3. Substituting the slope and the given point into the equation $y = m(x - x_1) + y_1$, we have

$y = -\dfrac{7}{5}(x + 5) + 6 \Rightarrow y = -\dfrac{7}{5}x - 7 + 6 \Rightarrow y = -\dfrac{7}{5}x - 1$. The function is $f(x) = -\dfrac{7}{5}x - 1$.

5. The slope of the line is $m = \dfrac{-2 - 0}{0 - 5} = \dfrac{-2}{-5} = \dfrac{2}{5}$, and then we substitute the slope and the point $(5, 0)$ into the

equation $y = m(x - x_1) + y_1$ to find $y = \dfrac{2}{5}(x - 5) + 0 \Rightarrow y = \dfrac{2}{5}x - 2$. The function is $f(x) = \dfrac{2}{5}x - 2$.

7. Using point-slope form $y = m(x - x_1) + y_1$, we get $y = -2(x + 2) + 3 \Rightarrow f(x) = -2x - 1$

9. $y = 7(x + 3) + 9 \Rightarrow y = 7x + 21 + 9 \Rightarrow y = 7x + 30$

11. Let $m = -3$. Then, $y = -3(x - 1) - 1 \Rightarrow y = -3x + 2$.

13. The line segment has slope $m = \dfrac{0 - 3.1}{5.7 - 0} = -\dfrac{31}{57}$; the parallel line has slope $m = -\dfrac{31}{57}$;

$y = -\dfrac{31}{57}(x - 1) - 7 \Rightarrow y = -\dfrac{31}{57}x - \dfrac{368}{57}$

15. The line is vertical passing through $(6, -7)$, so the equation is $x = 6$.

17. The line is horizontal passing through $(1, 3)$, so the equation is $y = 3$.

19. The equation of the vertical line with x-intercept 2.7 is $x = 2.7$.

21. For x-intercept: $y = 0 \Rightarrow 5x - 4(0) = 20 \Rightarrow 5x = 20 \Rightarrow x = 4$; for y-intercept: $x = 0 \Rightarrow$

    $5(0) - 4y = 20 \Rightarrow -4y = 20 \Rightarrow y = -5$; use $(4, 0)$ and $(0, -5)$ to graph the equation. See Figure 21.

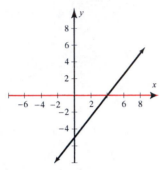

Figure 21

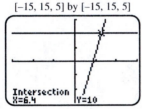

$[-15, 15, 5]$ by $[-15, 15, 5]$

Figure 23

23. Graphical: Graph $Y_1 = 5X - 22$ and $Y_2 = 10$. Their graphs intersect at $(6.4, 10)$ as shown in Figure 23. The

    solution is $x = 6.4$.

    Symbolic: $5x - 22 = 10 \Rightarrow 5x = 32 \Rightarrow x = \dfrac{32}{5} = 6.4$

25. Graphical: Graph $Y_1 = -2(3X - 7) + X$ and $Y_2 = 2X - 1$. Their graphs intersect near $(2.143, 3.286)$ as

    shown in Figure 25. The solution is approximately 2.143.

    Symbolic: $-2(3x - 7) + x = 2x - 1 \Rightarrow -6x + 14 + x = 2x - 1 \Rightarrow -7x = -15 \Rightarrow x = \dfrac{15}{7} \approx 2.143$

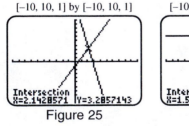

$[-10, 10, 1]$ by $[-10, 10, 1]$ $[-10, 10, 1]$ by $[-10, 10, 1]$

Figure 25   Figure 27

27. Graphical: Graph $Y_1 = \pi X + 1$ and $Y_2 = 6$. Their graphs intersect near $(1.592, 6)$ as shown in Figure 27. The

    solution is approximately 1.592.

    Symbolic: $\pi x + 1 = 6 \Rightarrow \pi x = 5 \Rightarrow x = \dfrac{5}{\pi} \approx 1.592$

29. Let $Y_1 = 3.1X - 0.2 - 2(X - 1.7)$ and approximate where $Y_1 = 0$. From Figure 29 this occurs when

   $x \approx -2.9$.

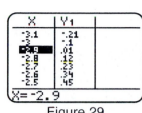

[-10, 10, 1] by [-10, 10, 1]

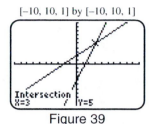

      Figure 29   Figure 39

31. (a) $4(6 - x) = -4x + 24 \Rightarrow 24 - 4x = -4x + 24 \Rightarrow 0 = 0 \Rightarrow$ all real numbers are solutions.

   (b) Because all real numbers are solutions, the equation is an identity.

33. (a) $5 - 2(4 - 3x) + x = 4(x - 3) \Rightarrow 5 - 8 + 6x + x = 4x - 12 \Rightarrow 7x - 3 = 4x - 12 \Rightarrow$

   $3x = -9 \Rightarrow x = -3$

   (b) Because there are finitely many solutions, the equation in condtional.

35. $(-3, \infty)$

37. $\left[-2, \dfrac{3}{4}\right)$

39. Graphical: Graph $Y_1 = 3X - 4$ and $Y_2 = 2 + X$. Their graphs intersect at $(3, 5)$. The graph of Y_1 is below the

   graph of Y_2 to the left of the point of intersection. Thus, $3x - 4 \leq 2 + x$ holds when $x \leq 3$ or $(-\infty, 3]$.

   See Figure 39. Symbolic: $3x - 4 \leq 2 + x \Rightarrow 2x \leq 6 \Rightarrow x \leq 3$ or $(-\infty, 3]$. In set-builder notation, the

   interval is $\{x \mid x \leq 3\}$.

41. Graphical: Graph $Y_1 = (2X - 5)/2$ and $Y_2 = (5X + 1)/5$. Their graphs are parallel and never intersect. The

   graph of Y_1 is always below the graph of Y_2, so $Y_1 < Y_2$ for all values of x; the inequality $\dfrac{2x - 5}{2} < \dfrac{5x + 1}{5}$

   holds when $-\infty < x < \infty$, or $(-\infty, \infty)$. See Figure 41. In set-builder notation the interval is

   $\{x \mid -\infty < x < \infty\}$.

[-10, 10, 1] by [-10, 10, 1]   [-10, 10, 1] by [-10, 10, 1]

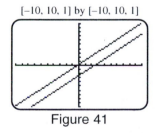 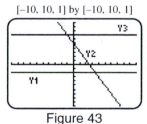

      Figure 41   Figure 43

43. Graphical: Graph $Y_1 = -2$, $Y_2 = 5 - 2X$ and $Y_3 = 7$. See Figure 43. Their graphs intersect at the points

   $(-1, 7)$ and $(3.5, -2)$. The graph of Y_2 is between the graphs of Y_1 and Y_3 when $-1 < x \leq 3.5$. In interval

   notation the solution is $(-1, 3.5]$.

Symbolic: $-2 \le 5 - 2x < 7 \implies -7 \le -2x < 2 \implies \dfrac{7}{2} \ge x > -1 \implies -1 < x \le \dfrac{7}{2}$ or $\left(-1, \dfrac{7}{2}\right]$

In set-builder notation, the interval is $\left\{ x \,\middle|\, -1 < x \le \dfrac{7}{2} \right\}$.

45. Graph $Y_1 = 2x$ and $Y_2 = x - 1$. See Figure 45. The lines intersect at $(-1, -2)$. $Y_1 > Y_2$ when the graph of Y_1 is above the graph of Y_2; this happens when $x > -1 \implies (-1, \infty)$. In set-builder notation the interval is $\{x \mid x > -1\}$.

[-10, 10, 1] by [-10, 10, 1]

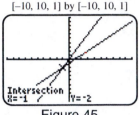

Figure 45

47. (a) The graphs intersect at $(2, 1)$. The solution to $f(x) = g(x)$ is 2.

(b) The graph of f is below the graph of g to the right of $(2, 1)$.

Thus, $f(x) < g(x)$ when $x > 2$ or on $(2, \infty)$.

(c) The graph of f is above the graph of g to the left of $(2, 1)$. Thus, $f(x) > g(x)$ when $x < 2$ or on $(-\infty, 2)$.

49. (a) $f(-2) = 8 + 2(-2) = 4$; $f(-1) = 8 + 2(-1) = 6$; $f(2) = 5 - 2 = 3$; $f(3) = 3 + 1 = 4$.

(b) The graph of f is shown in Figure 49. It is essentially a piecewise line graph with the points $(-3, 2)$, $(-1, 6)$, $(2, 3)$, and $(5, 6)$. Since there are no breaks in the graph, f is continuous.

(c) From the graph we can see that there are two x-values where $f(x) = 3$. They occur when

$8 + 2x = 3 \implies x = -2.5$ and when $5 - x = 3 \implies x = 2$. The solutions are $x = -2.5$ or 2.

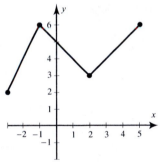

Figure 49

51. $|2x - 5| - 1 = 8 \implies |2x - 5| = 9 \implies 2x - 5 = -9$ or $2x - 5 = 9$; $2x - 5 = -9 \implies 2x = -4 \implies$ $x = -2$; $2x - 5 = 9 \implies 2x = 14 \implies x = 7 \implies -2, 7$

53. $|6 - 4x| = -2$ has no solutions since the absolute value of any quantity is always greater than or equal to 0.

55. $|x| = 3 \Rightarrow x = \pm 3$. The solutions to $|x| > 3$ lie to the left of -3 and to the right of 3. That is, $x < -3$ or $x > 3$. This can be supported by graphing $Y_1 = |x|$ and $Y_2 = 3$ and determining where the graph of Y_1 is above the graph of Y_2. To support this result numerically, table $Y_1 = \text{abs}(X)$ starting at -9 and incrementing by 3.

57. $|3x - 7| = 10 \Rightarrow 3x - 7 = 10$ or $3x - 7 = -10 \Rightarrow x = \dfrac{17}{3}$ or $x = -1$. The solutions to $|3x - 7| > 10$ lie to the left of -1 or to the right of $x = \dfrac{17}{3}$; that is, $x < -1$ or $x > \dfrac{17}{3}$. This can be supported by graphing $Y_1 = |3x - 7|$ and $Y_2 = 10$ and determining where the graph of Y_1 is above the graph of Y_2. To support this result numerically, table $Y_1 = \text{abs}(3X - 7)$ starting at -3 and incrementing by $\dfrac{1}{3}$.

59. The solutions to $|3 - 2x| < 9$ satisfy $s_1 < x < s_2$ where s_1 and s_2, are the solutions to $|3 - 2x| = 9$. $|3 - 2x| = 9$ is equivalent to $3 - 2x = -9 \Rightarrow x = 6$ or $3 - 2x = 9 \Rightarrow x = -3$. The solutions are $-3 < x < 6$ or $(-3, 6)$.

61. The solutions to $\left|\dfrac{1}{3}x - \dfrac{1}{6}\right| \geq 1$ satisfy $x \leq s_1$ or $x \geq s_2$ where s_1 and s_2, are the solutions to $\left|\dfrac{1}{3}x + \dfrac{1}{6}\right| = 1$. $\left|\dfrac{1}{3}x - \dfrac{1}{6}\right| = 1$ is equivalent to $\dfrac{1}{3}x - \dfrac{1}{6} = -1 \Rightarrow x = -\dfrac{5}{2}$ or $\dfrac{1}{3}x - \dfrac{1}{6} = 1 \Rightarrow x = \dfrac{7}{2}$.
The solutions are $x \leq -\dfrac{5}{2}$ or $x \geq \dfrac{7}{2}$ or $\left(-\infty, -\dfrac{5}{2}\right] \cup \left[\dfrac{7}{2}, \infty\right)$.

63. (a) Using the points (1980, 17,700) and (2010, 49,500), $m = \dfrac{49,500 - 17,700}{2010 - 1980} = \dfrac{31,800}{30} = 1060$. Since x represents the number of years after 1980, we have a y-intercept of 17,700, and the function is $f(x) = 1060x + 17,700$.

(b) The median income increased, on average, by about $1060 per year. In 1980 median income was $17,700.

(c) Since 1992 is 12 years after 1980, we will let $x = 12$ and $f(12) = 1060(12) + 17,700 = 30,420$. The calculated result of $30,420 and the true value of $30,600 are approximately equal.

(d) $60,000 = 1060x + 17,700 \Rightarrow 42,300 = 1060x \Rightarrow x \approx 40$, Since x represents the number of years after 1980 the result is about 2020. The calculation involved extrapolation.

65. (a) Using the points (2010, 524) and (2020, 949), $m = \dfrac{524 - 949}{2010 - 2020} = \dfrac{-425}{-10} = 42.5$. Since x represents the number of years after 2010, we have a y-intercept of 524, and the function is $f(x) = 42.5x + 524$.

(b) The spending increased, on average, by about $42.5 billion per year. In 2010 spending was $524 billion.

(c) Since 2016 is 6 years after 2010, we will let $x = 6$ and $f(6) = 42.5(6) + 524 = 779$. The result is $779 billion. The calculation involved interpolation.

(d) $694 \leq 42.5x + 524 \leq 864 \Rightarrow 170 \leq 42.5x \leq 340 \Rightarrow 4 \leq x \leq 8$, since x represents the number of years after 2010 the result is about from 2014 to 2018.

67. Since the graph is piecewise linear, the slope each line segment represents a constant speed. Initially, the car is

home. After 1 hour it is 30 miles from home and has traveled at a constant speed of 30 mph. After 2 hours it is 50 miles away. During the second hour the car travels 20 mph. During the third hour the car travels toward home at 30 mph until it is 20 miles away. During the fourth hour the car travels away from home at 40 mph until it is 60 miles away from home. The last hour the car travels 60 miles at 60 mph until it arrives back at home.

69. The midpoint is computed by $\left(\dfrac{2012 + 2008}{2}, \dfrac{167{,}933 + 143{,}247}{2} \right) = (2010,\ 155{,}590)$.

71. Let $x = $ time it takes for both working together; the first worker can shovel $\dfrac{1}{50}$ of the sidewalk in 1 minute, and the second worker can shovel $\dfrac{1}{30}$ of the sidewalk in 1 minute; for the entire job, we get the equation

$\dfrac{x}{50} + \dfrac{x}{30} = 1 \Rightarrow 3x + 5x = 150 \Rightarrow 8x = 150 \Rightarrow x = 18.75$; it takes the two workers 18.75 minutes to shovel the sidewalk together.

73. Let $t = $ time spent jogging at 7 mph; then $1.8 - t = $ time spent jogging at 8 mph; since $d = rt$ and the total distance jogged is 13.5 miles, we get the equation $7t + 8(1.8 - t) = 13.5 \Rightarrow 7t + 14.4 - 8t = 13.5 \Rightarrow -t = -0.9 \Rightarrow t = 0.9$ and $1.8 - t = 0.9$; the runner jogged 0.9 hour at 7 mph and 0.9 hour at 8 mph.

75. (a) Begin by selecting any two points to determine the equation of the line. For example, if we use $(-1, 4.2)$ and $(2, 0.6)$, then $m = \dfrac{4.2 - 0.6}{-1 - 2} = \dfrac{3.6}{-3} = -1.2$. $y - y_1 = m(x - x_1) \Rightarrow y - 0.6 = -1.2(x - 2) \Rightarrow$

$y - 0.6 = -1.2x + 2.4 \Rightarrow y = -1.2x + 3$.

(b) When $x = -1.5$, then $y = -1.2(-1.5) + 3 = 4.8$. This involves interpolation.

When $x = 3.5$, then $y = -1.2(3.5) + 3 = -1.2$. This involves extrapolation.

(c) $1.3 = -1.2x + 3 \Rightarrow -1.7 = -1.2x \Rightarrow x = \dfrac{17}{12}$.

77. The tank is initially empty. When $0 \le x \le 3$, the slope is 5. The inlet pipe is open; the outlet pipe is closed. When $3 < x \le 5$, the slope is 2. Both pipes are open. When $5 < x \le 8$, the slope is 0. Both pipes are closed. When $8 < x \le 10$, the slope is -3. The inlet pipe is closed; the outlet pipe is open.

79. Let x represent the distance above the ground and let y represent the temperature. Since the ground temperature is $25\,°C$, the point $(0, 25)$ is on the graph of the function which models the situation. Since the rate of change is a constant $-6\,°C$ per kilometer, the model is linear with a slope of $m = -6$. Therefore, the equation of the linear model is $y = -6x + 25$.

Graphically: Graph $Y_1 = 15$, $Y_2 = -6x + 25$, and $Y_3 = 5$ in $[0, 4, 1]$ by $[0, 30, 5]$ See Figure 79. The intersection points are $\left(1\dfrac{2}{3}, 15 \right)$ and $\left(3\dfrac{1}{3}, 5 \right)$. The distance above the ground is between $1\dfrac{2}{3}$ km and $3\dfrac{1}{3}$ km.

Symbolically: Solve $5 \le -6x + 25 \le 15 \Rightarrow -20 \le -6x \le -10 \Rightarrow \dfrac{20}{6} \ge x \ge \dfrac{10}{6} \Rightarrow 1\dfrac{2}{3} \le x \le 3\dfrac{1}{3}$.

The solution interval is the same for either method, $\left[1\dfrac{2}{3}, 3\dfrac{1}{3}\right]$. The distance above the ground is between

$1\dfrac{2}{3}$ km and $3\dfrac{1}{3}$ km.

[0, 4, 1] by [0, 30, 5]

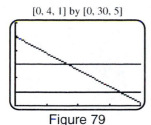

Figure 79

81. $\left|\dfrac{C - A}{A}\right| \leq 0.003 \Rightarrow -0.003 \leq \dfrac{C - 52.3}{52.3} \leq 0.003 \Rightarrow -0.1569 \leq C - 52.3 \leq 0.1569 \Rightarrow$

$52.1431 \leq C \leq 52.4569 \Rightarrow$ between 52.1431 and 52.4569 ft.

Extended and Discovery Exercises for Chapter 2

1. (a) 62.8 inches

   (b) The (x, y) pairs for females are plotted in Figure 1a and for males in Figure 1b. Both sets of data appear to

      be linear.

   (c) Female: 3.1 inches; male: 3.0 inches

   (d) $f(x) = 3.1(x - 8) + 50.4$; $g(x) = 3.0(x - 8) + 53$

   (e) $f(9.7) = 55.67$ and $f(10.1) = 56.91$. For a female, the height could vary between 55.67 and 56.91 inches.

      $g(9.7) = 58.1$ and $g(10.1) = 59.3$. For a male, the height could vary between 58.1 and 59.3 inches.

[7, 15, 1] by [45, 75, 5]   [7, 15, 1] by [45, 75, 5]

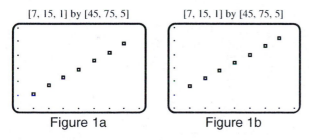

Figure 1a   Figure 1b

3. Let x represent the distance walked by the 1st person. Let z represent the distance walked by the 2nd person.

   Let y represent the distance the car travels between dropping off the 2nd person and picking up the 1st person.

   Refer to Figure 3.

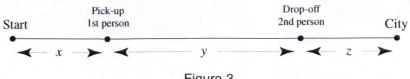

Figure 3

Using the formula time $= \dfrac{\text{distance}}{\text{rate}}$ we obtain the following results:

(1) (time for 1st person to walk distance x) $=$ (time for car to drive distance $x + 2y$) $\Rightarrow \dfrac{x}{4} = \dfrac{x + 2y}{28} \Rightarrow$

$28x = 4x + 8y \Rightarrow 24x = 8y \Rightarrow 3x = y$.

(3) The total distance is 15 miles. Thus, $x + y + z = 15$.

Solving these three equations simultaneously results in $x = 3, y = 9, z = 3$. Each person walked 3 miles.

Chapters 1-2 Cumulative Review Exercises

1. Move the decimal point five places to the left; $123{,}000 = 1.23 \times 10^5$

   Move the decimal point three places to the right; $0.005 = 5.1 \times 10^{-3}$

3. $\dfrac{4 + \sqrt{2}}{4 - \sqrt{2}} \approx 2.09$

5. The standard equation of a circle must fit the form $(x - h)^2 + (y - k)^2 = r^2$, where (h, k) is the center and the

   radius r. The equation of the circle with center $(-2, 3)$ and radius 7 is $(x + 2)^2 + (y - 3)^2 = 49$.

7. $d = \sqrt{[2 - (-3)]^2 + ((-3) - 5)^2} = \sqrt{25 + 64} = \sqrt{89}$

9. (a) $D =$ all real numbers $\Rightarrow \{x \mid -\infty < x < \infty\}$; $R = \{y \mid y \geq -2\}$; $f(-1) = -1$

   (b) $D = \{x \mid -3 \leq x \leq 3\}$; $R = \{y \mid -3 \leq y \leq 2\}$; $f(-1) = -\dfrac{1}{2}$

11. (a) $f(2) = 5(2) - 3 = 7$; $f(a - 1) = 5(a - 1) - 3 = 5a - 5 - 3 = 5a - 8$

    (b) The domain of f includes all real numbers $\Rightarrow D = \{x \mid -\infty \leq x \leq \infty\}$

13. No, this is not a graph of a function because some vertical lines intersect the graph twice.

15. $f(1) = (1)^2 - 2(1) + 1 = 1 - 2 + 1 = 0 \Rightarrow (1, 0)$; $f(2) = (2)^2 - 2(2) + 1 = 4 - 4 + 1 = 1 \Rightarrow$

    $(2, 1)$. The slope $m = \dfrac{1 - 0}{2 - 1} = \dfrac{1}{1} = 1$, so the average rate of change is 1.

17. (a) $m = \dfrac{2}{3}$; y-intercept: -2, x-intercept: 3

   (b) $f(x) = mx + b \Rightarrow f(x) = \dfrac{2}{3}x - 2$

   (c) 3

19. $m = \dfrac{\frac{1}{2} - (-5)}{-3 - 1} = \dfrac{\frac{11}{2}}{-4} = -\dfrac{11}{8}$; using $(1, -5)$ and point-slope form: $y = -\dfrac{11}{8}(x - 1) - 5 \Rightarrow$

   $y = -\dfrac{11}{8}x + \dfrac{11}{8} - 5 \Rightarrow y = -\dfrac{11}{8}x - \dfrac{29}{8}$

21. All lines parallel to the y-axis have undefined slope $\Rightarrow$ y changes but x remains constant $\Rightarrow$ $x = -1$.

23. For the points $(2.4, 5.6)$ and $(3.9, 8.6)$ we get $m = \dfrac{8.6 - 5.6}{3.9 - 2.4} = \dfrac{3}{1.5} = 2$. A line parallel to this has

   the same slope. Using point-slope form: $y = 2(x + 3) + 5 \Rightarrow y = 2x + 11$.

25. For $-2x + 3y = 6$: x-intercept, then $y = 0 \Rightarrow -2x + 3(0) = 6 \Rightarrow -2x = 6 \Rightarrow x = -3$;

   y-intercept, then $x = 0 \Rightarrow -2(0) + 3y = 6 \Rightarrow 3y = 6 \Rightarrow y = 2$. See Figure 25.

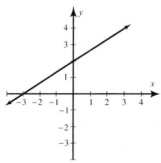

Figure 25

27. $4x - 5 = 1 - 2x \Rightarrow 6x = 6 \Rightarrow x = 1$; 1

29. $\dfrac{2}{3}(x - 2) - \dfrac{4}{5}x = \dfrac{4}{15} + x \Rightarrow \dfrac{2}{3}x - \dfrac{4}{3} - \dfrac{4}{5}x = \dfrac{4}{15} + x \Rightarrow 15\left(\dfrac{2}{3}x - \dfrac{4}{3} - \dfrac{4}{5}x = \dfrac{4}{15} + x\right) \Rightarrow$

   $10x - 20 - 12x = 4 + 15x \Rightarrow -17x = 24 \Rightarrow x = -\dfrac{24}{17}$; $-\dfrac{24}{17}$

31. Graph $Y_1 = X + 1$ and $Y_2 = 2X - 2$. See Figure 31a. The lines intersect at point $(3, 4) \Rightarrow x = 3$. Make a

   table of $Y_1 = X + 1$ and $Y_2 = 2X - 2$ for x values from 0 to 5. See Figure 31b.

   Both equations have y-value 4 at $x = 3$.

$[-10, 10, 1]$ by $[-10, 10, 1]$

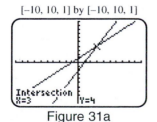

Figure 31a   Figure 31b

33. $(-\infty, 5)$

35. $(-\infty, -2)\cup(2, \infty)$

37. $-3(1 - 2x) + x \le 4 - (x + 2) \Rightarrow -3 + 6x + x \le 4 - x - 2 \Rightarrow 7x - 3 \le -x + 2 \Rightarrow 8x \le 5 \Rightarrow$

    $x \le \dfrac{5}{8} \Rightarrow \left(-\infty, \dfrac{5}{8}\right]$. In set builder notation the interval is $\left\{x \mid x \le \dfrac{5}{8}\right\}$.

39. (a) 2

    (b) The graph of $f(x)$ is above the graph of $g(x)$ to the left of $x = 2 \Rightarrow x < 2$

    (c) The graph of $f(x)$ intersects or is below the graph of $g(x)$ to the right of $x = 2 \Rightarrow f(x) \le g(x)$ when $x \ge 2$.

41. $|d + 1| = 5 \Rightarrow d + 1 = 5$ or $d + 1 = -5$. If $d + 1 = -5, d = -6$; if $d + 1 = 5, d = 4 \Rightarrow -6, 4$.

43. $|2t| - 4 = 10 \Rightarrow |2t| = 14 \Rightarrow 2t = 14$ or $2t = -14$. If $2t = -14, t = -\dfrac{14}{2} \Rightarrow t = -7$; if

    $2t = 14, t = 7 \Rightarrow -7, 7$.

45. The solutions to $|2t - 5| \le 5$ satisfy $s_1 \le t \le s_2$ where s_1 and s_2, are the solutions to $|2t - 5| = 5$.

    $|2t - 5| = 5$ is equivalent to $2t - 5 = -5 \Rightarrow t = 0$ or $2t - 5 = 5 \Rightarrow t = 5$.

    The interval is $[0, 5]$. In set-builder notation the interval is $\{t \mid 0 \le t \le 5\}$.

47. (a) $C(1500) = 500(1500) + 20,000 = 770,000$; it costs \$770,000 to manufacture 1500 computers.

    (b) 500; each additional computer costs \$500 to manufacture and fixed costs are \$20,000.

49. (a) $T(2) = 70 + \dfrac{3}{2}(2)^2 = 70 + 6 = 76$; $T(4) = 70 + \dfrac{3}{2}(4)^2 = 70 + 24 = 94$

    Using (2, 76) and (4, 94): $m = \dfrac{94 - 76}{4 - 2} = \dfrac{18}{2} = 9°F$ increase per hour.

    (b) On average the temperature increased by $9°F$ per hour over this 2-hour period.

51. Let $t =$ time for the two to mow the lawn together. Then the first person mows $\dfrac{1}{5}t$ of the lawn and the second

    person mows $\dfrac{1}{12}t$ of the lawn $\Rightarrow \dfrac{1}{5}t + \dfrac{1}{12}t = 1 \Rightarrow \dfrac{12}{60}t + \dfrac{5}{60}t = 1 \Rightarrow \dfrac{17}{60}t = 1 \Rightarrow t = \dfrac{60}{17} = 3.53$ hours.

53. (a) Using (2001, 56) and (2010, 84), $m = \dfrac{84 - 56}{2010 - 2001} = \dfrac{28}{9}$; $f(x) = \dfrac{28}{9}(x - 2001) + 56$ or

    $f(x) = \dfrac{28}{9}(x - 2010) + 84$.

    (b) $f(2015) = \dfrac{28}{9}(2015 - 2010) + 84 = \dfrac{140}{9} + 84 \approx 100$ pounds.

Chapter 3: Quadratic Functions and Equations

3.1: Quadratic Functions and Models

1. $f(x) = 1 - 2x + 3x^2$ is quadratic; $a = 3$; $f(-2) = 1 - 2(-2) + 3(-2)^2 = 17$.

3. $f(x) = \dfrac{1}{x^2 - 1}$ is neither linear nor quadratic.

5. $f(x) = \dfrac{1}{2} - \dfrac{3}{10}x$ is linear.

7. (a) $a > 0$

   (b) vertex: $(1, 0)$

   (c) axis of symmetry: $x = 1$

   (d) f is increasing for $x > 1$ and decreasing for $x < 1$.

   (e) $D = (-\infty, \infty)$; $R = [0, \infty)$

9. (a) $a < 0$

   (b) vertex: $(-3, -2)$

   (c) axis of symmetry: $x = -3$

   (d) f is increasing for $x < -3$ and decreasing for $x > -3$.

   (e) $D = (-\infty, \infty)$; $R = (-\infty, -2]$

11. The graph of g is narrower than the graph of f.

13. The graph of g is wider than the graph of f and opens downward rather than upward.

15. $f(x) = -3(x - 1)^2 + 2 \Rightarrow$ vertex: $(1, 2)$; leading coefficient: -3; $f(x) = -3x^2 + 6x - 1$

17. $f(x) = 5 - 2(x - 4)^2 \Rightarrow$ vertex: $(4, 5)$; leading coefficient: -2; $f(x) = -2x^2 + 16x - 27$

19. $f(x) = \dfrac{3}{4}(x + 5)^2 - \dfrac{7}{4} \Rightarrow$ vertex: $\left(-5, -\dfrac{7}{4}\right)$; leading coefficient: $\dfrac{3}{4}$; $f(x) = \dfrac{3}{4}x^2 + \dfrac{15}{2}x + 17$

21. The vertex of the parabola is $(2, -2)$, so $f(x) = a(x - h)^2 + k \Rightarrow f(x) = a(x - 2)^2 - 2$;   since $(0, 2)$ is a point on the parabola, $f(0) = a(0 - 2)^2 - 2 \Rightarrow 2 = a(-2)^2 - 2 \Rightarrow 4 = 4a \Rightarrow a = 1$; $f(x) = (x - 2)^2 - 2$.

23. The vertex of the parabola is $(2, -3)$, so $f(x) = a(x - h)^2 + k \Rightarrow f(x) = a(x - 2)^2 - 3$;   since $(0, -1)$ is a point on the parabola, $f(0) = a(0 - 2)^2 - 3 \Rightarrow -1 = a(4) - 3 \Rightarrow -1 = 4a - 3 \Rightarrow 2 = 4a \Rightarrow a = \dfrac{1}{2}$; $f(x) = \dfrac{1}{2}(x - 2)^2 - 3$.

25. The vertex of the parabola is $(-1, 3)$, so $f(x) = a(x - h)^2 + k \Rightarrow f(x) = a(x + 1)^2 + 3$;   since $(0, 1)$ is a point on the parabola, $f(0) = a(0 + 1)^2 + 3 \Rightarrow 1 = a(1) + 3 \Rightarrow 1 = a + 3 \Rightarrow a = -2$; $f(x) = -2(x + 1)^2 + 3$.

27. The vertex of the parabola is $(2, 6)$, so $f(x) = a(x - h)^2 + k \Rightarrow f(x) = a(x - 2)^2 + 6$;   since $(0, -6)$ is a point on the parabola, $f(0) = a(0 - 2)^2 + 6 \Rightarrow -6 = 4a + 6 \Rightarrow -12 = 4a \Rightarrow a = -3$; $f(x) = -3(x - 2)^2 + 6$.

29. $f(x) = x^2 + 4x - 5 \Rightarrow x = -\dfrac{b}{2a} = -\dfrac{4}{2(1)} = -2$ and

$f(-2) = (-2)^2 + 4(-2) - 5 = -9 \Rightarrow$ vertex: $(-2, -9)$; since $a = 1, f(x) = 1(x - (-2))^2 - 9$, or

$f(x) = (x + 2)^2 - 9$

31. $f(x) = x^2 - 3x \Rightarrow x = -\dfrac{b}{2a} = -\dfrac{(-3)}{2(1)} = \dfrac{3}{2}$ and

$f\left(\dfrac{3}{2}\right) = \left(\dfrac{3}{2}\right)^2 - 3\left(\dfrac{3}{2}\right) = -\dfrac{9}{4} \Rightarrow$ vertex: $\left(\dfrac{3}{2}, -\dfrac{9}{4}\right)$; since $a = 1, f(x) = 1\left(x - \dfrac{3}{2}\right)^2 - \dfrac{9}{4}$, or

$f(x) = \left(x - \dfrac{3}{2}\right)^2 - \dfrac{9}{4}$

33. $f(x) = 2x^2 - 5x + 3 \Rightarrow x = -\dfrac{b}{2a} = -\dfrac{(-5)}{2(2)} = \dfrac{5}{4}$ and

$f\left(\dfrac{5}{4}\right) = 2\left(\dfrac{5}{4}\right)^2 - 5\left(\dfrac{5}{4}\right) + 3 = -\dfrac{1}{8} \Rightarrow$ vertex: $\left(\dfrac{5}{4}, -\dfrac{1}{8}\right)$; since $a = 2, f(x) = 2\left(x - \dfrac{5}{4}\right)^2 - \dfrac{1}{8}$

35. $f(x) = \dfrac{1}{3}x^2 + x + 1 \Rightarrow x = -\dfrac{b}{2a} = -\dfrac{1}{2(\frac{1}{3})} = -\dfrac{1}{\frac{2}{3}} = -\dfrac{3}{2}$ and

$f\left(-\dfrac{3}{2}\right) = \dfrac{1}{3}\left(-\dfrac{3}{2}\right)^2 - \dfrac{3}{2} + 1 = \dfrac{1}{4} \Rightarrow$ vertex: $\left(-\dfrac{3}{2}, \dfrac{1}{4}\right)$; since $a = \dfrac{1}{3}, f(x) = \dfrac{1}{3}\left(x + \dfrac{3}{2}\right)^2 + \dfrac{1}{4}$

37. $f(x) = 2x^2 - 8x - 1 \Rightarrow x = -\dfrac{b}{2a} = -\dfrac{(-8)}{2(2)} = 2$ and

$f(2) = 2(2)^2 - 8(2) - 1 = -9 \Rightarrow$ vertex: $(2, -9)$; since $a = 2, f(x) = 2(x - 2)^2 - 9$

39. $f(x) = 2 - 9x - 3x^2 \Rightarrow x = -\dfrac{b}{2a} = -\dfrac{(-9)}{2(-3)} = -1.5$ and

$f(-1.5) = 2 - 9(-1.5) - 3(-1.5)^2 = 8.75 \Rightarrow$ vertex: $(-1.5, 8.75)$; since $a = -3$,

$f(x) = -3(x - (-1.5))^2 + 8.75$, or $f(x) = -3(x + 1.5)^2 + 8.75$

41. (a) It may be helpful to write f in standard form as follows: $f(x) = -x^2 + 0x + 6$.

To find the vertex symbolically, use the vertex formula with $a = -1$ and $b = 0$.

$x = -\dfrac{b}{2a} = -\dfrac{0}{2(-1)} = \dfrac{0}{2} = 0$. The x-coordinate of the vertex is 0.

$y = f\left(-\dfrac{b}{2a}\right) = f(0) = 6 - (0)^2 = 6$.   The y-coordinate of the vertex is 6. Thus, the vertex is $(0, 6)$.

The graph of $Y_1 = 6 - X^2$ shows graphical support for this vertex. See Figure 41.

(b) f is increasing on $x \le 0$ or $(-\infty, 0]$ and decreasing on $x \ge 0$ or $[0, \infty)$.

[−10, 10, 1] by [−10, 10, 1]   [−10, 10, 1] by [−15, 5, 1]

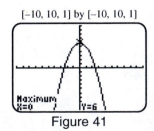
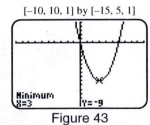

Figure 41   Figure 43

43. (a) It may be helpful to write f in standard form as follows: $f(x) = x^2 - 6x + 0$.

To find the vertex symbolically, use the vertex formula with $a = 1$ and $b = -6$.

$x = -\dfrac{b}{2a} = -\dfrac{(-6)}{2(1)} = \dfrac{6}{2} = 3$. The x-coordinate of the vertex is 3.

$y = f\left(-\dfrac{b}{2a}\right) = f(3) = (3)^2 - 6(3) = -9$. The y-coordinate of the vertex is –9. Thus, the vertex is

$(3, -9)$. The graph of $Y_1 = X^2 - 6X$ shows graphical support for this vertex. See Figure 43.

(b) f is increasing on $x \geq 3$ or $[3, \infty)$ and decreasing on $x \leq 3$ or $(-\infty, 3]$.

45. (a) The function is already in standard form: $f(x) = 2x^2 - 4x + 1$.

To find the vertex symbolically, use the vertex formula with $a = 2$ and $b = -4$.

$x = -\dfrac{b}{2a} = -\dfrac{(-4)}{2(2)} = \dfrac{4}{4} = 1$. The x-coordinate of the vertex is 1.

$y = f\left(-\dfrac{b}{2a}\right) = f(1) = 2(1)^2 - 4(1) + 1 = -1$. The y-coordinate of the vertex is -1 Thus, the vertex

is $(1, -1)$. The graph of $Y_1 = 2X^2 - 4X + 1$ shows graphical support for this vertex. See Figure 45.

(b) f is increasing on $x \geq 1$ or $[1, \infty)$ and decreasing on $x \leq 1$ or $(-\infty, 1]$.

[–10, 10, 1] by [–10, 10, 1]   [–40, 40, 5] by [–40, 40, 5]

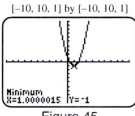
Figure 45

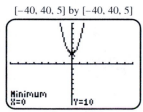
Figure 47

47. (a) It may be helpful to write f in standard form as follows: $f(x) = \dfrac{1}{2}x^2 + 0x + 10$.

To find the vertex symbolically, use the vertex formula with $a = \dfrac{1}{2}$ and $b = 0$.

$x = -\dfrac{b}{2a} = -\dfrac{0}{2(\frac{1}{2})} = 0$. The x-coordinate of the vertex is 0.

$y = f\left(-\dfrac{b}{2a}\right) = f(0) = \dfrac{1}{2}(0)^2 + 0(0) + 10 = 10$. The y-coordinate of the vertex is 10. Thus, the

vertex is $(0, 10)$. The graph of $Y_1 = (1/2)X^2 + 10$ shows graphical support for this vertex. See Figure 47.

(b) f is increasing on $x \geq 0$ or $[0, \infty)$ and decreasing on $x \leq 0$ or $(-\infty, 0]$.

49. (a) The function is already in standard form: $f(x) = -\dfrac{3}{4}x^2 + \dfrac{1}{2}x - 3$.

   To find the vertex symbolically, use the vertex formula with $a = -\dfrac{3}{4}$ and $b = \dfrac{1}{2}$.

   $x = -\dfrac{b}{2a} = -\dfrac{\frac{1}{2}}{2\left(-\frac{3}{4}\right)} = \dfrac{1}{3}$. The x-coordinate of the vertex is $\dfrac{1}{3}$.

   $y = f\left(-\dfrac{b}{2a}\right) = f\left(\dfrac{1}{3}\right) = -\dfrac{3}{4}\left(\dfrac{1}{3}\right)^2 + \dfrac{1}{2}\left(\dfrac{1}{3}\right) - 3 = -\dfrac{35}{12}$. The y-coordinate of the vertex is $-\dfrac{35}{12}$.

   Thus, the vertex is $\left(\dfrac{1}{3}, -\dfrac{35}{12}\right)$. The graph of $Y_1 = (-3/4)X^2 + (1/2)X - 3$ shows graphical support for this vertex. See Figure 49.

   (b) f is increasing on $x \le \dfrac{1}{3}$ or $\left(-\infty, \dfrac{1}{3}\right]$ and decreasing on $x \ge \dfrac{1}{3}$ or $\left[\dfrac{1}{3}, \infty\right)$.

   [−10, 10, 1] by [−10, 10, 1]

   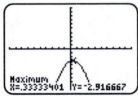

   Figure 49

51. (a) It may be helpful to write f in standard form as follows: $f(x) = -6x^2 - 3x + 1.5$.

   To find the vertex symbolically, use the vertex formula with $a = -6$ and $b = -3$.

   $x = -\dfrac{b}{2a} = -\dfrac{(-3)}{2(-6)} = -\dfrac{3}{12} = -\dfrac{1}{4}$. The x-coordinate of the vertex is $-\dfrac{1}{4}$.

   $y = f\left(-\dfrac{b}{2a}\right) = f\left(-\dfrac{1}{4}\right) = 1.5 - 3\left(-\dfrac{1}{4}\right) - 6\left(-\dfrac{1}{4}\right)^2 = 1\dfrac{7}{8}$. The y-coordinate of the vertex

   is $1\dfrac{7}{8}$. Thus, the vertex is $\left(-\dfrac{1}{4}, 1\dfrac{7}{8}\right)$. The graph of $Y_1 = 1.5 - 3X - 6X^2$ shows graphical support for this vertex. See Figure 51.

   (b) f is increasing on $x \le -\dfrac{1}{4}$ or $\left(-\infty, -\dfrac{1}{4}\right]$ and decreasing on $x \ge -\dfrac{1}{4}$ or $\left[-\dfrac{1}{4}, \infty\right)$.

   [−3, 3, 1] by [−5, 5, 1]]

   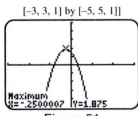

   Figure 51

53. Because $a > 0$ the graph of f is a parabola opening upward. The vertex is the lowest point on the graph, with

    an x-coordinate of $x = \dfrac{-4}{2(1)} = -2$. The corresponding y-coordinate is $f(-2) = (-2)^2 + 4(-2) - 2 = -6$.

    Thus the vertex is $-2, -6$ and the minimum y-value is -6.

55. Because $a > 0$ he graph of f is a parabola opening upward. The vertex is the lowest point on the graph, with an

    x-coordinate of $x = \dfrac{4}{2(3)} = \dfrac{2}{3}$. The corresponding y-coordinate is $f\left(\dfrac{2}{3}\right) = 3\left(\dfrac{2}{3}\right)^2 - 4\left(\dfrac{2}{3}\right) + 2 = \dfrac{2}{3}$. Thus

    the vertex is $\left(\dfrac{2}{3}, \dfrac{2}{3}\right)$ and the minimum y-value is $\dfrac{2}{3}$.

57. Because $a > 0$ the graph of f is a parabola opening upward. The vertex is the lowest point on the graph, with an

    x-coordinate of $x = \dfrac{-3}{2(1)} = -\dfrac{3}{2}$. The corresponding y-coordinate is $f\left(-\dfrac{3}{2}\right) = \left(-\dfrac{3}{2}\right)^2 + 3\left(-\dfrac{3}{2}\right) + 5 = \dfrac{11}{4}$.

    Thus the vertex is $\left(-\dfrac{3}{2}, \dfrac{11}{4}\right)$ and the minimum y-value is $\dfrac{11}{4}$.

59. Because $a > 0$ the graph of f is a parabola opening upward. The vertex is the lowest point on the graph, with

    an x-coordinate of $x = \dfrac{-3}{2(-1)} = \dfrac{3}{2}$. The corresponding y-coordinate is $f\left(\dfrac{3}{2}\right) = -\left(\dfrac{3}{2}\right)^2 + 3\left(\dfrac{3}{2}\right) - 2 = \dfrac{1}{4}$.

    Thus the vertex is $\left(\dfrac{3}{2}, \dfrac{1}{4}\right)$ and the maximum y-value is $\dfrac{1}{4}$.

61. Because $a < 0$ the graph of f is a parabola opening downward. The vertex is the highest point on the graph,

    with an x-coordinate of $x = \dfrac{-5}{2(-1)} = \dfrac{5}{2}$. The corresponding y-coordinate is $f\left(\dfrac{5}{2}\right) = 5\left(\dfrac{5}{2}\right) - \left(\dfrac{5}{2}\right)^2 = \dfrac{25}{4}$.

    Thus the vertex is $\left(\dfrac{5}{2}, \dfrac{25}{4}\right)$ and the maximum y-value is $\dfrac{25}{4}$.

63. Because $a < 0$ the graph of f is a parabola opening downward. The vertex is the highest point on the graph,

    with an x-coordinate of $x = \dfrac{-2}{2(-3)} = \dfrac{1}{3}$. The corresponding y-coordinate is $f\left(\dfrac{1}{3}\right) = 2\left(\dfrac{1}{3}\right) - 3\left(\dfrac{1}{3}\right)^2 = \dfrac{1}{3}$.

    and the maximum y-value is $\dfrac{1}{3}$.

65. The graph of $f(x) = x^2$ is shown in Figure 65.

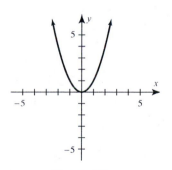

Figure 65

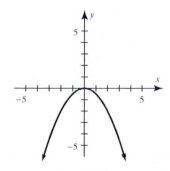

Figure 67

67. The graph of $f(x) = -\dfrac{1}{2}x^2$ is shown in Figure 67.

69. The graph of $f(x) = x^2 - 3$ is shown in Figure 69.

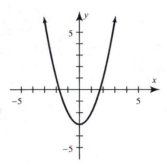

Figure 69

71. The graph of $f(x) = (x - 2)^2 + 1$ is shown in Figure 71.

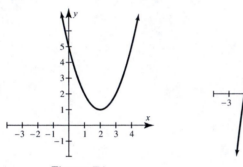

Figure 71   Figure 73

73. The graph of $f(x) = -3(x + 1)^2 + 3$ is shown in Figure 73.

75. The graph of $f(x) = x^2 - 2x - 2$ is shown in Figure 75.

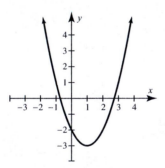

Figure 75

77. The graph of $f(x) = -x^2 + 4x - 2$ is shown in Figure 77.

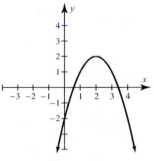

Figure 77

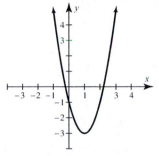

Figure 79

79. The graph of $f(x) = 2x^2 - 4x - 1$ is shown in Figure 79.

81. The graph of $f(x) = -3x^2 - 6x + 1$ is shown in Figure 81.

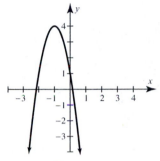

Figure 81

83. The graph of $f(x) = -\dfrac{1}{2}x^2 + x + 1$ is shown in Figure 83.

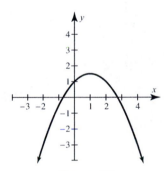

Figure 83

85. The average rate of change from 1 to 3 for $f(x) = -3x^2 + 5x$ is $\dfrac{f(3) - f(1)}{3 - 1} = \dfrac{-14}{2} = -7.$

87. $\dfrac{f(x + h) - f(x)}{h} = \dfrac{3(x + h)^2 - 2(x + h) - [3x^2 - 2x]}{h} =$

 $\dfrac{3x^2 + 6xh + 3h^2 - 2x - 2h - 3x^2 + 2x}{h} = \dfrac{6xh + 3h^2 - 2h}{h} = 6x + 3h - 2$

89. Since the axis of symmetry is $x = 3$ and the graph passes through the point $(3, 1)$, the vertex is $(3, 1)$.

 $f(x) = a(x - h)^2 + k \Rightarrow f(x) = a(x - 3)^2 + 1.$ Using the point $(1, 9)$ to find a gives

 $9 = a(1 - 3)^2 + 1 \Rightarrow a = 2; \; f(x) = 2(x - 3)^2 + 1.$

91. A stone thrown from ground level would first rise to some maximum height and then fall back to the ground. This is represented in figure d.

93. When the furnace first fails to work, the temperature begins to drop. Then, after the furnace is repaired, the temperature begins to rise. This is represented in figure a.

95. Perimeter of fence $= 2l + 2w = 1000 \Rightarrow l = \dfrac{1000 - 2w}{2} \Rightarrow l = 500 - w$.

   If $A = lw$ then $A = (500 - w)w \Rightarrow A = 500w - w^2 \Rightarrow A = -w^2 + 500w$.   This is a parabola opening downward and by the vertex formula, the maximum area occurs when $w = -\dfrac{b}{2a} = -\dfrac{500}{2(-1)} = 250$. The dimensions that maximize area are 250 ft. by 250 ft.

97. (a) $R(x) = x(40 - 2x) \Rightarrow R(2) = 2(40 - 2(2)) = 2(36) = 72$;   the company receives \$72,000 for producing 2000 CD players.

   (b) $R(x) = 40x - 2x^2$ is a quadratic function; to find the value at which the maximum value occurs, we need to find the vertex: $x = -\dfrac{b}{2a} = -\dfrac{40}{2(-2)} = 10$ and $R(10) = 40(10) - 2(10)^2 = 200 \Rightarrow (10, 200)$ is the vertex, thus, the company needs to produce 10,000 CD players to maximize its revenue.

   (c) Since $R(10) = 200$, the maximum revenue for the company is \$200,000.

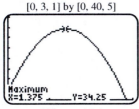

[0, 3, 1] by [0, 40, 5]

Figure 101

99. (a) Because $a > 0$ the graph of f is a parabola opening upward.

   (b) The vertex is the lowest point on the graph, with an x-coordinate of $x = \dfrac{0.0145}{2(0.00000093)} \approx 7796$. The corresponding y-coordinate is $f(7796) = 0.00000093(7796)^2 - 0.0145(7796) + 60 \approx 3.5$. Thus the vertex is approximately $(7796, 3.5)$. The minimum average cost per copy is 3.5 cents when 7796 copies are made.

101.(a) $s(1) = -16(1)^2 + 44(1) + 4 = 32$; the baseball is 32 feet high after 1 second.

   (b) For $s(t) = -16t^2 + 44t + 4$, the vertex formula gives $t = -\dfrac{b}{2a} = -\dfrac{44}{2(-16)} = 1.375$ and $f(1.375) = -16(1.375)^2 + 44(1.375) + 4 = 34.25$; the maximum height 34.25 feet. See Figure 101.

103.(a) The initial velocity $v_0 = -66$ and the initial height is $h_0 = 120$, so $s(t) = -16t^2 + v_0 t + h_0 \Rightarrow s(t) = -16t^2 - 66t + 120$.

   (b) When the stone hits the water, $s(t) = 0$; $0 = -16t^2 - 66t + 120 \Rightarrow 0 = -2(8t^2 + 33t - 60) \Rightarrow 0 = 8t^2 + 33t - 60$; using the quadratic formula or graphing the parabola, we find that $s(t) = 0$ when $t \approx 1.37$ seconds, so the stone hits the water within the first 2 seconds.

105. Using the wall of the barn for one side gives us $W + L + W = 160$ for the three sides $\Rightarrow L = 160 - 2W$.

If $A = LW$ then $A = (160 - 2W)W \Rightarrow A = 160W - 2W^2$.   Since this is a parabola opening downward,

maximum area occurs when $W = -\dfrac{b}{2a} \Rightarrow W = \dfrac{-160}{2(-2)} \Rightarrow W = 40$.   Then $L = 160 - 2(40) = 80$.   Thus,

the dimensions that yield maximum area are 40 feet by 80 feet.

107. (a) When $g = 32$, $v_0 = 88$, and $h_0 = 25$, then $f(x) = -\dfrac{1}{2}(32)x^2 + 88x + 25 = -16x^2 + 88x + 25$.

Graph $Y_1 = -16X^2 + 88X + 25$. By using the calculator, the maximum height is found to be

approximately 146 feet. The maximum height of 146 ft occurs when $x = 2.75$ seconds. See Figure 107.

(b) To find the maximum height symbolically, use the vertex formula with $a = -16$ and $b = 88$.

$x = -\dfrac{b}{2a} = -\dfrac{88}{2(-16)} = \dfrac{88}{32} = 2.75$. The maximum height occurs at $x = 2.75$ seconds.

$y = f\left(-\dfrac{b}{2a}\right) = f(2.75) = -16(2.75)^2 + 88(2.75) + 25 = 146$. The maximum height is 146 feet.

[0, 6, 1] by [0, 160, 10]   [0, 15, 5] by [0, 400, 100]

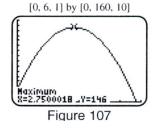

Maximum
X=2.7500018 Y=146

Maximum
X=6.7692285 Y=322.84615

Figure 107   Figure 109

109. (a) When $g = 13$, $v_0 = 88$, and $h_0 = 25$, then $f(x) = -\dfrac{1}{2}(13)x^2 + 88x + 25 = -6.5x^2 + 88x + 25$.

Graph $Y_1 = -6.5X^2 + 88X + 25$. By using the calculator, the maximum height is found to be

approximately 323 feet. The maximum height of about 323 ft occurs when $x \approx 6.77$ seconds. See Figure 109.

(b) To find the maximum height symbolically, use the vertex formula with $a = -6.5$ and $b = 88$.

$x = -\dfrac{b}{2a} = -\dfrac{88}{2(-6.5)} = \dfrac{88}{13} \approx 6.77$. The maximum height occurs at $x \approx 6.77$ seconds.

$y = f\left(-\dfrac{b}{2a}\right) = f(6.77) = -6.5(6.77)^2 + 88(6.77) + 25 \approx 323$. The maximum height is about 323 feet.

111. The cables drawn as a parabola can be placed in the coordinate plane. Use $(0, 20)$ as the vertex.

$f(x) = a(x - h)^2 + k \Rightarrow f(x) = a(x - 0)^2 + 20 = f(x) = ax^2 + 20$; since the parabola passes through

$(150, 120)$, $f(150) = 120$; $f(150) = a(150)^2 + 20 \Rightarrow 120 = 22{,}500a + 20 \Rightarrow 100 = 22{,}500a \Rightarrow$

$a = \dfrac{1}{225}$; $f(x) = \dfrac{1}{225}x^2 + 20$, or $f(x) \approx 0.0044x^2 + 20$.

113. The smallest y-value is -3 when $x = 1$; the symmetry in the y-values about $x = 1$ indicates that the axis of

symmetry is $x = 1$ and the vertex is $(1, -3)$, so $f(x) = a(x - 1)^2 - 3$. Since $(0, -1)$ is a data point,

$f(0) = -1$; $f(0) = a(0 - 1)^2 - 3 \Rightarrow -1 = a - 3 \Rightarrow a = 2$; the function $f(x) = 2(x - 1)^2 - 3$

models the data exactly.

115. (a) Since the minimum occurs at $t = 4$, let $(4, 90)$ be the vertex and write $H(t) = a(t - 4)^2 + 90$. Use

the data point $(0, 122)$ to find a: $H(0) = a(0 - 4)^2 + 90 \Rightarrow 122 = a(0 - 4)^2 + 90 \Rightarrow$

$122 = 16a + 90 \Rightarrow 16a = 32 \Rightarrow a = 2$; $H(t) = 2(t - 4)^2 + 90$; Domain of $H = \{0 \le t \le 4\}$

   (b) $H(1.5) = 2(1.5 - 4)^2 + 90 \Rightarrow H(1.5) = 2(-2.5)^2 + 90 \Rightarrow H(1.5) = 12.5 + 90 \Rightarrow$

$H(1.5) = 102.5$ beats per minute.

117. (a) Since the minimum number of visitors for the selected years is 1.5 million in 2007, let $(0, 1.5)$ be the ver-

tex and write $f(x) = a(x - 0)^2 + 1.5$. We choose the point $(4, 60.1)$ to substitute into the equation and

solve for a as follows, $60.1 = a(4 - 0)^2 + 1.5 \Rightarrow 60.1 = 16a + 1.5 \Rightarrow 58.6 = 16a \Rightarrow a \approx 3.66$ Thus

$f(x) = 3.66x^2 + 1.5$ can be used to model the number of Slideshare visitors since 2007.

   (b) Since x represents the number of years since 2007 let $x = 5$. $f(5) = 3.66(5)^2 + 1.5 = 93$ million.

119. (a) Using the coordinates $(1982, 1.6)$, $f(x) = a(x - 1982)^2 + 1.6$.   To find a, use $(1994, 442) \Rightarrow$

$442 = a(1994 - 1982)^2 + 1.6 \Rightarrow 440.4 = a(144) \Rightarrow a \approx 3.06$; $f(x) = 3.06(x - 1982)^2 + 1.6$.

   Answers may vary.

   (b) See Figure 119.

   (c) $f(1991) = 3.06(1991 - 1982)^2 + 1.6 = 3.06(81) + 1.6 = 249.46$. By 1991, a total of about 250,000

AIDS cases had been reported.

[1980, 1996, 2] by [−50, 500, 100]

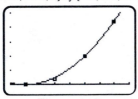

Figure 119

121. Enter the data into your calculator. See Figure 121a. Select quadratic regression from the STAT menu. See

Figure 121b. In Figure 121c, the modeling function is given (approximately) by

$f(x) = 3.125x^2 + 2.05x - 0.9$. $f(3.5) = 3.125(3.5)^2 + 2.05(3.5) - 0.9 \approx 44.56$

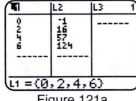

Figure 121a

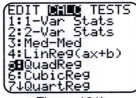

Figure 121b

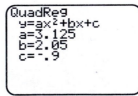

Figure 121c

123. (a) Use the Quadratic regression function on the calculator to find the function

$f(x) = 0.055562x^2 - 220.05x + 217886$. See Figure 123.

   (b) $f(x) = 0.055562(2006)^2 - 220.05(2006) + 217886 \approx 49$ thousand, The estimate is about 2 thousand

lower than the actual value of 51 thousand.

125. (a) Enter data into your calculator. See Figure 125a. Select quadratic regression from the STAT menu. See

Figure 125b. In Figure 125c, the modeling function is given by approximately

$f(x) = 0.59462x^2 - 2350.82x + 2,323,895.$

(b) $f(1985) = 0.59462(1985)^2 - 2350.82(1985) + 2,323,895 = 453.8895.$ In 1985, the approximate

enrollment was 454 thousand.

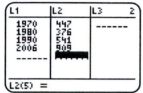
Figure 125a

Figure 125b

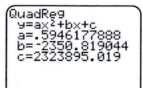
Figure 125c

Extended and Discovery Exercises for Section 3.1

1. (a) See Figure 1.

(b) For $(1, -2)$ and $(2, 1)$, $m = \dfrac{1 - (-2)}{2 - 1} = 3$; for $(2, 1)$ and $(3, 6)$, $m = \dfrac{6 - 1}{3 - 2} = 5$;

for $(3, 6)$ and $(4, 13)$, $m = \dfrac{13 - 6}{4 - 3} = 7$; for $(4, 13)$ and $(5, 22)$, $m = \dfrac{22 - 13}{5 - 4} = 9 \Rightarrow 3, 5, 7, 9.$

(c) $f(x + h) = (x + h)^2 - 3 = x^2 + 2xh + h^2 - 3$;

the difference quotient $= \dfrac{f(x + h) - f(x)}{h} = \dfrac{x^2 + 2xh + h^2 - 3 - (x^2 - 3)}{h} = \dfrac{2xh + h^2}{h} = 2x + h;$

when $h = 1$, the difference quotient is $2x + 1$.

(d) $x = 1, 2(1) + 1 = 3$; $x = 2, 2(2) + 1 = 5$; $x = 3, 2(3) + 1 = 7$; $x = 4, 2(4) + 1 = 9$;

the results are the same.

x	1	2	3	4	5
$f(x)$	-2	1	6	13	22

Figure 1

3. (a) See Figure 3.

(b) For $(1, 0)$ and $(2, -3)$, $m = \dfrac{-3 - 0}{2 - 1} = -3$; for $(2, -3)$ and $(3, -10)$, $m = \dfrac{-10 - (-3)}{3 - 2} = -7$; for

$(3, -10)$ and $(4, -21)$, $m = \dfrac{-21 - (-10)}{4 - 3} = -11$; for $(4, -21)$ and $(5, -36)$,

$m = \dfrac{-36 - (-21)}{5 - 4} = -15 \Rightarrow -3, -7, -11, -15.$

(c) $f(x) = -2x^2 + 3x - 1, f(x + h) = -2(x + h)^2 + 3(x + h) - 1 =$

$-2x^2 - 4xh - 2h^2 + 3x + 3h - 1;$ the difference quotient $= \dfrac{f(x + h) - f(x)}{h} =$

$$\frac{-2x^2 - 4xh - 2h^2 + 3x + 3h - 1 - (-2x^2 + 3x - 1)}{h} =$$

$$\frac{-4xh - 2h^2 + 3h}{h} = -4x - 2h + 3; \text{ when } h = 1, \text{ the difference quotient is } -4x + 1.$$

(d) $x = 1, -4(1) + 1 = -3; \; x = 2, -4(2) + 1 = -7; \; x = 3, -4(3) + 1 = -11;$

$x = 4, -4(4) + 1 = -15;$ the results are the same.

x	1	2	3	4	5
$f(x)$	0	-3	-10	-21	-36

Figure 3

3.2: Quadratic Equations and Problem Solving

1. $x^2 + x - 11 = 1 \Rightarrow x^2 + x - 12 = 0$. Factoring this we get: $(x + 4)(x - 3) = 0$. Then $x + 4 = 0 \Rightarrow$

    $x = -4$ or $x - 3 = 0 \Rightarrow x = 3$. Therefore $x = -4, 3$.

    Check: $(-4)^2 + (-4) - 11 = 1 \Rightarrow 16 - 4 - 11 = 1 \Rightarrow 12 - 11 = 1 \Rightarrow 1 = 1;$

    $(3)^2 + (3) - 11 = 1 \Rightarrow 9 + 3 - 11 = 1 \Rightarrow 12 - 11 = 1 \Rightarrow 1 = 1$

3. $t^2 = 2t \Rightarrow t^2 - 2t = 0$. Factoring this we get: $t(t - 2) = 0$. Then $t = 0$ or $t - 2 = 0 \Rightarrow t = 2$.

    Therefore $t = 0, 2$. Check: $0^2 = 2(0) \Rightarrow 0 = 0; \; 2^2 = 2(2) \Rightarrow 4 = 4$

5. $3x^2 - 7x = 0$. Factoring we get $x(3x - 7) = 0$. Then $x = 0$ or $3x - 7 = 0 \Rightarrow x = \frac{7}{3}$. Therefore $x = 0, \frac{7}{3}$.

    Check: $3(0)^2 - 7(0) = 0 \Rightarrow 0 = 0; \; 3\left(\frac{7}{3}\right)^2 - 7\left(\frac{7}{3}\right) = 0 \Rightarrow \frac{49}{3} - \frac{49}{3} = 0 \Rightarrow 0 = 0$

7. $2z^2 = 13z + 15 \Rightarrow 2z^2 - 13z - 15 = 0$. Factoring we get $(2z - 15)(z + 1) = 0$. Then $2z - 15 = 0 \Rightarrow$

    $z = \frac{15}{2}$ or $z + 1 = 0 \Rightarrow z = -1$. Check: $2\left(\frac{15}{2}\right)^2 = 13\left(\frac{15}{2}\right) + 15 \Rightarrow \frac{225}{2} = \frac{195}{2} + \frac{30}{2} \Rightarrow \frac{225}{2} = \frac{225}{2};$

    $2(-1)^2 = 13(-1) + 15 \Rightarrow 2 = -13 + 15 \Rightarrow 2 = 2.$

9. $x^2 + 6x + 9 = 0$ Factoring this we get $(x + 3)^2 = 0 \Rightarrow x + 3 = 0 \Rightarrow x = -3$. Therefore, $x = -3$. Check:

    $(-3)^2 + 6(-3) + 9 = 0 \Rightarrow 9 - 18 + 0 = 0 \Rightarrow 0 = 0.$

11. $4x^2 + 1 = 4x \Rightarrow 4x^2 - 4x + 1 = 0$ factoring this we get

    $(2x - 1)^2 = 0 \Rightarrow 2x - 1 = 0 \Rightarrow 2x - 1 = 0 \Rightarrow x = \frac{1}{2}$. Therefore, $x = \frac{1}{2}$. Check:

    $4\left(\frac{1}{2}\right)^2 + 1 = 4\left(\frac{1}{2}\right) \Rightarrow 1 + 1 = 2 \Rightarrow 2$

13. $x(3x + 14) = 5 \Rightarrow 3x^2 + 14x - 5 = 0$. Factoring this we get: $(3x - 1)(x + 5) = 0$.

Then $3x - 1 = 0 \Rightarrow 3x = 1 \Rightarrow x = \dfrac{1}{3}$ or $x + 5 = 0 \Rightarrow x = -5$. Therefore $x = -5, \dfrac{1}{3}$.

Check: $-5(3(-5) + 14) = 5 \Rightarrow -5(-1) = 5 \Rightarrow 5 = 5$; $\dfrac{1}{3}\left(3\left(\dfrac{1}{3}\right) + 14\right) = 5 \Rightarrow \dfrac{1}{3}\left(15\right) = 5 \Rightarrow 5 = 5$

15. $6x^2 + \dfrac{5}{2} = 8x \Rightarrow 6x^2 - 8x + \dfrac{5}{2} = 0 \Rightarrow 12x^2 - 16x + 5 = 0$. Factoring this we get:

$(2x - 1)(6x - 5) = 0$.

Then $2x - 1 = 0 \Rightarrow 2x = 1 \Rightarrow x = \dfrac{1}{2}$ or $6x - 5 = 0 \Rightarrow 6x = 5 \Rightarrow x = \dfrac{5}{6}$. Therefore $x = \dfrac{1}{2}, \dfrac{5}{6}$.

Check: $6\left(\dfrac{1}{2}\right)^2 + \dfrac{5}{2} = 8\left(\dfrac{1}{2}\right) \Rightarrow \dfrac{3}{2} + \dfrac{5}{2} = 4 \Rightarrow 4 = 4$;

$6\left(\dfrac{5}{6}\right)^2 + \dfrac{5}{2} = 8\left(\dfrac{5}{6}\right) \Rightarrow \dfrac{25}{6} + \dfrac{15}{6} = \dfrac{40}{6} \Rightarrow \dfrac{40}{6} = \dfrac{40}{6}$

17. $(t + 3)^2 = 5 \Rightarrow t^2 + 6t + 4 = 0$. Using the quadratic formula: $t = \dfrac{-b \pm \sqrt{b^2 - 4ac}}{2a} \Rightarrow$

$t = \dfrac{-6 \pm \sqrt{6^2 - 4(1)(4)}}{2(1)} \Rightarrow t = \dfrac{-6 \pm \sqrt{20}}{2} \Rightarrow t = \dfrac{-6 \pm 2\sqrt{5}}{2} \Rightarrow t = -3 \pm \sqrt{5}$

19. $4x^2 - 13 = 0 \Rightarrow 4x^2 + 0x - 13 = 0$. Using the quadratic formula: $x = \dfrac{-b \pm \sqrt{b^2 - 4ac}}{2a} \Rightarrow$

$x = \dfrac{0 \pm \sqrt{0^2 - 4(4)(-13)}}{2(4)} \Rightarrow x = \dfrac{\pm\sqrt{208}}{8} \Rightarrow x = \dfrac{\pm 4\sqrt{13}}{8} \Rightarrow x = \dfrac{\pm\sqrt{13}}{2}$

21. $2(x - 1)^2 + 4 = 0 \Rightarrow 2(x^2 - 2x + 1) + 4 = 0 \Rightarrow 2x^2 - 4x + 6 = 0$. Using the quadratic formula:

$x = \dfrac{-b \pm \sqrt{b^2 - 4ac}}{2a} \Rightarrow x = \dfrac{-(-4) \pm \sqrt{(-4)^2 - 4(2)(6)}}{2(2)} \Rightarrow x = \dfrac{4 \pm \sqrt{-32}}{4} = \dfrac{4 \pm 4\sqrt{-2}}{4} =$

$1 \pm \sqrt{-2}$. Since $\sqrt{-2}$ is not real, there is no real solution to the equation.

23. $\dfrac{1}{2}x^2 - 3x + \dfrac{1}{2} = 0 \Rightarrow x^2 - 6x + 1 = 0$. Using the quadratic formula: $x = \dfrac{-b \pm \sqrt{b^2 - 4ac}}{2a} \Rightarrow$

$x = \dfrac{-(-6) \pm \sqrt{(-6)^2 - 4(1)(1)}}{2(1)} \Rightarrow x = \dfrac{6 \pm \sqrt{32}}{2} \Rightarrow x = \dfrac{6 \pm 4\sqrt{2}}{2} \Rightarrow x = 3 \pm 2\sqrt{2}$

25. $-3z^2 - 2z + 4 = 0$. Using the quadratic formula: $x = \dfrac{-b \pm \sqrt{b^2 - 4ac}}{2a} \Rightarrow$

$x = \dfrac{-(-2) \pm \sqrt{(-2)^2 - 4(-3)(4)}}{2(-3)} \Rightarrow x = \dfrac{2 \pm \sqrt{52}}{-6} \Rightarrow x = \dfrac{-2 \pm 2\sqrt{13}}{6} \Rightarrow x = \dfrac{-1 \pm \sqrt{13}}{3}$

27. $25k^2 + 1 = 10k \Rightarrow 25k^2 - 10k + 1 = 0$. Using the quadratic formula: $x = \dfrac{-b \pm \sqrt{b^2 - 4ac}}{2a} \Rightarrow$

$x = \dfrac{-(-10) \pm \sqrt{(-10)^2 - 4(25)(1)}}{2(25)} \Rightarrow x = \dfrac{10 \pm \sqrt{0}}{50} \Rightarrow x = \dfrac{1}{5}$

29. $-0.3x^2 + 0.1x = -0.02 \Rightarrow -30x^2 + 10x + 2 = 0$.

Using the quadratic formula: $x = \dfrac{-b \pm \sqrt{b^2 - 4ac}}{2a} \Rightarrow$

$x = \dfrac{-10 \pm \sqrt{10^2 - 4(-30)(2)}}{2(-30)} \Rightarrow x = \dfrac{-10 \pm \sqrt{340}}{-60} \Rightarrow x = \dfrac{10 \pm 2\sqrt{85}}{60} \Rightarrow x = \dfrac{5 \pm \sqrt{85}}{30}$

31. $2x(x + 2) = (x - 1)(x + 2) \Rightarrow 2x^2 + 4x = x^2 + x - 2 \Rightarrow x^2 + 3x + 2 = 0$.   Factoring we get

$(x + 2)(x + 1) = 0$. Then $x + 2 = 0 \Rightarrow x = -2$ or $x + 1 = 0 \Rightarrow x = -1$.

33. For the x-intercepts, $y = 0 \Rightarrow 6x^2 + 13x - 5 = 0$. Factoring this we get $(2x + 5)(3x - 1) = 0$. Then

$2x + 5 = 0 \Rightarrow 2x = -5 \Rightarrow x = -\dfrac{5}{2}$ or $3x - 1 = 0 \Rightarrow 3x = 1 \Rightarrow x = \dfrac{1}{3}$. Therefore $x = -\dfrac{5}{2}, \dfrac{1}{3}$. For the

y-intercept, $x = 0 \Rightarrow y = 6(0)^2 + 13(0) - 5 \Rightarrow y = -5$

35. For the x-intercept, $y = 0 \Rightarrow -4x^2 + 12x - 9 = 0$. Using the quadratic formula:

$x = \dfrac{-12 \pm \sqrt{12^2 - 4(-4)(-9)}}{2(-4)} \Rightarrow x = \dfrac{-12 \pm \sqrt{144 - 144}}{-8} \Rightarrow x = \dfrac{-12 \pm \sqrt{0}}{-8} \Rightarrow x = \dfrac{3}{2}$. For the y-

intercept, $x = 0 \Rightarrow y = -4(0)^2 + 12(0) - 9 \Rightarrow y = -9$

37. For the x-intercepts, $y = 0 \Rightarrow -3x^2 + 11x - 6 = 0$. Using the quadratic formula:

$x = \dfrac{-11 \pm \sqrt{11^2 - 4(-3)(-6)}}{2(-3)} \Rightarrow x = \dfrac{-11 \pm \sqrt{121 - 72}}{-6} \Rightarrow x = \dfrac{-11 \pm \sqrt{49}}{-6} \Rightarrow x = \dfrac{11 \pm 7}{6} \Rightarrow$

$x = \dfrac{2}{3}, 3$. For the y-intercept, $x = 0 \Rightarrow y = -3(0)^2 + 11(0) - 6 \Rightarrow y = -6$

39. (a) Graphical: Graph $Y_1 = X^2 + 2X$ and locate the x-intercepts. See Figure 39a and Figure 39b; the

    solutions to the equation are -2 and 0.

    (b) Numerical: Table $Y_1 = X^2 + 2X$ starting at $x = -4$, incrementing by 1. $Y_1 = 0$ when $x = -2$ or 0.

    See Figure 39c.

    (c) Symbolic: $x^2 + 2x = 0 \Rightarrow x(x + 2) = 0$. Then $x = 0$ or $x + 2 = 0 \Rightarrow x = -2$. Therefore $x = -2, 0$.

[–10, 10, 1] by [–10, 10, 1]   [–10, 10, 1] by [–10, 10, 1]

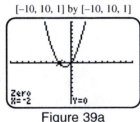
Figure 39a

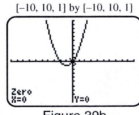
Figure 39b

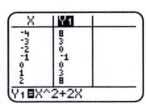
Figure 39c

41. (a) Graphical: Graph $Y_1 = X^2 - X - 6$ and locate the x-intercepts. See Figures 41a and 41b; the solutions to

    the equation are -2 and 3.

(b) Numerical: Table $Y_1 = X{\wedge}2 - X - 6$ starting at $x = -3$, incrementing by 1. $Y_1 = 0$ when $x = -2$ or 3.

See Figure 41c.

(c) Symbolic: $x^2 - x - 6 = 0 \Rightarrow (x - 3)(x + 2) = 0 \Rightarrow x = 3$ or -2 .

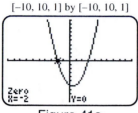
[-10, 10, 1] by [-10, 10, 1]

Figure 41a

[-10, 10, 1] by [-10, 10, 1]

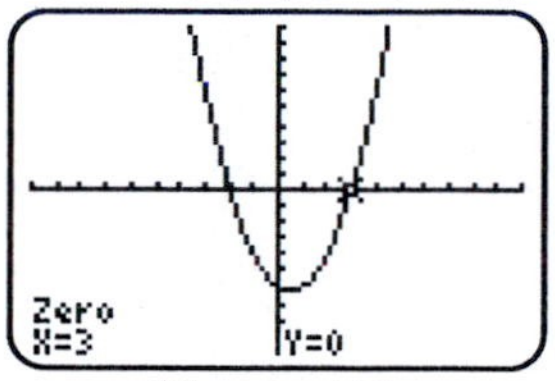

Figure 41b

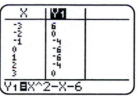

Figure 41c

43. (a) Graphical: Graph $Y_1 = 2X{\wedge}2 - 6$ and locate the *x*-intercepts. See Figures 43a and 43b; the solutions to the

equation are $x \approx \pm 1.7$.

(b) Numerical: Table $Y_1 = 2X{\wedge}2 - 6$ starting at $x = -2\sqrt{3}$, incrementing by $\sqrt{3}$. $Y_1 \approx 0$ when $x \approx \pm 1.7$.

See Figure 43c.

(c) Symbolic: $2x^2 = 6 \Rightarrow x^2 = 3 \Rightarrow x = \pm\sqrt{3} \approx \pm 1.7$.

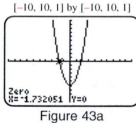
[-10, 10, 1] by [-10, 10, 1]

Figure 43a

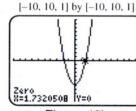
[-10, 10, 1] by [-10, 10, 1]

Figure 43b

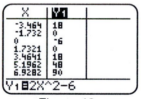

Figure 43c

45. (a) Graphical: Graph $Y_1 = 4X{\wedge}2 - 12X + 9$ and locate the *x*-intercept. See Figure 45a; the solution to

the equation is $x = 1.5$.

(b) Numerical: Table $Y_1 = 4X{\wedge}2 - 12X + 9$ starting at $x = 0$, incrementing by 0.5. $Y_1 = 0$ when $x = 1.5$.

See Figure 45b.

(c) Symbolic: $4x^2 - 12x + 9 = 0 \Rightarrow (2x - 3)(2x - 3) = 0 \Rightarrow x = \dfrac{3}{2}$.

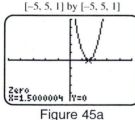
[-5, 5, 1] by [-5, 5, 1]

Figure 45a

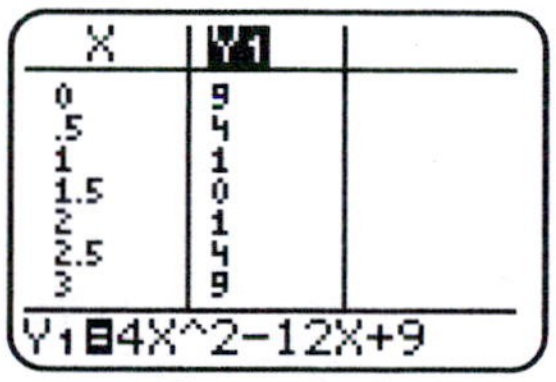

Figure 45b

47. $20x^2 + 11x = 3 \Rightarrow 20x^2 + 11x - 3 = 0$. Graph $Y_1 = 20X{\wedge}2 + 11X - 3$ and locate the *x*-intercepts.

See Figures 47a and 47b; the solutions to the equation are $x = -0.75$ or 0.2.

[0, 2, 0.1] by [−1, 1, 0.1]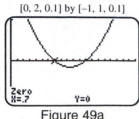

[0, 2, 0.1] by [−1, 1, 0.1]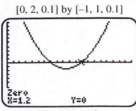

Figure 49a   Figure 49b

49. $2.5x^2 = 4.75x - 2.1 \Rightarrow 2.5x^2 - 4.75x + 2.1 = 0$. Graph $Y_1 = 2.5X^2 - 4.75X + 2.1$ and locate the

x-intercepts. See Figures 49a and 49b; the solutions to the equation are $x = 0.7$ or 1.2.

51. $x^2 + 4x - 6 = 0 \Rightarrow x^2 + 4x = 6 \Rightarrow x^2 + 4x + 4 = 6 + 4 \Rightarrow (x + 2)^2 = 10 \Rightarrow$

$x + 2 = \pm\sqrt{10} \Rightarrow x = -2 \pm \sqrt{10}$

53. $x^2 + 5x = 4 \Rightarrow x^2 + 5x + \dfrac{25}{4} = 4 + \dfrac{25}{4} \Rightarrow \left(x + \dfrac{5}{2}\right)^2 = \dfrac{41}{4} \Rightarrow$

$x + \dfrac{5}{2} = \pm\sqrt{\dfrac{41}{4}} \Rightarrow x = -\dfrac{5}{2} \pm \dfrac{\sqrt{41}}{2} \Rightarrow x = -\dfrac{5}{2} \pm \dfrac{1}{2}\sqrt{41}$

55. $3x^2 - 6x = 2 \Rightarrow x^2 - 2x = \dfrac{2}{3} \Rightarrow x^2 - 2x + 1 = \dfrac{2}{3} + 1 \Rightarrow (x - 1)^2 = \dfrac{5}{3} \Rightarrow$

$x - 1 = \pm\sqrt{\dfrac{5}{3}} \Rightarrow x = 1 \pm \sqrt{\dfrac{5}{3}} \Rightarrow 1 \pm \dfrac{\sqrt{15}}{3}$

57. $x^2 - 8x = 10 \Rightarrow x^2 - 8x + 16 = 10 + 16 \Rightarrow (x - 4)^2 = 26 \Rightarrow x - 4 = \pm\sqrt{26} \Rightarrow x = 4 \pm \sqrt{26}$

59. $\dfrac{1}{2}t^2 - \dfrac{3}{2}t = 1 \Rightarrow t^2 - 3t = 2 \Rightarrow t^2 - 3t + \dfrac{9}{4} = 2 + \dfrac{9}{4} \Rightarrow \left(t - \dfrac{3}{2}\right)^2 = \dfrac{17}{4} \Rightarrow t - \dfrac{3}{2} = \pm\sqrt{\dfrac{17}{4}} \Rightarrow$

$t = \dfrac{3}{2} \pm \dfrac{\sqrt{17}}{2} \Rightarrow t = \dfrac{3 \pm \sqrt{17}}{2}$

61. $-2z^2 + 3z + 1 = 0 \Rightarrow z^2 - \dfrac{3}{2}z = \dfrac{1}{2} \Rightarrow z^2 - \dfrac{3}{2}z + \dfrac{9}{16} = \dfrac{1}{2} + \dfrac{9}{16} \Rightarrow \left(z - \dfrac{3}{4}\right)^2 = \dfrac{17}{16} \Rightarrow$

$z - \dfrac{3}{4} = \pm\sqrt{\dfrac{17}{16}} \Rightarrow z = \dfrac{3}{4} \pm \dfrac{\sqrt{7}}{4} \Rightarrow z = \dfrac{3 \pm \sqrt{17}}{4}$

63. $-\dfrac{3}{2}z^2 - \dfrac{1}{4}z + 1 = 0 \Rightarrow z^2 + \dfrac{1}{6}z = \dfrac{2}{3} \Rightarrow z^2 + \dfrac{1}{6}z + \dfrac{1}{144} = \dfrac{2}{3} + \dfrac{1}{144} \Rightarrow \left(z + \dfrac{1}{12}\right)^2 = \dfrac{97}{144} \Rightarrow$

$z + \dfrac{1}{12} = \pm\sqrt{\dfrac{97}{144}} \Rightarrow z = -\dfrac{1}{12} \pm \dfrac{\sqrt{97}}{12} \Rightarrow z = \dfrac{-1 \pm \sqrt{97}}{12}$

65. $D =$ all real numbers except when the denominator $x^2 - 5 = 0 \Rightarrow x^2 = 5 \Rightarrow x = \pm\sqrt{5} \Rightarrow$

$D = \{x \mid x \neq \sqrt{5}, x \neq -\sqrt{5}\}$

67. $D =$ all real numbers except when the denominator

$t^2 - t - 2 = 0 \Rightarrow (t - 2)(t + 1) = 0 \Rightarrow t = -1, 2 \Rightarrow D = \{t \mid t \neq -1, t \neq 2\}$

69. $4x^2 + 3y = \dfrac{y + 1}{3} \Rightarrow 3y - \dfrac{y + 1}{3} = -4x^2 \Rightarrow 9y - y - 1 = -12x^2 \Rightarrow 8y - 1 = -12x^2 \Rightarrow$

$8y = -12x^2 + 1 \Rightarrow y = \dfrac{-12x^2 + 1}{8}$; yes, y is a function of x since each x-input produces only one y-output.

71. $3y = \dfrac{2x - y}{3} \Rightarrow 9y = 2x - y \Rightarrow 10y = 2x \Rightarrow y = \dfrac{x}{5}$;   yes, y is a function of x since each x-input produces

only one y-output.

73. $x^2 + (y - 3)^2 = 9 \Rightarrow (y - 3)^2 = 9 - x^2 \Rightarrow y - 3 = \pm\sqrt{9 - x^2} \Rightarrow y = 3 \pm \sqrt{9 - x^2}$; no, y is not a

function of x since some x-inputs produce two y-outputs.

75. $3x^2 + 4y^2 = 12 \Rightarrow 4y^2 = 12 - 3x^2 \Rightarrow 2y = \pm\sqrt{12 - 3x^2} \Rightarrow y = \pm\dfrac{\sqrt{12 - 3x^2}}{2}$;

no, y is not a function of x because some x-inputs produce two y-outputs.

77. $V = \dfrac{1}{3}\pi r^2 h$ for $r \Rightarrow \dfrac{3V}{\pi h} = r^2 \Rightarrow r = \pm\sqrt{\dfrac{3V}{\pi h}}$

79. $K = \dfrac{1}{2}mv^2$ for $v \Rightarrow \dfrac{2K}{m} = v^2 \Rightarrow v = \pm\sqrt{\dfrac{2K}{m}}$

81. $a^2 + b^2 = c^2$ for $b \Rightarrow b^2 = c^2 - a^2 \Rightarrow b = \pm\sqrt{c^2 - a^2}$

83. $s = -16t^2 + 100t$ for $t \Rightarrow -s = 16t^2 - 100t \Rightarrow \dfrac{-s}{16} = t^2 - \dfrac{25}{4}t \Rightarrow \dfrac{-s}{16} + \dfrac{625}{64} = t^2 - \dfrac{25}{4}t + \dfrac{625}{64} \Rightarrow$

$\left(t - \dfrac{25}{8}\right)^2 = \dfrac{625 - 4s}{64} \Rightarrow t - \dfrac{25}{8} = \pm\sqrt{\dfrac{625 - 4s}{64}} \Rightarrow t = \dfrac{25}{8} \pm \sqrt{\dfrac{625 - 4s}{64}} \Rightarrow$

$t = \dfrac{25 \pm \sqrt{625 - 4s}}{8}$

85. (a) $3x^2 = 12 \Rightarrow 3x^2 - 12 = 0$

(b) $b^2 - 4ac = 0^2 - 4(3)(-12) = 144 > 0$. There are two real solutions.

(c) $3x^2 = 12 \Rightarrow x^2 = 4 \Rightarrow x = \pm 2$

87. (a) $x^2 - 2x = -1 \Rightarrow x^2 - 2x + 1 = 0$

(b) $b^2 - 4ac = (-2)^2 - 4(1)(1) = 0$. There is one real solution.

(c) $x^2 - 2x + 1 = 0 \Rightarrow (x - 1)^2 = 0 \Rightarrow x = 1$

89. (a) $4x = x^2 \Rightarrow x^2 - 4x = 0$

(b) $b^2 - 4ac = (-4)^2 - 4(1)(0) = 16 > 0$. There are two real solutions.

(c) $x^2 - 4x = 0 \Rightarrow x(x - 4) = 0 \Rightarrow x = 0$ or $x - 4 = 0 \Rightarrow x = 4$.   Therefore, $x = 0, 4$.

91. (a) $x^2 + 1 = x \Rightarrow x^2 - x + 1 = 0$

(b) $b^2 - 4ac = (-1)^2 - 4(1)(1) = -3 < 0$. There are no real solutions.

(c) There are no real solutions.

93. (a) $2x^2 + 3x = 12 - 2x \Rightarrow 2x^2 + 5x - 12 = 0$.

(b) $b^2 - 4ac = (5)^2 - 4(2)(-12) = 121 > 0$. There are two real solutions.

(c) $2x^2 + 5x - 12 = 0 \Rightarrow (2x - 3)(x + 4) = 0 \Rightarrow 2x = 3 \Rightarrow x = 1.5$ or $x = -4$. So, $x = 1.5, -4$.

95. (a) $9x(x - 4) = -36 \Rightarrow 9x^2 - 36x + 36 = 0$

    (b) $b^2 - 4ac = (-36)^2 - 4(9)(36) = 0.$ There is one real solution.

    (c) $9x^2 - 36x + 36 = 0 \Rightarrow x^2 - 4x + 4 = 0 \Rightarrow (x - 2)^2 = 0 \Rightarrow x = 2$

97. (a) $x\left(\dfrac{1}{2}x + 1\right) = -\dfrac{13}{2} \Rightarrow \dfrac{1}{2}x^2 + x + \dfrac{13}{2} = 0$

    (b) $b^2 - 4ac = (1)^2 - 4\left(\dfrac{1}{2}\right)\left(\dfrac{13}{2}\right) = -12 < 0.$ There are no real solutions.

    (c) There are no real solutions.

99. (a) $3x^2 = 1 - x \Rightarrow 3x^2 + x - 1 = 0$

    (b) $b^2 - 4ac = 1^2 - 4(3)(-1) = 13 > 0.$ There are two real solutions.

    (c) $x = \dfrac{-1 \pm \sqrt{1^2 - 4(3)(-1)}}{2(3)} = \dfrac{-1 \pm \sqrt{13}}{2(3)} = \dfrac{-1 \pm \sqrt{13}}{6}$

101.(a) Since the parabola opens upward, $a > 0$.

    (b) Since the zeros of f are -6 and 2, the solutions to $ax^2 + bx + c = 0$ are also -6 and 2.

    (c) Since there are two real solutions, the discriminant is positive.

103.(a) Since the parabola opens upward, $a > 0$.

    (b) Since the only zero of f is -4, the solution is also -4.

    (c) Since there is one real solutions, the discriminant is equal to zero.

105. The height when it hits the ground will be $0 \Rightarrow 75 - 16t^2 = 0 \Rightarrow -16t^2 = -75 \Rightarrow t^2 = \dfrac{75}{16} \Rightarrow$

    $t = \sqrt{\dfrac{75}{16}} \Rightarrow t = \dfrac{\sqrt{75}}{4} \Rightarrow t \approx 2.2$ seconds.

107. $55 = 16x^2 + 7x + 32 \Rightarrow 16x^2 + 7x - 23 = 0$ Factoring this we get $(x - 1)(16x + 23) = 0 \Rightarrow$

    $x - 1 = 0 \Rightarrow x = 1$ or $16x + 23 = 0 \Rightarrow x = -\dfrac{23}{16}.$ Since x represents the number of years after 2008 the

    result cannot be negative and the only answer is $x = 1$. Therefore, the year is 2009.

109. Graphical: First, we must determine a formula for the area. Let x represent the height of the computer screen.

    Then, $x + 2.5$ is the width. The area of the screen is height times width, computed $A(x) = x(x + 2.5)$. We

    must solve the quadratic equation $x(x + 2.5) = 93.5$ or $x^2 + 2.5x - 93.5 = 0$. Graph

    $Y_1 = X{\wedge}2 + 2.5X - 93.5$and determine any zeros. Figure 109a shows that the equation has two zeros, one

    negative and one positive. The positive zero is located at $x = 8.5$. The height is 8.5 inches and the width is

    $8.5 + 2.5 = 11$ inches.

    Symbolic: The quadratic equation $x^2 + 2.5x - 93.5 = 0$ can be solved by the quadratic formula.

    $x = \dfrac{-b \pm \sqrt{b^2 - 4ac}}{2a} = \dfrac{-2.5 \pm \sqrt{2.5^2 - 4(1)(-93.5)}}{2(1)} = \dfrac{-2.5 \pm 19.5}{2} = 8.5, -11.$ The positive

    answer gives a height of 8.5 inches. It follows that the width is 11 inches.

    The numerical solution is shown in Figure 109b. Yes, the symbolic, graphical and numerical answers agree.

[−15, 15, 1] by [−100, 100, 10]

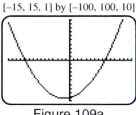

Figure 109a

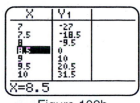

Figure 109b

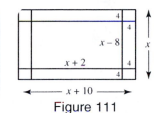

Figure 111

111. Let $x =$ width of the metal sheet in inches and $x + 10 =$ length of the metal sheet in inches. Make a sketch to find expressions for the dimensions of the box. See Figure 111. The width of the box is $x - 8$ inches, the length of the box is $x + 2$ inches, and the height of the box is 4 inches; the volume of the box, which is given as 476 cubic inches, is determined by the length times the width times the height: $4(x - 8)(x + 2) = 476 \Rightarrow$ $(x - 8)(x + 2) = 119 \Rightarrow x^2 - 6x - 16 = 119 \Rightarrow x^2 - 6x - 135 = 0 \Rightarrow (x + 9)(x - 15) = 0 \Rightarrow$ $x = -9$ or $x = 15$. Since width cannot be negative, $x = 15$ inches; thus, the dimensions of the metal sheet is 15 inches by 25 inches.

113. $V = \pi r^2 h$, $V = 28$ cubic inches, and $h = 4$ inches $\Rightarrow 28 = \pi r^2 (4) \Rightarrow r^2 = \dfrac{28}{4\pi} \Rightarrow r = \sqrt{\dfrac{28}{4\pi}} \approx 1.49$; the radius of the cylinder is approximately 1.49 inches.

115. Since the diameter of the semicircle is x, the radius is $\dfrac{x}{2}$; the area of the semicircle

$= \dfrac{1}{2}\pi \left(\dfrac{x}{2}\right)^2 = \dfrac{1}{2}\pi \left(\dfrac{x^2}{4}\right) = \dfrac{1}{8}\pi x^2$. The area of the square $= x^2$; thus, the total area of the window, which

is 463 square inches, is $x^2 + \dfrac{1}{8}\pi x^2$; $463 = x^2 + \dfrac{1}{8}\pi x^2 \Rightarrow 463 = \left(1 + \dfrac{\pi}{8}\right)x^2 \Rightarrow x^2 = \dfrac{463}{(1 + \frac{\pi}{8})} \Rightarrow$

$x = \sqrt{\dfrac{463}{1 + \frac{\pi}{8}}} \approx 18.23$ inches.

117. Let x be the number of shirts ordered. Revenue equals the number of shirts sold times the price of each shirt. If x shirts are sold, then the price in dollars of each shirt is $20 - 0.10(x - 1)$. The revenue $R(x)$ is given by $R(x) = x(20 - 0.10(x - 1))$.   We must solve the equation $989 = x(20 - 0.10(x - 1)) \Rightarrow$ $-0.10x^2 + 20.1x - 989 = 0$. Let $y_1 = -0.10x^2 + 20.1x - 989$ and use the graphing calculator to find the x-intercept. Because the discount applies to orders up to 100 shirts, we find that $x = 86$ shirts. See Figure 117.

[0, 150, 10] by [−20, 50, 10]

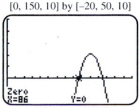

Figure 117

119. (a) Since $s(0) = 32$, $s(t) = -16t^2 + v_0 t + h_0 \Rightarrow 32 = -16(0)^2 + v_0(0) + h_0 \Rightarrow h_0 = 32$ and so $s(t) = -16t^2 + v_0 t + 32$; since $s(1) = 176$, $176 = -16(1)^2 + v_0(1) + 32 \Rightarrow v_0 = 160$; thus, $s(t) = -16t^2 + 160t + 32$ models the data.

(b) When the projectile strikes the ground, $s(t) = 0$; $0 = -16t^2 + 160t + 32 \Rightarrow 0 = t^2 - 10t - 2 \Rightarrow$

$$t = \frac{10 \pm \sqrt{10^2 - 4(1)(-2)}}{2} \approx 10.2 \text{ or } -0.2; \quad \text{since } t \text{ cannot be negative, } t \approx 10.2; \text{ the projectile strikes}$$

the ground after about 10.2 seconds.

121. (a) The vertex of (1995, 180) substituted into the function will be written as $B(x) = a(x - 1985)^2 + 180$.

   We now us the point (2009, 1200) to substitute into the equation and solve for a as follows,

   $1200 = a(2009 - 1995)^2 + 180 \Rightarrow 1200 = 196a + 180 \Rightarrow 1020 = 196a \Rightarrow a \approx 5.2$. Thus

   $B(x) = 5.2(x - 1995)^2 + 180$ can be used to model the budget for pedestrian and bicycle programs in

   selected years.

   (b) $700 = 5.2(x - 1995)^2 + 180 \Rightarrow 520 = 5.2(x - 1995)^2 \Rightarrow 100 = (x - 1995) \Rightarrow$

   $10 = x - 1995 \Rightarrow x = 2005$

123. (a) Use the Quadratic regression function on the calculator to find the function

   $I(x) = -2.277x^2 + 11.71x + 40.4$.

   (b) $28 = -2.27x^2 + 11.71x + 40.4 \Rightarrow -2.277x^2 + 11.71x + 12.4 = 0$, using the quadratic formula:

   $$\frac{-11.71 \pm \sqrt{(11.71)^2 - 4(-2.277)(12.4)}}{2(-2.277)} \Rightarrow x \approx -1 \text{ and } 6. \text{ Since } x = 0 \text{ corresponds to 2006 the results}$$
   are 2005 and 2012.

125. (a) $E(15) = 1.4$, $1987 + 15 = 2002$; in 2002 there were 1.4 million Wal-Mart employees.

   (b) Use the quadratic regression function on your graphing calculator to find

   $f(x) = 0.00474x^2 + 0.00554x + 0.205$.   *Answers may vary.*

   (c) See Figure 125c.

   (d) Let $Y_1 = (0.00474)X^2 + 0.00554X + 0.205$ and $Y_2 = 3$. Use the CALC function on your calculator to

   find the intersection points of the Y_1 and Y_2. See Figure 125d. When $x \approx 24$, $f(x) = 3$.   Therefore, in

   the year 2011 the number of employees may reach 3 million. *Answers may vary.*

[0, 25, 5] by [0, 2.6, 0.2]   [0, 25, 5] by [0, 3.6, 0.2]

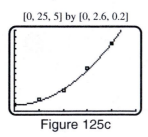   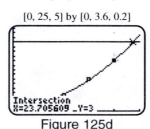

Figure 125c   Figure 125d

Extended and Discovery Exercises for Section 3.2

1. $b^2 - 4ac = 14^2 - 4(8)(-15) = 676 = 26^2$.   Since 676 is a perfect square the equation can be solved by

    factoring. $8x^2 + 14x - 15 = 0 \Rightarrow (4x - 3)(2x + 5) = 0$.   Then $4x - 3 = 0 \Rightarrow x = \dfrac{3}{4}$ or

    $2x + 5 = 0 \Rightarrow x = -\dfrac{5}{2}$.

3. $b^2 - 4ac = (-3)^2 - 4(5)(-3) = 69$.   Since 69 is not a perfect square the equation cannot be solved by

    factoring. $x = \dfrac{-b \pm \sqrt{b^2 - 4ac}}{2a} \Rightarrow x = \dfrac{3 \pm \sqrt{(-3)^2 - 4(5)(-3)}}{2(5)} \Rightarrow x = \dfrac{3 \pm \sqrt{69}}{10}$.

5. (a) Follow the steps for completing the square in this equation:

    $$ax^2 + bx + c = 0 \Rightarrow x^2 + \frac{b}{a}x + \frac{c}{a} = 0 \Rightarrow x^2 + \frac{b}{a}x = -\frac{c}{a}$$

    (b) Add $\left(\dfrac{b}{2a}\right)^2 = \dfrac{b^2}{4a^2}$ to both sides to obtain $x^2 + \dfrac{b}{a}x + \dfrac{b^2}{4a^2} = -\dfrac{c}{a} + \dfrac{b^2}{4a^2}$ or $\left(x + \dfrac{b}{2a}\right)^2 = \dfrac{b^2 - 4ac}{4a^2}$.

    (c) In order to derive the quadratic formula we must solve this second equation for x.

    $$\left(x + \frac{b}{2a}\right)^2 = \frac{b^2 - 4ac}{4a^2} \Rightarrow \left(x + \frac{b}{2a}\right) = \pm\sqrt{\frac{b^2 - 4ac}{4a^2}} \Rightarrow \qquad \{\text{Square root property}\}$$

    $$x = -\frac{b}{2a} \pm \sqrt{\frac{b^2 - 4ac}{4a^2}} \Rightarrow x = -\frac{b}{2a} \pm \frac{\sqrt{b^2 - 4ac}}{|2a|} \Rightarrow \qquad \{\sqrt{4a^2} = |2a|\}$$

    $$x = -\frac{b}{2a} \pm \frac{\sqrt{b^2 - 4ac}}{2a} \Rightarrow x = \frac{-b \pm \sqrt{b^2 - 4ac}}{2a}$$

    {Since there is a $\pm$ in front of the expression, the absolute value does not matter.}

7. $x^2 = k \Rightarrow \sqrt{x^2} = \sqrt{k} \Rightarrow |x| = \sqrt{k} \Rightarrow x = \pm\sqrt{k}$

Checking Basic Concepts for Sections 3.1 and 3.2

1. Vertex: $(1, -4)$; axis of symmetry: $x = 1$; x-intercepts: $-1, 3$. See Figure 1.

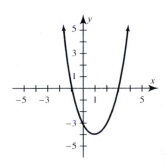

Figure 1

3. Since 3 is the smallest y-value and there is symmetry of the y-values around $x = -1$, the vertex is

    $(-1, 3)$; $f(x) = a(x - h)^2 + k \Rightarrow f(x) = a(x + 1)^2 + 3$;   since $(0, 5)$ is a data point, $f(0) = 5$;

    $5 = a(0 + 1)^2 + 3 \Rightarrow a = 2$; $f(x) = 2(x + 1)^2 + 3$ models the data exactly.

5. Find the vertex:

$$x = -\frac{b}{2a} \Rightarrow x = -\frac{4}{2(1)} \Rightarrow x = -2.\text{ Since } f(-2) = (-2)2 + 4(-2) - 3 \Rightarrow f(-2) = -7,$$

$(-2, -7)$ is the vertex. The leading coefficient is $a = 1$. Converting to $f(x) = a(x - h)^2 + k$, we get

$$f(x) = 1(x - (-2))^2 + (-7) \Rightarrow f(x) = (x + 2)^2 - 7. \quad \text{The minimum value is at the vertex or } -7.$$

7. Let $x =$ width of rectangle, then $x + 4 =$ length of rectangle; since area $= 165$ square inches,

$$x(x + 4) = 165 \Rightarrow x^2 + 4x - 165 = 0 \Rightarrow (x - 11)(x + 15) = 0 \Rightarrow x = 11, -15; \text{ since } x = -15 \text{ has}$$

no physical meaning for the width, the width of the rectangle is 11 inches and the length is $11 + 4 = 15$ inches.

3.3: Complex Numbers

1. $\sqrt{-4} = i\sqrt{4} = 2i$

3. $\sqrt{-100} = i\sqrt{100} = 10i$

5. $\sqrt{-23} = i\sqrt{23}$

7. $\sqrt{-12} = i\sqrt{12} = i\sqrt{4}\sqrt{3} = 2i\sqrt{3}$

9. $\sqrt{-54} = \sqrt{(9)(6)(-1)} = \sqrt{9(-1)}\sqrt{6} = 3i\sqrt{6}$

11. $\dfrac{4 \pm \sqrt{-16}}{2} = \dfrac{4 \pm \sqrt{(16)(-1)}}{2} = \dfrac{4 \pm 4i}{2} = 2 \pm 2i$

13. $\dfrac{-6 \pm \sqrt{-72}}{3} = \dfrac{-6 \pm \sqrt{(36)(-1)}\sqrt{2}}{3} = \dfrac{-6 \pm 6i\sqrt{2}}{3} = -2 \pm 2i\sqrt{2}$

15. $\sqrt{-5} \cdot \sqrt{-5} = \sqrt{5}\sqrt{-1} \cdot \sqrt{5}\sqrt{-1} = \sqrt{5}i \cdot \sqrt{5}i = \sqrt{25}i^2 = 5(-1) = -5.$

17. $\sqrt{-18} \cdot \sqrt{-2} = (\sqrt{9}\sqrt{2}\sqrt{-1})(\sqrt{2}\sqrt{-1}) = (3\sqrt{2}\,i)(\sqrt{2}\,i) = 3\sqrt{4}i^2 = (3)(2)(-1) = -6$

19. $\sqrt{-3} \cdot \sqrt{-6} = (\sqrt{3}\sqrt{-1})(\sqrt{6}\sqrt{-1}) = (\sqrt{3}i)(\sqrt{6}i) = \sqrt{18}i^2 = \sqrt{9}\sqrt{2}(-1) = -3\sqrt{2}$

21. $3i + 5i = (3 + 5)i = 8i$

23. $(3 + i) + (-5 - 2i) = (3 + (-5)) + (1 - 2)i = -2 - i$

25. $2i - (-5 + 23i) = (-5) + (2 - 23)i = 5 - 21i$

27. $3 - (4 - 6i) = (3 - 4) + i = -1 + 6i$

29. $(2)(2 + 4i) = 4 + 8i$

31. $(1 + i)(2 - 3i) = (1)(2) + (1)(-3i) + (i)(2) + (i)(-3i) = 2 - i - 3i^2 = 2 - i - 3(-1) = 5 - i$

33. $(-3 + 2i)(-2 + i) = (-3)(-2) + (-3)(i) + (2i)(-2) + (2i)(i) = 6 - 7i + 2i^2 =$

   $6 - 7i + 2(-1) = 4 - 7i$

35. $(-2 + 3i)^2 = (-2 + 3i)(-2 + 3i) = (-2)(-2) + (-2)(3i) + (-2)(3i) + (3i)(3i) =$

   $4 + (-12i) + (9i^2) = 4 - 12i + (-9) = -5 - 12i$

37. $2i(1 - i)^2 = 2i(1 - i)(1 - i) = 2i[(1)(1) + (1)(-i) + (1)(-i) + (-i)(-i)] =$

   $2i[1 - 2i + (-1)] = 2i(-2i) = -4i^2 = 4$

39. $\dfrac{1}{1+i} = \dfrac{1}{1+i} \cdot \dfrac{1-i}{1-i} = \dfrac{1-i}{(1+i)(1-i)} = \dfrac{1-i}{1-i^2} = \dfrac{1-i}{2} = \dfrac{1}{2} - \dfrac{1}{2}i$

41. $\dfrac{4+i}{5-i} = \dfrac{4+i}{5-i} \cdot \dfrac{5+i}{5+i} = \dfrac{(4+i)(5+i)}{(5-i)(5+i)} = \dfrac{20+9i+i^2}{25-i^2} = \dfrac{19+9i}{26} = \dfrac{19}{26} + \dfrac{9}{26}i$

43. $\dfrac{2i}{10-5i} = \dfrac{2i}{10-5i} \cdot \dfrac{10+5i}{10+5i} = \dfrac{20i+10i^2}{(10-5i)(10+5i)} = \dfrac{-10+20i}{100-25i^2} = \dfrac{-10+20i}{125} = -\dfrac{2}{25} + \dfrac{4}{25}i$

45. $\dfrac{3}{-i} \cdot \dfrac{i}{i} = \dfrac{3i}{-i^2} = \dfrac{3i}{-i^2} = \dfrac{3i}{-(-1)} = 3i$

47. $\dfrac{-2+i}{(1+i)^2} = \dfrac{-2+i}{(1+i)(1+i)} = \dfrac{-2+i}{(1)(1)+(1)(i)+(1)(i)+(i)(i)} = \dfrac{-2+i}{1+2i+i^2} = \dfrac{-2+i}{1+2i-1} =$

$\dfrac{-2+i}{2i} \cdot \dfrac{i}{i} = \dfrac{-2i+i^2}{2i^2} = \dfrac{-2i-1}{2(-1)} = \dfrac{-2i-1}{-2} = \dfrac{1}{2} + i$

49. $(23 - 5.6i) + (-41.5 + 93i) = -18.5 + 87.4i$

51. $(17.1 - 6i) - (8.4 + 0.7i) = 8.7 - 6.7i$

53. $(-12.6 - 5.7i)(5.1 - 9.3i) = -117.27 + 88.11i$

55. $\dfrac{17 - 135i}{18 + 142i} \approx -0.921 - 0.236i$

57. $i^{50} = \left(i^4\right)^{12} \cdot i^2 = (1)^{12} \cdot (-1) = -1$

59. $i^{32} = \left(i^4\right)^{7} \cdot i^3 = (1)^{7} \cdot (-i) = -i$

61. $i^{12} = \left(i^4\right)^{3} = (1)^{3} = 1$

63. $i^{57} = \left(i^4\right)^{14} \cdot i = (1)^{14} \cdot (i) = i$

65. $x^2 + 5 = 0 \Rightarrow x^2 = -5 \Rightarrow x = \pm\sqrt{-5} = \pm i\sqrt{5}$

67. $5x^2 + 1 = 3x^2 \Rightarrow 2x^2 = -1 \Rightarrow x^2 = -\dfrac{1}{2} \Rightarrow x = \pm i\sqrt{\dfrac{1}{2}}$

69. $3x = 5x^2 + 1 \Rightarrow 5x^2 - 3x + 1 = 0$; use the quadratic formula with $a = 5$, $b = -3$, and $c = 1$.

$x = \dfrac{-(-3) \pm \sqrt{(-3)^2 - 4(5)(1)}}{2(5)} = \dfrac{3 \pm \sqrt{9-20}}{10} = \dfrac{3 \pm \sqrt{-11}}{10} = \dfrac{3 \pm i\sqrt{11}}{10} = \dfrac{3}{10} \pm \dfrac{i\sqrt{11}}{10}$

71. $x(x-4) = -5 \Rightarrow x^2 - 4x = -5 \Rightarrow x^2 - 4x + 5 = 0$; use the quadratic formula to solve

$x^2 - 4x + 5 = 0$ with $a = 1$, $b = -4$, and $c = 5$.

$x = \dfrac{-(-4) \pm \sqrt{(-4)^2 - 4(1)(5)}}{2(1)} = \dfrac{4 \pm \sqrt{16-20}}{2} = \dfrac{4 \pm \sqrt{-4}}{2} = \dfrac{4 \pm 2i}{2} = 2 \pm i$

73. Use the quadratic formula to solve $x^2 - 3x + 5 = 0$ with $a = 1$, $b = -3$, and $c = 5$.

$x = \dfrac{-(-3) \pm \sqrt{(-3)^2 - 4(1)(5)}}{2(1)} = \dfrac{3 \pm \sqrt{9-20}}{2} = \dfrac{3 \pm \sqrt{-11}}{2} = \dfrac{3 \pm i\sqrt{11}}{2} = \dfrac{3}{2} \pm \dfrac{i\sqrt{11}}{2}$

75. Use the quadratic formula to solve $x^2 + 2x + 4 = 0$ with $a = 1$, $b = 2$, and $c = 4$.

$$x = \frac{-2 \pm \sqrt{2^2 - 4(1)(4)}}{2(1)} = \frac{-2 \pm \sqrt{4 - 16}}{2} = \frac{-2 \pm \sqrt{-12}}{2} = \frac{-2 \pm 2i\sqrt{3}}{2} = -1 \pm i\sqrt{3}$$

77. $3x^2 - 4x = x^2 - 3 \Rightarrow 2x^2 - 4x + 3 = 0$.   Using the quadratic formula we get:

$$\frac{4 \pm \sqrt{16 - 4(2)(3)}}{2(2)} = \frac{4 \pm \sqrt{-8}}{4} = \frac{4 \pm 2\sqrt{-2}}{4} = 1 + \frac{\sqrt{-2}}{2} = 1 \pm \frac{i\sqrt{2}}{2}$$

79. $2x(x - 2) = x - 4 \Rightarrow 2x^2 - 4x = x - 4 \Rightarrow 2x^2 - 5x + 4 = 0$.

Using the quadratic formula we get: $\dfrac{5 \pm \sqrt{(5)^2 - 4(2)(4)}}{2(2)} = \dfrac{5 \pm \sqrt{-7}}{4} = \dfrac{5}{4} \pm \dfrac{i\sqrt{7}}{4}$

81. $3x(3 - x) - 8 = x(x - 2) \Rightarrow 9x - 3x^2 - 8 = x^2 - 2x \Rightarrow -4x^2 + 11x - 8 = 0$.

Using the quadratic formula we get: $\dfrac{-11 \pm \sqrt{(11)^2 - 4(-4)(-8)}}{2(-4)} = \dfrac{-11 \pm \sqrt{-7}}{-8} = \dfrac{11}{8} \pm \dfrac{i\sqrt{7}}{8}$

83. (a) The graph of $y = 2x^2 - x - 3$ intersects the x-axis twice, so there are two real zeros.

(b) $x = \dfrac{-b \pm \sqrt{b^2 - 4ac}}{2a} = \dfrac{1 \pm \sqrt{(-1)^2 - 4(2)(-3)}}{2(2)} = \dfrac{1 \pm 5}{4} = \dfrac{3}{2}, -1$

85. (a) The graph of $f(x) = x^2 + x + 2$ does not intersect the x-axis. Both of its zeros are imaginary.

(b) $x = \dfrac{-b \pm \sqrt{b^2 - 4ac}}{2a} = \dfrac{-1 \pm \sqrt{1^2 - 4(1)(2)}}{2(1)} = \dfrac{-1 \pm \sqrt{-7}}{2} = -\dfrac{1}{2} \pm \dfrac{i\sqrt{7}}{2}$

87. (a) The graph of $y = -x^2 - 2$ does not intersect the x-axis. Both of its zeros are imaginary.

(b) $x = \dfrac{-b \pm \sqrt{b^2 - 4ac}}{2a} = \dfrac{0 \pm \sqrt{0^2 - 4(-1)(-2)}}{2(-1)} = \dfrac{\pm\sqrt{-8}}{-2} = \dfrac{\pm 2i\sqrt{2}}{-2} = \pm i\sqrt{2}$

89. $Z = \dfrac{V}{I} \Rightarrow Z = \dfrac{50 + 98i}{8 + 5i} \Rightarrow Z = \dfrac{50 + 98i}{8 + 5i} \cdot \dfrac{8 - 5i}{8 - 5i} \Rightarrow Z = \dfrac{400 - 250i + 784i - 490i^2}{64 - 25i^2} \Rightarrow$

$Z = \dfrac{400 + 534i + 490}{64 + 25} \Rightarrow Z = \dfrac{890 + 534i}{89} \Rightarrow Z = 10 + 6i$

91. $Z = \dfrac{V}{I} \Rightarrow 3 - 4i = \dfrac{V}{1 + 2i} \Rightarrow V = (3 - 4i)(1 + 2i) \Rightarrow V = 3 + 6i - 4i - 8i^2 \Rightarrow$

$V = 3 + 2i + 8 \Rightarrow V = 11 + 2i$

93. $Z = \dfrac{V}{I} \Rightarrow 22 - 5i = \dfrac{27 + 17i}{I} \Rightarrow (22 - 5i)V = 27 + 17i \Rightarrow V = \dfrac{27 + 17i}{22 - 5i} \Rightarrow V = \dfrac{27 + 17i}{22 - 5i} \cdot \dfrac{22 + 5i}{22 + 5i} \Rightarrow$

$V = \dfrac{594 + 135i + 374i + 85i^2}{484 - 25i^2} \Rightarrow V = \dfrac{594 + 509i - 85}{484 + 25} \Rightarrow V = \dfrac{509 + 509i}{509} \Rightarrow V = 1 + i$

Extended and Discovery Exercise for Section 3.3

1. (a) $i^1 = i$, $i^2 = -1$, $i^3 = -i$, $i^4 = 1$, $i^5 = i$, $i^6 = -1$, $i^7 = -i$, $i^8 = 1$, and so on.

(b) Divide n by 4. If the remainder is r, then $i^n = i^r$, where $i^0 = 1$, $i^1 = i$, $i^2 = -1$ and $i^3 = -i$.

3.4: Quadratic Inequalities

1. (a) To solve the inequality $x^2 - 1 > 0$, we will first solve the quadratic equation by factoring.

    $x^2 - 1 = 0 \Rightarrow (x - 1)(x + 1) = 0 \Rightarrow x = \pm 1$ These two boundary numbers separate the number line

    into three intervals $(-\infty, -1)$, $(-1, 1)$, and $(1, \infty)$. We choose the test values $x = -2, 0$, and 2. The

    expression $x^2 - 1$ is positive (or greater than zero) when x is in the interval $(-\infty, -1) \cup (1, \infty)$. See table

    1 below.

    (b) To solve the inequality $x^2 - 1 < 0$, we will first solve the quadratic equation by factoring.

    $x^2 - 1 = 0 \Rightarrow (x - 1)(x + 1) = 0 \Rightarrow x = \pm 1$ These two boundary numbers separate the number line

    into three intervals $(-\infty, -1)$, $(-1, 1)$, and $(1, \infty)$. We choose the test values $x = -2, 0$, and 2. The

    expression $x^2 - 1$ is negative (or less than zero) when x is in the interval $(-1, 1)$. See Table 1 below.

Table 1

Interval	Test Value	Positive (+) or Negative (−) Result
$-\infty, -1$	-2	+
$-1, 1$	0	−
$1, \infty$	2	+

3. (a) To solve the inequality $x^2 - 16 \le 0$, we will first solve the quadratic equation by factoring.

    $x^2 - 16 = 0 \Rightarrow (x - 4)(x + 4) = 0 \Rightarrow x = \pm 4$ These two boundary numbers separate the number line

    into three intervals $(-\infty, -4]$, $[-4, 4]$, and $[4, \infty)$. We choose the test values $x = -5, 0$, and 5. the xpres-

    sion $x^2 - 16$ is negative (or less than or equal to zero) when x is in the interval $[-4, 4]$. See Table 3 below.

    (b) To solve the inequality $x^2 - 16 \ge 0$, we will first solve the quadratic equation by factoring.

    $x^2 - 16 = 0 \Rightarrow (x - 4)(x + 4) = 0 \Rightarrow x = \pm 4$ These two boundary numbers separate the number line

    into three intervals $(-\infty, -4]$, $[-4, 4]$, and $[4, \infty)$. We choose test values $x = -5, 0$, and 5. The expres-

    sion $x^2 - 16$ is positive (or greater than or equal to zero) when x is in the interval $(-\infty, -4] \cup [4, \infty)$. See

    Table 3 below.

Table 3

Interval	Test Value	Positive (+) or Negative (−) Result
$-\infty, -4$	-5	+
$-4, 4$	0	−
$4, \infty$	5	+

5. (a) To solve the inequality $4 - x^2 > 0$, we will first solve the quadratic equation by factoring.

$4 - x^2 = 0 \Rightarrow (2 - x)(2 + x) = 0 \Rightarrow x = \pm 2$ These two boundary numbers separate the number line into three intervals $(-\infty, -2)$, $(-2, 2)$, and $(2, \infty)$. We choose test values $x = -3, 0$, and 3. The expression $4 - x^2$ is positive (or greater than zero) when x is in the interval $(-2, 2)$. See Table 5 below.

(b) To solve the inequality $4 - x^2 < 0$, we will first solve the quadratic equation by factoring.

$4 - x^2 = 0 \Rightarrow (2 - x)(2 + x) = 0 \Rightarrow x = \pm 2$ These two boundary numbers separate the number line into three intervals $(-\infty, -2)$, $(-2, 2)$, and $(2, \infty)$. We choose test values $x = -3, 0$, and 3. The expression $4 - x^2$ is negative (or less than zero) when x is in the interval $(-\infty, -2) \cup (2, \infty)$. See Table 5 below.

Table 5

Interval	Test Value	Positive (+) or Negative (−) Result
$-\infty, -2$	-3	$-$
$-2, 2$	0	$+$
$2, \infty$	3	$-$

7. (a) $x^2 - x - 12 = 0 \Rightarrow (x - 4)(x + 3) = 0 \Rightarrow x = -3, 4$.

(b) Using test intervals, $x^2 - x - 12 < 0$ on $(-3, 4)$ or $\{x \mid -3 < x < 4\}$.

(c) Using test intervals, $x^2 - x - 12 > 0$ on $(-\infty, -3) \cup (4, \infty)$ or $\{x \mid x < -3 \text{ or } x > 4\}$.

9. (a) $k^2 - 5 = 0 \Rightarrow k^2 = 5 \Rightarrow k = \pm\sqrt{5}$.

(b) Using test intervals, $k^2 - 5 \leq 0$ on $\left[-\sqrt{5}, \sqrt{5}\right]$ or $\{k \mid -\sqrt{5} \leq k \leq \sqrt{5}\}$.

(c) Using test intervals, $k^2 - 5 \geq 0$ on $(-\infty, -\sqrt{5}] \cup [\sqrt{5}, \infty)$ or $\{k \mid k \leq -\sqrt{5} \text{ or } k \geq \sqrt{5}\}$.

11. (a) $3x^2 + 8x = 0 \Rightarrow x(3x + 8) = 0 \Rightarrow x = -\dfrac{8}{3}, 0$. Using $x = -1$, which is inbetween $-\dfrac{8}{3}$ and 0 produces $3(-1)^2 + 8(-1) = -5 \Rightarrow$ less than 0.

(b) Using test intervals, $3x^2 + 8x \leq 0$ on $\left[-\dfrac{8}{3}, 0\right]$ or $\left\{x \mid -\dfrac{8}{3} \leq x \leq 0\right\}$.

(c) Using test intervals, $3x^2 + 8x \geq 0$ on $\left(-\infty, -\dfrac{8}{3}\right] \cup [0, \infty)$ or $\left\{x \mid x \leq -\dfrac{8}{3} \text{ or } x \geq 0\right\}$.

13. (a) $-4x^2 + 12x - 9 = 0 \Rightarrow (2x - 3)(-2x + 3) = 0 \Rightarrow x = \dfrac{3}{2}$.

(b) Using test intervals, $-4x^2 + 12x - 9 < 0$ on $x \neq \dfrac{3}{2}$ on $\left(-\infty, \dfrac{3}{2}\right) \cup \left(\dfrac{3}{2}, \infty\right)$ or $\left\{x \mid x < \dfrac{3}{2} \text{ or } x > \dfrac{3}{2}\right\}$

(c) Using test intervals, $-4x^2 + 12x - 9 > 0$ has no solution

15. (a) $12z^2 - 23z + 10 = 0 \Rightarrow (3z - 2)(4z - 5) = 0 \Rightarrow z = \dfrac{2}{3}, \dfrac{5}{4}$.

(b) Using test intervals, $12z^2 - 23z + 10 \leq 0$ on $\left[\dfrac{2}{3}, \dfrac{5}{4}\right]$ or $\left\{z \mid \dfrac{2}{3} \leq z \leq \dfrac{5}{4}\right\}$

(c) Using test intervals, $12z^2 - 23z + 10 \geq 0$ on $\left(-\infty, \dfrac{2}{3}\right] \cup \left[\dfrac{5}{4}, \infty\right)$ or $\left\{z \mid z \leq \dfrac{2}{3} \text{ or } z \geq \dfrac{5}{4}\right\}$

17. (a) $x^2 + 2x - 1 = 0$, using the quadratic formula we get: $x = \dfrac{-2 \pm \sqrt{(2)^2 - 4(1)(-1)}}{2(1)} \Rightarrow$

$x = \dfrac{-2 \pm \sqrt{8}}{2} \Rightarrow x = \dfrac{-2 \pm 2\sqrt{2}}{2} \Rightarrow x = -1 \pm \sqrt{2}$.

(b) Using test intervals, $x^2 + 2x - 1 < 0$ on $(-1 - \sqrt{2}, -1 + \sqrt{2})$ or $\{x \mid -1 - \sqrt{2} < x < -1 + \sqrt{2}\}$

(c) Using test intervals, $x^2 + 2x - 1 > 0$ on $(-\infty, -1 - \sqrt{2}) \cup (-1 + \sqrt{2}, \infty)$ or

$\{x \mid x < -1 - \sqrt{2} \text{ or } x > -1 + \sqrt{2}\}$

19. (a) The inequality $f(x) < 0$ is satisfied when the graph of f is below the x-axis. This occurs when

$-3 < x < 2$.

(b) The inequality $f(x) \geq 0$ is satisfied when the graph of f is above the x-axis or intersects it. This occurs

when $x \leq -3$ or $x \geq 2$.

21. (a) The inequality $f(x) \leq 0$ is satisfied when the graph of f is below the x-axis or intersects it. This occurs

only when $x = -2$.

(b) The inequality $f(x) > 0$ is satisfied when the graph of f is above the x-axis. This occurs when $x \neq -2$.

23. (a) The inequality $f(x) > 0$ is satisfied when the graph of f is above the x-axis. Since the graph of f is always

below the x-axis, this inequality has no solutions.

(b) The inequality $f(x) < 0$ is satisfied when the graph of f is below the x-axis. This occurs for all real

values of x.

25. (a) The x-intercepts are the solutions to $f(x) = 0$: $0 = 2x^2 + 6x + \dfrac{5}{2} \Rightarrow 4x^2 + 12x + 5 = 0$.

$x = \dfrac{-12 \pm \sqrt{(12)^2 - 4(4)(5)}}{2(4)} \Rightarrow x = \dfrac{-12 \pm \sqrt{64}}{8} \Rightarrow x = \dfrac{-12 \pm 8}{8} \Rightarrow x = \dfrac{-4}{8} \Rightarrow$

$x = -\dfrac{1}{2}$ or $x = -\dfrac{20}{8} \Rightarrow x = -2\dfrac{1}{2}, x = -2\dfrac{1}{2}, -\dfrac{1}{2}$

(b) The inequality $f(x) < 0$ is satisfied when the graph of f is above the x-axis. This occurs in the interval

$\left(-\dfrac{5}{2}, -\dfrac{1}{2}\right)$ or $\left\{x \mid -\dfrac{5}{2} < x < -\dfrac{1}{2}\right\}$.

(c) The inequality $f(x) > 0$ is satisfied when the graph of f is above the x-axis. This occurs in the interval

$\left(-\infty, -\dfrac{5}{2}\right) \cup \left(-\dfrac{1}{2}, \infty\right)$ or $\left\{x \mid x < -\dfrac{5}{2} \text{ or } x > -\dfrac{1}{2}\right\}$.

27. (a) The x-intercepts are the solutions to $f(x) = 0$: $\quad 0 = -5x^2 + 2x + 7$.

$$x = \frac{-2 \pm \sqrt{(2)^2 - 4(-5)(7)}}{2(-5)} \Rightarrow x = \frac{-2 \pm \sqrt{144}}{-10} \Rightarrow x = \frac{-2 \pm 12}{-10} \Rightarrow x = -1, \frac{7}{5}$$

(b) The inequality $f(x) < 0 \quad$ is satisfied when the graph of f is above the x-axis. This occurs in the interval

$$(-\infty, -1) \cup \left(\frac{7}{5}, \infty\right) \text{ or } \left\{x \mid x < -1 \text{ or } x > \frac{7}{5}\right\}$$

(c) The inequality $f(x) > 0 \quad$ is satisfied when the graph of f is above the x-axis. This occurs in the interval

$$\left(-1, \frac{7}{5}\right) \text{ or } \left\{x \mid -1 < x < \frac{7}{5}\right\}$$

29. (a) $f(x) > 0$ when $x < -1$ or $x > 1$

    (b) $f(x) \leq 0$ when $-1 \leq x \leq 1$

31. (a) $f(x) > 0$ when $-6 < x < -2$

    (b) $f(x) \leq 0$ when $x \leq -6$ or $x \geq -2$

33. Start by solving $2x^2 + 5x + 2 = 0 \Rightarrow (2x + 1)(x + 2) = 0 \Rightarrow x = -2$ or -0.5; thus,

    $2x^2 + 5x + 2 \leq 0$ when $-2 \leq x \leq -0.5$.

35. Start by solving $x^2 + x - 6 = 0 \Rightarrow (x + 3)(x - 2) = 0 \Rightarrow x = -3$ or 2; thus,

    $x^2 + x - 6 > 0$ when $x < -3$ or $x > 2$.

37. Start by solving the equation $x^2 = 4 \Rightarrow x = \pm 2$. Next write the quadratic inequality with $a > 0$.

    $x^2 \leq 4 \Rightarrow x^2 - 4 \leq 0$. The graph of $y = x^2 - 4$ is a parabola opening up. It will be less than or equal to 0

    between (and including) the solutions to the equation. The solutions are $-2 \leq x \leq 2$.

39. Since $x(x - 4) \geq 4 \Rightarrow x^2 - 4x + 4 \geq 0 \Rightarrow (x - 2)^2 \geq 0$, the solutions are all real numbers.

41. Start by solving the equation $-x^2 + x + 6 = 0 \Rightarrow x^2 - x - 6 = 0 \Rightarrow (x + 2)(x - 3) = 0 \Rightarrow$

    $x = -2, 3$. The graph of $y = -x^2 + x + 6$ is a parabola opening downward. It will be below the x-axis or

    intersect it outside of the solutions of the equation. The solutions are $x \leq -2$ or $x \geq 3$.

43. Start by solving the equation $6x^2 - x = 1 \Rightarrow 6x^2 - x - 1 = 0 \Rightarrow (3x + 1)(2x - 1) = 0 \Rightarrow$

    $x = -\frac{1}{3}, \frac{1}{2}$. The graph of $y = 6x^2 - x - 1$ is a parabola opening up. It will be below the x-axis between the

    solutions of the equation. The solutions to $6x^2 - x < 1$ are $-\frac{1}{3} < x < \frac{1}{2}$.

45. Start by solving the equation $(x + 4)(x - 10) = 0 \Rightarrow x = -4, 10$. The graph of

    $y = (x + 4)(x - 10) = x^2 - 6x - 40$ is a parabola opening up. It will intersect or be below the x-axis

    between (and including) the solutions of the equation. The solutions to $(x + 4)(x - 10) \leq 0$ are $-4 \leq x \leq 10$.

47. Since $2x^2 + 4x + 3 > 0 \quad$ for all values of x, there are no solutions to $2x^2 + 4x + 3 < 0$.

49. Start by solving the equation $9x^2 - 12x + 4 = 0 \Rightarrow (3x - 2)(3x - 2) = 0 \Rightarrow x = \frac{2}{3}$. The graph of

    $y = 9x^2 - 12x + 4 = (3x - 2)^2$ is a parabola opening up and intersecting the x-axis at $x = \frac{2}{3}$. It will always

    be above the x-axis except at $x = \frac{2}{3}$. The solutions to $9x^2 + 4 > 12x$ are all real numbers except $\frac{2}{3}$.

51. Start by solving the equation $x^2 = x \Rightarrow x^2 - x = 0 \Rightarrow x(x - 1) = 0 \Rightarrow x = 0, 1$.   The graph of $y = x^2 - x$ is a parabola opening up. It will be above or intersect the x-axis outside (and including) the solutions of the equation. The solutions to $x^2 \geq x$ are $x \leq 0$ or $x \geq 1$.

53. Start by solving the equation $x(x - 1) - 6 = 0 \Rightarrow x^2 - x - 6 = 0 \Rightarrow (x + 2)(x - 3) = 0 \Rightarrow x = -2, 3$. The graph of $y = x^2 - x - 6$ is a parabola opening up. It will be above or intersect the x-axis outside (and including) the solutions of the equation. The solutions to $x(x - 1) \geq 6$ are $x \leq -2$ or $x \geq 3$.

55. Start by solving the equation $x^2 - 5 = 0 \Rightarrow x^2 = 5 \Rightarrow x = \pm\sqrt{5}$. The graph of $y = x^2 - 5$ is a parabola opening up. It will intersect or be below the x-axis between (and including) the solutions of the equation. The solutions are $-\sqrt{5} \leq x \leq \sqrt{5}$.

57. Start by graphing $Y_1 = 7X^2 - 179.8X + 515.2$ in $[-4, 30, 2]$ by $[-750, 250, 50]$. The inequality given by $7x^2 - 179.8x + 515.2 \geq 0$ is satisfied when the graph Y_1 is above or intersects the x-axis. This occurs when $x \leq k$ or $x \geq 22.4$, where $k = \dfrac{23}{7}$.

59. For $x^2 - 9x + 14 \leq 0$, first solve $x^2 - 9x + 14 = 0 \Rightarrow (x - 7)(x - 2) = 0 \Rightarrow x = 2, 7$. These boundary numbers give us the disjoint intervals: $(-\infty, 2), (2, 7), (7, \infty)$. See Figure 59. $x^2 - 9x + 14 \leq 0$ when $2 \leq x \leq 7$.

Interval	Test Value x	$x^2 - 9x + 14$	Positive or Negative?
$(-\infty, 2)$	0	14	Positive
$(2, 7)$	4	-6	Negative
$(7, \infty)$	10	24	Positive

Figure 59

61. For $x^2 \geq 3x + 10 \Rightarrow x^2 - 3x - 10 \geq 0$, first solve $x^2 - 3x - 10 = 0 \Rightarrow (x - 5)(x + 2) = 0 \Rightarrow x = -2, 5$. These boundary numbers give us the disjoint intervals: $(-\infty, -2), (-2, 5), (5, \infty)$. See Figure 61. $x^2 - 3x - 10 \geq 0$ when $x \leq -2$ or $x \geq 5$.

Interval	Test Value x	$x^2 - 3x - 10$	Positive or Negative?
$(-\infty, -2)$	-3	8	Positive
$(-2, 5)$	0	-10	Negative
$(5, \infty)$	6	8	Positive

Figure 61

63. For $\dfrac{1}{8}x^2 + x + 2 \geq 0 \Rightarrow x^2 + 8x + 16 \geq 0$, first solve $x^2 + 8x + 16 = 0 \Rightarrow (x + 4)(x + 4) = 0 \Rightarrow x = -4$. This boundary gives us the disjoint intervals: $(-\infty, -4), (-4, \infty)$. See Figure 63. $\dfrac{1}{8}x^2 + x + 2 \geq 0$ for all real numbers.

Interval	Test Value x	$1/8x^2 + x + 2$	Positive or Negative?
$(-\infty, -4)$	-8	2	Positive
$(-4, \infty)$	0	2	Positive

Figure 63

65. For $x^2 > 3 - 4x \Rightarrow x^2 + 4x - 3 > 0$, first solve $x^2 + 4x - 3 = 0$. Using the quadratic formula we get:

$$x = \frac{-4 \pm \sqrt{(4)^2 - 4(1)(-3)}}{2(1)} \Rightarrow x = \frac{-4 \pm \sqrt{28}}{2} \Rightarrow x = -2 + \sqrt{7} \text{ or } x = -2 - \sqrt{7}.$$ These

boundary numbers give us the disjoint intervals:

$(-\infty, -2 - \sqrt{7}), (-2 - \sqrt{7}, -2 + \sqrt{7}), (-2 + \sqrt{7}, \infty)$. See Figure 65. $x^2 + 4x - 3 > 0$ when

$x < -2 - \sqrt{7}$ or $x > -2 + \sqrt{7}$.

Interval	Test Value x	$x^2 + 4x - 3$	Positive or Negative?
$(-\infty, -4.6)$	-5	2	Positive
$(-4.6, 0.6)$	0	-3	Negative
$(0.6, \infty)$	1	2	Positive

Figure 65

67. Graph $Y_1 = (1/9)X^2 + (11/3)X$, $Y_2 = 300$ in $[30, 40, 1]$ by $[0, 400, 100]$ (Not Shown). The intersection is

(38.018, 300). Thus, the safe stopping speed is to be no more than 38 mph. The speed limit might be 35 mph.

69. To find the possible values of the radius of the can, solve the following inequality: $24\pi \le \pi(r^2)6 \le 54\pi \Rightarrow$

$24 \le 6r^2 \le 54 \Rightarrow 4 \le r^2 \le 9 \Rightarrow 2 \le r \le 3$. The possible values of r range between 2 and 3 inches.

71. (a) $f(0) = \frac{4}{5}(0 - 10)^2 + 80 = 160$ and $f(2) = \frac{4}{5}(2 - 10)^2 + 80 = 131.2$.   Initially when the person stops

    exercising the heart rate is 160 beats per minute, and after 2 minutes the heart rate has dropped to about 131

    beats per minute.

    (b) Graph $Y_1 = 100$, $Y_2 = 0.8(X - 10)^2 + 80$ and $Y_3 = 120$ (Not shown). To find the points of intersection,

    use the calc function on the calculator. The person's heart rate is between 100 and 120 when the graph of

    Y_2 is between the graphs of Y_1 and Y_3. This occurs between approximately 2.9 minutes and 5 minutes after

    the person stops exercising.

73. Graph $Y_1 = 2375X^2 + 5134X - 5020$, $Y_2 = 90,000$ and $Y_3 = 200,000$ in $[-1, 11, 1]$ by

$[5000, 210000, 10000]$ (Not shown). The intersection points are approximately (4.9977492, 90000) and

(8.044127, 200,000). From the graphs, we see that $90,000 \le Y_1 \le 200,000$ when $4.998 \le X \le 8.044$. Since

1984 corresponds to $x = 0$, the number of AIDS deaths was from 90,000 to 200,000 for the years from 1989 to 1992.

[-25, 200, 25] by [-5, 20, 5]   [-25, 200, 25] by [-5, 20, 5]

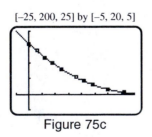   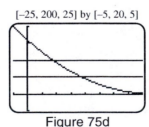

Figure 75c   Figure 75d

75. To find when the sales were between 50 and 55 million iPods, graph the function.

$Y_1 = 50$, $Y_2 = -2.277x^2 + 11.71x + 40.4$, $Y_3 = 55$ in the same window. The intersection points are approximately (1.02, 50), (2.12, 55), and (4.11, 50). Since x represents the number of years 2006, the result is from 2007 to 2008 and 2009 to 2010. See Figures 75c and 75d.

Checking Basic Concepts for Sections 3.3 and 3.4

1. (a) $\sqrt{-25} = \sqrt{25}\sqrt{-1} = 5i$

    (b) $\sqrt{-3} \cdot \sqrt{-18} = (\sqrt{3}\sqrt{-1})(\sqrt{9}\sqrt{2}\sqrt{-1}) = (i\sqrt{3})(3i\sqrt{2}) = 3i^2\sqrt{6} = -3\sqrt{6}$

    (c) $\dfrac{7 + \sqrt{-98}}{14} = \dfrac{7 \pm \sqrt{49}\sqrt{2}\sqrt{-1}}{14} = \dfrac{7 \pm 7i\sqrt{2}}{14} = \dfrac{1}{2} \pm \dfrac{i\sqrt{2}}{2}$

3. (a) The x-intercepts are -3 and 0. The graph of $y = f(x)$ is below the x-axis between these values, so

    $f(x) \leq 0$ on the interval $[-3, 0]$. The graph of $y = f(x)$ is above the x-axis outside of these values, so

    $f(x) > 0$ on the intervals $(-\infty, -3) \cup (0, \infty)$ or $\{x \mid x < -3$ or $x > 0\}$.

    (b) The x-intercept is -2. The graph of $y = f(x)$ is always below the x-axis for all real numbers except

    at $x = -2$. That is, $f(x) \leq 0$ on $(-\infty, \infty)$. The graph of $y = f(x)$ is never

    above the x-axis, so there are no solutions to $f(x) > 0$.

5. (a) Start by solving the equation $x^2 - 36 = 0 \Rightarrow x^2 = 36 \Rightarrow x = \pm 6$. The graph of $y = x^2 - 36$ is a parabola

    opening up and y is positive when x is greater than 6 or less than negative 6. The solution is

    $(-\infty, -6] \cup [6, \infty)$ or $\{x \mid x \leq -6$ or $x \geq 6\}$.

    (b) Start by solving the equation $4x^2 + 9 = 9x \Rightarrow 4x^2 - 9x + 9 = 0$. Using the quadratic formula we get:

    $x = \dfrac{9 \pm \sqrt{81 - 4(4)(9)}}{2(4)} \Rightarrow x = \dfrac{9 \pm \sqrt{-63}}{8} \Rightarrow$ no real solutions $\Rightarrow$ no intersection with the

    x-axis. This is a parabola opening up completely above the x-axis $\Rightarrow$ the solutions to $4x^2 + 9 > 9x$ are

    all real numbers.

    (c) Start by solving the equation $2x(x - 1) = 2 \Rightarrow 2x^2 - 2x = 2 \Rightarrow 2x^2 - 2x - 2 = 0$.

    Using the quadratic formula we get: $x = \dfrac{2 \pm \sqrt{4 - 4(2)(-2)}}{2(2)} \Rightarrow x = \dfrac{2 \pm \sqrt{20}}{4} \Rightarrow \dfrac{1 \pm \sqrt{5}}{2}$.

    The graph of $y = 2x^2 - 2x - 2$ is a parabola opening up. It will be below the x-axis between

    the solutions of the equation. The solutions to $2x(x - 1) \leq 2$ is $\left[\dfrac{1 - \sqrt{5}}{2}, \dfrac{1 + \sqrt{5}}{2} \right]$ or

    $\left\{ x \mid \dfrac{1 - \sqrt{5}}{2} \leq x \leq \dfrac{1 + \sqrt{5}}{2} \right\}$.

3.5: Transformations of Graphs

1. The vertex is (–2, 0). The parabola $y = x^2$ has been shifted 2 units to the left. The equation is $y = (x + 2)^2$.

3. The endpoint is (–3, 0). The curve $y = \sqrt{x}$ has been shifted 3 units to the left. The equation is $y = \sqrt{x + 3}$.

5. The vertex is (–2, –1). The graph of $y = |x|$ has been shifted 2 units to the left and 1 unit down.

    The equation is $y = |x + 2| - 1$.

7. The endpoint is (–2, –3). The curve $y = \sqrt{x}$ has been shifted 2 units to the left and 3 units down.

    The equation is $y = \sqrt{x + 2} - 3$.

9. To shift the graph of $f(x) = x^2$ to the right 2 units, replace x with $(x - 2)$ in the formula for $f(x)$.

    This results in $y = f(x - 2) = (x - 2)^2$. To shift the graph of this new equation down 3 units, subtract 3

    from the formula to obtain $y = f(x - 2) - 3 = (x - 2)^2 - 3$. See Figure 9.

[–10, 10, 1] by [–10, 10, 1]   [–10, 10, 1] by [–10, 10, 1]

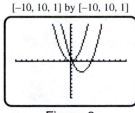
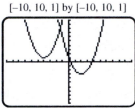

Figure 9   Figure 11

11. To shift the graph of $f(x) = x^2 - 4x + 1$ to the left 6 units, replace x with $(x + 6)$ in the formula for $f(x)$.

    This results in $y = f(x + 6) = (x + 6)^2 - 4(x + 6) + 1$. To shift the graph of this new equation up 4

    units, add 4 to the formula to obtain

    $y = f(x + 6) + 4 = (x + 6)^2 - 4(x + 6) + 1 + 4 = (x + 6)^2 - 4(x + 6) + 5$. See Figure 11.

13. To shift the graph of $f(x) = \dfrac{1}{2}x^2 + 2x - 1$ to the left 3 units, replace x with $(x + 3)$ in the formula for $f(x)$.

    The result is $y = f(x + 3) = \dfrac{1}{2}(x + 3)^2 + 2(x + 3) - 1$.   To shift the graph of this new equation down 2

    units, subtract 2 to obtain $y = f(x + 3) - 2 = \dfrac{1}{2}(x + 3)^2 + 2(x + 3) - 1 - 2 \Rightarrow$

    $y = \dfrac{1}{2}(x + 3)^2 + 2(x + 3) - 3$. See Figure 13.

[–10, 10, 1] by [–10, 10, 1]

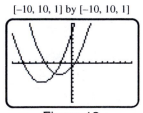

Figure 13

15. (a) $g(x) = 3(x + 3)^2 + 2(x + 3) - 5$

    (b) $g(x) = 3x^2 + 2x - 9$

17. (a) $g(x) = 2(x - 2)^2 + 4$

    (b) $g(x) = 2(x + 8)^2 - 5$

19. (a) $g(x) = 3(x - 2000)^2 - 3(x - 2000) + 72$

    (b) $g(x) = 3(x + 300)^2 - 3(x + 300) - 28$

21. (a) $g(x) = -\sqrt{x} - 4$

    (b) $g(x) = \sqrt{-x} + 2$

23. $(x - 3)^2 + (y + 4)^2 = 4$; Center: $(3, -4)$; Radius $= 2$

25. $(x + 5)^2 + (y - 3)^2 = 5$; Center: $(-5, 3)$; Radius $= \sqrt{5}$

27. (a) See Figure 27a.

    (b) See Figure 27b.

    (c) See Figure 27c.

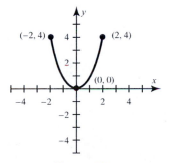

Figure 27a

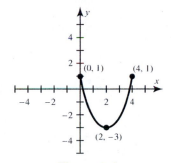

Figure 27b

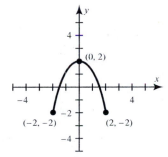

Figure 27c

29. (a) See Figure 29a.

    (b) See Figure 29b.

    (c) See Figure 29c.

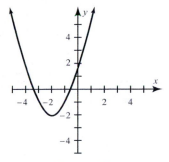

Figure 29a

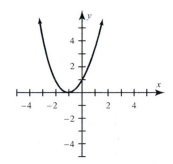

Figure 29b

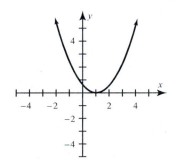

Figure 29c

31. (a) See Figure 31a.

    (b) See Figure 31b.

    (c) See Figure 31c.

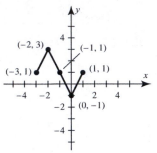

Figure 31a

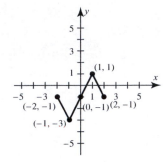

Figure 31b

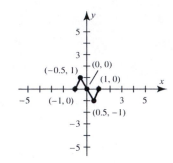

Figure 31c

33. (a) See Figure 33a.

    (b) See Figure 33b.

    (c) See Figure 33c.

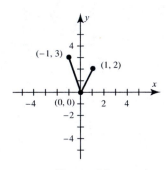

Figure 33a

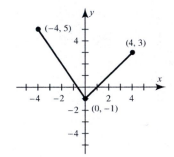

Figure 33b

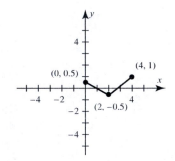

Figure 33c

35. The graph of $y = -f(x)$ is a reflection of the graph of $y = f(x)$ across the x-axis. Therefore the new equation

    is $y = -\sqrt{x} - 1$. See Figure 35a and Figure 35b.

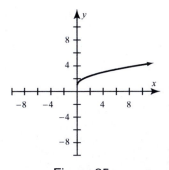

Figure 35a

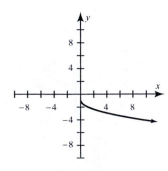

Figure 35b

37. The graph of $y = f(-x)$ is a reflection of the graph of $y = f(x)$ across the x-axis. Therefore the new equation

    is $y = x^2 + x$. See Figure Figure 37a and Figure 37b.

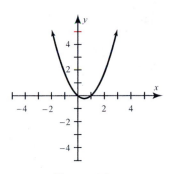

Figure 37a   Figure 37b

39. To reflect this function across the x-axis, graph $y = -f(x) = -(x^2 - 2x - 3) = -x^2 + 2x + 3$.
    See Figure 39a.

    To reflect this function across the y-axis, graph $y = f(-x) = (-x)^2 - 2(-x) - 3 = x^2 + 2x - 3$.
    See Figure 39b.

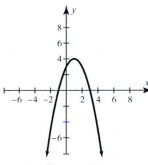 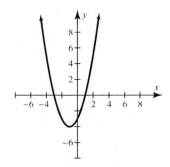

Figure 39a   Figure 39b

41. To reflect this function across the x-axis, graph $y = -f(x) = -(|x + 1| - 1) = -|x + 1| + 1$.
    See Figure 41a.

    To reflect this function across the y-axis, graph $y = f(-x) = |-x + 1| - 1 = |-x + 1| - 1$.
    See Figure 41b.

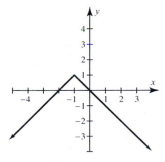 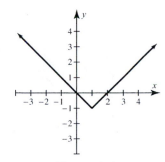

Figure 41a   Figure 41b

43. (a) To reflect this line graph across the x-axis, make a table of values for $y = -f(x)$; that is, change each (x, y) to $(x, -y)$. Plot the points $(-3, -2), (-1, -3), (1, 1)$, and $(2, 2)$. See Figure 43a.

   (b) To reflect this line graph across the y-axis, make a table of values for $y = f(-x)$; that is, change each (x, y) to $(-x, y)$. Plot the points $(3, 2), (1, 3), (-1, -1)$, and $(-2, -2)$. See Figure 43b.

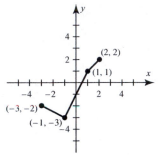

Figure 43a

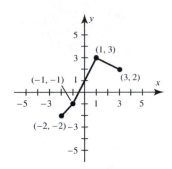

Figure 43b

45. Shift the graph of $y = x^2$ right 3 units and upward 1 unit.

47. Shift the graph of $y = x^2$ left 1 unit and vertically shrink it with factor $\frac{1}{4}$.

49. Reflect the graph of $y = \sqrt{x}$ across the x-axis and shift it left 5 units.

51. Reflect the graph of $y = \sqrt{x}$ across the y-axis and vertically stretch it with factor 2.

53. Reflect the graph of $y = |x|$ across the y-axis and shift it left 1 unit.

55. Shift the graph of $y = x^2$ down 3 units. The vertex is located at $(0, -3)$. See Figure 55.

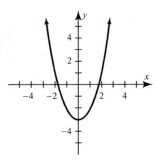

Figure 55

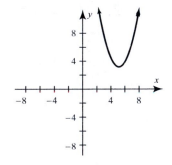

Figure 57

57. Shift the graph of $y = x^2$ to the right 5 units and up 3 units. The vertex is located at $(5, 3)$. See Figure 57.

59. Reflect the graph of $y = \sqrt{x}$ across the x-axis. See Figure 59.

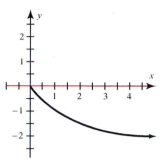

Figure 59

61. Reflect the graph of $y = x^2$ across the x-axis and shift the graph 4 units up. See Figure 61.

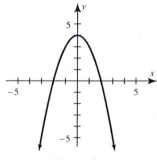

Figure 61

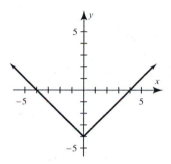

Figure 63

63. Shift the graph of $y = |x|$ down 4 units. See Figure 63.

65. Shift the graph of $y = \sqrt{x}$ to the right 3 units and up 2 units. See Figure 65.

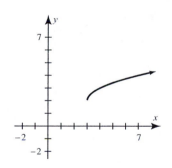

Figure 65

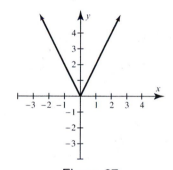

Figure 67

67. The graph is like $f(x) = |x|$ but narrower by a factor of 2, vertex $(0, 0)$. See Figure 67.

69. Reflect the graph of $f(x) = \sqrt{x}$ across the x-axis and then shift up 1 unit, the vertex is $(0, 1)$. See Figure 69.

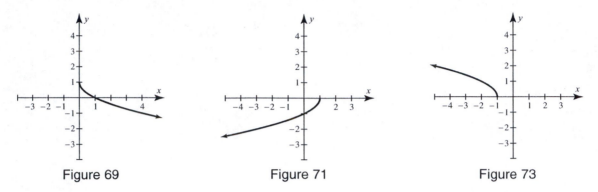

Figure 69   Figure 71   Figure 73

71. Shift the graph of $f(x) = \sqrt{x}$ left 1 unit, then reflect this graph across the y-axis and then reflect this graph across the x-axis. See Figure 71.

73. Reflect the graph of $f(x) = \sqrt{x}$ across the y-axis, then shift left 1 unit. See Figure 73.

75. Shift the graph of $f(x) = x^3$ right 1 unit. See Figure 75.

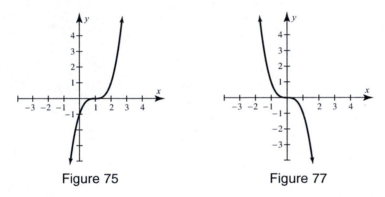

Figure 75   Figure 77

77. Reflect the graph of $f(x) = x^3$ across the x-axis. See Figure 77.

79. Since $g(x) = f(x) + 7$, the output for $g(x)$ can be determined by adding 7 to the output of $f(x)$.
See Figure 79.

x	1	2	3	4	5	6
$g(x)$	12	8	13	9	14	16

Figure 79

81. Since $g(x) = f(x - 2)$, each input value of g is found by shifting each input value of f two units right.

For example, $g(4) = f(4 - 2) = f(2) = -5$ and $g(2) = f(2 - 2) = f(0) = -3$.

However, the value of $g(-4) = f(-4 - 2) = f(-6)$ is undefined because -6 is not in the domain of f.

See Figure 81.

x	-2	0	2	4	6
$g(x)$	5	2	-3	-5	-9

Figure 81

83. Since $g(x) = f(x + 1) - 2$, each input value of g is found by shifting each input value of f one unit left. Then each output of g is found by shifting the cooresponding output of f two units down.

   For example, $g(1) = f(1 + 1) - 2 = f(2) - 2 = 4 - 2 = 2$ and

   $g(1) = f(1 + 1) - 2 = f(2) - 2 = 4 - 2 = 2$ and $g(2) = f(2 + 1) - 2 = f(3) - 2 = 3 - 2 = 1$.

   $g(5) = f(6) - 2 = 10 - 2 = 8$; $g(6) = f(7) - 2$ is undefined because 7 is not in the domain of f.

   See Figure 83.

x	0	1	2	3	4	5
$g(x)$	0	2	1	5	6	8

   Figure 83

85. Since $g(x) = f(-x) + 1$, each input value of g is the opposite of each input value of f .

   Then each output of g is found by shifting the cooresponding output of f one unit up.

   For example, $g(-2) = f(-(-2)) + 1 = f(2) + 1 = -1 + 1 = 0$ and

   $g(-1) = f(-(-1)) + 1 = f(1) + 1 = 2 + 1 = 3$. See Figure 85.

x	-2	-1	0	1	2
$g(x)$	0	3	6	9	12

   Figure 85

87. The graph of $f(x)$ is shifted up 2 units $\Rightarrow (-12, 8), (0, 10)$, and $(8, -2)$.

89. The graph of $f(x)$ is shifted right 2 units and up 1 unit $\Rightarrow (-10, 7), (2, 9)$, and $(10, -3)$.

91. The graph is reflected across the x-axis $\Rightarrow$ the y-coordinate is $-\frac{1}{2}$ times the given y-coordinate $\Rightarrow$ $(-12, -3), (0, -4)$, and $(8, 2)$.

93. The graph is reflected across the y-axis $\Rightarrow$ the x-coordinate is the reciprocal of -2 or $-\frac{1}{2}$ times the given x-coordinate $\Rightarrow (6, 6), (0, 8)$, and $(-4, -4)$.

95. Let $(2008, 11.6)$ be the vertex (h, k). Translate the graph of $y = x^2$ to the right 2008 units up 11.6 units to find

   $f(x) = a(x - 2008)^2 + 11.6$. Choose the point $(2011, 72.3)$ to determine the value for a as follows:

   $72.3 = a(2011 - 2008)^2 + 11.6 \Rightarrow 60.7 = 9a \Rightarrow a \approx 6.7$. The function is $f(x) = 6.7(x - 2008)^2 + 11.6$.

97. Let $(2008, 22)$ be the vertex (h, k). Translate the graph of $y = x^2$ to the right 2009 units up 22 units to find

   $f(x) = a(x - 2008)^2 + 22$. Choose the point $(2011, 38)$ to determine the value for a as follows:

   $38 = a(2011 - 2008)^2 + 22 \Rightarrow 16 = 9a \Rightarrow a \approx 1.8$. The function is $f(x) = 1.8(x - 2008)^2 + 22$.

99. We must determine a function g such that $g(1991) = P(1), g(1992) = P(2)...$.   Thus, the relationship

   $g(x) = P(x - 1990)$ holds for $x = 1990, 1991, 1992...$. It follows that the representation for $g(x)$ is

   $g(x) = 0.00075(x - 1990)^2 + 0.17(x - 1990) + 44$.

101.In 15 seconds, the plane has moved $15(0.2) = 3$ kilometers to the left. To show this movement, translate the graph of the mountain 3 kilometers (units) to the right; $y = -0.4(x - 3)^2 + 4$. See Figure 101.

[-4, 4, 1] by [0, 6, 1]

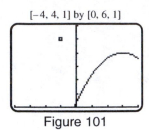

Figure 101

103. (a) In 4 hours, the cold front has moved $4(40) = 160$ miles, which is $\dfrac{160}{100} = 1.6$ units on the graph. To show this movement, shift the graph 1.6 units down; $y = \dfrac{1}{20}x^2 - 1.6$. See Figure 103a.

(b) The graph of the cold front has shifted $\dfrac{250}{100} = 2.5$ units down and $\dfrac{210}{100} = 2.1$ units to the right. To show this movement, the new equation should be $y = \dfrac{1}{20}(x - 2.1)^2 - 2.5$. The location of Columbus, Ohio is at $(5.5, -0.8)$ relative to the location of Des Moines at $(0, 0)$. Since the parabola representing the cold front at midnight has past the point $(5.5, -0.8)$, the front reaches Columbus by midnight. See Figure 103b.

[-15, 15, 1] by [-10, 10, 1]   [-15, 15, 1] by [-10, 10, 1]

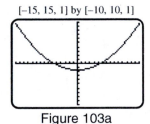

Figure 103a   Figure 103b

Checking Basic Concepts for Section 3.5

1. (a) The graph of $y = (x + 4)^2$ is the graph of $f(x) = x^2$ shifted 4 units to the left.

    (b) The graph of $y = x^2 - 3$ is the graph of $f(x) = x^2$ shifted 3 units down.

    (c) The graph of $y = (x - 5)^2 + 3$ is the graph of $f(x) = x^2$ shifted 5 units to the right and 3 units up.

3. (a) To shift $y = x^2$ to the right 3 units replace x with $(x - 3)$. To shift the graph of the new function down 4 units subtract 4 from y. The new equation is $y = (x - 3)^2 - 4$.

    (b) To reflect $y = x^2$ about the y-axis, replace $f(x)$ with $-f(x)$. The new equation is $y = -x^2$.

    (c) To shift $y = x^2$ to the left 6 units replace x with $(x + 6)$. To reflect the graph of the new function about the y-axis, replace $f(x)$ with $f(-x)$. The new equation is $y = (-x + 6)^2$.

    (d) To reflect the graph of $f(x) = x^2 - 4x + 1$ about the y-axis, replace x with $-x$. This results in $y = f(-x) = (-x)^2 - 4(-x) + 1$. To shift the new equation to the left 6 units, replace x with $(x + 6)$. This results in $f(-(x + 6)) = (-(x + 6))^2 - 4(-(x + 6)) + 1$.

5. (a) Since $g(x) = f(x - 2) + 3$, each input value of g is found by shifting each input value of f two units

    right. Then each output of g is found by shifting the cooresponding output of f three units up. For example,

    $g(-2) = f(-2 - 2) + 3 = f(-4) + 3 = 1 + 3 = 4.$ See Figure 5a.

x	-2	0	2	4	6
$g(x)$	4	6	9	11	12

    Figure 5a

    (b) Since $h(x) = -2(f(x + 1))$, each input value of g is found by shifting each input value of f one unit left.

    Then each output of g is twice the opposite of the cooresponding output of f. For example,

    $h(-3) = -2(f(-3 + 1)) = -2(f(-2)) = -2(3) = -6.$ See Figure 5b.

x	-5	-3	-1	1	3
$h(x)$	-2	-6	-12	-16	-18

    Figure 5b

Chapter 3 Review Exercises

1. (a) The parabola opens downward, so $a < 0$.

    (b) Vertex: (2, 4)

    (c) Axis of symmetry: $x = 2$.

    (d) f is increasing for $x < 2$ and decreasing for $x > 2$.

3. $f(x) = -2(x - 5)^2 + 1 \Rightarrow f(x) = -2(x^2 - 10x + 25) + 1 \Rightarrow f(x) = -2x^2 + 20x - 49$

    The leading coefficient is –2.

5. The graph of $y = x^2$ has been shifted 1 unit left, reflected across the x-axis, and shifed 2 units up $\Rightarrow$

    $f(x) = -(x + 1)^2 + 2.$

7. $f(x) = x^2 + 6x - 1 \Rightarrow f(x) = x^2 + 6x + 9 - 9 - 1 \Rightarrow f(x) = (x^2 + 6x + 9) - 10 \Rightarrow$

    $f(x) = (x + 3)^2 - 10$; the vertex is (–3, –10).

9. $-\dfrac{b}{2a} = -\dfrac{2}{2(-3)} = \dfrac{1}{3}$ and $f\left(\dfrac{1}{3}\right) = -3\left(\dfrac{1}{3}\right)^2 + 2\left(\dfrac{1}{3}\right) - 4 = -\dfrac{11}{3}$; the vertex is $\left(\dfrac{1}{3}, -\dfrac{11}{3}\right)$.

11. Shift the graph of $f(x) = x^2$ up 3 units, reflect it across the x-axis, and make it narrower by the scale factor 3.

    The vertex is located at (0, 3). See Figure 11.

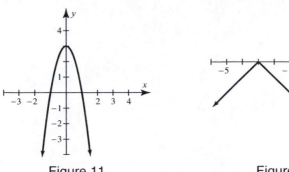

Figure 11   Figure 13

13. Shift the graph of $f(x) = |x|$ left 3 units, and reflect it across the x-axis. The vertex is located at $(-3, 0)$. See Figure 13.

15. Average rate of change $= \dfrac{f(b) - f(a)}{b - a} \Rightarrow \dfrac{f(4) - f(2)}{4 - 2} = \dfrac{-63 - (-5)}{2} = \dfrac{-58}{2} = -29$

17. $x^2 - x - 20 = 0 \Rightarrow (x - 5)(x + 4) = 0 \Rightarrow x = -4, 5$

19. $4z^2 - 7 = 0 \Rightarrow 4z^2 = 7 \Rightarrow z^2 = \dfrac{7}{4} \Rightarrow z = \pm\sqrt{\dfrac{7}{4}} \Rightarrow z = \pm\dfrac{\sqrt{7}}{2}$

21. $-2t^2 - 3t + 14 = 0 \Rightarrow (2t + 7)(-t + 2) = 0 \Rightarrow 2t + 7 = 0 \Rightarrow 2t = -7 \Rightarrow$

    $t = -\dfrac{7}{2}$ or $-t + 2 = 0 \Rightarrow -t = -2 \Rightarrow t = 2 \Rightarrow t = -\dfrac{7}{2}, 2$

23. $0.1x^2 - 0.3x = 1 \Rightarrow x^2 - 3x - 10 = 0 \Rightarrow (x - 5)(x + 2) = 0 \Rightarrow x = -2, 5$

25. $x^2 + 2x = 5 \Rightarrow x^2 + 2x + 1 = 5 + 1 \Rightarrow (x + 1)^2 = 6 \Rightarrow x + 1 = \pm\sqrt{6} \Rightarrow x = -1 \pm \sqrt{6}$

27. $2z^2 - 6z - 1 = 0 \Rightarrow z^2 - 3z = \dfrac{1}{2} \Rightarrow z^2 - 3z + \dfrac{9}{4} = \dfrac{1}{2} + \dfrac{9}{4} \Rightarrow \left(z - \dfrac{3}{2}\right)^2 = \dfrac{11}{4} \Rightarrow$

    $z - \dfrac{3}{2} = \pm\sqrt{\dfrac{11}{4}} \Rightarrow z = \dfrac{3}{2} \pm \dfrac{\sqrt{11}}{2} \Rightarrow z = \dfrac{3 \pm \sqrt{11}}{2}$

29. $2x^2 - 3y^2 = 6 \Rightarrow -3y^2 = -2x^2 + 6 \Rightarrow y^2 = \dfrac{2x^2 - 6}{3} \Rightarrow y = \pm\sqrt{\dfrac{2x^2 - 6}{3}}$

    Since there are two output values for each input, y is not a function of x.

31. (a) $\sqrt{-16} = \sqrt{16}\sqrt{-1} = 4i$

    (b) $\sqrt{-48} = \sqrt{16}\sqrt{-1}\sqrt{3} = 4i\sqrt{3}$

    (c) $\sqrt{-5} \cdot \sqrt{-15} = i\sqrt{5} \cdot i\sqrt{5}\sqrt{3} = i^2(5)\sqrt{3} = 5(-1)\sqrt{3} = -5\sqrt{3}$

33. (a) The x-intercepts of f are $-\dfrac{5}{2}$ and $\dfrac{1}{2}$.

    (b) The complex zeros of f are $-\dfrac{5}{2}$ and $\dfrac{1}{2}$.

35. $4x^2 + 9 = 0 \Rightarrow 4x^2 = -9 \Rightarrow x^2 = -\dfrac{9}{4} \Rightarrow x = \pm\sqrt{-\dfrac{9}{4}} = \pm\dfrac{3}{2}i$

37. (a) The graph of $y = f(x)$ has x-intercepts at $x = -3, 2$. It is above the x-axis between these values, which

are solutions to $f(x) = 0$, so $f(x) > 0$ when $-3 < x < 2$ or on $(-3, 2)$.

(b) The graph of $y = f(x)$ is below or intersects the x-axis outside of and including the values where $f(x) = 0$,

so $f(x) \leq 0$ when $x \leq -3$ or $x \geq 2$ or on $(-\infty, -3] \cup [2, \infty)$.

39. Start by solving $x^2 - 3x + 2 = 0 \Rightarrow (x - 1)(x - 2) = 0 \Rightarrow x = 1$ or 2; thus,

$x^2 - 3x + 2 \leq 0$ on $[1, 2]$ or $\{x \mid 1 \leq x \leq 2\}$. Note that a graph of $y = x^2 - 3x + 2$ intersects or is below

the x-axis for values of x between the x-intercepts.

41. Start by solving $n(n - 2) = 15 \Rightarrow n^2 - 2n - 15 = 0 \Rightarrow (n - 5)(n + 3) = 0 \Rightarrow n = -3$ or 5; thus

$n(n - 2) \geq 15$ on $(-\infty, -3] \cup [5, \infty)$ or $\{x \mid x \leq -3$ or $x \geq 5\}$. Note that a graph of $y = n^2 - 2n - 15$

intersects or is above the n-axis for values of n outside of the n-intercepts.

43. $y = -f(x) = -(2x^2 - 3x + 1) = -2x^2 + 3x - 1$. See Figure 43a.

$y = f(-x) = 2(-x)^2 - 3(-x) + 1 = 2x^2 + 3x + 1$. See Figure 43b.

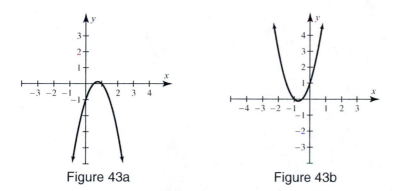

Figure 43a   Figure 43b

45. Shift the graph of $y = x^2$ four units down. See Figure 45.

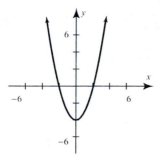

Figure 45

47. Reflect the graph of $y = x^2$ across the x-axis, shift it two units right and three unnits up, then stretch it vertical-

ly. See Figure 47.

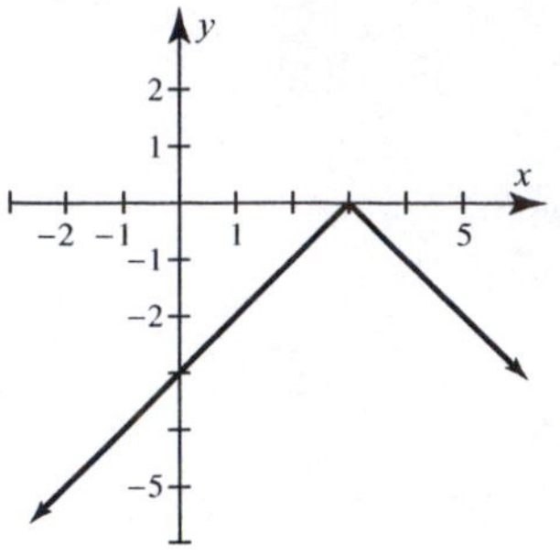

Figure 47

49. Let $W + L + W = 44 \Rightarrow L = 44 - 2W$. If $A = L \cdot W$ then $A = (44 - 2W)(W) \Rightarrow A = 44W - 2W^2$ or

$A = -2W^2 + 44W$. Now use the vertex formula: $W = -\dfrac{44}{2(-2)} \Rightarrow W = 11$ and $L = 44 - 2(11) \Rightarrow$

$L = 22$. So the dimensions are 11 feet by 22 feet.

51. (a) $h(t) = -16t^2 + 88t + 5 \Rightarrow h(0) = -16(0)^2 + 88(0) + 5 \Rightarrow h(0) = 5$

The stone was 5 feet above the ground when it was released.

(b) $h(2) = -16(2)^2 + 88(2) + 5 = 117$, the stone was 117 feet high after 2 seconds.

(c) $t = -\dfrac{b}{2a} = -\dfrac{88}{2(-16)} = 2.75$ and $h(2.75) = -16(2.75)^2 + 88(2.75) + 5 = 126$; the maximum height of

the stone was 126 feet.

(d) $h(t) = 117 \Rightarrow 117 = -16t^2 + 88t + 5 \Rightarrow 16t^2 - 88t + 112 = 0 \Rightarrow$

$t = \dfrac{88 \pm \sqrt{(-88)^2 - 4(16)(112)}}{2(16)} \Rightarrow t = 3.5 \text{ or } t = 2$

The stone was 117 feet high after 2 seconds and after 3.5 seconds.

53. Use the sketch shown in Figure 53 to determine the dimensions of the box; let $x = $ width of the metal sheet.

The dimensions of the box are 3 inches by $(x - 6)$ inches by $(x - 2)$ inches, and the volume is 135 cubic

inches.

$135 = 3(x - 6)(x - 2) \Rightarrow 135 = 3x^2 - 24x + 36 \Rightarrow 0 = 3x^2 - 24x - 99 \Rightarrow 0 = x^2 - 8x - 33 \Rightarrow$

$0 = (x - 11)(x + 3) \Rightarrow x = 11 \text{ or } x = -3$. Since $x = -3$ is impossible, the width of the metal sheet is

11 inches and the length is $11 + 4 = 15$ inches.

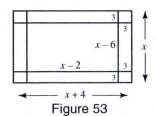
Figure 53

55. Let $(2007, 60)$ be the vertex (h, k). Translate the graph of $y = x^2$ to the right 2006 units and up 60 units to find

$f(x) = a(x - 2007)^2 + 60$. Choose the point $(2009, 180)$ to determine the value for a as follows:

$180 = a(2009 - 2007)^2 + 60 \Rightarrow 120 = 4a \Rightarrow a = 30$. The function is $f(x) = 30(x - 2007)^2 + 60$.

57. (a) No, the change in debt is not constant for each 4-year period.

(b) Using $(1980, 82)$ as the vertex, let $h = 1980$ and $k = 82$ in $f(x) = a(x - h)^2 + k$. Then the function can

be written $f(x) = a(x - 1980)^2 + 82$. To find the value of a, note that $f(1996) = 444$.

By substitution, $444 = a(1996 - 1980)^2 + 82 \Rightarrow 362 = 256a \Rightarrow a = \dfrac{362}{256} \approx 1.4$ When the data is

plotted along with the graph of $f(x) = 1.4(x - 1980)^2 + 82$, it may be reasonable to adjust the value of a.

A good fit appears to be obtained when $f(x) = 1.3(x - 1980)^2 + 82$. To solve $f(x) = 212$, substitute

212 for $f(x)$. $212 = 1.3(x - 1980)^2 + 82 \Rightarrow 130 = 1.3(x - 1980)^2 \Rightarrow 100 = (x - 1980)^2 \Rightarrow$

$\sqrt{100} = x - 1980 \Rightarrow x = 1980 + \sqrt{100} = 1990$. Note that only the positive square root of 100 is

necessary for dates after 1980. Visa and Mastercard debt reached \$212 billion in 1990.

Extended and Discovery Exercises for Chapter 3

1. (a) For $y = 10$ and $x = 15$, $y = \dfrac{-16x^2}{0.434v^2} + 1.15x + 8 \Rightarrow 10 = \dfrac{-16(15)^2}{0.434v^2} + 1.15(15) + 8 \Rightarrow$

$10 = \dfrac{-3600}{0.434v^2} + 17.25 + 8 \Rightarrow -15.25 = \dfrac{-3600}{0.434v^2} \Rightarrow v^2 \approx 543.93 \Rightarrow v \approx 23.32$feet per second.

(b) Graph $Y_1 = (-16X^2/(0.434*543.93)) + 1.15X + 8$ along with the points $(0, 8)$ and $(15, 10)$ in the

window $[-1, 16, 1]$ by $[-1, 16, 1]$. See Figure 1. The graph passes through both points.

(c) Using the maximum finder feature on the graphing calculator, we see that the maximum point is

approximately $(8.48, 12.88)$. The maximum height of the basketball is about 12.88 feet.

$[-1, 16, 1]$ by $[-1, 16, 1]$

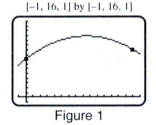

Figure 1

3. (a) See Figure 3.

(b) The graph of $y = f(2k - x) = f(4 - x)$ is a reflection of $y = f(x)$ across the line $x = 2$.

$[-1, 8, 1]$ by $[-4, 4, 1]$ $[-15, 3, 1]$ by $[-3, 9, 1]$

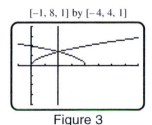   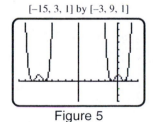

Figure 3   Figure 5

5. (a) See Figure 5.

    (b) The graph of $y = f(2k - x) = f(-12 - x)$ is a reflection of $y = f(x)$ across the line $x = -6$.

7. (a) The front reached St. Louis, but not Nashville. See Figure 7a.

    (b) $g(x) = -\sqrt{750^2 - (x - 160)^2} - 110$

    (c) The front reached both cities in less than 12 hours. See Figure 7c.

[0, 1200, 100] by [−800, 0, 100]   [0, 1200, 100] by [−800, 0, 100]

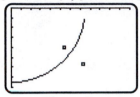 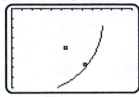

Figure 7a   Figure 7c

Chapter 4: More Nonlinear Functions and Equations

4.1: More Nonlinear Functions and Their Graphs

1. $f(x) = 2x^3 - x + 5$ is a polynomial; degree: 3; leading coefficient: 2.

3. $f(x) = \sqrt{x}$ is not a polynomial.

5. $f(x) = 1 - 4x - 5x^4$ is a polynomial; degree: 4; leading coefficient: –5.

7. $g(t) = \dfrac{1}{t^2 + 3t - 1}$ is not a polynomial.

9. $g(t) = 22$ is a polynomial; degree: 0; leading coefficient: 22.

11. (a) A local maximum of approximately 5.5 occurs when $x \approx -2$. A local minimum of approximately –5.5 occurs when $x \approx 2$. *Answers may vary slightly.*

    (b) There are no absolute extrema.

13. (a) Local maxima of approximately 17 and 27 occur when $x \approx -3$ and 2, respectively. Local minima of approximately –10 and 24 occur when $x \approx -1$ and 3, respectively. *Answers may vary slightly.*

    (b) There are no absolute extrema.

15. (a) Local maxima of approximately 0.5 and 2.8 occur when $x \approx 1$ and -2, respectively. A local minimum of approximately 0 occur when $x = 0$. *Answers may vary slightly.*

    (b) The absolute maximum is 2.8 when $x \approx -2$. There is no absolute minimum.

17. (a) A Local maximum of 0 occurs when $x = 0$. A local minimum of $-1{,}000$ occurs when $x = \pm 8$.

    (b) There is no absolute maximum. The absolute minimum of $-1{,}000$ occurs when $x = \pm 8$.

19. (a) Each local minimum is –1. Each local maximum is 1.

    (b) The absolute minimum is –1 and the absolute maximum is 1.

21. (a) A local minimum of approximately –3.2 occurs when $x \approx 0.5$. There are no local maxima. *Answers may vary slightly.*

    (b) The absolute minimum is approximately –3.2. The absolute maximum is 3 and occurs when $x = -2$.

23. (a) Local minima are approximately -0.5 and –2. Local maxima are approximately 0.5 and 2.

    (b) The absolute minimum is –2 and the absolute maximum is 2.

25. (a) There is no local maximum. A local minimum of -2 occurs when $x \approx 1$.

    (b) There is no absolute maximum. An absolute minimum of -2 occurs when $x \approx 1$.

27. (a) The graph of g is linear $\Rightarrow$ no local extrema.

    (b) The graph of g is linear $\Rightarrow$ no absolute extrema.

29. (a) The graph of g is a parabola opening up with a vertex $(0, 1) \Rightarrow$ it has a local minimum: 1; and no local maximum.

    (b) The graph of g is a parabola opening up with a vertex $(0, 1) \Rightarrow$ it has an absolute minimum: 1; and no absolute maximum.

31. (a) The graph of g is a parabola opening down with a vertex $(-3, 4) \Rightarrow$ it has a local maximum: 4; and no local minimum.

    (b) The graph of g is a parabola opening down with a vertex $(-3, 4) \Rightarrow$ it has an absolute maximum: 4; and no absolute minimum.

33. (a) The graph of g is a parabola opening up with a vertex $\left(\frac{3}{4}, -\frac{1}{8}\right) \Rightarrow$ it has a local minimum: $-\frac{1}{8}$; and no local maximum.

    (b) The graph of g is a parabola opening up with a vertex $\left(\frac{3}{4}, -\frac{1}{8}\right) \Rightarrow$ it has an absolute minimum: $-\frac{1}{8}$; and no absolute maximum.

35. (a) The absolute value graph opening up has a vertex $(-3, 0) \Rightarrow$ it has a local minimum: 0; and no local maximum.

    (b) The absolute value graph opening up has a vertex $(-3, 0) \Rightarrow$ it has an absolute minimum: 0; and no absolute maximum.

37. (a) The graph of g is the basic $\sqrt[3]{x}$ graph and has no local extrema.

    (b) The graph of g is the basic $\sqrt[3]{x}$ graph and has no absolute extrema.

39. (a) The cubic function graph of g has turning points $(-1, -2)$ and $(1, 2) \Rightarrow$ it has a local minimum: -2; and a local maximum: 2.

    (b) The cubic function graph of g has turning points $(-1, -2)$ and $(1, 2)$ but then continues on to $-\infty$ and $\infty \Rightarrow$ it has no absolute extrema.

41. (a) A graph of $f(x) = -3x^4 + 8x^3 + 6x^2 - 24x$ is shown in Figure 41. Local maxima of 19 and -8 occur at $x \approx -1$ and $x \approx 2$ respectively. A local minima of -13 occurs when $x \approx 1$.

    (b) The absolute maximum of 19 occurs at $x \approx -1$, and there is no absolute minimum.

[-5, 5, 1] by [-30, 30, 5]

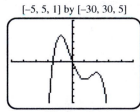

Figure 41

43. (a) A graph of $f(x) = 0.5x^4 - 5x^2 + 4.5$ is shown in Figure 43. A local maximum of 4.5 occurs at $x = 0$. A local minimum of -8 occurs when $x \approx \pm 2.236$.

    (b) There is no absolute maximum. An absolute minimum of -8 occurs at $x \approx \pm 2.236$.

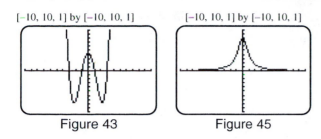

[−10, 10, 1] by [−10, 10, 1]   [−10, 10, 1] by [−10, 10, 1]

Figure 43   Figure 45

45. (a) A graph of $f(x) = \dfrac{8}{1 + x^2}$ is shown in Figure 45. A local maximum of 8 occurs at $x = 0$. There are no local minima.

   (b) An absolute maximum of 8 occurs at $x = 0$. There is no absolute minimum.

47. From the graph we can see that the function f is neither even or odd.

49. From the graph we can see that the function is even. Since the point $(1, -1)$ is on the graph of f, the point $(-1, -1)$ must also be located on the graph.

51. The function f is a monomial with only an odd power $\Rightarrow$ it is odd.

53. The function f is a polynomial with an odd power and a constant term, which is an even power $\Rightarrow$ it is neither.

55. The function f is a polynomial with an even power and a constant term, which is also even $\Rightarrow$ it is even.

57. The function f is a polynomial with all even powers and a constant term, which is also even $\Rightarrow$ it is even.

59. The function f is a polynomial with all odd powers $\Rightarrow$ it is odd.

61. The function f is a polynomial with both an odd powered term and an even powered term $\Rightarrow$ it is neither.

63. If $f(x) = \sqrt[3]{x^2}$, then $f(-x) = \sqrt[3]{(-x)^2} = \sqrt[3]{x^2}$, which is equal to $f(x) \Rightarrow$ it is even.

65. If $f(x) = \sqrt{1 - x^2}$, then $f(-x) = \sqrt{1 - (-x)^2} = \sqrt{1 - x^2}$, which is equal to $f(x) \Rightarrow$ it is even.

67. If $f(x) = \dfrac{1}{1 + x^2}$, then $f(-x) = \dfrac{1}{1 + (-x)^2} = \dfrac{1}{1 + x^2}$, which is equal to $f(x) \Rightarrow$ it is even.

69. If $f(x) = |\,x + 2\,|$, then $f(-x) = |\,-x + 2\,|$, and $-f(x) = -|\,x + 2\,|$. Since $f(x) \neq f(-x)$ and $f(-x) \neq -f(x)$. It is neither.

71. Notice that $f(-100) = 56$ and $f(100) = -56$. In general, since $f(-x) = -f(x)$ holds for each x in the table, f is odd.

73. Since f is even, the condition $f(-x) = f(x)$ must hold. For example, since $f(-3) = 21$, $f(3)$ must also equal 21. The value assigned to $f(0)$ makes no difference. See Figure 73.

x	-3	-2	-1	0	1	2	3
$f(x)$	21	-12	-25	1	-25	-12	21

Figure 73

75. Since f is an odd function and $(-5, -6)$ and $(-3, 4)$ are on the graph of f, then $(5, 6)$ and $(3, -4)$ are on the graph. Thus, $f(5) = 6$ and $f(3) = -4$.

77. Sketch any linear function that passes through the origin will work. For example, the graph of $y = x$ is shown in Figure 77.

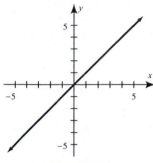

Figure 77

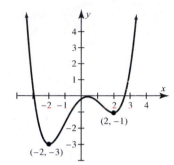

Figure 81

79. No. If $(2, 5)$ is on the graph of an odd function f, then so is $(-2, -5)$. Since f would pass through $(-3, -4)$ and then $(-2, -5)$, it could not always be increasing.

81. Plot points $(-2, -3)$ and $(2, -1)$ and make them minima. See Figure 81.

83. Plot point $(2, 3)$ and make it a maxima. See Figure 83. Yes, it could be quadratic, but it does not have to be quadratic.

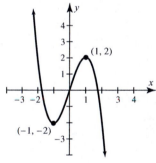

Figure 82

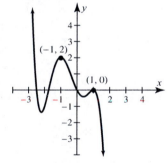

Figure 84

85. Shift the graph of $y = f(x) = 4x - \dfrac{x^3}{3}$ to the left 1 unit to get the graph of

$$y = f(x + 1) = 4(x + 1) - \dfrac{(x + 1)^3}{3}.$$ See Figure 85.

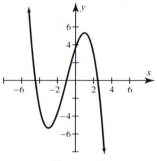

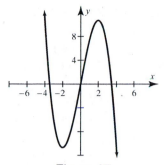

Figure 85   Figure 87

87. Sketch the graph of $y = f(x) = 4x - \dfrac{x^3}{3}$ vertically by multiplying each y-value by 2, that is, each (x, y)

   point becomes $(x, 2y)$, to get the graph of $y = 2f(x) = 2\left(4x - \dfrac{x^3}{3}\right) = 8x - \dfrac{2}{3}x^3$. See Figure 87.

89. The first equation graph is transformed 1 unit left $\Rightarrow y = f(x + 1) - 2$ increases on $(0, 3)$. The second

   equation graph is transformed across the x-axis and transformed 2 units right $\Rightarrow y = -f(x - 2)$ decreases on

   $(3, 6)$.

91. (a) An absolute maximum of approximately 84°F occurs when $x \approx 5$. An absolute minumum of approximately 63°F

   occurs when $x \approx 1.6$. These are the high and low temperatures between 1 p.m. and 6 p.m. *Answers may vary slightly.*

   (b) Local maxima of approximately 78°F and 84°F occur when $x \approx 2.9$ and 5, respectively. Local mimima of

   approximately 63°F and 72°F occur when $x \approx 1.6$ and 3.8. *Answers may vary slightly.*

   (c) The temperature is increasing on the intervals $1.6 < x < 2.9$ and $3.8 < x < 5$. *Answers may vary slightly.*

93. (a) $F(1) = 0.0484(1) - 1.504(1)^3 + 17.7(1)^2 + 53 = 0.0484 - 1.504 + 17.7 + 53 \approx 69$

   $F(12) = 0.0484(12)^3 - 1.504(12)^2 + 17.7(12) + 53 \approx 132$

   In January 2009, Facebook had about 69 million unique monthly users. In December 2009 this number

   increased to 132 million.

   (b) See Figure 93. One can see that F does not have any local extrema for the given interval.

   (c) According to the graph one can see that F is increasing on the interval $(1, 12)$.

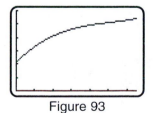

Figure 93

95. (a) One might expect an absolute maximum to occur in January since it is the coldest month and requires the

   heating. An absolute minimum might occur in July since it is the warmest month.

   (b) The graph f is shown in both Figures 95a & 95b. One can see from the graph that there is one local

   maximum and one local minimum where $1 \leq x \leq 12$. The peak is near the point $(1.46, 140.06)$ and the

   valley is located near $(7.12, 14.78)$. This means that the peak heating expense of $140 occurs in January

   $(x \approx 1)$. The minimum heating cost is during July $(x \approx 7)$ when it is roughly $15.

[1, 12, 1] by [0, 150, 10]   [1, 12, 1] by [0, 150, 10]

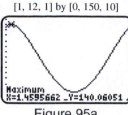

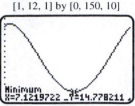

Figure 95a   Figure 95b

97. (a) Since the graph of f is symmetric with respect to the y-axis, it is a graphical representation of an even function.

(b) If f represents the average temperature in the month x, $f(1) = f(-1)$. Thus, the average monthly temperature in August is also 83°F.

(c) The average monthly temperatures in March and November are equal, since $f(-4) = f(4)$.

(d) In Austin the average monthly temperature is symmetric about July. In July the highest average monthly temperature occurs. In January the lowest temperature occurs. The pairs June-August, May-September, April-October, March-November, and February-December have approximately the same average temperatures.

Extended and Discovery Exercise for Section 4.1

1. The semicircle with radius 3 has the equation $y = \sqrt{9 - x^2}$. (Recall that the equation for a circle with center $(0, 0)$ and radius 3 is $x^2 + y^2 = 9 \Rightarrow y = \pm\sqrt{9 - x^2}$. So any point (x, y) on the semicircle would have coordinates $(x, \sqrt{9 - x^2})$. Since the rectangle is symmetric about the y-axis, the length of the rectangle $= 2x$, as shown in Figure 1a. The width of the rectangle $= y$.

The area of the rectangle is: $A = lw = 2x(y) = 2x(\sqrt{9 - x^2})$. To find the maximum area, graph the function $A(x) = 2x\sqrt{9 - x^2}$ on a graphing calculator, and then find the maximum value. See Figure 1b. The maximum area of 9 occurs when $x \approx 2.12$. Thus, the length of the rectangle is $2x \approx 2(2.12) \approx 4.24$, and the height is $y = \sqrt{9 - x^2} \approx \sqrt{9 - (2.12)^2} \approx 2.12$.

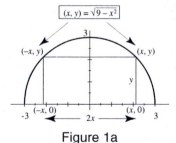

$(x, y) = \sqrt{9 - x^2}$

                                    [0, 5, 1] by [0, 10, 1]

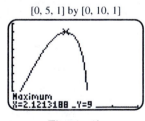

Figure 1a   Figure 1b

3. (a) Since $r \cdot t = D$ the time to shore is $4 \cdot t = 3 \Rightarrow t = \dfrac{3}{4}$ hour or 45 minutes; and the time to the cabin is $7 \cdot t = 8 \Rightarrow t = \dfrac{8}{7}$ hours or about 69 minutes. Therefore the combined time is $45 + 69 = 114$ minutes or 1 hour and 54 minutes.

(b) Using pythagorean theorem; the distance to the cabin is $3^2 + 8^2 = d^2 \Rightarrow 9 + 64 = d^2 \Rightarrow d^2 = 73 \Rightarrow$

$d \approx 8.5$. Now the time is $4 \cdot t = 8.5 \Rightarrow t = \dfrac{8.5}{4} \Rightarrow t \approx 128$ minutes or 2 hours and 8 minutes.

(c) Let $x =$ the distance jogged, then the distance rowed by using pythagorean theorem is;

$3^2 + (8 - x)^2 = d^2$ or $d = \sqrt{x^2 - 16x + 73} \Rightarrow$ the time needed to reach the cabin if $t = \dfrac{d}{r}$ is

$t = \dfrac{d(\text{jogged})}{7} + \dfrac{d(\text{rowed})}{4}$ or $t = \dfrac{x}{7} + \dfrac{\sqrt{x^2 - 16x + 73}}{4}$. Graph this equation to find the minimum

time. See Figure 3. The minimum time is $t \approx 1.76$ hours or about 1 hour 46 minutes.

4.2: Polynomial Functions and Models

1. (a) Turning points occur near (1.6, 3.6), (3, 1.2), and (4.4, 3.6).

   (b) The turning points represent the times and distances where the runner turned around and jogged the other

   direction. In this example, after 1.6 minutes the runner is 360 feet from the starting line. The runner turns

   and jogs toward the starting line. After 3 minutes the runner is 120 feet from the starting line, turns, and

   jogs away from the starting line. After 4.4 minutes the runner is again 360 feet from the starting line. The

   runner turns and jogs back to the starting line.

3. (a) It has no turning points and one x-intercept of 0.5.

   (b) The leading coefficient is positive, since the slope of the line is positive.

   (c) Since the graph appears to be a line that is not horizontal, its minimum degree is 1.

5. (a) It has three turning points and three x-intercepts of $-6, -1, 6$.

   (b) Since the graph goes down for large values of x, the leading coefficient is negative.

   (c) Since the graph has three turning points, its minimum degree is 4.

7. (a) It has four turning points and five x-intercepts of $-3, -1, 0, 1,$ and 2.

   (b) Since the graph goes up for large values of x, the leading coefficient is positive.

   (c) Since the graph has four turning points, its minimum degree is 5.

9. (a) It has two turning points and one x-intercept of -3.

   (b) Since the graph goes down to the left and up to the right, the leading coefficient is positive.

   (c) The graph has two turning points. The minimum degree of f is 3.

11. (a) It has one turning point and two x-intercepts of -1, and 2.

    (b) Since the graph is a parabola opening up, the leading coefficient is positive.

    (c) The graph has one turning point. The minimum degree of f is 2.

13. (a) Graph d.

    (b) There is one turning point at (1, 0).

    (c) It has one x-intercept: $x = 1$.

(d) The only local minimum is 0.

(e) The absolute minimum is 0. There is no absolute maximum.

15. (a) Graph b.

(b) The turning points are at $(-3, 27)$ and $(1, -5)$.

(c) The three x-intercepts are $x \approx -4.9$, $x = 0$, and $x \approx 1.9$.

(d) The local minimum is -5, and the local maximum is 27.

(e) There are no absolute extrema.

17. (a) Graph a.

(b) The turning points are at $(-2, 16)$, $(0, 0)$, and $(2, 16)$.

(c) The three x-intercepts are $x \approx -2.8$, $x = 0$, and $x \approx 2.8$.

(d) The local minimum is 0, and the local maximum is 16.

(e) The absolute maximum is 16. There is no absolute minimum.

19. (a) The graph of $f(x) = \dfrac{1}{9}x^3 - 3x$ is shown in Figure 19.

(b) There are two turning points located at $(-3, 6)$ and $(3, -6)$.

(c) At $x = -3$ there is a local maximum of 6 and at $x = 3$ there is a local minimum of -6.

[−10, 10, 1] by [−10, 10, 1]	[−10, 10, 1] by [−10, 10, 1]	[−10, 10, 1] by [−10, 10, 1]
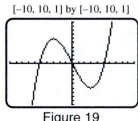	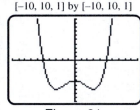	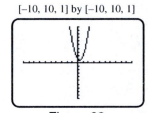
Figure 19	Figure 21	Figure 23

21. (a) The graph of $f(x) = 0.025x^4 - 0.45x^2 - 3$ is shown in Figure 21.

(b) There are three turning points located at $(-3, -7.025)$, $(0, -5)$, and $(3, -7.025)$.

(c) At $x = \pm 3$ there is a local minimum of -7.025. At $x = 0$ there is a local maximum of -5.

23. (a) The graph of $f(x) = 1 - 2x + 3x^2$ is shown in Figure 23.

(b) There is one turning point located at $\left(\dfrac{1}{3}, \dfrac{2}{3} \right) \approx (0.333, 0.667)$.

(c) At $x = \dfrac{1}{3}$ there is a local minimum of $\dfrac{2}{3} \approx 0.667$.

25. (a) The graph of $f(x) = \dfrac{1}{3}x^3 + \dfrac{1}{2}x^2 - 2x$ is shown in Figure 25.

(b) The turning points are at $\left(-2, \dfrac{10}{3} \right) \approx (-2, 3.333)$ and $\left(1, -\dfrac{7}{6} \right) \approx (1, -1.167)$.

(c) There is a local minimum of $-\dfrac{7}{6} \approx -1.167$, and there is a local maximum of $\dfrac{10}{3} \approx 3.333$.

[-10, 10, 1] by [-10, 10, 1]

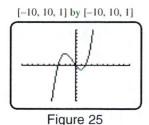

Figure 25

27. (a) The degree is 1 and the leading coefficient is −2.

(b) Since the degree of *f* is 1 and the lead coefficient is negative the graph is a line with a negative slope ⟹ it goes up on the left end and down on the right.

29. (a) The degree is 2 and the leading coefficient is 1.

(b) Since the degree of *f* is 2 and the lead coefficient is positive the graph is a parabola that opens up on each side.

31. (a) The degree is 3 and the leading coefficient is −2.

(b) Since the degree of *f* is odd and the lead coefficient is negative its graph will go up on the left and down on the right.

33. (a) The degree is 3 and the leading coefficient is –1.

(b) Since the degree of *f* is odd and the leading coefficient is negative, its graph will go up on the left and down on the right.

35. (a) The degree is 5 and the leading coefficient is 0.1.

(b) Since the degree of *f* is odd and the leading coefficient is positive, its graph will go down on the left and up on the right.

37. (a) The degree is 2 and the leading coefficient is $-\dfrac{1}{2}$.

(b) Since the degree of *f* is even and the leading coefficient is negative, its graph will go down on both sides.

39. (a) A line graph of the data is shown in Figure 39.

(b) The data appears to decrease, increase, and decrease. The line graph crosses the *x*-axis three times. Neither a linear or quadratic function could do this. Thus, *f* must be degree 3.

(c) Since *f* rises to the left and falls to the right $a < 0$.

(d) Using the cubic regession function on the calculator we find $f(x) = -x^3 + 8x$.

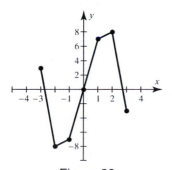

Figure 39

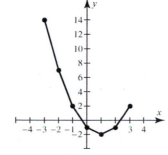

Figure 41

41. (a) A line graph of the data is shown in Figure 41.

   (b) The data decreases and then increases. This could not be a linear function. The data appears to be parabolic and the line graph crosses the x-axis twice. Since all zeros are real and between –3 and 3, f must be degree 2 with two real zeros.

   (c) Since f is opening upward $a > 0$.

   (d) Using the quadratic regession function on the calclator we find $f(x) = x^2 - 2x - 1$.

43. (a) See Figure 43.

   (b) The data increases then decreases then increases and the decreases. The graph crosses the x-axis 4 times. Therefore, f is degree 4.

   (c) Since both ends of f go down $a < 0$.

   (d) Using the quartic regession function on the calculator we find $f(x) = -x^4 + 3x^2 - 1$.

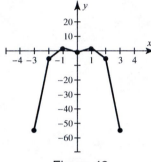

Figure 43

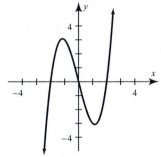

Figure 45

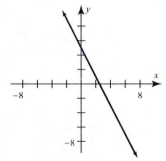

Figure 47

45. One possible graph is shown in Figure 45.

47. One possible graph is shown in Figure 47.

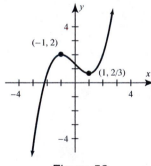

Figure 53

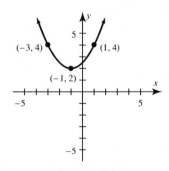

Figure 55

49. No such graph is possible.

51. No such graph is possible.

53. Start by plotting the points $(-1, 2)$ and $\left(1, \dfrac{2}{3}\right)$. Then sketch a cubic polynomial having these two turning points. One possible graph is shown in Figure 53.

55. Since a quadratic polynomial has only one turning point, $(-1, 2)$ must be the vertex. Plot the points $(-1, 2)$, $(-3, 4)$, and $(1, 4)$. Sketch a parabola passing through these points with a vertex of $(-1, 2)$. The parabola must open up. See Figure 55.

57. The graphs of $f(x) = 2x^4$, $g(x) = 2x^4 - 5x^2 + 1$, and $h(x) = 2x^4 + 3x^2 - x - 2$ are shown in Figures 57a-c. As the viewing rectangle increases in size, the graphs begin to look alike. Each formula contains the term $2x^4$. This term determines the end behavior of the graph for large values of $|x|$. The other terms are relatively small in absolute value by comparison when $|x| \geq 100$. End behavior is determined by the leading term.

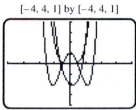

[−4, 4, 1] by [−4, 4, 1]

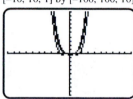

[−10, 10, 1] by [−100, 100, 10]

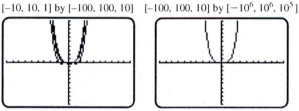

[−100, 100, 10] by [−10⁶, 10⁶, 10⁵]

Figure 57a   Figure 57b   Figure 57c

59. For $f(x) = x$ we get $(0, 0)$ and $\left(\dfrac{1}{2}, \dfrac{1}{2}\right)$ $\Rightarrow$ the average rate of change is $\dfrac{\frac{1}{2} - 0}{\frac{1}{2} - 0} = \dfrac{\frac{1}{2}}{\frac{1}{2}} = 1$

For $g(x) = x^2$ we get $(0, 0)$ and $\left(\dfrac{1}{2}, \dfrac{1}{4}\right)$ $\Rightarrow$ the average rate of change is $\dfrac{\frac{1}{4} - 0}{\frac{1}{2} - 0} = \dfrac{\frac{1}{4}}{\frac{1}{2}} = 0.5$

For $h(x) = x^3$ we get $(0, 0)$ and $\left(\dfrac{1}{2}, \dfrac{1}{8}\right)$ $\Rightarrow$ the average rate of change is $\dfrac{\frac{1}{8} - 0}{\frac{1}{2} - 0} = \dfrac{\frac{1}{8}}{\frac{1}{2}} = 0.25$

It decreases with higher degree.

61. (a) For $[1.9, 2.1]$ we get: $(1.9, 6.859)$ and $(2.1, 9.261)$ $\Rightarrow$ the average rate of change is $\dfrac{9.261 - 6.859}{2.1 - 1.9} = 12.01$

    (b) For $[1.99, 2.01]$ we get: $(1.99, 7.880599)$ and $(2.01, 8.120601)$ $\Rightarrow$ the average rate of change is

    $\dfrac{8.120601 - 7.880599}{2.01 - 1.99} = 12.0001$

    (c) For $[1.999, 2.001]$ we get: $(1.999, 7.988005999)$ and $(2.001, 8.012006001)$ $\Rightarrow$ the average rate of change is

    $\dfrac{8.012006001 - 7.988005999}{2.001 - 1.999} = 12.000001$

    As the interval decreases in length the average rate of change is approaching 12.

63. (a) For $[1.9, 2.1]$ we get: $\dfrac{1}{4}(1.9)^4 - \dfrac{1}{3}(1.9)^3 = 0.971691\overline{6}$ or $(1.9, 0.971691\overline{6})$ and $\dfrac{1}{4}(2.1)^4 - \dfrac{1}{3}(2.1)^3 =$

    1.775025 or $(2.1, 1.775025)$ $\Rightarrow$ the average rate of change is: $\dfrac{1.775025 - 0.971691\overline{6}}{2.1 - 1.9} = 4.01\overline{6}$

    (b) For $[1.99, 2.01]$ we get: $\dfrac{1}{4}(1.99)^4 - \dfrac{1}{3}(1.99)^3 = 1.293731\overline{6}$ or $(1.99, 1.293731\overline{6})$ and

$\frac{1}{4}(2.01)^4 - \frac{1}{3}(2.01)^3 = 1.373735003$ or $(2.01, 1.373735003) \Rightarrow$ the average rate of change is:

$$\frac{1.373735003 - 1.293731\overline{6}}{2.01 - 1.99} = 4.0001\overline{6}$$

(c) For $[1.999, 2.001]$ we get: $\frac{1}{4}(1.999)^4 - \frac{1}{3}(1.999)^3 = 1.329337332$ or $(1.999, 1.329337332)$ and

$\frac{1}{4}(2.001)^4 - \frac{1}{3}(2.001)^3 = 1.337337335$ or $(2.001, 1.337337335) \Rightarrow$ the average rate of change is:

$$\frac{1.337337335 - 1.329337332}{2.001 - 1.999} = 4.000001\overline{6}$$

As the interval decreases in length the average rate of change is approaching 4.

65. Use $\frac{f(x+h) - f(x)}{h}$, then $\frac{3(x+h)^3 - 3x^3}{h} = \frac{3x^3 + 9x^2h + 9xh^2 + 3h^3 - 3x^3}{h} = 9x^2 + 9xh + 3h^2$

67. Use $\frac{f(x+h) - f(x)}{h}$, then $\frac{1 + (x+h) - (x+h)^3 - (1 + x - x^3)}{h} =$

$\frac{1 + x + h - x^3 - 3x^2h - 3xh^2 - h^3 - 1 - x + x^3}{h} = -3x^2 - 3xh - h^2 + 1$

69. $f(-2) \approx 5$ and $f(1) \approx 0$

71. $f(-1) \approx -1$, $f(1) \approx 1$, and $f(2) \approx -2$

73. $f(-3) = (-3)^3 - 4(-3)^2 = -27 - 36 = -63$, $f(1) = 3(1)^2 = 3$, and $f(4) = (4)^3 - 54 =$

64 - 54 = 10

75. $f(-2) = (-2)^2 + 2(-2) + 6 = 6$, $f(1) = 1 + 6 = 7$, and $f(2) = 2^3 + 1 = 9$

77. (a) The graph of f is shown in Figure 77.

(b) The function f is discontinuous at $x = 0$.

(c) First solve: $4 - x^2 = 0 \Rightarrow -x^2 = -4 \Rightarrow x^2 = 4 \Rightarrow x = -2, 2$; only $x = -2$ is in the domain for the

equation. Now solve: $x^2 - 4 = 0 \Rightarrow x^2 = 4 \Rightarrow x = -2, 2$; only $x = 2$ is in the domain for the

equation $\Rightarrow x = -2, 2$.

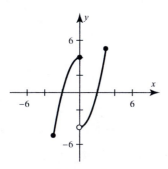

Figure 77

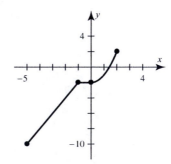

Figure 79

79. (a) The graph of f is shown in Figure 79.

(b) f is continuous.

(c) First solve: $2x = 0 \Rightarrow x = 0$; this is not in the domain for the equation.

$-2 \neq 0 \Rightarrow$ no solutions $-1 \leq x \leq 0$. Finally solve: $x^2 - 2 = 0 \Rightarrow x^2 = 2 \Rightarrow x = \pm\sqrt{2}$; only $\sqrt{2}$

is in the domain for the equation $\Rightarrow x = \sqrt{2}$.

81. (a) The graph of f is shown in Figure 81.

(b) f is continuous on its domain.

(c) First solve: $x^3 + 3 = 0 \Rightarrow x^3 = -3 \Rightarrow x = \sqrt[3]{-3}$ or $-\sqrt[3]{3}$; this is in the domain for the equation.

Now solve: $x + 3 = 0 \Rightarrow x = -3$; this is not in the domain for the equation.

Finally solve: $4 + x - x^2 = 0 \Rightarrow$ using quadratic formula for $-x^2 + x + 4 = 0$ we get:

$$\frac{-1 \pm \sqrt{1 - 4(-1)(4)}}{2(-1)} = \frac{-1 \pm \sqrt{17}}{-2}; \text{ only } \frac{1 + \sqrt{17}}{2} \text{ is in the domain of the equation} \Rightarrow$$

$$x = -\sqrt[3]{3}, \frac{\sqrt{17} + 1}{2}.$$

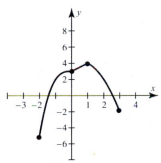

Figure 81

83. (a) To evaluate $F(3)$ we use the formula $F(x) = 2x$ since 3 is in the interval $0 \leq x \leq 4$. $F(3) = 2(3) = 6$, 300

boats are able to harvest 6 thousand tons of fish.

(b) The maximum of $F(x) = 2x$ is 8 when $x = 4$. The maximum of $F(x) = -\frac{1}{4}x^2 + 4x - 4$ is the y-coordi-

nate of the vertex. The x-coordinate of the vertex is $x = \dfrac{-4}{2\left(-\dfrac{1}{4}\right)} = 8$ and the y-coordinate is

$F(8) = -\frac{1}{4}(8)^2 + 4(8) - 4 = -16 + 32 - 4 = 12$. 12 thousand tons of fish are the maximum that can

be caught.

85. (a) $H(-2) = 0$, because $t < 0$; $H(0) = 1$, because $t \geq 0$; $H(3.5) = 1$, because $t \geq 0$

(b) See Figure 85.

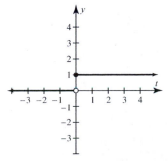

Figure 85

87. (a) There are two turning points that occur near $(1, 13)$ and $(7, 72)$.

   (b) The point $(1, 13)$ means that in January the average temperature is 13°F. This is the minimum average monthly temperature in Minneapolis. After this the average temperature begins to increase. The point $(7, 72)$ means that in July the average temperature is 72°F. This is the maximum average monthly temperature in Minneapolis. After this the average temperature begins to decrease.

89. (a) The height begins to decrease after 4 seconds. The object appears to have dropped after 4 seconds.

   (b) From 0 to 4 seconds, when the object is being lifted, the height could be modeled with a linear function. From 4 to 7 seconds, when the object begins to fall, the height can be modeled with a nonlinear function.

   (c) The altitude increases by 36 feet each second, during the first four seconds. Therefore, let $m = 36$. When $x = 4$, the height is 144 feet. Let $b = 144$. When $x = 7$, the height was 0,
   $a(7 - 4)^2 + 144 = 0 \Rightarrow a = -16$. Thus, $f(x) = \begin{cases} 36x, & \text{if } 0 \le x \le 4 \\ -16(x - 4)^2 + 144, & \text{if } 4 < x \le 7 \end{cases}$

   (d) Using $f(x) = 36x$, $f(x) = 100 \Rightarrow 100 = 36x \Rightarrow x \approx 2.8$;
   using $f(x) = -16(x - 4)^2 + 144$, $f(x) = 100 \Rightarrow 100 = -16(x - 4)^2 + 144 \Rightarrow$
   $-44 = -16(x - 4)^2 \Rightarrow x \approx 5.7$; the object is 100 feet high at about 2.8 seconds and at about 5.7 second

91. (a) Using the cubic regression function on the calculator we find

   $$f(x) \approx -0.0006296x^3 + 0.06544x^2 - 0.368x + 2.8.$$

   (b) See Figure 91.

   (c) Since x represents the number of years after 1960 we will let $x = 34$ and 60.

   $$f(34) \approx -0.0006296(34)^3 + 0.06544(34)^2 - 0.368(34) + 2.8 \approx 41.2$$

   $$f(60) \approx -0.0006296(60)^3 + 0.06544(60)^2 - 0.368(60) + 2.8 \approx 80.3$$

   (d) The computation for 1994 involved interpolation and the computation for 2020 involved extrapolation.

Extended and Discovery Exercise for Section 4.2

1. (a) $D = [0, 10]$

   (b) $A(1) = 500\left(1 - \dfrac{1}{10}\right)^2 \Rightarrow A(1) = 500\left(\dfrac{9}{10}\right)^2 \Rightarrow A(1) = 500\left(\dfrac{81}{100}\right) \Rightarrow A(1) = 405$

   After one minute of draining the tank it contains 405 gallons of water.

   (c) $500\left(1 - \dfrac{t}{10}\right)^2 = 500\left(1 - \dfrac{2t}{10} + \dfrac{t^2}{100}\right) = 5t^2 - 100t + 500 \Rightarrow$ the degree is: 2.

   The leading coefficient is: 5.

   (d) $A(5) = 500\left(1 - \dfrac{5}{10}\right)^2 \Rightarrow A(5) = 500\left(\dfrac{1}{2}\right)^2 \Rightarrow A(5) = 500\left(\dfrac{1}{4}\right) \Rightarrow A(5) = \dfrac{500}{4} \Rightarrow$

   $A(5) = 125$ gallons remaining. No more than half is drained. This is reasonable because the water will drain faster at first.

3. *f* is concave down on $(-\infty, \infty)$ and is not concave up.

5. *f* is concave up on $(1, \infty)$ and is concave down on $(-\infty, 1)$.

7. *f* is concave up on $(-2, 2)$ and is concave down on $(-\infty, -2); (2, \infty)$.

Checking Basic Concepts for Section 4.1 and 4.2

1. (a) In the set builder notation, these intervals are $\{x \mid -2 < x < 1 \text{ or } x > 3\}$ and $\{x \mid x < -2 \text{ or } 1 < x < 3\}$ respectively.

    (b) Local maximum: approximately 3; local minimum; approximately $-13, -2$.

    (c) Absolute minimum: approximately -13; no absolute maximum.

    (d) From the graph they are approximately $-3.1, 0, 2.2$, and 3.6; they are the same values.

3. (a) Cubic graphs have a range from $-\infty$ to $\infty \Rightarrow$ must have an *x*-intercept $\Rightarrow$ not possible.

    (b) See Figure 3b.

    (c) See Figure 3c.

    (d) Cubic graphs have at most two turning points $\Rightarrow$ have at most 3 *x*-intercept $\Rightarrow$ not possible.

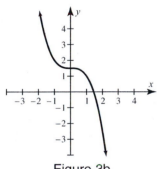

Figure 3b

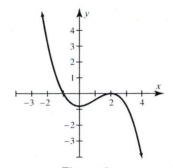

Figure 3c

5. $f(x) \approx -1.01725x^4 + 10.319x^2 - 10$

4.3: Division of Polynomials

1. $\dfrac{5x^4 - 15}{10x} = \dfrac{5x^4}{10x} - \dfrac{15}{10x} = \dfrac{x^3}{2} - \dfrac{3}{2x}$

3. $\dfrac{3x^4 - 2x^2 - 1}{3x^3} = \dfrac{3x^4}{3x^3} - \dfrac{2x^2}{3x^3} - \dfrac{1}{3x^3} = x - \dfrac{2}{3x} - \dfrac{1}{3x^3}$

5. $\dfrac{x^3 - 4}{4x^3} = \dfrac{x^3}{4x^3} - \dfrac{4}{4x^3} = \dfrac{1}{4} - \dfrac{1}{x^3}$

7. $\dfrac{5x(3x^2 - 6x + 1)}{3x^2} = \dfrac{15x^3 - 30x^2 + 5x}{3x^2} = \dfrac{15x^3}{3x^2} - \dfrac{30x^2}{3x^2} + \dfrac{5x}{3x^2} = 5x - 10 + \dfrac{5}{3x}$

9. $x^3 - 2x^2 - 5x + 6$ divided by $x - 3$ can be performed using synthetic division.

    $$\begin{array}{r|rrrr} 3 & 1 & -2 & -5 & 6 \\ & & 3 & 3 & -6 \\ \hline & 1 & 1 & -2 & 0 \end{array}$$

    The quotient is $x^2 + x - 2$ and the remainder is 0.

11. $2x^4 - 7x^3 - 5x^2 - 19x + 17$ divided by $x + 1$ can be performed using synthetic division.

$$
\begin{array}{r|rrrrr}
-1 & 2 & -7 & -5 & -19 & 17 \\
   &   & -2 & 9 & -4 & 23 \\
\hline
   & 2 & -9 & 4 & -23 & 40
\end{array}
$$

The quotient is $2x^3 - 9x^2 + 4x - 23$ and the remainder is 40.

13. $3x^3 - 7x + 10$ divided by $x - 1$ can be performed using synthetic division.

$$
\begin{array}{r|rrrr}
1 & 3 & 0 & -7 & 10 \\
  &   & 3 & 3 & -4 \\
\hline
  & 3 & 3 & -4 & 6
\end{array}
$$

The quotient is $3x^2 + 3x - 4$ and the remainder is 6.

This can also be found using long division as shown.

$$
\begin{array}{r}
3x^2 + 3x - 4 \\
x - 1 \overline{)3x^3 - 0x^2 - 7x + 10} \\
\underline{3x^3 - 3x^2} \\
3x^2 - 7x \\
\underline{3x^2 - 3x} \\
-4x + 10 \\
\underline{-4x + 4} \\
6
\end{array}
$$

15. We can use synthetic division to divide $x^4 - 3x^3 - x + 3$ by $x - 3$.

$$
\begin{array}{r|rrrrr}
3 & 1 & -3 & 0 & -1 & 3 \\
  &   & 3 & 0 & 0 & -3 \\
\hline
  & 1 & 0 & 0 & -1 & 0
\end{array}
$$

The quotient is $x^3 - 1$ and the remainder is 0.

17. We can use long division to divide $4x^3 - x^2 - 5x + 6$ by $x - 1$.

$$
\begin{array}{r}
4x^2 + 3x - 2 \\
x - 1 \overline{)4x^3 - x^2 - 5x + 6} \\
\underline{4x^3 - 4x^2} \\
3x^2 - 5x \\
\underline{3x^2 - 3x} \\
-2x + 6 \\
\underline{-2x + 2} \\
4
\end{array}
$$

The quotient is $4x^2 + 3x - 2 + \dfrac{4}{x - 1}$.

19. We can use synthetic division to divide $x^3 + 1$ by $x + 1$.

$$
\begin{array}{r|rrrr}
-1 & 1 & 0 & 0 & 1 \\
   &   & -1 & 1 & -1 \\
\hline
   & 1 & -1 & 1 & 0
\end{array}
$$

The quotient is $x^2 - x + 1$ and the remainder is 0.

21. We can use long division to divide $6x^3 + 5x^2 - 8x + 4$ by $2x - 1$.

$$\begin{array}{r} 3x^2 + 4x - 2 \\ 2x - 1\overline{)6x^3 + 5x^2 - 8x + 4} \\ \underline{6x^3 - 3x^2} \\ 8x^2 - 8x + 4 \\ \underline{8x^2 - 4x} \\ -4x + 4 \\ \underline{-4x + 2} \\ 2 \end{array}$$

The quotient is $3x^2 + 4x - 2 + \dfrac{2}{2x - 1}$.

23. We can use long division to divide $3x^4 - 7x^3 + 6x - 16$ by $3x - 7$.

$$\begin{array}{r} x^3 + 2 \\ 3x - 7\overline{)3x^4 - 7x^3 + 0x^2 + 6x - 16} \\ \underline{3x^4 - 7x^3} \\ 6x - 16 \\ \underline{6x - 14} \\ -2 \end{array}$$

The quotient is $x^3 + 2 + \dfrac{-2}{3x - 7}$.

25. We can use long division to divide $5x^4 - 2x^2 + 6$ by $x^2 + 2$.

$$\begin{array}{r} 5x^2 - 12 \\ x^2 + 2\overline{)5x^4 - 2x^2 + 6} \\ \underline{5x^4 + 10x^2} \\ -12x^2 + 6 \\ \underline{-12x^2 - 24} \\ 30 \end{array}$$

The quotient is $5x^2 - 12 + \dfrac{30}{x^2 + 2}$.

27. We can use long division to divide $8x^3 + 10x^2 - 12x - 15$ by $2x^2 - 3$.

$$\begin{array}{r} 4x + 5 \\ 2x^2 - 3\overline{)8x^3 + 10x^2 - 12x - 15} \\ \underline{8x^3 \qquad - 12x} \\ 10x^2 \qquad - 15 \\ \underline{10x^2 \qquad - 15} \\ 0 \end{array}$$

The quotient is $4x + 5$.

29. We can use long division to divide $2x^4 - x^3 + 4x^2 + 8x + 7$ by $2x^2 + 3x + 2$.

$$\begin{array}{r} x^2 - 2x + 4 \\ 2x^2 + 3x + 2\overline{)2x^4 - x^3 + 4x^2 + 8x + 7} \\ \underline{2x^4 + 3x^3 + 2x^2} \\ -4x^3 + 2x^2 + 8x \\ \underline{-4x^3 - 6x^2 - 4x} \\ 8x^2 + 12x + 7 \\ \underline{8x^2 + 12x + 8} \\ -1 \end{array}$$

The quotient is $x^2 - 2x + 4 + \dfrac{-1}{2x^3 + 3x + 2}$.

31. The divisor times the quotient will be equal to the dividend, $(x - 2)(x^2 - 6x + 3) = x^3 - 8x^2 + 15x - 6$.

33.
$$
\begin{array}{r}
x - 1 \\
x - 2\overline{\smash{)}x^2 - 3x + 1} \\
\underline{x^2 - 2x} \\
-x + 1 \\
\underline{-x + 2} \\
-1 \quad (x - 2)(x - 1) - 1
\end{array}
$$

35.
$$
\begin{array}{r}
x^2 - 1 \\
2x + 1\overline{\smash{)}2x^3 - x^2 - 2x} \\
\underline{2x^3 + x^2} \\
-2x + 0 \\
\underline{-2x - 1} \\
1 \quad (2x + 1)(x^2 - 1) + 1
\end{array}
$$

37.
$$
\begin{array}{r}
x - 1 \\
x^2 + 0x + 1\overline{\smash{)}x^3 - x^2 + x + 1} \\
\underline{x^3 + 0x^2 + x} \\
-x^2 \quad + 1 \\
\underline{-x^2 \quad - 1} \\
2 \quad (x^2 + 1)(x - 1) + 2
\end{array}
$$

39.
$$
\begin{array}{r|rrrr}
-5 & 1 & 2 & -17 & -10 \\
& & -5 & 15 & 10 \\
\hline
& 1 & -3 & -2 & 0
\end{array}
$$
The quotient is $x^2 - 3x - 2$.

41.
$$
\begin{array}{r|rrrr}
5 & 3 & -11 & -20 & 3 \\
& & 15 & 20 & 0 \\
\hline
& 3 & 4 & 0 & 3
\end{array}
$$
The quotient is $3x^2 + 4x + \dfrac{3}{x - 5}$.

43.
$$
\begin{array}{r|rrrrr}
2 & 1 & -3 & -4 & 12 & 0 \\
& & 2 & -2 & -12 & 0 \\
\hline
& 1 & -1 & -6 & 0 &
\end{array}
$$
The quotient is $x^3 + x^2 - 6x$.

45.
$$
\begin{array}{r|rrrrrr}
-\frac{1}{2} & 2 & -1 & -1 & 0 & 4 & 3 \\
& & -1 & 1 & 0 & 0 & -2 \\
\hline
& 2 & -2 & 0 & 0 & 4 & 1
\end{array}
$$
The quotient is $2x^4 - 2x^3 + 4 + \dfrac{1}{x + 0.5}$.

47. Using the remainder theorem we find: $f(1), \Rightarrow 5(1)^2 - 3(1) + 1 = 5 - 3 + 1 = 3$

49. Using the remainder theorem we find:

$$f(-2), \Rightarrow 4(-2)^3 - (-2)^2 + 4(-2) + 2 = -32 - 4 - 8 + 2 = -42$$

51. If we divide the Area by the Width we will find the Length:

$$
\begin{array}{r}
4x + 3 \\
3x + 1\overline{\smash{)}12x^2 + 13x + 3} \\
\underline{12x^2 + 4x} \\
9x + 3 \\
\underline{9x + 3} \\
0
\end{array}
$$

The length is $4x + 3$. When $x = 10$, the Length is $4(10) + 3 = 43$ feet.

4.4: Real Zeros of Polynomial Functions

1. The *x*-intercepts of *f* are -1, 1, and 2. Since $f(-1) = 0$, the factor theorem states that $(x + 2)$ is a factor of $f(x)$. Similarly, $f(-1) = 0$ implies that $(x + 1)$ is a factor, and $f(1) = 0$ implies that $(x - 1)$ is a factor.

3. The *x*-intercepts of *f* are -2, -1, 1 and 2. Since $f(-2) = 0$, the factor theorem states that $(x + 2)$ is a factor of $f(x)$. Similarly, $f(-1) = 0$ implies that $(x + 1)$ is a factor, $f(1) = 0$ implies that $(x - 1)$ is a factor and $f(2) = 0$ implies that $(x - 2)$ is a factor.

5. $f(x) = 2x^2 - 25x + 77$ and zeros: $\dfrac{11}{2}$ and $7 \Rightarrow f(x) = 2\left(x - \dfrac{11}{2}\right)(x - 7)$

7. $f(x) = x^3 - 2x^2 - 5x + 6$ and zeros: -2, 1, and $3 \Rightarrow f(x) = (x + 2)(x - 1)(x - 3)$

9. $f(x) = -2x^3 + 3x^2 + 59x - 30$ and zeros: $-5, \dfrac{1}{2}$, and $6 \Rightarrow f(x) = -2(x + 5)\left(x - \dfrac{1}{2}\right)(x - 6)$

11. If $f(-3) = 0$ then the quadratic equation has a factor of $x - (-3)$ or $x + 3$, likewise if $f(2) = 0$ then the quadratic equation has a factor $x - 2$. If this quadratic equation has a leading coefficient 7, the complete factored form of $f(x)$ is $f(x) = 7(x + 3)(x - 2)$.

13. To factor $f(x)$ we need to determine the leading coefficient and zeros of *f*. The leading coefficient is -2 and the zeros are $-1, 0$, and 1. The complete factorization is $f(x) = -2x(x + 1)(x - 1)$.

15. From the graph the zeros are -4, 2, and 8. $f(x)$ is a cubic with a positive leading coefficient. Therefore, $f(x) = (x + 4)(x - 2)(x - 8)$.

17. From the graph the zeros are $-8, -4, -2$, and 4. $f(x)$ is a quartic polynomial with a negative leading coefficient. Therefore, $f(x) = -1(x + 8)(x + 4)(x + 2)(x - 4)$.

19. Since the polynomial has zeros of $-1, 2$, and 3, it has factors $(x + 1)(x - 2)(x - 3)$.
    If *f* passes through $(0, 3)$ then $a(0 + 1)(0 - 2)(0 - 3) = 3 \Rightarrow a(1)(-2)(-3) = 3 \Rightarrow 6a = 3 \Rightarrow a = \dfrac{1}{2}$.
    The complete factored form is: $f(x) = \dfrac{1}{2}(x + 1)(x - 2)(x - 3)$.

21. Since *f* has zeros -1, 1, and 2, it has factors $(x + 1)(x - 1)(x - 2)$. If $f(0) = 1$ then $a(1)(-1)(-2) = 1$ or $2a = 1 \Rightarrow a = \dfrac{1}{2}$. The complete factored form is: $f(x) = \dfrac{1}{2}(x + 1)(x - 1)(x - 2)$.

23. Since *f* has zeros $-2, -1$, 1, and 2, it has factors $(x + 2)(x + 1)(x - 1)(x - 2)$. If $f(0) = -8$ then $a(2)(1)(-1)(-2) = -8 \Rightarrow 4a = -8 \Rightarrow a = -2$.
    The complete factored form is: $f(x) = -2(x + 2)(x + 1)(x - 1)(x - 2)$.

25. A graph of $Y_1 = 10X^2 + 17X - 6$ is shown in Figure 25. Its zeros are -2 and 0.3. Since the leading coefficient is 10, the complete factorization is $f(x) = 10(x + 2)\left(x - \dfrac{3}{10}\right)$.

[−5, 5, 1] by [−20, 20, 5]   [−5, 5, 1] by [−40, 40, 5]

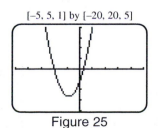 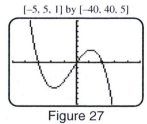

Figure 25   Figure 27

27. A graph of $Y_1 = -3X^3 - 3X^2 + 18X$ is shown in Figure 27. Its zeros are $-3, 0,$ and 2. Since the leading coefficient is -3, the complete factorization is $f(x) = -3(x - 0)(x - 2)(x + 3) = -3x(x - 2)(x + 3)$.

29. A graph of $Y_1 = X^4 + (5/2)X^3 - 3X^2 - (9/2)X$ is shown in Figure 29. Its zeros are $-3, -1, 0,$ and $\dfrac{3}{2}$.

    Since the leading coefficient is 1, the complete factorization is

    $$f(x) = (x + 3)(x + 1)(x - 0)\left(x - \frac{3}{2}\right) = x(x + 3)(x + 1)\left(x - \frac{3}{2}\right).$$

    [−5, 5, 1] by [−10, 10, 1]

    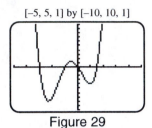

    Figure 29

31. By the factor theorem, since 1 is a zero, $(x - 1)$ is a factor. $x^3 - 9x^2 + 23x - 15$ divided by $x - 1$ can be performed using synthetic division.

    $$\begin{array}{r|rrrr} 1 & 1 & -9 & 23 & -15 \\ & & 1 & -8 & 15 \\ \hline & 1 & -8 & 15 & 0 \end{array}$$

    The quotient is $x^2 - 8x + 15$ and the remainder is 0.

    Thus, $x^3 - 9x^2 + 23x - 15 = (x - 1)(x^2 - 8x + 15) = (x - 1)(x - 3)(x - 5)$.

    The complete factored form of $f(x) = x^3 - 9x^2 + 23x - 15$ is $f(x) = (x - 1)(x - 3)(x - 5)$.

33. By the factor theorem, since -4 is a zero, $(x + 4)$ is a factor. $-4x^3 - x^2 + 51x - 36$ divided by $x + 4$ can be performed using synthetic division.

    $$\begin{array}{r|rrrr} -4 & -4 & -1 & 51 & -36 \\ & & 16 & -60 & 36 \\ \hline & -4 & 15 & -9 & 0 \end{array}$$

    The quotient is $-4x^2 + 15x - 9$ and the remainder is 0.

    Thus, $-4x^3 - x^2 + 51x - 36 = (x + 4)(-4x^2 + 15x - 9) = (x + 4)(-4x + 3)(x - 3)$.

    The complete factored form of $f(x) = -4x^3 - x^2 + 51x - 36$ is $f(x) = -4(x + 4)\left(x - \frac{3}{4}\right)(x - 3)$.

35. By the factor theorem, since -2 is a zero, $(x + 2)$ is a factor. $2x^4 - x^3 - 13x^2 - 6x$ divided by $x + 2$ can be performed using synthetic division.

    $$\begin{array}{r|rrrrr} -2 & 2 & -1 & -13 & -6 & 0 \\ & & -4 & 10 & 6 & 0 \\ \hline & 2 & -5 & -3 & 0 & 0 \end{array}$$

    The quotient is $2x^3 - 5x^2 - 3x$ and the remainder is 0. Thus,

    $$2x^4 - x^3 - 13x^2 - 6x = (x + 2)(2x^3 - 5x^2 - 3x) = (x + 2)x(2x^2 - 5x - 3) =$$

    $$x(x + 2)(2x + 1)(x - 3).$$

    The complete factored form of $f(x) = 2x^4 - x^3 - 13x^2 - 6x$ is $f(x) = 2x(x + 2)\left(x + \frac{1}{2}\right)(x - 3)$.

37. $f(2) = (2)^3 - 6(2)^2 + 11(2) - 6 = 0$; since $f(2) = 0$, by the factor theorem,

    $x - 2$ is a factor of $f(x) = x^3 - 6x^2 + 11x - 6$.

39. $f(3) = (3)^4 - 2(3)^3 - 13(3)^2 - 10(3) = -120$; since $f(-3) \neq 0$, by the factor theorem,

    $x - 3$ is not a factor of $x^4 - 2x^3 - 13x^2 - 10x$.

41. The zeros of $f(x)$ are 4 and –2. Since the graph does not cross the x-axis at $x = 4$, the zero of 4 has even

    multiplicity. The graph crosses the x-axis at $x = -2$. The zero of –2 has odd multiplicity. Since the graph

    levels off, crossing the x-axis at $x = -2$, this zero has at least multiplicity 3, and the zero of 4 has at least

    multiplicity 2. Thus, the minimum degree of $f(x)$ is $3 + 2$, or 5.

43. Degree: 3; zeros: –1 with multiplicity 2 and 6 with multiplicity 1. $f(x) = (x + 1)^2(x - 6)$

45. Degree: 4; zeros: 2 with multiplicity 3 and 6 with multiplicity 1. $f(x) = (x - 2)^3(x - 6)$

47. The graph shows a cubic polynomial with a positive leading coefficient and zeros of 4 and –2. The zero of 4

    has odd multiplicity, whereas the zero of –2 has even multiplicity. Since the graph has a degree three, its

    complete factored form is $f(x) = (x - 4)(x + 2)^2$.

49. The graph shows a quartic polynomial with a negative leading coefficient and zeros of –3 and 3. Both zeros

    have even multiplicity. Since the graph has a degree four polynomial, its complete factored form is

    $f(x) = -1(x + 3)^2(x - 3)^2$.

51. The graph shows a fifth degree polynomial with a positive leading coefficient and zeros of -1 and 1. The -1

    has an even multiplicity, whereas the zero of 1 has an odd multiplicity. Since the graph shows a fifth degree

    polynomial the factors are $(x + 1)^2(x - 1)^3$. To find the leading coefficient we use

    $f(0) = -2 \Rightarrow a(1)^2(-1)^3 = -2 \Rightarrow -a = -2 \Rightarrow a = 2$. Its factored form is $f(x) = 2(x + 1)^2(x - 1)^3$.

53. (a) From the factors the x-intercepts are: $-2, -1$. To find the y-intercept set $x = 0 \Rightarrow$

    $2(0 + 2)(0 + 1)^2 = 2(2)(1)^2 = 4$

    (b) The zero -2 has multiplicity 1. The zero -1 has multiplicity 2.

    (c) See Figure 53.

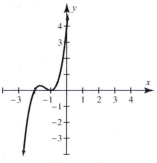

Figure 53

55. (a) From the factors the x-intercepts are: $-2, 0,$ and $2.$ To find the y-intercept set $x = 0 \Rightarrow$

$$0(0 + 2)(0 - 2) = 0(2)(-2) = 0$$

(b) The zero -2 has multiplicity 1. The zero 0 has multiplicity 2. The zero 2 has multiplicity 1.

(c) See Figure 55.

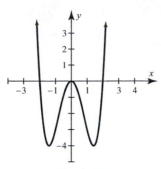

Figure 55

57. $f(x) = 2x^3 + 3x^2 - 8x + 3$

(a) If $\frac{p}{q}$ is a rational zero, then p is a factor of 3, which are ± 1 and ± 3 and q is a factor of 2, which are

± 1 or $\pm 2.$ Thus, any rational zero must be in the list $\pm \frac{1}{2}, \pm 1, \pm \frac{3}{2},$ or $\pm 3.$ From Figure 57 we see that

there are three rational zeros of $\frac{1}{2}, 1,$ and $-3.$

(b) The complete factored form is $f(x) = 2 \left(x - \frac{1}{2} \right)(x - 1)(x + 3).$

x	$\frac{1}{2}$	$-\frac{1}{2}$	1	-1	$\frac{3}{2}$	$-\frac{3}{2}$	3	-3
$f(x)$	0	$\frac{15}{2}$	0	12	$\frac{9}{2}$	15	60	0

Figure 57

59. $f(x) = 2x^4 + x^3 - 8x^2 - x + 6$

(a) If $\frac{p}{q}$ is a rational zero, then p is a factor of 6, which are $\pm 1, \pm 2, \pm 3,$ and ± 6 and q is a factor of 2, which

are ± 1 and $\pm 2.$ Thus, any rational zero must be in the list $\pm \frac{1}{2}, \pm 1, \pm \frac{3}{2}, \pm 2, \pm 3,$ or $\pm 6.$ By evaluating $f(x)$

at each of these values, we find that the zeros are $-2, -1, 1,$ and $\frac{3}{2}.$

(b) The complete factored form is $f(x) = 2(x + 2)(x + 1)(x - 1) \left(x - \frac{3}{2} \right).$

61. $f(x) = 3x^3 - 16x^2 + 17x - 4$

(a) If $\frac{p}{q}$ is a rational zero, then p is a factor of 4, which are $\pm 1, \pm 2,$ and ± 4 and q is a factor of 3, which are

± 1 and ± 3. Thus, any rational zero must be in the list $\pm\dfrac{1}{3}$, ± 1, $\pm\dfrac{2}{3}$, ± 2, $\pm\dfrac{4}{3}$, or ± 4. By evaluating $f(x)$ at

each of these values, we find that the zeros are $\dfrac{1}{3}$, 1, and 4.

(b) The complete factored form is $f(x) = 3\left(x - \dfrac{1}{3}\right)(x - 1)(x - 4)$.

63. $f(x) = x^3 - x^2 - 7x + 7$

(a) If $\dfrac{p}{q}$ is a rational zero, then p is a factor of 7, which are ± 1, and ± 7 and q is a factor of 1, which are ± 1.

Thus, any rational zero must be in the list ± 1 or ± 7. By evaluating $f(x)$ at each of these values, we find

that the only rational zero is 1.

(b) In order to find the complete factored form of $f(x)$ we need to divide the factor $(x - 1)$ into

$x^3 - x^2 - 7x + 7$ using synthetic division.

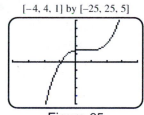

Thus $x^3 - x^2 - 7x + 7 = (x - 1)(x^2 - 7)$.

The complete factored form is $f(x) = (x - 1)(x - \sqrt{7})(x + \sqrt{7})$.

65. $P(x) = 2x^3 - 4x^2 + 2x + 7$, $P(x)$ has two sign changes, therefore are 2 or $2 - 2 = 0$ possible positive

zeros. $P(-x) = 2(-x)^3 - 4(-x)^2 + 2(-x) + 7 = -2(x)^3 - 4x^2 - 2x + 7$, $P(-x)$ has one sign change,

therefore there is one possible negative zero. From the graph of $P(x)$ in Figure 65, we see that the actual num-

ber of positive and negative zeros are 0 and 1 repectively.

[−4, 4, 1] by [−25, 25, 5]

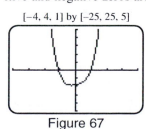

Figure 65

67. $P(x) = 5x^4 + 3x^2 + 2x - 9$, $P(x)$ has one sign change, therefore there is one possible positive zero.

$P(-x) = 5(-x)^4 + 3(-x)^2 + 2(-x) - 9 = 5x^4 + 3x - 2x - 9$, $P(-x)$ has one sign change, therefore

there is one possible negative zero. From the graph of $P(x)$ in Figure 67, we see that the actual number of pos-

itive and negative zeros are 1 and 1 respectively.

[−4, 4, 1] by [−25, 25, 5]   [−4, 4, 1] by [−50, 50, 10]

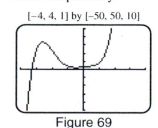

Figure 67   Figure 69

69. $P(x) = x^5 + 3x^4 - x^3 + 2x + 3$, $P(x)$ has two sign changes, therefore there are 2 or $2 - 2 = 0$ possible positive zeros. $P(-x) = (-x)^5 + 3(-x)^4 - (-x)^3 + 2(-x) + 3 = -x^5 + 3x + x^3 - 2x + 3$, $P(-x)$ has three sign changes, therefore there are 3 or $3 - 2 = 1$ possible negative zeros. From the graph of $P(x)$ in Figure 69, we see that the actual number of positive and negative zeros are 0 and 1 respectively.

71. (a) $f(x) = x^3 + x^2 - 6x = 0 \Rightarrow x(x^2 + x - 6) = x(x + 3)(x - 2) = 0 \Rightarrow x = 0, -3,$ or 2.

    (b) Graph $Y_1 = X^3 + X^2 - 6X$ in $[-5, 5, 1]$ by $[-10, 10, 1]$. The x-intercepts are -3, 0, and 2.

    (c) Table $Y_1 = X^3 + X^2 - 6X$ starting at $x = -4$, incrementing by 1. The zeros or x-intercepts are -3, 0, and 2.

73. (a) $f(x) = x^4 - 1 = 0 \Rightarrow x^4 = 1 \Rightarrow x = \pm 1$.

    (b) Graph $Y_1 = X^4 - 1$ in $[-2, 2, 1]$ by $[-10, 10, 1]$. The x-intercepts are -1 and 1.

    (c) Table $Y_1 = X^4 - 1$ starting at $x = -2$, incrementing by 1. The zeros or x-intercepts are ± 1.

75. (a) $f(x) = -x^3 + 4x = 0 \Rightarrow -x(x^2 - 4) = x(x + 2)(x - 2) = 0 \Rightarrow x = 0,$ or ± 2.

    (b) Graph $Y_1 = -X^3 + 4X$ in $[-5, 5, 1]$ by $[-5, 5, 1]$. The x-intercepts are -2, 0, and 2.

    (c) Table $Y_1 = -X^3 + 4X$ starting at $x = -3$, incrementing by 1. The zeros or x-intercepts are -2, 0, and 2.

77. $x^3 - 25x = 0 \Rightarrow x(x^2 - 25) = 0 \Rightarrow x(x - 5)(x + 5) = 0 \Rightarrow x = 0,$ or ± 5

79. $x^4 - x^2 = 2x^2 + 4 \Rightarrow x^4 - 3x^2 - 4 = 0 \Rightarrow (x^2 - 4)(x^2 + 1) = 0 \Rightarrow$
    $(x - 2)(x + 2)(x^2 + 1) = 0 \Rightarrow x = \pm 2$

81. $x^3 - 3x^2 - 18x = 0 \Rightarrow x(x^2 - 3x - 18) = 0 \Rightarrow x(x - 6)(x + 3) = 0 \Rightarrow x = 0, 6,$ or -3

83. $2x^3 = 4x^2 - 2x \Rightarrow 2x^3 - 4x^2 + 2x = 0 \Rightarrow 2x(x^2 - 2x + 1) = 0 \Rightarrow$
    $2x(x - 1)(x - 1) = 0 \Rightarrow x = 0$ or 1

85. $12x^3 = 17x^2 + 5x \Rightarrow 12x^3 - 17x^2 - 5x = 0 \Rightarrow x(12x^2 - 17x - 5) = 0 \Rightarrow$
    $x(4x + 1)(3x - 5) = 0 \Rightarrow x = 0, -\dfrac{1}{4},$ or $\dfrac{5}{3}$

87. $9x^4 + 4 = 13x^2 \Rightarrow 9x^4 - 13x^2 + 4 = 0 \Rightarrow (9x^2 - 4)(x^2 - 1) = 0 \Rightarrow x = \pm\dfrac{2}{3}, x = \pm 1$

89. $4x^3 + 4x^2 - 3x - 3 = 0 \Rightarrow (4x^3 + 4x^2) - (3x + 3) = 0 \Rightarrow 4x^2(x + 1) - 3(x + 1) = 0 \Rightarrow$
    $(4x^2 - 3)(x + 1) = 0$. Set $4x^2 - 3 = 0 \Rightarrow 4x^2 = 3 \Rightarrow x^2 = \dfrac{3}{4} \Rightarrow x = \pm\dfrac{\sqrt{3}}{2}$.
    The solutions are; $x = -1, \pm\dfrac{\sqrt{3}}{2}$.

91. $2x^3 + 4 = x(x + 8) \Rightarrow 2x^3 + 4 = x^2 + 8x \Rightarrow 2x^3 - x^2 - 8x + 4 = 0 \Rightarrow$
    $(2x^3 - x^2) - (8x - 4) = 0 \Rightarrow x^2(2x - 1) - 4(2x - 1) = 0 \Rightarrow (x^2 - 4)(2x - 1) = 0 \Rightarrow$
    $(x + 2)(x - 2)(2x - 1) = 0 \Rightarrow x = -2, 2, \dfrac{1}{2}$

93. $8x^4 - 30x^2 + 27 = 0 \Rightarrow (4x^2 - 9)(2x^2 - 3) = 0 \Rightarrow (2x + 3)(2x - 3)(2x^2 - 3) = 0.$

Set $2x^2 - 3 = 0 \Rightarrow 2x^2 = 3 \Rightarrow x^2 = \dfrac{3}{2} \Rightarrow x = \pm\sqrt{\dfrac{3}{2}} \Rightarrow x = \pm\dfrac{\sqrt{6}}{2}.$ The solutions are;

$x = \pm\dfrac{3}{2}, \pm\dfrac{\sqrt{6}}{2}$

95. $x^6 - 19x^3 - 216 = 0 \Rightarrow (x^3 + 8)(x^3 - 27) = 0.$ set $x^3 + 8 = 0 \Rightarrow x^3 = -8 \Rightarrow x = -2.$

Also set $x^3 - 27 = 0 \Rightarrow x^3 = 27 \Rightarrow x = 3.$ The solutions are; $x = -2, 3.$

97. The graph of $f(x) = x^3 - 1.1x^2 - 5.9x + 0.7$ is shown in Figure 97. Its zeros are approximately $-2.0095,$

0.11639, and 2.9931. The solutions are $x \approx -2.01, 0.12,$ or $2.99.$

[−10, 10, 1] by [−10, 10, 1]	[−10, 10, 1] by [−10, 10, 1]	[−10, 10, 1] by [−120, 120, 20]
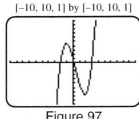	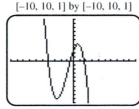	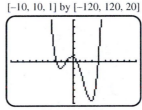
Figure 97	Figure 99	Figure 101

99. The graph of $f(x) = -0.7x^3 - 2x^2 + 4x + 2.5$ is shown in Figure 99. Its zeros are approximately $-4.0503,$

$-0.51594,$ and $1.7091.$ The solutions are $x \approx -4.05, -0.52,$ or $1.71.$

101. The graph of $f(x) = 2x^4 - 1.5x^3 - 24x^2 - 10x + 13$ is shown in Figure 101. Its zeros are approximately

$-2.6878, -1.0957, 0.55475,$ and $3.9787.$ The solutions are $x \approx -2.69, -1.10, 0.55$ or $3.98.$

103. $f(x) = x^2 - 5 \Rightarrow f(2) = 2^2 - 5 = -1$ and $f(3) = 3^2 - 5 = 4;$ because $\lceil(2) < 0$ and $\lceil(3) > 0,$ by the

intermediate value property, there exists an x-value between 2 and 3 such that $\lceil(x) = 0.$

105. $f(x) = 2x^3 - 1 \Rightarrow f(0) = 2(0)^3 - 1 = -1$ and $f(1) = 2(1)^3 - 1 = 1;$ because

$\lceil(0) < 0$ and $\lceil(1) > 0,$ by the intermediate value property, there exists an x-value between 0 and 1 such that

$\lceil(x) = 0.$

107. $f(x) = x^5 - x^2 + 4 \Rightarrow f(1) = 1^5 - 1^2 + 4 = 4$ and $f(2) = 2^5 - 2^2 + 4 = 32.$ Because

$f(1) < 20$ and $f(2) > 20,$ by the intermediate value property, there exists a number K such that $f(K) = 20.$

[−20, 40, 5] by [−2500, 2500, 500]

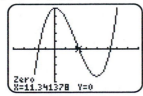

Figure 111

109. $T(x) = x^3 - 6x^2 + 8x$ when $0 \leq x \leq 4$. Graph $Y_1 = X^3 - 6X^2 + 8X$ in the window

    $[0, 4, 1]$ by $[-5, 5, 1]$. The zeros are $x = 0$, $x = 2$, and $x = 4$. Since $x = 0, 2$, and 4 correspond to the hours

    after midnight, then the temperature was 0°F at 12 am, 2 am and 4 am.

111. The graph of $Y_1 = (\pi/3) X^3 - 10 \pi X^2 + ((4000\pi)(0.6))/3$ and the smallest positive zero are shown in

    Figure 111. The ball with a 20-centimeter diameter will sink approximately 11.34 centimeters into the water.

113. $f(x) = x^3 - 66x^2 + 1052x + 652$ and $f(x) = 2500 \Rightarrow 2500 = x^3 - 66x^2 + 1052x + 652 \Rightarrow$

    $x^3 - 66x^2 + 1052x - 1848 = 0$; graph $Y_1 = X^3 - 66X^2 + 1052X - 1848$ in the window $[0, 45, 5]$ by

    $[-5000, 5000, 1000]$. The zeros are at $x = 2$, $x = 22$, and $x = 42$. Since $x = 1$ corresponds to June 1, there

    were 2500 birds on approximately June 2, June 22, and July 12.

115. (a) Graph f in $[-10, 15, 1]$ by $[-70, 70, 10]$. It has three zeros of approximately -6.01, 2.15, and 11.7. The

    approximate complete factored form is $-0.184(x + 6.01)(x - 2.15)(x - 11.7)$.

    (b) The zeros represent the months when the average temperature is 0°F. The zero of -6.01 has no

    significance since it does not correspond to a month. The zeros of 2.15 and 11.7 mean that in

    approximately February and November the average temperature in Trout Lake is 0°F.

[0, 70, 10] by [0, 22, 5]   [0, 70, 10] by [0, 22, 5]

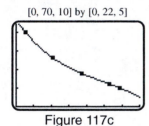   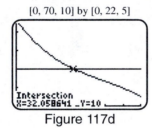

Figure 117c   Figure 117d

117. (a) The greater the distance downstream from the plant the less the concentration of copper. This agrees with

    intuition.

    (b) Using the cubic regression function on your calculator we find

    $f(x) \approx -0.000068x^3 + 0.0099x^2 - 0.653x + 23$.

    (c) See Figure 117c.

    (d) We must approximate the distance where the concentration of copper first drops to 10. Graph $Y_1 = C(x)$

    and $Y_2 = 10$. The point of intersection is near $(32.1, 10)$ as shown in Figure 117d. Mussels would not be

    expected to live between the plant and approximately 32.1 miles downstream, that is when $0 \leq x < 32.1$

    (approximately).

Extended and Discovery Exercise for Section 4.4

1. $P(x) = x^4 - x^3 + 3x^2 - 8x + 8; \; c = 2$

$$
\begin{array}{r|rrrrr}
2 & 1 & -1 & 3 & -8 & 8 \\
& & 2 & 2 & 10 & 4 \\
\hline
& 1 & 1 & 5 & 2 & 12
\end{array}
$$

Since the bottom row of the synthetic division is all non-negative and $c > 0$, $P(x)$ has no real zero greater than 2.

3. $P(x) = x^4 + x^3 - x^2 + 3; \; c = -2$

$$
\begin{array}{r|rrrrr}
-2 & 1 & 1 & -1 & 0 & 3 \\
& & -2 & 2 & -2 & 4 \\
\hline
& 1 & -1 & 1 & -2 & 7
\end{array}
$$

Since the bottom row of the synthetic division alternates in sign and $c < 0$, $P(x)$ has no real zero less than -2.

5. $P(x) = 3x^4 + 2x^3 - 4x^2 + x - 1; \; c = 1$

$$
\begin{array}{r|rrrrr}
1 & 3 & 2 & -4 & 1 & -1 \\
& & 3 & 5 & 1 & 2 \\
\hline
& 3 & 5 & 1 & 2 & 1
\end{array}
$$

Since the bottom row of the synthetic division are all non-negative and $c > 0$, $P(x)$ has no real zero greater than 1.

Checking Basic Concepts for Sections 4.3 and 4.4

1. $\dfrac{5x^4 - 10x^3 + 5x^2}{5x^2} = \dfrac{5x^4}{5x^2} - \dfrac{10x^3}{5x^2} + \dfrac{5x^2}{5x^2} = x^2 - 2x + 1$

3. Since the graph of the cubic polynomial has zeros -2 and 1, the multiplicities of the degree 3 equation is the zero -2 is even $\Rightarrow 2$ and the zero 1 is odd $\Rightarrow 1$. The factors are $(x + 2)(x + 2)(x - 1)$. To find the leading coefficient use:

$$f(0) = 2 \Rightarrow a(0 + 2)(0 + 2)(0 - 1) = 2 \Rightarrow a(2)(2)(-1) = 2 \Rightarrow -4a = 2 \Rightarrow a = -\frac{1}{2}.$$

The complete factored form is: $f(x) = -\dfrac{1}{2}(x + 2)^2(x - 1)$.

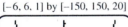

$[-6, 6, 1]$ by $[-150, 150, 20]$

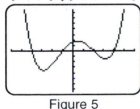

Figure 5

5. Graph $Y_1 = X^4 - X^3 - 18X^2 + 16X + 32$ in $[-6, 6, 1]$ by $[-150, 150, 20]$. The x-intercepts are $-4, -1,$ 2, and 4. See Figure 5. The factored form of f is $f(x) = (x + 4)(x + 1)(x - 2)(x - 4)$.

4.5: The Fundamental Theorem of Algebra

1. The graph of $f(x)$ does not intersect the x-axis. Therefore, f has no real zeros. Since f is degree 2, there must be two imaginary zeros.

3. The graph of $f(x)$ intersects the x-axis once. Therefore, f has one real zero. Since f is degree 3, there must be two imaginary zeros.

5. The graph of $f(x)$ intersects the x-axis twice. Therefore, f has two real zeros. Since f is degree 4, there must be two imaginary zeros.

7. The graph of $f(x)$ intersects the x-axis three times. Therefore, f has three real zeros. Since f is degree 5, there must be two imaginary zeros.

9. Degree: 2; leading coefficient: 1; zeros: $6i$ and $-6i$

    (a) $f(x) = (x - 6i)(x + 6i)$

    (b) $(x - 6i)(x + 6i) = x^2 - 36i^2 = x^2 + 36$. Thus, $f(x) = x^2 + 36$.

11. Degree: 3; leading coefficient: -1; zeros: $-1, 2i,$ and $-2i$

    (a) $f(x) = -1(x + 1)(x - 2i)(x + 2i)$

    (b) $-1(x + 1)(x - 2i)(x + 2i) = -1(x + 1)(x^2 - 4i^2) = -1(x + 1)(x^2 + 4) =$
    $-(x^3 + 4x + x^2 + 4) = -x^3 - x^2 - 4x - 4$. Thus, $f(x) = -x^3 - x^2 - 4x - 4$.

13. Degree: 4; leading coefficient: 10; zeros: $1, -1, 3i,$ and $-3i$

    (a) $f(x) = 10(x - 1)(x + 1)(x - 3i)(x + 3i)$

    (b) $10(x - 1)(x + 1)(x - 3i)(x + 3i) = 10(x^2 - 1)(x^2 + 9) = 10(x^4 + 8x^2 - 9) = 10x^4 + 80x^2 - 90$.
    Thus, $f(x) = 10x^4 + 80x^2 - 90$.

15. Degree: 4; leading coefficient: $\dfrac{1}{2}$; zeros: $-i$ and $2i$

    (a) Since $f(x)$ has real coefficients, it must also have a third and fourth zero of i and $-2i$ the conjugate of

    $-i$ and $2i$. Therefore the complete factored form is: $f(x) = \dfrac{1}{2}(x + i)(x - i)(x + 2i)(x - 2i)$

    (b) $\dfrac{1}{2}(x + i)(x - i)(x + 2i)(x - 2i) = \dfrac{1}{2}(x^2 - i^2)(x^2 - 4i^2) = \dfrac{1}{2}(x^2 + 1)(x^2 + 4) =$

    $\dfrac{1}{2}(x^4 + 4x^2 + x^2 + 4) = \dfrac{1}{2}x^4 + \dfrac{5}{2}x^2 + 2$. Thus, $f(x) = \dfrac{1}{2}x^4 + \dfrac{5}{2}x^2 + 2$.

17. Degree: 3; leading coefficient: -2; zeros: $1 - i$ and 3

    (a) Since $f(x)$ has real coefficients, it must also have a third zero of $1 + i$ the conjugate of

    $1 - i$. Therefore the complete factored form is: $f(x) = -2(x - (1 + i))(x - (1 - i))(x - 3)$

    (b) $-2(x - (1 + i))(x - (1 - i))(x - 3) = -2(x^2 - x + xi - x - xi + 1 - i^2)(x - 3) =$

    $-2(x^2 - 2x + 2)(x - 3) = -2(x^3 - 2x^2 + 2x - 3x^2 + 6x - 6) = -2(x^3 - 5x^2 + 8x - 6) =$

    $-2x^3 + 10x^2 - 16x + 12$. Thus, $f(x) = -2x^3 + 10x^2 - 16x + 12$.

19. First divide:

$$
\begin{array}{r}
3x^2 + 75 \\
x - \tfrac{5}{3}\overline{)\,3x^3 - 5x^2 + 75x - 125} \\
\underline{3x^3 - 5x^2} \\
75x - 125 \\
\underline{75x - 125} \\
0
\end{array}
$$

Now set $3x^2 + 75 = 0$ and solve. $3x^2 + 75 = 0 \Rightarrow 3x^2 = -75 \Rightarrow x^2 = -25 \Rightarrow x = \pm\sqrt{-25} \Rightarrow$

$x = \pm 5i$. The solutions are $x = \dfrac{5}{3}, \pm 5i$.

21. If $-3i$ is a zero then $3i$ is also a zero $\Rightarrow (x + 3i)(x - 3i) = x^2 - 9i^2 = x^2 + 9$. Now divide this into the

equation:

$$
\begin{array}{r}
2x^2 - x + 1 \\
x^2 + 9\overline{)\,2x^4 - x^3 + 19x^2 - 9x + 9} \\
\underline{2x^4 + 18x^2} \\
-x^3 + x^2 - 9x + 9 \\
\underline{-x^3 - 9x} \\
x^2 + 9 \\
\underline{x^2 + 9} \\
0
\end{array}
$$

Then use the quadratic formula to solve $2x^2 - x + 1$. $\dfrac{1 \pm \sqrt{1 - 4(2)(1)}}{2(2)} = \dfrac{1 \pm \sqrt{-7}}{4} =$

$\dfrac{1}{4} \pm \dfrac{i\sqrt{7}}{4}$. Therefore the solutions are: $x = \pm 3i, \dfrac{1}{4} \pm \dfrac{i\sqrt{7}}{4}$.

23. $x^2 + 25 = 0 \Rightarrow x^2 = -25 \Rightarrow x = \pm 5i$. Thus, $f(x) = (x - 5i)(x + 5i)$.

25. $3x^3 + 3x = 0 \Rightarrow 3x(x^2 + 1) = 0 \Rightarrow x = 0$ or $\pm i$. Thus, $f(x) = 3(x - 0)(x - i)(x + i)$.

27. $x^4 + 5x^2 + 4 = 0 \Rightarrow (x^2 + 1)(x^2 + 4) = 0 \Rightarrow x = \pm i$ or $\pm 2i$.

    Thus, $f(x) = (x - i)(x + i)(x - 2i)(x + 2i)$.

29. The graph of $y = x^3 + 2x^2 + 16x + 32$ is shown in Figure 29. The x-intercept appears to be -2. We can use

    synthetic division to help factor f.

$$
\begin{array}{r|rrrr}
-2 & 1 & 2 & 16 & 32 \\
   &   & -2 & 0 & -32 \\
\hline
   & 1 & 0 & 16 & 0
\end{array}
$$

$x^3 + 2x^2 + 16x + 32 = (x + 2)(x^2 + 16) = (x + 2)(x + 4i)(x - 4i)$.

Thus, $f(x) = (x + 2)(x + 4i)(x - 4i)$.

[−10, 10, 1] by [−50, 50, 10]

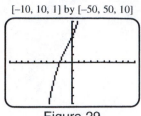

Figure 29

31. $x^3 + x = 0 \Rightarrow x(x^2 + 1) = 0 \Rightarrow x(x + i)(x - i) = 0 \Rightarrow x = 0, \pm i$

33. Factor or find the zero of 2 graphically.

$x^3 = 2x^2 - 7x + 14 \Rightarrow x^3 - 2x^2 + 7x - 14 = 0 \Rightarrow x^2(x - 2) + 7(x - 2) = 0 \Rightarrow$

$(x - 2)(x^2 + 7) = 0 \Rightarrow (x - 2)(x + i\sqrt{7})(x - i\sqrt{7}) = 0 \Rightarrow x = 2, \pm i\sqrt{7}$

35. $x^4 + 5x^2 = 0 \Rightarrow x^2(x^2 + 5) = x^2(x + i\sqrt{5})(x - i\sqrt{5}) = 0 \Rightarrow x = 0, \pm i\sqrt{5}$

37. $x^4 = x^3 - 4x^2 \Rightarrow x^4 - x^3 + 4x^2 = 0 \Rightarrow x^2(x^2 - x + 4) = 0$

Use the quadratic formula to find the zeros of $x^2 - x + 4$. The solutions are $x = 0, \dfrac{1}{2} \pm \dfrac{i\sqrt{15}}{2}$.

39. Find the zeros of 1 and 2 using the rational zero test or a graph. See Figures 39a & 39b.

$x^4 + x^3 = 16 - 8x - 6x^2 \Rightarrow x^4 + x^3 + 6x^2 + 8x - 16 = 0$

The graph of $x^4 + x^3 + 6x^2 + 8x - 16$ shows that –2 and 1 are zeros, so $(x + 2)$ and $(x - 1)$ are factors;

using synthetic division twice gives the missing factor.

$$
\begin{array}{r|rrrrr}
-2 & 1 & 1 & 6 & 8 & -16 \\
   &   & -2 & 2 & -16 & 16 \\
\hline
   & 1 & -1 & 8 & -8 & 0
\end{array}
$$

$$
\begin{array}{r|rrrr}
1 & 1 & -1 & 8 & -8 \\
  &   & 1 & 0 & 8 \\
\hline
  & 1 & 0 & 8 & 0
\end{array}
$$

$x^4 + x^3 + 6x^2 + 8x - 16 = (x + 2)(x - 1)(x^2 + 8) = (x + 2)(x - 1)(x + i\sqrt{8})(x - i\sqrt{8})$

The solutions are $x = -2, 1, \pm i\sqrt{8}$.

[−4, 4, 1] by [−20, 20, 5]   [−4, 4, 1] by [−20, 20, 5]

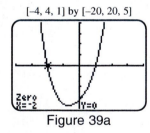   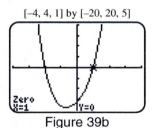

Figure 39a   Figure 39b

41. Find the zero of –2 using the rational zero test or a graph of $y = 3x^3 + 4x^2 - x + 6$. Use synthetic division

to factor $3x^3 + 4x^2 - x + 6$.

$$
\begin{array}{r|rrrr}
-2 & 3 & 4 & -1 & 6 \\
   &   & -6 & 4 & -6 \\
\hline
   & 3 & -2 & 3 & 0
\end{array}
$$

$3x^3 + 4x^2 - x + 6 = (x + 2)(3x^2 - 2x + 3)$

Use the quadratic formula to find the zeros of $3x^2 - 2x + 3$. The solutions are $x = -2, \dfrac{1}{3} \pm \dfrac{i\sqrt{8}}{3}$.

4.6: Rational Functions and Models

1. Yes, since the numerator and denominator are both polynomials. Since $4x - 5 \neq 0, D = \left\{ x \mid x \neq \dfrac{5}{4} \right\}$.

3. Yes, since f can be written as $f(x) = \dfrac{x^2 - x - 2}{1}$, and $g(x) = 1$ is a polynomial. $D =$ all real numbers.

5. No, since the numerator is not a polynomial. $D = \{x \mid x \neq -1\}$

7. Yes, since the numerator and denominator are both polynomials. Since $x^2 + 1 \neq 0, D =$ all real numbers.

9. No, since the numerator is not a polynomial. Since $x^2 + x \neq 0 \Rightarrow x(x + 1) \neq 0, D = \{x \mid x \neq -1, x \neq 0\}$.

11. Yes, since f can be written as $f(x) = \dfrac{4(x + 1) - 3}{x + 1} \Rightarrow f(x) = \dfrac{4x + 1}{x + 1}$.

    Since $x + 1 \neq 0, D = \{x \mid x \neq -1\}$.

13. There is a horizontal asymptote of $y = 4$ and a vertical asymptote of $x = 2$; $D = \{x \mid x \neq 2\}$.

15. There is a horizontal asymptote of $y = -4$ and a vertical asymptote of $x = \pm 2$; $D = \{x \mid x \neq \pm 2\}$.

17. There is a horizontal asymptote of $y = 0$ and there is no vertical asymptote; $D =$ all real numbers.

19. Since the output of Y_1 gets closer to 3 as the input x gets larger, it is reasonable to conjecture that the equation for the horizontal asymptote is $y = 3$.

21. Horizontal Asymptotes: The degree of the numerator is equal to the degree of the denominator and the ratio of the leading coefficients is $\dfrac{4}{2} = 2$. Therefore, $y = 2$ is a horizontal asymptote.

    Vertical Asymptotes: Find the zeros of the denominator, $2x - 6 = 0 \Rightarrow 2x = 6 \Rightarrow x = 3$. Therefore $x = 3$ is a vertical asymptote. (Note that 3 is not a zero of the numerator.)

23. Horizontal Asymptotes: The degree of the numerator is less than the degree of the denominator, therefore, the x-axis, $y = 0$, is a horizontal asymptote.

    Vertical Asymptotes: Find the zeros of the denominator, $x^2 - 5 = 0 \Rightarrow x^2 = 5 \Rightarrow x = \pm\sqrt{5}$. Therefore $x = \pm\sqrt{5}$ are vertical asymptotes. (Note that $\pm\sqrt{5}$ are not zeros of the numerator.)

25. Horizontal Asymptotes: The degree of the numerator is greater than the degree of the denominator, therefore there is no horizontal asymptote.

    Vertical Asymptotes: Find the zeros of the denominator,
    $x^2 + 3x - 10 = 0 \Rightarrow (x + 5)(x - 2) = 0 \Rightarrow x = -5, 2$. Therefore $x = -5$ and $x = 2$ are the vertical asymptotes. (Note that –5 and 2 are not a zeros of the numerator.)

27. Horizontal Asymptotes: The degree of the numerator is equal to the degree of the denominator and the ratio of the leading coefficients is $\dfrac{1}{2}$. Therefore, $y = \dfrac{1}{2}$ is a horizontal asymptote.

    Vertical Asymptotes: Find the zeros of the denominator, $(2x - 5)(x + 1) = 0 \Rightarrow x = \dfrac{5}{2}$ and $x = -1$.

    Therefore $x = \dfrac{5}{2}$ is a vertical asymptote. (Note that $\dfrac{5}{2}$ is not a zero of the numerator and $x = -1$ is a zero of the numerator).

29. Horizontal Asymptotes: The degree of the numerator is equal to the degree of the denominator and the ratio of the leading coefficients is $\frac{3}{1} = 3$. Therefore, $y = 3$ is a horizontal asymptote.

    Vertical Asymptotes: Find the zeros of the denominator, $(x + 2)(x - 1) = 0 \Rightarrow x = -2, 1$. Here, only $x = 1$ is a vertical asymptote. (Note that –2 is a zero of the numerator.)

31. Horizontal Asymptotes: The degree of the numerator is greater than the degree of the denominator, therefore there is no horizontal asymptote.

    Vertical Asymptotes: Find the zeros of the denominator, $x + 3 = 0 \Rightarrow x = -3$. But $x = -3$ is not a vertical asymptote since –3 is a zero of the numerator and $f(x) = x - 3$ for $x \neq -3$. There are no vertical asymptotes.

33. $f(x) = \dfrac{a}{x - 1}$ has a vertical asymptote of $x = 1$. It has a horizontal asymptote of $y = 0$, since the degree of the numerator is less than the degree of the denominator. The best choice is graph *b*.

35. $f(x) = \dfrac{x - a}{x + 2}$ has a vertical asymptote of $x = -2$ and a horizontal asymptote of $y = 1$, since the degree of the numerator equals the degree of the denominator and the ratio of the leading coefficients is $\dfrac{1}{1}$.

    The best choice is graph *d*.

37. One example of a symbolic representation of a rational function with a vertical asymptote of $x = -3$ and a horizontal asymptote of $y = 1$ is $f(x) = \dfrac{x + 1}{x + 3}$. *Answers may vary.*

39. One example of a symbolic representation of a rational function with vertical asymptotes of $x = \pm 3$ and a horizontal asymptote of $y = 0$ is $f(x) = \dfrac{1}{x^2 - 9}$. *Answers may vary.*

41. See Figure 41. Since the degree of the numerator is less than the degree of the denominator, the horizontal asymptote is $y = 0$. To find the vertical asymptote we find the zero of the denominator, $x^2 = 0 \Rightarrow x = 0$.

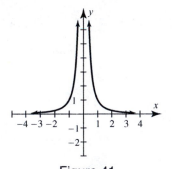

Figure 41

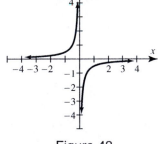

Figure 43

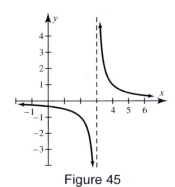

Figure 45

43. See Figure 43. Since the degree of the numerator is less than the degree of the denominator, the horizontal asymptote is $y = 0$. To find the vertical asymptote we find the zero of the denominator, $2x = 0 \Rightarrow x = 0$.

45. $g(x) = \dfrac{1}{x - 3}$ is $y = \dfrac{1}{x}$ transformed 3 units right $\Rightarrow$ it has a vertical asymptote of $x = 3$. It still has a horizontal asymptote of $y = 0$. See Figure 45. Then $g(x) = f(x - 3)$.

47. $g(x) = \dfrac{1}{x} + 2$ is $y = \dfrac{1}{x}$ transformed 2 units up $\Rightarrow$ it has a horizontal asymptote of $y = 2$. It still has a

vertical asymptote of $x = 0$. See Figure 47. Then $g(x) = f(x) + 2$.

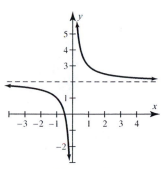

Figure 47

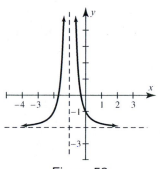

Figure 49

49. $g(x) = \dfrac{1}{x + 1} - 2$ is $y = \dfrac{1}{x}$ transformed 1 unit left and 2 units down. It now has a vertical asymptote of

$x = -1$, and a horizontal asymptote of $y = -2$. See Figure 49. Then $g(x) = f(x + 1) - 2$.

51. $g(x) = -\dfrac{2}{(x - 1)^2}$ is $y = \dfrac{1}{x^2}$ transformed by reflecting it across the x-axis, then shifting 1 unit right, and

vertically stretching by a factor of 2. It now has a vertical asymptote of $x = 1$, and it still has a horizontal

asymptote of $y = 0$. See Figure 51. Then $g(x) = -2h(x - 1)$.

53. $g(x) = \dfrac{1}{(x + 1)^2} - 2$ is $y = \dfrac{1}{x^2}$ transformed 1 unit left and 2 units down. It now has a vertical asymptote of

$x = -1$, and a horizontal asymptote of $y = -2$. See Figure 53. Then $g(x) = h(x + 1) - 2$.

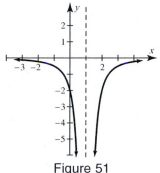

Figure 51

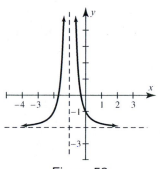

Figure 53

55. (a) $x - 2 = 0 \Rightarrow x = 2; D = \{x \mid x \neq 2\}$

(b) The graph of f using dot mode is shown in Figure 55b.

(c) Since the degree of the numerator equals the degree of the denominator and the ratio of the leading

coefficients is $\dfrac{1}{1} = 1$, the horizontal asymptote is $y = 1$. There is a vertical asymptote at $x = 2$.

(d) First, sketch the vertical and horizontal asymptotes found in part (c). Then use Figure 55b as a guide to a more complete graph of f. The sketch is shown in Figure 55d.

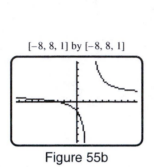

[−8, 8, 1] by [−8, 8, 1]

Figure 55b

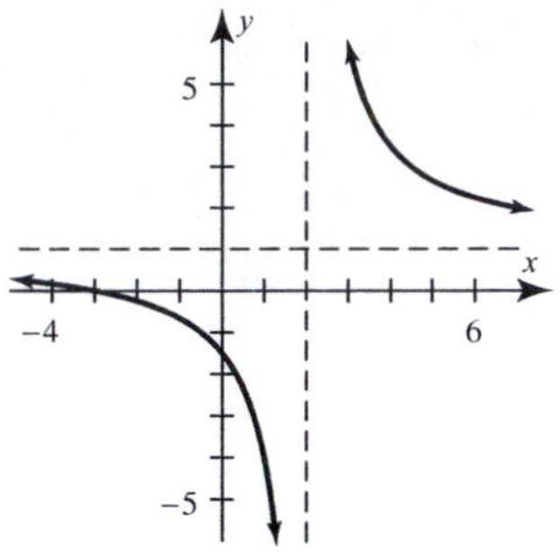

Figure 55d

57. (a) $x^2 - 4 = 0 \Rightarrow x^2 = 4 \Rightarrow x = \pm 2; D = \{x \mid x \neq \pm 2\}$

(b) The graph of f using dot mode is shown in Figure 57b.

(c) Since the degree of the numerator is less than the degree of the denominator there is a horizontal asymptote at $y = 0$. There are vertical asymptotes at $x = \pm 2$.

(d) First, sketch the vertical and horizontal asymptotes found in part (c). Then use Figure 57b as a guide to a more complete graph of f. The sketch is shown in Figure 57d.

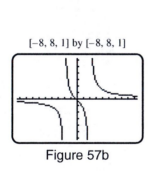

[−8, 8, 1] by [−8, 8, 1]

Figure 57b

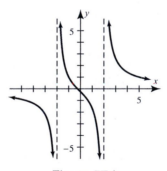

Figure 57d

59. (a) $1 - 0.25x^2 = 0 \Rightarrow 0.25x^2 = 1 \Rightarrow x^2 = 4 \Rightarrow x = \pm 2; D = \{x \mid x \neq \pm 2\}$

(b) The graph of f using dot mode is shown in Figure 59b.

(c) Since the degree of the numerator is less than the degree of the denominator there is a horizontal asymptote at $y = 0$. There are vertical asymptotes at $x = \pm 2$.

(d) First, sketch the vertical and horizontal asymptotes found in part (c). Then use Figure 59b as a guide to a more complete graph of f. The sketch is shown in Figure 59d.

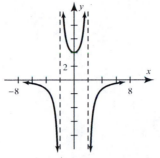

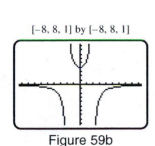

[–8, 8, 1] by [–8, 8, 1]

Figure 59b   **Figure 59d**

61. (a) $x - 2 = 0 \Rightarrow x = 2; D = \{x \mid x \neq 2\}$

   (b) The graph of f using dot mode is shown in Figure 61b.

   (c) Since the degree of the numerator is greater than the degree of the denominator there is no horizontal
       asymptote. There are no vertical asymptotes. (The numerator equals 2 when $x = 2$).

   (d) First, sketch the vertical and horizontal asymptotes found in part (c). Then use Figure 61b as a guide to a
       more complete graph of f. The sketch is shown in Figure 61d.

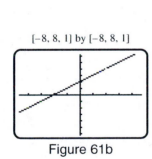

[–8, 8, 1] by [–8, 8, 1]

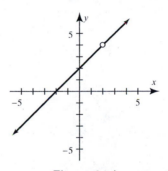

Figure 61b   **Figure 61d**

63. First divide:

$$
\require{enclose}
\begin{array}{r}
x - 1 \\
x - 1 \enclose{longdiv}{x^2 - 2x + 1} \\
\underline{x^2 - x} \\
-x + 1 \\
\underline{-x + 1} \\
0
\end{array}
$$

Now graph $x - 1$; since the denominator $x - 1 \neq 0$ there is a hole at $x = 1$ in the graph. See Figure 63.

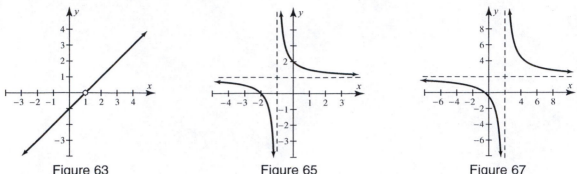

Figure 63   Figure 65   Figure 67

65. First we find the asymptotes. Since the degree of the numerator and denominator are the same, the ratio of the leading coefficients $\frac{1}{1}$ or $y = 1$ is the horizontal asymptote. To find the vertical asymptote we find the zero of the denominator. $x + 1 = 0 \Rightarrow x = -1$. Since $x = -1$ is a vertical asymptote it has no holes.

See Figure 65.

67. First divide: $g(x) = \frac{(2x + 1)(x - 2)}{(x - 2)(x - 2)} = \frac{2x + 1}{x - 2}$. Now graph $\frac{2x + 1}{x - 2}$ by finding it's asymptotes. The ratio of the leading coefficients $\frac{2}{1}$ or $y = 2$ is the horizontal asymptote. To find the vertical asymptote we find the zero of the denominator. $x - 2 = 0 \Rightarrow x = 2$. Since $x = 2$ is an asymptote there are no holes.

See Figure 67.

69. Factor the numerator and denominator, then divide: $f(x) = \frac{2x^2 + 9x + 9}{2x^2 + 7x + 6} = \frac{(2x + 3)(x + 3)}{(2x + 3)(x + 2)} = \frac{x + 3}{x + 2}$.

Now graph $\frac{x + 3}{x + 2}$ by finding it's asymptotes. The ratio of the leading coefficients $\frac{1}{1}$ or $y = 1$ is the horizontal asymptote. To find the vertical asymptote we find the zero of the denominator. $x + 2 = 0 \Rightarrow x = -2$.

Since $x = -2$ is an asymptote there is only a hole is at $2x + 3 = 0$ or $x = \frac{-3}{2}$. See Figure 69.

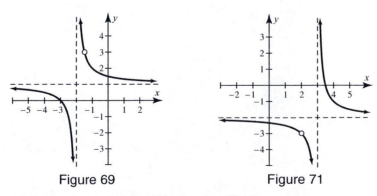

Figure 69   Figure 71

71. Factor the numerator and denominator, then divide:

$f(x) = \frac{-2x^2 + 11x - 14}{x^2 - 5x + 6} = \frac{(-2x + 7)(x - 2)}{(x - 3)(x - 2)} = \frac{-2x + 7}{x - 3}$.

Now graph $\frac{-2x + 7}{x - 3}$ by finding it's asymptotes. The ratio of the leading coefficients $\frac{-2}{1}$ or $y = -2$ is the horizontal asymptote. To find the vertical asymptote we find the zero of the denominator.

$x - 3 = 0 \Rightarrow x = 3$. Since $x = 3$ is an asymptote there is only a hole is at $x = 2$. See Figure 71.

73. There is a vertical asymptote at $x = -1$ since -1 is a zero of the denominator but is not a zero of the numerator.

To find any slant asymptote we will divide the numerator by the denominator using synthetic division:

$$
\begin{array}{r|rrr}
-1 & 1 & 0 & 1 \\
   &   & -1 & 1 \\
\hline
   & 1 & -1 & 2 \\
\end{array}
$$

Therefore, $f(x) = x - 1 + \dfrac{2}{x + 1}$ and as $|x|$ becomes large, $f(x)$ approaches $y = x - 1$. There is a slant asymptote at $y = x - 1$. The graph of f using dot mode is shown in Figure 73a. To sketch this graph, first sketch the vertical and slant asymptotes found above. Then use Figure 73a as a guide to a more complete graph of f. The sketch is shown in Figure 73b.

[−8, 8, 1] by [−8, 8, 1]

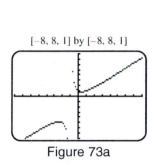

Figure 73a

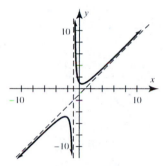

Figure 73b

75. There is a vertical asymptote at $x = -2$ since -2 is a zero of the denominator but is not a zero of the numerator.

To find any slant asymptote we will divide the numerator by the denominator using synthetic division:

$$
\begin{array}{r|rrr}
-2 & 0.5 & -2 & 2 \\
   &   & -1 & 6 \\
\hline
   & 0.5 & -3 & 8 \\
\end{array}
$$

Therefore, $f(x) = 0.5x - 3 + \dfrac{8}{x + 2}$ and as $|x|$ becomes large, $f(x)$ approaches $y = 0.5x - 3$. There is a slant asymptote at $y = 0.5x - 3$. The graph of f using dot mode is shown in Figure 75a. To sketch this graph, first sketch the vertical and slant asymptotes found above. Then use Figure 75a as a guide to a more complete graph of f. The sketch is shown in Figure 75b.

[−14, 14, 2] by [−14, 14, 2]

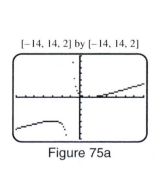

Figure 75a

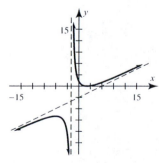

Figure 75b

77. There is a vertical asymptote at $x = 1$ since 1 is a zero of the denominator but is not a zero of the numerator.

    To find any slant asymptote we will divide the numerator by the denominator using synthetic division:

    $$\begin{array}{r|rrr} 1 & 1 & 2 & 1 \\ & & 1 & 3 \\ \hline & 1 & 3 & 4 \end{array}$$

    Therefore, $f(x) = x + 3 + \dfrac{4}{x - 1}$ and as $|x|$ becomes large, $f(x)$ approaches $y = x + 3$. There is a slant

    asymptote at $y = x + 3$. The graph of f using dot mode is shown in Figure 77a. To sketch this graph, first

    sketch the vertical and slant asymptotes found above. Then use Figure 77a as a guide to a more complete graph

    of f. The sketch is shown in Figure 77b.

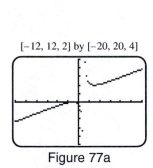

[−12, 12, 2] by [−20, 20, 4]

Figure 77a

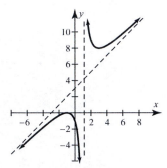

Figure 77b

79. There is a vertical asymptote at $x = \dfrac{1}{2}$ since $\dfrac{1}{2}$ is a zero of the denominator but is not a zero of the

    numerator. To find any slant asymptote we will divide the numerator by the denominator.

    $$\begin{array}{r} 2x + 1 + \frac{1}{2x-1} \\ 2x - 1 \overline{\smash{)}4x^2 + 0x + 0} \\ \underline{4x^2 - 2x} \\ 2x + 0 \\ \underline{2x - 1} \\ 1 \end{array}$$

    Therefore, $f(x) = 2x + 1 + \dfrac{1}{2x - 1}$ and as $|x|$ becomes large, $f(x)$ approaches $y = 2x + 1$. There is a slant

    asymptote at $y = 2x + 1$. The graph of f using dot mode is shown in Figure 79a. To sketch this graph, first

    sketch the vertical and slant asymptotes found above. Then use Figure 79a as a guide to a more complete

    graph of f. The sketch is shown in Figure 79b.

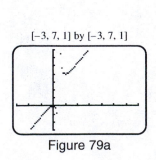

[−3, 7, 1] by [−3, 7, 1]

Figure 79a

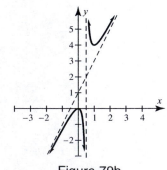

Figure 79b

81. $\dfrac{4}{x+2} = -4 \Rightarrow 4 = -4(x+2) \Rightarrow 4 = -4x - 8 \Rightarrow 12 = -4x \Rightarrow x = -3$

83. $\dfrac{x+1}{x} = 2 \Rightarrow x + 1 = 2x \Rightarrow x = 1$

85. $\dfrac{1-x}{3x-1} = -\dfrac{3}{5} \Rightarrow -3(3x-1) = 5(1-x) \Rightarrow -9x + 3 = 5 - 5x \Rightarrow -2 = 4x \Rightarrow x = -\dfrac{1}{2}$

87. $f(x) = \dfrac{2x-4}{x-1}$ is in lowest terms.

   1) To find the vertical asymptote we find the zero of the denominator, $x - 1 = 0 \Rightarrow x = 1$.

   2) Since the degree of the numerator is equal to the degree of the denominator the equation of the horizontal asymptote is $y = \dfrac{2}{1} = 2$.

   3) $f(0) = \dfrac{2(0) - 4}{0 - 1} = 4;\ (0, 4)$

   4) $\dfrac{2x-4}{x-1} = 0 \Rightarrow 2x - 4 = 0 \Rightarrow 2x = 4 \Rightarrow x = 2;\ (2, 0)$

   5) $\dfrac{2x-4}{x-1} = 2 \Rightarrow 2x - 4 = 2(x - 1) \Rightarrow 2x - 4 = 2x - 2 \Rightarrow 0 = 2 \Rightarrow f$ does not cross the horizontal asymptote.

   6) See Figure 87.

   7) See Figure 87.

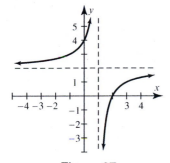

Figure 87

89. $f(x) = \dfrac{x^2 - 2x}{x^2 + 6x + 9}$ is in lowest terms.

   1) To find the vertical asymptote we find the zero of the denominator, $x^2 + 6x + 9 = 0 \Rightarrow$

       $(x+3)^2 = 0 \Rightarrow x + 3 = 0 \Rightarrow x = -3$.

   2) Since the degree of the numerator is equal to the degree of the denominator the equation of the horizontal asymptote is $y = \dfrac{1}{1} = 1$.

   3) $f(0) = \dfrac{0^2 - 2(0)}{0^2 + 6(0) + 9} = \dfrac{0}{9} = 0;\ (0, 0)$

4) $\dfrac{x^2 - 2x}{x^2 + 6x + 9} = 0 \Rightarrow x^2 - 2x = 0 \Rightarrow x(x - 2) = 0 \Rightarrow x = 0, x = 2;\ (0, 0), (2, 0)$

5) $\dfrac{x^2 - 2x}{x^2 + 6x + 9} = 1 \Rightarrow x^2 - 2x = x^2 + 6x + 9 \Rightarrow -8x = 9 \Rightarrow x = -\dfrac{9}{8} \Rightarrow f$ crosses the horizontal

asymptote at $\left(-\dfrac{9}{8}, 1\right)$

6) See Figure 89.

7) See Figure 89.

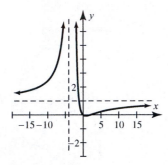

Figure 89

91. $f(x) = \dfrac{x^2 + 2x + 1}{x^2 - x - 6}$ is in lowest terms.

1) To find the vertical asymptotes we find the zeros of the denominator, $x^2 - x - 6 = 0 \Rightarrow$

$(x - 3)(x + 2) = 0 \Rightarrow x = 3, -2.$

2) Since the degree of the numerator is equal to the degree of the denominator the equation of the horizontal

asymptote is $y = \dfrac{1}{1} = 1.$

3) $f(0) = \dfrac{0^2 + 2(0) + 1}{0^2 - 0 - 6} = -\dfrac{1}{6} = 0;\ \left(0, -\dfrac{1}{6}\right)$

4) $\dfrac{x^2 + 2x + 1}{x^2 - x - 6} = 0 \Rightarrow x^2 + 2x + 1 = 0 \Rightarrow (x + 1)^2 = 0 \Rightarrow x = -1;\ (-1, 0)$

5) $\dfrac{x^2 + 2x + 1}{x^2 - x - 6} = 1 \Rightarrow x^2 + 2x + 1 = x^2 - x - 6 \Rightarrow 3x = -7 \Rightarrow x = -\dfrac{7}{3} \Rightarrow f$ crosses the

horizontal asymptote at $-\dfrac{7}{3}.$

6) See Figure 91.

7) See Figure 91.

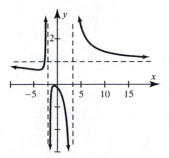

Figure 91

93. (a) $f(98) = \dfrac{100}{101 - 98} = \dfrac{100}{3} \approx 33\%$, the least active 98% post about 33% of the postings.

(b) $f(100) = \dfrac{100}{101 - 100} = \dfrac{100}{1} = 100\%$, from part (a) we found $f(98) \approx 33\%$, so

$f(100) - f(98) = 100 - 33 = 67\%$. The most active 2% post about 67% of the postings.

95. (a) $P(1) = \dfrac{5}{0.03(1) + 0.97} = \dfrac{5}{1} = 5\%$, at 1 second there is a 5% chance that the visitor is abandoning the

website.

(b) $P(60) = \dfrac{5}{0.03(60) + 0.97} = \dfrac{5}{2.77} \approx 1.8\%$, at 60 seconds there is a 1.8% chance the visitor is abandoning

the website.

97. (a) $T(4) = -\dfrac{1}{4 - 8} \Rightarrow T(4) = \dfrac{1}{4} \Rightarrow T(4) = 0.25$; when vehicles leave the ramp at an average rate

of 4 vehicles per minute, the wait is 0.25 minutes or 15 seconds.

$T(7.5) = \dfrac{1}{7.5 - 8} \Rightarrow T(7.5) = \dfrac{1}{0.5} \Rightarrow T(7.5) = 2.0$; when vehicles leave the ramp at an average rate of

7.5 vehicles per minute, the wait is 2 minutes.

(b) The wait increases dramatically.

99. (a) $N(20) = \dfrac{20^2}{1600 - 40(20)} = \dfrac{400}{800} = \dfrac{1}{2}$; $N(39) = \dfrac{39^2}{1600 - 40(39)} = \dfrac{1521}{40} = 38.025$

(b) The wait increases dramatically.

(c) We find the vertical asymptote by finding the zero of the denominator,

$1600 - 40x = 0 \Rightarrow 1600 = 40x \Rightarrow x = 40.$

101.(a) Graph $Y_1 = (10X + 1)/(X + 1)$ and $Y_2 = 10$ in [0, 14, 1] by [0, 14, 1]. Since the degree of the

numerator and denominator are equal, there is a horizontal asymptote at $y = \dfrac{10}{1} = 10$. See Figure 101.

(b) The initial population would be $f(0) = 1$ million insects.

(c) After many months the population starts to level off at 10 million.

(d) The horizontal asymptote $y = 10$ represents the limiting population after a very long time.

[0, 14, 1] by [0, 14, 1]   [0, 600, 100] by [0, 50, 5]

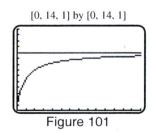
Figure 101

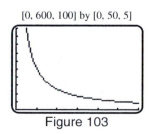
Figure 103

103.(a) $f(x) = \dfrac{2540}{x} \Rightarrow f(400) = \dfrac{2540}{400} = 6.35$ inches. A curve designed for 60 mph with a radius of 400 feet

should have the outer rail elevated 6.35 inches.

(b) Figure 103 shows the graph of $Y_1 = 2540/X$. This means that a sharper curve must be banked more. As

the radius increases, the curve is not as sharp and the elevation of the outer rail decreases.

(c) Since the degree of the numerator is less than the degree of the denominator, the graph of $f(x) = \dfrac{2540}{x}$

has a horizontal asymptote of $y = 0$. As the radius of the curve increases without bound $(x \to \infty)$, the

tracks become straight and no elevation or banking $(y \to 0)$ of the outer rail is necessary.

(d) $f(x) = 12.7 \Rightarrow 12.7 = \dfrac{2540}{x} \Rightarrow x = \dfrac{2540}{12.7} = 200$; a radius of 200 feet requires an elevation of 12.7

inches.

Extended and Discovery Exercise for Section 4.6

1. $f(x) = \dfrac{1}{x}$

$\dfrac{f(3) - f(1)}{3 - 1} = \dfrac{\frac{1}{3} - 1}{2} = -\dfrac{1}{3}, \quad \dfrac{f(x + h) - f(x)}{h} = \dfrac{\frac{1}{x+h} - \frac{1}{x}}{h} = \dfrac{\frac{x - x - h}{(x+h)(x)}}{h} = \dfrac{\frac{-h}{x(x+h)}}{h} = -\dfrac{1}{x(x+h)}$

3. $f(x) = \dfrac{3}{2x}$

$\dfrac{f(3) - f(1)}{3 - 1} = \dfrac{\frac{3}{6} - \frac{3}{2}}{2} = -\dfrac{1}{2}, \quad \dfrac{f(x + h) - f(x)}{h} = \dfrac{\frac{3}{2(x+h)} - \frac{3}{2x}}{h} = \dfrac{\frac{3x - 3x - 3h}{2x(x+h)}}{h} =$

$\dfrac{x}{x} \cdot \dfrac{3}{2(x+h)} - \dfrac{3}{2x} \cdot \dfrac{x+h}{x+h} = \dfrac{3x}{2x(x+h)} - \dfrac{3(x+h)}{2x(x+h)} = -\dfrac{3}{2x(x+h)}$

Checking Basic Concepts for Sections 4.5 and 4.6

1. Since the polynomial is quadratic, it can have at most two zeros, which are given as $\pm 4i$. Thus, there no real

    zeros; since the leading coefficient is 3, $f(x) = 3(x + 4i)(x - 4i) = 3(x^2 + 16) = 3x^2 + 48$.

3. Use a graph of $y = x^3 - x^2 + 4x - 4$ or use the rational zero test to find the zero of 1. See Figure 3. The

    graph indicates that 1 is a zero, so $(x - 1)$ is a factor. Use synthetic division to find the other factor.

$$
\begin{array}{r|rrrr}
1 & 1 & -1 & 4 & -4 \\
  &   & 1 & 0 & 4 \\
\hline
  & 1 & 0 & 4 & 0 \\
\end{array}
$$

$f(x) = (x - 1)(x^2 + 4) = (x - 1)(x + 2i)(x - 2i)$

5. (a) The denominator $x - 1 \neq 0 \Rightarrow D\{x \mid x \neq 1\}$

    (b) $f(x) = \dfrac{1}{x-1} + 2 \Rightarrow f(x) = \dfrac{1}{x-1} + \dfrac{2(x-1)}{x-1} \Rightarrow f(x) = \dfrac{2x-1}{x-1}$. To find the vertical asymptote

    we find the zero of the denominator. $x - 1 = 0 \Rightarrow x = 1$ is the vertical asymptote. Since the degree of

    the numerator and denominator is the same, we use the ratio of the leading coefficients $\dfrac{2}{1} \Rightarrow y = 2$ to find

    the horizontal asymptote.

    (c) See Figure 5.

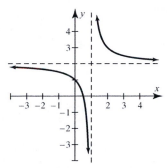

Figure 5

7. (a) $f(x) = \dfrac{3x-1}{2x-2}$

    Since $2x - 2 = 0 \Rightarrow x = 1$, the vertical asymptote is the line $x = 1$. Since the degree of the

    numerator is equal to the degree of the denominator the equation of the horizontal asymptote is $y = \dfrac{3}{2}$.

    See Figure 7a.

    (b) $f(x) = \dfrac{1}{(x+1)^2}$

    Since $(x + 1)^2 = 0 \Rightarrow x = -1$, the vertical asymptote is the line $x = -1$. Since the degree of the

    denominator is greater than the degree of the numerator the equation of the horizontal asymptote is $y = 0$.

    See Figure 7b.

    (c) $f(x) = \dfrac{x+2}{x^2-4} = \dfrac{(x+2)}{(x+2)(x-2)} \Rightarrow f(x)$ has a common factor of $(x+2)$ shows that $f(x)$ has a hole

    at $x = -2$. Since $x - 2 = 0 \Rightarrow x = 2$, the equation of the vertical asymptote is the line $x = 2$. Since

    the degree of the numerator is less than the degree of the denominator the equation of the horizontal

    asymptote is $y = 0$. See Figure 7c.

    (d) $f(x) = \dfrac{(x^2+1)}{x^2-1}$

    Since $x^2 - 1 = 0 \Rightarrow x = \pm 1$, the equations of the vertical asymptotes are the lines $x = \pm 1$. Since the

    degree of the numerator is equal to the degree of the denominator the equation of the horizontal asymptote

    is $y = 1$. See Figure 7d.

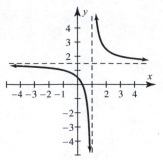

Figure 7a

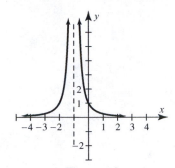

Figure 7b

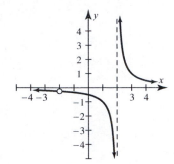

Figure 7c

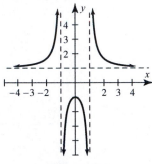

Figure 7d

4.7: More Equations and Inequalities

1. (a) $\dfrac{2x}{x + 2} = 6 \Rightarrow 2x = 6(x + 2) \Rightarrow 2x = 6x + 12 \Rightarrow 4x = -12 \Rightarrow x = -3$

    (b) Graph $Y_1 = (2X)/(X + 2)$ and $Y_2 = 6$ in $[-10, 10, 1]$ by $[-10, 10, 1]$ using dot mode. See Figure 1b.

    The intersection point is $(-3, 6)$. The solution is $x = -3$.

    (c) Table $Y_1 = (2X)/(X + 2)$ starting at $x = -5$ and incrementing by 1. See Figure 1c. The solution is

    $x = -3$.

$[-10, 10, 1]$ by $[-10, 10, 1]$

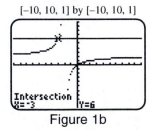

Figure 1b

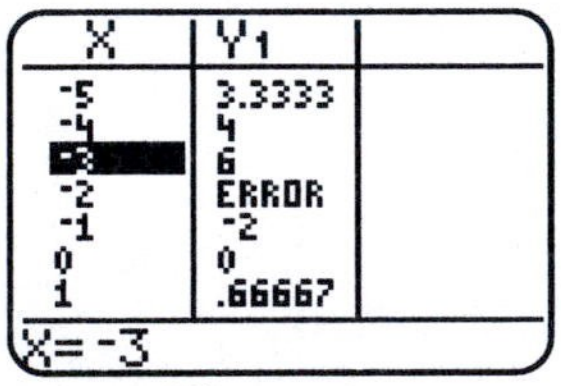

Figure 1c

3. (a) $2 - \dfrac{5}{x} + \dfrac{2}{x^2} = 0 \Rightarrow \dfrac{2x^2}{x^2} - \dfrac{5x}{x^2} + \dfrac{2}{x^2} = 0 \Rightarrow \dfrac{2x^2 - 5x + 2}{x^2} = 0 \Rightarrow 2x^2 - 5x + 2 = 0 \Rightarrow$

    $(2x - 1)(x - 2) = 0 \Rightarrow x = \dfrac{1}{2}, 2$

    (b) Graph $Y_1 = 2 - 5/X + 2/X^2$ in $[-10, 10, 1]$ by $[-10, 10, 1]$ using dot mode. The x-intercepts are the

    points is $(0.5, 0)$ and $(2, 0)$. The solutions are $x = \dfrac{1}{2}, 2$.

    (c) Table $Y_1 = 2 - 5/X + 2/X^2$ starting at $x = -0.5$ and incrementing by 0.5. The solutions are $x = \dfrac{1}{2}, 2$.

5. (a) $\dfrac{1}{x+1} + \dfrac{1}{x-1} = \dfrac{1}{x^2-1} \Rightarrow \dfrac{x-1}{x^2-1} + \dfrac{x+1}{x^2-1} = \dfrac{1}{x^2-1} \Rightarrow \dfrac{2x}{x^2-1} = \dfrac{1}{x^2-1} \Rightarrow 2x = 1 \Rightarrow$

$x = \dfrac{1}{2}$

(b) Graph $Y_1 = 1/(X+1) + 1/(X-1)$ and $Y_2 = 1/(X^2-1)$ in $[-2, 2, 1]$ by $[-5, 5, 1]$ using dot mode.

The intersection point is $\left(\dfrac{1}{2}, -\dfrac{4}{3}\right)$. The solution is $x = \dfrac{1}{2}$.

(c) Table $Y_1 = 1/(X+1) + 1/(X-1) - 1/(X^2-1)$ starting at $x = -1$ and incrementing by 0.5. Find the

x-value where $Y_1 = 0$. The solution is $x = \dfrac{1}{2}$.

7. $\dfrac{x+1}{x-5} = 0 \Rightarrow x+1 = 0(x-5) \Rightarrow x+1 = 0 \Rightarrow x = -1$; Check: $\dfrac{-1+1}{-1-5} = \dfrac{0}{-6} = 0$

9. $\dfrac{6(1-2x)}{x-5} = 4 \Rightarrow 6(1-2x) = 4(x-5) \Rightarrow 6 - 12x = 4x - 20 \Rightarrow -16x = -26 \Rightarrow x = \dfrac{13}{8}$

Check: $\dfrac{6(1 - 2(\frac{13}{8}))}{\frac{13}{8} - 5} = \dfrac{6(-\frac{9}{4})}{-\frac{27}{8}} = \dfrac{-\frac{27}{2}}{-\frac{27}{8}} = -\dfrac{27}{2} \cdot \left(-\dfrac{8}{27}\right) = 4$

11. $\dfrac{1}{x+2} + \dfrac{1}{x} = 1 \Rightarrow x + (x+2) = x(x+2) \Rightarrow 2x + 2 = x^2 + 2x \Rightarrow x^2 = 2 \Rightarrow x = \pm\sqrt{2}$

Check: $\dfrac{1}{\sqrt{2}+2} + \dfrac{1}{\sqrt{2}} = \dfrac{\sqrt{2}}{\sqrt{2}(\sqrt{2}+2)} + \dfrac{\sqrt{2}+2}{\sqrt{2}(\sqrt{2}+2)} = \dfrac{2\sqrt{2}+2}{2+2\sqrt{2}} = 1$

Check: $\dfrac{1}{-\sqrt{2}+2} + \dfrac{1}{\sqrt{2}} = \dfrac{-\sqrt{2}}{-\sqrt{2}(-\sqrt{2}+2)} + \dfrac{-\sqrt{2}+2}{-\sqrt{2}(-\sqrt{2}+2)} = \dfrac{2-2\sqrt{2}}{2-2\sqrt{2}} = 1$

13. $\dfrac{1}{x} - \dfrac{2}{x^2} = 5 \Rightarrow x - 2 = 5x^2 \Rightarrow 5x^2 - x + 2 = 0$; This quadratic equation has no solutions since the

discriminant is negative: $(-1)^2 - 4(5)(2) = 1 - 40 = -39 < 0$. No real solution.

15. $\dfrac{x^3 - 4x}{x^2 + 1} = 0 \Rightarrow x^3 - 4x = 0(x^2+1) \Rightarrow x^3 - 4x = 0 \Rightarrow x(x+2)(x-2) = 0 \Rightarrow x = 0, -2, 2$

Check: $\dfrac{(0)^3 - 4(0)}{(0)^2 + 1} = \dfrac{0}{1} = 0$; Check: $\dfrac{(-2)^3 - 4(-2)}{(-2)^2 + 1} = \dfrac{0}{5} = 0$; Check: $\dfrac{(2)^3 - 4(2)}{(2)^2 + 1} = \dfrac{0}{5} = 0$

17. $\dfrac{35}{x^2} = \dfrac{4}{x} + 15 \Rightarrow 35 = 4x + 15x^2 \Rightarrow 15x^2 + 4x - 35 = 0 \Rightarrow (5x-7)(3x+5) = 0 \Rightarrow x = \dfrac{-5}{3}, \dfrac{7}{5}$

Check: $\dfrac{35}{(\frac{-5}{3})^2} = \dfrac{4}{\frac{-5}{3}} + 15 \Rightarrow \dfrac{63}{5} = \dfrac{-12}{5} + \dfrac{75}{5} \Rightarrow \dfrac{63}{5} = \dfrac{63}{5}$;

Check: $\dfrac{35}{(\frac{7}{5})^2} = \dfrac{4}{\frac{7}{5}} + 15 \Rightarrow \dfrac{125}{7} = \dfrac{20}{7} + \dfrac{105}{7} \Rightarrow \dfrac{125}{7} = \dfrac{125}{7}$

19. $\dfrac{x+5}{x+2} = \dfrac{x-4}{x-10} \Rightarrow (x+5)(x-10) = (x-4)(x+2) \Rightarrow x^2 - 5x - 50 = x^2 - 2x - 8 \Rightarrow$

$3x + 42 = 0 \Rightarrow x = -14$; Check: $\dfrac{(-14)+5}{(-14)+2} = \dfrac{(-14)-4}{(-14)-10} \Rightarrow \dfrac{-9}{-12} = \dfrac{-18}{-24} \Rightarrow \dfrac{3}{4} = \dfrac{3}{4}$

21. $\dfrac{1}{x-2} - \dfrac{2}{x-3} = \dfrac{-1}{x^2 - 5x + 6} \Rightarrow \dfrac{1}{x-2} - \dfrac{2}{x-3} = \dfrac{-1}{(x-2)(x-3)} \Rightarrow$

    $x - 3 - 2(x-2) = -1 \Rightarrow -x + 1 = -1 \Rightarrow x = 2$; No solution. Check: There is no solution since

    $x = 2$ is not definrd in the original equation.

23. $\dfrac{2}{x-1} + 1 = \dfrac{4}{x^2 - 1} \Rightarrow \dfrac{2}{x-1} + 1 = \dfrac{4}{(x+1)(x-1)} \Rightarrow 2(x+1) + (x+1)(x-1) = 4 \Rightarrow$

    $2x + 2 + x^2 - 1 = 4 \Rightarrow x^2 + 2x - 3 = 0 \Rightarrow (x+3)(x-1) = 0 \Rightarrow x = -3, 1$. Since $x = 1$ is not

    defined in the original equation $x = -3$. Check: $\dfrac{2}{-3-1} + 1 = \dfrac{4}{(-3)^2 - 1} \Rightarrow -\dfrac{1}{2} + 1 = \dfrac{4}{8} \Rightarrow \dfrac{1}{2} = \dfrac{1}{2}$

25. $\dfrac{1}{x+2} = \dfrac{4}{4-x^2} - 1 \Rightarrow \dfrac{1}{x+2} = \dfrac{4}{(2+x)(2-x)} - 1 \Rightarrow 2 - x = 4 - (2+x)(2-x) \Rightarrow$

    $2 - x = 4 - (4 - x^2) \Rightarrow 2 - x = 4 - 4 + x^2 \Rightarrow x^2 + x - 2 = 0 \Rightarrow (x+2)(x-1) = 0 \Rightarrow$

    $x = -2, 1$. Since $x = -2$ is not defined in the original equation $x = 1$.

    Check: $\dfrac{1}{1+2} = \dfrac{4}{4-(1)^2} - 1 \Rightarrow \dfrac{1}{3} = \dfrac{4}{3} - 1 \Rightarrow \dfrac{1}{3} = \dfrac{1}{3}$

27. $\dfrac{1}{x-1} + \dfrac{1}{x+1} = \dfrac{2}{x^2 - 1} \Rightarrow \dfrac{1}{x-1} + \dfrac{1}{x+1} = \dfrac{2}{(x+1)(x-1)} \Rightarrow x + 1 + x - 1 = 2 \Rightarrow$

    $2x = 2 \Rightarrow x = 1$. Since $x = 1$ is not defined in the original equation, there is no solution to the equation.

29. (a) The boundary numbers, the x-values for which $f(x) = 0$, are $-4, -2$, and 2.

    (b) $f(x) > 0$ on the interval for which the graph of f is above the x-axis. That is $(-4, -2) \cup (2, \infty)$.

    In set builder notation the intervals are $\{x \mid -4 < x < -2 \text{ or } x > 2\}$.

    (c) $f(x) < 0$ on the interval for which the graph of f is below the x-axis. That is $(-\infty, -4) \cup (-2, 2)$.

    In set builder notation the intervals are $\{x \mid x < -4 \text{ or } -2 < x < 2\}$.

31. (a) The boundary numbers, the x-values for which $f(x) = 0$, are $-4, -2, 0$, and 2.

    (b) $f(x) > 0$ on the interval for which the graph of f is above the x-axis. That is $(-4, -2) \cup (0, 2)$.

    In set builder notation the intervals are $\{x \mid -4 < x < -2 \text{ or } 0 < x < 2\}$.

    (c) $f(x) < 0$ on the interval for which the graph of f is below the x-axis. That is $(-\infty, -4) \cup (-2, 0) \cup (2, \infty)$.

    In set builder notation the intervals are $\{x \mid x < -4 \text{ or } -2 < x < 0 \text{ or } x > 2\}$.

33. (a) The boundary numbers, the x-values for which $f(x) = 0$, are $-2, 1$, and 2.

    (b) $f(x) > 0$ on the interval for which the graph of f is above the x-axis. That is $(-\infty, -2) \cup (-2, 1)$.

    In set builder notation the intervals are $\{x \mid x < -2 \text{ or } -2 < x < 1\}$.

    (c) $f(x) < 0$ on the interval for which the graph of f is below the x-axis. That is $(1, 2) \cup (2, \infty)$.

    In set builder notation the intervals are $\{x \mid 1 < x < 2 \text{ or } x > 2\}$.

35. (a) $f(x)$ is undefined at $x = 0$.

    (b) $f(x) > 0$ on the interval for which the graph of f is above the x-axis. That is $(-\infty, 0) \cup (0, \infty)$.

    In set builder notation the intervals are $\{x \mid x < 0 \text{ or } x > 0\}$.

    (c) $f(x) < 0$ on the interval for which the graph of f is below the x-axis. There are no solutions.

37. (a) $f(x)$ is undefined at $x = 1$, $f(x) = 0$ at $x = 0$.

(b) $f(x) > 0$ on the interval for which the graph of f is above the x-axis. That is $(-\infty, 0)\cup(1, \infty)$.

In set builder notation the intervals are $\{x \mid x < 0 \text{ or } x > 1\}$.

(c) $f(x) < 0$ on the interval for which the graph of f is below the x-axis. That is $(0, 1)$.

In set builder notation the interval is $\{x \mid 0 < x < 1\}$.

39. (a) $f(x)$ is undefined at $x = \pm 2$, $f(x) = 0$ at $x = 0$.

(b) $f(x) > 0$ on the interval for which the graph of f is above the x-axis. That is $(-\infty, -2)\cup(2, \infty)$.

In set builder notation the intervals are $\{x \mid x < -2 \text{ or } x > 2\}$.

(c) $f(x) < 0$ on the interval for which the graph of f is below the x-axis. That is $(-2, 0)\cup(0, 2)$.

In set builder notation the intervals are $\{x \mid -2 < x < 0 \text{ or } 0 < x < 2\}$.

41. (a) First find the boundary numbers by solving $x^3 - x = 0$.

$$x^3 - x = 0 \Rightarrow x(x^2 - 1) = 0 \Rightarrow x(x + 1)(x - 1) = 0 \Rightarrow x = 0, -1, \text{ or } 1$$

The boundary values divide the number line into four intervals. Choose a test value from each of these

intervals and evaluate $x^3 - x$ for these values. See Figure 41a. The solution is $(-1, 0)\cup(1, \infty)$.

In set builder notation the intervals are $\{x \mid -1 < x < 0 \text{ or } x > 1\}$.

(b) Graph $Y_1 = X^{\wedge}3 - X$ as shown in Figure 41b. The x-intercepts are -1, 0, and 1. The graph is above the

x-axis on the interval $(-1, 0)\cup(1, \infty)$.

Interval	Test Value x	$x^3 - x$	Positive or Negative?
$(-\infty, -1)$	-2	-6	Negative
$(-1, 0)$	-0.5	0.375	Positive
$(0, 1)$	0.5	-0.375	Negative
$(1, \infty)$	2	6	Positive

Figure 41a

[-3, 3, 0.5] by [-3, 3, 0.5]

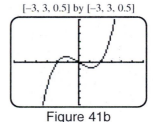

Figure 41b

43. (a) Write the inequality as $x^3 + x^2 - 2x \geq 0$. Then find the boundary numbers by solving $x^3 + x^2 - 2x = 0$.

$$x^3 + x^2 - 2x = 0 \Rightarrow x(x^2 + x - 2) = 0 \Rightarrow x(x + 2)(x - 1) = 0 \Rightarrow x = 0, -2, \text{ or } 1$$

The boundary values divide the number line into four intervals. Choose a test value from each of these

intervals and evaluate $x^3 + x^2 - 2x$ for these values. See Figure 43a. The solution is $[-2, 0]\cup[1, \infty)$.

In set builder notation the intervals are $\{x \mid -2 \leq x \leq 0 \text{ or } x \geq 1\}$.

(b) Graph $Y_1 = X^{\wedge}3 + X^{\wedge}2 - 2X$ as shown in Figure 43b. The x-intercepts are -1, 0, and 1. The graph

intersects or is above the x-axis on the interval $[-2, 0]\cup[1, \infty)$.

Interval	Test Value x	$x^3 + x^2 - 2x$	Positive or Negative?
$(-\infty, -2)$	-3	-12	Negative
$(-2, 0)$	-1	2	Positive
$(0, 1)$	0.5	-0.625	Negative
$(1, \infty)$	2	8	Positive

Figure 43a

[−3, 3, 1] by [−5, 5, 1]

Figure 43b

45. (a) First find the boundary numbers by solving $x^4 - 13x^2 + 36 = 0$.

$$x^4 - 13x^2 + 36 = 0 \Rightarrow (x^2 - 4)(x^2 - 9) = 0 \Rightarrow (x + 2)(x - 2)(x + 3)(x - 3) \Rightarrow$$

$x = -3, -2, 2$ or 3 The boundary values divide the number line into five intervals. Choose a test value

from each of these intervals and evaluate $x^4 - 13x^2 + 36$ for these values. See Figure 45a. The solution is

$(-3, -2)\cup(2, 3)$. In set builder notation the intervals are $\{x\,|\,-3 < x < -2 \text{ or } 2 < x < 3\}$.

(b) Graph $Y_1 = X^4 - 13X^2 + 36$ as shown in Figure 45b. The x-intercepts are –3, –2, 2 and 3. The graph

is below the x-axis on the interval $(-3, -2)\cup(2, 3)$.

Interval	Test Value x	$x^4 - 13x^2 + 36$	Positive or Negative?
$(-\infty, -3)$	-4	84	Positive
$(-3, -2)$	-2.5	-6.1875	Negative
$(-2, 2)$	0	36	Positive
$(2, 3)$	2.5	-6.1875	Negative
$(3, \infty)$	4	84	Positive

Figure 45a

[−5, 5, 1] by [−10, 100, 10]

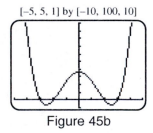

Figure 45b

47. Write the inequality as $7x^4 - 14x^2 \geq 0$. Then find the boundary numbers by solving $7x^4 - 14x^2 = 0$.
$7x^4 - 14x^2 = 0 \Rightarrow 7x^2(x^2 - 2) = 0 \Rightarrow x = 0$ or $\pm\sqrt{2}$. The boundary values divide the number line into
four intervals. Choose a test value from each of these intervals and evaluate $7x^4 - 14x^2$ for these values. See
Figure 47. The solution is $(-\infty, -\sqrt{2})\cup(\sqrt{2}, \infty)$. In set builder notation the intervals are
$\{x\,|\,x < -\sqrt{2} \text{ or } x > \sqrt{2}\}$.

Interval	Test Value x	$7x^4 - 14x^2$	Positive or Negative?
$(-\infty, -1\overline{2})$	-2	56	Positive
$(-1\overline{2}, 0)$	-1	-7	Negative
$(0, 1\overline{2})$	1	-7	Negative
$(1\overline{2}, \infty)$	2	56	Positive

Figure 47

49. First find the boundary numbers by solving $(x - 1)(x - 2)(x + 2) = 0$.

    $(x - 1)(x - 2)(x + 2) = 0 \Rightarrow x = -2, 1 \text{ or } 2$

    The boundary values divide the number line into four intervals. Choose a test value from each of these intervals and evaluate $(x - 1)(x - 2)(x + 2)$ for these values. See Figure 49. The solution is $[-2, 1] \cup [2, \infty)$.

    In set builder notation the intervals are $\{x \mid 2 \le x \le 1 \text{ or } x \ge 2\}$.

Interval	Test Value x	$(x - 1)(x - 2)(x + 2)$	Positive or Negative?
$(-\infty, -2)$	-3	-20	Negative
$(-2, 1)$	0	4	Positive
$(1, 2)$	1.5	-0.875	Negative
$(2, \infty)$	3	10	Positive

Figure 49

51. Write the inequality as $2x^4 + 2x^3 - 12x^2 \le 0$. Find the boundary numbers by solving $2x^4 + 2x^3 - 12x^2 = 0$.

    $2x^4 + 2x^3 - 12x^2 = 0 \Rightarrow 2x^2(x^2 + x - 6) = 0 \Rightarrow 2x^2(x + 3)(x - 2) = 0 \Rightarrow x = -3, 0 \text{ or } 2$

    The boundary values divide the number line into four intervals. Choose a test value from each of these intervals and evaluate $2x^4 + 2x^3 - 12x^2$ for these values. See Figure 51. The solution is $[-3, 2]$.

    In set builder notation the interval is $\{x \mid -3 \le x \le 2\}$.

Interval	Test Value x	$2x^4 + 2x^3 - 12x^2$	Positive or Negative?
$(-\infty, -3)$	-4	192	Positive
$(-3, 0)$	-1	-12	Negative
$(0, 2)$	1	-8	Negative
$(2, \infty)$	3	108	Positive

Figure 51

53. Write the inequality as $x^3 - 7x^2 + 14x - 8 \le 0$.

    Then Graph $Y_1 = X^3 - 7X^2 + 14X - 8$ as shown in Figure 53. The x-intercepts are 1, 2, and 4.

    The graph intersects or is below the x-axis on the interval $(-\infty, 1] \cup [2, 4]$. In set builder notation the intervals are $\{x \mid x \le 1 \text{ or } 2 \le x \le 4\}$.

[-2, 5, 1] by [-10, 10, 1]   [-4, 4, 1] by [-5, 5, 1]

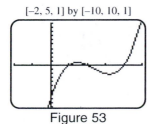
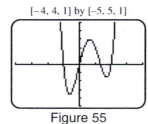

Figure 53   **Figure 55**

55. Graph $Y_1 = 3X^4 - 7X^3 - 2X^2 + 8X$ as shown in Figure 55. The *x*-intercepts are $-1, 0, \dfrac{4}{3}$ and 2.

The graph is above the *x*-axis on the interval $(-\infty, -1) \cup \left(0, \dfrac{4}{3}\right) \cup (2, \infty)$. In set builder notation the

intervals are $\left\{ x \mid x < -1 \text{ or } 0 < x < \dfrac{4}{3} \text{ or } x > 2 \right\}$.

57. (a) Find the zeros of the numerator and the denominator.

Numerator: No zero;   Denominator: $x = 0$

This boundary number divides the number line into two intervals. Choose a test value from each of these

intervals and evaluate $\dfrac{1}{x}$ for these values. See Figure 57a. The solution is $(-\infty, 0)$. In set builder notation

the interval is $\{x \mid x < 0\}$.

(b) Graph $Y_1 = 1/X$ as shown in Figure 57b. The graph has a vertical asymptote at $x = 0$ and no *x*-intercepts.

The graph is below the *x*-axis on the interval $(-\infty, 0)$.

$[-4, 4, 1]$ by $[-2, 2, 0.5]$

Interval	Test Value x	$1/x$	Positive or Negative?
$(-\infty, 0)$	-1	-1	Negative
$(0, \infty)$	1	1	Positive

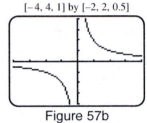

Figure 57a   Figure 57b

59. (a) Find the zeros of the numerator and the denominator.

Numerator: No zero;   Denominator: $x + 3 = 0 \Rightarrow x = -3$

This boundary number divides the number line into two intervals. Choose a test value from each of these

intervals and evaluate $\dfrac{4}{x + 3}$ for these values. See Figure 59a. The solution is $(-3, \infty)$. In set builder

notation the interval is $\{x \mid x > -3\}$.

(b) Graph $Y_1 = 4/(X + 3)$ using dot mode as shown in Figure 59b. The graph has a vertical asymptote at

$x = -3$ and no *x*-intercepts. The graph is above the *x*-axis on the interval $(-3, \infty)$. Note that the value

$x = -3$ can not be included in the solution set since $\dfrac{4}{x + 3}$ is undefined for $x = -3$.

$[-6, 1, 1]$ by $[-4, 4, 1]$

Interval	Test Value x	$4/(x + 3)$	Positive or Negative?
$(-\infty, -3)$	-4	-4	Negative
$(-3, \infty)$	1	1	Positive

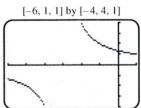

Figure 59a   Figure 59b

61. (a) Find the zeros of the numerator and the denominator.

    Numerator: No zero; Denominator: $x^2 - 4 = 0 \Rightarrow x = \pm 2$

    These boundary numbers divide the number line into three intervals. Choose a test value from each of these

    intervals and evaluate $\dfrac{5}{x^2 - 4}$ for these values. See Figure 61a. The solution is $(-2, 2)$. In set builder

    notation the interval is $\{x \mid -2 < x < 2\}$.

    (b) Graph $Y_1 = 5/(X^2 - 4)$ using dot mode as shown in Figure 61b. The graph has vertical asymptotes

    at $x = \pm 2$ and no x-intercepts. The graph is below the x-axis on the interval $(-2, 2)$.

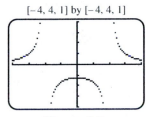

$[-4, 4, 1]$ by $[-4, 4, 1]$

Interval	Test Value x	$5/(x^2 - 4)$	Positive or Negative?
$(-\infty, -2)$	-3	1	Positive
$(-2, 2)$	0	-1.25	Negative
$(2, \infty)$	3	1	Positive

Figure 61a   Figure 61b

63. The graph of $Y_1 = (X + 1)^2/(X - 2)$ has a vertical asymptote at $x = 2$ and an x-intercept at $x = -1$.

    The graph of Y_1 intersects or is below the x-axis on the interval $(-\infty, 2)$. Note that 2 can not be included.

    In set builder notation the interval is $\{x \mid x > 2\}$.

65. The graph of $Y_1 = (3 - 2X)/(1 + X)$ has a vertical asymptote at $x = -1$ and an x-intercept at $x = \dfrac{3}{2}$.

    The graph of Y_1 is below the x-axis on the interval $(-\infty, -1) \cup \left(\dfrac{3}{2}, \infty\right)$.

    In set builder notation the intervals are $\left\{x \mid x < -1 \text{ or } x > \dfrac{3}{2}\right\}$.

67. The graph of $Y_1 = (X + 1)(X - 2)/(X + 3)$ has a vertical asymptote of at $x = -3$ and x-intercepts at

    $x = -1$ and $x = 2$. The graph of Y_1 is below the x-axis on the interval $(-\infty, -3) \cup (-1, 2)$.

    In set builder notation the intervals are $\{x \mid x < -3 \text{ or } -1 < x < 2\}$.

69. The graph of $Y_1 = 2X - 5/((X + 1)(X - 1))$ has vertical asymptotes at $x = -1$ and $x = 1$ and an x-intercept

    at $x = \dfrac{5}{2}$. The graph of Y_1 is on or above the x-axis on the interval $(-1, 1) \cup \left[\dfrac{5}{2}, \infty\right)$.

    In set builder notation the intervals are $\left\{x \mid -1 < x < 1 \text{ or } x \geq \dfrac{5}{2}\right\}$.

71. First, rewrite the inequality:

    $$\frac{1}{x - 3} \leq \frac{5}{x - 3} \Rightarrow \frac{1}{x - 3} - \frac{5}{x - 3} \leq 0 \Rightarrow \frac{-4}{x - 3} \leq 0$$

    The graph of $Y_1 = -4/(X - 3)$ has a vertical asymptote at $x = 3$ and no x-intercept.

The graph of Y_1 intersects or is below the x-axis on the interval $(3, \infty)$. Note that 3 can not be included.

In set builder notation the interval is $\{x \mid x > 3\}$.

73. First, rewrite the inequality:

$$2 - \frac{5}{x} + \frac{2}{x^2} \geq 0 \Rightarrow \frac{2x^2 - 5x + 2}{x^2} \geq 0 \Rightarrow \frac{(2x - 1)(x - 2)}{x^2} \geq 0$$

The graph of $Y_1 = (2X - 1)(X - 2)/X^2$ has a vertical asymptote at $x = 0$ and x-intercepts

$x = \dfrac{1}{2}$ and $x = 2$. The graph of Y_1 intersects or is above the x-axis on the interval $(-\infty, 0) \cup \left(0, \dfrac{1}{2}\right] \cup [2, \infty)$.

In set builder notation the intervals are $\left\{ x \mid x < 2 \text{ or } 0 < x \leq \dfrac{1}{2} \text{ or } x \geq 2 \right\}$.

75. First, rewrite the inequality:

$$\frac{1}{x} \leq \frac{2}{x + 2} \Rightarrow \frac{1}{x} - \frac{2}{x + 2} \leq 0 \Rightarrow \frac{x + 2 - 2x}{x(x + 2)} \leq 0 \Rightarrow \frac{2 - x}{x(x + 2)} \leq 0. \text{ The graph of}$$

$Y_1 = (2 - X)/((X(+ 2))$ has vertical asymptotes at $x = 0$ and $x = -2$, and x-intercept of $x = 2$. The graph of Y_1 is below the x-axis on the interval $(-2, 0) \cup [2, \infty)$. In set builder notation the intervals are

$\left\{ x \mid -2 < x < 0 \text{ or } x \geq 2 \right\}$.

77. (a) Since 15 seconds equals 0.25 minute, graph $Y_1 = (X - 5)/(X^2 - 10X)$ and $Y_2 = 0.25$ as shown in

Figure 77. The intersection point is near $(12.4, 0.25)$. The gate should admit 12.4 cars per minute on

average to keep the wait less than 15 seconds. Note: The reason the answer is greater than 10 cars per

minute is because cars are arriving randomly. For some minutes, more than 10 vehicles might arrive at

the gate.

(b) $\dfrac{12.4}{5} = 2.48$ or 3 parking attendants must be on duty to keep the average wait less than 15 seconds.

[10, 15, 1] by [0, 1, 0.1]

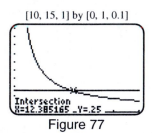

Figure 77

79. Let $y =$ height, $x =$ width, and $2x =$ length of the box. To relate the variables, we use the volume formula for

a box and the fact that $V = 196$ cubic inches. $V = (2x)xy = 2x^2 y \Rightarrow y = \dfrac{V}{2x^2} = \dfrac{196}{2x^2} = \dfrac{98}{x^2}$.

The surface area of the box is the sum of the areas of the 6 rectangular sides. We are given that $A = 280 \text{ in}^2$.

$A = 2(x \cdot y) + 2(2x \cdot y) + 2(x \cdot 2x) \Rightarrow A = 6xy + 4x^2 = 280$; and since $y = \dfrac{98}{x^2}$ we get the equation

$6x\left(\dfrac{98}{x^2}\right) + 4x^2 = 280 \Rightarrow \dfrac{588}{x} + 4x^2 = 280$.

Figures 79a and 79b show the intersection points when graphing $Y_1 = 588/X + 4X^2$ and $Y_2 = 280$

There are two possible solutions (in inches):

width $= 7$, length $= 14$, height $= \dfrac{98}{7^2} = 2$ or width ≈ 2.266, length ≈ 4.532, height $\approx \dfrac{98}{2.266^2} \approx 19.086.$

[0, 10, 2] by [100, 400, 20]   [0, 10, 2] by [100, 400, 20]

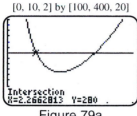   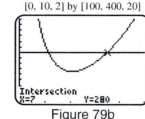

Figure 79a   Figure 79b

81. (a) $V = L \cdot W \cdot H \Rightarrow V = x \cdot x \cdot h \Rightarrow 108 = x^2 h \Rightarrow \dfrac{108}{x^2} = h.$ The surface area $A(x) = x^2 + 4xh \Rightarrow$

$A(x) = x^2 + 4x\left(\dfrac{108}{x^2}\right) \Rightarrow A(x) = \left(x^2 + \dfrac{432}{x}\right).$ This finds surface area in square inches. To find

square feet we must divide by 144 (the number of square inches in a square foot) $\Rightarrow$

$A(x) = \left(x^2 + \dfrac{432}{x}\right) \div 144 \Rightarrow A(x) = \dfrac{x^2}{144} + \dfrac{3}{x}.$

(b) $C = 0.10\left(\dfrac{x^2}{144} + \dfrac{3}{x}\right)$

(c) Graph the equation: $C = 0.10\left(\dfrac{x^2}{144} + \dfrac{3}{x}\right).$ The minimum cost is when $x \approx 6 \Rightarrow h = \dfrac{108}{(6)^2} \Rightarrow h = 3.$

The dimensions are $6 \times 6 \times 3$ inches.

83. (a) $D(0.05) = \dfrac{2500}{30(0.3 + 0.05)} \approx 238$

The braking distance far a car traveling at 50 mph on a 5% uphill grade is about 238 feet.

(b) Table $Y_1 = 2500/(30(0.3 + X))$ starting at $x = 0$ and incrementing by 0.05 as shown in Figure 83.

The table indicates that as the grade x increases, the braking distance decreases. This agrees with driving experience.

(c) $220 = \dfrac{2500}{30(0.3 + x)} \Rightarrow 0.3 + x = \dfrac{2500}{30(220)} \Rightarrow x = \dfrac{2500}{30(220)} - 0.3 \approx 0.079,$ or 7.9%

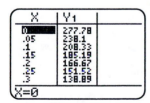

Figure 83

85. (a) Graph $Y_1 = X^2/(1600 - 40X)$ and $Y_2 = 8$(not shown). The graphs intersect approximately at (36, 8).

The graph of Y_1 is equal to or is below the graph of Y_2 when $x \le 36$ (approximately).

(b) The average line length is less than or equal to 8 cars when the average arrival rate is 36 cars per hour or less.

87. (a) The graph of $D(x) = \dfrac{120}{x}$ increases as x decreases, which means the braking distance increases as the

coefficient of friction becomes smaller.

(b) $\dfrac{120}{x} \geq 400 \Rightarrow \dfrac{120}{400} \geq x \Rightarrow 0.3 \geq x$; the braking distance is 400 feet or more when $0 < x \leq 0.3$.

89. The volume of a cube is $V = x^3$, where x is the length of a side.

$212.8 \leq V \leq 213.2 \Rightarrow 212.8 \leq x^3 \leq 213.2 \Rightarrow \sqrt[3]{212.8} \leq x \leq \sqrt[3]{213.2}$ inches.

This is approximately $5.97022 \leq x \leq 5.97396$ inches.

91. If $y = \dfrac{k}{x}$ and $y = 2$ when $x = 3$, then $2 = \dfrac{k}{3} \Rightarrow k = 6$.

93. If $y = kx^3$ and $y = 64$ when $x = 2$, then $64 = k(8) \Rightarrow 8k = 64 \Rightarrow k = 8$.

95. If $T = kx^{3/2}$ and $T = 20$ when $x = 4$, then $20 = k(4)^{3/2} \Rightarrow 20 = 8k \Rightarrow k = 2.5$.

The variation equation becomes $T = 2.5x^{3/2}$. When $x = 16$, $T = 2.5(16)^{3/2} = 2.5(64) = 160$.

97. If $y = \dfrac{k}{x}$ and $y = 5$ when $x = 6$, then $5 = \dfrac{k}{6} \Rightarrow k = 30$.

The variation equation becomes $y = \dfrac{30}{x}$. When $x = 15$, $y = \dfrac{30}{15} = 2$.

99. If y is inversely proportional to x, then $y = \dfrac{k}{x}$ must hold. So if the value of x is doubled, the right side of this

equation becomes $\dfrac{k}{2x} = \dfrac{1}{2} \cdot \dfrac{k}{x} = \dfrac{1}{2}y$. Therefore y becomes half its original value.

101. If y is directly proportional to x^3, then $y = kx^3$ must hold. So if the value of x is tripled, the right side of this

equation becomes $k(3x)^3 = 27 \cdot kx^3 = 27y$. Therefore y becomes 27 times its original value.

103. Since $y = kx^n$, we know that $k = \dfrac{y}{x^n}$ where k is a constant. Using trial and error, let $n = 1$ while calculating

various values for k using x- and y-values from the table. For example, when $x = 2$ and $y = 2$, the value of k is

1. But for $x = 4$ and $y = 8$, the value of k is 2. Since the value of k did not remain constant we know that the

value of n is not 1. Repeat this process for $n = 2$. For each x and y pair in the table, the value of k is 0.5.

Therefore $k = 0.5$ and $n = 2$.

105. Since $y = \dfrac{k}{x^n}$, we know that $k = yx^n$ where k is a constant. Using trial and error, let $n = 1$ while calculating

various values for k using x- and y-values from the table. For example, when $x = 2$ and $y = 1.5$, the value of

k is 3. And for $x = 3$ and $y = 1$, the value of k is also 3. Since the value of k remained constant we know

that the value of n is 1. For each x and y pair in the table, the value of k is 3. Therefore $k = 3$ and $n = 1$.

107. If y is directly proportional to $x^{1.25}$, then $y = kx^{1.25}$ must hold. Given that $y = 1.9$ when $x = 1.1$, we may

calculate the value of $k = \dfrac{y}{x^{1.25}} = \dfrac{1.9}{1.1^{1.25}} \approx 1.69$. The variation equation can be written as $y \approx 1.69x^{1.25}$.

When a fiddler crab has claws weighing 0.75 grams, its body weight will be $y \approx 1.69(0.75)^{1.25} \approx 1.18$ grams.

109. Let I = the intensity of the light, and let d = the distance from a star to the earth. If I is inversely

proportional to d^2, then $I = \dfrac{k}{d^2}$ must hold. Suppose a ground-based telescope can see a star with intensity I_g.

Then the Hubble Telescope can see stars of intensity $\dfrac{1}{50} \cdot I_g = \dfrac{1}{50} \cdot \dfrac{k}{d^2} = \dfrac{k}{(d\sqrt{50})^2}$. That is, the Hubble

Telescope can see $\sqrt{50} \approx 7$ times as far as ground-based telescopes.

111. Since the resistance varies inversely as the square of the diameter of the wire, increasing the diameter of the

wire by a factor of 1.5 will decrease the resistance by a factor of $\dfrac{1}{1.5^2} = \dfrac{4}{9}$. Hence, a 25 foot wire with a

diameter of 3 millimeters (1.5 times 2 millimeters) will have a resistance of $\dfrac{4}{9}$ (0.5 ohm) $= \dfrac{2}{9}$ ohm.

113. In this exercise $F = \dfrac{K\sqrt{T}}{L}$. If both T and L are doubled,

$$\dfrac{K\sqrt{2T}}{2L} = \dfrac{\sqrt{2}}{2} \cdot \dfrac{K\sqrt{T}}{L} = \dfrac{\sqrt{2}}{2} F. \text{ Therefore } F$$

decreases by a factor of $\dfrac{\sqrt{2}}{2}$.

4.8: Radical Equations and Power Functions

1. $8^{2/3} = (8^{1/3})^2 = 2^2 = 4$

3. $16^{-3/4} = (16^{1/4})^{-3} = (2)^{-3} = \dfrac{1}{2^3} = \dfrac{1}{8}$

5. $-81^{0.5} = -81^{1/2} = -\sqrt{81} = -9$

7. $(9^{3/4})^2 = 9^{3/2} = (\sqrt{9})^3 = 3^3 = 27$

9. $\dfrac{8^{5/6}}{8^{1/2}} = 8^{5/6-1/2} = 8^{1/3} = 2$

11. $27^{5/6} \cdot 27^{-1/6} = 27^{5/6+(-1/6)} = 27^{2/3} = \sqrt[3]{27^2} = 3^2 = 9$

13. $(-27)^{-5/3} = \dfrac{1}{(-27)^{5/3}} = \dfrac{1}{\sqrt[3]{(-27)^5}} = \dfrac{1}{(-3)^5} = -\dfrac{1}{243}$

15. $(0.5^{-2})^2 = \left(\dfrac{1}{2}\right)^{-4} = 2^4 = 16$

17. $\left(\dfrac{2}{3}\right)^{-2} = \left(\dfrac{3}{2}\right)^2 = \dfrac{9}{4}$

19. $\sqrt{2x} = (2x)^{1/2}$

21. $\sqrt[3]{z^5} = z^{5/3}$

23. $(\sqrt[4]{y})^{-3} = (y^{1/4})^{-3} = y^{-3/4} = \dfrac{1}{y^{3/4}}$

25. $\sqrt{x} \cdot \sqrt[3]{x} = x^{1/2} \cdot x^{1/3} = x^{1/2+1/3} = x^{5/6}$

27. $\sqrt{y \cdot \sqrt{y}} = (y \cdot y^{1/2})^{1/2} = (y^{3/2})^{1/2} = y^{3/4}$

29. $a^{-3/4}b^{1/2} = \dfrac{b^{1/2}}{a^{3/4}} = \dfrac{\sqrt{b}}{\sqrt[4]{a^3}}$

31. $(a^{1/2} + b^{1/2})^{1/2} = \sqrt{\sqrt{a} + \sqrt{b}}$

33. $f(32) = \sqrt[3]{2(32)} = \sqrt[3]{64} = 4$

35. $f(4) = 2\sqrt{(4)^5} = 2\sqrt{1024} = 2(32) = 64$

37. $f(125) = \sqrt[4]{5(125)} = \sqrt[4]{625} = 5$

39. $f(-1) = \sqrt[5]{32(-1)} = \sqrt[5]{-32} = -2$

41. Reflect the graph of $y = \sqrt[3]{x}$ through the y-axis to get the graph of $f(x) = \sqrt[3]{-x}$. See Figure 41.

43. Shift the graph of $y = \sqrt[3]{x}$ right 1 unit and up 1 unit to get the graph of $f(x) = \sqrt[3]{x-1} + 1$. See Figure 43.

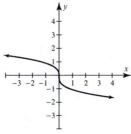

Figure 41

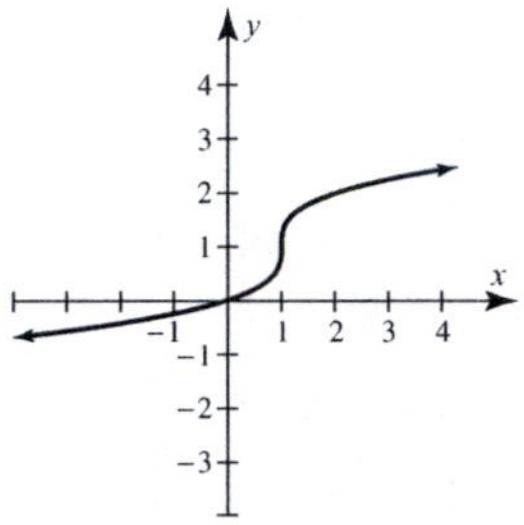

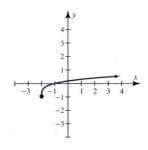

Figure 43

Figure 45

45. Shift the graph of $y = \sqrt[4]{x}$ left 2 units and down 1 unit to get the graph of $f(x) = \sqrt[4]{x+2} - 1$. See Figure 45.

47. Shift the graph of $y = \sqrt[4]{x}$ right 1 unit and vertically stretch by a multiple of 2 units to get the graph of $f(x) = 2\sqrt[4]{x-1}$. See Figure 47.

49. Shift the graph of $y = \sqrt{x}$ left 3 units and up 2 units to get the graph of $f(x) = \sqrt{x+3} + 2$. See Figure 49.

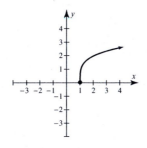

Figure 47

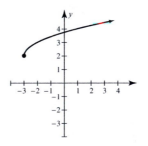

Figure 49

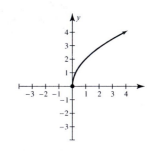

Figure 51

51. Vertically stretch the graph of $y = \sqrt{x}$ by a multiple of 2 units to get the graph of $f(x) = 2\sqrt{x}$. See Figure 51.

53. $\sqrt{x+2} = x - 4 \Rightarrow x + 2 = x^2 - 8x + 16 \Rightarrow x^2 - 9x + 14 = 0 \Rightarrow (x-2)(x-7) = 0 \Rightarrow$
    $x = 2$ or 7.

Check: $\sqrt{2+2} = 2 \neq 2 - 4$ (not a solution); $\sqrt{7+2} = 3 = 7 - 4$. The only solution is $x = 7$.

55. $\sqrt{3x+7} = 3x+5 \Rightarrow 3x+7 = 9x^2 + 30x + 25 \Rightarrow 9x^2 + 27x + 18 = 0 \Rightarrow 9(x^2 + 3x + 2) = 0 \Rightarrow$

    $9(x+2)(x+1) = 0 \Rightarrow x = -2 \text{ or } -1$

    Check: $\sqrt{3(-2)+7} \neq 3(-2)+5$ (not a solution); $\sqrt{3(-1)+7} = 2 = 3(-1)+5$. The only solution is

    $x = -1$.

57. $\sqrt{5x-6} = x \Rightarrow 5x-6 = x^2 \Rightarrow x^2 - 5x + 6 = 0 \Rightarrow (x-2)(x-3) = 0 \Rightarrow x = 2 \text{ or } 3$

    Check: $\sqrt{5(2)-6} = 2$; $\sqrt{5(3)-6} = 3$

59. $\sqrt{x+5} + 1 = x \Rightarrow \sqrt{x+5} = x-1 \Rightarrow x+5 = (x-1)^2 \Rightarrow x+5 = x^2 - 2x + 1 \Rightarrow$

    $x^2 - 3x - 4 = 0 \Rightarrow (x-4)(x+1) = 0 \Rightarrow x = -1 \text{ or } x = 4$

    Check: $\sqrt{-1+5} + 1 = -1 \Rightarrow \sqrt{4} + 1 = -1 \Rightarrow 3 \neq -1$ (not a solution).

    Check: $\sqrt{4+5} + 1 = 4 \Rightarrow \sqrt{9} + 1 = 4 \Rightarrow 4 = 4$. The only solution is $x = 4$.

61. $\sqrt{x+1} + 3 = \sqrt{3x+4} \Rightarrow (\sqrt{x+1} + 3)^2 = 3x + 4 \Rightarrow x + 1 + 6\sqrt{x+1} + 9 = 3x + 4 \Rightarrow$

    $6\sqrt{x+1} = 2x - 6 \Rightarrow 3\sqrt{x+1} = x - 3 \Rightarrow (3\sqrt{x+1})^2 = (x-3)^2 \Rightarrow$

    $9(x+1) = x^2 - 6x + 9 \Rightarrow x^2 - 15x = 0 \Rightarrow x(x-15) = 0 \Rightarrow x = 0 \text{ or } 15$

    Check: $\sqrt{(0)+1} + 3 \neq \sqrt{3(0)+4}$ (not a solution); $\sqrt{(15)+1} + 3 = 7 = \sqrt{3(15)+4}$.

    The only solution is $x = 15$.

63. $\sqrt{2x} - \sqrt{x+1} = 1 \Rightarrow \sqrt{2x} = 1 + \sqrt{x+1} \Rightarrow 2x = (1 + \sqrt{x+1})^2 \Rightarrow$

    $2x = 1 + 2\sqrt{x+1} + x + 1 \Rightarrow x - 2 = 2\sqrt{x+1} \Rightarrow x^2 - 4x + 4 = 4(x+1) \Rightarrow$

    $x^2 - 4x + 4 = 4x + 4 \Rightarrow x^2 - 8x = 0 \Rightarrow x(x-8) = 0 \Rightarrow x = 0 \text{ or } 8.$

    Check: $\sqrt{2(0)} - \sqrt{0+1} = 1 \Rightarrow \sqrt{0} - \sqrt{1} = 1 \Rightarrow -1 = 1$ (not a solution);

    $\sqrt{2(8)} - \sqrt{8+1} = 1 \Rightarrow \sqrt{16} - \sqrt{9} = 1 \Rightarrow 4 - 3 = 1 \Rightarrow 1 = 1$ The only solution is 8.

65. $\sqrt[3]{z+1} = -3 \Rightarrow z+1 = (-3)^3 \Rightarrow z+1 = -27 \Rightarrow z = -28$

    Check: $\sqrt[3]{-28-1} = \sqrt[3]{-27} = -3$

67. $\sqrt[3]{x+1} = \sqrt[3]{2x-1} \Rightarrow x+1 = 2x-1 \Rightarrow x = 2$

    Check: $\sqrt[3]{2+1} = \sqrt[3]{3} = \sqrt[3]{2(2)-1}$

69. $\sqrt[4]{x-2} + 4 = 20 \Rightarrow \sqrt[4]{x-2} = 16 \Rightarrow x-2 = (16)^4 \Rightarrow x-2 = 65,536 \Rightarrow x = 65,538$

    Check: $\sqrt[4]{65,538-2} + 4 = 20 \Rightarrow \sqrt[4]{65,536} + 4 = 20 \Rightarrow 20 = 20$

71. $f(x) = x^{3/2} - x^{1/2} \Rightarrow f(50) = 50^{3/2} - 50^{1/2} \approx 346.48$

73. The graph of $f(x) = x^a$ is increasing, since $a > 0$. However, since $a < b$ its graph increases more slowly

    than the graph of $f(x) = x^b$. The best choice is graph b.

75. Shift the graph of $f(x) = x^{1/2}$ up 1 unit to get the graph of $f(x) = x^{1/2} + 1$. See Figure 75.

77. Shift the graph of $f(x) = x^{2/3}$ down 1 unit to get the graph of $f(x) = x^{2/3} - 1$. See Figure 77.

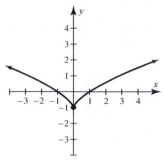

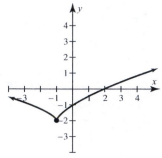

Figure 75   Figure 77   Figure 79

79. Shift the graph of $f(x) = x^{2/3}$ to the left 1 unit and down 2 units to get the graph of $f(x) = (x + 1)^{2/3} - 2$.
    See Figure 79.

81. $x^3 = 8 \Rightarrow x = \sqrt[3]{8} \Rightarrow x = 2$; Check: $2^3 = 8$

83. $x^{1/4} = 3 \Rightarrow x = 3^4 \Rightarrow x = 81$; Check: $81^{1/4} = \sqrt[4]{81} = 3$

85. $x^{2/5} = 4 \Rightarrow (x^{2/5})^{5/2} = 4^{5/2} \Rightarrow x = (\sqrt{4})^5 \Rightarrow x = (\pm 2)^5 \Rightarrow x = \pm 32$

    Check: $(\pm 32)^{2/5} = (\sqrt[5]{\pm 32})^2 = (\pm 2)^2 = 4$

87. $2(x^{1/5} - 2) = 0 \Rightarrow x^{1/5} - 2 = 0 \Rightarrow x^{1/5} = 2 \Rightarrow (x^{1/5})^5 = 2^5 \Rightarrow x = 32$

    Check: $2(32^{1/5} - 2) = 2(2 - 2) = 2(0) = 0$

89. $4x^{3/2} + 5 = 21 \Rightarrow 4x^{3/2} = 16 \Rightarrow x^{3/2} = 4 \Rightarrow (x^{3/2})^{2/3} = 4^{2/3} \Rightarrow x = \sqrt[3]{4^2} \Rightarrow x = \sqrt[3]{16}$

    Check: $4(\sqrt[3]{16})^{3/2} + 5 = 4(16^{1/3})^{3/2} + 5 = 4(16^{1/2}) + 5 = 4(\sqrt{16}) + 5 = 4(4) + 5 = 16 + 5 = 21$

91. $n^{-2} + 3n^{-1} + 2 = 0 \Rightarrow (n^{-1})^2 + 3(n^{-1}) + 2 = 0$, let $u = n^{-1}$, then $u^2 + 3u + 2 = 0 \Rightarrow$

    $(u + 2)(u + 1) = 0 \Rightarrow u = -2$ or $u = -1$. Because $u = n^{-1}$, it follows that $n = u^{-1}$.

    Thus $n = (-2)^{-1} \Rightarrow n = \dfrac{1}{(-2)} \Rightarrow n = -\dfrac{1}{2}$ or $n = (-1)^{-1} \Rightarrow n = \dfrac{1}{(-1)} = -1$. Therefore, $n = -1, -\dfrac{1}{2}$.

93. $5n^{-2} + 13n^{-1} = 28 \Rightarrow 5(n^{-1})^2 + 13(n^{-1}) - 28 = 0$, let $u = n^{-1}$, then $5u^2 + 13u - 28 = 0 \Rightarrow$

    $(5u - 7)(u + 4) = 0 \Rightarrow u = \dfrac{7}{5}$ or $u = -4$. Because $u = n^{-1}$, it follows that $n = u^{-1}$.

    Thus $n = \left(\dfrac{7}{5}\right)^{-1} \Rightarrow n = \dfrac{5}{7}$ or $n = (-4)^{-1} \Rightarrow n = -\dfrac{1}{4}$. Therefore, $n = -\dfrac{1}{4}, \dfrac{5}{7}$.

95. $x^{2/3} - x^{1/3} - 6 = 0 \Rightarrow (x^{1/3})^2 - x^{1/3} - 6 = 0$, let $u = x^{1/3}$, then $u^2 - u - 6 = 0 \Rightarrow$

    $(u - 3)(u + 2) = 0 \Rightarrow u = -2$ or $u = 3$. Because $u = x^{1/3}$, it follows that $x = u^3$.

    Thus $x = (-2)^3 \Rightarrow x = -8$ or $x = 3^3 \Rightarrow x = 27$. Therefore, $x = -8, 27$.

97. $6x^{2/3} - 11x^{1/3} + 4 = 0 \Rightarrow 6(x^{1/3})^2 - 11(x^{1/3}) + 4 = 0$, let $u = x^{1/3}$, then $6u^2 - 11u + 4 = 0$.

    Using the quadratic formula to solve we get: $u = \dfrac{11 \pm \sqrt{121 - 4(6)(4)}}{2(6)} = \dfrac{11 \pm \sqrt{25}}{12} = \dfrac{11 \pm 5}{12} \Rightarrow$

    $u = \dfrac{16}{12} \Rightarrow u = \dfrac{4}{3}$ or $u = \dfrac{6}{12} \Rightarrow u = \dfrac{1}{2}$. Because $u = x^{1/3}$, it follows that $x = u^3$.

    Thus $x = \left(\dfrac{4}{3}\right)^3 \Rightarrow x = \dfrac{64}{27}$ or $x = \left(\dfrac{1}{2}\right)^3 \Rightarrow x = \dfrac{1}{8}$. Therefore, $x = \dfrac{1}{8}, \dfrac{64}{27}$.

99. $x^{3/4} - x^{1/2} - x^{1/4} + 1 = 0 \Rightarrow (x^{1/4})^3 - (x^{1/4})^2 - (x^{1/4}) + 1 = 0$, let $u = x^{1/4}$,

then $u^3 - u^2 - u + 1 = 0 \Rightarrow (u^3 - u^2) - (u - 1) = 0 \Rightarrow u^2(u - 1) - 1(u - 1) = 0 \Rightarrow$

$(u^2 - 1)(u - 1) = 0 \Rightarrow (u + 1)(u - 1)(u - 1) = 0 \Rightarrow u = -1$ or $u = 1$.

Because $u = x^{1/4}$ it follows that $x = u^4$.

Thus $x = (-1)^4 \Rightarrow x = 1$ or $x = (1)^4 \Rightarrow x = 1$. Therefore, $x = 1$.

101. $x^{-2/3} - 2x^{-1/3} - 3 = 0 \Rightarrow (x^{-1/3})^2 - 2x^{-1/3} - 3 = 0$, let $u = x^{-1/3}$, then $u^2 - 2u - 3 = 0 \Rightarrow$

$(u - 3)(u + 1) = 0 \Rightarrow u = 3$ or -1. Because $u = x^{-1/3}$ it follows that $x = u^{-3}$.

Thus $x = 3^{-3} = \dfrac{1}{27}, x = -1^{-3} = -1$.

103. Average rate of change $= \dfrac{s(\frac{9}{2}) - s(\frac{1}{2})}{\frac{9}{2} - \frac{1}{2}} = \dfrac{\sqrt{96(\frac{9}{2})} - \sqrt{96(\frac{1}{2})}}{4} = \dfrac{\sqrt{432} - \sqrt{48}}{4} \approx 3.5$. The average speed

over the time interval is about 3.5 mph.

105. $S(w) = 3 \Rightarrow 1.27w^{2/3} = 3 \Rightarrow w^{2/3} = \dfrac{3}{1.27} \Rightarrow w = \left(\dfrac{3}{1.27}\right)^{3/2} \Rightarrow w \approx 3.63$

The bird weighs about 3.63 pounds

107. $f(15) = 15^{1.5} \approx 58.1$; The planet would take about 58.1 years to orbit the sun.

109. (a) Following the hint, $f(1) = a(1)^b = 1960 \Rightarrow a = 1960$

(b) Plot the four data points (0.5, 4500), (1, 1960), (2, 850), and (3, 525) together with $f(x) = 1960x^b$ for

different values of b. Through trial and error, a value of $b \approx -1.2$ can be found. Figure 109 shows a graph

of $Y_1 = 1960X^{\wedge}(-1.2)$ along with the data.

(c) $f(x) = 1960x^{-1.2} \Rightarrow f(4) = 1960(4)^{-1.2} \approx 371$. This means that if the zinc ion concentration in

the water reaches 371 milligrams per liter, the rainbow trout will live only 4 minutes on average.

[0, 5, 1] by [0, 5000, 1000]]

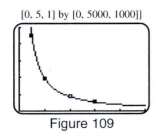

Figure 109

111. (a) $f(2) = 0.445(2)^{5/4} \approx 1.06$ grams

(b) We must solve the equation $0.445x^{5/4} = 0.5$ for x. Graph $Y_1 = 0.445X^{\wedge}1.25$ and $Y_2 = 0.5$ Their graphs

intersect near (1.1, 0.5). See Figure 111. When the claw weighs 0.5 grams, the crab weighs about 1.1 grams.

(c) $0.445x^{1.25} = 0.5 \Rightarrow 0.445x^{5/4} = 0.5 \Rightarrow x^{5/4} = \dfrac{0.5}{0.445} \Rightarrow x = \left(\dfrac{0.5}{0.445}\right)^{4/5} \Rightarrow x \approx 1.1$ grams

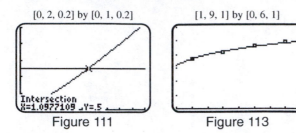

[0, 2, 0.2] by [0, 1, 0.2]   [1, 9, 1] by [0, 6, 1]

Figure 111   Figure 113

113. Use the power regression feature of your graphing calculator to find the values of a and b in the equation

$f(x) = ax^b$. For this exercise, $a \approx 3.20$ and $b \approx 0.20$.

The graph of $Y_1 = 3.20X{}^{\wedge}0.20$ is shown together with the data in Figure 113.

115. (a) Using the power regression function on the graphing calculator we see that $f(x) \approx 0.005192x^{1.7902}$.

Answers may vary.

(b) The year 2012 is 32 years after 1980. $f(32) \approx 0.005192(32)^{1.7902} \approx 2.6$ million. It appears that the value

is about a half million to high. The estimate involved extrapolation.

(c) See Figure 115. We can see from the graph of $f(x)$ that the number of employees reached 1.million in

about 1999.

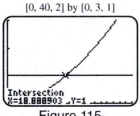

[0, 40, 2] by [0, 3, 1]

Figure 115

117. Use the power regression feature of your graphing calculator to find the values of a and b in the equation

$f(x) = ax^b$. For this exercise, $a \approx 874.54$ and $b \approx -0.49789$.

Extended and Discovery Exercise for Section 4.8

1. The graph of an odd root function is always increasing; the function is negative for $x < 0$, positive for $x > 0$,

and zero at $x = 0$.

3. $\dfrac{f(x + h) - f(x)}{h} = \dfrac{\sqrt{x + h} - \sqrt{x}}{h} = \dfrac{\sqrt{x + h} - \sqrt{x}}{h} \cdot \dfrac{\sqrt{x + h} + \sqrt{x}}{\sqrt{x + h} + \sqrt{x}} = \dfrac{x + h - x}{h(\sqrt{x + h} + \sqrt{x})} =$

$\dfrac{1}{\sqrt{x + h} + \sqrt{x}}$

5. $\dfrac{x^{-2/3} + x^{1/3}}{x} = \dfrac{x^{-2/3}(1 + x)}{x} = \dfrac{1 + x}{x^{2/3}(x)} = \dfrac{1 + x}{x^{5/3}}$

7. $\dfrac{\frac{2}{3}(x + 1)x^{-1/3} - x^{2/3}}{(x + 1)^2} = \dfrac{x^{-1/3}[\frac{2}{3}x + \frac{2}{3} - x]}{(x + 1)^2} = \dfrac{-\frac{1}{3}x + \frac{2}{3}}{x^{1/3}(x + 1)^2} = \dfrac{\frac{-x + 2}{3}}{x^{1/3}(x + 1)^2} = \dfrac{2 - x}{3x^{1/3}(x + 1)^2}$

9. (a) Since the distance from C to D is 20 feet, the distance from P to C is $20 - x$

(b) The value must be between 0 and 20. That is, $0 < x < 20$.

(c) $(AP)^2 = (DP)^2 + (AD)^2 \Rightarrow (AP)^2 = x^2 + 12^2 \Rightarrow AP = \sqrt{x^2 + 12^2}$;

$(BP)^2 = (CP)^2 + (BC)^2 \Rightarrow (BP)^2 = (20 - x)^2 + 16^2 \Rightarrow BP = \sqrt{(20 - x)^2 + 16^2}$

(d) $f(x) = \sqrt{x^2 + 12^2} + \sqrt{(20 - x)^2 + 16^2}$, $0 < x < 20$.

(e) The graph of $y_1 = \sqrt{x^2 + 12^2} + \sqrt{(20 - x)^2 + 16^2}$ is shown in Figure 73. Here $f(4) \approx 35.28$. When the

stake is 4 feet from the 12-foot pole, approximately 35.28 feet of wire will be required.

(f) Using the calculator, $f(x)$ is a minimum (about 34.41 feet) when $x \approx 8.57$ feet.

(g) This problem examined how the total amount of wire used can be expressed in terms of the distance from

the stake at P to the base of the 12-foot pole. We find that the amount of wire used can be minimized when

the stake is approximately 8.57 feet from the 12-foot pole.

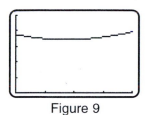

Figure 9

Checking Basic Concepts for Sections 4.7 and 4.8

1. (a) $\dfrac{3x - 1}{1 - x} = 1 \Rightarrow 3x - 1 = 1 - x \Rightarrow 4x = 2 \Rightarrow x = \dfrac{1}{2}$; Check: $\dfrac{3(\frac{1}{2}) - 1}{1 - (\frac{1}{2})} = 1 \Rightarrow \dfrac{\frac{1}{2}}{\frac{1}{2}} = 1 \Rightarrow 1 = 1$

(b) $3 + \dfrac{8}{x} = \dfrac{35}{x^2} \Rightarrow 3x^2 + 8x = 35 \Rightarrow 3x^2 + 8x - 35 = 0 \Rightarrow (3x - 7)(x + 5) = 0 \Rightarrow x = -5, \dfrac{7}{3}$

Check: $3 + \dfrac{8}{(-5)} = \dfrac{35}{(-5)^2} \Rightarrow 3 - \dfrac{8}{5} = \dfrac{35}{25} \Rightarrow \dfrac{75}{25} - \dfrac{40}{25} = \dfrac{35}{25} \Rightarrow \dfrac{35}{25} = \dfrac{35}{25}$

Check: $3 + \dfrac{8}{(\frac{7}{3})} = \dfrac{35}{(\frac{7}{3})^2} \Rightarrow 3 + \dfrac{24}{7} = \dfrac{45}{7} \Rightarrow \dfrac{21}{7} + \dfrac{24}{7} = \dfrac{45}{7} \Rightarrow \dfrac{45}{7} = \dfrac{45}{7}$

(c) $\dfrac{1}{x - 1} - \dfrac{1}{3(x + 2)} = \dfrac{1}{(x + 2)(x - 1)} \Rightarrow 3(x + 2) - (x - 1) = 3 \Rightarrow 3x + 6 - x + 1 = 3 \Rightarrow$

$2x = -4 \Rightarrow x = -2$; Check: $\dfrac{1}{(-2) - 1} - \dfrac{1}{3(-2 + 2)} = \dfrac{1}{(-2 + 2)(-2 - 1)} \Rightarrow \dfrac{1}{-3} - \dfrac{1}{0} = \dfrac{1}{0}$

Since -2 is not defined in the original equation there is no solution.

3. The graph of $Y_1 = (X^2 - 1)/(X + 2)$ has a vertical asymptote at $x = -2$ and x-intercepts at $x = \pm 1$.

The graph of Y_1 intersects or is above the x-axis on the interval $(-2, -1] \cup [1, \infty)$.

In set builder notation the interval is $\{x \mid -2 < x \le -1 \text{ or } x \ge 1\}$.

5. (a) $-4^{3/2} = -(\sqrt{4})^3 = -(2)^3 = -8$

(b) $(8^{-2})^{1/3} = 8^{-2/3} = \dfrac{1}{8^{2/3}} = \dfrac{1}{(\sqrt[3]{8})^2} = \dfrac{1}{2^2} = \dfrac{1}{4}$

(c) $\sqrt[3]{27^2} = (\sqrt[3]{27})^2 = 3^2 = 9$

7. $\sqrt{5x-4} = x-2 \Rightarrow 5x-4 = x^2 - 4x + 4 \Rightarrow x^2 - 9x + 8 = 0 \Rightarrow (x-1)(x-8) = 0 \Rightarrow$

    $x = 1$ or 8

    Check: $\sqrt{5(1)-4} \neq 1-2$ (not a solution); $\sqrt{5(8)-4} = 6 = 8-2$. The only solution is $x = 8$.

9. Use the power regression feature of your graphing calculator to find the values of a and b in the equation

    $f(x) = ax^b$. For this exercise, $a = 2$ and $b = 0.5$. That is, $f(x) = 2x^{1/2}$ or $f(x) = 2\sqrt{x}$.

Chapter 4 Review Exercises

1. First put in standard order:

    $f(x) = -7x^3 - 2x^2 + x + 4$, now the degree is: 3, and the leading coefficient is: -7.

3. (a) There is a local minimum of –2. There is a local maximum of 4.

    (b) There is neither an absolute minimum nor an absolute maximum.

5. $f(x) = 2x^6 - 5x^4 - x^2 \Rightarrow f$ is even since each power of x is even.

7. $f(x) = 7x^5 + 3x^3 - x \Rightarrow f$ is odd since each power of x is odd.

9. f is odd since $f(-x) = -f(x)$ for all x-values in the table.

11. See Figure 11. *Answers may vary.*

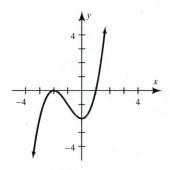

Figure 11

13. (a) 2 turning points; 3 x-intercepts at: $x \approx -2, 0, 1$.

    (b) This is a an odd degreed function going down to the right $\Rightarrow$ negative.

    (c) Since it has 2 turning points, it is a minimum of 3rd degree.

15. A negative cubic function $\Rightarrow$ up on the left end and down on the right end;

    $f(x) \to \infty$ as $x \to -\infty$; $f(x) \to -\infty$ as $x \to \infty$.

17. $f(-2) = (-2)^3 + 1 \Rightarrow f(-2) = -7 \Rightarrow (-2, -7)$ and $f(-1) = (-1)^3 + 1 \Rightarrow f(-1) = 0 \Rightarrow (-1, 0)$.

    To find the average rate of change we use: $\dfrac{y_2 - y_1}{x_2 - x_1} \Rightarrow \dfrac{0 - (-7)}{(-1) - (-2)} \Rightarrow \dfrac{7}{1} \Rightarrow 7$

19. (a) The graph of f has of two different pieces: $y = 2x$ when $0 \leq x < 2$ and $y = 8 - x^2$ when $2 \leq x \leq 4$.

The graph is shown in Figure 19. The graph of f is continuous.

(b) To evaluate $f(1)$ we must use the first piece of the function: $f(1) = 2(1) = 2$

To evaluate $f(3)$ we must use the second piece of the function: $f(3) = 8 - (3)^2 = 8 - 9 = -1$

(c) We must solve the equation $f(x) = 2$ for each piece of the piecewise-defined function f.

$2x = 2 \Rightarrow x = 1$ and $8 - x^2 = 2 \Rightarrow x^2 = 6 \Rightarrow x = \pm\sqrt{6}$

Since $x = -\sqrt{6}$ is not in the domain of f, the solutions are $x = 1$ or $\sqrt{6}$.

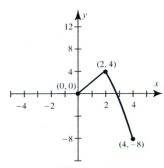

Figure 19

21. $\dfrac{14x^3 - 21x^2 - 7x}{7x} = 2x^2 - 3x - 1$

23.
$$
\begin{array}{r}
2x^2 - 3x + 1 \\
2x + 3 \overline{)\;4x^3 \qquad\quad - 7x + 4} \\
\underline{4x^3 + 6x^2} \\
-6x^2 - 7x + 4 \\
\underline{-6x^2 - 9x} \\
2x + 4 \\
\underline{2x + 3} \\
1
\end{array}
$$

Therefore the quotient is: $2x^2 - 3x + 1 + \dfrac{1}{2x + 3}$

25. Since the leading coefficient of $f(x)$ is $\dfrac{1}{2}$ and the degree of the polynomial is 3, all of the zeros are given, thus:

$f(x) = \dfrac{1}{2}(x - 1)(x - 2)(x - 3)$.

27. The graph has x-intercepts or zeros of $x = -2, 1, 3 \Rightarrow$ it has factors $(x + 2)(x - 1)(x - 3)$.

Since $f(0) = 3$ we can solve for the leading coefficient a by setting $a(0 + 2)(0 - 1)(0 - 3) = 3$

and solving $\Rightarrow a(2)(-1)(-3) = 3 \Rightarrow 6a = 3 \Rightarrow a = \dfrac{1}{2}$.

The complete factored form is: $f(x) = \dfrac{1}{2}(x + 2)(x - 1)(x - 3)$.

29. By the rational zeros test, any rational zero of $2x^3 + x^2 - 13x + 6$ must be one of the following:

$\pm 6, \pm 3, \pm 2, \pm 1, \pm\dfrac{3}{2}, \pm\dfrac{1}{2}$. By evaluating f at each of these values we find that the three rational zeros are

$-3, \dfrac{1}{2}$, and 2.

31. $9x = 3x^3 \Rightarrow 3x^3 - 9x = 0 \Rightarrow 3x(x^2 - 3) = 0 \Rightarrow x = 0, \pm\sqrt{3}$

33. The x-intercepts on the graph of $Y_1 = X^3 - 3X + 1$ are approximately $-1.88, 0.35$ and 1.53. See Figure 33.

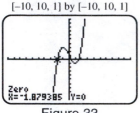

[−10, 10, 1] by [−10, 10, 1]

Figure 33

35. $x^3 + x = 0 \Rightarrow x(x^2 + 1) = 0 \Rightarrow x = 0$ or $x = \pm i$

37. $f(x) = 4(x - 1)(x - 3i)(x + 3i)$; the expanded form is $f(x) = 4x^3 - 4x^2 + 36x - 36$.

39. If i is a zero then $-i$ is also a zero $\Rightarrow (x + i)(x - i)$ or $x^2 + 1$ is a factor. By polynomial long division:

$$
\begin{array}{r}
x^2 + x + 1 \\
x^2 + 1 \overline{)\ x^4 + x^3 + 2x^2 + x + 1} \\
\underline{x^4 \qquad\ \ + x^2} \\
x^3 + x^2 + x + 1 \\
\underline{x^3 \qquad\ \ + x} \\
x^2 \qquad + 1 \\
\underline{x^2 \qquad + 1} \\
0
\end{array}
$$

Solve for $x^2 + x + 1 = 0$ using the quadratic formula: $x = \dfrac{-1 \pm \sqrt{1 - 4(1)(1)}}{2(1)} = \dfrac{-1 \pm \sqrt{-3}}{2} =$

$-\dfrac{1}{2} \pm \dfrac{i\sqrt{3}}{2}$. Therefore the zeros are $\pm i, -\dfrac{1}{2} \pm \dfrac{i\sqrt{3}}{2}$ and the complete factored form is:

$$f(x) = (x + i)(x - i)\left(x - \left(-\dfrac{1}{2} + \dfrac{i\sqrt{3}}{2}\right)\right)\left(x - \left(-\dfrac{1}{2} - \dfrac{i\sqrt{3}}{2}\right)\right)$$

41. Horizontal asymptote: since the degree of the numerator and denominator is the same the horizontal asymptote

is the leading coefficient ratio $y = \dfrac{2}{3}$. Vertical asymptotes: they can be found by finding the zero's of the

denominator that are not zeros of the numerator

$\Rightarrow$ solve: $3x^2 + 8x - 3 \Rightarrow (3x - 1)(x + 3) = 0 \Rightarrow x = -3, \dfrac{1}{3}$. However, $x = -3$ is not a vertical

asymptote because -3 is a zero of the numerator.

The horizontal asymptote is $y = \dfrac{2}{3}$ and the vertical asymptote is $x = \dfrac{1}{3}$.

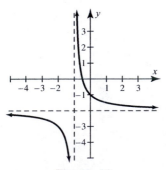

Figure 43

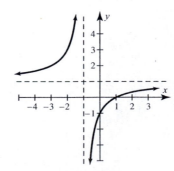

Figure 45

43. $g(x)$ is the function $f(x) = \dfrac{1}{x}$ shifted 1 left and 2 down. See Figure 43.

45. $g(x) = \dfrac{x^2 - 1}{x^2 + 2x + 1} \Rightarrow g(x) = \dfrac{(x + 1)(x - 1)}{(x + 1)(x + 1)} \Rightarrow g(x) = \dfrac{(x - 1)}{(x + 1)} \Rightarrow g(x) = \dfrac{x - 1}{x + 1}$. This is similar

  to $f(x) = \dfrac{1}{x}$ with a horizontal asymptote of $y = 1$ and a vertical asymptote of $x = -1$. See Figure 45.

47. One example of a function that has a vertical asymptote at $x = -2$ and a horizontal asymptote at $y = 2$ is

  shown in Figure 47.

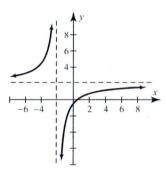

Figure 47

49. $\dfrac{5x + 1}{x + 3} = 3 \Rightarrow 5x + 1 = 3(x + 3) \Rightarrow 5x + 1 = 3x + 9 \Rightarrow 2x = 8 \Rightarrow x = 4$

  This answer is supported graphically by graphing $Y_1 = (5X + 1)/(X + 3)$ and $Y_2 = 3$ The intersection point

  is $(4, 3)$. Making a table of Y_1 starting at 0 and incrementing by 1 will support this result numerically.

51. $\dfrac{1}{x + 2} + \dfrac{1}{x - 2} = \dfrac{4}{x^2 - 4} \Rightarrow x - 2 + x + 2 = 4 \Rightarrow 2x - 4 \Rightarrow x = 2$. Since 2 is defined in the

  original equation, there is no solution.

53. (a) $(-\infty, -4) \cup (-2, 3)$; In set builder notation the interval is $\{x \mid x < -4 \text{ or } -2 < x < 3\}$.

  (b) $(-4, -2) \cup (3, \infty)$; In set builder notation the interval is $\{x \mid -4 < x < -2 \text{ or } x > 3\}$.

55. Find the boundary numbers by solving $x^3 + x^2 - 6x = 0$.

  $x^3 + x^2 - 6x = 0 \Rightarrow x(x^2 + x - 6) = 0 \Rightarrow x(x + 3)(x - 2) = 0 \Rightarrow x = -3, 0 \text{ or } 2$.

  The boundary values divide the number line into four intervals. Choose a test value from each of these intervals

  and evaluate $x^3 + x^2 - 6x$ for these values. See Figure 55. The solution is $(-3, 0) \cup (2, \infty)$; In set builder

  notation the interval is $\{x \mid -3 < x < 0 \text{ or } x > 2\}$.

Interval	Test Value x	$x^3 + x^2 - 6x$	Positive or Negative?
$(-\infty, -3)$	-4	-24	Negative
$(-3, 0)$	-1	6	Positive
$(0, 2)$	1	-4	Negative
$(2, \infty)$	3	18	Positive

Figure 55

57. Graph $Y_1 = (2X - 1)/(X + 2)$ using dot mode as shown in Figure 57. The graph has a vertical asymptote at $x = -2$ and an x-intercept at $x = \dfrac{1}{2}$. The graph is above the x-axis on the interval $(-\infty, -2) \cup \left(\dfrac{1}{2}, \infty\right)$;

In set builder notation the interval is $\left\{ x \mid x < -2 \text{ or } x > \dfrac{1}{2} \right\}$

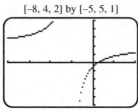

[−8, 4, 2] by [−5, 5, 1]

Figure 57

59. $(36^{3/4})^2 = 36^{3/2} = (\sqrt{36}\,)^3 = 6^3 = 216$

61. $(2^{-3/2} \cdot 2^{1/2})^{-3} = (2^{-3/2+1/2})^{-3} = (2^{-1})^{-3} = 2^3 = 8$

63. $\sqrt[3]{x^4} = x^{4/3}$

65. $\sqrt[3]{y} \cdot \sqrt{y} = (y \cdot y^{1/2})^{1/3} = (y^{3/2})^{1/3} = y^{1/2}$

67. $D = \{x \mid x \geq 0\}; f(3) = 3^{5/2} \approx 15.59$

69. $x^5 = 1024 \Rightarrow x = \sqrt[5]{1024} = 4$ Check: $4^5 = 1024$

71. $\sqrt{x - 2} = x - 4 \Rightarrow x - 2 = (x - 4)^2 \Rightarrow x - 2 = x^2 - 8x + 16 \Rightarrow x^2 - 9x + 18 = 0 \Rightarrow$

$(x - 6)(x - 3) = 0 \Rightarrow x = 3 \text{ or } 6$

Check: $\sqrt{3 - 2} \neq 3 - 4$ (not a solution); $\sqrt{6 - 2} = 2 = 6 - 4$. The only solution is $x = 6$.

73. $2x^{1/4} + 3 = 6 \Rightarrow 2x^{1/4} = 3 \Rightarrow x^{1/4} = \dfrac{3}{2} \Rightarrow x = \left(\dfrac{3}{2}\right)^4 \Rightarrow x = \dfrac{81}{16}$

Check: $2\left(\dfrac{81}{16}\right)^{1/4} + 3 = 2\left(\dfrac{3}{2}\right) + 3 = 3 + 3 = 6$

75. $\sqrt[3]{2x - 3} + 1 = 4 \Rightarrow \sqrt[3]{2x - 3} = 3 \Rightarrow 2x - 3 = 3^3 \Rightarrow 2x - 3 = 27 \Rightarrow 2x = 30 \Rightarrow x = 15$

Check: $\sqrt[3]{2(15) - 3} + 1 = \sqrt[3]{27} + 1 = 3 + 1 = 4$

77. $2n^{-2} - 5n^{-1} = 3 \Rightarrow 2(n^{-1})^2 - 5(n^{-1}) = 3$, let $u = n^{-1}$, then $2u^2 - 5u - 3 = 0 \Rightarrow$

$(2u + 1)(u - 3) = 0 \Rightarrow u = -\dfrac{1}{2}, 3$. If $u = n^{-1}$, then it follows that $n = u^{-1}$, thus $n = \left(-\dfrac{1}{2}\right)^{-1} \Rightarrow$

$n = -2 \text{ or } n = (3)^{-1} \Rightarrow n = \dfrac{1}{3}$

Check: $2(-2)^{-2} - 5(-2)^{-1} = 3 \Rightarrow 2\left(\dfrac{1}{4}\right) - 5\left(-\dfrac{1}{2}\right) = 3 \Rightarrow \dfrac{1}{2} + \dfrac{5}{2} = 3 \Rightarrow 3 = 3$

Check: $2\left(\dfrac{1}{3}\right)^{-2} - 5\left(\dfrac{1}{3}\right)^{-1} = 3 \Rightarrow 2(9) - 5(3) = 3 \Rightarrow 18 - 15 = 3 \Rightarrow 3 = 3$

The solution are: $n = -2, \dfrac{1}{3}$.

79. $k^{2/3} - 4k^{1/3} - 5 = 0 \Rightarrow (k^{1/3})^2 - 4(k^{1/3}) - 5 = 0$, let $u = k^{1/3}$, then $u^2 - 4u - 5 = 0 \Rightarrow$

$(u - 5)(u + 1) = 0 \Rightarrow u = -1, 5$. If $u = k^{1/3}$, then it follows that $k = u^3$, thus $k = (-1)^3 \Rightarrow$

$k = -1$ or $k = (5)^3 \Rightarrow k = 125$.

Check: $(-1)^{2/3} - 4(-1)^{1/3} - 5 = 0 \Rightarrow 1 - (-4) - 5 = 0 \Rightarrow 0 = 0$

Check: $(125)^{2/3} - 4(125)^{1/3} - 5 = 0 \Rightarrow 25 - 4(5) - 5 = 0 \Rightarrow 0 = 0$. The solutions are: $k = -1, 125$.

81. (a) Dog: $f(24) = 1607(24)^{-0.75} \approx 148$ bpm; Person: $f(66) = 1607(66)^{-0.75} \approx 69$ bpm

(b) We must approximate the length of an animal with a heart rate of 400 bpm. To do this, solve the

equation $f x) = 400$ or $1607x^{-0.75} = 400$ bpm.

$$1607x^{-0.75} = 400 \Rightarrow x^{-0.75} = \frac{400}{1607} \Rightarrow x^{0.75} = \frac{1607}{400} \Rightarrow x^{3/4} = \frac{1607}{400} \Rightarrow x = \left(\frac{1607}{400}\right)^{4/3} \approx 6.4$$

A six-inch animal such as a large bird or smaller rodent might have a heart rate of 400 bpm.

[0, 13, 1] by [0, 100, 10]

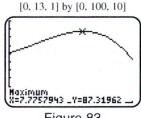

Maximum
X=7.7757943   Y=87.31962

Figure 83

83. (a) May is $T(5) \Rightarrow T(5) = -0.064(5)^3 + 0.56(5)^2 + 2.9(5) + 61 \Rightarrow$

$T(5) = -8 + 14 + 14.5 + 61 \Rightarrow T(5) = 81.5$. The temperature in May is 81.5°F.

(b) Graph the function. See Figure 83. From the graph the ocean reaches a maximum temperature of about

87.3°F in July.

85.   Since the time is directly proportional to the square root of the height, the equation $t = k\sqrt{h}$ must hold. If it

takes 1 second to drop 16 feet then $1 = k\sqrt{16} \Rightarrow k = 0.25$. Therefore, from a height of 256 feet the object

will take $t = 0.25\sqrt{256} = 4$ seconds to strike the ground.

Extended and Discovery Exercise for Chapter 4

1.   (a) See Figure 1.

(b) The velocity of the bike rider is 20 ft/sec at 10 seconds.

$f(t) = t^2$	$t_1 = 10$ $t_2 = 11$	$t_1 = 10$ $t_2 = 10.1$	$t_1 = 10$ $t_2 = 10.01$	$t_1 = 10$ $t_2 = 10.001$
average velocity (ft/sec)	21	20.1	20.01	20.001

Figure 1

3. (a) See Figures 3a-d.

    (b) Neither the graph of the linear function nor the graph of its average rate of change have any turning points.

    (c) For any linear function, the graph of it average rate of change is a constant function whose value is equal to the slope of the graph of the linear function.

[−10, 10, 1] by [−10, 10, 1]   [−10, 10, 1] by [−10, 10, 1]   [−10, 10, 1] by [−10, 10, 1]   [−10, 10, 1] by [−10, 10, 1]

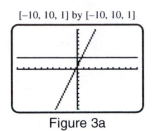   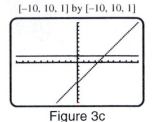   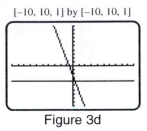

Figure 3a   Figure 3b   Figure 3c   Figure 3d

5. (a) See Figures 5a-d.

    (b) The graph of each cubic function has two turning points or none, whereas the graph of its average rate of change has one turning point.

    (c) For any cubic function, the graph of its average rate of change is a quadratic function. The leading coefficient of the cubic function and the leading coefficient of its average rate of change have the same sign.

[−10, 10, 1] by [−10, 10, 1]   [−10, 10, 1] by [−10, 10, 1]   [−10, 10, 1] by [−10, 10, 1]   [−10, 10, 1] by [−10, 10, 1]

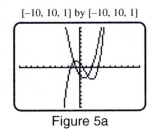   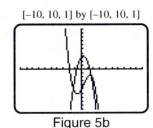   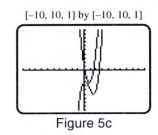   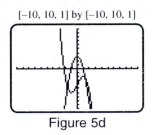

Figure 5a   Figure 5b   Figure 5c   Figure 5d

Chapters 1-4 Cumulative Review Exercises

1. (a) $D = \{-3, -1, 0, 1\}$; $R = \{4, -2, 5\}$

    (b) No, all input must have unique outputs and -1 has an output of -2 and 5.

3. $D = \{x \mid -2 \le x \le 2\}$; $R = \{x \mid 0 \le x \le 3\}$

5. $c(x) = 0.25x + 200$; $c(2000) = 0.25(2000) + 200 \Rightarrow c(2000) = 700$. The monthly cost of driving and maintaining a car driven 2000 miles is \$700.

7. $f(-3) = (-3)^3 - (-3) \Rightarrow f(-3) = -24$ or $(-3, -24)$; $f(-2) = (-2)^3 - (-2) \Rightarrow$

    $f(-2) = -6$ or $(-2, -6)$. The average rate of change $= \dfrac{-6 - (-24)}{-2 - (-3)} = \dfrac{18}{1} = 18$.

9. Slope $= \dfrac{(-4) - 5}{3 - (-2)} = \dfrac{-9}{5}$. Using point slope form $y = m(x - x_1) + y$ and point $(-2, 5)$ we get:

$$y = \frac{-9}{5}(x + 2) + 5 \Rightarrow y = \frac{-9}{5}x - \frac{18}{5} + 5 \Rightarrow y = \frac{-9}{5}x + \frac{7}{5}$$

11. All lines parallel to the x-axis have 0 slope and will have graphs of $y =$ (the y-coordinate) $\Rightarrow y = -5$.

13. Each radio costs \$15 to manufacture. The fixed cost for this company to produce radios is \$2000.

15. $-3(2 - 3x) - (-x - 1) = 1 \Rightarrow -6 + 9x + x + 1 = 1 \Rightarrow 10x - 5 = 1 \Rightarrow 10x = 6 \Rightarrow x = \dfrac{3}{5}$

17. $|3x - 4| + 1 = 5 \Rightarrow |3x - 4| = 4 \Rightarrow 3x - 4 = 4 \Rightarrow 3x = 8 \Rightarrow x = \dfrac{8}{3}$ or

$3x - 4 = -4 \Rightarrow 3x = 0 \Rightarrow x = 0$. Therefore $x = 0, \dfrac{8}{3}$.

19. $7x^2 + 9x = 10 \Rightarrow 7x^2 + 9x - 10 = 0 \Rightarrow (7x - 5)(x + 2) = 0 \Rightarrow x = -2, \dfrac{5}{7}$

21. $3x^{2/3} + 5x^{1/3} - 2 = 0 \Rightarrow 3(x^{1/3})^2 + 5(x^{1/3}) - 2 = 0$. Let $u = x^{1/3}$, then $3u^2 + 5u - 2 = 0 \Rightarrow$

$(3u - 1)(u + 2) = 0 \Rightarrow u = -2, \dfrac{1}{3}$, If $u = x^{1/3}$ then it follows $x = u^3$, thus $x = (-2)^3 \Rightarrow x = -8$ or

$x = \left(\dfrac{1}{3}\right)^3 \Rightarrow x = \dfrac{1}{27}$. Therefore $x = -8, \dfrac{1}{27}$

23. $\dfrac{2x - 3}{5 - x} = \dfrac{4x - 3}{1 - 2x} \Rightarrow (2x - 3)(1 - 2x) = (4x - 3)(5 - x) \Rightarrow$

$2x - 4x^2 - 3 + 6x = 20x - 4x^2 - 15 + 3x \Rightarrow 15x = 12 \Rightarrow x = \dfrac{4}{5}$

25. $\dfrac{1}{2}x - (4 - x) + 1 = \dfrac{3}{2}x - 5 \Rightarrow \dfrac{1}{2}x - 4 + x + 1 = \dfrac{3}{2}x - 5 \Rightarrow \dfrac{3}{2}x - 3 = \dfrac{3}{2}x - 5 \Rightarrow$

$-3 = -5$, since this is false the equation has no solutions and is a contradiction.

27. $-\dfrac{1}{3}x - (1 + x) > \dfrac{2}{3}x \Rightarrow -\dfrac{1}{3}x - 1 - x > \dfrac{2}{3}x \Rightarrow -\dfrac{4}{3}x - 1 > \dfrac{2}{3}x \Rightarrow -2x > 1 \Rightarrow x < -\dfrac{1}{2} \Rightarrow$

$\left(-\infty, -\dfrac{1}{2}\right)$; In set builder notation the interval is $\left\{ x \,|\, x < -\dfrac{1}{2} \right\}$.

29. The solutions to $|5x - 7| \geq 3$ satisfy $x \leq s_1$ or $x \geq s_2$ where s_1 and s_2 are the solutions to $|5x - 7| = 3$.

$|5x - 7| = 3$ is equivalent to $5x - 7 = -3 \Rightarrow x = \dfrac{4}{5}$ and $5x - 7 = 3 \Rightarrow x = 2$.

The interval is $\left(-\infty, \dfrac{4}{5}\right] \cup [2, \infty)$; In set builder notation the interval is $\left\{ x \,|\, x \leq \dfrac{4}{5} \text{ or } x \geq 2 \right\}$.

31. For $x^3 - 9x \leq 0$, first set $x^3 - 9x = 0 \Rightarrow x(x^2 - 9) = 0 \Rightarrow x(x + 3)(x - 3) = 0 \Rightarrow x = -3, 0, 3$.

These boundary numbers separate the number line into these intervals: $(-\infty, -3), (-3, 0), (0, 3)$ and $(3, \infty)$.

Checking an x value in each interval we get:

for $x = -5, (-5)^3 - 9(-5) \leq 0 \Rightarrow -125 + 45 \leq 0 \Rightarrow -80 \leq 0$, which is true $\Rightarrow$ a solution;

for $x = -1, (-1)^3 - 9(-1) \leq 0 \Rightarrow -1 + 9 \leq 0 \Rightarrow 8 \leq 0$, which is false $\Rightarrow$

not a solution; for $x = 1, (1)^3 - 9(1) \leq 0 \Rightarrow 1 - 9 \leq 0 \Rightarrow -8 \leq 0$, which is true $\Rightarrow$ a solution.

for $x = 5, (5)^3 - 9(5) \leq 0 \Rightarrow 125 - 45 \leq 0 \Rightarrow 80 \leq 0$, which is false $\Rightarrow$ not a solution.

Therefore $(-\infty, -3] \cup [0, 3]$. In set builder notation the intervals are $\{x \mid x \leq -3 \text{ or } 0 \leq x \leq 3\}$.

33. (a) $x = -3, -1, 1, 2$

(b) $(-\infty, -3) \cup (-1, 1) \cup (2, \infty)$; In set builder notation the intervals are

$\{x \mid x < -3 \text{ or } -1 < x < 1 \text{ or } x > 2\}$.

(c) $[-3, -1] \cup [1, 2]$; In set builder notation the intervals are $\{x \mid -3 \leq x \leq -1 \text{ or } 1 \leq x \leq 2\}$.

35. (a) Shift $f(x)$ 2 units left and 1 unit down. See Figure 35a.

(b) Multiply the y-coordinate of each point by -2 and plot. The new graph will open up. See Figure 35b.

(c) Shift $f(x)$ up one unit and reflect across the y-axis. See Figure 35c.

(d) Multiply the x-coordinate of each point by 2 and plot. See Figure 35d.

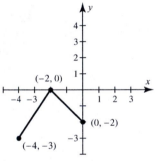

Figure 35a

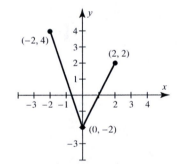

Figure 35b

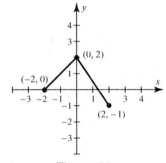

Figure 35c

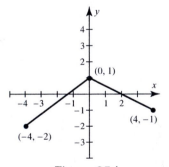

Figure 35d

37. (a) It is increasing: $(-\infty, -2)$ and $(1, \infty)$; it is decreasing: $(-2, 1)$. In set builder notation the intervals are

$\{x \mid x < -2 \text{ or } x > 1\}$ and $\{x \mid -2 < x < 1\}$.

(b) The zeros are approximately $x = -3.3, 0$ and 1.8.

(c) The turning points are: $(-2, 3)$ and $(1, -1)$.

(d) It has local extrema, maximum: 3, and minimum: -1.

39. Answers may vary. See Figure 39.

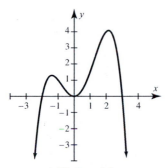

Figure 39

41. (a)

$$a - 2 + \frac{12}{4a^2}$$
$$4a^2 \overline{)4a^3 - 8a^2 + 12}$$
$$\underline{4a^3}$$
$$-8a^2 + 12$$
$$\underline{-8a^2}$$
$$12$$

Therefore the quotient is: $a - 2 + \dfrac{3}{a^2}$

(b)

$$2x^2 + 2x - 2 + \frac{-1}{x-1}$$
$$x - 1 \overline{)2x^3 \qquad - 4x + 1}$$
$$\underline{2x^3 - 2x^2}$$
$$2x^2 - 4x + 1$$
$$\underline{2x^2 - 2x}$$
$$-2x + 1$$
$$\underline{-2x + 2}$$
$$-1$$

Therefore the quotient is: $2x^2 + 2x - 2 + \dfrac{-1}{x-1}$

43. $f(x) = 4(x + 3)(x - 1)^2(x - 4)^3$

45. $\dfrac{3 + 4i}{1 - i} \cdot \dfrac{1 + i}{1 + i} = \dfrac{3 + 3i + 4i + 4i^2}{1 - (i)^2} = \dfrac{3 + 7i + 4(-1)}{1 - (-1)} = \dfrac{-1 + 7i}{2}$ or $\dfrac{-1}{2} + \dfrac{7i}{2}$

47. For the domain the denominator cannot equal zero $\Rightarrow$ set $x^2 - 3x - 4 = 0 \Rightarrow (x + 1)(x - 4) = 0$.

Then $D = \{x \mid x \neq -1, x \neq 4\}$. Since the degree of the numerator is less then the degree of the denominator

the horizontal asymptote is $y = 0$. The vertical asymptotes are values when the demoninator equals $0 \Rightarrow$

the vertical asymptotes are: $x = -1, x = 4$.

49. For m_1 use $(0, 0)$ and $(2, 40,000)$ $\Rightarrow \dfrac{40,000 - 0}{2 - 0} \Rightarrow \dfrac{40,000}{2} \Rightarrow m_1 = 20,000$; the pool is being filled at a

    rate of 20,000 gallons per hour.

    For m_2 use $(2, 40,000)$ and $(3, 50,000)$ $\Rightarrow \dfrac{50,000 - 40,000}{3 - 2} \Rightarrow \dfrac{10,000}{1} \Rightarrow m_2 = 10,000$; now the pool is

    being filled at a rate of 10,000 gallons per hour.

    For m_3 use $(3, 50,000)$ and $(4, 50,000)$ $\Rightarrow \dfrac{50,000 - 50,000}{4 - 3} \Rightarrow \dfrac{0}{1} \Rightarrow m_3 = 0$; the amount of water in the

    pool is remaining constant.

    For m_4 use $(4, 50,000)$ and $(6, 20,000)$ $\Rightarrow \dfrac{20,000 - 50,000}{6 - 4} \Rightarrow \dfrac{-30,000}{2} \Rightarrow m_4 = -15,000$; the pool is

    being drained at a rate of 15,000 gallons per hour.

51. $0.35(2) + 0.12x = 0.20(x + 2) \Rightarrow 0.70 + 0.12x = 0.20x + 0.40 \Rightarrow 0.30 = 0.08x \Rightarrow x = 3.75$ liters

53. Graph $R(x) = x(800 - x)$ with x by the 100's from 0 to 800. See Figure 53. The maximum revenue is made

    when 400 toy figures are sold.

[0, 1000, 100] by [0, 200,000, 20,000] [0, 300, 100] by [0, 50,000, 10,000]

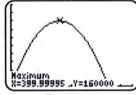 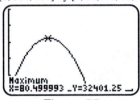

Figure 53   Figure 55

55. (a) $C(t) = t(805 - 5t)$

    (b) $C(t) = 17,000 \Rightarrow 17,000 = t(805 - 5t) \Rightarrow 17,000 = 805t - 5t^2 \Rightarrow 5t^2 - 805t + 17,000 = 0 \Rightarrow$
       $5(t^2 - 161t + 3400) = 0 \Rightarrow 5(t - 25)(t - 136) = 0 \Rightarrow t = 25, 136$. The cost is $17,000 when either
       25 or 136 tickets are purchased.

    (c) Graph $C(t) = t(805 - 5t)$. See Figure 55. The maximum cost of \$32,400 is reached at 80 to 81 tickets
       are sold.

57. Since volume of a cylinder is: $V = \pi r^2 h$. Then $10\pi = \pi r^2 h \Rightarrow h = \dfrac{10}{r^2}$.

    The surface area of the can is: $S(r) = 2\pi r^2 + 2\pi r h \Rightarrow S(r) = 2\pi r^2 + 2\pi r \cdot \dfrac{10}{r^2} \Rightarrow$

    $S(r) = 2\pi r^2 + \dfrac{20\pi}{r}$. Graph this equation. See Figure 57. The minimum surface area occurs when

    $r \approx 1.7$ inches $\Rightarrow$ then $h = \dfrac{10}{(1.7)^2}$ or $h \approx 3.5$ inches.

[0, 5, 1] by [0, 100, 10]

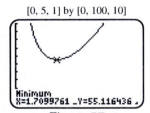

Figure 57

Chapter 5: Exponential and Logarithmic Functions

5.1: Combining Functions

1. $(f + g)(3) = f(3) + g(3) = 2 + 5 = 7$

3. $(fg)(x) = f(x)g(x) = (x^2)(4x) = 4x^3$

5. $(g \circ f)(x) = g(f(x))$. Therefore, $(g \circ f)(x)$ calculates the cost of x square yards of carpet.

7. (a) $f(3) = 2(3) - 3 = 3$ and $g(3) = 1 - 3^2 = -8$. Thus, $(f + g)(3) = f(3) + g(3) = 3 + (-8) = -5$.

   (b) $f(-1) = 2(-1) - 3 = -5$ and $g(-1) = 1 - (-1)^2 = 0$.

   Thus, $(f - g)(-1) = f(-1) - g(-1) = -5 - 0 = -5$.

   (c) $f(0) = 2(0) - 3 = -3$ and $g(0) = 1 - 0^2 = 1$. Thus, $(fg)(0) = f(0) \cdot g(0) = (-3)(1) = -3$.

   (d) $f(2) = 2(2) - 3 = 1$ and $g(2) = 1 - 2^2 = -3$. Thus, $(f/g)(2) = \dfrac{f(2)}{g(2)} = \dfrac{1}{-3} = -\dfrac{1}{3}$.

9. (a) $f(2) = 2(2) + 1 = 5$ and $g(2) = \dfrac{1}{2}$. Thus, $(f + g)(2) = f(2) + g(2) = 5 + \dfrac{1}{2} = \dfrac{11}{2}$.

   (b) $f\left(\dfrac{1}{2}\right) = 2\left(\dfrac{1}{2}\right) + 1 = 2$ and $g\left(\dfrac{1}{2}\right) = \dfrac{1}{\frac{1}{2}} = 2$. Thus, $(f - g)\left(\dfrac{1}{2}\right) = f\left(\dfrac{1}{2}\right) - g\left(\dfrac{1}{2}\right) = 2 - 2 = 0$.

   (c) $f(4) = 2(4) + 1 = 9$ and $g(4) = \dfrac{1}{4}$. Thus, $(fg)(4) = f(4) \cdot g(4) = (9)\left(\dfrac{1}{4}\right) = \dfrac{9}{4}$.

   (d) $f(0) = 2(0) + 1 = 1$ and $g(0) = \dfrac{1}{0} \Rightarrow$ undefined. Thus, $(f/g)(0) \Rightarrow$ undefined.

11. (a) $(f + g)(x) = f(x) + g(x) = 2x + x^2$; Domain: all real numbers.

    (b) $(f - g)(x) = f(x) - g(x) = 2x - x^2$; Domain: all real numbers.

    (c) $(fg)(x) = f(x) \cdot g(x) = (2x)(x^2) = 2x^3$; Domain: all real numbers.

    (d) $(f/g)(x) = \dfrac{f(x)}{g(x)} = \dfrac{2x}{x^2} = \dfrac{2}{x}$; Domain: $\{x \mid x \neq 0\}$.

13. (a) $(f + g)(x) = f(x) + g(x) = (x^2 - 1) + (x^2 + 1) = 2x^2$; Domain: all real numbers.

    (b) $(f - g)(x) = f(x) - g(x) = (x^2 - 1) - (x^2 + 1) = -2$; Domain: all real numbers.

    (c) $(fg)(x) = f(x) \cdot g(x) = (x^2 - 1)(x^2 + 1) = x^4 - 1$; Domain: all real numbers.

    (d) $(f/g)(x) = \dfrac{f(x)}{g(x)} = \dfrac{x^2 - 1}{x^2 + 1}$; Domain: all real numbers.

15. (a) $(f + g)(x) = f(x) + g(x) = (x - \sqrt{x - 1}) + (x + \sqrt{x - 1}) = 2x$; Domain: $\{x \mid x \geq 1\}$.

    (b) $(f - g)(x) = f(x) - g(x) = (x - \sqrt{x - 1}) - (x + \sqrt{x - 1}) = -2\sqrt{x - 1}$; Domain: $\{x \mid x \geq 1\}$.

    (c) $(fg)(x) = f(x) \cdot g(x) = (x - \sqrt{x - 1})(x + \sqrt{x - 1}) = x^2 - (x - 1) = x^2 - x + 1$;

    Domain: $\{x \mid x \geq 1\}$.

    (d) $(f/g)(x) = \dfrac{f(x)}{g(x)} = \dfrac{x - \sqrt{x - 1}}{x + \sqrt{x - 1}}$; Domain: $\{x \mid x \geq 1\}$.

17. (a) $(f + g)(x) = f(x) + g(x) = (\sqrt{x} - 1) + (\sqrt{x} + 1) = 2\sqrt{x}$; Domain: $\{x \mid x \geq 0\}$.

    (b) $(f - g)(x) = f(x) - g(x) = (\sqrt{x} - 1) - (\sqrt{x} + 1) = -2$; Domain: $\{x \mid x \geq 0\}$.

    (c) $(fg)(x) = f(x) \cdot g(x) = (\sqrt{x} - 1)(\sqrt{x} + 1) = x - 1$; Domain: $\{x \mid x \geq 0\}$.

    (d) $(f/g)(x) = \dfrac{f(x)}{g(x)} = \dfrac{\sqrt{x} - 1}{\sqrt{x} + 1}$; Domain: $\{x \mid x \geq 0\}$.

19. (a) $(f + g)(x) = f(x) + g(x) = \dfrac{1}{x + 1} + \dfrac{3}{x + 1} = \dfrac{4}{x + 1}$; Domain: $\{x \mid x \neq -1\}$.

    (b) $(f - g)(x) = f(x) - g(x) = \dfrac{1}{x + 1} - \dfrac{3}{x + 1} = -\dfrac{2}{x + 1}$; Domain: $\{x \mid x \neq -1\}$.

    (c) $(fg)(x) = f(x) \cdot g(x) = \left(\dfrac{1}{x + 1}\right)\left(\dfrac{3}{x + 1}\right) = \dfrac{3}{(x + 1)^2}$; Domain: $\{x \mid x \neq -1\}$.

    (d) $(f/g)(x) = \dfrac{f(x)}{g(x)} = \dfrac{\frac{1}{x+1}}{\frac{3}{x+1}} = \dfrac{1}{x + 1} \cdot \dfrac{x + 1}{3} = \dfrac{1}{3}$; Domain: $\{x \mid x \neq -1\}$.

21. (a) $(f + g)(x) = f(x) + g(x) = \dfrac{1}{2x - 4} + \dfrac{x}{2x - 4} = \dfrac{1 + x}{2x - 4}$; Domain: $\{x \mid x \neq 2\}$.

    (b) $(f - g)(x) = f(x) - g(x) = \dfrac{1}{2x - 4} - \dfrac{x}{2x - 4} = \dfrac{1 - x}{2x - 4}$; Domain: $\{x \mid x \neq 2\}$.

    (c) $(fg)(x) = f(x) \cdot g(x) = \left(\dfrac{1}{2x - 4}\right)\left(\dfrac{x}{2x - 4}\right) = \dfrac{x}{(2x - 4)^2}$; Domain: $\{x \mid x \neq 2\}$.

    (d) $(f/g)(x) = \dfrac{f(x)}{g(x)} = \dfrac{\frac{1}{2x-4}}{\frac{x}{2x-4}} = \dfrac{1}{2x - 4} \cdot \dfrac{2x - 4}{x} = \dfrac{1}{x}$; Domain: $\{x \mid x \neq 0 \text{ and } x \neq 2\}$.

23. (a) $(f + g)(x) = f(x) + g(x) = (x^2 - 1) + (|x + 1|) = x^2 - 1 + |x + 1|$; Domain: all real numbers.

    (b) $(f - g)(x) = f(x) - g(x) = (x^2 - 1) - (|x + 1|) = x^2 - 1 - |x + 1|$; Domain: all real numbers.

    (c) $(fg)(x) = f(x) \cdot g(x) = (x^2 - 1)(|x + 1|) = (x^2 - 1)(|x + 1|)$; Domain: all real numbers.

    (d) $(f/g)(x) = \dfrac{f(x)}{g(x)} = \dfrac{x^2 - 1}{|x + 1|}$; Domain: $\{x \mid x \neq -1\}$.

25. (a) $(f + g)(x) = f(x) + g(x) = \dfrac{(x - 1)(x - 2)}{x + 1} + \dfrac{(x + 1)(x - 1)}{x - 2} =$

    $\dfrac{(x - 1)(x - 2)(x - 2)}{(x + 1)(x - 2)} + \dfrac{(x - 1)(x + 1)(x + 1)}{(x + 1)(x - 2)} =$

    $\dfrac{(x - 1)(x^2 - 4x + 4)}{(x + 1)(x - 2)} + \dfrac{(x - 1)(x^2 + 2x + 1)}{(x + 1)(x - 2)} = \dfrac{(x - 1)(2x^2 - 2x + 5)}{(x + 1)(x - 2)}$;

    Domain: $\{x \mid x \neq -1, x \neq 2\}$

    (b) $(f - g)(x) = f(x) - g(x) = \dfrac{(x - 1)(x - 2)}{x + 1} - \dfrac{(x + 1)(x - 1)}{x - 2} =$

    $\dfrac{(x - 1)(x - 2)(x - 2)}{(x + 1)(x - 2)} - \dfrac{(x - 1)(x + 1)(x + 1)}{(x + 1)(x - 2)} =$

    $\dfrac{(x - 1)(x^2 - 4x + 4)}{(x + 1)(x - 2)} - \dfrac{(x - 1)(x^2 + 2x + 1)}{(x + 1)(x - 2)} = \dfrac{(x - 1)(-6x + 3)}{(x + 1)(x - 2)} = \dfrac{-3(x - 1)(2x - 1)}{(x + 1)(x - 2)}$;

    Domain: $\{x \mid x \neq -1, x \neq 2\}$

    (c) $(fg)(x) = f(x) \cdot g(x) = \dfrac{(x - 1)(x - 2)}{x + 1} \cdot \dfrac{(x + 1)(x - 1)}{x - 2} =$

    $\dfrac{(x - 2)(x - 1)(x + 1)(x - 1)}{(x + 1)(x - 2)} = (x - 1)(x - 1) = (x - 1)^2$; Domain: $\{x \mid x \neq -1, x \neq 2\}$

(d) $(f/g)(x) = \dfrac{f(x)}{g(x)} = \dfrac{(x-2)(x-1)}{x+1} \div \dfrac{(x+1)(x-1)}{x-2} = \dfrac{(x-2)(x-1)}{x+1} \cdot \dfrac{x-2}{(x+1)(x-1)} =$

$\dfrac{(x-2)^2}{(x+1)^2}$; Domain: $\{x \mid x \neq 1, x \neq -1 \text{ and } x \neq 2\}$

27. (a) $(f+g)(x) = f(x) + g(x) = \dfrac{2}{(x+1)(x-1)} + \dfrac{(x+1)}{(x-1)(x-1)} =$

$\dfrac{2(x-1)}{(x+1)(x-1)^2} + \dfrac{(x+1)^2}{(x+1)(x-1)^2} = \dfrac{2x-2}{(x+1)(x-1)^2} + \dfrac{x^2+2x+1}{(x+1)(x-1)^2} = \dfrac{x^2+4x-1}{(x+1)(x-1)^2}$;

Domain: $\{x \mid x \neq -1, x \neq 1\}$

(b) $(f-g)(x) = f(x) - g(x) = \dfrac{2}{(x+1)(x-1)} - \dfrac{(x+1)}{(x-1)(x-1)} =$

$\dfrac{2(x-1)}{(x+1)(x-1)^2} - \dfrac{(x+1)^2}{(x+1)(x-1)} = \dfrac{2x-2}{(x+1)(x-1)^2} - \dfrac{x^2+2x+1}{(x+1)(x-1)^2} = \dfrac{-x^2-3}{(x+1)(x-1)^2}$;

Domain: $\{x \mid x \neq -1, x \neq 1\}$

(c) $(fg)(x) = f(x) \cdot g(x) = \dfrac{2}{(x+1)(x-1)} \cdot \dfrac{(x+1)}{(x-1)(x-1)} = \dfrac{2}{(x-1)^3}$; Domain: $\{x \mid x \neq -1, x \neq 1\}$

(d) $(f/g)(x) = f(x) \div g(x) = \dfrac{2}{(x+1)(x-1)} \div \dfrac{(x+1)}{(x-1)(x-1)} =$

$\dfrac{2}{(x+1)(x-1)} \cdot \dfrac{(x-1)(x-1)}{(x+1)} = \dfrac{2(x-1)}{(x+1)^2}$; Domain: $\{x \mid x \neq -1, x \neq 1\}$

29. (a) $(f+g)(x) = f(x) + g(x) = (x^{5/2} - x^{3/2}) + (x^{1/2}) = x^{5/2} - x^{3/2} + x^{1/2} = x^{1/2}(x^2 - x + 1)$;

Domain: $\{x \mid x \geq 0\}$

(b) $(f-g)(x) = f(x) - g(x) = (x^{5/2} - x^{3/2}) - (x^{1/2}) = x^{5/2} - x^{3/2} - x^{1/2} = x^{1/2}(x^2 - x - 1)$;

Domain: $\{x \mid x \geq 0\}$

(c) $(fg)(x) = f(x) \cdot g(x) = (x^{5/2} - x^{3/2})(x^{1/2}) = x^3 - x^2 = x^2(x-1)$; Domain: $\{x \mid x \geq 0\}$

(d) $(f/g)(x) = f(x) \div g(x) = \left(\dfrac{x^{5/2} - x^{3/2}}{x^{1/2}} \right) = x^2 - x = x(x-1)$; Domain: $\{x \mid x > 0\}$

31. (a) From the graph $f(2) = 4$ and $g(2) = -2$. Thus, $(f+g)(2) = f(2) + g(2) = 4 + (-2) = 2$.

(b) From the graph $f(1) = 1$ and $g(1) = -3$. Thus, $(f-g)(1) = f(1) - g(1) = 1 - (-3) = 4$.

(c) From the graph $f(0) = 0$ and $g(0) = -4$. Thus, $(fg)(0) = f(0) \cdot g(0) = (0)(-4) = 0$.

(d) From the graph $f(1) = 1$ and $g(1) = -3$. Thus, $(f/g)(1) = \dfrac{f(1)}{g(1)} = \dfrac{1}{-3} = -\dfrac{1}{3}$.

33. (a) From the graph $f(0) = 0$ and $g(0) = 2$. Thus, $(f+g)(0) = f(0) + g(0) = 0 + 2 = 2$.

(b) From the graph $f(-1) = -2$ and $g(-1) = 1$. Thus, $(f-g)(-1) = f(-1) - g(-1) = -2 - 1 = -3$.

(c) From the graph $f(1) = 2$ and $g(1) = 1$. Thus, $(fg)(1) = f(1) \cdot g(1) = (2)(1) = 2$.

(d) From the graph $f(2) = 4$ and $g(2) = -2$. Thus, $(f/g)(2) = \dfrac{f(2)}{g(2)} = \dfrac{4}{-2} = -2$.

35. (a) $(f + g)(-1) = f(-1) + g(-1) = -3 + -2 = -5.$

    (b) $(g - f)(0) = g(0) - f(0) = 3 - 5 = -2.$

    (c) $(gf)(2) = g(2) \cdot f(2) = 0 \cdot 1 = 0.$

    (d) $(f/g)(2) = \dfrac{f(2)}{g(2)} = \dfrac{1}{0} \Rightarrow$ undefined.

37. (a) $(f + g)(2) = f(2) + g(2) = 7 + (-2) = 5$

    (b) $(f - g)(4) = f(4) - g(4) = 10 - 5 = 5$

    (c) $(fg)(-2) = f(-2) \cdot g(-2) = (0)(6) = 0$

    (d) $(f/g)(0) = \dfrac{f(0)}{g(0)} = \dfrac{5}{0} \Rightarrow$ undefined

39. For example: $(f - g)(2) = f(2) - g(2) = 7 - (-2) = 9.$ See Figure 39. A dash (—) indicates that the value of the function is undefined.

x	-2	0	2	4
$(f + g)(x)$	6	5	5	15
$(f - g)(x)$	-6	5	9	5
$(fg)(x)$	0	0	-14	50
$(f/g)(x)$	0	—	-3.5	2

Figure 39

41. (a) $g(x) = 2x + 1 \Rightarrow g(-3) = 2(-3) + 1 \Rightarrow g(-3) = -5$

    (b) $g(x) = 2x + 1 \Rightarrow g(b) = 2(b) + 1 \Rightarrow g(b) = 2b + 1$

    (c) $g(x) = 2x + 1 \Rightarrow g(x^3) = 2(x^3) + 1 \Rightarrow g(x^3) = 2x^3 + 1$

    (d) $g(x) = 2x + 1 \Rightarrow g(2x - 3) = 2(2x - 3) + 1 \Rightarrow g(2x - 3) = 4x - 6 + 1 \Rightarrow$
    $g(2x - 3) = 4x - 5$

43. (a) $g(x) = 2(x + 3)^2 - 4 \Rightarrow g(-3) = 2((-3) + 3)^2 - 4 \Rightarrow g(-3) = 2(0) - 4 \Rightarrow g(-3) = -4$

    (b) $g(x) = 2(x + 3)^2 - 4 \Rightarrow g(b) = 2((b) + 3)^2 - 4 \Rightarrow g(b) = 2(b + 3)^2 - 4$

    (c) $g(x) = 2(x + 3)^2 - 4 \Rightarrow g(x^3) = 2((x^3) + 3)^2 - 4 \Rightarrow g(x^3) = 2(x^3 + 3)^2 - 4$

    (d) $g(x) = 2(x + 3)^2 - 4 \Rightarrow g(2x - 3) = 2((2x - 3) + 3)^2 - 4 \Rightarrow g(2x - 3) = 2(2x)^2 - 4 \Rightarrow$
    $g(2x - 3) = 2(4x^2) - 4 \Rightarrow g(2x - 3) = 8x^2 - 4$

45. (a) $g(x) = \dfrac{1}{2}x^2 + 3x - 1 \Rightarrow g(-3) = \dfrac{1}{2}(-3)^2 + 3(-3) - 1 \Rightarrow g(-3) = \dfrac{9}{2} - 9 - 1 \Rightarrow$
    $g(-3) = \dfrac{9}{2} - \dfrac{20}{2} \Rightarrow g(-3) = -\dfrac{11}{2}$

    (b) $g(x) = \dfrac{1}{2}x^2 + 3x - 1 \Rightarrow g(b) = \dfrac{1}{2}(b)^2 + 3(b) - 1 \Rightarrow g(b) = \dfrac{1}{2}b^2 + 3b - 1$

    (c) $g(x) = \dfrac{1}{2}x^2 + 3x - 1 \Rightarrow g(x^3) = \dfrac{1}{2}(x^3)^2 + 3(x^3) - 1 \Rightarrow g(x^3) = \dfrac{1}{2}x^6 + 3x^3 - 1$

    (d) $g(x) = \dfrac{1}{2}x^2 + 3x - 1 \Rightarrow g(2x - 3) = \dfrac{1}{2}(2x - 3)^2 + 3(2x - 3) - 1 \Rightarrow$
    $g(2x - 3) = \dfrac{4x^2 - 12x + 9}{2} + 6x - 9 - 1 \Rightarrow g(2x - 3) = 2x^2 - 6x + \dfrac{9}{2} + 6x - \dfrac{20}{2} \Rightarrow$
    $g(2x - 3) = 2x^2 - \dfrac{11}{2}$

47. (a) $g(x) = \sqrt{x+4} \Rightarrow g(-3) = \sqrt{(-3)+4} \Rightarrow g(-3) = \sqrt{1} \Rightarrow g(-3) = 1$

    (b) $g(x) = \sqrt{x+4} \Rightarrow g(b) = \sqrt{(b)+4} \Rightarrow g(b) = \sqrt{b+4}$

    (c) $g(x) = \sqrt{x+4} \Rightarrow g(x^3) = \sqrt{(x^3)+4} \Rightarrow g(x^3) = \sqrt{x^3+4}$

    (d) $g(x) = \sqrt{x+4} \Rightarrow g(2x-3) = \sqrt{(2x-3)+4} \Rightarrow g(2x-3) = \sqrt{2x+1}$

49. (a) $g(x) = |3x-1|+4 \Rightarrow g(-3) = |3(-3)-1|+4 \Rightarrow g(-3) = |-10|+4 \Rightarrow g(-3) = 14$

    (b) $g(x) = |3x-1|+4 \Rightarrow g(b) = |3(b)-1|+4 \Rightarrow g(b) = |3b-1|+4$

    (c) $g(x) = |3x-1|+4 \Rightarrow g(x^3) = |3(x^3)-1|+4 \Rightarrow g(x^3) = |3x^3-1|+4$

    (d) $g(x) = |3x-1|+4 \Rightarrow g(2x-3) = |3(2x-3)-1|+4 \Rightarrow g(2x-3) = |6x-10|+4$

51. (a) $g(x) = \dfrac{4x}{x+3} \Rightarrow g(-3) = \dfrac{4(-3)}{(-3)+3} \Rightarrow g(-3) = \dfrac{-12}{0} \Rightarrow g(-3) =$ undefined

    (b) $g(x) = \dfrac{4x}{x+3} \Rightarrow g(b) = \dfrac{4(b)}{b+3} \Rightarrow g(b) = \dfrac{4b}{b+3}$

    (c) $g(x) = \dfrac{4x}{x+3} \Rightarrow g(x^3) = \dfrac{4(x^3)}{x^3+3} \Rightarrow g(x^3) = \dfrac{4x^3}{x^3+3}$

    (d) $g(x) = \dfrac{4x}{x+3} \Rightarrow g(2x-3) = \dfrac{4(2x-3)}{(2x-3)+3} \Rightarrow g(2x-3) = \dfrac{8x-12}{2x} \Rightarrow$

    $g(2x-3) = \dfrac{4x-6}{x} \Rightarrow g(2x-3) = \dfrac{2(2x-3)}{x}$

53. (a) $(f \circ g)(2) = f(g(2)) = f(2^2) = f(4) = \sqrt{4+5} = \sqrt{9} = 3$

    (b) $(g \circ f)(-1) = g(f(-1)) = g(\sqrt{-1+5}) = g(2) = 2^2 = 4$

55. (a) $(f \circ g)(-4) = f(g(-4)) = f(|-4|) = f(4) = 5(4) - 2 = 18$

    (b) $(g \circ f)(5) = g(f(5)) = g(5(5)-2) = g(23) = |23| = 23$

57. (a) $(f \circ g)(x) = f(g(x)) = f(x^2+3x-1) = (x^2+3x-1)^3$; D: all real numbers

    (b) $(g \circ f)(x) = g(f(x)) = g(x^3) = (x^3)^2 + 3(x^3) - 1 = x^6 + 3x^3 - 1$; D: all real numbers

    (c) $(f \circ f)(x) = f(f(x)) = f(x^3) = (x^3)^3 = x^9$; D: all real numbers

59. (a) $(f \circ g)(x) = f(g(x)) = f(x^4+x^2-3x-4) = x^4+x^2-3x-2$; D: all real numbers

    (b) $(g \circ f)(x) = g(f(x)) = g(x+2) = (x+2)^4 + (x+2)^2 - 3(x+2) - 4$; D: all real numbers

    (c) $(f \circ f)(x) = f(f(x)) = f(x+2) = (x+2) + 2 = x+4$; D: all real numbers

61. (a) $(f \circ g)(x) = f(g(x)) = f(x^3) = 2 - 3x^3$; D: all real numbers

    (b) $(g \circ f)(x) = g(f(x)) = g(2-3x) = (2-3x)^3$; D: all real numbers

    (c) $(f \circ f)(x) = f(f(x)) = f(2-3x) = 2 - 3(2-3x) = 9x - 4$; D: all real numbers

63. (a) $(f \circ g)(x) = f(g(x)) = f(5x) = \dfrac{1}{5x+1}$; $D = \left\{ x \,\middle|\, x \neq -\dfrac{1}{5} \right\}$

    (b) $(g \circ f)(x) = g(f(x)) = g\!\left(\dfrac{1}{x+1}\right) = 5\!\left(\dfrac{1}{x+1}\right) = \dfrac{5}{x+1}$; $D = \{ x \mid x \neq -1 \}$

    (c) $(f \circ f)(x) = f(f(x)) = f\!\left(\dfrac{1}{x+1}\right) = \dfrac{1}{\frac{1}{x+1}+1} = \dfrac{1}{\frac{1+x+1}{x+1}} = \dfrac{x+1}{x+2}$;

    $D = \{ x \mid x \neq -1 \text{ and } x \neq -2 \}$

65. (a) $(f \circ g)(x) = f(g(x)) = f(\sqrt{4 - x^2}) = \sqrt{4 - x^2} + 4; \ D = \{x | -2 \le x \le 2\}$

    (b) $(g \circ f)(x) = g(f(x)) = g(x + 4) = \sqrt{4 - (x + 4)^2}; D = \{x | -6 \le x \le -2\}$

    (c) $(f \circ f)(x) = f(f(x)) = f(x + 4) = (x + 4) + 4 = x + 8$; D: all real numbers

67. (a) $(f \circ g)(x) = f(g(x)) = f(3x) = \sqrt{3x - 1}$; $D = \left\{x \mid x \ge \dfrac{1}{3}\right\}$

    (b) $(g \circ f)(x) = g(f(x)) = g(\sqrt{x - 1}) = 3\sqrt{x - 1}$; $D = \{x | x \ge 1\}$

    (c) $(f \circ f)(x) = f(f(x)) = f(\sqrt{x - 1}) = \sqrt{\sqrt{x - 1} - 1}; D = \{x | x \ge 2\}$

69. (a) $(f \circ g)(x) = f(g(x)) = f\left(\dfrac{1 - x}{5}\right) = 1 - 5\left(\dfrac{1 - x}{5}\right) = 1 - (1 - x) = x; \ D$: all real numbers

    (b) $(g \circ f)(x) = g(f(x)) = g(1 - 5x) = \dfrac{1 - (1 - 5x)}{5} = \dfrac{5x}{5} = x; \ D$: all real numbers

    (c) $(f \circ f)(x) = f(f(x)) = f(1 - 5x) = 1 - 5(1 - 5x) = 1 - 5 + 25x = 25x - 4;$

    D: all real numbers

71. (a) $(f \circ g)(x) = f(g(x)) = f\left(\dfrac{1}{kx}\right) = \dfrac{1}{k\left(\frac{1}{kx}\right)} = \dfrac{1}{\frac{1}{x}} = x; \ D: \{x | x \ne 0\}$

    (b) $(g \circ f)(x) = g(f(x)) = g\left(\dfrac{1}{kx}\right) = \dfrac{1}{k\left(\frac{1}{kx}\right)} = \dfrac{1}{\frac{1}{x}} = x; \ D: \{x | x \ne 0\}$

    (c) $(f \circ f)(x) = f(f(x)) = f\left(\dfrac{1}{kx}\right) = \dfrac{1}{k\left(\frac{1}{kx}\right)} = \dfrac{1}{\frac{1}{x}} = x; \ D: \{x | x \ne 0\}$

73. (a) $(f \circ g)(4) = f(g(4)) = f(0) = -4$

    (b) $(g \circ f)(3) = g(f(3)) = g(2) = 2$

    (c) $(f \circ f)(2) = f(f(2)) = f(0) = -4$

75. (a) $(f \circ g)(1) = f(g(1)) = f(2) = -3$

    (b) $(g \circ f)(-2) = g(f(-2)) = g(-3) = -2$

    (c) $(g \circ g)(-2) = g(g(-2)) = g(-1) = 0$

77. (a) $(g \circ f)(1) = g(f(1)) = g(4) = 5$

    (b) $(f \circ g)(4) = f(g(4)) = f(5) \Rightarrow$ undefined

    (c) $(f \circ f)(3) = f(f(3)) = f(1) = 4$

79. $g(3) = 4; \ f(4) = 2$

81. $h(x) = \sqrt{x - 2} \Rightarrow g(x) = \sqrt{x}$ and $f(x) = x - 2$. **Answers may vary.**

83. $h(x) = \dfrac{1}{x + 2} \Rightarrow g(x) = \dfrac{1}{x}$ and $f(x) = x + 2$. **Answers may vary.**

85. $h(x) = 4(2x + 1)^3 \Rightarrow g(x) = 4x^3$ and $f(x) = 2x + 1$. **Answers may vary.**

87. $h(x) = (x^3 - 1)^2 \Rightarrow g(x) = x^2$ and $f(x) = x^3 - 1$. **Answers may vary.**

89. $h(x) = -4|x + 2| - 3 \Rightarrow g(x) = -4|x| - 3$ and $f(x) = x + 2$. **Answers may vary.**

91. $h(x) = \dfrac{1}{(x - 1)^2} \Rightarrow g(x) = \dfrac{1}{x^2}$ and $f(x) = x - 1$. **Answers may vary.**

93. $h(x) = x^{3/4} - x^{1/4} \Rightarrow g(x) = x^3 - x$ and $f(x) = x^{1/4}$. *Answers may vary.*

95. Since each album sells for \$15 each, the revenue function would be $R(x) = 15x$. Therefore,

$$P(x) = R(x) - C(x) = 15x - (5x + 12,000) = 10x - 12,000; P(3000) = 10(3000) - 12,000 = \$18,000$$

97. (a) $I(x) = 36x$

   (b) $C(x) = 2.54x$

   (c) $F(x) = (C \circ I)(x)$

   (d) $F(x) = 36(2.54x) = 91.44x$

99. (a) $d(x) = f(x) + g(x) = \dfrac{11}{6}x + \dfrac{1}{9}x^2$

   (b) $d(60) = \dfrac{11}{6}(60) + \dfrac{1}{9}(60)^2 = 510$; this car requires 510 feet to stop at 60 mph.

101. (a) $(g \circ f)(1) = g(f(1)) = g(1.5) = 5.25$;

   A 1% decrease in the ozone layer could result in a 5.25% increase in skin cancer.

   (b) $(f \circ g)(21) = f(g(21))$, but $g(21)$ is not given by the table. So $(f \circ g)(21)$ is not possible using these tables.

103. (a) $(g \circ f)(1975) = g(f(1975)) = g(3) = 4.5\%$

   (b) $(g \circ f)(x)$ computes the percent increase in peak demand during year x.

105. (a) $(g \circ f)(1960) = g(f(1960)) = g(0.11(1960 - 1948)) = g(1.32) = 1.5(1.32) = 1.98$

   In 1960 the temperature had risen 1.32°C which resulted in a 1.98% increase in peak-demand for electricity.

   (b) $(g \circ f)(x) = g(f(x)) = g(0.11(x - 1948)) = 1.5(0.11(x - 1948)) = 0.165(x - 1948)$

   (c) f, g, and $f \circ g$ are all linear functions.

107. (a) $(g \circ f)(2) = g(f(2)) = g(77) \approx 25°C$. After 2 hours the temperature is about 25°C.

   Answers may vary.

   (b) $(g \circ f)(x)$ computes the Celsius temperature after x hours.

109. $A = \pi r^2 \Rightarrow A(t) = \pi(6t)^2 = 36\pi t^2$

111. (a) For $A(s) = \dfrac{\sqrt{3}}{4}s^2$, $A(4s) = \dfrac{\sqrt{3}}{4}(4s)^2 = 16\dfrac{\sqrt{3}}{4}s^2 = 16A(s)$; if the length of a side is quadrupled, the

   area increases by a factor of 16.

   (b) For $A(s) = \dfrac{\sqrt{3}}{4}s^2$, $A(s + 2) = \dfrac{\sqrt{3}}{4}(s + 2)^2 = \dfrac{\sqrt{3}}{4}(s^2 + 4s + 4) = \dfrac{\sqrt{3}}{4}s^2 + \sqrt{3}s + \sqrt{3} =$

   $A(s) + \sqrt{3}(s + 1)$; if the length of a side increases by 2, the area increases by $\sqrt{3}(s + 1)$.

113. (a) To find the total emissions we must add the developed and developing countries' emissions for each year.

   The resulting table is shown in Figure 113.

   (b) $h(x) = f(x) + g(x)$

x	1990	2000	2010	2020	2030
$h(x)$	32	35.5	39	42.5	46

Figure 113

115. $h(x) = f(x) + g(x) = (0.1x - 172) + (0.25x - 492.5) = 0.35x - 664.5$

117. (a) $M(20) = 0.1(20) + 48 = 50, N(20) = -0.15(20)^2 + 4(20) + 120 = 140$; the median household income

was $50 thousand in 2010, while the 90th percentile was $140 thousand.

(b) $D(x) = -0.15x^2 + 4x + 120 - (0.1x + 48) = -0.15x^2 + 3.9x + 72$; The resulting formula gives the

household income difference between the 90th percentile and the median.

(c) $D(20) = -0.15(20)^2 + 3.9(20) + 72 = 90$; The income difference between the 90th percentile and the

median in 2010 was $90 thousand.

119. Let $f(x) = ax + b$ and $g(x) = cx + d$. Then $f(x) + g(x) = (ax + b) + (cx + d) =$
$(a + c)x + (b + d)$, which is linear.

121. (a) $f(x) = k$ and $g(x) = ax + b$. Thus, $(f \circ g)(x) = f(g(x)) = f(ax + b) = k$; $f \circ g$ is a constant
function.

(b) $f(x) = k$ and $g(x) = ax + b$. Thus, $(g \circ f)(x) = g(f(x)) = g(k) = ak + b$; $g \circ f$ is a constant
function.

5.2: Inverse Functions and Their Representations

1. Closing a window.

3. Closing a book, standing up, and walking out of the classroom.

5. Subtract 2 from x; $x + 2$ and $x - 2$

7. Divide x by 3 and add 2; $3(x - 2)$ and $\dfrac{x}{3} + 2$

9. Subtract 1 from x and cube the result; $\sqrt[3]{x} + 1$ and $(x - 1)^3$

11. Take the reciprocal of x; $\dfrac{1}{x}$ and $\dfrac{1}{x}$

13. Since a horizontal line will intersect the graph at most once, f is one-to-one.

15. Since a horizontal line will intersect the graph two times, f is not one-to-one.

17. Since a horizontal line will intersect the graph two times, f is not one-to-one.

19. Since $f(2) = 3$ and $f(3) = 3$, different inputs result in the same output. Therefore f is not one-to-one. Does

not have an inverse.

21. Since different inputs always result in different outputs, f is one-to-one. Does have an inverse.

23. Since the graph of f is a line sloping upward from left to right, a horizontal line can intersect it at most once.

Therefore, f is one-to-one.

25. Since the graph of f is a parabola, a horizontal line can intersect it more than once. Therefore, f is not one-to-one.

27. Since the graph of f is a parabola, a horizontal line can intersect it more than once. Therefore, f is not one-to-one.

29. Since the graph of f is a V-shape, a horizontal line can intersect it more than once. Therefore, f is not one-to-one.

31. Since the graph of f is both increasing and decreasing, a horizontal line can intersect it more than once. Therefore, f is not one-to-one.

33. Since the graph of f is both increasing and decreasing, a horizontal line can intersect it more than once. Therefore, f is not one-to-one.

35. Since the graph of f is a line sloping upward from left to right, a horizontal line can intersect it at most once. Therefore f is one-to-one.

37. Since the person goes up and down, there would be several times when the person attained a particular height above the ground. A one-to-one function would not model this situation.

39. Since the population of the United States increased each of these years, a horizontal line would intersect a graph of this population at most once. A one-to-one function would model this situation.

41. The inverse operation of taking the cube root of x is cubing x. Therefore, $f^{-1}(x) = x^3$.

43. The inverse operations of multiplying by -2 and adding 10 are subtracting 10 and dividing by -2.
    Therefore, $f^{-1}(x) = \dfrac{x - 10}{-2} = -\dfrac{1}{2}x + 5$.

45. The inverse operations of multiplying by 3 and subtracting 1 are adding 1 and dividing by 3.
    Therefore, $f^{-1}(x) = \dfrac{x + 1}{3}$.

47. The inverse operations of cubing a number, multiplying by 2 and subtracting 5 are adding 5, dividing by 2 and taking the cube root of the result. Therefore, $f^{-1}(x) = \sqrt[3]{\dfrac{x + 5}{2}}$.

49. The inverse operations of squaring a number and subtracting 1 are adding 1 and taking the square root.
    Therefore, $f^{-1}(x) = \sqrt{x + 1}$.

51. The inverse operations of multiplying by 2 and taking the reciprocal are taking the reciprocal and dividing by 2.
    Therefore, $f^{-1}(x) = \dfrac{1}{x} \div 2 = \dfrac{1}{2x}$.

53. The inverse operations of multiplying by -5, adding 4, multiplying by $\dfrac{1}{2}$, and adding 1 are subtracting 1,

    dividing by $\dfrac{1}{2}$ (or multipying by 2), subtracting 4, and dividing by $-5 \Rightarrow$

    $f^{-1}(x) = \dfrac{2(x - 1) - 4}{-5} = \dfrac{2x - 6}{-5} = -\dfrac{2(x - 3)}{5}$.

55. $y = \dfrac{x}{x + 2} \Rightarrow y(x + 2) = x \Rightarrow xy + 2y = x \Rightarrow 2y = x - xy \Rightarrow 2y = x(1 - y) \Rightarrow x = \dfrac{2y}{1 - y} \Rightarrow$

    $f^{-1}(x) = \dfrac{2x}{1 - x} \Rightarrow f^{-1}(x) = -\dfrac{2x}{x - 1}$

57. $y = \dfrac{2x + 1}{x - 1} \Rightarrow y(x - 1) = 2x + 1 \Rightarrow xy - y = 2x + 1 \Rightarrow xy - 2x = y + 1 \Rightarrow$

    $x(y - 2) = y + 1 \Rightarrow x = \dfrac{y + 1}{y - 2} \Rightarrow f^{-1}(x) = \dfrac{x + 1}{x - 2}$

59. $y = \dfrac{1}{x} - 3 \Rightarrow y + 3 = \dfrac{1}{x} \Rightarrow x = \dfrac{1}{y + 3} \Rightarrow f^{-1}(x) = \dfrac{1}{x + 3}$

61. $y = \dfrac{1}{x^3 - 1} \Rightarrow \dfrac{1}{y} = x^3 - 1 \Rightarrow \dfrac{1}{y} + 1 = x^3 \Rightarrow \dfrac{1}{y} + \dfrac{y}{y} = x^3 \Rightarrow \dfrac{y + 1}{y} = x^3 \Rightarrow f^{-1}(x) = \sqrt[3]{\dfrac{x + 1}{x}}$

63. $y = 4 - x^2, x \geq 0 \Rightarrow x^2 = 4 - y \Rightarrow x = \sqrt{4 - y}$; therefore $f^{-1}(x) = \sqrt{4 - x}$

65. $y = (x - 2)^2 + 4, x \geq 2 \Rightarrow y - 4 = (x - 2)^2 \Rightarrow \sqrt{y - 4} = x - 2 \Rightarrow x = \sqrt{y - 4} + 2$;

    therefore $f^{-1}(x) = \sqrt{x - 4} + 2$.

67. $y = x^{2/3} + 1, x \geq 0 \Rightarrow x^{2/3} = y - 1 \Rightarrow x = (y - 1)^{3/2}; \Rightarrow f^{-1}(x) = (x - 1)^{3/2}$

69. $y = \sqrt{9 - 2x^2}, -\dfrac{3}{12} \leq x \leq \dfrac{3}{\sqrt{2}} \Rightarrow y^2 = 9 - 2x^2 \Rightarrow y^2 - 9 = -2x^2 \Rightarrow \dfrac{9 - y^2}{2} = x^2 \Rightarrow$

    $\sqrt{\dfrac{9 - y^2}{2}} = x; \Rightarrow f^{-1}(x) = \sqrt{\dfrac{9 - x^2}{2}}$

71. $y = 5x - 15 \Rightarrow y + 15 = 5x \Rightarrow \dfrac{y + 15}{5} = x \Rightarrow f^{-1}(x) = \dfrac{x + 15}{5}$; D and R are all real numbers.

73. $y = \sqrt[3]{x - 5} \Rightarrow y^3 = x - 5 \Rightarrow y^3 + 5 = x \Rightarrow f^{-1}(x) = x^3 + 5$; D and R are all real numbers

75. $y = \dfrac{x - 5}{4} \Rightarrow 4y = x - 5 \Rightarrow 4y + 5 = x \Rightarrow f^{-1}(x) = 4x + 5$; D and R are all real numbers.

77. $y = \sqrt{x - 5}, x \geq 5 \Rightarrow y^2 = x - 5 \Rightarrow y^2 + 5 = x \Rightarrow f^{-1}(x) = x^2 + 5$;

    $D = \{x \mid x \geq 0\}$ and $R = \{y \mid y \geq 5\}$

79. $y = \dfrac{1}{x + 3} \Rightarrow \dfrac{1}{y} = x + 3 \Rightarrow \dfrac{1}{y} - 3 = x \Rightarrow f^{-1}(x) = \dfrac{1}{x} - 3$; $D = \{x \mid x \neq 0\}$ and $D = \{y \mid y \neq -3\}$

81. $y = 2x^3 \Rightarrow \dfrac{y}{2} = x^3 \Rightarrow \sqrt[3]{\dfrac{y}{2}} = x \Rightarrow f^{-1}(x) = \sqrt[3]{\dfrac{x}{2}}$; D and R are all real numbers.

83. $y = x^2, x \geq 0 \Rightarrow \sqrt{y} = x \Rightarrow f^{-1}(x) = \sqrt{x}$; $D = \{x \mid x \geq 0\}$ and $D = \{y \mid y \geq 0\}$

85. The domain and range of f are $D = \{1, 2, 3\}$ and $R = \{5, 7, 9\}$. Interchange these to get the domain and range

    of f^{-1}; $D = \{5, 7, 9\}$ and $R = \{1, 2, 3\}$. See Figure 85.

x	5	7	9
$f^{-1}(x)$	1	2	3

Figure 85

87. The domain and range of f are $D = \{0, 2, 4\}$ and $R = \{0, 4, 16\}$. Interchange these to get the domain and

    range of f^{-1}; $D = \{0, 4, 16\}$ and $R = \{0, 2, 4\}$. See Figure 87.

x	0	4	16
$f^{-1}(x)$	0	2	4

Figure 87

89. Since f multiplies x by 4, f^{-1} divides x by 4. See Figure 89.

x	0	2	4	6
$f^{-1}(x)$	0	$\frac{1}{2}$	1	$\frac{3}{2}$

Figure 89

91. $f(1) = 3 \Rightarrow f^{-1}(3) = 1$

93. $g(3) = 4 \Rightarrow g^{-1}(4) = 3$

95. $(f \circ g^{-1})(1) = f(g^{-1}(1)) = f(2) = 5$. Note that $g(2) = 1 \Rightarrow g^{-1}(1) = 2$.

97. $(g \circ f^{-1})(5) = g(f^{-1}(5)) = g(2) = 1$. Note that $f(2) = 5 \Rightarrow f^{-1}(5) = 2$.

99. (a) $f(1) \approx \$110$

   (b) $f^{-1}(110) \approx 1$ year

   (c) $f^{-1}(160) \approx 5$ years

   The function f^{-1} computes the number of years it takes for this savings account to accumulate x dollars.

101. (a) $f(-1) = 2$

   (b) $f^{-1}(-2) = 3$

   (c) $f^{-1}(0) = 1$

   (d) $(f^{-1} \circ f)(3) = f^{-1}(f(3)) = f^{-1}(-2) = 3$

103. (a) $f(4) = 4$

   (b) $f^{-1}(0) = 0$

   (c) $f^{-1}(6) = 9$

   (d) $(f^{-1} \circ f)(4) = f^{-1}(f(4)) = f^{-1}(4) = 4$

105. The graph of f passes through the points $(-2, -4)$ and $(1, 2)$. Thus, the graph of f^{-1} passes through the points $(-4, -2)$ and $(2, 1)$. Plot these points and sketch the reflection of f in the line $y = x$ to obtain the graph of f^{-1}. Notice that since the graph of f is a line, its reflection will also be a line. See Figure 105.

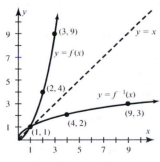

Figure 106

107. The graph of f passes through the points $\left(-2, \dfrac{1}{4}\right)$, $(0, 1)$, $(1, 2)$ and $(2, 4)$. Thus, the graph of f^{-1} passes through the points $\left(\dfrac{1}{4}, -2\right)$, $(1, 0)$, $(2, 1)$ and $(4, 2)$. Plot these points and sketch the reflection of f in the line $y = x$ to obtain the graph of f^{-1}. See Figure 107.

109. The graph of f passes through the points $(-3, -2)$, $(-1, 1)$ and $(2, 2)$. Thus, the graph of f^{-1} passes through the points $(-2, -3)$, $(1, -1)$ and $(2, 2)$. Plot these points and sketch the reflection of f in the line $y = x$ to obtain the graph of f^{-1}. See Figure 109.

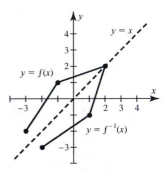

Figure 109

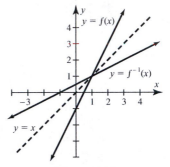

Figure 111

111. The graphs of $y = 2x - 1$, $y = \dfrac{x + 1}{2}$, and $y = x$ are shown in Figure 111.

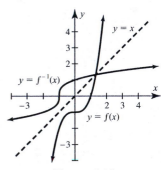

Figure 113

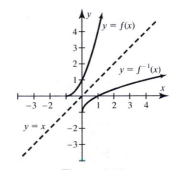

Figure 115

113. The graphs of $y = x^3 - 1$, $y = \sqrt[3]{x + 1}$, and $y = x$ are shown in Figure 113.

115. The graphs of $y = (x + 1)^2$, $y = \sqrt{x} - 1$, and $y = x$ are shown in Figure 115.

117. The graphs of $Y_1 = 3X - 1$, $Y_2 = (X + 1)/3$ and $Y_3 = X$ are shown in Figure 117.

[−4.7, 4.7, 1] by [−3.1, 3.1, 1]   [−4.7, 4.7, 1] by [−3.1, 3.1, 1]

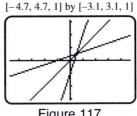

Figure 117

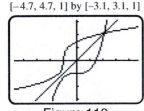

Figure 119

119. The graphs of $Y_1 = X^3/3 - 1$, $Y_2 = \sqrt[3]{(3X + 3)}$ and $Y_3 = X$ are shown in Figure 119.

121. (a) Since each volume value is the result of exactly one radius value, V represents a one-to-one function.

(b) The inverse of V computes the radius r of a sphere with volume V.

(c) $V = \dfrac{4}{3}\pi r^3 \Rightarrow r^3 = \dfrac{3V}{4\pi} \Rightarrow r = \sqrt[3]{\dfrac{3V}{4\pi}}$

(d) No; if V and r were interchanged, then r would represent the volume and V would represent the radius.

123. (a) $W = \dfrac{25}{7}(70) - \dfrac{800}{7} = \dfrac{950}{7} \approx 135.7$ pounds

(b) Yes, since no weight value corresponds to more than one height value.

(c) $W = \dfrac{25}{7}h - \dfrac{800}{7} \Rightarrow \dfrac{25}{7}h = W + \dfrac{800}{7} \Rightarrow h = \dfrac{7}{25}\left(W + \dfrac{800}{7}\right) = \dfrac{7}{25}W + 32 \Rightarrow W^{-1} = \dfrac{7}{25}W + 32$

(d) $W^{-1}(150) = \dfrac{7}{25}(150) + 32 \Rightarrow W^{-1}(150) = 74$; the maximum recommended height for a person

weighing 150 pounds is 74 inches.

(e) The inverse computes the maximum recommended height of a person with a given weight.

125. (a) $(F \circ Y)(2) = F(Y(2)) = F(3520) = 10{,}560.$ $(F \circ Y)(2)$ computes the number of feet in 2 miles.

(b) $F^{-1}(26{,}400) = 8800.$ F^{-1} converts feet to yards. There are 8800 yards in 26,400 feet.

(c) $(Y^{-1} \circ F^{-1})(21{,}120) = Y^{-1}(F^{-1}(21{,}120)) = Y^{-1}(7040) = 4.$ $(Y^{-1} \circ F^{-1})(21{,}120)$ computes the

number of miles in 21,120 feet.

127. (a) $(Q \circ C)(96) = Q(C(96)) = Q(6) = 1.5.$ $(Q \circ C)(96)$ computes the number of quarts in 96 tablespoons.

(b) $Q^{-1}(2) = 8.$ Q^{-1} converts quarts into cups. There are 8 cups in 2 quarts.

(c) $(C^{-1} \circ Q^{-1})(1.5) = C^{-1}(Q^{-1}(1.5)) = C^{-1}(6) = 96.$ $(C^{-1} \circ Q^{-1})(1.5)$ computes the number of

tablespoons in 1.5 quarts.

Extended and Discovery Exercises for Section 5.2

1. (a) f^{-1} computes the elapsed time in seconds when the rocket was x feet above the ground.

(b) The solution to the equation $f(x) = 5000$ is the time in seconds when the rocket was 5000 feet above the

ground.

(c) Evaluate $f^{-1}(5000)$.

Checking Basic Concepts for Sections 5.1 and 5.2

1. (a) $(f + g)(1) = f(1) + g(1) = -1 + 2 = 1$

(b) $(f - g)(-1) = f(-1) - g(-1) = 1 - (-2) = 3$

(c) $(fg)(0) = f(0) \cdot g(0) = (-2)(-1) = 2$

(d) $(f/g)(2) = \dfrac{f(2)}{g(2)} = \dfrac{2}{0} \Rightarrow$ undefined

(e) $(f \circ g)(2) = f(g(2)) = f(0) = -2$

(f) $(g \circ f)(-2) = g(f(-2)) = g(0) = -1$

3. (a) $(f + g)(x) = f(x) + g(x) = (x^2 + 3x - 2) + (3x - 1) = x^2 + 6x - 3$

(b) $(f/g)(x) = \dfrac{f(x)}{g(x)} = \dfrac{x^2 + 3x - 2}{3x - 1}, x \neq \dfrac{1}{3}$

(c) $(f \circ g)(x) = f(g(x)) = f(3x - 1) = (3x - 1)^2 + 3(3x - 1) - 2 = 9x^2 + 3x - 4$

5. (a) All horizontal lines intersect once $\Rightarrow$ yes; yes; $y = x + 1 \Rightarrow y - 1 = x$, therefore $f^{-1}(x) = x - 1$.

(b) A parabola $\Rightarrow$ no; no.

7. (a) Since $f(0) = -2$, then $f^{-1}(-2) = 0$.

(b) $(f^{-1} \circ g)(1) \Rightarrow f^{-1}(g(1)) \Rightarrow f^{-1}(2) = 2$

5.3: Exponential Functions and Models

1. $2^{-3} = \dfrac{1}{2^3} = \dfrac{1}{8}$

3. $3(4)^{1/2} = 3\sqrt{4} = 3(2) = 6$

5. $-2(27)^{2/3} = -2(\sqrt[3]{27})^2 = -2(3)^2 = -2(9) = -18$

7. $4^{1/6}4^{1/3} = 4^{1/6+1/3} = 4^{1/2} = \sqrt{4} = 2$

9. $3^0 = 1$

11. $(5^{101})^{1/101} = 5^1 = 5$

13. For each unit increase in x, the y-values decrease by 1.2, so the data is linear. Since $y = 2$ when $x = 0$, the function $f(x) = -1.2x + 2$ ca

15. For each unit increase in x, the y-values are multiplied by $\dfrac{1}{2}$, so the data is exponential. Since the initial value is $C = 8$ and $a = \dfrac{1}{2}$, the function $f(x) = 8\left(\dfrac{1}{2}\right)^x$ can model the data.

17. For each 2-unit increase in x, the y-values are multiplied by 4. That is, for each unit increase in x, the y-values are multiplied by 2, so the data is exponential. Since the initial value is $C = 5$ and $a = 2$, the function $f(x) = 5(2^x)$ can model the data.

19. For each unit increase in x, the y-values increase by a fixed amount, so the data are linear. The slope is $m = \dfrac{19 - (-23)}{6 - (-6)} = \dfrac{42}{12} = 3.5$. We have $y = 3.5x + b \Rightarrow 5 = 3.5(2) + b \Rightarrow 5 = 7 + b \Rightarrow b = -2$, it follows that the data can be modeled by $f(x) = 3.5x - 2$.

21. Since the data do not change a fixed amount for each unit increase in x, the data are not linear. Because we are not given $f(0)$, we cannot immediately determine C. Instead, find a by evaluating the following ratios.

$\dfrac{f(5)}{f(2)} = \dfrac{\frac{1}{3}}{9} = \dfrac{1}{27}$ It follows that $a^3 = \dfrac{1}{27} \Rightarrow a = \dfrac{1}{3}$. Next find C by using the fact that $f(2) = 9 = C\left(\dfrac{1}{3}\right)^2 \Rightarrow \dfrac{1}{9}C = 9 \Rightarrow C = 81$. Thus $f(x) = 81\left(\dfrac{1}{3}\right)^x$.

23. For $x > 4$, $f(x) > g(x)$; for example $f(10) = 1024$ whereas $g(10) = 100$. That is, $f(x) = 2^x$ becomes larger.

25. $f(x) = 2x + 1$, $f(0) = 1$, $f(10) = 21$. $g(x) = 2^{-x}$, $g(0) = 2$, $g(10) = 9.76 \times 10^{-4}$.

   Therefore, $f(x)$ becomes larger for $0 \le x \le 10$.

27. $f(0) = 5 \Rightarrow C = 5$ and $a = 1.5$

29. $f(0) = 10$ and $f(1) = 20 \Rightarrow C = 10$ and $a = \dfrac{20}{10} = 2$

31. $f(1) = 9$ and $f(2) = 27 \Rightarrow a = \dfrac{27}{9} = 3$; $C = f(0) = 9 \div 3 = 3$

33. $f(-2) = \dfrac{9}{2}$ and $f(2) = \dfrac{1}{18} \Rightarrow Ca^{-2} = \dfrac{9}{2} \Rightarrow \dfrac{C}{a^2} = \dfrac{9}{2} \Rightarrow C = \dfrac{9a^2}{2}$; $Ca^2 = \dfrac{1}{18} \Rightarrow C = \dfrac{1}{18a^2}$ thus

   $\dfrac{1}{18a^2} = \dfrac{9a^2}{2} \Rightarrow 162a^4 = 2 \Rightarrow a^4 = \dfrac{1}{81} \Rightarrow a = \dfrac{1}{3}$; then $C = \dfrac{1}{18(\frac{1}{3})^2} \Rightarrow C = \dfrac{1}{18(\frac{1}{9})} \Rightarrow C = \dfrac{1}{2}$.

35. Linear: the slope is $m = \dfrac{4 - 8}{0 - 1} = \dfrac{-4}{-1} = 4$. Since the y-intercept is given as $(0, 4)$ it follows that the data can be modeled by $f(x) = 4x + 4$.

   Exponential: Using the point $(0, 4)$ and the exponential model $g(x) = Ca^x$ we have $4 = Ca^0 \Rightarrow C = 4$. Now using the point $(1, 8)$ we have $8 = 4a^1 \Rightarrow a = 2$. The function is $g(x) = 4(2)^x$.

37. Linear: The slope is $m = \dfrac{12 - 1.5}{-2 - 1} = \dfrac{10.5}{-3} = -\dfrac{7}{2}$. We have $y = -\dfrac{7}{2}x + b \Rightarrow 12 = -\dfrac{7}{2}(-2) + b \Rightarrow$,

   $12 = 7 + b \Rightarrow b = 5$ it follows that the data can be modeled by $f(x) = -\dfrac{7}{2}x + 5$.

   Exponential: Find a by evaluating the following ratios. $\dfrac{g(1)}{g(-2)} = \dfrac{1.5}{12} = \dfrac{1}{8}$ and $\dfrac{g(1)}{g(-2)} = \dfrac{Ca^1}{Ca^{-2}} = a^3$. it follows

   that $a^3 = \dfrac{1}{8} \Rightarrow a = \dfrac{1}{2}$. Thus $g(x) = C\left(\dfrac{1}{2}\right)^1 \Rightarrow \dfrac{1}{2}C = 1.5 \Rightarrow C = 3$. Thus, $g(x) = 3\left(\dfrac{1}{2}\right)^x$.

39. $C = 5000$, $a = 2$; x represents time in hours

41. $C = 200{,}000$, $a = 0.95$; x represents the number of years after 2008

43. $f(9.5) = 30(0.9)^{9.5} \approx 11$; the tire's pressure is about 11 pounds per square inch after 9.5 minutes.

45. $f(x) = 4e^{-1.2x} \Rightarrow f(-2.4) = 4e^{-1.2(-2.4)} = 4e^{2.88} \approx 71.2571$

47. $f(x) = \dfrac{e^x - e^{-x}}{2} \Rightarrow f(-0.7) = \dfrac{e^{-0.7} - e^{0.7}}{2} \approx -0.7586$

49. See Figure 49.

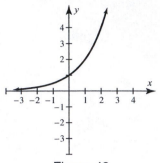

Figure 49

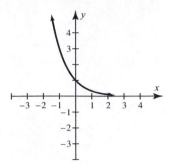

Figure 51

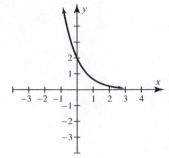

Figure 53

51. See Figure 51.

53. See Figure 53.

55. See Figure 55.

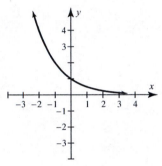

Figure 55

57. Since $y = 1$ when $x = 0$, $C = 1$ and so $y = a^x$.

Since $y = 4$ when $x = -2$, $4 = a^{-2} \Rightarrow \dfrac{1}{a^2} = 4 \Rightarrow a^2 = \dfrac{1}{4} \Rightarrow a = \dfrac{1}{2}$. That is $C = 1$ and $a = \dfrac{1}{2}$.

59. Since $y = \dfrac{1}{2}$ when $x = 0$, $C = \dfrac{1}{2}$ and so $y = \dfrac{1}{2}a^x$.

Since $y = 8$ when $x = 2$, $8 = \dfrac{1}{2}a^2 \Rightarrow a^2 = 16 \Rightarrow a = 4$. That is $C = \dfrac{1}{2}$ and $a = 4$.

61. (a) D: $(-\infty, \infty)$; R: $(0, \infty)$

   (b) Decreasing, as x increases $\left(\dfrac{1}{8}\right)^x$ decreases.

   (c) $y = 0$ as x increases, $7\left(\dfrac{1}{8}\right)^x$ gets closer and closer to 0.

   (d) y-intercept: 7; no x-intercept.

   (e) All horizontal lines intersect once $\Rightarrow$ yes; yes.

63. (i) The graph of $y = e^x$ increases faster than the graph of $y = 1.5^x$. The best choice is graph b.

   (ii) The graph of $y = 3^{-x}$ decreases faster than the graph of $y = 0.99^x$. The best choice is graph d.

   (iii) The graph of $y = 1.5^x$ increases slower than the graph of $y = e^x$. The best choice is graph a.

   (iv) The graph of $y = 0.99^x$ is almost a horizontal line since $y = 1^x$ is horizontal. The best choice is graph c.

65. (a) To graph $y = 2^x - 2$, translate the graph of $y = 2^x$ down 2 units. See Figure 65a.

   (b) To graph $y = 2^{x-1}$, translate the graph of $y = 2^x$ right 1 unit. See Figure 65b.

   (c) To graph $y = 2^{-x}$, reflect the graph of $y = 2^x$ about the *y*-axis. See Figure 65c.

   (d) To graph $y = -2^x$, reflect the graph of $y = 2^x$ about the *x*-axis. See Figure 65d.

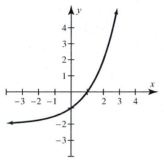

Figure 65a

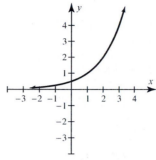

Figure 65b

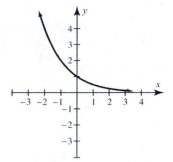

Figure 65c

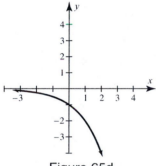

Figure 65d

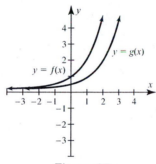

Figure 67

67. See Figure 67.

69. See Figure 69

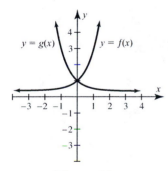

Figure 69

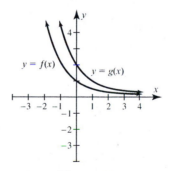

Figure 71

71. See Figure 71

73. $A_n = A_0(1 + r)^n \Rightarrow A_5 = 600(1 + 0.07)^5 \approx \$841.53$

75. $A_n = A_0\left(1 + \dfrac{r}{m}\right)^{mn} \Rightarrow A_{20} = 950\left(1 + \dfrac{0.03}{365}\right)^{365(20)} \approx \$1730.97$

77. $A_n = A_0\,e^{rn} \Rightarrow A_8 = 2000\,e^{0.10\,(8)} \approx \$4451.08$

79. $A_n = A_0\left(1 + \dfrac{r}{m}\right)^{mn} \Rightarrow A_{2.5} = 1600\left(1 + \dfrac{0.013}{12}\right)^{12(2.5)} \approx \$1652.83$

81. $A_{10} = 8000(1 + 0.06)^{10} \approx \$14{,}326.78$

83. We will use the graph of the graph of $y = e^{0.06x}$ and $y = (1.063)^x$ to represent the continuous and annual

growth respectively. From the graph we see that the $y = (1.063)^x$ grows at a faster rate making the annual

investment better. See Figure 83.

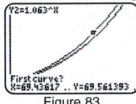
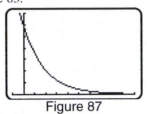

Figure 83   Figure 87

85. (a) The exponential model is given as $B(x) = Ca^x$, where C is the initial concentration, a is the growth factor,
    and x is the number of weeks. therefore, we have $B(x) = 2.5(3)^x$.

    (b) $B(1.5) = 2.5(3)^{1.5} \approx 13$ million per millimeter.

87. (a) The exponential model is given as $I(x) = Ca^x$, where C is the intensity at the surface, a is the

    growth/decline factor, and x is the number of feet. Therefore, we have $I(x) = 300\left(\dfrac{9}{10}\right)^x$ and

    $I(50) = 300\left(\dfrac{9}{10}\right)^{50} \approx 1.5$ watts per square meter.

    (b) See Figure 87. At the surface the intensity is 300 watts.

89. The exponential model is given as $G(x) = Pe^{rt}$, where P is the initial population, r is the rate and t is the num-

    ber of years. Therefore, we have $G(x) = 6.6e^{0.0144(x)}$ and $G(x) = 6.6e^{0.0144(6)} \approx 7.2$ million people.

91. (a) $f(0) = 0.72 \Rightarrow C \approx 0.72$, so $f(x) = 0.72a^x$. One possible way to determine the value of a is to let the

    graph of f pass through the point $(20, 1.60)$. $f(20) = 1.60 \Rightarrow 1.60 = 0.72a^{20} \Rightarrow a^{20} = \dfrac{1.60}{0.72} \Rightarrow$

    $a = \left(\dfrac{1.60}{0.72}\right)^{1/20} \Rightarrow a \approx 1.041$; Thus $f(x) = 0.72(1.041)^x$. *Answers may vary slightly.*

    (b) Let $x = 13$ correspond to 2013, so ; $f(13) = 0.72(1.041)^{13} \approx 1.21$ the CFC-12 concentration in 2013 is
    about 1.21 ppb.   *Answers may vary slightly.*

93. (a) The number of E. coli was modeled by $N(x) = N_0\, e^{0.014x}$, where x is in minutes and $N_0 = 500{,}000$.
    Since 3 hours is 180 minutes, $N(180) = 500{,}000\, e^{0.014(180)} \approx 6{,}214{,}000$ bacteria per milliter.

    (b) Solve the equation $500{,}000\, e^{0.014x} = 10{,}000{,}000$. Graph $Y_1 = 500000e^{\wedge}(0.014X)$ and $Y_2 = 10E6$
    as shown in Figure 93. The point of intersection is near $(214, 10{,}000{,}000)$. Thus, there will be 10
    million E. coli after about 214 minutes, or about 3.6 hours.

[0, 300, 100] by [0, 20,000,000, 5,000,000]   [0, 5, 1] by [0, 1, 0.1]

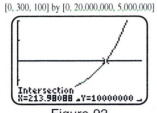   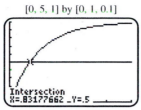

Figure 93   **Figure 95**

95. (a) $p(x) = 1 - e^{-5x/6} \Rightarrow p(3) = 1 - e^{-5(3)/6} \Rightarrow p(3) = 1 - e^{-15/6} \approx 0.92$, or 92%

(b) Find x when $p(x) = 0.5$. That is, solve $0.5 = 1 - e^{-5x/6}$ for x.

Graph $Y_1 = 0.5$ and $Y_2 = 1 - e^{\wedge}(-5X/6)$ as shown in Figure 95. There is a 50-50 chance of at least one car entering the intersection during an interval of about 0.83 minutes.

97. Let $T(t) = Ca^t$, where t is the number of hours. Initially, $T(0) = 100$, so $C = 100$ and $T(t) = 100a^t$. Next we

will find the value of a. Because the half-life is 2.8 hours, 50% of its hits remain after 2.8 hours, so

$T(2.8) = 50$. $100a^{2.8} = 50 \Rightarrow a^{2.8} = \dfrac{1}{2} \Rightarrow a = \left(\dfrac{1}{2}\right)^{1/2.8}$ Thus we have $T(t) = 100\left(\left(\dfrac{1}{2}\right)^{1/2.8}\right)^t = 100\left(\dfrac{1}{2}\right)^{t/2.8}$

and $T(5.5) = 100\left(\dfrac{1}{2}\right)^{5.5/2.8} \approx 25.6\%$.

99. (a) The exponential model is given as $P(x) = Ca^x$, where C is the inital percentage of putts made, a is the

growth/decline factor, and x is the number of feet. Therefore, we have $P(x) = 95(0.9)^x$.

(b) $P(20) = 95(0.9)^{20} \approx 11.5\%$

(c) Since the percentage decreases by a factor of 0.9 we use the graph of $y = 0.9^x$ to see that with every

increase in distance of 6.6 feet the percentage of putts made will decrease by half. See Figure 99.

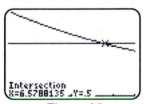

Figure 99

101. To find the age of the fossil solve $0.10 = 1\left(\dfrac{1}{2}\right)^{x/5700}$. Graph $Y_1 = 0.10$ and $Y_2 = 0.5^{\wedge}(X/5700)$. The graphs

intersect near (18934.99, 0.1), so the fossil is about 18,935 years old.

103. $P = 1\left(\dfrac{1}{2}\right)^{x/1600} \Rightarrow P = 1\left(\dfrac{1}{2}\right)^{3000/1600} \Rightarrow P \approx 0.273$, or 27.3%

105. (a) Since the initial amount is 2.5 ppm, $C = 2.5$. Since 30% of the chlorine dissipates each day, 70% remains.

So, $a = 0.7$.

(b) $f(2) = 2.5(0.7)^2 = 1.225$ parts per million.

107.(a) $H(30) = 0.157(1.033)^{30} \approx 0.42$. This means that approximately 0.42 horsepower are required for each ton that the locomotive is pulling at 30 mph.

(b) To pull a 5000-ton train at 30 mph, approximately a $0.42(5000) = 2100$ horsepower engine is needed.

(c) $\dfrac{2100}{1350} \approx 1.56$. This value must be rounded up. Two locomotives having 1350 horsepower would move a 5000-ton train at 30 mph.

Extended and Discovery Exercises for Section 5.3

1. $P = A(1 + r/n)^{-nt} \Rightarrow P = \dfrac{A}{(1 + r/n)^{nt}} \Rightarrow A = P(1 + r/n)^{nt}$

3. $P = 15{,}000(1 + 0.05/12)^{-12\,(3)} = 12{,}914.64367$; The present value should be \$12,914.64.

5. (a) $x = 0 \Rightarrow e^0 = 1 \Rightarrow e^{0+0.001} = 1.0010005 \Rightarrow \dfrac{1 + 1.0010005}{2} \approx 1.0005$

(b) $e^0 = 1$

(c) They are very similar.

7. (a) $x = -0.5 \Rightarrow e^{-0.5} = 0.6065 \Rightarrow e^{-0.5+0.001} = 0.6071 \Rightarrow \dfrac{0.6065 + 0.6071}{2} \approx 0.6068$

(b) $e^{-0.5} \approx 0.6065$

(c) They are very similar.

9. The average rate of change near x and the value of the function at x are approximately equal.

5.4: Logarithmic Functions and Models

1. See Figure 1.

x	10^0	10^4	10^{-8}	$10^{1.26}$
$\log x$	0	4	-8	1.26

Figure 1

3. (a) $\log(-3)$ is undefined

(b) $\log\dfrac{1}{100} = \log\dfrac{1}{10^2} = \log 1^{-2} = -2$

(c) $\log\sqrt{0.1} = \log 10^{-1/2} = -\dfrac{1}{2}$

(d) $\log 5^0 = 0$

5. (a) $\log 10 = \log 10^1 = 1$

(b) $\log 10{,}000 = \log 10^4 = 4$

(c) $20 \log 0.1 = 20 \log 10^{-1} = 20(-1) = -20$

(d) $\log 10 + \log 0.001 = \log 10^1 + \log 10^{-3} = 1 + (-3) = -2$

7.   (a) $2 \log 0.1 + 4 = 2 \log 10^{-1} + 4 = 2(-1) + 4 = -2 + 4 = 2$

   (b) $\log 10^{1/2} = \dfrac{1}{2}$

   (c) $3 \log 100 - \log 1000 = 3 \log 10^2 - \log 10^3 = 3(2) - 3 = 6 - 3 = 3$

   (d) $\log(-10)$ is undefined

9.   (a) Since $10^1 \le 79 \le 10^2, \log 10^1 \le \log 79 \le \log 10^2 \Rightarrow 1 \le \log 79 \le 2 \Rightarrow n = 1; \; \log 79 \approx 1.898$

   (b) Since $10^2 \le 500 \le 10^3, \log 10^2 \le \log 500 \le \log 10^3 \Rightarrow 2 \le \log 500 \le 3 \Rightarrow n = 2; \; \log 500 \approx 2.699$

   (c) Since $10^0 \le 5 \le 10^1, \log 10^0 \le \log 5 \le \log 10^1 \Rightarrow 0 \le \log 5 \le 1 \Rightarrow n = 0; \; \log 5 \approx 0.6990$

   (d) Since $10^{-1} \le 0.5 \le 10^0, \log 10^{-1} \le \log 0.5 \le \log 10^0 \Rightarrow -1 \le \log 0.5 \le 0 \Rightarrow n = -1;$
   $\log 0.5 \approx -0.3010$

11. (a) $\dfrac{3}{2}$ since $\sqrt{1000} = 1000^{1/2} = (10^3)^{1/2} = 10^{3/2}$

   (b) $\dfrac{1}{3}$ since $\log \sqrt[3]{10} = \log 10^{1/3}$

   (c) $\log \sqrt[5]{0.1} = \log (10^{-1})^{1/5} = \log (10)^{-1/5} = -\dfrac{1}{5}$

   (d) $\log \sqrt{0.01} = \log (10^{-2})^{1/2} = \log 10^{-1} = -1$

13. The input to a logarithmic function must be positive. Thus, any element of the domain of f must satisfy
   $x + 3 > 0$, or equivalently, $x > -3$. Thus D: $(-3, \infty)$, or $\{x \mid x > -3\}$.

15. The input to a logarithmic function must be positive. Thus, any element of the domain of f must satisfy
   $x^2 - 1 > 0 \Rightarrow x^2 > 1 \Rightarrow x < -1$ or $x > 1$ Thus D: $(-\infty, -1) \cup (1, \infty)$, or $\{x \mid x < -1, \text{ or } x > 1\}$.

17. The input to a logarithmic function must be positive. Thus, any element of the domain of f must satisfy
   $4^x > 0 \Rightarrow x$ can be any real number. Thus D: $(-\infty, \infty)$, or $\{x \mid -\infty < x < \infty\}$.

19. The input to a logarithmic function must be positive. Thus, any element of the domain of f must satisfy
   $\sqrt{3 - x} - 1 > 0 \Rightarrow \sqrt{3 - x} > 1 \Rightarrow x < 2$. Thus D: $(-\infty, 2)$, or $\{x \mid x < 2\}$.

21. $\log_8 8^{-5.7} = -5.7$

23. $7^{\log_7 2x} = 2x$ for $x > 0$

25. $\log_{1/3} \left(\dfrac{1}{3}\right)^{64} = 64$

27. $\ln e^{-4} = -4$

29. $\log_5 5^{\pi} = \pi$

31. $3^{\log_3 (x-1)} = x - 1$, for $x > 1$

33. $\log_2 64 = \log_2 2^6 = 6$

35. $\log_4 2 = \log_4 \sqrt{4} = \log_4 4^{1/2} = \dfrac{1}{2}$

37. $\ln e^{-3} = -3$

39. $\log_8 64 = \log_8 8^2 = 2$

41. $\log_{1/2}\left(\dfrac{1}{4}\right) = \log_{1/2}\left(\dfrac{1}{2}\right)^2 = 2$

43. $\log_{1/6} 36 = \log_{1/6}\left(\dfrac{1}{6}\right)^{-2} = -2$

45. $\log_a \dfrac{1}{a} = \log_a a^{-1} = -1$

47. $\log_5 5^0 = 0$

49. $\log_2 \dfrac{1}{16} = \log_2 2^{-4} = -4$

51. See Figure 51.

x	6	7	21
$f(x)$	0	2	8

Figure 51

53. (a) $7^{4x} = 4 \Rightarrow \log_7 4 = 4x$

    (b) $e^x = 7 \Rightarrow \ln(7) = x$

    (c) $c^x = b \Rightarrow \log_c(b) = x$

55. (a) $\log_8(x) = 3 \Rightarrow x = 8^3$

    (b) $\log_9(2+x) = 5 \Rightarrow 2+x = 9^5$

    (c) $\log_k b = c \Rightarrow b = k^c$

57. (a) $10^x = 0.01 \Rightarrow \log 10^{-2} = x \Rightarrow x = -2$

    (b) $10^x = 7 \Rightarrow \log 7 = x \Rightarrow x \approx 0.85$

    (c) $10^x = -4 \Rightarrow \log(-4) = x \Rightarrow$ no solution

59. (a) $4^x = \dfrac{1}{16} \Rightarrow 4^x = 4^{-2} \Rightarrow x = -2$

    (b) $e^x = 2 \Rightarrow \ln(2) = x \Rightarrow x \approx 0.69$

    (c) $5^x = 125 \Rightarrow 5^x = 5^3 \Rightarrow x = 3$

61. (a) $9^x = 1 \Rightarrow 9^x = 9^0 \Rightarrow x = 0$

    (b) $10^x = \sqrt{10} \Rightarrow 10^x = 10^{1/2} \Rightarrow x = \dfrac{1}{2}$

    (c) $4^x = \sqrt[3]{4} \Rightarrow 4^x = 4^{1/3} \Rightarrow x = \dfrac{1}{3}$

63. $e^{-x} = 3 \Rightarrow \ln e^{-x} = \ln 3 \Rightarrow -x = \ln 3 \Rightarrow x = -\ln 3 \approx -1.10$

65. $10^x - 5 = 95 \Rightarrow 10^x = 100 \Rightarrow 10^x = 10^2 \Rightarrow x = 2$

67. $10^{3x} = 100 \Rightarrow 10^{3x} = 10^2 \Rightarrow 3x = 2 \Rightarrow x = \dfrac{2}{3} \approx 0.67$

69. $5(10^{4x}) = 65 \Rightarrow 10^{4x} = 13 \Rightarrow \log(13) = 4x \Rightarrow x = \dfrac{\log(13)}{4} \approx 0.28$

71. $4(3^x) - 3 = 13 \Rightarrow 4(3^x) = 16 \Rightarrow 3^x = 4 \Rightarrow \log_3(4) = \dfrac{\log(4)}{\log(3)} \approx 1.26$

73. $e^x + 1 = 24 \Rightarrow e^x = 23 \Rightarrow \ln e^x = \ln 23 \Rightarrow x = \ln 23 \approx 3.14$

75. $2^x + 1 = 15 \Rightarrow 2^x = 14 \Rightarrow \log_2 2^x = \log_2 14 \Rightarrow x = \log_2 14$ or $x \approx 3.81$

77. $5e^x + 2 = 20 \Rightarrow 5e^x = 18 \Rightarrow e^x = \dfrac{18}{5} \Rightarrow \ln e^x = \ln \dfrac{18}{5} \Rightarrow x = \ln\left(\dfrac{18}{5}\right) \approx 1.28$

79. $8 - 3(2)^{0.5x} = -40 \Rightarrow -3(2)^{0.5x} = -48 \Rightarrow 2^{0.5x} = 16 \Rightarrow 2^{0.5x} = 2^4 \Rightarrow 0.5x = 4 \Rightarrow x = \dfrac{4}{0.5} \Rightarrow$

$x = 8$

81. (a) $\log x = 2 \Rightarrow 10^2 = x \Rightarrow x = 100$

 (b) $\log x = -3 \Rightarrow 10^{-3} = x \Rightarrow x = \dfrac{1}{1000}$

 (c) $\log x = 1.2 \Rightarrow 10^{1.2} = x \Rightarrow x \approx 15.8489$

83. (a) $\log_2 x = 6 \Rightarrow x = 2^6 = x \Rightarrow 64$

 (b) $\log_3 x = -2 \Rightarrow x = 3^{-2} = x \Rightarrow \dfrac{1}{9}$

 (c) $\ln x = 2 \Rightarrow x = e^2 \Rightarrow x \approx 7.3891$

85. $\log_2 x = 1.2 \Rightarrow 2^{\log_2 x} = 2^{1.2} \Rightarrow x = 2^{1.2} \approx 2.2974$

87. $5\log_7(2x) = 10 \Rightarrow \log_7(2x) = 2 \Rightarrow 2x = 7^2 \Rightarrow 2x = 49 \Rightarrow x = \dfrac{49}{2}$

89. $2\log x = 6 \Rightarrow \log x = 3 \Rightarrow 10^{\log x} = 10^3 \Rightarrow x = 10^3 = 1000$

91. $2\log 5x = 4 \Rightarrow \log 5x = 2 \Rightarrow 10^{\log 5x} = 10^2 \Rightarrow 5x = 100 \Rightarrow x = 20$

93. $4\ln x = 3 \Rightarrow \ln x = \dfrac{3}{4} \Rightarrow e^{\ln x} = e^{3/4} \Rightarrow x = e^{3/4} \approx 2.1170$

95. $5\ln x - 1 = 6 \Rightarrow 5\ln x = 7 \Rightarrow \ln x = \dfrac{7}{5} \Rightarrow e^{\ln x} = e^{7/5} \Rightarrow x = e^{7/5} \approx 4.0552$

97. $4\log_2 x = 16 \Rightarrow \log_2 x = 4 \Rightarrow 2^{\log_2 x} = 2^4 \Rightarrow x = 2^4 = 16$

99. $5\ln(2x) + 6 = 12 \Rightarrow 5\ln(2x) = 6 \Rightarrow \ln(2x) = \dfrac{6}{5} \Rightarrow e^{\ln 2x} = e^{6/5} \Rightarrow 2x = e^{6/5} \Rightarrow$

 $x = \dfrac{e^{6/5}}{2} \approx 1.6601$

101. $9 - 3\log_4 2x = 3 \Rightarrow -3\log_4 2x = -6 \Rightarrow \log_4 2x = 2 \Rightarrow 4^{\log_4 2x} = 4^2 \Rightarrow 2x = 16 \Rightarrow x = 8$

103. $f(x) = a + b\log x$ and $f(1) = 5 \Rightarrow 5 = a + b\log 1 \Rightarrow 5 = a + b(0) \Rightarrow a = 5$

 $f(x) = 5 + b\log x$ and $f(10) = 7 \Rightarrow 7 = 5 + b\log 10 \Rightarrow 7 = 5 + b(1) \Rightarrow b = 2$

 The function is $f(x) = 5 + 2\log x$.

105. Since $f(x) = e^x$, $f^{-1}(x) = \ln x$. See Figure 105.

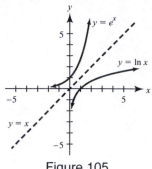

Figure 105

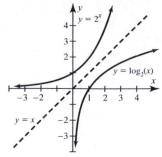

Figure 107

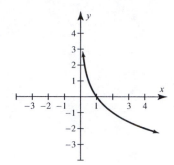

Figure 109

107. See Figure 107.

109. Decreasing. See Figure 109.

111. (a) See Figure 111.

    (b) f is increasing on $(0, \infty)$; f^{-1} is increasing on $(-\infty, \infty)$.

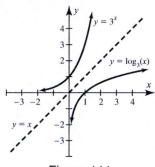

Figure 111

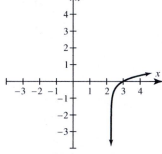

Figure 113

113. The graph of $g(x) = \log(x - 2)$ is similar to the graph of $f(x) = \log x$, except it is shifted to the right 2 units. The graph has a vertical asymptote of $x = 2$. See Figure 113.

115. The graph of $g(x) = 3 \log x$ is similar to the graph of $f(x) = \log x$, except it is stretched by a factor of 3. The graph has a vertical asymptote of $x = 0$. See Figure 115.

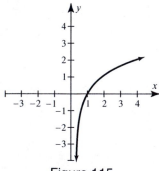

Figure 115

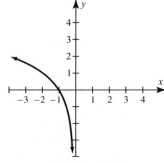

Figure 117

117. The graph of $g(x) = \log_2(-x)$ is similar to the graph of $f(x) = \log_2 x$, except it is reflected through the y-axis. The graph has a vertical asymptote of $x = 0$. See Figure 117.

119. The graph of $g(x) = 2 + \ln(x - 1)$ is similar to the graph of $f(x) = \ln x$, except it is shifted to the right one unit and up two units. The graph has a vertical asymptote of $x = 1$. See Figure 119.

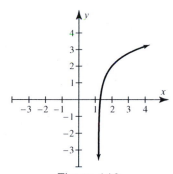

Figure 119

121. The graph is shown in Figure 121. $D = \{x \mid x > -1\}$

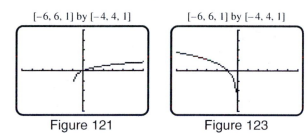

[−6, 6, 1] by [−4, 4, 1]   [−6, 6, 1] by [−4, 4, 1]

Figure 121   Figure 123

123. The graph is shown in Figure 123. $D = \{x \mid x < 0\}$

125. $D\left(10^{-4}\right) = 10 \log\left(10^{16} \cdot 10^{-4}\right) = 10 \log\left(10^{12}\right) = 10 \cdot 12 = 120$ db

127. (a) $L(100) = 3 \log (100) = 3 \cdot 2 = 6$, a 100-thousand pound airplanee needs 6000 feet of runway.

   (b) Since the weight is measured in 1000-pound units, start by evaluating $L(10)$ and $L(100)$.

   $L(10) = 3 \log 10 = 3$ and $L(100) = 3 \log 100 = 6$. Thus, a 10,000-pound plane requires approximately 3000 feet of runway, whereas as 100,000-pound plane requires approximately 6000 feet. The distance does not increase by a factor of 10.

   (c) If the weight increases tenfold, the runway length increases by 3000 feet.

129. (a) $\text{pH} = -\log (x) \Rightarrow \text{pH} = -\log (10^{-4.7}) \Rightarrow \text{pH} = -(-4.7) = 4.7$

   (b) $-\log (x) = 8.2 \Rightarrow \log (x) = -8.2 \Rightarrow 10^{\log (x)} = 10^{-8.2} \Rightarrow x = 10^{-8.2}$. Since $\dfrac{10^{-4.7}}{10^{-8.2}} = 10^{3.5}$ the

   hydrogen concentration in the rainwater is $10^{3.5} \approx 3162$ times greater than it is in seawater.

131. (a) Since $I_0 = 1$, $R(x) = 6.0 \Rightarrow \log x = 6.0 \Rightarrow 10^{\log x} = 10^{6.0} \Rightarrow x = 10^6 = 1{,}000{,}000$

   Similarly $R(x) = 8.0 \Rightarrow \log x = 8.0 \Rightarrow 10^{\log x} = 10^{8.0} \Rightarrow x = 10^8 = 100{,}000{,}000$

   (b) $\dfrac{10^8}{10^6} = 10^{8-6} = 10^2 = 100$. The Indonesian earthquake was 100 times more intense (powerful) than the

   Yugoslovian earthquake.

133. (a) $f(x) = 0.48 \ln (x + 1) + 27 \Rightarrow f(0) = 0.48 \ln (0 + 1) + 27 = 0.48 \ln 1 + 27 =$

   $0.48(0) + 27 = 27$ and $f(100) = 0.48 \ln (100 + 1) + 27 \approx 29.2$ inches. At the center, or eye, of the hurricane the pressure is 27 inches of mercury, while 100 miles from the eye the air pressure has risen to 29.2 inches of mercury.

   (b) Graph $Y_1 = 0.48 \ln (X + 1) + 27$ as shown in Figure 133. At first, the air pressure rises rapidly as one moves away from the eye. Then, the air pressure starts to level off and does not increase significantly for distances greater than 200 miles.

   (c) $f(x) = 28 \Rightarrow 28 = 0.48 \ln (x + 1) + 27 \Rightarrow \ln (x + 1) = \dfrac{1}{0.48} \Rightarrow e^{\ln (x+1)} = e^{1/0.48} \Rightarrow$

   $x + 1 = e^{1/0.48} \Rightarrow x = e^{1/0.48} - 1 \approx 7.03$; the air pressure is 28 inches of mercury about 7 miles from the eye of the hurricane.

   [0, 250, 50] by [25, 30, 1]

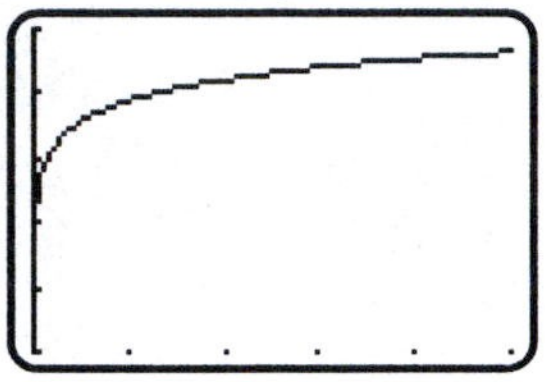

Figure 133

135. (a) $T(x) = 20 + 80e^{-x} \Rightarrow T(1) = 20 + 80e^{-1} \approx 49.4$; the temperature is about 49.4°C after 1 hour.

   (b) $T(x) = 60 \Rightarrow 60 = 20 + 80e^{-x} \Rightarrow 40 = 80e^{-x} \Rightarrow e^{-x} = \dfrac{1}{2} \Rightarrow \ln e^{-x} = \ln \left(\dfrac{1}{2}\right) \Rightarrow$

   $-x = \ln \left(\dfrac{1}{2}\right) \Rightarrow x = -\ln \left(\dfrac{1}{2}\right) \approx 0.693$; the water took about 0.69 hours, or 41.4 minutes to cool to 60°C.

137. (a) $f(x) = e^{-x/3} \Rightarrow f(5) = e^{-5/3} \approx 0.189$

   The probability that no car enters the intersection during a 5-minute period is about 0.189 or 18.9%.

   (b) $f(x) = 0.30 \Rightarrow 0.30 = e^{-x/3} \Rightarrow \ln e^{-x/3} = \ln 0.3 \Rightarrow -\dfrac{x}{3} = \ln 0.3 \Rightarrow x = -3 \ln 0.3 \approx 3.6$

   The probability that no car enters the intersection will be 30% during a period of about 3.6 minutes.

139. $f(x) = a + b \log x$ and $f(1) = 7 \Rightarrow 7 = a + b \log 1 \Rightarrow 7 = a + b(0) \Rightarrow a = 7$

   $f(x) = 7 + b \log x$ and $f(10) = 11 \Rightarrow 11 = 7 + b \log 10 \Rightarrow 11 = 7 + b(1) \Rightarrow b = 4$

   The function that models the given data is $f(x) = 7 + 4 \log x$.

   $f(x) = 16 \Rightarrow 16 = 7 + 4 \log x \Rightarrow 9 = 4 \log x \Rightarrow \log x = \dfrac{9}{4} \Rightarrow 10^{\log x} = 10^{9/4} \Rightarrow$

   $x = 10^{9/4} \approx 178$. The island would be about 178 square kilometers.

141. (a) $f(x) = Ca^x$ and $f(0) = 3 \Rightarrow 3 = Ca^0 \Rightarrow 3 = C(1) \Rightarrow C = 3$

   $f(x) = 3a^x$ and $f(1) = 6 \Rightarrow 6 = 3a^1 \Rightarrow 2 = a^1 \Rightarrow a = 2$; thus $f(x) = 3(2^x)$ models the data.

   (b) $f(x) = 16 \Rightarrow 16 = 3(2^x) \Rightarrow 2^x = \dfrac{16}{3}$. Graph $Y_1 = 2\text{^}X$ and $Y_2 = 16/3$ in [0, 5, 1] by [4, 20, 5].

   The graphs (not shown) intersect near (2.41, 5.33), so there were 16 million bacteria after about 2.4 days.

Extended and Discovery Exercises for Section 5.4

1. (a) If $x = 1$ then $\ln(1) = 0$ and if $x = 1.001$ then $\ln(1.001) = 0.0009995 \Rightarrow$

    $(1, 0)$ and $(1.001, 0.0009995)$, the average rate of change is $\dfrac{0.0009995 - 0}{1.001 - 1} \approx 1.00$.

    (b) If $x = 2$ then $\ln(2) = 0.693147$ and if $x = 2.001$ then $\ln(2.001) = 0.693647 \Rightarrow$

    $(2, 0.693147)$ and $(2.001, 0.693647)$, the average rate of change is $\dfrac{0.693647 - 0.693147}{2.001 - 2} \approx 0.50$.

    (c) If $x = 3$ then $\ln(3) = 1.098612$ and if $x = 3.001$ then $\ln(3.001) = 1.098946 \Rightarrow$

    $(3, 1.098612)$ and $(3.001, 1.098946)$, the average rate of change is $\dfrac{1.098946 - 1.098612}{3.001 - 2} \approx 0.33$.

    (d) If $x = 4$ then $\ln(4) = 1.386294$ and if $x = 4.001$ then $\ln(4.001) = 1.386544 \Rightarrow$

    $(4, 1.386294)$ and $(4.001, 1.386544)$, the average rate of change is $\dfrac{1.386544 - 1.386294}{4.001 - 4} \approx 0.25$.

3. (a) $T(C(x)) = 6.5 \ln\left(\dfrac{364(1.005)^x}{280}\right) = 6.5 \ln(1.3 \cdot 1.005^x);\ T(100) = 6.5 \ln\left(\dfrac{364(1.005)^{100}}{280}\right) \approx 4.95$

    This model predicts an average global temperature increase of about 5°F in the year 2100.

    (b) Graph $Y_1 = 364(1.005)\wedge X$ in $[0, 200, 50]$ by $[0, 1000, 100]$ and $Y_2 = 6.5 \ln(364(1.005)\wedge X/280)$ in

    $[0, 200, 50]$ by $[0, 10, 1]$. The graph of Y_1 is exponential while the graph of Y_2 appears linear over this

    time interval. See Figures 3a and 3b.

    (c) C is an exponential function and T is approximately linear over the same time period. While the carbon

    dioxide levels in the atmosphere increase exponentially, the average global temperature rises at nearly a

    constant rate each year.

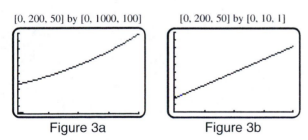

[0, 200, 50] by [0, 1000, 100]   [0, 200, 50] by [0, 10, 1]

Figure 3a   Figure 3b

Checking Basic Concepts for Sections 5.3 and 5.4

1. $A = 1200\left(1 + \dfrac{0.095}{12}\right)^{12(4)} \approx 1752.12$; After four years the account balance will be \$1752.12. If

    compounding continuously, the account balance will be $A = 1200e^{0.095(4)} \approx \$1754.74$.

3. $\log_2 15$ represents the power of 2 resulting in 15. That is, $2^{\log_2 15} = 15$ by definition. There is no integer k such

    that $2^k = 15$, so $\log_2 15$ does not equal an integer. Since $2^4 = 16$, the value of $\log_2 15$ is slightly less than 4.

5. (a) $e^x = 5 \Rightarrow \ln e^x = \ln 5 \Rightarrow x = \ln 5 \approx 1.609$

    (b) $10^x = 25 \Rightarrow \log 10^x = \log 25 \Rightarrow x = \log 25 \approx 1.398$

    (c) $\log x = 1.5 \Rightarrow 10^{\log x} = 10^{1.5} \Rightarrow x = 10^{1.5} \approx 31.623$

7. (a) $P(2) = 37.3e^{0.01(2)} \approx 38.1$ million, in 2012 California's population was about 38.1 million.

    (b) The y-intercept is found by letting $x = 0$, $P(x) = 37.3e^{0.01(0)} = 37.3$, in 2010 California's population was about 37.3 million.

    (c) The function $P(x) = 37.3e^{0.01(x)}$ is in the form $A = Pe^{rt}$ where r is the rate of increase in the population. The annual increase is calculated as $e^r - 1 \Rightarrow e^{001} - 1 \approx 0.01$.

    (d) $40 = 37.3e^{0.01(x)} \Rightarrow \dfrac{40}{37.3} = e^{0.01x} \Rightarrow \ln\left(\dfrac{40}{37.3}\right) = 0.01x \Rightarrow x = \dfrac{\ln\left(\dfrac{40}{37.3}\right)}{0.01} \approx 7$, since x represents the number of years after 2010 the result is 2017.

5.5: Properties of Logarithms

1. $\log 4 + \log 7 \approx 1.447$; $\log 28 \approx 1.447$; $\log 4 + \log 7 = \log 28 = \log (4 \cdot 7)$; Property 2

3. $\ln 72 - \ln 8 \approx 2.197$; $\ln 9 \approx 2.197$; $\ln 72 - \ln 8 = \ln 9 = \ln\left(\dfrac{72}{8}\right)$; Property 3

5. $10 \log 2 \approx 3.010$; $\log 1024 \approx 3.010$; $10 \log 2 = \log 1024 = \log 2^{10}$; Property 4

7. $\log_2 ab = \log_2 a + \log_2 b$

9. $\ln 7a^4 = \ln 7 + 4 \ln a$

11. $\log \dfrac{6}{z} = \log 6 - \log z$

13. $\log \dfrac{x^2}{3} = \log x^2 - \log 3 = 2 \log x - \log 3$

15. $\ln \dfrac{2x^7}{3k} = \ln 2x^7 - \ln 3k = \ln 2 + \ln x^7 - (\ln 3 + \ln k) = \ln 2 + 7 \ln x - \ln 3 - \ln k$

17. $\log_2 4k^2 x^3 = \log_2 4k^2 + \log_2 x^3 = \log_2 4 + \log_2 k^2 + \log_2 x^3 = 2 + 2 \log_2 k + 3 \log_2 x$

19. $\log_5 \dfrac{25x^3}{y^4} = \log_5 25x^3 - \log_5 y^4 = 2 + 3 \log_5 x - 4 \log_5 y$

21. $\ln \dfrac{x^4}{y^2\sqrt{z^3}} = \ln (x^4) - (\ln y^2 + \ln (z^{3/2})) = 4 \ln (x) - 2 \ln (y) - \dfrac{3}{2} \ln (z)$

23. $\log_4 0.25(x + 2)^3 = \log_4 0.25 + \log_4 (x + 2)^3 = -1 + 3 \log_4 (x + 2)$

25. $\log_5 \dfrac{x^3}{(x - 4)^4} = \log_5 x^3 - \log_5 (x - 4)^4 = 3 \log_5 x - 4 \log_5 (x - 4)$

27. $\log_2 \dfrac{\sqrt{x}}{z^2} = \log_2 \sqrt{x} - \log_2 z^2 = \log_2 x^{1/2} - \log_2 z^2 = \dfrac{1}{2} \log_2 x - 2 \log_2 z$

29. $\ln \sqrt[3]{\dfrac{2x + 6}{(x + 1)^5}} = \ln \left(\dfrac{2x + 6}{(x + 1)^5}\right)^{1/3} = \dfrac{1}{3} \left[\ln (2x + 6) - 5 \ln (x + 1)\right] = \dfrac{1}{3} \ln (2x + 6) - \dfrac{5}{3} \ln (x + 1)$

31. $\log_2 \dfrac{\sqrt[3]{x^2 - 1}}{\sqrt{1 + x^2}} = \log_2 \sqrt[3]{x^2 - 1} - \log_2 \sqrt{1 + x^2} = \log_2 (x^2 - 1)^{1/3} - \log_2 (1 + x^2)^{1/2} =$

$\dfrac{1}{3} \log_2 (x^2 - 1) - \dfrac{1}{2} \log_2 (1 + x^2)$

33. $\log 2 + \log 3 = \log (2 \cdot 3) = \log 6$

35. $\ln \sqrt{5} - \ln 25 = \ln \left(\dfrac{5^{1/2}}{5^2} \right) = \ln 5^{-3/2} = -\dfrac{3}{2} \ln 5$

37. $\log 20 + \log \dfrac{1}{10} = \log (20) \left(\dfrac{1}{10} \right) = \log \dfrac{20}{10} = \log 2$

39. $\log 4 + \log 3 - \log 2 = \log \dfrac{4(3)}{2} = \log \dfrac{12}{2} = \log 6$

41. $\log_7 5 + \log_7 k^2 = \log_7 (5)(k^2) = \log_7 5k^2$

43. $\ln x^6 - \ln x^3 = \ln \dfrac{x^6}{x^3} = \ln x^{6-3} = \ln x^3$

45. $\log \sqrt{x} + \log x^2 - \log x = \log \left(\dfrac{x^{1/2} \cdot x^2}{x} \right) = \log x^{3/2} = \dfrac{3}{2} \log x$

47. $3 \ln (x) - \dfrac{3}{2} \ln (y) + 4 \ln (z) = \ln (x^3) - \ln (\sqrt{y^3}) + \ln (z^2) = \ln \left(\dfrac{x^3 z^4}{\sqrt{y^3}} \right)$

49. $\ln \dfrac{1}{e^2} + \ln 2e = \ln \left(\dfrac{1}{e^2} \cdot 2e \right) = \ln \dfrac{2}{e}$

51. $2 \ln x - 4 \ln y + \dfrac{1}{2} \ln z = \ln x^2 - \ln y^4 + \ln z^{1/2} = \ln \dfrac{x^2}{y^4} + \ln \sqrt{z} = \ln \dfrac{x^2 \sqrt{z}}{y^4}$

53. $\log 4 - \log x + 7 \log \sqrt{x} = \log \dfrac{4}{x} + \log (\sqrt{x})^7 = \log \left(\dfrac{4x^{7/2}}{x} \right) = \log 4x^{5/2} = \log 4\sqrt{x^5}$

55. $2 \log (x^2 - 1) + 4 \log (x - 2) - \dfrac{1}{2} \log y = \log (x^2 - 1)^2 + \log (x - 2)^4 - \log y^{1/2} =$

$\log \dfrac{(x^2 - 1)^2 (x - 2)^4}{\sqrt{y}}$

57. (a) Table $Y_1 = \log (3X) + \log (2X)$ and $Y_2 = \log (6X^2)$ starting at $x = 1$, incrementing by 1.

See Figure 57. From the table we see that $Y_1 = Y_2$, so $f(x) = g(x)$.

(b) Product Rule: $\log 3x + \log 2x = \log (3x \cdot 2x) = \log 6x^2$

X	Y1	Y2
1	.77815	.77815
2	1.3802	1.3802
3	1.7324	1.7324
4	1.9823	1.9823
5	2.1761	2.1761
6	2.3345	2.3345
7	2.4683	2.4683

X=1

Figure 57

X	Y1	Y2
1	.69315	.69315
2	1.3863	1.3863
3	1.7918	1.7918
4	2.0794	2.0794
5	2.3026	2.3026
6	2.4849	2.4849
7	2.6391	2.6391

X=1

Figure 59

59. (a) Table $Y_1 = \ln(2X^2) - \ln(X)$ and $Y_2 = \ln(2X)$ starting at $x = 1$, incrementing by 1. See Figure 59.

    From the table we see that $Y_1 = Y_2$, so $f(x) = g(x)$.

    (b) Quotient Rule: $\ln 2x^2 - \ln x = \ln\left(\dfrac{2x^2}{x}\right) = \ln 2x$

61. (a) Table $Y_1 = \ln(X^4) - \ln(X^2)$ and $Y_2 = 2\ln(X)$ starting at $x = 1$, incrementing by 1. See Figure 61.

    From the table we see that $Y_1 = Y_2$, so $f(x) = g(x)$.

    (b) Power Rule: $\ln x^4 - \ln x^2 = 4\ln x - 2\ln x = 2\ln x$

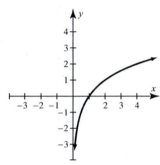

Figure 61

63. See Figure 63.

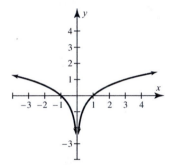

Figure 63   Figure 65

65. See Figure 65.

67. $\log_2 25 = \dfrac{\log 25}{\log 2} \approx 4.644$

69. $\log_5 130 = \dfrac{\log 130}{\log 5} \approx 3.024$

71. $\log_2 5 + \log_2 7 = \dfrac{\log 5}{\log 2} + \dfrac{\log 7}{\log 2} \approx 5.129$

73. $\sqrt{\log_4 46} = \sqrt{\dfrac{\log 46}{\log 4}} \approx 1.662$

75. $\dfrac{\log_2 12}{\log_2 3} = \dfrac{\frac{\log 12}{\log 2}}{\frac{\log 3}{\log 2}} = \dfrac{\log 12}{\log 2} \cdot \dfrac{\log 2}{\log 3} = \dfrac{\log 12}{\log 3} \approx 2.262$

77. Graph $Y_1 = \log(X^3 + X^2 + 1)/\log 2$ and $Y_2 = 7$ in $[0, 10, 2]$ by $[0, 10, 2]$. The graphs intersect near the point $(4.714, 7)$, so the solution to $\log_2(x^3 + x^2 + 1) = 7$ is $x \approx 4.714$.

79. Graph $Y_1 = \log(X^2 + 1)/\log 2$ and $Y_2 = 5 - \log(X^4 + 1)/\log 3$ in $[-5, 5, 2]$ by $[-5, 10, 2]$. The graphs intersect near the points $(-2.035, 2.362)$ and $(2.035, 2.362)$, so the solutions to the equation $\log_2(x^2 + 1) = 5 - \log_3(x^4 + 1)$ are $x \approx \pm 2.035$.

81. Change base 10 to base e: $L(x) = 3 \log x = \dfrac{3 \cdot \ln x}{\ln 10}$;

    $L(50) = 3 \log 50 \approx 5.097$ and $L(50) = \dfrac{3 \ln 50}{\ln 10} \approx 5.097$. Yes, the answers agree.

83. $f(x) = 160 + 10 \log x \Rightarrow f(10x) = 160 + 10 \log (10x) = 160 + 10 \, (\log 10 + \log x) =$

    $160 + 10(1 + \log x) = 160 + 10 + 10 \log x = 160 + 10 \log x + 10$, thus $f(10x) = f(x) + 10$.

    In other words, the decibel level increases by 10 decibels.

85. $\ln I - \ln I_0 = -kx \Rightarrow \ln I = \ln I_0 - kx \Rightarrow e^{\ln I} = e^{\ln I_0 - kx} \Rightarrow e^{\ln I} = e^{\ln I_0} \cdot e^{-kx} \Rightarrow I = I_0 \, e^{-kx}$

87. (a) $P = 37.3 e^{0.01x} \Rightarrow \dfrac{P}{37.3} = e^{0.01x} \Rightarrow \ln \left(\dfrac{P}{37.3} \right) = 0.01x \Rightarrow x = 100 \ln \left(\dfrac{P}{37.3} \right)$

    (b) $x = 100 \ln \dfrac{40}{37.3} \Rightarrow x \approx 7$, The population is expected to reach 40 million during 2017.

89. $A = Pe^{rt} \Rightarrow \dfrac{A}{P} = e^{rt} \Rightarrow \ln \dfrac{A}{P} = \ln e^{rt} \Rightarrow \ln \dfrac{A}{P} = rt \Rightarrow \dfrac{\ln \frac{A}{P}}{r} = t \Rightarrow t = \dfrac{\ln \frac{A}{P}}{r}$

91. $\log 1 + 2 \log 2 + 3 \log 3 + 4 \log 4 + 5 \log 5 = \log 1 + \log 2^2 + \log 3^3 + \log 4^4 + \log 5^5 =$

    $\log (1 \cdot 2^2 \cdot 3^3 \cdot 4^4 \cdot 5^5) = \log 86{,}400{,}000$

5.6: Exponential and Logarithmic Equations

1. (a) The graphs appear to intersect near the point $(2, 7.5)$. Thus, the solution is $x \approx 2$.

    (b) $f(x) = g(x) \Rightarrow e^x = 7.5 \Rightarrow \ln e^x = \ln 7.5 \Rightarrow x = \ln 7.5$ about 2.015

3. (a) The graphs appear to intersect near the point $(2, 2.5)$. Thus, the solution is $x \approx 2$.

    (b) $f(x) = g(x) \Rightarrow 10^{0.2x} = 2.5 \Rightarrow \log 10^{0.2x} = \log 2.5 \Rightarrow 0.2x = \log 2.5 \Rightarrow x = \dfrac{\log 2.5}{0.2}$ about 1.990

5. $4e^x = 5 \Rightarrow e^x = \dfrac{5}{4} \Rightarrow \ln e^x = \ln \dfrac{5}{4} \Rightarrow x = \ln \dfrac{5}{4} \approx 0.2231$

7. $2(10^x) + 5 = 45 \Rightarrow 2(10^x) = 40 \Rightarrow 10^x = 20 \Rightarrow \log 10^x = \log 20 \Rightarrow x = \log 20 \approx 1.301$

9. $2.5 e^{-1.2x} = 1 \Rightarrow e^{-1.2x} = \dfrac{1}{2.5} \Rightarrow \ln e^{-1.2x} = \ln \dfrac{1}{2.5} \Rightarrow -1.2x = \ln \dfrac{1}{2.5} \Rightarrow x = \dfrac{\ln \frac{1}{2.5}}{-1.2} \approx 0.7636$

11. $1.2(0.9^x) = 0.6 \Rightarrow 0.9^x = 0.5 \Rightarrow \ln 0.9^x = \ln 0.5 \Rightarrow x \ln 0.9 = \ln 0.5 \Rightarrow x = \dfrac{\ln 0.5}{\ln 0.9} \approx 6.579$

13. $4(1.1^{x-1}) = 16 \Rightarrow 1.1^{x-1} = 4 \Rightarrow \ln 1.1^{x-1} = \ln 4 \Rightarrow (x - 1) \ln 1.1 = \ln 4 \Rightarrow x - 1 = \dfrac{\ln 4}{\ln 1.1} \Rightarrow$

    $x = \dfrac{\ln 4}{\ln 1.1} + 1 \approx 15.55$

15. $5(1.2)^{3x-2} + 94 = 100 \Rightarrow 5(1.2)^{3x-2} = 6 \Rightarrow (1.2)^{3x-2} = 1.2 \Rightarrow \ln 1.2^{3x-2} = \ln 1.2 \Rightarrow$

    $(3x - 2) \ln 1.2 = \ln 1.2 \Rightarrow 3x - 2 = \dfrac{\ln 1.2}{\ln 1.2} \Rightarrow 3x - 2 = 1 \Rightarrow 3x = 3 \Rightarrow x = 1$

17. $5^{3x} = 5^{1-2x} \Rightarrow \log_5 5^{3x} = \log_5 5^{1-2x} \Rightarrow 3x = 1 - 2x \Rightarrow 5x = 1 \Rightarrow x = \dfrac{1}{5}$

19. $10^{x^2} = 10^{3x-2} \Rightarrow \log 10^{x^2} = \log 10^{3x-2} \Rightarrow x^2 = 3x - 2 \Rightarrow x^2 - 3x + 2 = 0 \Rightarrow$

$(x - 1)(x - 2) = 0 \Rightarrow x = 1 \text{ or } x = 2$

21. No solution since no power of $\dfrac{1}{5}$ will result in a negative value.

23. $\left(\dfrac{2}{5}\right)^{x-2} = \dfrac{1}{3} \Rightarrow \log\left(\dfrac{2}{5}\right)^{x-2} = \log\left(\dfrac{1}{3}\right) \Rightarrow (x-2)\log\left(\dfrac{2}{5}\right) = \log\left(\dfrac{1}{3}\right) \Rightarrow x - 2 = \dfrac{\log\left(\frac{1}{3}\right)}{\log\left(\frac{2}{5}\right)} \Rightarrow$

$x = 2 + \dfrac{\log\left(\frac{1}{3}\right)}{\log\left(\frac{2}{5}\right)} \approx 3.199$

25. $4^{x-1} = 3^{2x} \Rightarrow \log 4^{x-1} = \log 3^{2x} \Rightarrow (x-1)\log 4 = 2x \log 3 \Rightarrow x\log 4 - \log 4 = 2x \log 3 \Rightarrow$

$x\log 4 - 2x\log 3 = \log 4 \Rightarrow x(\log 4 - 2\log 3) = \log 4 \Rightarrow x = \dfrac{\log 4}{\log 4 - 2\log 3} \approx -1.710$

27. $e^{x-3} = 2^{3x} \Rightarrow \ln e^{x-3} = \ln 2^{3x} \Rightarrow x - 3 = 3x\ln 2 \Rightarrow -3 = -x + 3x\ln 2 \Rightarrow$

$3 = x - 3x\ln 2 \Rightarrow 3 = x(1 - 3\ln 2) \Rightarrow x = \dfrac{3}{1 - 3\ln 2} \Rightarrow x = \dfrac{3}{1 - \ln 8} \Rightarrow x \approx -2.779$

29. $3(1.4)^x - 4 = 60 \Rightarrow 3(1.4)^x = 64 \Rightarrow 1.4^x = \dfrac{64}{3} \Rightarrow \log 1.4^x = \log \dfrac{64}{3} \Rightarrow x\log 1.4 = \log \dfrac{64}{3} \Rightarrow$

$x = \dfrac{\log \frac{64}{3}}{\log 1.4} \approx 9.095$

31. $5(1.015)^{x-1980} = 8 \Rightarrow 1.015^{x-1980} = \dfrac{8}{5} \Rightarrow \log 1.015^{x-1980} = \log \dfrac{8}{5} \Rightarrow (x - 1980)\log 1.015 = \log \dfrac{8}{5} \Rightarrow$

$x\log 1.015 - 1980\log 1.015 = \log \dfrac{8}{5} \Rightarrow x\log 1.015 = \log \dfrac{8}{5} + 1980\log 1.015 \Rightarrow$

$x = \dfrac{\log \frac{8}{5} + 1980\log 1.015}{\log 1.015} = \dfrac{\log \frac{8}{5}}{\log 1.015} + 1980 \approx 2012$

33. $5^{2x} = 5^{x-3} \Rightarrow 2x = x - 3 \Rightarrow x = -3$

35. $e^{-x} = e^{x^2} \Rightarrow -x = x^2 \Rightarrow 0 = x^2 + x \Rightarrow 0 = x(x + 1) \Rightarrow x = 0 \text{ or } x + 1 = 0 \Rightarrow x = 0 \text{ or } x = -1$

37. $2^{3x} = 8^{-x+2} \Rightarrow 2^{3x} = \left(2^3\right)^{-x+2} \Rightarrow 2^{3x} = 2^{-3x+6} \Rightarrow 3x = -3x + 6 \Rightarrow 6x = 6 \Rightarrow x = 1$

39. $25^{2x} = 125^{2-x} \Rightarrow \left(5^2\right)^{2x} = \left(5^3\right)^{2-x} \Rightarrow 5^{4x} = 5^{6-3x} \Rightarrow 4x = 6 - 3x \Rightarrow 7x = 6 \Rightarrow x = \dfrac{6}{7}$

41. $32^{3x} = 16^{5x+3} \Rightarrow \left(2^5\right)^{3x} = \left(2^4\right)^{5x+3} \Rightarrow 2^{15x} = 2^{20x+12} \Rightarrow 15x = 20x + 12 \Rightarrow -5x = 12 \Rightarrow x = -\dfrac{12}{5}$

43. $3\log x = 2 \Rightarrow \log x = \dfrac{2}{3} \Rightarrow 10^{\log x} = 10^{2/3} \Rightarrow x = 10^{2/3} \approx 4.642$

45. $\ln 2x = 5 \Rightarrow e^{\ln 2x} = e^5 \Rightarrow 2x = e^5 \Rightarrow x = \dfrac{e^5}{2} \approx 74.207$

47. $\log 2x^2 = 2 \Rightarrow 10^{\log 2x^2} = 10^2 \Rightarrow 2x^2 = 100 \Rightarrow x^2 = 50 \Rightarrow x = \pm\sqrt{50} \approx \pm 7.071$

49. $\log_2 (3x - 2) = 4 \Rightarrow 2^{\log_2 (3x-2)} = 2^4 \Rightarrow 3x - 2 = 16 \Rightarrow 3x = 18 \Rightarrow x = 6$

51. $\log_5 (8 - 3x) = 3 \Rightarrow 5^{\log_5 (8-3x)} = 5^3 \Rightarrow 8 - 3x = 125 \Rightarrow -3x = 117 \Rightarrow x = -39$

53. $160 + 10 \log x = 50 \Rightarrow 10 \log x = -110 \Rightarrow \log x = -11 \Rightarrow 10^{\log x} = 10^{-11} \Rightarrow x = 10^{-11}$

55. $\ln x + \ln x^2 = 3 \Rightarrow \ln x + 2 \ln x = 3 \Rightarrow 3 \ln x = 3 \Rightarrow \ln x = 1 \Rightarrow e^{\ln x} = e^1 \Rightarrow x = e \approx 2.718$

57. $2 \log_2 x = 4.2 \Rightarrow \log_2 x = 2.1 \Rightarrow 2^{\log_2 x} = 2^{2.1} \Rightarrow x = 2^{2.1} \approx 4.287$

59. $\log x + \log 2x = 2 \Rightarrow \log (x \cdot 2x) = 2 \Rightarrow \log 2x^2 = 2 \Rightarrow 10^{\log 2x^2} = 10^2 \Rightarrow 2x^2 = 100 \Rightarrow$

    $x^2 = 50 \Rightarrow x = \pm\sqrt{50}$. When $x = -\sqrt{50} < 0$, $\log x$ is undefined. Therefore, the only solution is

    $x = \sqrt{50} \approx 7.071$.

61. $\log (2 - 3x) = 3 \Rightarrow 10^3 = 2 - 3x \Rightarrow 1000 = 2 - 3x \Rightarrow 998 = -3x \Rightarrow x = -\dfrac{998}{3}$

63. $\ln (x) + \ln (3x - 1) = \ln (10) \Rightarrow \ln (x(3x - 1)) = \ln (10) \Rightarrow \ln (3x^2 - x) = \ln (10) \Rightarrow$

    $3x^2 - x = 10 \Rightarrow 3x^2 - x - 10 = 0 \Rightarrow (3x + 5)(x - 2) = 0 \Rightarrow 3x + 5 = 0 \text{ or } x - 2 = 0 \Rightarrow$

    $x = -\dfrac{5}{3}$ or $x = 2$. When $x = -\dfrac{5}{3}$, both $\ln (3x - 1)$ and $\ln (x)$ are undefined, therefore the only solution is $x = 2$.

65. $2 \ln x = \ln (2x + 1) \Rightarrow \ln x^2 = \ln (2x + 1) \Rightarrow e^{\ln x^2} = e^{\ln (2x+1)} \Rightarrow x^2 = 2x + 1 \Rightarrow$

    $x^2 - 2x - 1 = 0 \Rightarrow x = \dfrac{2 \pm \sqrt{2^2 - 4(1)(-1)}}{2(1)} = \dfrac{2 \pm \sqrt{8}}{2} = \dfrac{2 \pm 2\sqrt{2}}{2} = 1 \pm \sqrt{2} \Rightarrow$

    $x = 1 + \sqrt{2}$ or $x = 1 - \sqrt{2}$.

    When $x = 1 - \sqrt{2} \approx -0.414$, $\ln x$ is undefined. The only solution is $x = 1 + \sqrt{2}$.

67. $\log (x + 1) + \log (x - 1) = \log 3 \Rightarrow \log [(x + 1)(x - 1)] = \log 3 \Rightarrow \log (x^2 - 1) = \log 3 \Rightarrow$

    $10^{\log (x^2-1)} = 10^{\log 3} \Rightarrow x^2 - 1 = 3 \Rightarrow x^2 = 4 \Rightarrow x = -2 \text{ or } x = 2$

    When $x = -2$ both $\log (x - 1)$ and $\log (x + 1)$ are undefined. The only solution is $x = 2$.

69. $\log_2 2x = 4 - \log_2 (x + 2) \Rightarrow \log_2 2x + \log_2 (x + 2) = 4 \Rightarrow \log_2 [2x(x + 2)] = 4 \Rightarrow$

    $\log_2 (2x^2 + 4x) = 4 \Rightarrow 2^{\log_2 (2x^2+4x)} = 4 \Rightarrow 2x^2 + 4x - 16 = 0 \Rightarrow 2(x^2 + 2x - 8) = 0 \Rightarrow$

    $2(x + 4)(x - 2) = 0 \Rightarrow x = -4, x = 2$, since we cannot take the log of a negative number, $x = 2$.

71. $\log_5 (x + 1) + \log_5 (x - 1) = \log_5 15 \Rightarrow \log_5 (x + 1)(x - 1) = \log_5 15 \Rightarrow$

    $\log_5 (x^2 - 1) = \log_5 15 \Rightarrow 5^{\log_5 (x^2-1)} = 5^{\log_5 15} \Rightarrow x^2 - 1 = 15 \Rightarrow x^2 - 16 = 0 \Rightarrow$

    $(x + 4)(x - 4) = 0$, since we cannot take the log of a negative number, $x = 4$.

73. Graph $Y_1 = 2X + e^{\wedge}(X)$ and $Y_2 = 2$ as shown in Figure 73.

    The graphs intersect near the point $(0.31, 2)$, so the solution to $2x + e^x = 2$ is $x \approx 0.31$.

    The graph intersects the x-axis near the point $(0.57, 0)$, so the solution to $xe^x - 1 = 0$ is $x \approx 0.57$.

[-5, 5, 1] by [-5, 5, 1]   [-5, 5, 1] by [-5, 5, 1]

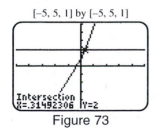

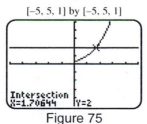

Figure 73   Figure 75

75. Graph $Y_1 = X^2 + X \ln (X)$ and $Y_2 = 2$ as shown in Figure 75.

The graphs intersect near the point $(1.71, 2)$, so the solution to $x^2 - x \ln x = 2$ is $x \approx 1.71$.

77. Graph $Y_1 = Xe^{\wedge}(-X) + \ln (X)$ and $Y_2 = 1$ as shown in Figure 77.

The graphs intersect near the point $(2.10, 2)$, so the solution to $xe^{-x} + \ln x = 1$ is $x \approx 2.10$.

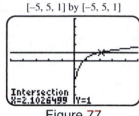

[–5, 5, 1] by [–5, 5, 1]

Intersection
X=2.1026499 Y=1

Figure 77

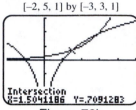

[–2, 5, 1] by [–3, 3, 1]

Intersection
X=1.5041186 Y=.7091283

Figure 78b

79. $7.5 = 7(1.01)^{x-2011} \Rightarrow \dfrac{7.5}{7} = 1.01^{x-2011} \Rightarrow \ln\left(\dfrac{7.5}{7}\right) = \ln(1.01)^{x-2011} \Rightarrow$

$$\ln\left(\dfrac{7.5}{7}\right) = (x - 2011) \ln (1.01) \Rightarrow \dfrac{\ln\left(\dfrac{7.5}{7}\right)}{\ln(1.01)} = x - 2011 \Rightarrow x = 2011 + \dfrac{\ln\left(\dfrac{7.5}{7}\right)}{\ln(1.01)} \Rightarrow x \approx 2018$$

81. $250 = 1000e^{-0.12x} \Rightarrow 0.25 = e^{-0.12x} \Rightarrow \ln (0.25) = -0.12x \Rightarrow x = \dfrac{\ln (0.25)}{-0.12} \approx 11.55$ ft.

83. (a) Let $T(x) = Ca^x$, where x is the number of transistors. Initially, $T(0) = 2300$, so $C = 2300$ and

$T(x) = 2300a^x$. Next we will find the value of a. Because the number of transistors doubles every 2 years

there will be 4600 transistors 2 years after 1971, so $T(2) = 4600$. $2300a^2 = 4600 \Rightarrow a^2 = 2 \Rightarrow a = (2)^{1/2}$.

Thus, $T(x) = 2300\left((2)^{1/2}\right)^x = 2300(2)^{x/2}$

(b) $T(40) = 2300(2)^{40/2} = 2300(2)^{20} = 2,411,724,800$. In 2011 there were about 2.4 billion transistors on an

integrated circuit.

(c) $10,000,000 = 2300(2)^{x/2} \Rightarrow \dfrac{10,000,000}{2300} = 2^{x/2} \Rightarrow \ln\left(\dfrac{10,000,000}{2300}\right) = \ln\left(2^{x/2}\right) \Rightarrow$

$$\ln\left(\dfrac{10,000,000}{2300}\right) = \dfrac{x}{2} \ln(2) \Rightarrow \dfrac{\ln\left(\dfrac{10,000,000}{2300}\right)}{\ln(2)} = \dfrac{x}{2} \Rightarrow$$

$$2\left(\dfrac{\ln\left(\dfrac{10,000,000}{2300}\right)}{\ln(2)}\right) = x \Rightarrow x \approx 24.$$ Since x represents the number of years after 1971 the year is about 1995.

85. (a) $H(8) = 0.3 + 0.28 \ln(8) \approx 0.88$. When x is 8 the Human Development Index is 0.88.

(b) $0.68 = 0.3 + 0.28 \ln(x) \Rightarrow 0.38 = 0.28 \ln(x) \Rightarrow \dfrac{0.38}{0.28} = \ln x \Rightarrow e^{0.38/0.28} = x \Rightarrow x \approx 3.9$

87. We must solve the equation $f(x) = 95$.

$$230(0.881)^x = 95 \Rightarrow 0.881^x = \frac{95}{230} \Rightarrow \log 0.881^x = \log \frac{95}{230} \Rightarrow x \log 0.881 = \log \frac{95}{230}$$

$$x = \frac{\log \frac{95}{230}}{\log 0.881} \approx 7$$

Since $x = 0$ corresponds to 1974, $x \approx 7$ represents $1974 + 7 = 1981$, rounded to the nearest year.

89. (a) $P(x) = Ca^{x-2000} \Rightarrow P(2000) = Ca^{2000-2000} \Rightarrow 1 - Ca^0 \Rightarrow C = 1$, so $P(x) = a^{x-2000}$

$$P(2025) = a^{2025-2000} \Rightarrow 1.4 = a^{25} \Rightarrow a = 1.4^{1/25} \approx 1.01355$$

(b) $P(x) = (1.01355)^{x-2000} \Rightarrow P(2010) = (1.01355)^{10} \approx 1.144$;

in 2010 the population if India will be about 1.14 billion.

(c) $P(x) = 1.5 \Rightarrow 1.01355^{x-2000} = 1.5 \Rightarrow \ln 1.01355^{x-2000} = \ln 1.5 \Rightarrow (x - 2000) \ln 1.01355 = \ln 1.5 \Rightarrow$

$x \ln 1.01355 - 2000 \ln 1.01355 = \ln 1.5 \Rightarrow x \ln 1.01355 = \ln 1.5 + 2000 \ln 1.01355 \Rightarrow$

$$x = \frac{\ln 1.5 + 2000 \ln 1.01355}{\ln 1.01355} = \frac{\ln 1.5}{\ln 1.01355} + 2000 \approx 2030.13;$$

India's population might reach 1.5 billion in 2030.

91. (a) $T_0 = 32, D = 212 - 32 \Rightarrow D = 180$. Solving for a use $70 = 32 + 180a^{1/2} \Rightarrow 38 = 180a^{1/2} \Rightarrow$

$$\frac{38}{180} = a^{1/2} \Rightarrow a = 0.045$$

(b) $T(t) = 32 + 180(0.045)^t \Rightarrow T\left(\frac{1}{6}\right) = 32 + 180(0.045)^{1/6} \Rightarrow T\left(\frac{1}{6}\right) \approx 139°F$

(c) $40 = 32 + 180(0.045)^t \Rightarrow 8 = 180(0.045)^t \Rightarrow \frac{8}{180} = 0.045^t \Rightarrow \log \frac{8}{180} = \log 0.045^t \Rightarrow$

$$\log \frac{8}{180} = t \log 0.045 \Rightarrow t = \frac{\log \frac{8}{180}}{\log 0.045} \Rightarrow t \approx 1$$

Graph $Y_1 = 32 + 180(0.045)^t$ and $Y_2 = 40$. The lines intersect at $t \approx 1$. See Figure 91.

[0,5,1] by [10,50,1]

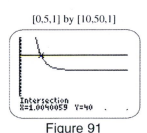

Figure 91

93. (a) $f(1.5) = 20 - 15(0.365)^{1.5} \approx 16.7°C$

(b) $15 = 20 - 15(0.365)^t \Rightarrow 15(0.365)^t = 5 \Rightarrow 0.365^t = \frac{1}{3} \Rightarrow \ln 0.365^t = \ln \frac{1}{3} \Rightarrow t \ln 0.365 = \ln \frac{1}{3} \Rightarrow$

$$t = \frac{\ln \frac{1}{3}}{\ln 0.365} \approx 1.09;$$ the soda can warmed to 15°C after about 1.09 minutes or 1 hour 5.4 minutes.

95. $280 \ln (x + 1) + 1925 = 2300 \Rightarrow 280 \ln (x + 1) = 375 \Rightarrow \ln (x + 1) = \frac{375}{280} \Rightarrow e^{\ln (x+1)} = e^{375/280} \Rightarrow$

$x + 1 = e^{375/280} \Rightarrow x = e^{375/280} - 1 \approx 2.8$; a person who consumes 2300 calories daily owns about 2.8 acres.

97. $\dfrac{2 - \log(100 - x)}{0.42} = 2 \Rightarrow 2 - \log(100 - x) = 0.84 \Rightarrow \log(100 - x) = 1.16 \Rightarrow$

$10^{\log(100-x)} = 10^{1.16} \Rightarrow 100 - x = 10^{1.16} \Rightarrow x = 100 - 10^{1.16} \approx 85.546$; after 2 years approximately 85.5% of the robins have died.

99. (a) A concentration increase of 15% implies $B(t) = 1.15$ when $B_0 = 1$.

$$1.15 = e^{6k} \Rightarrow \ln(1.15) = 6k \Rightarrow k = \frac{\ln(1.15)}{6} \approx 0.0233.$$

(b) $B(t) = 1.2e^{0.0233(8.2)} \approx 1.45$ billion per liter.

(c) $B(t) = 1e^{0.0233(1)} \approx 1.0236 \Rightarrow$ the concentration increases by about 2.36% per hour.

101. $A = P\left(1 + \dfrac{r}{n}\right)^{nt} \Rightarrow 2000 = 1000\left(1 + \dfrac{0.085}{4}\right)^{0.4t} \Rightarrow 2 = (1.02125)^{4t} \Rightarrow \ln(2) = \ln(1.02125)^{4t} \Rightarrow$

$\dfrac{\ln(2)}{\ln(1.02125)} = 4t \Rightarrow t = \dfrac{\frac{\ln(2)}{\ln(1.02125)}}{4} \approx 8.25$ yrs.

103. (a) $750 = 500e^{0.03t} \Rightarrow \dfrac{750}{500} = e^{0.03t} \Rightarrow \ln(1.5) = 0.03t \Rightarrow \dfrac{\ln 1.5}{0.03} = t \Rightarrow t \approx 13.5$

(b) Five hundred dollars invested at 3% compounded continuously results in $\$750$ after 13.5 years.

105. $P = 100\left(\dfrac{1}{2}\right)^{t/5700} \Rightarrow 35 = 100\left(\dfrac{1}{2}\right)^{t/5700} \Rightarrow \left(\dfrac{1}{2}\right)^{t/5700} = 0.35 \Rightarrow \ln\left(\dfrac{1}{2}\right)^{t/5700} = \ln 0.35 \Rightarrow$

$\dfrac{t}{5700} \ln \dfrac{1}{2} = \ln 0.35 \Rightarrow \dfrac{t}{5700} = \dfrac{\ln 0.35}{\ln \frac{1}{2}} \Rightarrow t = 5700\left(\dfrac{\ln 0.35}{\ln \frac{1}{2}}\right) \approx 8633$. The fossil is about 8633 years old.

107. $0.5 = 1 - e^{-0.5x} \Rightarrow -0.5 = -e^{-0.5x} \Rightarrow 0.5 = e^{-0.5x} \Rightarrow \ln(0.5) = -0.5x \Rightarrow x = \dfrac{\ln(0.5)}{-0.5} \Rightarrow$

$x = -2\ln(0.5) \approx 1.39$ min.

109. $N(t) = 100{,}000e^{rt} \Rightarrow 200{,}000 = 100{,}000e^{r(2)} \Rightarrow 2 = e^{2r} \Rightarrow \ln 2 = \ln e^{2r} \Rightarrow \ln 2 = 2r \Rightarrow$

$\dfrac{\ln 2}{2} = r \Rightarrow r = 0.3466.$ Now $350{,}000 = 100{,}000e^{0.3466t} \Rightarrow 3.5 = e^{0.3466t} \Rightarrow \ln 3.5 = \ln e^{0.3466t} \Rightarrow$

$\ln 3.5 = 0.3466t \Rightarrow t = \dfrac{\ln 3.5}{0.3466} \Rightarrow t \approx 3.6$ hrs.

111. $A_n = A_0 e^{rn} \Rightarrow 2300 = 2000e^{r(4)} \Rightarrow 1.15 = e^{4r} \Rightarrow \ln 1.15 = \ln e^{4r} \Rightarrow \ln 1.15 = 4r \Rightarrow$

$r = \dfrac{\ln 1.15}{4} \Rightarrow r = 0.03494.$ Now $3200 = 2000e^{0.03494t} \Rightarrow 1.6 = e^{0.03494t} \Rightarrow \ln 1.6 = \ln e^{0.03494t} \Rightarrow$

$\ln 1.6 = 0.03494t \Rightarrow t = \dfrac{\ln 1.6}{0.03494} \Rightarrow t \approx 13.5$ years.

113. (a) The initial concentration of the drug is 11 milligrams per liter since $C(0) = 11(0.72)^0 = 11$ milligrams per liter.

(b) 50% of $11 = 5.5$ milligrams per liter.

$11(0.72)^t = 5.5 \Rightarrow 0.72^t = 0.5 \Rightarrow \ln 0.72^t = \ln 0.5 \Rightarrow t\ln 0.72 = \ln 0.5 \Rightarrow t = \dfrac{\ln 0.5}{\ln 0.72} \approx 2.11$

The drug concentration decreases to 50% of its initial level after about 2 hours.

115. $P\left(1 + \dfrac{r}{n}\right)^{nt} = A \implies \left(1 + \dfrac{r}{n}\right)^{nt} = \dfrac{A}{P} \implies \log\left(1 + \dfrac{r}{n}\right)^{nt} = \log\left(\dfrac{A}{P}\right) \implies$

$nt \ \log\left(1 + \dfrac{r}{n}\right) = \log\left(\dfrac{A}{P}\right) \implies t = \dfrac{\log\left(A/P\right)}{n \log\left(1 + r/n\right)}$

Extended and Discovery Exercises for Section 5.6

1. $f(x) = Ca^x = Ce^{\ln\left(a^x\right)} = Ce^{x \ln\left(a\right)}$, that is $k = \ln a;\ g(x) = 2^x = e^{\ln\left(2^x\right)} = e^{x \ln\left(2\right)} \implies k = \ln\left(2\right)$

Checking Basic Concepts for Sections 5.5 and 5.6

1. $\log \dfrac{x^2 y^3}{\sqrt[3]{z}} = \log x^2 + \log y^3 - \log z^{1/3} = 2 \log x + 3 \log y - \dfrac{1}{3} \log z$

3. (a) $5(1.4)^x - 4 = 25 \implies 5(1.4)^x = 29 \implies 1.4^x = \dfrac{29}{5} \implies \ln 1.4^x = \ln \dfrac{29}{5} \implies x \ln 1.4 = \ln \dfrac{29}{5} \implies$

$x = \dfrac{\ln \frac{29}{5}}{\ln 1.4} \approx 5.224$

(b) $4^{2-x} = 4^{2x+1} \implies \log_4 4^{2-x} = \log_4 4^{2x+1} \implies 2 - x = 2x + 1 \implies 3x = 1 \implies x = \dfrac{1}{3}$

5. (a) Looking at the graph of $y = 80 + 120(0.9)^x$ we see that eventually the values stay around 80. After a long

time, the temperature of the object stays around 80°F

(b) $80 + 120(0.9)^x = 100 \implies 120(0.9)^x = 20 \implies 0.9^x = \dfrac{1}{6} \implies \ln 0.9^x = \ln \dfrac{1}{6} \implies x \ln 0.9 = \ln \dfrac{1}{6} \implies$

$x = \dfrac{\ln \frac{1}{6}}{\ln 0.9} \approx 17$; the object's temperature is 100°F after about 17 minutes.

5.7: Constructing Nonlinear Models

1. Growth slows down as x increases; logarithmic.

3. Growth rate increases as x increases; exponential.

5. Exponential; least-squares regression gives $f(x) = 1.2(1.7)^x$.

7. Logarithmic; least-squares regression gives $f(x) = 1.088 + 2.937 \ln x$.

9. Logistic; least-squares regression gives $f(x) = \dfrac{9.96}{1 + 30.6e^{-1.51x}}$.

11. (a) The best fit model for the data set is a quadratic function. Using the regression function on your calculator

yields $V(x) = 3.55x^2 + 0.21x + 1.9$.

(b) Since x represents the number of years after 2007 we will let $x = 5$.

$V(5) = 3.55(5)^2 + 0.21(5) + 1.9 = 91.7$ million

13. $a \approx 9.02, b \approx 1.03$, or $f(x) = 9.02 + 1.03 \ln x$.

15. (a) $a \approx 1.4734, b \approx 0.99986$, or $f(x) = 1.4734(0.99986)^x$.

    (b) $f(7000) = 1.4734(0.99986)^{7000} \approx 0.55$; approximately 0.55 kg/m^3.

17. (a) $f(x) = \dfrac{4.9955}{1 + 49.7081e^{-0.6998x}}$

    (b) For large x, $f(x) \approx \dfrac{4.9955}{1} \approx 5$; the density after a long time is about 5 thousand per acre.

    [0, 20, 2] by [150, 700, 50]

    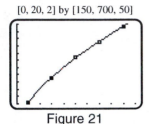
    Figure 21

19. (a) Using the logsitic regression function on your calculator we have $C(x) = \dfrac{0.94}{1 + 17.97e^{-0.862x}}$.

    (b) Since x represents the number of years after 2005 we will let $x = 9$.

    $$C(9) = \dfrac{0.94}{1 + 17.97e^{-0.862(9)}} \approx 0.93 \text{ billion}.$$

21. (a) Power: least-squares regession gives $A(w) \approx 101x^{0.662}$.

    (b) See Figure 21.

    (c) $500 = 101w^{0.662} \Rightarrow \dfrac{500}{101} = w^{0.662} \Rightarrow \left(\dfrac{500}{101}\right)^{1/0.662} = (w^{0.662})^{1/0.662} \Rightarrow w \approx 11.2.$ lbs. A weight of

    about 11.2 lbs. corresponds to a bird with a wing area of 500 in^2.

    [0, 45, 5] by [0, 55, 5]

    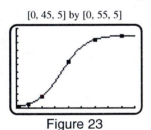
    Figure 23

23. (a) From the table we see that $H(5) = 3$. After 5 years the tree is 3 feet tall.

    (b) Logistic: least-squares regressions gives $H(x) \approx \dfrac{50.1}{1 + 47.4e^{-0.221x}}$.

    (c) See Figure 23.

    (d) $25 = \dfrac{50.1}{1 + 47.4e^{-0.221x}} \Rightarrow \dfrac{50.1}{25} = 1 + 47.4e^{-0.221x} \Rightarrow \dfrac{50.1}{25} - 1 = 47.4e^{-0.221x} \Rightarrow$

    $\dfrac{\frac{50.1}{25} - 1}{47.4} = e^{-0.221x} \Rightarrow \ln\left(\dfrac{\frac{50.1}{25} - 1}{47.4}\right) = -0.221x \Rightarrow x = \dfrac{\ln\left(\dfrac{\frac{50.1}{25} - 1}{47.4}\right)}{-0.221} \approx 17.4$ years. After about 17.4

    years the tree is 25 feet tall.

    (e) The answer involved interpolation.

25. (a) The data is not linear. See Figure 25.

    (b) $a \approx 12.42$, $b \approx 1.066$, or $f(x) = 12.42(1.066)^x$.

    (c) $f(39) = 12.42(1.066)^{39} \approx 150$; 150 kilograms per hectare. Chemical fertilizer use increased, but at a

       slower rate than predicted by f.

[–2, 32, 5] by [0, 80, 10]

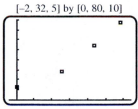

Figure 25

Extended and Discovery Exercises for Section 5.7

1.   (a) Least-squares regression gives $f(x) \approx 0.09(0.844)^x$. A better fit by trial and error is

      $f(x) \approx 0.128(0.777)^x$.

    (b) 30% of 0.133 is $0.0399 \Rightarrow 0.128(0.777)^x = 0.0399 \Rightarrow 0.777^x = \dfrac{0.0399}{0.128} \Rightarrow$

      $\ln 0.777^x = \ln \dfrac{0.0399}{0.128} \Rightarrow x \ln 0.777 = \ln \dfrac{0.0399}{0.128} \Rightarrow x = \dfrac{\ln \frac{0.0399}{0.128}}{\ln 0.777} \approx 4.6$; after about 4.6 minutes.

    (c) *Answers may vary.*

Checking Basic Concepts for Section 5.7

1.   Exponential; least-squares regression gives $f(x) \approx 0.5(1.2)^x$.

3.   Logistic; least-squares regression gives $f(x) = \dfrac{4.5}{1 + 277e^{-1.4x}}$.

Chapter 5 Review Exercises

1.   (a) $(f + g)(1) = f(1) + g(1) = 7 + 1 = 8$

    (b) $(f - g)(3) = f(3) - g(3) = 9 - 9 = 0$

    (c) $(fg)(-1) = f(-1)g(-1) = 3(-2) = -6$

    (d) $(f/g)(0) = \dfrac{f(0)}{g(0)} = \dfrac{5}{0}$; undefined.

3. (a) $f(x) = x^2 \Rightarrow f(3) = 9$ and $g(x) = 1 - x \Rightarrow g(3) = 1 - 3 = -2$.

   Thus, $(f + g)(3) = f(3) + g(3) = 9 + (-2) = 7$.

   (b) $f(-2) = 4$ and $g(-2) = 1 - (-2) = 3$. Thus, $(f - g)(-2) = f(-2) - g(-2) = 4 - 3 = 1$.

   (c) $f(1) = 1$ and $g(1) = 1 - 1 = 0$. Thus, $(fg)(1) = f(1)g(1) = (1)(0) = 0$.

   (d) $f(3) = 9$ and $g(3) = 1 - 3 = -2$. Thus, $(f/g)(3) = \dfrac{f(3)}{g(3)} = \dfrac{9}{-2} = -\dfrac{9}{2}$.

5. (a) $(g \circ f)(-2) = g(f(-2)) = g(1) = 2$

   (b) $(f \circ g)(3) = f(g(3)) = f(-2) = 1$

   (c) $f^{-1}(3) = 2$ since $f(2) = 3$

7. (a) $f(x) = \sqrt{x}$ and $g(x) = x^2 + x \Rightarrow (f \circ g)(2) = f(g(2)) = f(6) = \sqrt{6}$

   (b) $(g \circ f)(9) = g(f(9)) = g(\sqrt{9}) = g(3) = 3^2 + 3 = 12$

9. $(f \circ g) = f(g(x)) = f\left(\dfrac{1}{x}\right) = \left(\dfrac{1}{x}\right)^3 - \left(\dfrac{1}{x}\right)^2 + 3\left(\dfrac{1}{x}\right) - 2;\ D = \{x \mid x \neq 0\}$

11. $(f \circ g) = f(g(x)) = f\left(\dfrac{1}{2}x^3 + \dfrac{1}{2}\right) = \sqrt[3]{2\left(\dfrac{1}{2}x^3 + \dfrac{1}{2}\right) - 1} = \sqrt[3]{x^3 + 1 - 1} = \sqrt[3]{x^3} = x;$

   $D =$ all real numbers

13. $h(x) = (g \circ f)(x) \Rightarrow h(x) = \sqrt{x^2 + 3} \Rightarrow f(x) = x^2 + 3, g(x) = \sqrt{x}$. *Answers may vary.*

15. Subtract 6 from x and then multiply the results by 10. $\dfrac{x}{10} + 6$ and $10(x - 6)$.

17. Since the graph of $f(x) = 3x - 1$ is a line sloping upward from left to right, a horizontal line can intersect it at most once. Therefore, f is one-to-one.

19. f is not one-to-one. It does not pass the Horizontal line test.

21. See Figure 21. The domain of f is $D = \{-1, 0, 4, 6\}$ and its range is $R = \{1, 3, 4, 6\}$. The domain and range of f^{-1} are $D = \{1, 3, 4, 6\}$ and $R = \{-1, 0, 4, 6\}$, repectively.

x	6	4	3	1
$f^{-1}(x)$	-1	0	4	6

Figure 21

23. The inverse operations are add 5 to x and divide by 3. Therefore, $f^{-1}(x) = \dfrac{x + 5}{3}$.

25. $(f \circ f^{-1})(x) = f(f^{-1}(x)) = f\left(\dfrac{x + 1}{2}\right) = 2\left(\dfrac{x + 1}{2}\right) - 1 = (x + 1) - 1 = x$

   $(f^{-1} \circ f)(x) = f^{-1}(f(x)) = f^{-1}(2x - 1) = \dfrac{(2x - 1) + 1}{2} = \dfrac{2x}{2} = x$

27. $(f \circ g^{-1})(4) = f(g^{-1}(4)) = f(3) = 1$

29. $y = \sqrt{x + 1} \Rightarrow y^2 = x + 1 \Rightarrow y^2 - 1 = x; \Rightarrow f^{-1}(x) = x^2 - 1, x \geq 0$

   For f: $D = \{x \mid x \geq -1\}$ and $R = \{y \mid y \geq 0\}$; for f^{-1}: $D = \{x \mid x \geq 0\}$ and $R = \{y \mid y \geq -1\}$

31. If $f(0) = 3 \Rightarrow Ca^x = 3 \Rightarrow Ca^0 = 3 \Rightarrow a^0 = 1; C(1) = 3 \Rightarrow C = 3$; then $f(3) = 24 \Rightarrow$

   $Ca^x = 24 \Rightarrow 3a^3 = 24 \Rightarrow a^3 = 8 \Rightarrow a = 2$. Therefore $C = 3$ and $a = 2$.

33. See Figure 33. D = all real numbers.

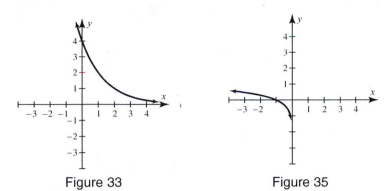

Figure 33   Figure 35

35. The graph of $f(x) = \log(-x)$ is similar to the graph of $f(x) = \log x$, except it is refelcted through the y-axis. The

domain is all real numbers less than 0. See Figure 35.

37. $y = Ca^x \Rightarrow y = 2(2)^x \Rightarrow C = 2;\ a = 2$

39. $A = 1200\left(1 + \dfrac{0.09}{2}\right)^{2(3)} \approx \$1562.71$

41. $e^x = 19 \Rightarrow \ln e^x = \ln 19 \Rightarrow x = \ln 19 \approx 2.9444$; the result can be supported graphically using the

intersection method with the graphs $Y_1 = e{\wedge}X$ and $Y_2 = 19$. The intersection point is near $(2.9444, 19)$. The

result may also be supported numerically with a table of $Y_1 = e{\wedge}X$ starting at 2.9 and incrementing by 0.01.

Here $Y_1 = 19$ when $x \approx 2.94$.

43. $\log 1000 = \log 10^3 = 3$

45. $10 \log 0.01 + \log \dfrac{1}{10} = 10 \log 10^{-2} + \log 10^{-1} = 10(-2) + (-1) = -21$

47. $\log_3 9 = \log_3 3^2 = 2$

49. $\ln e = \ln e^1 = 1$

51. $\log_3 18 = \dfrac{\log 18}{\log 3} \approx 2.631$

53. $10^x = 125 \Rightarrow \log 10^x = \log 125 \Rightarrow x = \log 125 \approx 2.097$

55. $e^{0.1x} = 5.2 \Rightarrow \ln e^{0.1x} = \ln 5.2 \Rightarrow 0.1x = \ln 5.2 \Rightarrow x = 10 \ln 5.2 \approx 16.49$

57. $5^{-x} = 10 \Rightarrow \log 5^{-x} = \log 10 \Rightarrow -x \log 5 = 1 \Rightarrow -x = \dfrac{1}{\log 5} \Rightarrow x = -\dfrac{1}{\log 5} \approx -1.431$

59. $50 - 3(0.78)^{x-10} = 21 \Rightarrow -3(0.78)^{x-10} = -29 \Rightarrow (0.78)^{x-10} = \dfrac{29}{3} \Rightarrow$

$(x - 10) \log (0.78) = \log\left(\dfrac{29}{3}\right) \Rightarrow x - 10 = \dfrac{\log\left(\frac{29}{3}\right)}{\log (0.78)} \Rightarrow x = 10 + \dfrac{\log\left(\frac{29}{3}\right)}{\log (0.78)} \approx 0.869$

61. For each unit increase in x, the y-values are multiplied by 2, so the data are exponential. Since $y = 1.5$ when $x = 0$, the initial value is 1.5, so $f(x) = Ca^x \Rightarrow f(x) = 1.5a^x$. Since $y = 3$ when $x = 1$, $f(1) = 1.5a^1 \Rightarrow 3 = 1.5a \Rightarrow a = 2$. Thus the function $f(x) = 1.5(2^x)$ can model the data.

63. $\log x = 1.5 \Rightarrow 10^{\log x} = 10^{1.5} \Rightarrow x = 10^{1.5} \approx 31.62$

65. $\ln x = 3.4 \Rightarrow e^{\ln x} = e^{3.4} \Rightarrow x = e^{3.4} \approx 29.96$

67. $\log 6 + \log 5x = \log (6 \cdot 5x) = \log 30x$

69. $\ln \dfrac{y}{x^2} = \ln (y) - \ln (x^2) = \ln (y) - 2 \ln (x)$

71. $8 \log x = 2 \Rightarrow \log x = \dfrac{1}{4} \Rightarrow 10^{\log x} = 10^{1/4} \Rightarrow x = 10^{1/4} \Rightarrow x = \sqrt[4]{10} \approx 1.778$

73. $2 \log 3x + 5 = 15 \Rightarrow 2 \log 3x = 10 \Rightarrow \log 3x = 5 \Rightarrow 10^{\log 3x} = 10^5 \Rightarrow 3x = 100,000 \Rightarrow$
    $x = \dfrac{100,000}{3} \approx 33,333$

75. $2 \log_5 (x + 2) = \log_5 (x + 8) \Rightarrow \log_5 (x + 2)^2 = \log_5 (x + 8) \Rightarrow 10^{\log_5 (x+2)^2} = 10^{\log_5 (x+8)} \Rightarrow$
    $(x + 2)^2 = (x + 8) \Rightarrow x^2 + 4x + 4 = x + 8 \Rightarrow x^2 + 3x - 4 = 0 \Rightarrow (x + 4)(x - 1) = 0 \Rightarrow$
    $x = -4$ or $x = 1$; since $x = -4$ is undefined in $\log (x + 2)$, the only solution is $x = 1$.

77. If b is the y-intercept, then the point $(0, b)$ is on the graph of f. Therefore, the point $(b, 0)$ is on the graph of f^{-1}. The point $(b, 0)$ lies on the x-axis and is the x-intercept of the graph of f^{-1}.

79. (a) $N(t) = N_0 e^{rt} \Rightarrow 6000 = 4000e^{r(1)} \Rightarrow 1.5 = e^r \Rightarrow \ln 1.5 = \ln e^r \Rightarrow \ln 1.5 = r \Rightarrow r = 0.4055$

    Now $N(2.5) = 4000e^{0.4055(2.5)} \Rightarrow N(2.5) \approx 11,022$

    (b) $8500 = 4000e^{0.4055t} \Rightarrow 2.125 = e^{0.4055t} \Rightarrow \ln 2.125 = \ln e^{0.4055t} \Rightarrow \ln 2.125 = 0.4055t \Rightarrow$

    $\dfrac{\ln 2.125}{0.4055} = t \Rightarrow t \approx 1.86$ hours.

81. $h(x) = f(x) + g(x) = 10x + 5x = 15x$

83. Use the intersection method by graphing $Y_1 = 175.6(1 - 0.66e^{\wedge}(-0.24X))^{\wedge}3$ and $Y_2 = 50$ in the viewing rectangle $[0, 14, 1]$ by $[0, 175, 25]$. The intersection is near the point $(2.74, 50)$. See Figure 83. This means that at approximately 3 weeks the fish weighed 50 milligrams.

[0, 14, 1] by [0, 175, 25]

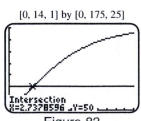

Figure 83

85. $15 = 32e^{-0.2t} \Rightarrow \dfrac{15}{32} = e^{-0.2t} \Rightarrow -0.2t = \ln\left(\dfrac{15}{32}\right) \Rightarrow t = \left(-\dfrac{1}{0.2}\right)\ln\left(\dfrac{15}{32}\right) \approx (-5)(-0.7577) \approx 3.8;$

     after about 3.8 minutes.

87. Logistic: least-squares regression gives $f(x) \approx \dfrac{171.4}{1 + 18.4e^{-0.0744x}}$

89. See Figure 89. $a \approx 3.50, b \approx 0.74,$ or $f(x) = 3.50(0.74)^x$

[0, 5, 1] by [0, 3, 1]

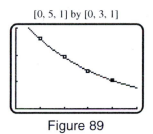

Figure 89

Extended and Discovery Exercises for Chapter 5

1. STEP 1: Using $w = \ln x$ and $z = \ln y$, the data points (w, z) are approximately $(4.85, -1.24)$, $(5.20, -0.69)$, $(5.40, -0.40)$, $(6.05, 0.57)$, $(6.51, 1.27)$, $(6.98, 1.97)$, and $(7.54, 2.81)$. Plot these points as shown in Figure 1a.

    STEP 2: Use the linear regression feature of the graphing calculator to find the values b and d. See Figure 1b.

    STEP 3: Here the slope of the line is approximately 1.505, so $b \approx 1.505$ in the equation $y = ax^b$. The y-intercept of the regression line is approximately -8.513, so $a \approx e^{-8.513} \approx 0.0002$ in the equation $y = ax^b$.

    The equation $y = 0.0002x^{1.5}$ models the data. A graph of $y = 0.0002x^{1.5}$ along with the original data is shown in Figure 1c.

[4, 8, 1] by [−2, 4, 1]   [0, 2000, 500] by [0, 20, 2]

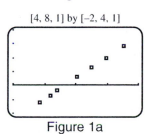 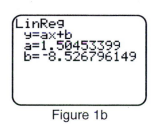 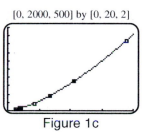

Figure 1a   Figure 1b   Figure 1c

3. $\left(\dfrac{1}{\pi}\ln\left(640{,}320^3 + 744\right)\right)^2 \approx 163.0000000000000000000000000000000232$

    That is, $\left(\dfrac{1}{\pi}\ln\left(640{,}320^3 + 744\right)\right)^2 - 163 \approx 2.32 \times 10^{-30}$. It is not an integer.

Chapter 6: Trigonometric Functions

6.1: Angles and Their Measure

1. (a) See Figure 1a.

   (b) See Figure 1b.

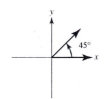

Figure 1a

Figure 1b

Figure 1c

Figure 1d

   (c) See Figure 1c.

   (d) See Figure 1d.

3. Any angle with degree measure between $0°$ and $90°$. See Figure 3. *Answers may vary.*

5. See Figure 5.

7. Any angle with degree measure between $180°$ and $270°$. See Figure 7. *Answers may vary.*

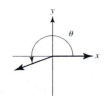

Figure 7

Figure 9

9. Any angle with degree measure of the form $90° - n(360°)$ where $n = 1, 2, 3, \ldots$. See Figure 9. *Answers may vary.*

11. (a) $\dfrac{90°}{360°} = \dfrac{1}{4}$

    (b) $\dfrac{30°}{360°} = \dfrac{1}{12}$

    (c) $\dfrac{\frac{\pi}{3}}{2\pi} = \dfrac{1}{6}$

    (d) $\dfrac{\frac{\pi}{4}}{2\pi} = \dfrac{1}{8}$

13. $150° + 360° = 510°$, $150° - 360° = -210°$ *Answers may vary.*

15. $-72° + 360° = 288°$, $-72° - 360° = -432°$ *Answers may vary.*

17. $\dfrac{\pi}{2} + 2\pi = \dfrac{\pi}{2} + \dfrac{4\pi}{2} = \dfrac{5\pi}{2}$, $\dfrac{\pi}{2} - 2\pi = \dfrac{\pi}{2} - \dfrac{4\pi}{2} = -\dfrac{3\pi}{2}$ *Answers may vary.*

19. $-\dfrac{\pi}{5} + 2\pi = -\dfrac{\pi}{5} + \dfrac{10\pi}{5} = \dfrac{9\pi}{5}$, $-\dfrac{\pi}{5} - 2\pi = -\dfrac{\pi}{5} - \dfrac{10\pi}{5} = -\dfrac{11\pi}{5}$ *Answers may vary.*

21. $125°15' = 125° + \left(\dfrac{15}{60}\right)^{\circ} = 125.25°$

23. $108°45'36'' = 108° + \left(\dfrac{45}{60}\right)^{\circ} + \left(\dfrac{36}{3600}\right)^{\circ} = 108.76°$

25. $125.3° = 125° + \left(\dfrac{3}{10}\right)^{\circ} = 125° + \left(\dfrac{18}{60}\right)^{\circ} = 125°18'$

27. $51.36° = 51° + \left(\dfrac{36}{100}\right)^{\circ} = 51° + \left(\dfrac{35}{100}\right)^{\circ} + \left(\dfrac{1}{100}\right)^{\circ} = 51° + \left(\dfrac{21}{60}\right)^{\circ} + \left(\dfrac{36}{3600}\right)^{\circ} = 51°21'36''$

29. $\alpha = 90° - 55.9° = 34.1°, \beta = 180° - 55.9° = 124.1°$

31. $\alpha = 90° - 85°23'45'' = 89°59'60'' - 85°23'45'' = 4°36'15''$

    $\beta = 180° - 85°23'45'' = 179°59'60'' - 85°23'45'' = 94°36'15''$

33. $\alpha = 90° - 23°40'35'' = 89°59'60'' - 23°40'35'' = 66°19'25''$

    $\beta = 180° - 23°40'35'' = 179°59'60'' - 23°40'35'' = 156°19'25''$

35. $\theta = \dfrac{4}{2} = 2$ radians. The degree measure of θ is $2\left(\dfrac{180°}{\pi}\right) = \dfrac{360°}{\pi} \approx 114.6°.$

37. $\theta = \dfrac{13}{10} = 1.3$ radians. The degree measure of θ is $\dfrac{13}{10}\left(\dfrac{180°}{\pi}\right) = \dfrac{234°}{\pi} \approx 74.5°.$

39. (a) $45°\left(\dfrac{\pi}{180°}\right) = \dfrac{45\pi}{180} = \dfrac{\pi}{4}$

    (b) $135°\left(\dfrac{\pi}{180°}\right) = \dfrac{135\pi}{180} = \dfrac{3\pi}{4}$

    (c) $-120°\left(\dfrac{\pi}{180°}\right) = -\dfrac{120\pi}{180} = -\dfrac{2\pi}{3}$

    (d) $-210°\left(\dfrac{\pi}{180°}\right) = -\dfrac{210\pi}{180} = -\dfrac{7\pi}{6}$

41. (a) $37°\left(\dfrac{\pi}{180°}\right) = \dfrac{37\pi}{180}$

    (b) $123.4°\left(\dfrac{\pi}{180°}\right) = \dfrac{123.4\pi}{180} \approx 2.15$

    (c) $-92°25' = -\left[92° + \left(\dfrac{25}{60}\right)^{\circ}\right] \approx -92.42°\left(\dfrac{\pi}{180°}\right) = -\dfrac{92.42\pi}{180} \approx -1.61$

    (d) $230°17' = 230 + \left(\dfrac{17}{60}\right)^{\circ} \approx 230.28\left(\dfrac{\pi}{180°}\right) = \dfrac{230.28\pi}{180} \approx 4.02$

43. (a) $\dfrac{\pi}{6}\left(\dfrac{180°}{\pi}\right) = \dfrac{180°}{6} = 30°$

    (b) $\dfrac{\pi}{15}\left(\dfrac{180°}{\pi}\right) = \dfrac{180°}{15} = 12°$

    (c) $-\dfrac{5\pi}{3}\left(\dfrac{180°}{\pi}\right) = -\dfrac{900°}{3} = -300°$

    (d) $-\dfrac{7\pi}{6}\left(\dfrac{180°}{\pi}\right) = -\dfrac{1260°}{6} = -210°$

45. (a) $\dfrac{\pi}{4}\left(\dfrac{180°}{\pi}\right) = \dfrac{180°}{4} = 45°$

    (b) $\dfrac{\pi}{7}\left(\dfrac{180°}{\pi}\right) = \dfrac{180°}{7} \approx 25.71°$

    (c) $3.1\left(\dfrac{180°}{\pi}\right) = \dfrac{558°}{\pi} \approx 177.62°$

    (d) $-\dfrac{5}{2}\left(\dfrac{180°}{\pi}\right) = -\dfrac{900°}{2\pi} \approx -143.24°$

47. Convert 60° to radians. $60°\left(\dfrac{\pi}{180°}\right) = \dfrac{60\pi}{180} = \dfrac{\pi}{3} \Rightarrow s = 2\left(\dfrac{\pi}{3}\right) = \dfrac{2\pi}{3}$ in.

49. $12 = 5\theta \Rightarrow \theta = \dfrac{12}{5}$ radians

51. $5 = r\pi \Rightarrow r = \dfrac{5}{\pi}$ ft.

53. $s = 3\left(\dfrac{\pi}{12}\right) = \dfrac{3\pi}{12} = \dfrac{\pi}{4}$ m.

55. Convert 15° to radians. $15°\left(\dfrac{\pi}{180°}\right) = \dfrac{15\pi}{180} = \dfrac{\pi}{12} \Rightarrow s = 12\left(\dfrac{\pi}{12}\right) = \dfrac{12\pi}{12} = \pi$ ft.

57. Note that $1°45' = 1° + \left(\dfrac{45}{60}\right)° = 1.75°$.

    Convert 1.75° to radians. $1.75°\left(\dfrac{\pi}{180°}\right) = \dfrac{1.75\pi}{180} = \dfrac{7\pi}{720} \Rightarrow s = 2\left(\dfrac{7\pi}{720}\right) = \dfrac{14\pi}{720} = \dfrac{7\pi}{360}$ mi.

59. A total of 15 minutes passes between 10:15 A.M. and 10:30 A.M. The minute hand has moved through the

    angle of $\dfrac{15}{60} \cdot 2\pi = \dfrac{30\pi}{60} = \dfrac{\pi}{2}$ radians. Therefore, the tip of the minute hand travels

    $s = 4\left(\dfrac{\pi}{2}\right) = \dfrac{4\pi}{2} = 2\pi$ inches. The linear speed of the tip of the minute hand is $\dfrac{2\pi}{15}$ in./min.

61. A total of 75 minutes passes between 3:00 P.M. and 4:15 P.M. The minute hand has moved through the

    angle of $\dfrac{75}{60} \cdot 2\pi = \dfrac{150\pi}{60} = \dfrac{5\pi}{2}$ radians. Therefore, the tip of the minute hand travels

    $s = 4\left(\dfrac{5\pi}{2}\right) = 10\pi$ inches. The linear speed of the tip of the minute hand is $\dfrac{10\pi}{75} = \dfrac{2\pi}{15}$ in./min.

63. $A = \dfrac{1}{2} \cdot 3^2 \cdot \left(\dfrac{\pi}{3}\right) = \dfrac{3\pi}{2} = 1.5\pi \text{ in}^2$

65. Convert 45° to radians. $45°\left(\dfrac{\pi}{180°}\right) = \dfrac{\pi}{4}$. Then, $A = \dfrac{1}{2} \cdot 6^2 \cdot \left(\dfrac{\pi}{4}\right) = \dfrac{9\pi}{2} = 4.5\pi \text{ in}^2$

67. $A = \dfrac{1}{2} \cdot 13.1^2 \cdot \left(\dfrac{\pi}{15}\right) = \dfrac{17,161\pi}{3000} \approx 5.72\pi \text{ cm}^2$

69. Convert 30° to radians. $30°\left(\dfrac{\pi}{180°}\right) = \dfrac{\pi}{6}$. Then, $A = \dfrac{1}{2} \cdot 1.5^2 \cdot \left(\dfrac{\pi}{6}\right) = \dfrac{3\pi}{16} \approx 0.59 \text{ ft}^2$

71. When rotating from $-45°$ to 90° the arm swings through an angle of 135°. Convert 135° to radians:

    $135°\left(\dfrac{\pi}{180°}\right) = \dfrac{3\pi}{4}$. To find the area of the work space, we subtract the area of the inner circular sector from

    the area of the outer circular sector. $A = \dfrac{1}{2}\left(\dfrac{3\pi}{4}\right)(26^2 - 6^2) = \dfrac{3\pi}{8}(640) = 240\pi \text{ in}^2$

73. When rotating from 15° to 195° the arm swings through an angle of 180°. Convert 180° to radians:

    $180°\left(\dfrac{\pi}{180°}\right) = \pi$. To find the area of the work space, we subtract the area of the inner circular sector from

    the area of the outer circular sector. $A = \dfrac{1}{2}(\pi)(95^2 - 21^2) = \dfrac{\pi}{2}(8584) = 4292\pi \text{ cm}^2$

75. The angular velocity is $\omega = 15$ radians/sec. The radius is $r = \dfrac{26}{2} = 13$ inches.

    The linear speed is given by $v = r\omega = 13(15) = 195$ in/sec. That is $\dfrac{195 \text{ in}}{\text{sec}} \cdot \dfrac{1 \text{ ft}}{12 \text{ in}} = 16.25 \text{ ft/sec}$.

    And $\dfrac{16.25 \text{ ft}}{\text{sec}} \cdot \dfrac{1 \text{ mi}}{5280 \text{ ft}} \cdot \dfrac{3600 \text{ sec}}{1 \text{ hr}} \approx 11.1 \text{ mi/hr}$.

77. (a) One revolution is equal to 2π radians, therefore the angular velocity is

    $\dfrac{2\pi}{420 \text{ sec}} = \dfrac{\pi}{210} \approx 0.015 \text{ radian/sec}$.

    (b) The linear speed is $v = r\omega$; $r = \dfrac{140}{2} = 70$ ft and $\omega = \dfrac{\pi}{210}$ therefore $v = 70\left(\dfrac{\pi}{210}\right) \approx 1.05 \text{ ft/sec}$.

79. (a) The angular velocity is $\omega = 500(2\pi) = 1000\pi \approx 3141.6 \text{ radians/min}$.

    (b) The linear speed is $v = r\omega$; $r = \dfrac{30}{2} = 15$ in. and $\omega = 1000\pi$

    therefore $v = 15(1000\pi) \approx 47,123.89 \text{ in./min}$ or about 65.5 ft/sec.

81. Convert $40°55' - 29°11' = 11°44'$ to radians: $\left[11° + \left(\dfrac{44}{60}\right)^{°}\right]\left(\dfrac{\pi}{180°}\right) \approx 0.2048$.

    Then, $s = 3955(0.2048) \approx 809.98 \approx 810 \text{ mi}$.

83. Angular speed ω measures the speed of rotation and is measured by $\omega = \dfrac{\theta}{t}$, $85° = \dfrac{17}{36}\pi$ and $t = 0.05 \Rightarrow$

    $\omega = \dfrac{\frac{17\pi}{36}}{0.05} \approx 29.7 \text{ rad/sec}$. The linear speed v at which the tip of the rocket travels is given by $v = r\omega$.

    $r = 28$ inches or $2\dfrac{1}{3}$ ft and $\omega \approx 29.7$ rad/sec $\Rightarrow v = \left(2\dfrac{1}{3}\right)(29.7) = 69.3$ or about 69 ft/sec.

85. Convert 75.3° to radians: $75.3°\left(\dfrac{\pi}{180°}\right) \approx 1.314$. Then, $s = 11(1.314) \approx 14.5$ in.

87. (a) In one revolution of the pedals, the chain will travel $s = 3.75(2\pi) = 7.5\pi$ inches. The circumference of

the small gear is $C = 1.5(2\pi) = 3\pi$ inches. When the chain moves 7.5π inches, the smaller gear will

rotate $\dfrac{7.5\pi}{3\pi} = \dfrac{5}{2} = 2.5$ times. Thus, the bicycle tire will rotate 2.5 times for each revolution of the pedals.

(b) When the pedals turn through 2 revolutions per second, the angular velocity of the tire is

$\omega = 2.5(2)(2\pi) = 10\pi$ radians/sec. Thus, the tire of radius 13 inches (diameter = 26 inches) will travel

$v = 13(10\pi) = 130\pi$ in./sec. That is $\dfrac{130\pi \text{ in.}}{\text{sec}} \cdot \dfrac{1 \text{ ft}}{12 \text{ in.}} \approx 34$ ft/sec.

89. (a) $\omega = \dfrac{2\pi}{0.615}$ radians/yr $\Rightarrow \dfrac{\frac{2\pi}{0.615} \text{ rad}}{1 \text{ yr}} \cdot \dfrac{1 \text{ yr}}{365 \text{ days}} \cdot \dfrac{1 \text{ day}}{24 \text{ hr}} \approx 1.1662741 \times 10^{-3}$ radians/hr.

Thus, the orbital velocity of Venus is $v = 67.2 \times 10^6(1.1662741 \times 10^{-3}) \approx 78,370$ mi/hr.

(b) $\omega = \dfrac{2\pi}{1} = 2\pi$ radians/yr $\Rightarrow \dfrac{2\pi \text{ rad}}{1 \text{ yr}} \cdot \dfrac{1 \text{ yr}}{365 \text{ days}} \cdot \dfrac{1 \text{ day}}{24 \text{ hr}} \approx 7.172585967 \times 10^{-4}$ radians/hr.

Thus, the orbital velocity of Earth is $v = 92.9 \times 10^6(7.172585967 \times 10^{-4}) \approx 66,630$ mi/hr.

(c) $\omega = \dfrac{2\pi}{11.86}$ radians/yr $\Rightarrow \dfrac{\frac{2\pi}{11.86} \text{ rad}}{1 \text{ yr}} \cdot \dfrac{1 \text{ yr}}{365 \text{ days}} \cdot \dfrac{1 \text{ day}}{24 \text{ hr}} \approx 6.047711608 \times 10^{-5}$ radians/hr.

Thus, the orbital velocity of Jupiter is $v = 483.6 \times 10^6(6.047711608 \times 10^{-5}) \approx 29,250$ mi/hr.

(d) $\omega = \dfrac{2\pi}{164.8}$ radians/yr $\Rightarrow \dfrac{\frac{2\pi}{164.8} \text{ rad}}{1 \text{ yr}} \cdot \dfrac{1 \text{ yr}}{365 \text{ days}} \cdot \dfrac{1 \text{ day}}{24 \text{ hr}} \approx 4.35229731 \times 10^{-6}$ radians/hr.

Thus, the orbital velocity of Neptune is $v = 2794 \times 10^6(4.35229731 \times 10^{-6}) \approx 12,160$ mi/hr.

91. Since the bar is 2 meters in length, we may use 2 meters as an estimate of the arc length from P to Q.

Convert 0.835° to radians: $0.835°\left(\dfrac{\pi}{180°}\right) = \dfrac{167\pi}{36,000}$. Then, $2 = r\left(\dfrac{167\pi}{36,000}\right) \Rightarrow r = 2\left(\dfrac{36,000}{167\pi}\right) \approx 137.2$ m.

A reasonable approximation for d is 137.2 meters.

93. (a) $475,000 = r^2\pi \Rightarrow r = \sqrt{\dfrac{475,000}{\pi}} \approx 388.8$ m

(b) Convert 70° to radians: $70°\left(\dfrac{\pi}{180°}\right) = \dfrac{7\pi}{18}$. So

$475,000 = \dfrac{1}{2}\left(\dfrac{7\pi}{18}\right)r^2 \Rightarrow r = \sqrt{2(475,000)\left(\dfrac{18}{7\pi}\right)} \approx 881.8$ m

Extended and Discovery Exercises for Section 6.1

1. To convert degrees to radians we use $\theta°\left(\dfrac{\pi}{180°}\right)$. So the modified formula is $s = r\left[\theta\left(\dfrac{\pi}{180°}\right)\right] = r\theta\left(\dfrac{\pi}{180°}\right)$.

The radian measure formula is simpler.

6.2: Right Triangle Trigonometry

1. See Figure 1.

3. See Figure 3.

Figure 1   Figure 3   Figure 5

5. See Figure 5.

7. $\sin 60° = \dfrac{\text{side opposite}}{\text{hypotenuse}} = \dfrac{\sqrt{3}}{2}$

9. $\cos 30° = \dfrac{\text{side adjacent}}{\text{hypotenuse}} = \dfrac{\sqrt{3}}{2}$

11. $\sec 60° = \dfrac{\text{hypotenuse}}{\text{side adjacent}} = \dfrac{2}{1} = 2$

13. $\tan 45° = \dfrac{\text{side opposite}}{\text{side adjacent}} = \dfrac{1}{1} = 1$

15. $\cot 45° = \dfrac{\text{side adjacent}}{\text{side opposite}} = \dfrac{1}{1} = 1$

17. $\sin 45° = \dfrac{\text{side opposite}}{\text{hypotenuse}} = \dfrac{1}{\sqrt{2}}$

19. Use the Pythagorean theorem to calculate the length of the hypotenuse:

$$h^2 = 3^2 + 4^2 \Rightarrow h^2 = 9 + 16 \Rightarrow h^2 = 25 \Rightarrow h = 5$$

$\sin\theta = \dfrac{4}{5} \qquad \cos\theta = \dfrac{3}{5} \qquad \tan\theta = \dfrac{4}{3} \qquad \csc\theta = \dfrac{5}{4} \qquad \sec\theta = \dfrac{5}{3} \qquad \cot\theta = \dfrac{3}{4}$

21. Use the Pythagorean theorem to calculate the length of the missing leg:

$$13^2 = 12^2 + l^2 \Rightarrow l^2 = 169 - 144 \Rightarrow l^2 = 25 \Rightarrow l = 5$$

$\sin\theta = \dfrac{12}{13} \qquad \cos\theta = \dfrac{5}{13} \qquad \tan\theta = \dfrac{12}{5} \qquad \csc\theta = \dfrac{13}{12} \qquad \sec\theta = \dfrac{13}{5} \qquad \cot\theta = \dfrac{5}{12}$

23. $\sin 60° \approx 0.866 \qquad \cos 60° \approx 0.5 \qquad \tan 60° \approx 1.732$

    $\csc 60° \approx 1.155 \qquad \sec 60° \approx 2 \qquad \cot 60° \approx 0.577$

25. $\sin 25° \approx 0.423 \qquad \cos 25° \approx 0.906 \qquad \tan 25° \approx 0.466$

    $\csc 25° \approx 2.366 \qquad \sec 25° \approx 1.103 \qquad \cot 25° \approx 2.145$

27. $\sin 5°35' \approx 0.097 \qquad \cos 5°35' \approx 0.995 \qquad \tan 5°35' \approx 0.098$

    $\csc 5°35' \approx 10.278 \qquad \sec 5°35' \approx 1.005 \qquad \cot 5°35' \approx 10.229$

29. $\sin 13°45'30'' \approx 0.238 \qquad \cos 13°45'30'' \approx 0.971 \qquad \tan 13°45'30'' \approx 0.245$

    $\csc 13°45'30'' \approx 4.205 \qquad \sec 13°45'30'' \approx 1.030 \qquad \cot 13°45'30'' \approx 4.084$

31. $\sin 1.05° \approx 0.018$ $\cos 1.05° \approx 1.000$ $\tan 1.05° \approx 0.018$

     $\csc 1.05° \approx 54.570$ $\sec 1.05° \approx 1.000$ $\cot 1.05° \approx 54.561$

33. $\sec \theta = \dfrac{1}{\cos \theta} = \dfrac{1}{\frac{1}{3}} = 3$

35. $\csc \theta = \dfrac{1}{\sin \theta} = \dfrac{1}{\frac{12}{13}} = \dfrac{13}{12}$

37. $\tan \theta = \dfrac{1}{\cot \theta} = \dfrac{1}{\frac{7}{24}} = \dfrac{24}{7}$

39. Triangle ABC is a $30° \sim 60°$ right triangle, thus $b = 8$. Find a using the Pythagorean theorem:

$$8^2 + a^2 = 16^2 \Rightarrow a^2 = 256 - 64 \Rightarrow a^2 = 192 \Rightarrow a = 8\sqrt{3} \approx 13.86.$$

41. To find b note that $\tan 50° = \dfrac{6}{b} \Rightarrow b = \dfrac{6}{\tan 50°} \Rightarrow b \approx 5.03.$

     To find c note that $\sin 50° = \dfrac{6}{c} \Rightarrow c = \dfrac{6}{\sin 50°} \Rightarrow c \approx 7.83.$

43. To find a note that $\sin 41° = \dfrac{a}{8} \Rightarrow a = 8\sin 41° \Rightarrow a \approx 5.25.$

     To find b note that $\cos 41° = \dfrac{b}{8} \Rightarrow b = 8\cos 41° \Rightarrow b \approx 6.04.$

45. To find a note that $\tan 46°15' = \dfrac{a}{16.1} \Rightarrow a = 16.1\tan 46°15' \Rightarrow a \approx 16.82.$

     To find c note that $\cos 46°15' = \dfrac{16.1}{c} \Rightarrow c = \dfrac{16.1}{\cos 46°15'} \Rightarrow c \approx 23.28.$

47. To find a note that $\cos 60° = \dfrac{a}{24} \Rightarrow a = 24\cos 60° \Rightarrow a = 24\left(\dfrac{1}{2}\right) = 12.$

     To find b note that $\sin 60° = \dfrac{b}{24} \Rightarrow b = 24\sin 60° \Rightarrow b = 24\left(\dfrac{\sqrt{3}}{2}\right) = 12\sqrt{3}.$

     To find c note that $\cos 45° = \dfrac{24\sin 60°}{c} \Rightarrow c = \dfrac{24\sin 60°}{\cos 45°} \Rightarrow c = 24\sqrt{2}\sin 60° \Rightarrow c = 24\sqrt{2}\left(\dfrac{\sqrt{3}}{2}\right) \Rightarrow$

     $c = 12\sqrt{6}.$

     To find d note that $\sin 45° = \dfrac{d}{24\sqrt{2}\sin 60°} \Rightarrow d = \sin 45°(24\sqrt{2}\sin 60°) \Rightarrow d = \dfrac{1}{\sqrt{2}}(24\sqrt{2}\sin 60°) \Rightarrow$

     $d = 24\sin 60° \Rightarrow d = 24\left(\dfrac{\sqrt{3}}{2}\right) \Rightarrow d = 12\sqrt{3}.$

49. To find a note that $\sin 60° = \dfrac{7}{a} \Rightarrow a = \dfrac{7}{\sin 60°} \Rightarrow a = \dfrac{7}{\frac{\sqrt{3}}{2}} \Rightarrow a = \dfrac{14\sqrt{3}}{3}.$

     To find b note that $\tan 60° = \dfrac{7}{b} \Rightarrow b = \dfrac{7}{\tan 60°} \Rightarrow b = \dfrac{7}{\sqrt{3}} \Rightarrow b = \dfrac{7\sqrt{3}}{3}.$

     To find c note that $\tan 45° = \left(\dfrac{7}{\sin 60°} \div c\right) \Rightarrow \tan 45° = \dfrac{7}{c\sin 60°} \Rightarrow (c\sin 60°)(\tan 45°) = 7,$

     $\tan 45° = 1 \Rightarrow (c\sin 60°)(1) = 7 \Rightarrow c = \dfrac{7}{\sin 60°} \Rightarrow c = \dfrac{7}{\frac{\sqrt{3}}{2}} \Rightarrow c = \dfrac{14\sqrt{3}}{3}.$

To find d note that $\sin 45° = \left(\dfrac{7}{\sin 60°} \div d \right) \Rightarrow \sin 45° = \dfrac{7}{d \sin 60°} \Rightarrow (d \sin 60°)(\sin 45°) = 7 \Rightarrow$

$d = \dfrac{7}{(\sin 60°)(\sin 45°)} \Rightarrow d = \dfrac{7}{\left(\frac{\sqrt{3}}{2} \right) \left(\frac{\sqrt{2}}{2} \right)} \Rightarrow d = \dfrac{14\sqrt{6}}{3}.$

51. To find a note that $\tan 60° = \dfrac{a}{12} \Rightarrow a = 12 \tan 60° \Rightarrow a \approx 20.78.$

53. To find c note that $\cos 53°43' = \dfrac{100}{c} \Rightarrow c = \dfrac{100}{\cos 53°43'} \Rightarrow c \approx 168.98.$

55. (a) $\sin 70° = \cos (90° - 70°) = \cos 20° \approx 0.9397$

    (b) $\cos 40° = \sin (90° - 40°) = \sin 50° \approx 0.7660$

57. (a) $\csc 49° = \sec (90° - 49°) = \sec 41° \approx 1.3250$

    (b) $\sec 63° = \csc (90° - 63°) = \csc 27° \approx 2.2027$

59. $\tan 37°30' = \dfrac{h}{1500} \Rightarrow h = 1500 \tan 37°30' \Rightarrow h \approx 1151$ ft.

61. $\tan 34° = \dfrac{5 \text{ ft } 3 \text{ in.}}{a}$ or $\tan 34° = \dfrac{63 \text{ in.}}{a} \Rightarrow a = \dfrac{63 \text{ in.}}{\tan 34°} \Rightarrow a \approx 93.4$ in. or about 7.8 ft.

63. $\tan 37° = \dfrac{h}{52} \Rightarrow h = 52(\tan 37°) \approx 39.2$ ft.

65. Using the 13° angle of depression, we may find the distance between the airplane and the stadium as follows:

$\tan 13° = \dfrac{12,000}{a} \Rightarrow a = \dfrac{12,000}{\tan 13°} \Rightarrow a \approx 51,977.7$ or 52,000 ft.

67. Using the 20° angle of depression, we may find the distance between the buildings as follows:

$\tan 20° = \dfrac{30}{d} \Rightarrow d = \dfrac{30}{\tan 20°} \Rightarrow d \approx 82.4$ ft. Using this distance, we may now calculate the height h of the

upper part of the building on the right: $\tan 50° = \dfrac{h}{82.4} \Rightarrow h = 82.4 \tan 50° \Rightarrow h \approx 98.2$ ft. The total height

of the building is about $98.2 + 30 = 128.2$ ft.

69. From the diagram in the text we see that: $\tan 23.76° = \dfrac{PR}{PQ} \Rightarrow PQ = \dfrac{85.62}{\tan 23.76°} \Rightarrow PQ = 194.5$ ft.

71. See Figure 71. From right triangle TCA, $\tan 13.7° = \dfrac{h}{d} \Rightarrow d = \dfrac{h}{\tan 13.7°}$. From right triangle TCB,

$\tan 10.4° = \dfrac{h}{d + 5} \Rightarrow d = \dfrac{h}{\tan 10.4°} - 5.$ Thus, we calculate h as follows:

$\dfrac{h}{\tan 10.4°} - 5 = \dfrac{h}{\tan 13.7°} \Rightarrow h \tan 13.7° - 5 \tan 13.7° \tan 10.4° = h \tan 10.4° \Rightarrow$

$h(\tan 13.7° - \tan 10.4°) = 5 \tan 13.7° \tan 10.4° \Rightarrow h = \dfrac{5 \tan 13.7° \tan 10.4°}{\tan 13.7° - \tan 10.4°} \Rightarrow h = 3.71$ mi $\approx 19,600$ ft.

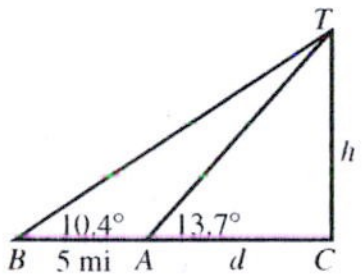

Figure 71

73. let y be the base of the smaller triangle, then $\tan (21°10') = \dfrac{x}{135 + y} \Rightarrow x = (135 + y) \tan 21°10'$

and $\tan 35°30' = \dfrac{x}{y} \Rightarrow x = y(\tan 35°30')$.

Now $(135 + y) \tan 21°10' = y(\tan 35°30') \Rightarrow 135(\tan 21°10') = y(\tan 35°30') - y(\tan 21°10') \Rightarrow$

$135(\tan 21°10') = y(\tan 35°30' - \tan 21°10') \Rightarrow y = \dfrac{135(\tan (21°10'))}{\tan (35°30') - \tan (21°10')} \Rightarrow y \approx 160.303 \Rightarrow$

Back substituting we find $x \approx 114.343$ or about 114 ft.

75. From the example we know that the distance d to a star with parallax θ is given by $d = \dfrac{93,000,000}{\sin \theta}$.

For Barnard's star: $d = \dfrac{93,000,000}{\sin 0.000152} \approx 3.5 \times 10^{13}$ mi or $\dfrac{3.5 \times 10^{13}}{5.9 \times 10^{12}} \approx 5.9$ light-years.

For Sirus: $d = \dfrac{93,000,000}{\sin 0.000105} \approx 5.1 \times 10^{13}$ mi or $\dfrac{5.1 \times 10^{13}}{5.9 \times 10^{12}} \approx 8.6$ light-years.

For 61 Cygni: $d = \dfrac{93,000,000}{\sin 0.0000811} \approx 6.6 \times 10^{13}$ mi or $\dfrac{6.6 \times 10^{13}}{5.9 \times 10^{12}} \approx 11.1$ light-years.

For Procyon: $d = \dfrac{93,000,000}{\sin 0.0000797} \approx 6.7 \times 10^{13}$ mi or $\dfrac{6.7 \times 10^{13}}{5.9 \times 10^{12}} \approx 11.3$ light-years.

77. Using the Earth - Sun distance of 93,000,000 miles, the minimum distance d is given by:

$\sin 18° = \dfrac{d}{93,000,000} \Rightarrow d = 93,000,000 \sin 18° \Rightarrow d \approx 2.9 \times 10^7$ mi.

The maximum distance is given by: $\sin 28° = \dfrac{d}{93,000,000} \Rightarrow d = 93,000,000 \sin 28° \Rightarrow d \approx 4.4 \times 10^7$ mi.

79. $d = r\left(\dfrac{1}{\cos \theta} - 1\right) = 3963\left(\dfrac{1}{\cos 76.1°} - 1\right) \approx 12,533.8 \approx 12,534$ mi.

81. (a) $r = \dfrac{66^2}{4.5 + 32.2 \tan 3°} \approx 704$ ft.

(b) $r = \dfrac{66^2}{4.5 + 32.2 \tan 5°} \approx 595$ ft.

(c) Larger values of θ correspond to a smaller safe radius. A table of $Y_1 = 66^2/(4.5 + 32.2 \tan (X))$ starting

at $x = 0$ and incrementing by 1 is shown in Figure 81.

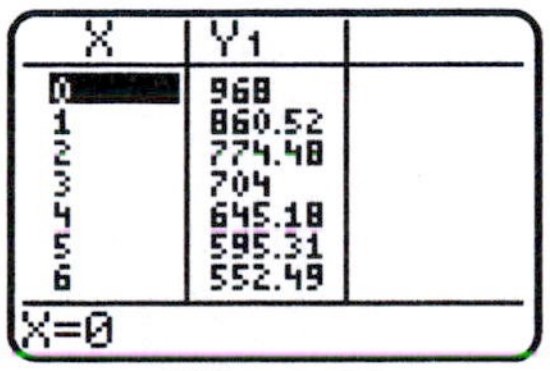

X	Y1
0	968
1	860.52
2	774.48
3	704
4	645.18
5	595.31
6	552.49

X=0

Figure 81

83. From the example $d = r\left(\dfrac{1}{\cos \theta} - 1\right)$. Thus, $r = 625$ and $\theta = 54° \Rightarrow d = 625\left(\dfrac{1}{\cos 54°} - 1\right) \approx 438$ ft.

85. The area of a triangle is $\dfrac{b \cdot h}{2}$ and the height is unknown. Height is found by using the following:

$$\cos 30° = \dfrac{h}{s} \Rightarrow s\cos 30° = h, \cos 30° = \dfrac{\sqrt{3}}{2} \Rightarrow s \cdot \dfrac{\sqrt{3}}{2} = \text{height or } h = \dfrac{s\sqrt{3}}{2}.$$

Since area equals $\dfrac{b \cdot h}{2}$ and $b = s$, then area $= \left(s \cdot \dfrac{s\sqrt{3}}{2}\right) \div 2$ therefore area $= \dfrac{\sqrt{3}}{4}s^2$.

Checking Basic Concepts for Sections 6.1 and 6.2

1. (a) $45°\left(\dfrac{\pi}{180°}\right) = \dfrac{\pi}{4}$

    (b) $75°\left(\dfrac{\pi}{180°}\right) = \dfrac{5\pi}{12}$

3. Convert 30° to radians: $30°\left(\dfrac{\pi}{180°}\right) = \dfrac{\pi}{6}$; $s = 12\left(\dfrac{\pi}{6}\right) = 2\pi$ in.; $A = \dfrac{1}{2}\left(\dfrac{\pi}{6}\right) \cdot 12^2 = \dfrac{144\pi}{12} = 12\pi$ in.^{2}

5. First find the value of c: $c^2 = 12^2 + 5^2 \Rightarrow c^2 = 144 + 25 \Rightarrow c^2 = 169 \Rightarrow c = 13$

$$\sin \theta = \dfrac{5}{13} \quad \cos \theta = \dfrac{12}{13} \quad \tan \theta = \dfrac{5}{12} \quad \csc \theta = \dfrac{13}{5} \quad \sec \theta = \dfrac{13}{12} \quad \cot \theta = \dfrac{12}{5}$$

6.3: The Sine and Cosine Functions and Their Graphs

1. (a) $r = \sqrt{12^2 + 5^2} = \sqrt{144 + 25} = \sqrt{169} = 13$

    (b) $\sin \theta = \dfrac{y}{r} = \dfrac{5}{13} \qquad \cos \theta = \dfrac{x}{r} = \dfrac{12}{13}$

3. (a) $r = \sqrt{(-15)^2 + 8^2} = \sqrt{225 + 64} = \sqrt{289} = 17$

    (b) $\sin \theta = \dfrac{y}{r} = \dfrac{8}{17} \qquad \cos \theta = \dfrac{x}{r} = -\dfrac{15}{17}$

5. $r = \sqrt{4^2 + 3^2} = \sqrt{16 + 9} = \sqrt{25} = 5, \quad \sin \theta = \dfrac{y}{r} = \dfrac{3}{5} \quad \cos \theta = \dfrac{x}{r} = \dfrac{4}{5}$

7. $r = \sqrt{1^2 + (-2)^2} = \sqrt{1 + 4} = \sqrt{5}, \quad \sin \theta = \dfrac{y}{r} = -\dfrac{2}{\sqrt{5}} \quad \cos \theta = \dfrac{x}{r} = \dfrac{1}{\sqrt{5}}$

9. The point $(1, 2)$ lies on the line $y = 2x$ in the first quadrant. This is shown in Figure 9.

So, $r = \sqrt{1^2 + 2^2} = \sqrt{1 + 4} = \sqrt{5}$. It follows that: $\sin \theta = \dfrac{2}{\sqrt{5}}$ and $\cos \theta = \dfrac{1}{\sqrt{5}}$.

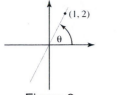

Figure 9

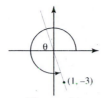

Figure 11

11. The point $(1, -3)$ lies on the line $y = -3x$ in the fourth quadrant. This is shown in Figure 11.

    So, $r = \sqrt{1^2 + (-3)^2} = \sqrt{1 + 9} = \sqrt{10}$. It follows that: $\sin\theta = -\dfrac{3}{\sqrt{10}}$ and $\cos\theta = \dfrac{1}{\sqrt{10}}$.

13. The length of the legs in a $45° \sim 45°$ right triangle are equal. For convenience we let $r = 1$ as shown in Figure 13a.

    Next, find x and y: $x^2 + y^2 = 1^2 \Rightarrow x^2 + x^2 = 1^2 \Rightarrow 2x^2 = 1 \Rightarrow x^2 = \dfrac{1}{2} \Rightarrow x = \dfrac{1}{\sqrt{2}}$. Thus,

    $x = y = \dfrac{1}{\sqrt{2}}$. $\sin 45° = \dfrac{1}{\sqrt{2}} \approx 0.7071$ $\cos 45° = \dfrac{1}{\sqrt{2}} \approx 0.7071$ The results are supported in Figure 13b.

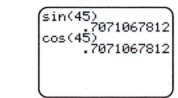

Figure 13a   Figure 13b

15. The length of the shorter leg in a $30° \sim 60°$ right triangle is equal to half the length of the hypotenuse. For convenience we let $r = 2$ as shown in Figure 15a. Note that y is negative so $y = -1$.

    Next, find x: $x^2 = r^2 - 1^2 \Rightarrow x^2 = 2^2 - 1^2 \Rightarrow x^2 = 3 \Rightarrow x = \sqrt{3}$.

    $\sin(-30°) = -\dfrac{1}{2}$ $\cos(-30°) = \dfrac{\sqrt{3}}{2} \approx 0.8660$ The results are supported in Figure 15b.

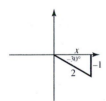

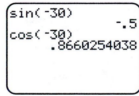

Figure 15a   Figure 15b

17. The lengths of the legs in a $45° \sim 45°$ right triangle are equal. For convenience we let $r = 1$ as shown in

    Figure 17a. Next, find x and y: $x^2 + y^2 = 1^2 \Rightarrow x^2 + x^2 = 1^2 \Rightarrow 2x^2 = 1 \Rightarrow x^2 = \dfrac{1}{2} \Rightarrow x = \pm\dfrac{1}{\sqrt{2}}$.

    Thus, $x = y = -\dfrac{1}{\sqrt{2}}$; $\sin(225°) = -\dfrac{1}{\sqrt{2}}$, $\cos(225°) = -\dfrac{1}{\sqrt{2}}$. The results are supported in Figure 17b.

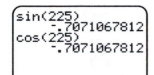

Figure 17a   Figure 17b

19. The length of the shorter leg in a 30° ~ 60° right triangle is equal to half the length of the hypotenuse. For convenience we let $r = 2$ as shown in Figure 19a. Note that x is positive, so $x = 1$.

$y^2 = r^2 - x^2 \Rightarrow y^2 = 2^2 - 1^2 \Rightarrow y^2 = 3 \Rightarrow y = -\sqrt{3}$. Thus, $\sin(-420°) = -\dfrac{\sqrt{3}}{2}$, $\cos(-420°) = \dfrac{1}{2}$.

The results are supported in Figure 19b.

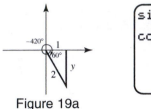   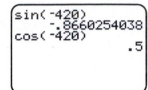

Figure 19a   Figure 19b

21. Note that $\dfrac{\pi}{3} = 60°$. The length of the shorter leg in a 30° ~ 60° right triangle is equal to half the length of the hypotenuse. For convenience we let $r = 2$ as shown in Figure 21a. Note that x is positive so $x = 1$.

Next, find y: $y^2 = r^2 - 1^2 \Rightarrow y^2 = 2^2 - 1^2 \Rightarrow y^2 = 3 \Rightarrow y = \sqrt{3}$.

$\sin\dfrac{\pi}{3} = \dfrac{\sqrt{3}}{2} \approx 0.8660 \qquad \cos\dfrac{\pi}{3} = \dfrac{1}{2}$ The results are supported in Figure 21b.

     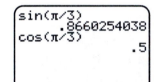

Figure 21a   Figure 21b

23. A $-\dfrac{\pi}{2}$ angle is quadrantal. Its terminal side lies on the y-axis. For convenience we let $r = 1$ as shown in Figure 23a. Note that if $r = 1$, then $x = 0$ and $y = -1$.

$\sin\left(-\dfrac{\pi}{2}\right) = -1 \qquad \cos\left(-\dfrac{\pi}{2}\right) = 0$ The results are supported in Figure 23b.

     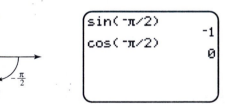

Figure 23a   Figure 23b

25. Note that $\dfrac{7\pi}{6} = 210°$ lies in the third quadrant and it forms a 30° ~ 60° right triangle with the x-axis.

For convenience we let $r = 2$ as shown in Figure 25a. Note that y is negative so $y = -1$.

Next find $x^2 = r^2 - y^2 \Rightarrow x^2 = 2^2 - (-1)^2 \Rightarrow x^2 = 3 \Rightarrow x = -\sqrt{3}$.

$\sin\dfrac{7\pi}{6} = -\dfrac{1}{2} \qquad \cos\dfrac{7\pi}{6} = -\dfrac{\sqrt{3}}{2} \approx -0.8660$ These results are supported in Figure 25b.

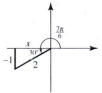
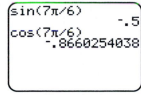

Figure 25a   Figure 25b

27. Note that $-\dfrac{9\pi}{4} = -405°$. The length of the legs in a $45° \sim 45°$ right triangle are equal. For convenience we

    let $r = 1$ as shown in Figure 27a. Next, find x and y:

    $$x^2 + y^2 = 1^2 \Rightarrow x^2 + x^2 = 1^2 \Rightarrow 2x^2 = 1 \Rightarrow x^2 = \dfrac{1}{2} \Rightarrow x = \pm\dfrac{1}{\sqrt{2}}.$$

    $$\sin\left(-\dfrac{9\pi}{4}\right) = -\dfrac{1}{\sqrt{2}} \quad \cos\left(-\dfrac{9\pi}{4}\right) = \dfrac{1}{\sqrt{2}} \qquad \text{The results are supported in Figure 27b.}$$

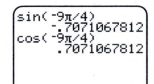

Figure 27a   Figure 27b

29. $\sin 93.2° \approx 0.9984$ $\cos 93.2° \approx -0.0558$

31. $\sin 123°50' \approx 0.8307$ $\cos 123°50' \approx -0.5568$

33. $\sin(-4) \approx 0.7568$ $\cos(-4) \approx -0.6536$

35. $\sin\dfrac{11\pi}{7} \approx -0.9749$ $\cos\dfrac{11\pi}{7} \approx 0.2225$

37. $\sin\theta = y = \dfrac{3}{5}$ $\cos\theta = x = \dfrac{4}{5}$

39. $\sin\theta = y = -\dfrac{5}{13}$ $\cos\theta = x = \dfrac{12}{13}$

41. An angle θ of $\dfrac{\pi}{2}$ radians in standard position has a terminal side that intersects the unit circle at the point $(0, 1)$.

    Therefore $\sin\dfrac{\pi}{2} = 1$ and $\cos\dfrac{\pi}{2} = 0$.

43. An angle θ of $\dfrac{7\pi}{6}$ radians in standard position has a terminal side that intersects the unit circle in the third quadrant.

    To find the point of intersection, the hypotenuse of a $30° \sim 60°$ right triangle has length 1 and the shorter leg has

    length $y = \dfrac{1}{2}$. Since $x^2 + y^2 = 1$, we can solve for x. $x^2 + \left(\dfrac{1}{2}\right)^2 = 1 \Rightarrow x^2 = \dfrac{3}{4} \Rightarrow x = \dfrac{\sqrt{3}}{2}$.

Since the point (x, y) is located in the third quadrant, the point becomes $\left(-\dfrac{\sqrt{3}}{2}, -\dfrac{1}{2}\right)$.

Therefore $\sin\left(\dfrac{7\pi}{6}\right) = -\dfrac{1}{2}$ and $\cos\left(\dfrac{7\pi}{6}\right) = -\dfrac{\sqrt{3}}{2}$.

45. An angle θ of $-\dfrac{3\pi}{4}$ radians in standard position has a terminal side that intersects the unit circle in the third

quadrant at the point $\left(-\dfrac{1}{\sqrt{2}}, -\dfrac{1}{\sqrt{2}}\right)$. Therefore $\sin\left(-\dfrac{3\pi}{4}\right) = -\dfrac{1}{\sqrt{2}}$ and $\cos\left(-\dfrac{3\pi}{4}\right) = -\dfrac{1}{\sqrt{2}}$.

47. An angle θ of $\dfrac{5\pi}{2}$ radians in standard position has a terminal side that intersects the unit circle at the point

$(0, 1)$. Therefore $\sin\dfrac{5\pi}{2} = 1$ and $\cos\dfrac{5\pi}{2} = 0$.

49. An angle θ of $-\dfrac{\pi}{3}$ radians in standard position has a terminal side that intersects the unit circle in the fourth

quadrant. To find the point of intersection, the hypotenuse of a $30°/60°$ right triangle has length 1 and the shorter

leg has length $x = \dfrac{1}{2}$. Since $x^2 + y^2 = 1$, we can solve for y. $\left(\dfrac{1}{2}\right)^2 + y^2 = 1 \Rightarrow y^2 = \dfrac{3}{4} \Rightarrow y = \dfrac{\sqrt{3}}{2}$.

Since the point (x, y) is located in the fourth quadrant, the point becomes $\left(\dfrac{1}{2}, -\dfrac{\sqrt{3}}{2}\right)$.

Therefore $\sin\left(-\dfrac{\pi}{3}\right) = -\dfrac{\sqrt{3}}{2}$ and $\cos\left(-\dfrac{\pi}{3}\right) = \dfrac{1}{2}$.

51. If a string of $s = 3\pi$ is wrapped counter-clockwise around the unit circle, it will make $1\dfrac{1}{2}$

revolutions and the terminal point will coincide with the point $(-1, 0)$. Thus, $W(3\pi) = (-1, 0)$.

53. If a string of $s = -2\pi$ is wrapped clockwise around the unit circle, it will make 1 complete

revolution and the terminal point will coincide with the point $(1, 0)$. Thus, $W(-2\pi) = (1, 0)$.

55. If a string of $s = \dfrac{3\pi}{2}$ is wrapped counter-clockwise around the unit circle, it will make $\dfrac{3}{4}$ of a revolution and the

terminal point will coincide with the point $(0, -1)$. Thus, $W\left(\dfrac{3\pi}{2}\right) = (0, -1)$.

57. If a string of $s = -\dfrac{5\pi}{2}$ is wrapped clockwise around the unit circle, it will make $1\dfrac{1}{4}$ revolutions and the

terminal point will coincide with the point $(0, -1)$. Thus, $W\left(-\dfrac{5\pi}{2}\right) = (0, -1)$.

59. A distance of $\dfrac{\pi}{4}$ represents an eighth of a revolution in the unit circle. The $45° \sim 45°$ triangle formed has a

hypotenuse with length 1 and legs both length a. $a^2 + a^2 = 1 \Rightarrow 2a^2 = 1 \Rightarrow a^2 = \dfrac{1}{2} \Rightarrow a = \pm\dfrac{1}{\sqrt{2}}$.

Since the terminal point of $\dfrac{5\pi}{4}$ lies in quadrant III, the x-coordinate is $-\dfrac{1}{\sqrt{2}}$, and the y-coordinate is $-\dfrac{1}{\sqrt{2}}$.

Thus, $W\left(\dfrac{5\pi}{4}\right) = \left(-\dfrac{1}{\sqrt{2}}, -\dfrac{1}{\sqrt{2}}\right)$.

61. A distance of $\dfrac{\pi}{4}$ represents an eighth of a revolution in the unit circle. The $45° \sim 45°$ triangle formed has a

hypotenuse with length 1 and legs both length a. $a^2 + a^2 = 1 \Rightarrow 2a^2 = 1 \Rightarrow a^2 = \dfrac{1}{2} \Rightarrow a = \pm\dfrac{1}{\sqrt{2}}$.

Since the terminal point of $\dfrac{-5\pi}{4}$ lies in quadrant II, the x-coordinate is $-\dfrac{1}{\sqrt{2}}$, and the y-coordinate is $\dfrac{1}{\sqrt{2}}$.

Thus, $W\left(-\dfrac{5\pi}{4}\right) = \left(-\dfrac{1}{\sqrt{2}}, \dfrac{1}{\sqrt{2}}\right)$.

63. $W\left(\dfrac{11\pi}{6}\right)$ is illustrated in Figure 63. $W\left(\dfrac{11\pi}{6}\right) = \left(\dfrac{\sqrt{3}}{2}, -\dfrac{1}{2}\right)$

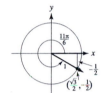

Figure 63

65. Use a similar procedure as is shown in solution 63. $W\left(-\dfrac{7\pi}{3}\right) = \left(\dfrac{1}{2}, -\dfrac{\sqrt{3}}{2}\right)$

67. Note that the values for s are the same as those used in exercises 51-66. Thus, if $W = (x, y)$, then we apply the fact that $\sin s = y$ and $\cos s = x$. $\sin 3\pi = 0 \quad \cos 3\pi = -1$

69. Note that the values for s are the same as those used in exercises 51-66. Thus, if $W = (x, y)$, then we apply the fact that $\sin s = y$ and $\cos s = x$. $\sin(-2\pi) = 0 \quad \cos(-2\pi) = 1$

71. Note that the values for s are the same as those used in exercises 51-66. Thus, if $W = (x, y)$, then we apply the fact that $\sin s = y$ and $\cos s = x$. $\sin\left(\dfrac{3\pi}{2}\right) = -1 \quad \cos\left(\dfrac{3\pi}{2}\right) = 0$

73. Note that the values for s are the same as those used in exercises 51-66. Thus, if $W = (x, y)$, then we apply the fact that $\sin s = y$ and $\cos s = x$. $\sin\left(-\dfrac{5\pi}{2}\right) = -1 \quad \cos\left(-\dfrac{5\pi}{2}\right) = 0$

75. Note that the values for s are the same as those used in exercises 51-66. Thus, if $W = (x, y)$, then we apply the fact that $\sin s = y$ and $\cos s = x$. $\sin\left(\dfrac{5\pi}{4}\right) = \cos\left(\dfrac{5\pi}{4}\right) = -\dfrac{1}{\sqrt{2}}$

77. Note that the values for s are the same as those used in exercises 51-66. Thus, if $W = (x, y)$, then we apply the fact that $\sin s = y$ and $\cos s = x$. $\sin\left(-\dfrac{5\pi}{4}\right) = \dfrac{1}{\sqrt{2}} \quad \cos\left(-\dfrac{5\pi}{4}\right) = -\dfrac{1}{\sqrt{2}}$

79. Note that the values for s are the same as those used in exercises 51-66. Thus, if $W = (x, y)$, then we apply the fact that $\sin s = y$ and $\cos s = x$. $\sin\left(\dfrac{11\pi}{6}\right) = -\dfrac{1}{2} \quad \cos\left(\dfrac{11\pi}{6}\right) = \dfrac{\sqrt{3}}{2}$

81. Note that the values for s are the same as those used in exercises 51-66. Thus, if $W = (x, y)$, then we apply the

fact that $\sin s = y$ and $\cos s = x$. $\sin \left(-\dfrac{7\pi}{3} \right) = -\dfrac{\sqrt{3}}{2}$ $\cos \left(-\dfrac{7\pi}{3} \right) = \dfrac{1}{2}$

83. See Figure 83.

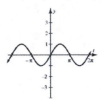

Figure 83

85. (a) $R = \{y \mid -3 \le y \le 3\}$. See Figure 85a. $f\left(\dfrac{3\pi}{2} \right) = 3 \sin \left(\dfrac{3\pi}{2} \right) = -3$

(b) $R = \{y \mid -1 \le y \le 1\}$. See Figure 85b. $f\left(\dfrac{3\pi}{2} \right) = \sin \left(\dfrac{9\pi}{2} \right) = 1$

$[-2\pi, 2\pi, \pi/2]$ by $[-4, 4, 1]$ $[-2\pi, 2\pi, \pi/2]$ by $[-4, 4, 1]$

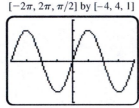

Figure 85a   Figure 85b

87. (a) $R = \{y \mid -1 \le y \le 3\}$. See Figure 87a. $f\left(\dfrac{3\pi}{2} \right) = 2 \cos \left(\dfrac{3\pi}{2} \right) + 1 = 1$

(b) $R = \{y \mid -2 \le y \le 0\}$. See Figure 87b. $f\left(\dfrac{3\pi}{2} \right) = \cos (3\pi) - 1 = -2$

$[-2\pi, 2\pi, \pi/2]$ by $[-4, 4, 1]$ $[-2\pi, 2\pi, \pi/2]$ by $[-4, 4, 1]$

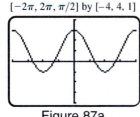 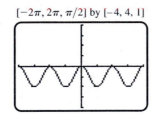

Figure 87a   Figure 87b

89. (a) The slope of the line is 0.03, so when x increases by 100 feet, y increases by 3 feet. The hill has a 3% grade.

(b) First we must find $\sin \theta$: $r = \sqrt{100^2 + 3^2} = \sqrt{10,009}$ and so $\sin \theta = \dfrac{y}{r} = \dfrac{3}{\sqrt{10,009}}$.

The grade resistance is $R = W \sin \theta = 25,000 \left(\dfrac{3}{\sqrt{10,009}} \right) \approx 750$ lb.

91. The first quarter phase corresponds to half the face on the moon being visible. This occurs when

$\theta = \dfrac{\pi}{2} + 2\pi n$ for $n = 0, 1, 2, 3, \ldots$.

93. (a) See Figure 93.

    (b) It appears from the graph the energy is the greatest when $x = 18$ or at 6 pm.

    (c) It appears from the graph the energy is the least when $x = 6$ or at 6 am.

95. (a) The graph is sinusoidal and varies between −310 volts and 310 volts. The voltage is changing direction.

    (b) $V\left(\dfrac{1}{120}\right) = 310 \sin\left(120\pi \cdot \dfrac{1}{120}\right) = 310 \sin(\pi) = 310(0) = 0$. After $\dfrac{1}{120}$ second the voltage is 0.

    (c) The maximum voltage is 310 volts. The root mean square voltage is $\dfrac{310}{\sqrt{2}} \approx 219$ volts. Note that the common electrical rating for electric ranges and ovens is 220 volts.

97. (a) $D = \dfrac{1.05(88^2 - 44^2)}{27 + 64.4 \sin 3^\circ} \approx 201$ ft

    (b) $D = \dfrac{1.05(88^2 - 44^2)}{27 + 64.4 \sin(-3^\circ)} \approx 258$ ft

    (c) When the slope is negative (down hill) the car will require a greater stopping distance.

99. (a) When $S = 390$ and $r = 600$, $\beta = \dfrac{S}{r} = \dfrac{390}{600} = \dfrac{13}{20}$ radians. Then, $d = 600\left(1 - \cos\dfrac{\frac{13}{20}}{2}\right) \approx 31.4$ ft.

    (b) When $S = 620$ and $r = 600$, $\beta = \dfrac{S}{r} = \dfrac{620}{600} = \dfrac{31}{30}$ radians. Then, $d = 600\left(1 - \cos\dfrac{\frac{31}{30}}{2}\right) \approx 78.3$ ft.

    (c) Higher speed limits require more land to be cleared on the inside of the curve.

101.(a) The scatterplot is shown in Figure 101a.

    (b) The graph of $Y_1 = 1.9 \sin(0.42X - 1.2) + 5.7$ in $[0, 13, 1]$ by $[3, 8, 1]$ is shown with the data in Figure 101b.

    From the close data fit shown, we may conclude that the flying squirrels become active near sunset.

$[0, 13, 1]$ by $[3, 8, 1]$ $[0, 13, 1]$ by $[3, 8, 1]$

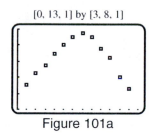   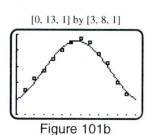

Figure 101a   Figure 101b

Extended and Discovery Exercises for Section 6.3

1. Graph $Y_1 = \sin(X)$ and $Y_2 = \cos(X)$ in $[-2\pi, 2\pi, \pi/2]$ by $[-2, 2, 1]$. See Figure 1a.

   Graph $Y_1 = \sin(X + \pi/2)$ and $Y_2 = \cos(X)$ in $[-2\pi, 2\pi, \pi/2]$ by $[-2, 2, 1]$. See Figure 1b.

   The two graphs are identical.

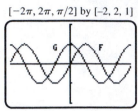

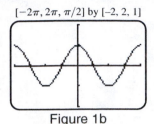

Figure 1a   Figure 1b

6.4: Other Trigonometric Functions and Their Graphs

1. $r = \sqrt{5^2 + 12^2} = \sqrt{25 + 144} = \sqrt{169} = 13$

$$\sin \theta = \frac{12}{13} \qquad \cos \theta = \frac{5}{13} \qquad \tan \theta = \frac{12}{5}$$

$$\csc \theta = \frac{13}{12} \qquad \sec \theta = \frac{13}{5} \qquad \cot \theta = \frac{5}{12}$$

3. $r = \sqrt{7^2 + (-24)^2} = \sqrt{49 + 576} = \sqrt{625} = 25$

$$\sin \theta = -\frac{24}{25} \qquad \cos \theta = \frac{7}{25} \qquad \tan \theta = -\frac{24}{7}$$

$$\csc \theta = -\frac{25}{24} \qquad \sec \theta = \frac{25}{7} \qquad \cot \theta = -\frac{7}{24}$$

5. The terminal side of a $90°$ angle intersects the unit circle at the point $(0, 1)$.

$$\sin 90° = 1 \qquad\qquad \cos 90° = 0 \qquad\qquad \tan 90° \to \text{undefined}$$

$$\csc 90° = 1 \qquad\qquad \sec 90° \to \text{undefined} \qquad\qquad \cot 90° = 0$$

7. The terminal side of a $-45°$ angle intersects the unit circle at the point $\left(\dfrac{1}{\sqrt{2}}, -\dfrac{1}{\sqrt{2}}\right)$.

$$\sin(-45°) = -\frac{1}{\sqrt{2}} \qquad \cos(-45°) = \frac{1}{\sqrt{2}} \qquad \tan(-45°) = -1$$

$$\csc(-45°) = -\sqrt{2} \qquad \sec(-45°) = \sqrt{2} \qquad \cot(-45°) = -1$$

9. The terminal side of a π radian angle intersects the unit circle at the point $(-1, 0)$.

$$\sin \pi = 0 \qquad\qquad \cos \pi = -1 \qquad\qquad \tan \pi = 0$$

$$\csc \pi \to \text{undefined} \qquad\qquad \sec \pi = -1 \qquad\qquad \cot \pi \to \text{undefined}$$

11. The terminal side of a $-\dfrac{\pi}{3}$ radian angle intersects the unit circle at the point $\left(\dfrac{1}{2}, -\dfrac{\sqrt{3}}{2}\right)$.

$$\sin\left(-\frac{\pi}{3}\right) = -\frac{\sqrt{3}}{2} \qquad \cos\left(-\frac{\pi}{3}\right) = \frac{1}{2} \qquad \tan\left(-\frac{\pi}{3}\right) = -\sqrt{3}$$

$$\csc\left(-\frac{\pi}{3}\right) = -\frac{2}{\sqrt{3}} \qquad \sec\left(-\frac{\pi}{3}\right) = 2 \qquad \cot\left(-\frac{\pi}{3}\right) = -\frac{1}{\sqrt{3}}$$

13. The terminal side of a $-\dfrac{\pi}{2}$ radian angle intersects the unit circle at the point $(0, -1)$.

$$\sin\left(-\frac{\pi}{2}\right) = -1 \qquad \cos\left(-\frac{\pi}{2}\right) = 0 \qquad \tan\left(-\frac{\pi}{2}\right) \rightarrow \text{undefined}$$

$$\csc\left(-\frac{\pi}{2}\right) = -1 \qquad \sec\left(-\frac{\pi}{2}\right) \rightarrow \text{undefined} \qquad \cot\left(-\frac{\pi}{2}\right) = 0$$

15. The terminal side of a $360°$ angle intersects the unit circle at the point $(1, 0)$.

$$\sin 360° = 0 \qquad \cos 360° = 1 \qquad \tan 360° = 0$$

$$\csc 360° \rightarrow \text{undefined} \qquad \sec 360° = 1 \qquad \cot 360° \rightarrow \text{undefined}$$

17. The terminal side of a $\dfrac{\pi}{6}$ radian angle intersects the unit circle at the point $\left(\dfrac{\sqrt{3}}{2}, \dfrac{1}{2}\right)$.

$$\sin\frac{\pi}{6} = \frac{1}{2} \qquad \cos\frac{\pi}{6} = \frac{\sqrt{3}}{2} \qquad \tan\frac{\pi}{6} = \frac{1}{\sqrt{3}}$$

$$\csc\frac{\pi}{6} = 2 \qquad \sec\frac{\pi}{6} = \frac{2}{\sqrt{3}} \qquad \cot\frac{\pi}{6} = \sqrt{3}$$

19. The terminal side of a $\dfrac{4\pi}{3}$ radian angle intersects the unit circle at the point $\left(-\dfrac{1}{2}, -\dfrac{\sqrt{3}}{2}\right)$.

$$\sin\left(\frac{4\pi}{3}\right) = -\frac{\sqrt{3}}{2} \qquad \cos\left(\frac{4\pi}{3}\right) = -\frac{1}{2} \qquad \tan\left(\frac{4\pi}{3}\right) = \sqrt{3}$$

$$\csc\left(\frac{4\pi}{3}\right) = -\frac{2}{\sqrt{3}} \qquad \sec\left(\frac{4\pi}{3}\right) = -2 \qquad \cot\left(\frac{4\pi}{3}\right) = \frac{1}{\sqrt{3}}$$

21. The terminal side of a $-225°$ radian angle intersects the unit circle at the point $\left(-\dfrac{1}{\sqrt{2}}, \dfrac{1}{\sqrt{2}}\right)$.

$$\sin(-225°) = \frac{1}{\sqrt{2}} \qquad \cos(-225°) = -\frac{1}{\sqrt{2}} \qquad \tan(-225°) = -1$$

$$\csc(-225°) = \sqrt{2} \qquad \sec(-225°) = -\sqrt{2} \qquad \cot(-225°) = -1$$

23. The terminal side of a $\dfrac{-13\pi}{6}$ radian angle intersects the unit circle at the point $\left(\dfrac{\sqrt{3}}{2}, -\dfrac{1}{2}\right)$.

$$\sin\left(\frac{-13\pi}{6}\right) = -\frac{1}{2} \qquad \cos\left(\frac{-13\pi}{6}\right) = \frac{\sqrt{3}}{2} \qquad \tan\left(\frac{-13\pi}{6}\right) = -\frac{1}{\sqrt{3}}$$

$$\csc\left(\frac{-13\pi}{6}\right) = -2 \qquad \sec\left(\frac{-13\pi}{6}\right) = \frac{2}{\sqrt{3}} \qquad \cot\left(\frac{-13\pi}{6}\right) = -\sqrt{3}$$

25. The point $(-1, 4)$ lies on the line $y = -4x$ in the second quadrant. This is shown in Figure 25.

So $r = \sqrt{(-1)^2 + 4^2} = \sqrt{1 + 16} = \sqrt{17}$.

$$\sin\theta = \frac{4}{\sqrt{17}} \qquad \cos\theta = -\frac{1}{\sqrt{17}} \qquad \tan\theta = -4$$

$$\csc\theta = \frac{\sqrt{17}}{4} \qquad \sec\theta = -\sqrt{17} \qquad \cot\theta = -\frac{1}{4}$$

The slope of the line is equal to $\tan\theta$.

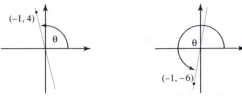

Figure 25   Figure 27

27. The point $(-1, -6)$ lies on the line $y = 6x$ in the third quadrant. This is shown in Figure 27.

So $r = \sqrt{(-1)^2 + (-6)^2} = \sqrt{1 + 36} = \sqrt{37}$.

$\sin \theta = -\dfrac{6}{\sqrt{37}}$ $\qquad$ $\cos \theta = -\dfrac{1}{\sqrt{37}}$ $\qquad$ $\tan \theta = 6$

$\csc \theta = -\dfrac{\sqrt{37}}{6}$ $\qquad$ $\sec \theta = -\sqrt{37}$ $\qquad$ $\cot \theta = \dfrac{1}{6}$

The slope of the line is equal to $\tan \theta$.

29. $\sin \theta = \dfrac{3}{5}$ $\qquad\qquad$ $\cos \theta = \dfrac{4}{5}$ $\qquad\qquad$ $\tan \theta = \dfrac{\sin \theta}{\cos \theta} = \dfrac{\frac{3}{5}}{\frac{4}{5}} = \dfrac{3}{4}$

$\csc \theta = \dfrac{1}{\sin \theta} = \dfrac{1}{\frac{3}{5}} = \dfrac{5}{3}$ $\qquad$ $\sec \theta = \dfrac{1}{\cos \theta} = \dfrac{1}{\frac{4}{5}} = \dfrac{5}{4}$ $\qquad$ $\cot \theta = \dfrac{\cos \theta}{\sin \theta} = \dfrac{\frac{4}{5}}{\frac{3}{5}} = \dfrac{4}{3}$

31. $\sin \theta = \dfrac{1}{\csc \theta} = \dfrac{1}{-\frac{17}{15}} = -\dfrac{15}{17}$ $\quad$ $\cos \theta = \dfrac{1}{\sec \theta} = \dfrac{1}{-\frac{17}{8}} = -\dfrac{8}{17}$ $\quad$ $\tan \theta = \dfrac{\sin \theta}{\cos \theta} = \dfrac{-\frac{15}{17}}{-\frac{8}{17}} = \dfrac{15}{8}$

$\csc \theta = -\dfrac{17}{15}$ $\qquad\qquad$ $\sec \theta = -\dfrac{17}{8}$ $\qquad$ $\cot \theta = \dfrac{\cos \theta}{\sin \theta} = \dfrac{-\frac{8}{17}}{-\frac{15}{17}} = \dfrac{8}{15}$

33. $\sin \theta = \tan \theta \cos \theta = \left(\dfrac{5}{12}\right)\left(\dfrac{12}{13}\right) = \dfrac{5}{13}$ $\quad$ $\cos \theta = \dfrac{12}{13}$ $\qquad\qquad$ $\tan \theta = \dfrac{5}{12}$

$\csc \theta = \dfrac{1}{\sin \theta} = \dfrac{1}{\frac{5}{13}} = \dfrac{13}{5}$ $\qquad$ $\sec \theta = \dfrac{1}{\cos \theta} = \dfrac{1}{\frac{12}{13}} = \dfrac{13}{12}$ $\quad$ $\cot \theta = \dfrac{\cos \theta}{\sin \theta} = \dfrac{\frac{12}{13}}{\frac{5}{13}} = \dfrac{12}{5}$

35. $\sin \theta = -\dfrac{3}{5}$ $\qquad\qquad$ $\cos \theta = \sqrt{1 - (-\frac{3}{5})^2} = \dfrac{4}{5}$ $\quad$ $\tan \theta = \dfrac{\sin \theta}{\cos \theta} = \dfrac{-\frac{3}{5}}{\frac{4}{5}} = -\dfrac{3}{4}$

$\csc \theta = \dfrac{1}{\sin \theta} = \dfrac{1}{-\frac{3}{5}} = -\dfrac{5}{3}$ $\qquad$ $\sec \theta = \dfrac{1}{\cos \theta} = \dfrac{1}{\frac{4}{5}} = \dfrac{5}{4}$ $\quad$ $\cot \theta = \dfrac{\cos \theta}{\sin \theta} = \dfrac{\frac{4}{5}}{-\frac{3}{5}} = -\dfrac{4}{3}$

37. $\sin \theta = -\sqrt{1 - (-\frac{4}{5})^2} = -\dfrac{3}{5}$ $\quad$ $\cos \theta = -\dfrac{4}{5}$ $\qquad\qquad$ $\tan \theta = \dfrac{\sin \theta}{\cos \theta} = \dfrac{-\frac{3}{5}}{-\frac{4}{5}} = \dfrac{3}{4}$

$\csc \theta = \dfrac{1}{\sin \theta} = \dfrac{1}{-\frac{3}{5}} = -\dfrac{5}{3}$ $\qquad$ $\sec \theta = \dfrac{1}{\cos \theta} = \dfrac{1}{-\frac{4}{5}} = -\dfrac{5}{4}$ $\quad$ $\cot \theta = \dfrac{\cos \theta}{\sin \theta} = \dfrac{-\frac{4}{5}}{-\frac{3}{5}} = \dfrac{4}{3}$

39. The terminal side of -7π corresponds to the point $(-1, 0)$. Therefore,

$\sin(-7\pi) = 0$ $\qquad\qquad$ $\cos(-7\pi) = -1$ $\qquad\qquad$ $\tan(-7\pi) = 0$

$\csc(-7\pi) \to$ undefined $\qquad$ $\sec(-7\pi) = -1$ $\qquad\qquad$ $\cot(-7\pi) \to$ undefined

41. The terminal side of $\dfrac{7\pi}{2}$ corresponds to the point $(0, -1)$.

$$\sin\left(\frac{7\pi}{2}\right) = -1 \qquad \cos\left(\frac{7\pi}{2}\right) = 0 \qquad \tan\left(\frac{7\pi}{2}\right) \rightarrow \text{undefined}$$

$$\csc\left(\frac{7\pi}{2}\right) = -1 \qquad \sec\left(\frac{7\pi}{2}\right) \rightarrow \text{undefined} \qquad \cot\left(\frac{7\pi}{2}\right) = 0$$

43. The terminal side of $-\dfrac{3\pi}{4}$ corresponds to the point $\left(-\dfrac{1}{\sqrt{2}}, -\dfrac{1}{\sqrt{2}}\right)$.

$$\sin\left(-\frac{3\pi}{4}\right) = -\frac{1}{\sqrt{2}} \qquad \cos\left(-\frac{3\pi}{4}\right) = -\frac{1}{\sqrt{2}} \qquad \tan\left(-\frac{3\pi}{4}\right) = 1$$

$$\csc\left(-\frac{3\pi}{4}\right) = -\sqrt{2} \qquad \sec\left(-\frac{3\pi}{4}\right) = -\sqrt{2} \qquad \cot\left(-\frac{3\pi}{4}\right) = 1$$

45. The terminal side of $\dfrac{7\pi}{6}$ corresponds to the point $\left(-\dfrac{\sqrt{3}}{2}, -\dfrac{1}{2}\right)$.

$$\sin\left(\frac{7\pi}{6}\right) = -\frac{1}{2} \qquad \cos\left(\frac{7\pi}{2}\right) = -\frac{\sqrt{3}}{2} \qquad \tan\left(\frac{7\pi}{6}\right) = \frac{1}{\sqrt{3}}$$

$$\csc\left(\frac{7\pi}{6}\right) = -2 \qquad \sec\left(\frac{7\pi}{6}\right) = -\frac{2}{\sqrt{3}} \qquad \cot\left(\frac{7\pi}{6}\right) = \sqrt{3}$$

47. $\sin\theta = \dfrac{1}{\sqrt{2}} \qquad \cos\theta = \dfrac{1}{\sqrt{2}} \qquad \tan\theta = 1 \qquad \csc\theta = \sqrt{2} \qquad \sec\theta = \sqrt{2} \qquad \cot\theta = 1$

49. $\sin\theta = -\dfrac{12}{13} \qquad \cos\theta = \dfrac{5}{13} \qquad \tan\theta = -\dfrac{12}{5} \qquad \csc\theta = -\dfrac{13}{12} \qquad \sec\theta = \dfrac{13}{5} \qquad \cot\theta = -\dfrac{5}{12}$

51. See Figure 51. $\tan\left(\dfrac{-5\pi}{4}\right) = \dfrac{\frac{1}{\sqrt{2}}}{-\frac{1}{\sqrt{2}}} = -1$

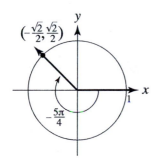

Figure 51

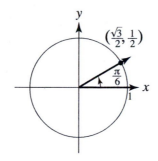

Figure 53

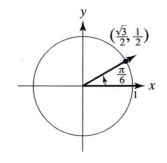

Figure 55

53. See Figure 53. $\csc\left(\dfrac{\pi}{6}\right) = \dfrac{1}{\frac{1}{2}} = 2$

55. See Figure 55. $\sec\left(\dfrac{\pi}{6}\right) = \dfrac{1}{\frac{\sqrt{3}}{2}} = \dfrac{2}{\sqrt{3}}$

57. See Figure 57.

$$\sin\left(\frac{9\pi}{2}\right) = 1 \qquad\qquad \cos\left(\frac{9\pi}{2}\right) = 0 \qquad\qquad \tan\left(\frac{9\pi}{2}\right) \to \text{undefined}$$

$$\csc\left(\frac{9\pi}{2}\right) = 1 \qquad\qquad \sec\left(\frac{9\pi}{2}\right) \to \text{undefined} \qquad \cot\left(\frac{9\pi}{2}\right) = 0$$

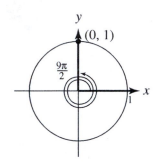

Figure 57

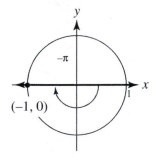

Figure 59

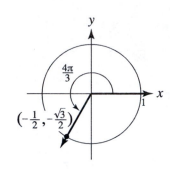

Figure 61

59. See Figure 59.

$$\sin\left(-\pi\right) = 0 \qquad\qquad \cos\left(-\pi\right) = -1 \qquad\qquad \tan\left(-\pi\right) = 0$$

$$\csc\left(-\pi\right) \to \text{undefined} \qquad \sec\left(-\pi\right) = -1 \qquad\qquad \cot\left(-\pi\right) \to \text{undefined}$$

61. See Figure 61.

$$\sin\left(\frac{4\pi}{3}\right) = -\frac{\sqrt{3}}{2} \qquad\qquad \cos\left(\frac{4\pi}{3}\right) = -\frac{1}{2} \qquad\qquad \tan\left(\frac{4\pi}{3}\right) = \sqrt{3}$$

$$\csc\left(\frac{4\pi}{3}\right) = -\frac{2}{\sqrt{3}} \qquad\qquad \sec\left(\frac{4\pi}{3}\right) = -2 \qquad\qquad \cot\left(\frac{4\pi}{3}\right) = \frac{1}{\sqrt{3}}$$

63. (a) $\sin 93.2° \approx 0.9984$

   (b) $\csc 93.2° = \dfrac{1}{\sin 93.2°} \approx 1.0016$

65. (a) $\tan 234°33' \approx 1.4045$

   (b) $\cot 234°33' = \dfrac{1}{\tan 234°33'} \approx 0.7120$

67. (a) $\cot\left(-4\right) = \dfrac{1}{\tan\left(-4\right)} \approx -0.8637$

   (b) $\tan\left(-4\right) \approx -1.1578$

69. (a) $\cos\dfrac{11\pi}{7} \approx 0.2225$

   (b) $\sec\dfrac{11\pi}{7} = \dfrac{1}{\cos\frac{11\pi}{7}} \approx 4.4940$

71. $\dfrac{\sin\left(\frac{\pi}{4}\right) - \sin\left(0\right)}{\frac{\pi}{4} - 0} \approx 0.900$

73. $\dfrac{\tan\left(\left(\frac{1}{2}\right)\left(\frac{\pi}{4}\right)\right) - \tan\left(0\right)}{\frac{\pi}{4} - 0} \approx 0.527$

75. See the sine graph in "Putting it All Together" at the end of section 6.4. $f(t) = \sin t$ has origin symmetry.

77. See the tangent graph in "Putting it All Together" at the end of section 6.4. $f(t) = \tan t$ has origin symmetry.

79. See the secant graph in "Putting it All Together" at the end of section 6.4. $f(t) = \sec t$ has y-axis symmetry.

81. The domain is all real numbers. The range is $\{y \,|\, -1 \le y \le 1\}$. The period is 2π.

83. The domain is $\left\{ t \,\middle|\, t \ne \pm\dfrac{\pi}{2}, \pm\dfrac{3\pi}{2}, \pm\dfrac{5\pi}{2}, \dots \right\}$. The range is all real numbers. The period is π.

85. The domain is $\left\{ t \,\middle|\, t \ne \pm\dfrac{\pi}{2}, \pm\dfrac{3\pi}{2}, \pm\dfrac{5\pi}{2}, \dots \right\}$. The range is $\{y \,|\, |y| \ge 1\}$. The period is 2π.

87. (a) See Figure 87.

(b) At sunrise $(t = 0)$ the stick would cast a very long shadow (longer than four meters). As time moves toward noon, the length of the shadow decreases until it has length zero at exactly noon $\left(t = \dfrac{\pi}{2} \right)$.

In the afternoon, the length of the shadow begins to increase again but now the shadow is on the other side of the stick (negative values). This continues until sunset $(t = \pi)$ when the shadow is again very long.

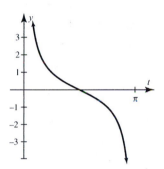

Figure 87

89. $L = \dfrac{2700}{h + 3\tan\alpha} = \dfrac{2700}{2 + 3\tan 1.5°} \approx 1300 \text{ ft.}$

91. $\theta = 57.35 \tan 17°23'43'' \approx 17.95''$; Because $\tan\alpha = \dfrac{\sin\alpha}{\cos\alpha}$, $\theta = 57.3\,\dfrac{\sin\alpha}{\cos\alpha}$.

93. At noon $\theta = 90°$. The sun moves $15°$ per hour. Thus, three hours later $\theta = 135°$. Since $\csc 90° = 1$ and $\csc 135° = \dfrac{1}{\sin 135°} \approx 1.414$, we may conclude that sunlight travels through about 41% more atmosphere at 3:00 P.M. than at noon.

95. Table $Y_1 = 1/\sin(X)$ and $Y_2 = 1/(\sin(X) + 0.5(6 + X)^{-1.64})$ starting at $x = 2$, incrementing by 1. See Figure 95. As θ increases, the values of Y_1 and Y_2 become closer together. When $\theta = 20°$ there is a difference of only about 0.02 or 2%.

X	Y1	Y2
2	28.654	19.449
3	19.107	15.163
4	14.336	12.314
5	11.474	10.314
6	9.5668	8.8478
7	8.2055	7.7328
8	7.1853	6.8601

Y1 ▤1/sin(X)

Figure 95

97. Graph $Y_1 = -16X^2/(750^2 (\cos (30))^2) + X \tan (30)$ in $[0, 20000, 5000]$ by $[0, 4000, 1000]$.

(a) The maximum height is shown to be about 2197 feet at $(7612, 2197)$. See Figure 97a.

(b) The total distance traveled is shown to be about 15,223 feet. See Figure 97b.

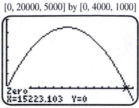

[0, 20000, 5000] by [0, 4000, 1000]   [0, 20000, 5000] by [0, 4000, 1000]

Figure 97a   Figure 97b

Extended and Discovery Exercise for Section 6.4

1. Using the small right triangle, $\sin \theta = \dfrac{\text{Opp.}}{\text{Hyp.}} = \dfrac{\text{Opp.}}{1} = \text{Opposite side.}$

Similarly, $\cos \theta = \dfrac{\text{Adj.}}{\text{Hyp.}} = \dfrac{\text{Adj.}}{1} = \text{Adjacent side.}$ Then using the large right triangle,

$\tan \theta = \dfrac{\text{Opp.}}{\text{Adj.}} = \dfrac{\text{Opp.}}{1} = \text{Opposite side.}$ Similarly, $\sec \theta = \dfrac{\text{Hyp.}}{\text{Adj.}} = \dfrac{\text{Hyp.}}{1} = \text{Hypotenuse.}$

Checking Basic Concepts for Sections 6.3 and 6.4

1. When the terminal side passes through $(-7, 6)$, $r = \sqrt{(-7)^2 + 6^2} = \sqrt{49 + 36} = \sqrt{85}$.

$\sin \theta = \dfrac{6}{\sqrt{85}}$ $\cos \theta = -\dfrac{7}{\sqrt{85}}$ $\tan \theta = -\dfrac{6}{7}$ $\csc \theta = \dfrac{\sqrt{85}}{6}$ $\sec \theta = -\dfrac{\sqrt{85}}{7}$ $\cot \theta = -\dfrac{7}{6}$

3. The graphs of $y = \sin x$, $y = \cos x$, and $y = \tan x$ are shown in Figures 3a, b, and c respectively.

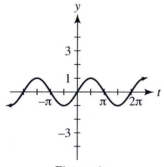

Figure 3a

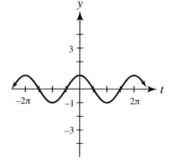

Figure 3b

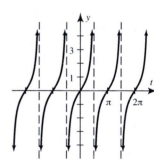

Figure 3c

6.5: Graphing Trigonometric Functions

1. Period is $\dfrac{2\pi}{\frac{1}{2}} = 4\pi$ and Amplitude is $|3| = 3$. See figure 1.

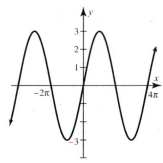

Figure 1

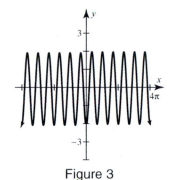

Figure 3

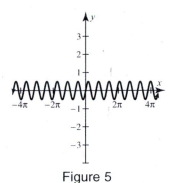

Figure 5

3. Period is $\dfrac{2\pi}{3}$ and Amplitude is $|-2| = 2$. See figure 3.

5. Period is $\dfrac{2\pi}{\pi} = 2$ and Amplitude is $\left|\dfrac{1}{2}\right| = \dfrac{1}{2}$. See figure 5.

7. Shorten the period of the sine graph to π, increase the amplitude to 3, and shift the graph downward 1 unit.

    See Figure 7.

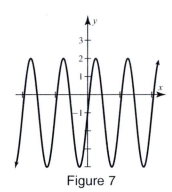

Figure 7

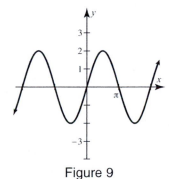

Figure 9

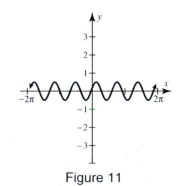

Figure 11

9. Shift the cosine graph $\dfrac{\pi}{2}$ units to the left, increase the amplitude to 2, and reflect the graph across the *x*-axis.

    See Figure 9.

11. Shift the cosine graph $\dfrac{1}{\pi}$ unit to the right, shorten the period to 2, and decrease the amplitude to $\dfrac{1}{2}$.

    See Figure 11.

13. Amplitude is 3; period is $\dfrac{2\pi}{4} = \dfrac{\pi}{2}$; phase shift is $\dfrac{\pi}{4}$; vertical shift is -4.

15. Amplitude is 4; period is $\dfrac{2\pi}{\frac{\pi}{2}} = 4$; phase shift is 1; vertical shift is 6.

17. Rewrite the equation as $y = \dfrac{2}{3} \sin\left(6\left(x + \dfrac{\pi}{2}\right)\right) - \dfrac{5}{2}$.

    Amplitude is $\dfrac{2}{3}$; period is $\dfrac{2\pi}{6} = \dfrac{\pi}{3}$; phase shift is $-\dfrac{\pi}{2}$; vertical shift is $-\dfrac{5}{2}$.

19. Half of the difference between the maximum and minimum values is $0.5(3 - (-3)) = 3$.

    Thus the amplitude is 3. The graph repeats every π units thus $\dfrac{2\pi}{b} = \pi \Rightarrow b = 2$. So $a = 3$ and $b = 2$.

21. Half of the difference between the maximum and minimum values is $0.5(3 - (-3)) = 3$. Thus the amplitude

    is 3. The graph repeats every 4π units, thus the period is 4π. There is no phase shift for this graph.

23. Graph c.   amplitude $= 2$, period $= \dfrac{2\pi}{0.5} = 4\pi$, no phase shift

25. Graph d.   amplitude $= 3$, period $= \dfrac{2\pi}{\pi} = 2$, no phase shift

27. Graph a.   amplitude $= 1$, period $= \dfrac{2\pi}{1} = 2\pi$, phase shift $= -\dfrac{\pi}{2}$

29. $a = 3, \dfrac{2\pi}{b} = 4\pi \Rightarrow b = \dfrac{1}{2}$, and $c = 0$. Thus $y = 3 \sin\left(\dfrac{1}{2}(x - 0)\right) \Rightarrow y = 3 \sin\left(\dfrac{1}{2}x\right)$.

31. Amplitude $= 2$, period $= 2\pi$, phase shift $= 0$. See Figure 31.

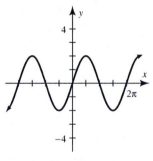

Figure 31

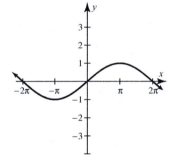

Figure 33

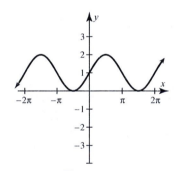

Figure 35

33. Amplitude is 1; period is $\dfrac{2\pi}{\frac{1}{2}} = 4\pi$; phase shift is 0. See figure 33.

35. Amplitude is 1; period is 2π; phase shift is 0. See figure 35.

37. Amplitude $= 1$, period $= \dfrac{2\pi}{\pi} = 2$, phase shift $= 0$, vertical shift up 2 units. See Figure 37.

39. Amplitude $= 1$, period $= \dfrac{2\pi}{2} = \pi$, phase shift $= -\pi$. See Figure 39.

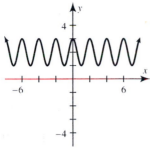

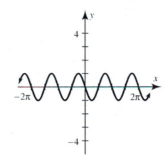

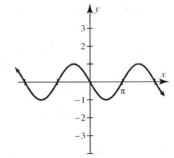

Figure 37   Figure 39   Figure 41

41. Amplitude is 1; period is 2π; phase shift is $\dfrac{\pi}{2}$. See figure 41.

43. Amplitude is $\left| -\dfrac{1}{2} \right| = \dfrac{1}{2}$; period is $\dfrac{2\pi}{2} = \pi$; phase shift is 0. See figure 43.

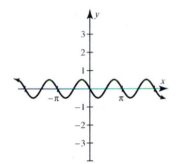

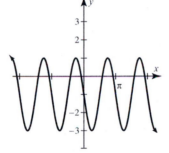

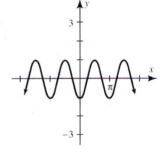

Figure 43   Figure 45   Figure 47

45. Rewrite the equation as $2\cos\left(2\left(x + \dfrac{\pi}{4} \right) \right) - 1$. Amplitude is 2; period is $\dfrac{2\pi}{2} = \pi$; phase shift is $-\dfrac{\pi}{4}$.
    See Figure 45.

47. Amplitude $= 1$, period $= \dfrac{2\pi}{2} = \pi$, phase shift $= \dfrac{\pi}{2}$. See Figure 47.

49. Graph $Y_1 = 2\sin(2X)$ in $[-2\pi, 2\pi, \pi/2]$ by $[-4, 4, 1]$. See Figure 49.

    amplitude $= 2$, period $= \pi$, phase shift $= 0$

$[-2\pi, 2\pi, \pi/2]$ by $[-4, 4, 1]$

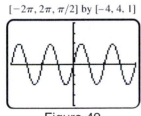

Figure 49

51. Graph $Y_1 = 0.5\cos(3(X + \pi/3))$ in $[-2\pi, 2\pi, \pi/2]$ by $[-4, 4, 1]$. See Figure 51.

    amplitude $= \dfrac{1}{2}$, period $= \dfrac{2\pi}{3}$, phase shift $= -\dfrac{\pi}{3}$

$[-2\pi, 2\pi, \pi/2]$ by $[-4, 4, 1]$ $[-2\pi, 2\pi, \pi/2]$ by $[-4, 4, 1]$ $[-2\pi, 2\pi, \pi/2]$ by $[-4, 4, 1]$

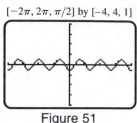

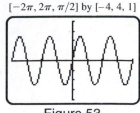

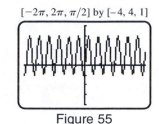

Figure 51   Figure 53   Figure 55

53. Graph $Y_1 = -2.5 \cos (2X + \pi/2))$ in $[-2\pi, 2\pi, \pi/2]$ by $[-4, 4, 1]$. See Figure 53.

amplitude $= 2.5$, period $= \dfrac{2\pi}{2} = \pi$, phase shift $= -\dfrac{\pi}{4}$

55. Rewrite the equation as $f(x) = -2 \cos \left(2\pi \left(x + \dfrac{1}{8} \right) \right) + 1$.

Graph $Y_1 = -2 \cos (2\pi(X + 1/8)) + 1$ in $[-2\pi, 2\pi, \pi/2]$ by $[-4, 4, 1]$. See Figure 55.

Amplitude is $|-2| = 2$; period is $\dfrac{2\pi}{2\pi} = 1$; phase shift is $-\dfrac{1}{8}$.

57. If $f(t) = \tan 2t$, then $b = 2$ and $c = 0$. Also, period $= \dfrac{\pi}{b} = \dfrac{\pi}{2}$ and phase shift $= 0$. See Figure 57.

The vertical asymptotes occur at $t = \pm\dfrac{\pi}{4}, \pm\dfrac{3\pi}{4}, \pm\dfrac{5\pi}{4}, \pm\dfrac{7\pi}{4}$.

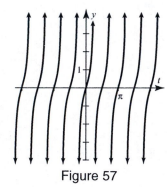

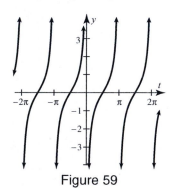

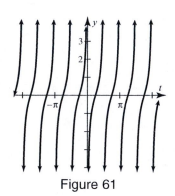

Figure 57   Figure 59   Figure 61

59. If $f(t) = \tan \left(t - \dfrac{\pi}{2} \right)$, then $b = 1$ and $c = \dfrac{\pi}{2}$. Also, period $= \dfrac{\pi}{b} = \dfrac{\pi}{1} = \pi$ and phase shift $= \dfrac{\pi}{2}$. See

Figure 59. The vertical asymptotes occur at $t = 0, \pm\pi, \pm 2\pi$.

61. If $f(t) = -\cot 2t$, then $b = 2$ and $c = 0$. Also, period $= \dfrac{\pi}{b} = \dfrac{\pi}{2}$ and phase shift $= 0$. See Figure 61.

The vertical asymptotes occur at $t = 0, \pm\dfrac{\pi}{2}, \pm\pi, \pm\dfrac{3\pi}{2}, \pm 2\pi$.

63. If $f(t) = \cot \left(2 \left(x - \dfrac{\pi}{4} \right) \right) - 1$, then $b = 2$ and $c = \dfrac{\pi}{4}$. Also, period $= \dfrac{\pi}{2}$ and phase shift $= \dfrac{\pi}{4}$.

The graph is shown in Figure 63. The vertical asymptotes occur at $x = \pm\dfrac{\pi}{4}, \pm\dfrac{3\pi}{4}, \pm\dfrac{5\pi}{4}, \pm\dfrac{7\pi}{4}$.

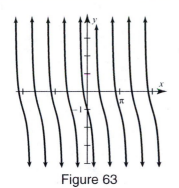

Figure 63

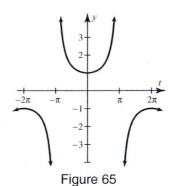

Figure 65

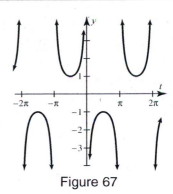

Figure 67

65. If $f(t) = \sec\left(\dfrac{1}{2}t\right)$, then $b = \dfrac{1}{2}$ and $c = 0$. Also, period $= \dfrac{2\pi}{\frac{1}{2}} = 4\pi$ and phase shift $= 0$. See Figure 65.

The vertical asymptotes occur at $t = \pm\pi$.

67. If $f(t) = \csc(t - \pi)$, then $b = 1$ and $c = \pi$. Also, period $= \dfrac{2\pi}{1} = 2\pi$ and phase shift $= \pi$. See Figure 67.

The vertical asymptotes occur at $t = 0, \pm\pi, \pm2\pi$.

69. If $f(x) = \sec\left(\dfrac{1}{3}\left(x - \dfrac{\pi}{6}\right)\right)$, then $b = \dfrac{1}{3}$ and $c = \dfrac{\pi}{6}$. Also, period $= \dfrac{2\pi}{\frac{1}{3}} = 6\pi$ and phase shift $= \dfrac{\pi}{6}$.

See Figure 69. The vertical asymptotes occur at $x = -\dfrac{4\pi}{3}, \dfrac{5\pi}{3}$.

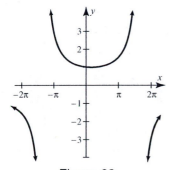

Figure 69

71. (a) Since $s(0) = 2$ inches, $a = 2$. Since the period is 0.5 seconds, the frequency is $F = \dfrac{1}{0.5} = 2$.

Thus $s(t) = a\cos(2\pi F t) = 2\cos(4\pi t)$.

(b) $s(1) = 2\cos(4\pi) = 2$. The spring is compressed 2 inches one second after the weight is released. The weight is moving neither upward nor downward.

73. (a) Since $s(0) = -3$ inches, $a = -3$. Since the period is 0.8 seconds, the frequency is $F = \dfrac{1}{0.8} = \dfrac{5}{4}$.

Thus $s(t) = a\cos(2\pi F t) = -3\cos(2.5\pi t)$.

(b) $s(1) = -3\cos(2.5\pi) = -3(0) = 0$. The spring is at its natural length one second after the weight is released. The weight is moving upward.

75. First note that $b = 2\pi F = 2\pi(27.5) = 55\pi$. Thus $s(t) = a \cos(55\pi t)$.

Hence $s(0) = a \cos(0) = 0.21 \Rightarrow a = 0.21$. The equation is $s(t) = 0.21 \cos(55\pi t)$. See Figure 75.

[0, 0.05, 0.01] by [−0.3, 0.3, 0.1]   [0, 0.05, 0.01] by [−0.3, 0.3, 0.1]   ]

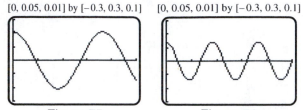

Figure 75   Figure 79

77. First note that $b = 2\pi F = 2\pi(55) = 110\pi$. Thus $s(t) = a \cos(110\pi t)$.

Hence $s(0) = a \cos(0) = 0.14 \Rightarrow a = 0.14$. The equation is $s(t) = 0.14 \cos(110\pi t)$. See Figure 79.

79. (a) $a = \dfrac{Max - Min}{2} = \dfrac{90 - 10}{2} = 40, b = \dfrac{2\pi}{P} = \dfrac{2\pi}{24} = \dfrac{\pi}{12}, d = \dfrac{Max + Min}{2} = \dfrac{90 + 10}{2} = 50$

and $P(x) = 40 \sin\left(\dfrac{\pi}{12}x\right) + 50$

(b) The graph of $P(x)$ is shown in Figure 79. The maximum will occur when $x = 6$ or at noon.

(c) The minimum will occur when $x = 18$ or at midnight.

81. (a) The maximum monthly average temperature is found at the peak of the graph: $y = 40°$ F.

The minimum monthly average temperature is found at the "valley" of the graph: $y = -40°$ F.

(b) The amplitude is $0.5(40 - (-40)) = 40$. This represents a total temperature swing from $-40°$ F to $40°$ F.

Since the average temperatures are cyclic, repeating every 12 months, the period is 12.

(c) The x-intercepts represent the months in which the average temperature is $0°$ F (April and October).

83. (a) Graph $Y_1 = 34 \sin((\pi/6)(X - 4.3))$ in $[0, 25, 2]$ by $[-50, 50, 10]$. See Figure 83a.

amplitude $= 34$, period $= \dfrac{2\pi}{\frac{\pi}{6}} = 12$, phase shift $= 4.3$

(b) May corresponds to $x = 5, y \approx 12.2°$ F. See Figure 83b.

December corresponds to $x = 12, y \approx -26.4°$ F. See Figure 83c.

(c) Since half of the months have average temperatures above zero and half have average temperatures below

zero, we would conjecture that the average yearly temperature is $0°$ F.

[0, 25, 2] by [−50, 50, 10]   [0, 25, 2] by [−50, 50, 10]   [0, 25, 2] by [−50, 50, 10]

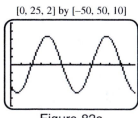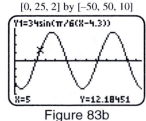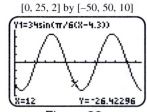

Figure 83a   Figure 83b   Figure 83c

85. (a) Plot the data is [0, 25, 2] by [0, 80, 10]. See Figure 85a.

   (b) The maximum monthly average temperature is 64° F and the minimum is 36° F. The midpoint of these values is $0.5(64 + 36) = 50$, thus $d = 50$. Half the difference between these temperatures is $0.5(64 - 36) = 14$, thus $a = 14$. Since the temperatures cycle every 12 months, $b = \dfrac{2\pi}{12} = \dfrac{\pi}{6}$. The maximum of the $y = \sin x$ graph occurs when $x = \dfrac{\pi}{2}$ while the maximum in the table occurs when $x = 7$.

   Thus $\dfrac{\pi}{6}(7 - c) = \dfrac{\pi}{2} \Rightarrow 7 - c = 3 \Rightarrow c = 4$. The function is $f(x) = 14 \sin\left(\dfrac{\pi}{6}(x - 4)\right) + 50$.

   (c) See Figure 85c.

[0, 25, 2] by [0, 80, 10]   [0, 25, 2] by [0, 80, 10]   [0, 25, 2] by [40, 100, 10]

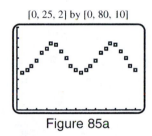   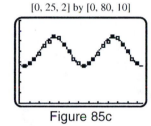   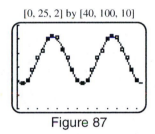

Figure 85a   Figure 85c   Figure 87

87. Refer to the solution for exercise 85.

   (a) $a = 0.5(92 - 58) = 17$, $b = \dfrac{\pi}{6}$, $c = 7$, $d = 0.5(92 + 58) = 75$.

   Thus $f(x) = 17 \cos\left(\dfrac{\pi}{6}(x - 7)\right) + 75$. The graph of f and the actual data are shown together in [0, 25, 2] by [40, 100, 10]. See Figure 87.

   (b) Yes, different values of c are possible of the form $c = 7 + 12n$ where n is an integer.

89. (a) The maximum number of daylight hours is about 18.5 hr. This occurs June 21st - summer solstice.

   (b) The minimum number of daylight hours is about 6 hr. This occurs December 21st - winter solstice.

   (c) The amplitude represents half the difference in daylight between the longest and shortest days. The period represents one year. *Answers may vary.*

91. (a) The maximum monthly average precipitation is about 8 inches.

   The minimum monthly average precipitation is about 0.5 inches.

   (b) The amplitude is $0.5(8 - 0.5) = 3.75$. The amplitude represents half of the difference between the maximum and minimum average monthly precipitation.

   (c) Since $a = 0.5(8 - 0.5) = 3.75$, $d = 0.5(8 + 0.5) = 4.25$. The maximum occurs at $x = 0$. Let $c = 0$.

   Since daylight hours cycle every 12 months, let $b = \dfrac{2\pi}{12} = \dfrac{\pi}{6}$. Thus, $f(x) = 3.75 \cos\left(\dfrac{\pi}{6}x\right) + 4.25$.

[0, 25, 2] by [60, 90, 5]   [0, 12, 3] by [0, 50, 10]   [0, 1/110, 1/880] by [−1.5, 1.5, 0.5]

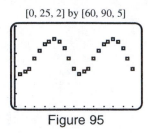

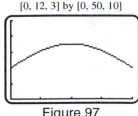

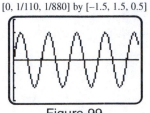

Figure 95   Figure 97   Figure 99

93. Refer to the solution for exercise 85.

(a) $a = 0.5(6 - 2) = 2, b = \dfrac{\pi}{6}, c = 1, d = 0.5(6 + 2) = 4.$ Thus $f(t) = 2 \cos\left(\dfrac{\pi}{6}(t - 1)\right) + 4.$

(b) $f(1) = 2 \cos\left(\dfrac{\pi}{6}(1 - 1)\right) + 4 = 2 \cos(0) + 4 = 2 + 4 = 6.$

$f(7) = 2 \cos\left(\dfrac{\pi}{6}(7 - 1)\right) + 4 = 2 \cos(\pi) + 4 = -2 + 4 = 2.$

95. Refer to the solution in exercise 85.

(a) Plot the data is [0, 25, 2] by [60, 90, 5]. See Figure 95.

(b) $a = 0.5(85 - 72) = 6.5, b = \dfrac{\pi}{6}, c = 4, d = 0.5(85 + 72) = 78.5.$

Thus $y = 6.5 \sin\left(\dfrac{\pi}{6}(x - 4)\right) + 78.5.$

97.   Graph $Y_1 = 20 + 15 \sin(\pi X/12)$ in [0, 12, 3] by [0, 50, 10]. See Figure 87.

At 9:00 A.M. the outdoor temperature is 20° F. The temperature increases to a maximum of 35° F at 3:00

P.M. Then the temperature begins to fall until it reaches 20° F again at 9:00 P.M.

99. (a) Graph $Y_1 = \sin(880\pi X)$ in [0, 1/110, 1/880] by [−1.5, 1.5, 0.5]. See Figure 99.

(b) The period can be found by finding the change in the x-coordinate from one peak to the next.

$x_1 \approx 0.00511371, x_2 \approx 0.00738592 \Rightarrow$

$x_2 - x_1 = 0.00738592 - 0.00511371 \approx 0.00227$ sec. $\left(\text{actual: } \dfrac{1}{440} \approx 0.00227 \text{ sec.}\right)$

(c) $F = \dfrac{1}{0.00227} \approx 440$ cycles per second $\left(\text{actual: } F = 440\right)$

101. $y \approx 13.2 \sin(0.524x - 2.18) + 49.7.$ See Figure 101a.

Graph $Y_1 = 13.2 \sin(0.524 X - 2.18) + 49.7$ together with the data in [0, 25, 2] by [30, 80, 10].

See Figure 101b.

[0, 25, 2] by [30, 80, 10]

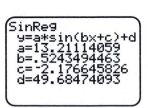

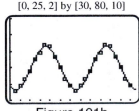

Figure 101a   Figure 101b

103. $y \approx 16.9 \sin (0.522x - 2.09) + 75.4$. See Figure 103a.

Graph $Y_1 = 16.9 \sin (0.522X - 2.09) + 75.4$ together with the data in [0, 25, 2] by [40, 100, 10].

See Figure 103b.

[0, 25, 2] by [40, 100, 10]

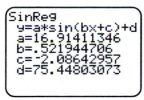   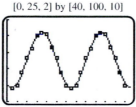

Figure 103a   Figure 103b

6.6: Inverse Trigonometric Functions

1.   one-to-one

3. π

5.   0

7. $\dfrac{\pi}{2}$

9. $-\dfrac{\pi}{4}$

11. $\dfrac{2\pi}{3}$

13. (a) The angle in $\left[-\dfrac{\pi}{2}, \dfrac{\pi}{2} \right]$ whose sine is 1 is $\dfrac{\pi}{2}$. Thus $\sin^{-1} 1 = \dfrac{\pi}{2}$ or 90°.

(b) No angle in $\left[-\dfrac{\pi}{2}, \dfrac{\pi}{2} \right]$ has a sine of 4. Thus, arcsin 4 is undefined.

(c) The angle in $\left[-\dfrac{\pi}{2}, \dfrac{\pi}{2} \right]$ whose sine is $-\dfrac{\sqrt{3}}{2}$ is $-\dfrac{\pi}{3}$. Thus $\arcsin \left(-\dfrac{\sqrt{3}}{2} \right) = -\dfrac{\pi}{3}$ or $-60°$.

15. (a) The angle in $[0, \pi]$ whose cosine is 0 is $\dfrac{\pi}{2}$. Thus $\cos^{-1} 0 = \dfrac{\pi}{2}$ or 90°.

(b) The angle in $[0, \pi]$ whose cosine is -1 is π. Thus $\arccos (-1) = \pi$ or 180°.

(c) The angle in $[0, \pi]$ whose cosine is $\dfrac{1}{2}$ is $\dfrac{\pi}{3}$. Thus $\cos^{-1} \left(\dfrac{1}{2} \right) = \dfrac{\pi}{3}$ or 60°.

17. (a) The angle in $\left(-\dfrac{\pi}{2}, \dfrac{\pi}{2} \right)$ whose tangent is 1 is $\dfrac{\pi}{4}$. Thus $\tan^{-1} 1 = \dfrac{\pi}{4}$ or 45°.

(b) The angle in $\left(-\dfrac{\pi}{2}, \dfrac{\pi}{2} \right)$ whose tangent is -1 is $-\dfrac{\pi}{4}$. Thus $\arctan (-1) = -\dfrac{\pi}{4}$ or $-45°$.

(c) The angle in $\left(-\dfrac{\pi}{2}, \dfrac{\pi}{2} \right)$ whose tangent is $\sqrt{3}$ is $\dfrac{\pi}{3}$. Thus $\tan^{-1} \sqrt{3} = \dfrac{\pi}{3}$ or 60°.

19. (a) $\sin^{-1} 1.5$ is undefined (Error). See Figure 19a.

    (b) $\tan^{-1} 10 \approx 1.47$ radians or $84.3°$. See Figure 19b.

    (c) $\arccos(-0.25) \approx 1.82$ radians or $104.5°$. See Figure 19c.

21. (a) Since $\tan \beta = \dfrac{4}{3}$, $\tan^{-1} \dfrac{4}{3} = \beta$.

    (b) Since $\sin \alpha = \dfrac{3}{5}$, $\sin^{-1} \dfrac{3}{5} = \alpha$.

    (c) Since $\cos \beta = \dfrac{3}{5}$, $\arccos \dfrac{3}{5} = \beta$.

23. $\sin(\sin^{-1} 1) = \sin\left(\dfrac{\pi}{2}\right) = 1$

25. $\cos^{-1}\left(\cos \dfrac{5\pi}{4}\right) = \cos^{-1}\left(-\dfrac{1}{\sqrt{2}}\right) = \dfrac{3\pi}{4}$

27. $\tan(\tan^{-1}(-3)) = -3$ since $-3 \in (-\infty, \infty)$

29. (a) $\sin^{-1} \dfrac{3}{5} + \cos^{-1} \dfrac{3}{5} = 90°$

    (b) $\sin^{-1} \dfrac{1}{3} + \cos^{-1} \dfrac{1}{3} = 90°$

    (c) $\sin^{-1} \dfrac{2}{7} + \cos^{-1} \dfrac{2}{7} = 90°$

    Conjecture:

31. Verbal: Determine the angle (or real number) θ such that $\sin \theta = 2x$ and $-\dfrac{\pi}{2} \le \theta \le \dfrac{\pi}{2}$.

    Numerical: See Figure 31a.

    Graphical: See Figure 31b.

[–1, 1, 1] by [–2, 2, 1]

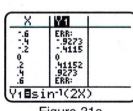
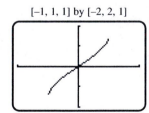

Figure 31a   Figure 31b

33. Verbal: Determine the angle (or real number) θ such that $\cos \theta = \dfrac{1}{2}x$ and $0 \le \theta \le \pi$.

    Numerical: See Figure 33a.

    Graphical: See Figure 33b.

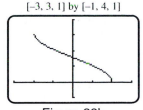

[−3, 3, 1] by [−1, 4, 1]

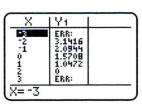

Figure 33a   **Figure 33b**

35. See Figure 35. By the Pythagorean theorem: $b^2 = 1^2 - x^2 \Rightarrow b = \sqrt{1 - x^2}$. Thus $\tan\theta = \dfrac{x}{\sqrt{1 - x^2}}$.

37. See Figure 43. By the Pythagorean theorem:

$$b^2 = \left(\sqrt{1 + x^2}\right)^2 - x^2 \Rightarrow b^2 = 1 + x^2 - x^2 \Rightarrow b^2 = 1 \Rightarrow b = 1. \text{ Thus } \cos\theta = \dfrac{1}{\sqrt{1 + x^2}}.$$

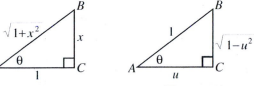

Figure 37   **Figure 39**

39. Let θ be the angle whose cosine is u. That is $\theta = \cos^{-1} u$. See Figure 39. We find the unknown side x by using

   the Pythagorean Theorem: $u^2 + x^2 = 1^2 \Rightarrow x^2 = 1 - u^2 \Rightarrow x = \sqrt{1 - u^2}$.

   Therefore $\sin\left(\cos^{-1} u\right) = \sin\theta = \dfrac{\sqrt{1 - u^2}}{1} = \sqrt{1 - u^2}$.

41. Let θ be the angle whose cosine is u. That is $\theta = \cos^{-1} u$. See Figure 41. We find the unknown side x by using

   the Pythagorean Theorem: $u^2 + x^2 = 1^2 \Rightarrow x^2 = 1 - u^2 \Rightarrow x = \sqrt{1 - u^2}$.

   Therefore $\tan\left(\cos^{-1} u\right) = \tan\theta = \dfrac{\sqrt{1 - u^2}}{u}$.

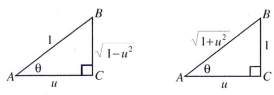

Figure 41   **Figure 43**

43. Let θ be the angle whose tangent is $\dfrac{1}{u}$. That is $\theta = \tan^{-1}\dfrac{1}{u}$. See Figure 43. We find the unknown side x by

   using the Pythagorean Theorem: $u^2 + 1^2 = x^2 \Rightarrow \sqrt{u^2 + 1^2} = x \Rightarrow x = \sqrt{u^2 + 1}$. Note that the

   unknown side is not needed for the requested expression. Therefore $\cot\left(\tan^{-1}\dfrac{1}{u}\right) = \cot\theta = \dfrac{1}{\tan\theta} = \dfrac{1}{\frac{1}{u}} = u$.

45. Since $\tan\alpha = \dfrac{7}{24}$, $\alpha = \tan^{-1}\dfrac{7}{24} \approx 16.3°$. Since $\tan\beta = \dfrac{24}{7}$, $\beta = \tan^{-1}\dfrac{24}{7} \approx 73.7°$.

   By the Pythagorean theorem: $c^2 = 7^2 + 24^2 \Rightarrow c^2 = 625 \Rightarrow c = 25$.

47. Since $\sin \alpha = \dfrac{6}{10}$, $\alpha = \sin^{-1} \dfrac{6}{10} \approx 36.9°$. Since $\cos \beta = \dfrac{6}{10}$, $\beta = \cos^{-1} \dfrac{6}{10} \approx 53.1°$.

    By the Pythagorean theorem: $b^2 = 10^2 - 6^2 \Rightarrow b^2 = 64 \Rightarrow b = 8$.

49. β and $55°$ are complimentary, so $\beta = 90° - 55° = 35°$

    Since $\tan 55° = \dfrac{5}{b}$, $b = \dfrac{5}{\tan 55°} \approx 3.5$. Since $\sin 55° = \dfrac{5}{c}$, $c = \dfrac{5}{\sin 55°} \approx 6.1$.

51. $\sin \theta = 1 \Rightarrow \theta = \sin^{-1} 1 = 90°$

53. $\tan \theta = 1 \Rightarrow \theta = \tan^{-1} 1 = 45°$

55. $\cos \theta = 0 \Rightarrow \theta = \cos^{-1} 0 = 90°$

57. $2 \cos \theta = \dfrac{1}{4} \Rightarrow \cos \theta = \dfrac{1}{8} \Rightarrow \theta = \cos^{-1} \dfrac{1}{8} \approx 82.8°$

59. $\tan \theta - 1 = 5 \Rightarrow \tan \theta = 6 \Rightarrow \theta = \tan^{-1} 6 \approx 80.5°$

61. $\sin^2 \theta = 0.87 \Rightarrow \sin \theta \approx 0.9327 \Rightarrow \theta \approx \sin^{-1} 0.9327 \approx 68.9°$

63. $\tan t = -\dfrac{1}{5} \Rightarrow t = \tan^{-1} \left(-\dfrac{1}{5} \right) \approx -0.197$. See Figure 63.

$[-\pi/2, \pi/2, \pi/4]$ by $[-2, 2, 1]$ $\quad$ $[0, \pi, \pi/4]$ by $[-2, 2, 1]$

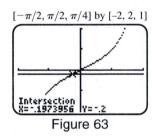
Figure 63
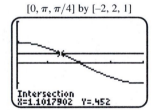
Figure 65

65. $\cos t = 0.452 \Rightarrow t = \cos^{-1} 0.452 \approx 1.102$. See Figure 65.

67. $2 \sin t = -0.557 \Rightarrow \sin t = -0.2785 \Rightarrow t = \sin^{-1} (-0.2785) \approx -0.282$. See Figure 67.

$[-\pi/2, \pi/2, \pi/4]$ by $[-2, 2, 1]$ $\quad$ $[0, \pi, \pi/4]$ by $[-1, 1, 1]$

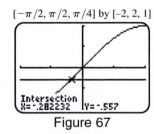
Figure 67
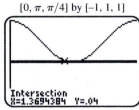
Figure 69

69. $\cos^2 t = \dfrac{1}{25} \Rightarrow \cos t = \pm \dfrac{1}{5} \Rightarrow t = \cos^{-1} \left(\pm \dfrac{1}{5} \right) \approx 1.369, 1.772$. See Figure 69.

71. $\sec^{-1}(-1) = \cos^{-1}\left(\dfrac{1}{-1}\right) = \pi$

73. $\csc^{-1}\left(-\sqrt{2}\right) = \sin^{-1}\left(\dfrac{1}{-\sqrt{2}}\right) = -\dfrac{\pi}{4}$

75. $\cot^{-1}(-1) = \dfrac{\pi}{2} - \tan^{-1}(-1) = \dfrac{\pi}{2} - \left(-\dfrac{\pi}{4}\right) = \dfrac{3\pi}{4}$

77. $\sec^{-1}(3) = \cos^{-1}\left(\dfrac{1}{3}\right) \approx 1.231$

79. $\csc^{-1}(5.1) = \sin^{-1}\left(\dfrac{1}{5.1}\right) \approx 0.197$

81. $\cot^{-1}(1.5) = \dfrac{\pi}{2} - \tan^{-1}(1.5) \approx \dfrac{\pi}{2} - 0.983 \approx 0.588$

83. $\tan\theta = \dfrac{50}{85}, \theta = \tan^{-1}\dfrac{50}{85} \approx 30.5°$

85. (a) $F = W\sin\theta \Rightarrow 400 = 5000\sin\theta \Rightarrow \sin\theta = \dfrac{2}{25} \Rightarrow \theta = \sin^{-1}\dfrac{2}{25} \approx 4.6°$

   (b) $F = W\sin\theta \Rightarrow 130 = 3500\sin\theta \Rightarrow \sin\theta = \dfrac{13}{350} \Rightarrow \theta = \sin^{-1}\dfrac{13}{350} \approx 2.1°$

   (c) $F = W\sin\theta \Rightarrow -200 = 4000\sin\theta \Rightarrow \sin\theta = -\dfrac{1}{20} \Rightarrow \theta = \sin^{-1}\left(-\dfrac{1}{20}\right) \approx -2.9°$

87. Since $\tan\theta = \dfrac{4}{10}, \theta = \tan^{-1}\dfrac{4}{10} \approx 21.8°$

89. (a) $\tan\theta = \dfrac{3}{12} \Rightarrow \theta = \tan^{-1}\dfrac{1}{4} \approx 14.0°$

   (b) $\tan\theta = \dfrac{4}{12} \Rightarrow \theta = \tan^{-1}\dfrac{1}{3} \approx 18.4°$

   (c) $\tan\theta = \dfrac{6}{12} \Rightarrow \theta = \tan^{-1}\dfrac{1}{2} \approx 26.6°$

   (d) $\tan\theta = \dfrac{12}{12} \Rightarrow \theta = \tan^{-1}1 = 45°$

91. $\theta = \sin^{-1}\left(\sqrt{\dfrac{43^2}{2(43^2) + 64.4(7)}}\right) \approx 41.9°$

93. If $\cos(0.1309H) = -0.4336\tan L$, then $0.1309H = \cos^{-1}(-0.4336\tan L) \Rightarrow H = \dfrac{\cos^{-1}(-0.4336\tan L)}{0.1309}$.

   (a) Convert $40°55'$ to radians: $\left(40 + \dfrac{55}{60}\right) \cdot \dfrac{\pi}{180} \approx 0.714$. Then $H \approx \dfrac{\cos^{-1}(-0.4336\tan 0.714)}{0.1309} \approx 14.9$ hr.

   (b) Convert $27°46'$ to radians: $\left(27 + \dfrac{46}{60}\right) \cdot \dfrac{\pi}{180} \approx 0.485$. Then $H \approx \dfrac{\cos^{-1}(-0.4336\tan 0.485)}{0.1309} \approx 13.8$ hr.

   (c) Convert $37°30'$ to radians: $\left(37 + \dfrac{30}{60}\right) \cdot \dfrac{\pi}{180} \approx 0.654$. Then $H \approx \dfrac{\cos^{-1}(-0.4336\tan 0.654)}{0.1309} \approx 14.6$ hr.

95. After 5 seconds, $\theta_1 \approx 35.8°$ and so $\theta_2 \approx 90° - 35.8° \approx 54.2°$. See Figure 95.

[0, 15, 5] by [0, 100, 10]

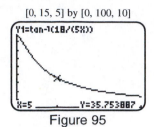

Figure 95

97. Let α and β represent the angles of elevation from the shrub to the shorter and taller buildings respectively. The distance from the shrub to the shorter building is $100 - x$, thus $\alpha = \arctan\left(\dfrac{75}{100 - x}\right)$. Similarly $\beta = \arctan\left(\dfrac{150}{x}\right)$. Because the angles α, θ, and β form a straight angle, $\theta = \pi - \alpha - \beta$. That is

$$\theta = \pi - \arctan\left(\frac{75}{100 - x}\right) - \arctan\left(\frac{150}{x}\right).$$

Extended and Discovery Exercise for Section 6.6

1. Graph $Y_1 = \tan^{-1}\left(10X/(X^2 + 39)\right)$ in $[0, 50, 10]$ by $[0°, 50°, 10°]$. See Figure 1.

   The maximum value of θ is $38.7°$ when $x \approx 6.24$.

   [0, 50, 10] by [0°, 50°, 10°]

   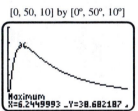

   Figure 1

Checking Basic Concepts for Sections 6.5 and 6.6

1. The graph is shown in Figure 1.

   The amplitude is 3. The period is $\dfrac{2\pi}{2} = \pi$. The phase shift is $\dfrac{\pi}{4}$.

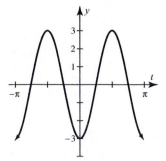

Figure 1

3. (a) $\sin^{-1} 0 = 0°$

(b) $\cos^{-1}(-1) = 180°$

(c) $\tan^{-1}(-1) = -45°$

(d) $\sin^{-1}\dfrac{1}{2} = 30°$

(e) $\tan^{-1}\sqrt{3} = 60°$

(f) The angle in $[0, \pi]$ whose cosine is $\dfrac{1}{2}$ is $60°$. Thus, $\cos^{-1}\left(\dfrac{1}{2}\right) = 60°$.

5. (a) $\sin t = 0.55 \Rightarrow t = \sin^{-1} 0.55 \approx 0.582$

(b) $\cos t = -0.35 \Rightarrow t = \cos^{-1}(-0.35) \approx 1.93$

(c) $\tan t = -2.9 \Rightarrow t = \tan^{-1}(-2.9) \approx -1.24$

Chapter 6 Review Exercises

1. (a) See Figure 1a.

(b) See Figure 1b.

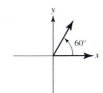

Figure 1a   Figure 1b

3. (a) $\dfrac{\pi}{3} \cdot \dfrac{180°}{\pi} = \dfrac{180°}{3} = 60°$

(b) $\dfrac{\pi}{36} \cdot \dfrac{180°}{\pi} = \dfrac{180°}{36} = 5°$

(c) $-\dfrac{5\pi}{6} \cdot \dfrac{180°}{\pi} = \dfrac{-5(180°)}{6} = -150°$

(d) $-\dfrac{7\pi}{4} \cdot \dfrac{180°}{\pi} = \dfrac{-7(180°)}{4} = -315°$

5. Note that $60° = \dfrac{\pi}{3}$ radians, thus $s = \theta r = \dfrac{\pi}{3}(6) = 2\pi$ ft.

7. $\sin 30° = \dfrac{\text{opposite}}{\text{hypotenuse}} = \dfrac{1}{2}$

9. $\cot 60° = \dfrac{\text{adjacent}}{\text{opposite}} = \dfrac{1}{\sqrt{3}}$

11. $\sec \dfrac{\pi}{4} = \dfrac{\text{hypotenuse}}{\text{adjacent}} = \dfrac{\sqrt{2}}{1} = \sqrt{2}$

13. $c = \sqrt{8^2 + 9^2} = \sqrt{145}$

$\sin\theta = \dfrac{8}{\sqrt{145}}$ $\qquad$ $\cos\theta = \dfrac{9}{\sqrt{45}}$ $\qquad$ $\tan\theta = \dfrac{8}{9}$ $\qquad$ $\csc\theta = \dfrac{\sqrt{145}}{8}$ $\qquad$ $\sec\theta = \dfrac{\sqrt{145}}{9}$ $\qquad$ $\cot\theta = \dfrac{9}{8}$

15. $\csc\theta = \dfrac{1}{\sin\theta} = \dfrac{1}{\frac{1}{3}} = 3$

17. $\sin 25° \approx 0.423$ $\cos 25° \approx 0.906$ $\tan 25° \approx 0.466$

    $\csc 25° \approx 2.366$ $\sec 25° \approx 1.103$ $\cot 25° \approx 2.145$

19. $r = \sqrt{1^2 + (-2)^2} = \sqrt{5}$

    $\sin\theta = -\dfrac{2}{\sqrt{5}}$ $\cos\theta = \dfrac{1}{\sqrt{5}}$ $\tan\theta = -\dfrac{2}{1} = -2$ $\csc\theta = -\dfrac{\sqrt{5}}{2}$ $\sec\theta = \dfrac{\sqrt{5}}{1} = \sqrt{5}$ $\cot\theta = -\dfrac{1}{2}$

21. $\sin\theta = \dfrac{\sqrt{3}}{2}$ $\cos\theta = -\dfrac{1}{2}$ $\tan\theta = -\sqrt{3}$ $\csc\theta = \dfrac{2}{\sqrt{3}}$ $\sec\theta = -2$ $\cot\theta = -\dfrac{1}{\sqrt{3}}$

23. An angle $\theta = -\dfrac{\pi}{2}$ radians in standard position has terminal side that intersects the unit

    circle at the point $(0, -1)$. Therefore $\sin\left(-\dfrac{\pi}{2}\right) = -1$.

25. An angle $\theta = -3\pi$ radians in standard position has terminal side that intersects the unit

    circle at the point $(-1, 0)$. Therefore $\tan(-3\pi) = 0$.

27. $\sin\theta = -\dfrac{4}{5}$ $\cos\theta = \dfrac{3}{5}$ $\tan\theta = \dfrac{-\frac{4}{5}}{\frac{3}{5}} = -\dfrac{4}{3}$ $\csc\theta = -\dfrac{5}{4}$ $\sec\theta = \dfrac{5}{3}$ $\cot\theta = -\dfrac{3}{4}$

29. See Figure 29. Amplitude $= 3$; Period $= \dfrac{2\pi}{2} = \pi$; Phase shift $= 0$

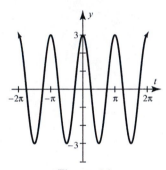

Figure 29

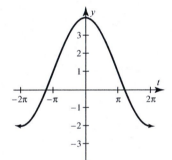

Figure 31

31. See Figure 31. Amplitude $= |-3| = 3$; Period $= \dfrac{2\pi}{\frac{1}{2}} = 4\pi$; Phase shift $= \pi$

33. Amplitude $= 2 \Rightarrow a = 2$ Period $= \dfrac{2\pi}{3} \Rightarrow \dfrac{2\pi}{3} = \dfrac{2\pi}{b} \Rightarrow b = 3$

35. See Figure 35. Period $= \dfrac{\pi}{b} = \dfrac{\pi}{2}$; Phase shift $= 0$

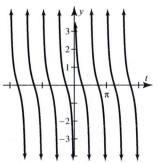

Figure 35

37. (a) $\sin^{-1}(-1) = -\dfrac{\pi}{2}$ or $-90°$

    (b) $\arccos\dfrac{1}{2} = \dfrac{\pi}{3}$ or $60°$

    (c) $\tan^{-1} 1 = \dfrac{\pi}{4}$ or $45°$

39. (a) $\sin^{-1}(-0.6) \approx -0.64$ radians or $-36.9°$

    (b) $\tan^{-1} 5 \approx 1.37$ radians or $78.7°$

    (c) $\arccos(0.12) \approx 1.45$ radians or $83.1°$

41. Since $\tan\alpha = \dfrac{5}{3}$, $\alpha = \tan^{-1}\dfrac{5}{3} \approx 59.0°$. And since $\tan\beta = \dfrac{3}{5}$, $\beta = \tan^{-1}\dfrac{3}{5} \approx 31.0°$.

    By the Pythagorean theorem: $c^2 = 5^2 + 3^2 \Rightarrow c^2 = 34 \Rightarrow c = \sqrt{34}$.

43. $\tan\theta = \dfrac{1}{\sqrt{3}} \Rightarrow \theta = \tan^{-1}\dfrac{1}{\sqrt{3}} \approx 30°$

45. $\cos\theta = \dfrac{1}{5} \Rightarrow \theta = \cos^{-1}\dfrac{1}{5} \approx 78.5°$

47. $\tan t = -\dfrac{3}{4} \Rightarrow t = \tan^{-1}\left(-\dfrac{3}{4}\right) \approx -0.6435$

    Graph $Y_1 = \tan(X)$ together with $Y_2 = -0.75$ in $[-\pi/2, \pi/2, \pi/4]$ by $[-4, 4, 1]$. See Figure 47.

$[-\pi/2, \pi/2, \pi/4]$ by $[-4, 4, 1]$

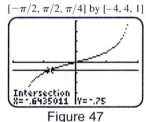

Figure 47

49. (a) Angular velocity $= \dfrac{\text{angle of rotation}}{\text{time}} = \dfrac{2\pi}{50 \text{ sec}} = \dfrac{\pi}{25}$ radians/sec.

    (b) Linear speed $= (\text{radius})(\text{angular velocity}) = (25)\left(\dfrac{\pi}{25}\right) = \pi$ ft/sec.

51. Let h represent the height of the tree. Then $\tan 48° = \dfrac{h}{80} \Rightarrow h = 80\tan 48° \approx 89$ ft.

53. (a) $350 = 6000\sin\theta \Rightarrow \sin\theta = \dfrac{7}{120} \Rightarrow \theta = \sin^{-1}\dfrac{7}{120} \approx 3.3°$

    (b) $160 = 4500\sin\theta \Rightarrow \sin\theta = \dfrac{8}{225} \Rightarrow \theta = \sin^{-1}\dfrac{8}{225} \approx 2.0°$

55. The two cities are $41°09' - 38°49' = 2°20'$ apart. Convert $2°20'$ to radians: $\left(2 + \dfrac{20}{60}\right)° \cdot \dfrac{\pi}{180°} \approx 0.0407$.

    Using the arc length formula: $s \approx 0.0407(3955) \approx 161$ mi.

57. (a) Plot the data in $[0, 25, 2]$ by $[0, 70, 10]$. See Figure 57a.

    (b) The maximum monthly average temperature is $58°F$ and the minimum is $6°F$. The midpoint of these values

       is $0.5(58 + 6) = 32$, thus $d = 32$. Half the difference between these temperatures is $0.5(58 - 6) = 26$,

       thus $a = 26$. Since the temperature cycles every 12 months, $b = \dfrac{2\pi}{12} = \dfrac{\pi}{6}$. The maximum of the

$y = \cos x$ graph occurs when $x = 0,$ while the maximum in the table occurs when $x = 7.$ Thus $c = 7.$

The function is $f(x) = 26 \cos\left(\dfrac{\pi}{6}(x - 7)\right) + 32.$

(c) Graph $Y_1 = 26 \cos((\pi/6)(X - 7)) + 32$ together with the data in $[0, 25, 2]$ by $[0, 70, 10].$

See Figure 57c.

[0, 25, 2] by [0, 70, 10]   [0, 25, 2] by [0, 70, 10]

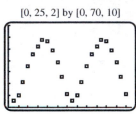

Figure 57a   Figure 57c

Extended and Discovery Exercises for Chapter 6

1. (a) Note that $\sin\theta = \dfrac{x_Q - x_P}{d}$ and $\cos\theta = \dfrac{y_Q - y_P}{d}$

    Solving for x_Q and y_Q we obtain $x_Q = d \sin\theta + x_P$ and $y_Q = d \cos\theta + y_P.$

    (b) $x_Q = 208 \sin 23.2° + 152 \approx 233.9$ and $y_Q = 208 \cos 23.2° + 186 \approx 377.2$

    The point is approximately $(233.9, 377.2).$

3. (a) $a = 0.5(74 - 49) = 12.5,\ b = \dfrac{2\pi}{12} = \dfrac{\pi}{6},\ c = 1,\ d = 0.5(74 + 49) = 61.5$

    The function is $f(x) = 12.5 \cos\left(\dfrac{\pi}{6}(x - 1)\right) + 61.5.$

    (b) Graph $Y_1 = 12.5 \cos((\pi/6)(X - 1)) + 61.5$ together with the data in $[0, 25, 2]$ by $[40, 80, 10].$

    See Figure 3.

    (c) The high temperatures in the Southern Hemisphere occur in January as opposed to July in the Northern Hemisphere. This affects the phase shift of $f.$

[0, 25, 2] by [40, 80, 10]

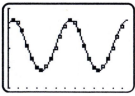

Figure 3

Chapters 1-6 Cumulative Review Exercises

1. Move the decimal point 5 place to the left, $125{,}000 = 1.25 \times 10^5$.

    Move the decimal point 3 places to the left, $4.67 \times 10^{-3} = 0.00467$

3. (a) See Figure 3a.

    (b) See Figure 3b.

    (c) See Figure 3c.

    (d) See Figure 3d.

    (f) See Figure 3f.

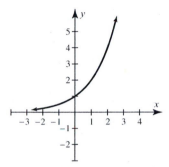

Figure 3a

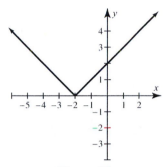

Figure 3b

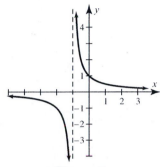

Figure 3c

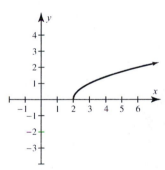

Figure 3d

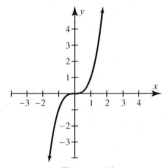

Figure 3f

5. (a) Since $4x^2 - 16 \neq 0 \Rightarrow 4x^2 \neq 16 \Rightarrow x^2 \neq 4 \Rightarrow x \neq -2, 2 \Rightarrow D = \{x \mid x \neq -2, x \neq 2\}$.

    (b) $f(-1) = \dfrac{-1 - 2}{4(-1)^2 - 16} = \dfrac{-3}{-12} = \dfrac{1}{4}$

    For $a \neq 1$, $f(2a) = \dfrac{2a - 2}{4(2a)^2 - 16} = \dfrac{2a - 2}{16a^2 - 16} = \dfrac{2(a - 1)}{16(a^2 - 1)} = \dfrac{2(a - 1)}{16(a - 1)(a + 1)} = \dfrac{1}{8(a + 1)}$

7. $f(1) = 10^1 = 10 \Rightarrow (1, 10)$, $f(2) = 10^2 = 100 \Rightarrow (2, 100)$, the average rate of change is $\dfrac{100 - 10}{2 - 1} =$

    $\dfrac{90}{1} = 90.$

9. The line $2x + 3y = 6 \Rightarrow 3y = -2x + 6 \Rightarrow y = \dfrac{-2}{3}x + 2$ has a slope $\dfrac{-2}{3}$. A line parallel to this line also

    has $m = \dfrac{-2}{3}$. Using point-slope form we get: $y = \dfrac{-2}{3}(x - 2) + (-3) \Rightarrow y = \dfrac{-2}{3}x - \dfrac{5}{3}$.

11. (a) $|4 - 5x| = 8 \Rightarrow 4 - 5x = -8$ or $4 - 5x = 8$; $4 - 5x = -8 \Rightarrow -5x = -12 \Rightarrow x = \dfrac{12}{5}$ or

$4 - 5x = 8 \Rightarrow -5x = 4 \Rightarrow x = -\dfrac{4}{5}$. Therefore $x = -\dfrac{4}{5}, \dfrac{12}{5}$.

(b) $2e^x - 1 = 27 \Rightarrow 2e^x = 28 \Rightarrow e^x = 14 \Rightarrow \ln e^x = \ln 14 \Rightarrow x = \ln 14 \Rightarrow x \approx 2.64$

(c) $x^3 - 3x^2 + 2x = 0 \Rightarrow x(x^2 - 3x + 2) = 0 \Rightarrow x(x - 2)(x - 1) = 0 \Rightarrow x = 0, 1, 2$

(e) $x^2 - x - 2 = 0 \Rightarrow (x - 2)(x + 1) = 0 \Rightarrow x = -1, 2$

(d) $\sqrt{2x - 1} = x - 2 \Rightarrow 2x - 1 = x^2 - 4x + 4 \Rightarrow x^2 - 6x + 5 = 0 \Rightarrow (x - 5)(x - 1) = 0 \Rightarrow$

$x = 1, 5$; Checking: $x = 1$ we get $\sqrt{2(1) - 1} = 1 - 2 \Rightarrow \sqrt{1} = -1$ which is false therefore $x = 5$.

(f) $\log_2(x + 1) = 16 \Rightarrow x + 1 = 2^{16} \Rightarrow x + 1 = 65{,}536 \Rightarrow x = 65{,}535$

13. (a) $-3(2 - x) < 4 - (2x + 1) \Rightarrow -6 + 3x < 3 - 2x \Rightarrow 5x < 9 \Rightarrow x < \dfrac{9}{5} \Rightarrow \left(-\infty, \dfrac{9}{5}\right)$

(b) $-3 \le 4 - 3x < 6 \Rightarrow -7 \le -3x < 2 \Rightarrow \dfrac{7}{3} \ge x > -\dfrac{2}{3} \Rightarrow \left(-\dfrac{2}{3}, \dfrac{7}{3}\right]$

(c) The solutions to $|4x - 3| \ge 9$ satisfy $x \le s_1$ or $x \ge s_2$ where s_1 and s_2 are the solutions to $|4x - 3| = 9$.

$|4x - 3| = 9$ is equivalent to $4x - 3 = -9 \Rightarrow x = -\dfrac{3}{2}$ and $4x - 3 = 9 \Rightarrow x = 3$.

The solution is $\left(-\infty, -\dfrac{3}{2}\right] \cup [3, \infty)$.

(d) For $x^2 - 5x + 4 \le 0$ first we solve $x^2 - 5x + 4 = 0 \Rightarrow (x - 4)(x - 1) = 0 \Rightarrow x = 1, 4$. These are

the boundary numbers and divide the number line into three sections: $(-\infty, 1], [1, 4]$, and $[4, \infty)$.

Testing a value in each section we get: $x = 0 \Rightarrow 0^2 - 5(0) + 4 \le 0 \Rightarrow 4 \le 0$, which is false;

$x = 2 \Rightarrow 2^2 - 5(2) + 4 \le 0 \Rightarrow 4 - 10 + 4 \le 0 \Rightarrow -2 \le 0$, which is true;

$x = 5 \Rightarrow 5^2 - 5(5) + 4 \le 0 \Rightarrow 25 - 25 + 4 \le 0 \Rightarrow 4 \le 0$, which is false.

Therefore the solution is; $[1, 4]$.

(e) For $t^3 - t > 0$ first solve $t^3 - t = 0 \Rightarrow t(t^2 - 1) = 0 \Rightarrow t(t + 1)(t - 1) = 0 \Rightarrow t = -1, 0, 1$. These

are the boundary numbers and divide the number line into four sections:

$(-\infty, -1), (-1, 0), (0, 1)$, and $(1, \infty)$.

Testing a value in each section we get: $t = -2 \Rightarrow -2^3 - (-2) > 0 \Rightarrow -6 > 0$, which is false;

$t = -\dfrac{1}{2} \Rightarrow \left(-\dfrac{1}{2}\right)^3 - \left(-\dfrac{1}{2}\right) > 0 \Rightarrow -\dfrac{1}{8} + \dfrac{1}{2} > 0 \Rightarrow \dfrac{3}{8} > 0$, which is true;

$t = \dfrac{1}{2} \Rightarrow \left(\dfrac{1}{2}\right)^3 - \dfrac{1}{2} > 0 \Rightarrow \dfrac{1}{8} - \dfrac{1}{2} > 0 \Rightarrow -\dfrac{3}{8} > 0$, which is false;

$t = 2 \Rightarrow 2^3 - 2 > 0 \Rightarrow 8 - 2 > 0 \Rightarrow 6 > 0$, which is true.

Therefore the solution is: $(-1, 0) \cup (1, \infty)$.

15. $2x^2 + 4x = 1 \Rightarrow 2(x^2 + 2x) = 1 \Rightarrow 2(x^2 + 2x + 1) = 1 + 2 \Rightarrow 2(x + 1)^2 = 3 \Rightarrow (x + 1)^2 = \dfrac{3}{2} \Rightarrow$

$x + 1 = \pm\sqrt{\dfrac{3}{2}} \Rightarrow x = -1 \pm \sqrt{\dfrac{3}{2}} \Rightarrow x = -1 \pm \dfrac{\sqrt{6}}{2} \Rightarrow \dfrac{-2 \pm \sqrt{6}}{2}$

17. (a) Increasing: $(-2, 0)$ and $(2, \infty)$; decreasing: $(-\infty, -2)$ and $(0, 2)$.

    (b) The graph crosses or touches the x-axis at $x \approx -2.8, 0, 2.8$.

    (c) The turning point coordinates: $(-2, -4), (0, 0)$, and $(2, -4)$.

    (d) Local minimum: -4; local maximum: 0.

19. (a) $\dfrac{5a^4 - 2a^2 + 4}{2a^2} = \dfrac{5a^4}{2a^2} - \dfrac{2a^2}{2a^2} + \dfrac{4}{2a^2} = \dfrac{5a^2}{2} - 1 + \dfrac{2}{a^2}$

    (b)

$$
\begin{array}{r}
x - 3 + \frac{4}{x^2+1} \\
x^2 + 1 \overline{) x^3 - 3x^2 + x + 1} \\
\underline{x^3 \qquad\quad + x} \\
-3x^2 \qquad + 1 \\
\underline{-3x^2 \qquad - 3} \\
4
\end{array}
$$

    The quotient is: $x - 3 + \dfrac{4}{x^2 + 1}$.

21. $3x - 7 \neq 0 \Rightarrow 3x \neq 7 \Rightarrow x \neq \dfrac{7}{3}$. Then $D = \left\{ x \,\middle|\, x \neq \dfrac{7}{3} \right\}$; the vertical asymptote is the denominator $=$

    $0 \Rightarrow x = \dfrac{7}{3}$; and the horizontal asymptote is the ratio of the lead coefficients $\Rightarrow y = \dfrac{2}{3}$.

23. (a) $(f + g)(2) = f(2) + g(2) = 2 + 3 = 5$

    (b) $(g/f)(4) = g(4)/f(4) = 1/0 \Rightarrow$ undefined

    (c) $(f \circ g)(3) = f(g(3)) = f(2) = 2$

    (d) $(f^{-1} \circ g)(1) = f^{-1}(g(1)) = f^{-1}(4) = 0$

25. (a) $f(2) = 2^2 + 3(2) - 2 = 4 + 6 - 2 = 8; \; g(2) = 2 - 2 = 0; \; (f + g)(2) = 8 + 0 = 8$

    (b) $(g \circ f)(1) = g(f(1)) \Rightarrow f(1) = 1^2 + 3(1) - 2 = 1 + 3 - 2 = 2$. Then $g(2) = 2 - 2 = 0 \Rightarrow$

      $(g \circ f)(1) = 0$

    (c) $f(x) = x^2 + 3x - 2$ and $g(x) = x - 2 \Rightarrow (f - g)(x) = x^2 + 3x - 2 - (x - 2) \Rightarrow$

      $(f - g)(x) = x^2 + 2x$

    (d) $(f \circ g)(x) = f(g(x)) \Rightarrow (f \circ g)(x) = (x - 2)^2 + 3(x - 2) - 2 \Rightarrow$

      $(f \circ g)(x) = x^2 - 4x + 4 + 3x - 6 - 2 \Rightarrow (f \circ g)(x) = x^2 - x - 4$

27. From the graph when $x = 0, y = \dfrac{1}{2}, \Rightarrow C = \dfrac{1}{2}$ and when $x = 1, y = 1 \Rightarrow$ using $y = Ca^x \Rightarrow 1 = \dfrac{1}{2}a^1 \Rightarrow$

    $a = 2$. So $C = \dfrac{1}{2}$ and $a = 2$.

29. (a) $\log 100 \Rightarrow 10^x = 100 \Rightarrow x = 2 \Rightarrow \log 100 = 2$

    (b) $\log_2 16 \Rightarrow 2^x = 16 \Rightarrow x = 4 \Rightarrow \log_2 16 = 4$

    (c) $\ln \dfrac{1}{e^2} \Rightarrow e^x = \dfrac{1}{e^2} \Rightarrow x = -2 \Rightarrow \ln \dfrac{1}{e^2} = -2$

    (d) $\log 4 + \log 25 = \log 4(25) = \log 100 \Rightarrow 10^x = 100 \Rightarrow x = 2 \Rightarrow \log 4 + \log 25 = 2$

31. $\log_3 125 = 4.395$

33. $150° \left(\dfrac{\pi}{180°} \right) = \dfrac{5\pi}{6}$

35. Convert 15° to radians $\Rightarrow$ $15° \left(\dfrac{\pi}{180°} \right) = \dfrac{15\pi}{180} = \dfrac{\pi}{12}$; $s = 3 \left(\dfrac{\pi}{12} \right) = \dfrac{3\pi}{12} = \dfrac{\pi}{4} \approx 0.79$ ft.

37. $\tan 30° = \dfrac{9}{a} \Rightarrow a \tan 30° = 9 \Rightarrow a = \dfrac{9}{\tan 30°} \Rightarrow a \approx 15.6$

39. $r = \sqrt{(-3)^2 + (-3)^2} = \sqrt{18} = 3\sqrt{2}$

$\sin \theta = \dfrac{-3}{3\sqrt{2}} = -\dfrac{1}{\sqrt{2}}$ $\qquad$ $\cos \theta = \dfrac{-3}{3\sqrt{2}} = -\dfrac{1}{\sqrt{2}}$ $\qquad$ $\tan \theta = \dfrac{-3}{-3} = 1$

$\csc \theta = \dfrac{3\sqrt{2}}{-3} = -\sqrt{2}$ $\qquad$ $\sec \theta = \dfrac{3\sqrt{2}}{-3} = -\sqrt{2}$ $\qquad$ $\cot \theta = \dfrac{-3}{-3} = 1$

41. Refer to Figure 41 and use the Pythagorean theorem to find the opposite side.

$5^2 + x^2 = 13^2, 25 + x^2 = 169, x^2 = 144, x = 12.$

$\sin \theta = -\dfrac{12}{13}$ $\qquad$ $\cos \theta = \dfrac{5}{13}$ $\qquad$ $\tan \theta = -\dfrac{12}{5}$

$\csc \theta = -\dfrac{13}{12}$ $\qquad$ $\sec \theta = \dfrac{13}{5}$ $\qquad$ $\cot \theta = -\dfrac{5}{12}$

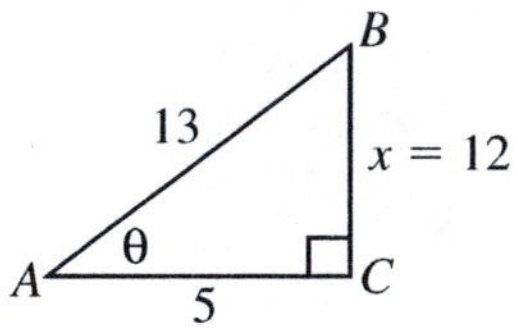

Figure 41

43. (a) See Figure 43a.

   (b) See Figure 43b.

   (c) See Figure 43c.

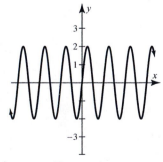

Figure 43a

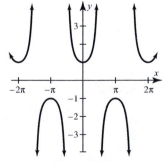

Figure 43b

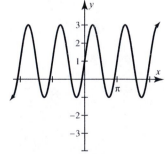

Figure 43c

45. $\sin \alpha = \dfrac{11}{20} \Rightarrow \alpha = \sin^{-1}\left(\dfrac{11}{20}\right) \Rightarrow \alpha \approx 33.4°$, $\cos \beta = \dfrac{11}{20} \Rightarrow \beta = \cos^{-1}\left(\dfrac{11}{20}\right) \Rightarrow \beta = 56.6$.

Using the Pythagorean theorem: $11^2 + b^2 = 20^2 \Rightarrow 121 + b^2 = 400 \Rightarrow b^2 = 279 \Rightarrow b \approx 16.7$.

47. $F = \dfrac{k}{x^2} \Rightarrow 150 = \dfrac{k}{(4000)^2} \Rightarrow k = 2{,}400{,}000{,}000$. Now, $F = \dfrac{2{,}400{,}000{,}000}{(10{,}000)^2} \Rightarrow F = \dfrac{2{,}400{,}000{,}000}{100{,}000{,}000} \Rightarrow$

$F = 24$ lbs.

49. (a) If $f(x) = \dfrac{5}{9}(x - 32)$, this is subtract 32 and multiply by $\dfrac{5}{9} \Rightarrow$ we divide by $\dfrac{5}{9}$, which is multiply by $\dfrac{9}{5}$,

and add 32 $\Rightarrow f(x)^{-1} = \dfrac{9}{5}x + 32$.

(b) Since $f(x)$ converted Fahrenheit to Celsius, $f(x)^{-1}$ converts Celsius to Fahrenheit.

51. (a) $300{,}000 = 200{,}000e^{k(3)} \Rightarrow \dfrac{3}{2} = e^{3k} \Rightarrow \ln\dfrac{3}{2} = \ln e^{3k} \Rightarrow \ln\dfrac{3}{2} = 3k \Rightarrow \dfrac{\ln\frac{3}{2}}{3} = k \Rightarrow k \approx 0.135$

The formula that models this is: $N(t) = 200{,}000e^{0.135t}$

(b) $N(5) = 200{,}000e^{0.135(5)} \Rightarrow N(5) = 200{,}000e^{0.675} \Rightarrow N(5) \approx 392{,}807$; after 5 hours there are about

393,000 bacteria per ml. in the sample.

(c) $500{,}000 = 200{,}000e^{0.135t} \Rightarrow \dfrac{5}{2} = e^{0.135t} \Rightarrow \ln\dfrac{5}{2} = \ln e^{0.135t} \Rightarrow \ln\dfrac{5}{2} = 0.135t \Rightarrow \dfrac{\ln\frac{5}{2}}{0.135} = t \Rightarrow$

$t \approx 6.8$ hrs.

53. The maximum monthly average temperature is 90° F and the minimum is 61° F. The midpoint of these values

is $0.5(90 + 61) = 75.5$, thus $d = 75.5$. Half the difference between these temperatures is

$0.5(90 - 61) = 14.5$, thus $a = 14.5$. Since the temperatures cycle every 12 months,

$b = \dfrac{2\pi}{12} = \dfrac{\pi}{6}$. The maximum of the $y = \sin x$ graph occurs when $x = \dfrac{\pi}{2}$, while the maximum in the table

occurs when $x = 7$. Thus $\dfrac{\pi}{6}(7 - c) = \dfrac{\pi}{2} \Rightarrow 7 - c = 3 \Rightarrow c = 4$.

The function $f(x) = 14.5\sin\left(\dfrac{\pi}{6}(x - 4)\right) + 75.5$.

Chapter 7: Trigonometric Identities and Equations

7.1: Fundamental Identities

1. $\cot \theta = \dfrac{1}{\tan \theta} = \dfrac{1}{\frac{1}{2}} = 2$

3. $\sec \theta = \dfrac{1}{\cos \theta} = \dfrac{1}{\frac{2}{7}} = \dfrac{7}{2}$

5. $\cos \theta = \dfrac{1}{\sec \theta} = \dfrac{1}{-4} = -\dfrac{1}{4}$

7. $\tan \theta = \dfrac{\sin \theta}{\cos \theta} = \dfrac{\sin \theta}{0} \Rightarrow$ undefined

9. $\dfrac{\tan \theta}{\cot \theta} = \dfrac{\tan \theta}{\frac{1}{\tan \theta}} = \tan^2 \theta = \dfrac{\sin^2 \theta}{\cos^2 \theta}$

11. $\dfrac{\cot^2 \theta}{\csc^2 \theta} = \dfrac{\frac{1}{\tan^2 \theta}}{\frac{1}{\sin^2 \theta}} = \dfrac{\frac{\cos^2 \theta}{\sin^2 \theta}}{\frac{1}{\sin^2 \theta}} = \cos^2 \theta$

13. $\dfrac{1}{\csc \theta} + \dfrac{1}{\sec \theta} = \sin \theta + \cos \theta$

15. $\tan \theta = \dfrac{\sin \theta}{\cos \theta} = \dfrac{\frac{3}{5}}{-\frac{4}{5}} = -\dfrac{3}{4}$ $\qquad$ $\csc \theta = \dfrac{1}{\sin \theta} = \dfrac{1}{\frac{3}{5}} = \dfrac{5}{3}$

$\quad$ $\sec \theta = \dfrac{1}{\cos \theta} = \dfrac{1}{-\frac{4}{5}} = -\dfrac{5}{4}$ $\qquad$ $\cot \theta = \dfrac{1}{\tan \theta} = \dfrac{1}{-\frac{3}{4}} = -\dfrac{4}{3}$

17. $\cos \theta = \sin \theta \cot \theta = -\dfrac{24}{25}\left(\dfrac{7}{24}\right) = -\dfrac{7}{25}$ $\qquad$ $\tan \theta = \dfrac{1}{\cot \theta} = \dfrac{1}{\frac{7}{24}} = \dfrac{24}{7}$

$\quad$ $\csc \theta = \dfrac{1}{\sin \theta} = \dfrac{1}{-\frac{24}{25}} = -\dfrac{25}{24}$ $\qquad$ $\sec \theta = \dfrac{1}{\cos \theta} = \dfrac{1}{-\frac{7}{25}} = -\dfrac{25}{7}$

19. $\tan \theta = \dfrac{\sin \theta}{\cos \theta} = \dfrac{-\frac{60}{61}}{-\frac{11}{61}} = \dfrac{60}{11}$ $\qquad$ $\cot \theta = \dfrac{\cos \theta}{\sin \theta} = \dfrac{-\frac{11}{61}}{-\frac{60}{61}} = \dfrac{11}{60}$

$\quad$ $\csc \theta = \dfrac{1}{\sin \theta} = \dfrac{1}{-\frac{60}{61}} = -\dfrac{61}{60}$ $\qquad$ $\sec \theta = \dfrac{1}{\cos \theta} = \dfrac{1}{-\frac{11}{61}} = -\dfrac{61}{11}$

21. $\sin \theta = \dfrac{1}{\csc \theta} = \dfrac{1}{\sqrt{2}}$ $\qquad$ $\cos \theta = \dfrac{1}{\sec \theta} = -\dfrac{1}{\sqrt{2}}$

$\quad$ $\tan \theta = \dfrac{\sin \theta}{\cos \theta} = \dfrac{\frac{1}{\sqrt{2}}}{-\frac{1}{\sqrt{2}}} = -\dfrac{1}{1} = -1$ $\qquad$ $\cot \theta = \dfrac{\cos \theta}{\sin \theta} = \dfrac{-\frac{1}{\sqrt{2}}}{\frac{1}{\sqrt{2}}} = -\dfrac{1}{1} = -1$

23. $\sec \theta \cos \theta = \dfrac{1}{\cos \theta} \cdot \cos \theta = 1$

25. $\sin \theta \csc \theta = \sin \theta \cdot \dfrac{1}{\sin \theta} = 1$

27. $\left(\sin^2 \theta + \cos^2 \theta\right)^3 = 1^3 = 1$

29. $1 - \sin^2 \theta = 1 - (1 - \cos^2 \theta) = \cos^2 \theta$

31. $\sec^2 \theta - 1 = (1 + \tan^2 \theta) - 1 = \tan^2 \theta$

33. $\dfrac{\sin(-\theta)}{\cos(-\theta)} = \dfrac{-\sin \theta}{\cos \theta} = -\tan \theta$

35. $\dfrac{\sin^2 \theta + \cos^2 \theta}{\cos \theta} = \dfrac{1}{\cos \theta} = \sec \theta$

37. $\dfrac{\sec^2(-\theta)}{\csc^2 \theta} = \dfrac{\sec^2 \theta}{\csc^2 \theta} = \left(\dfrac{\sec \theta}{\csc \theta}\right)^2 = \left(\dfrac{\frac{1}{\cos \theta}}{\frac{1}{\sin \theta}}\right)^2 = \left(\dfrac{1}{\cos \theta} \cdot \dfrac{\sin \theta}{1}\right)^2 = \left(\dfrac{\sin \theta}{\cos \theta}\right)^2 = \tan^2 \theta$

39. $\dfrac{\cot x}{\csc x} = \dfrac{\frac{\cos x}{\sin x}}{\frac{1}{\sin x}} = \dfrac{\cos x}{\sin x} \cdot \dfrac{\sin x}{1} = \cos x$

41. $(\sin^2 x)(1 + \cot^2 x) = (\sin^2 x)(\csc^2 x) = (\sin^2 x)\left(\dfrac{1}{\sin^2 x}\right) = 1$

43. $\sec(-x) + \csc(-x) = \sec x + (-\csc x) = \sec x - \csc x$

45. Yes, since $\sec \theta \cot \theta = \left(\dfrac{1}{\cos \theta}\right)\left(\dfrac{\cos \theta}{\sin \theta}\right) = \dfrac{1}{\sin \theta} = \csc \theta$. For numerical support, table the following in degree mode: $Y_1 = (1/\cos(X))(1/\tan(X))$ and $Y_2 = (1/\sin(X))$ starting at $x = 0$, incrementing by 50. See Figure 45.

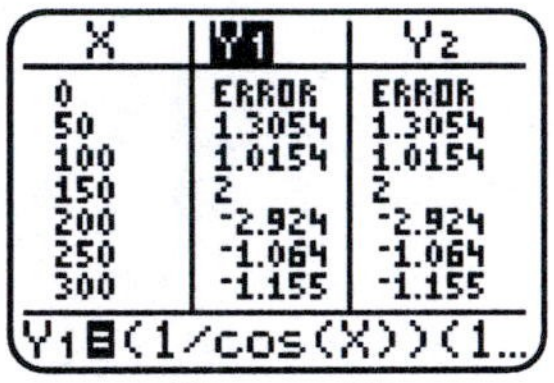

X	Y1	Y2
0	ERROR	ERROR
50	1.3054	1.3054
100	1.0154	1.0154
150	2	2
200	-2.924	-2.924
250	-1.064	-1.064
300	-1.155	-1.155

Y1 ＝ (1/cos(X))(1...

Figure 45

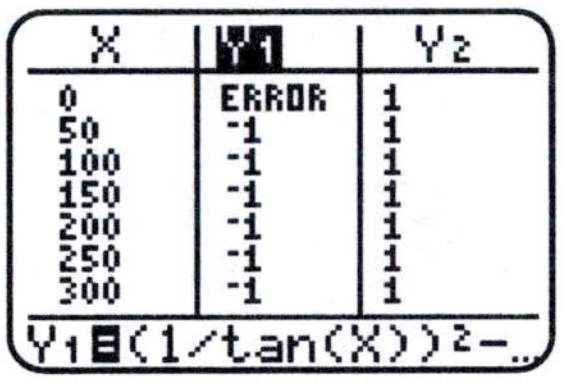

X	Y1	Y2
0	ERROR	1
50	-1	1
100	-1	1
150	-1	1
200	-1	1
250	-1	1
300	-1	1

Y1 ＝ (1/tan(X))²-...

Figure 47

47. No, since $1 + \cot^2 \theta = \csc^2 \theta \Rightarrow \cot^2 \theta - \csc^2 \theta = -1 \neq 1$. For numerical support, table the following in degree mode: $Y_1 = (1/\tan(X))^2 - (1/\sin(X))^2$ and $Y_2 = 1$ starting at $x = 0$, incrementing by 50. See Figure 47.

49. If $\sin \theta < 0$ and $\cos \theta > 0$ then any point (x, y) on the terminal side of θ must satisfy $y < 0$ and $x > 0$. Thus, θ is contained in Quadrant IV. To support this result numerically, table the following in degree mode: $Y_1 = \sin(X)$ and $Y_2 = \cos(X)$ starting at $x = 270$, incrementing by 15. See Figure 49.

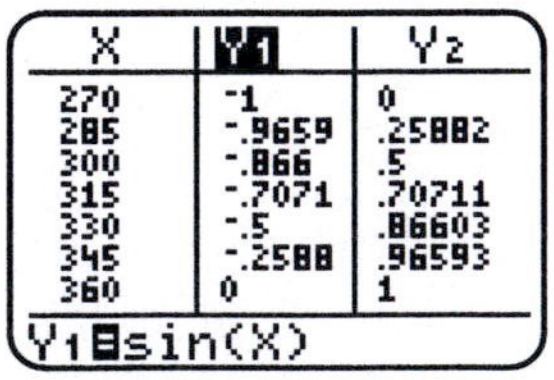

X	Y1	Y2
270	-1	0
285	-.9659	.25882
300	-.866	.5
315	-.7071	.70711
330	-.5	.86603
345	-.2588	.96593
360	0	1

Y1 ＝ sin(X)

Figure 49

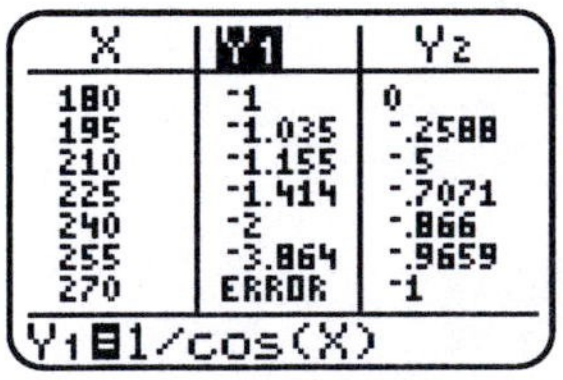

X	Y1	Y2
180	-1	0
195	-1.035	-.2588
210	-1.155	-.5
225	-1.414	-.7071
240	-2	-.866
255	-3.864	-.9659
270	ERROR	-1

Y1 ＝ 1/cos(X)

Figure 51

51. If $\sec\theta < 0$ and $\sin\theta < 0$ then any point (x, y) on the terminal side of θ must satisfy the following conditions:

$y < 0$ and $\dfrac{r}{x} < 0 \Rightarrow x < 0$. Thus, θ is contained in quadrant III. For numerical support, table the following

in degree mode: $Y_1 = 1/\cos(X)$ and $Y_2 = \sin(X)$ starting at $x = 180$, incrementing by 15. See Figure 51.

53. If $\cot\theta < 0$ and $\sin\theta > 0$ then any point (x, y) on the terminal side of θ must satisfy the following conditions:

$y > 0$ and $\dfrac{x}{y} < 0 \Rightarrow x < 0$. Thus, θ is contained in quadrant II. For numerical support, table the following

in degree mode: $Y_1 = 1/\tan(X)$ and $Y_2 = \sin(X)$ starting at $x = 90$, incrementing by 15. See Figure 53.

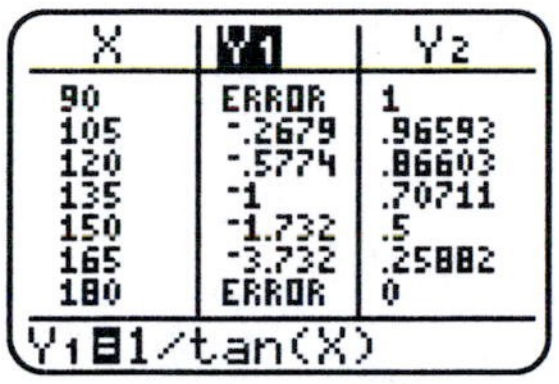

X	Y1	Y2
90	ERROR	1
105	-.2679	.96593
120	-.5774	.86603
135	-1	.70711
150	-1.732	.5
165	-3.732	.25882
180	ERROR	0

$Y_1 \blacksquare 1/\tan(X)$

Figure 53

55. $\sin^2\theta + \cos^2\theta = 1 \Rightarrow \sin\theta = \pm\sqrt{1 - \cos^2\theta}$ and since $\sin\theta < 0$ we know $\sin\theta = -\sqrt{1 - \cos^2\theta}$.

$$\sin\theta = -\sqrt{1 - \cos^2\theta} = -\sqrt{1 - \left(\frac{1}{2}\right)^2} = -\sqrt{1 - \frac{1}{4}} = -\sqrt{\frac{3}{4}} = -\frac{\sqrt{3}}{2} \qquad \tan\theta = \frac{\sin\theta}{\cos\theta} = \frac{-\frac{\sqrt{3}}{2}}{\frac{1}{2}} = -\sqrt{3}$$

$$\csc\theta = \frac{1}{\sin\theta} = \frac{1}{-\frac{\sqrt{3}}{2}} = -\frac{2}{\sqrt{3}} \qquad \sec\theta = \frac{1}{\cos\theta} = \frac{1}{\frac{1}{2}} = 2 \qquad \cot\theta = \frac{1}{\tan\theta} = \frac{1}{-\sqrt{3}} = -\frac{1}{\sqrt{3}}$$

57. $1 + \cot^2\theta = \csc^2\theta \Rightarrow \csc\theta = \pm\sqrt{1 + \cot^2\theta}$ and since $\csc\theta < 0$, $\csc\theta = -\sqrt{1 + \cot^2\theta}$

$$\cot\theta = \frac{1}{\tan\theta} = \frac{1}{-\frac{11}{60}} = -\frac{60}{11}$$

$$\csc\theta = -\sqrt{1 + \cot^2\theta} = -\sqrt{1 + \left(-\frac{60}{11}\right)^2} = -\sqrt{1 + \frac{3600}{121}} = -\sqrt{\frac{3721}{121}} = -\frac{61}{11}$$

$$\sin\theta = \frac{1}{\csc\theta} = \frac{1}{-\frac{61}{11}} = -\frac{11}{61} \qquad \cos\theta = \frac{\sin\theta}{\tan\theta} = \frac{-\frac{11}{61}}{-\frac{11}{60}} = \frac{60}{61} \qquad \sec\theta = \frac{1}{\cos\theta} = \frac{1}{\frac{60}{61}} = \frac{61}{60}$$

59. $\sin^2\theta + \cos^2\theta = 1 \Rightarrow \cos\theta = \pm\sqrt{1 - \sin^2\theta}$ and since $\cos\theta > 0$, $\cos\theta = \sqrt{1 - \sin^2\theta}$

$$\cos\theta = \sqrt{1 - \sin^2\theta} = \sqrt{1 - \left(\frac{7}{25}\right)^2} = \sqrt{1 - \frac{49}{625}} = \sqrt{\frac{576}{625}} = \frac{24}{25} \qquad \tan\theta = \frac{\sin\theta}{\cos\theta} = \frac{\frac{7}{25}}{\frac{24}{25}} = \frac{7}{24}$$

$$\csc\theta = \frac{1}{\sin\theta} = \frac{1}{\frac{7}{25}} = \frac{25}{7} \qquad \sec\theta = \frac{1}{\cos\theta} = \frac{1}{\frac{24}{25}} = \frac{25}{24} \qquad \cot\theta = \frac{1}{\tan\theta} = \frac{1}{\frac{7}{24}} = \frac{24}{7}$$

61. If $\sin\theta < 0$ and $\sec\theta < 0$, θ lies in quadrant III, $\sin^2\theta + \cos^2\theta = 1 \Rightarrow \left(-\frac{1}{3}\right)^2 + \cos^2\theta = 1 \Rightarrow$

$$\cos^2\theta = 1 - \frac{1}{9} = \frac{8}{9} \Rightarrow \cos\theta = \pm\frac{\sqrt{8}}{3} \Rightarrow \cos\theta = \pm\frac{2\sqrt{2}}{3} \Rightarrow \cos\theta = -\frac{2\sqrt{2}}{3} \text{ in quadrant III.}$$

$$\tan\theta = \frac{\sin\theta}{\cos\theta} = \frac{-\frac{1}{3}}{-\frac{2\sqrt{2}}{3}} = \frac{1}{2\sqrt{2}} \qquad \cot\theta = \frac{\cos\theta}{\sin\theta} = \frac{-\frac{2\sqrt{2}}{3}}{-\frac{1}{3}} = 2\sqrt{2} = \sqrt{8} \qquad \csc\theta = \frac{1}{\sin\theta} = \frac{1}{-\frac{1}{3}} = -3$$

$$\csc\theta = \frac{1}{\cos\theta} = \frac{1}{-\frac{2\sqrt{2}}{3}} = -\frac{3}{2\sqrt{2}} = -\frac{3}{\sqrt{8}}$$

63. $\sin^2\theta + \cos^2\theta = 1 \Rightarrow \sin\theta = \pm\sqrt{1 - \cos^2\theta}$ and since θ is in quadrant IV, $\sin\theta = -\sqrt{1 - \cos^2\theta}$.

$$\cos\theta = \frac{1}{\sec\theta} = \frac{12}{37} \qquad \sin\theta = -\sqrt{1 - \cos^2\theta} = -\sqrt{1 - \left(\frac{12}{37}\right)^2} = -\frac{35}{37}$$

$$\tan\theta = \frac{\sin\theta}{\cos\theta} = \frac{-\frac{35}{37}}{\frac{12}{37}} = -\frac{35}{12} \qquad \csc\theta = \frac{1}{\sin\theta} = \frac{1}{-\frac{35}{37}} = -\frac{37}{35} \qquad \cot\theta = \frac{\cos\theta}{\sin\theta} = \frac{\frac{12}{37}}{-\frac{35}{37}} = -\frac{12}{35}$$

65. $\sin^2\theta + \cos^2\theta = 1 \Rightarrow \cos\theta = \pm\sqrt{1 - \sin^2\theta}$ and since θ is in quadrant II, $\cos\theta = -\sqrt{1 - \sin^2\theta}$.

$$\sin\theta = \frac{1}{\csc\theta} = \frac{1}{\frac{7}{3}} = \frac{3}{7} \qquad \cos\theta = -\sqrt{1 - \sin^2\theta} = -\sqrt{1 - \left(\frac{3}{7}\right)^2} = -\sqrt{1 - \frac{9}{49}} = -\sqrt{\frac{40}{49}} = -\frac{\sqrt{40}}{7}$$

$$\tan\theta = \frac{\sin\theta}{\cos\theta} = \frac{\frac{3}{7}}{-\frac{\sqrt{40}}{7}} = -\frac{3}{\sqrt{40}} \qquad \sec\theta = \frac{1}{\cos\theta} = \frac{1}{-\frac{\sqrt{40}}{7}} = -\frac{7}{\sqrt{40}} \qquad \cot\theta = \frac{1}{\tan\theta} = \frac{1}{-\frac{3}{\sqrt{40}}} = -\frac{\sqrt{40}}{3}$$

67. If $\csc x > 0$, $\sin x > 0$. Then $\sin^2 x + \cos^2 x = 1 \Rightarrow \sin^2 x = 1 - \cos^2 x \Rightarrow \sin x = \sqrt{1 - \cos^2 x}$, thus

$$\tan x = \frac{\sin x}{\cos x} \Rightarrow \tan x = \frac{\sqrt{1 - \cos^2\theta}}{\cos x}.$$

69. If $\cos x < 0$, $\sec x < 0$. Then $\sec^2 x = 1 + \tan^2 x \Rightarrow \sec x = -\sqrt{1 + \tan^2 x}$ and

$$\sin x = \frac{\tan x}{\sec x} \Rightarrow \sin x = -\frac{\tan x}{\sqrt{1 + \tan^2 x}}.$$

71. If $\cot x < 0$, $\cot^2 x = \csc^2 x - 1 \Rightarrow \cot x = -\sqrt{\csc^2 x - 1}$ and

$$\cos x = \frac{\cot x}{\csc x} \Rightarrow \cos x = -\frac{\sqrt{\csc^2 x - 1}}{\csc x}.$$

73. $\cos^2\theta + \sin^2\theta = 1 \Rightarrow \cos\theta = \pm\sqrt{1 - \sin^2\theta}$ and since θ is acute, $\cos\theta = \sqrt{1 - \sin^2\theta}$,

thus, $\cos\theta = \sqrt{1 - x^2} = \sqrt{1 - 0.5126^2} \approx 0.8586$.

75. $1 + \cot^2\theta = \csc^2\theta \Rightarrow \frac{1}{\sin^2\theta} = 1 + \cot^2\theta \Rightarrow \sin^2\theta = \frac{1}{1 + \cot^2\theta} \Rightarrow \sin\theta = \pm\sqrt{\frac{1}{1 + \cot^2\theta}}$

Since θ is in quadrant III, $\sin\theta = -\sqrt{\frac{1}{1 + \cot^2\theta}}$, $\sin\theta = -\sqrt{\frac{1}{1 + x^2}} = -\sqrt{\frac{1}{1 + 0.5126^2}} \approx -0.8899$.

77. If θ is acute, θ lies in quadrant I, $\sin^2\theta + \cos^2\theta = 1 \Rightarrow \cos^2\theta = 1 - \sin^2\theta \Rightarrow \cos\theta = \sqrt{1 - \sin^2\theta}$,

$$\tan\theta = \frac{\sin\theta}{\sqrt{1 - \sin^2\theta}}. \text{ Since } \sin\theta = x \Rightarrow \tan\theta = \frac{x}{\sqrt{1 - x^2}}.$$

79. Since sine is an odd function, $\sin(-13°) = -\sin 13°$.

81. Since tangent is an odd function, $\tan\left(-\dfrac{\pi}{11}\right) = -\tan\dfrac{\pi}{11}$.

83.   Since secant is an even function, $\sec\left(-\dfrac{2\pi}{5}\right) = \sec\dfrac{2\pi}{5}$.

85. (a) Graph $Y_1 = (\cos X)^2$ in degree mode in $[-90°, 90°, 45°]$ by $[-1, 2, 1]$. See Figure 85.

   I is at maximum when $\theta = 0$.

   (b) $I = k\cos^2\theta = k\left(1 - \sin^2\theta\right)$

$[-90°, 90°, 45°]$ by $[-1, 2, 1]$ $[0, 10^{-6}, 10^{-7}]$ by $[-1, 4, 1]$

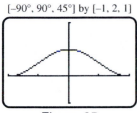
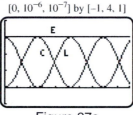
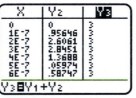

| | Figure 85 | | | Figure 87a | | | Figure 87b | |

87. (a) Graph $Y_1 = 3\,(\cos\,(6000000X))^2$, $Y_2 = 3\,(\sin\,(6000000X))^2$ and $Y_3 = Y_1 + Y_2$

   in $[0, 10^{-6}, 10^{-7}]$ by $[-1, 4, 1]$. See Figure 87a. The total energy E (the sum of L and C) is always 3.

   (b) Table Y_1, Y_2 and Y_3 starting at $x = 0$, incrementing by 10^{-7}. See Figure 87b.

   (c) $E(t) = 3\cos^2\,(6000000t) + 3\sin^2\,(6000000t) = 3\,(\cos^2\,(6000000t) + \sin^2\,(6000000t)) =$

   $3(1) = 3$

89. (a) In radian mode, the graph has y-axis symmetry. The monthly high temperatures x months before and x

   months after July are equal. See Figure 89a.

   (b) A table of Y_1 is shown in Figure 89b. f is an even function since the sign of the input does not affect the

   output.

   (c) Symbolically this symmetry can be expressed as $f(-x) = f(x)$.

$[-6, 6, 1]$ by $[0, 100, 10]$ $[0, 2, 0.5]$ by $[-1, 40, 8]$

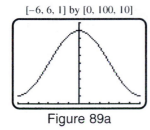
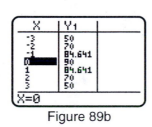
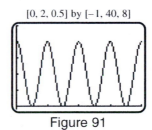

| | Figure 89a | | | Figure 89b | | | Figure 91 | |

91. (a) $P = ky^2 = k(4\cos\,(2\pi t))^2 = 16k\cos^2\,(2\pi t)$

   (b) Graph $Y_1 = 32\,(\cos\,(2\pi X))^2$ in $[0, 2, 0.5]$ by $[-1, 40, 8]$. See Figure 91. P has a maximum value of 32

   when $t = 0, 0.5, 1, 1.5, 2$ and P has a minimum value of 0 when $t = 0.25, 0.75, 1.25, 1.75$. The spring is

   either stretched or compressed the most when P is maximum.

   (c) $P = 16k\cos^2\,(2\pi t) = 16k\,(1 - \sin^2\,(2\pi t))$

7.2: Verifying Identities

1. (a) $(1 + x)(1 - x) = 1 - x^2$

    (b) $(1 + \sin \theta)(1 - \sin \theta) = 1 - \sin^2 \theta = \cos^2 \theta$

3. (a) $x(x - 1) = x^2 - x$

    (b) $\sec \theta \, (\sec \theta - 1) = \sec^2 \theta - \sec \theta$

5. $\dfrac{x}{1} \cdot \dfrac{y}{x} = \dfrac{y}{1} = y$

7. (a) $x^2 + 2x + 1 = (x + 1)(x + 1)$

    (b) $\cos^2 \theta + 2 \cos \theta + 1 = (\cos \theta + 1)(\cos \theta + 1)$

9. (a) $x^2 - 2x = x(x - 2)$

    (b) $\sec^2 t - 2 \sec t = \sec t \, (\sec t - 2)$

11. (a) $x + x^3 = x\left(1 + x^2\right)$

    (b) $\tan \theta + \tan^3 \theta = \tan \theta \left(1 + \tan^2 \theta\right) = \tan \theta \sec^2$

13. (a) $\dfrac{1}{1 - x} + \dfrac{1}{1 + x} = \dfrac{1 + x}{(1 - x)(1 + x)} + \dfrac{1 - x}{(1 - x)(1 + x)} = \dfrac{1 + x + 1 - x}{(1 - x)(1 + x)} = \dfrac{2}{1 - x^2}$

    (b) $\dfrac{1}{1 - \cos \theta} + \dfrac{1}{1 + \cos \theta} = \dfrac{1 + \cos \theta}{(1 - \cos \theta)(1 + \cos \theta)} + \dfrac{1 - \cos \theta}{(1 - \cos \theta)(1 + \cos \theta)} =$

    $\dfrac{1 + \cos \theta + 1 - \cos \theta}{(1 - \cos \theta)(1 + \cos \theta)} = \dfrac{2}{1 - \cos^2 \theta} = \dfrac{2}{\sin^2 \theta} = 2 \csc^2 \theta$

15. (a) $\dfrac{x}{y} + \dfrac{y}{x} = \dfrac{x^2}{xy} + \dfrac{y^2}{xy} = \dfrac{x^2 + y^2}{xy}$

    (b) $\dfrac{\cos t}{\sin t} + \dfrac{\sin t}{\cos t} = \dfrac{\cos^2 t}{\cos t \, \sin t} + \dfrac{\sin^2 t}{\cos t \, \sin t} = \dfrac{\cos^2 t + \sin^2 t}{\cos t \, \sin t} = \dfrac{1}{\cos t \, \sin t} = \sec t \, \csc t$

17. (a) $\dfrac{1}{\frac{1}{y^2}} + \dfrac{1}{\frac{1}{x^2}} = y^2 + x^2$

    (b) $\dfrac{1}{\csc^2 t} + \dfrac{1}{\sec^2 t} = \sin^2 t + \cos^2 t = 1$

19. (a) $\dfrac{\frac{x}{y}}{\frac{1}{y}} = \dfrac{x}{y} \cdot \dfrac{y}{1} = x$

    (b) $\dfrac{\cot \theta}{\csc \theta} = \dfrac{\frac{\cos \theta}{\sin \theta}}{\frac{1}{\sin \theta}} = \dfrac{\cos \theta}{\sin \theta} \cdot \dfrac{\sin \theta}{1} = \cos \theta$

21. $\cos \theta \, \tan \theta = \cos \theta \cdot \dfrac{\sin \theta}{\cos \theta} = \sin \theta$

23. $\tan \theta \, (\cos \theta - \csc \theta) = \dfrac{\sin \theta}{\cos \theta}\left(\cos \theta - \dfrac{1}{\sin \theta}\right) = \sin \theta - \dfrac{1}{\cos \theta} = \sin \theta - \sec \theta$

25. $(1 + \tan t)^2 = 1 + 2 \tan t + \tan^2 t = 2 \tan t + (1 + \tan^2 t) = 2 \tan t + \sec^2 t$

27. $\dfrac{\csc^2 \theta - 1}{\csc^2 \theta} = \dfrac{\csc^2 \theta}{\csc^2 \theta} - \dfrac{1}{\csc^2 \theta} = 1 - \sin^2 \theta = \cos^2 \theta$

29. $1 - \tan^2 \theta = (1 - \tan \theta)(1 + \tan \theta)$

31. $\sec^2 t - \sec t - 6 = (\sec t - 3)(\sec t + 2)$

33. $\tan^4 \theta + 3 \tan^2 \theta + 2 = (\tan^2 \theta + 1)(\tan^2 \theta + 2) = \sec^2 \theta \, (\tan^2 \theta + 2)$

35. $\csc^2 \theta - \cot^2 \theta = (1 + \cot^2 \theta) - \cot^2 \theta = 1$

37. $(1 - \sin t)^2 = (1 - \sin t)(1 - \sin t) = 1 - \sin t - \sin t + \sin^2 t = 1 - 2 \sin t + \sin^2 t$

39. $\dfrac{\sin t + \cos t}{\sin t} = \dfrac{\sin t}{\sin t} + \dfrac{\cos t}{\sin t} = 1 + \cot t$

41. $\sec^2 \theta - 1 = (1 + \tan^2 \theta) - 1 = \tan^2 \theta$

43. $\dfrac{\tan^2 t}{\sec t} = \dfrac{\sec^2 t - 1}{\sec t} = \dfrac{\sec^2 t}{\sec t} - \dfrac{1}{\sec t} = \sec t - \cos t$

45. $\cot x + 1 = \dfrac{\cos x}{\sin x} + \dfrac{\sin x}{\sin x} = \dfrac{1}{\sin x}(\cos x + \sin x) = \csc x \, (\cos x + \sin x)$

47. $\dfrac{\sec t}{1 + \sec t} = \dfrac{\sec t}{1 + \sec t} \cdot \dfrac{\cos t}{\cos t} = \dfrac{1}{\cos t + 1}$

49. $(\sec t - 1)(\sec t + 1) = \sec^2 t - 1 = \tan^2 t$

51. $\dfrac{1 - \sin^2 \theta}{\cos \theta} = \dfrac{\cos^2 \theta}{\cos \theta} = \cos \theta$

53. $\dfrac{\sec t}{\tan t} - \dfrac{\tan t}{\sec t} = \dfrac{\sec^2 t - \tan^2 t}{\sec t \tan t} = \dfrac{(1 + \tan^2 t) - \tan^2 t}{\sec t \tan t} = \dfrac{1}{\sec t \tan t} = \cos t \cot t$

55. $\dfrac{\cot^2 t}{\csc t + 1} = \dfrac{\csc^2 t - 1}{\csc t + 1} = \dfrac{(\csc t - 1)(\csc t + 1)}{\csc t + 1} = \csc t - 1$

57. $\dfrac{\cot t}{\cot t + 1} = \dfrac{\cot t}{\cot t + 1} \cdot \dfrac{\tan t}{\tan t} = \dfrac{\cot t \tan t}{\cot t \tan t + \tan t} = \dfrac{1}{1 + \tan t}$

59. $\dfrac{1}{1 - \sin t} + \dfrac{1}{1 + \sin t} = \dfrac{(1 + \sin t) + (1 - \sin t)}{(1 - \sin t)(1 + \sin t)} = \dfrac{2}{1 - \sin^2 t} = \dfrac{2}{\cos^2 t} = 2 \sec^2 t$

61. $\dfrac{\csc t + \cot t}{\csc t - \cot t} = \dfrac{\csc t + \cot t}{\csc t - \cot t} \cdot \dfrac{\csc t + \cot t}{\csc + \cot t} = \dfrac{(\csc t + \cot t)^2}{\csc^2 t - \cot^2 t} = \dfrac{(\csc t + \cot t)^2}{\csc^2 t - (\csc^2 t - 1)} = (\csc t + \cot t)^2$

63. $\dfrac{\cos^2 t}{1 - \sin t} = \dfrac{1 - \sin^2 t}{1 - \sin t} = \dfrac{(1 + \sin t)(1 - \sin t)}{1 - \sin t} = 1 + \sin t$

65. $\dfrac{1}{1 + \sin \theta} = \dfrac{1}{1 + \sin \theta} \cdot \dfrac{1 - \sin \theta}{1 - \sin \theta} = \dfrac{1 - \sin \theta}{1 - \sin^2 \theta} = \dfrac{1 - \sin \theta}{\cos^2 \theta}$

67. $\sqrt{1 - \sin^2 \theta} = \sqrt{\cos^2 \theta} = \pm \cos \theta = \cos \theta$, since θ is acute.

69. $\dfrac{1 + 2 \sin x + \sin^2 x}{\cos^2 x} = \dfrac{(1 + \sin x)^2}{1 - \sin^2 x} = \dfrac{(1 + \sin x)(1 + \sin x)}{(1 - \sin x)(1 + \sin x)} = \dfrac{1 + \sin x}{1 - \sin x}$

71. $(1 - \cos^2 x)(1 + \cos^2 x) = \sin^2 x \, (1 + (1 - \sin^2 x)) = \sin^2 x \, (2 - \sin^2 x) = 2 \sin^2 x - \sin^4 x$

73. $\cot \theta \sin \theta = \dfrac{\cos \theta}{\sin \theta} \cdot \sin \theta = \cos \theta$

Graph $Y_1 = (\cos (X)/\sin (X)) \sin (X)$ and $Y_2 = \cos (X)$ in $[-2\pi, 2\pi, \pi/2]$ by $[-4, 4, 1]$.

Table Y_1 and Y_2 together in degree mode starting at $x = 0$, incrementing by 50.

Graph Y_1 is shown in Figure 73a. Graph Y_2 is shown in Figure 73b. The table is shown in Figure 73c.

$[-2\pi, 2\pi, \pi/2]$ by $[-4, 4, 1]$ $[-2\pi, 2\pi, \pi/2]$ by $[-4, 4, 1]$

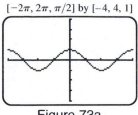   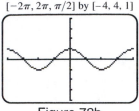   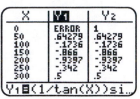

Figure 73a   Figure 73b   Figure 73c

75. $(1 - \cos^2 \theta)(1 + \tan^2 \theta) = \sin^2 \theta \ \sec^2 \theta = \sin^2 \theta \cdot \dfrac{1}{\cos^2 \theta} = \dfrac{\sin^2 \theta}{\cos^2 \theta} = \tan^2 \theta$

Graph $Y_1 = (1 - (\cos (X))^\wedge 2)(1 + (\tan (X))^\wedge 2)$ and $Y_2 = (\tan (X))^\wedge 2$ in $[-2\pi, 2\pi, \pi/2]$ by $[-4, 4, 1]$.

Table Y_1 and Y_2 together in degree mode starting at $x = 0$, incrementing by 50.

Graph Y_1 is shown in Figure 75a. Graph Y_2 is shown in Figure 75b. The table is shown in Figure 75c.

$[-2\pi, 2\pi, \pi/2]$ by $[-4, 4, 1]$ $[-2\pi, 2\pi, \pi/2]$ by $[-4, 4, 1]$

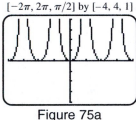   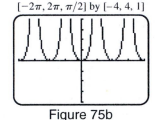   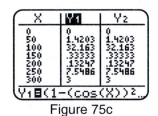

Figure 75a   Figure 75b   Figure 75c

77. $\cos t \left(\tan t - \sec t\right) = \cos t \left(\dfrac{\sin t}{\cos t} - \dfrac{1}{\cos t}\right) = \sin t - 1$

Graph $Y_1 = \cos (X)(\tan (X) - 1/\cos (X))$ and $Y_2 = \sin (X) - 1$ in $[-2\pi, 2\pi, \pi/2]$ by $[-4, 4, 1]$.

Table Y_1 and Y_2 together in degree mode starting at $x = 0$, incrementing by 50.

Graph Y_1 is shown in Figure 77a. Graph Y_2 is shown in Figure 77b. The table is shown in Figure 77c.

$[-2\pi, 2\pi, \pi/2]$ by $[-4, 4, 1]$ $[-2\pi, 2\pi, \pi/2]$ by $[-4, 4, 1]$

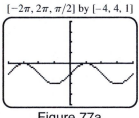   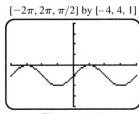   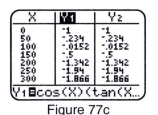

Figure 77a   Figure 77b   Figure 77c

79. $\dfrac{\tan (-\theta)}{\sin (-\theta)} = \dfrac{-\tan \theta}{-\sin \theta} = \dfrac{\frac{\sin \theta}{\cos \theta}}{\frac{\sin \theta}{1}} = \dfrac{\sin \theta}{\cos \theta} \cdot \dfrac{1}{\sin \theta} = \dfrac{1}{\cos \theta} = \sec \theta$

Graph $Y_1 = \tan (-X)/\sin (-X)$ and $Y_2 = 1/\cos (X)$ in $[-2\pi, 2\pi, \pi/2]$ by $[-4, 4, 1]$.

Table Y_1 and Y_2 together in degree mode starting at $x = 0$, incrementing by 50.

Graph Y_1 is shown in Figure 79a. Graph Y_2 is shown in Figure 79b. The table is shown in Figure 79c.

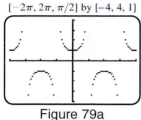

[$-2\pi, 2\pi, \pi/2$] by [$-4, 4, 1$]

Figure 79a

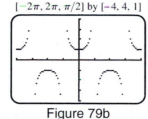

[$-2\pi, 2\pi, \pi/2$] by [$-4, 4, 1$]

Figure 79b

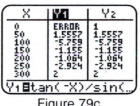

Figure 79c

81. (a) $W(t) = 5\cos^2(120\pi t) = 5(1 - \sin^2(120\pi t))$

$V = 25\sin(120\pi t)$ is maximum when $\sin(120\pi t) = 1$ and is minimum when $\sin(120\pi t) = -1$.

Thus, $W = 5(1 - (\pm 1)^2) = 0$.

(b) Graph $Y_1 = 5(\cos(120\pi X))^{\wedge}2$ and $Y_2 = 25\sin(120\pi X)$ in $[0, 1/15, 1/60]$ by $[-30, 30, 10]$.

See Figure 81. The wattage $Y_1 = 0$ whenever the voltage $Y_2 = \pm 25$.

Checking Basic Concepts for Sections 7.1 and 7.2

1.   If $\cot\theta > 0$ and $\sin\theta < 0$ then any point (x, y) on the terminal side of θ must satisfy the following:

$y < 0$ and $\dfrac{x}{y} > 0 \Rightarrow x < 0$. Thus, θ is contained in Quadrant III.

3.   (a) $(1 - \sin\theta)(1 + \sin\theta) = 1 - \sin^2\theta = \cos^2\theta$

(b) $\tan^2 t \csc^2 t - 1 = \tan^2 t\,(1 + \cot^2 t) - 1 = \tan^2 t + 1 - 1 = \tan^2 t$

5.   (a) $(1 - \sin^2\theta)(1 + \cot^2\theta) = \cos^2\theta \csc^2\theta = \dfrac{\cos^2\theta}{\sin^2\theta} = \cot^2\theta$

(b) $\dfrac{\cot^2 t}{\csc t} = \dfrac{\csc^2 t - 1}{\csc t} = \dfrac{\csc^2 t}{\csc t} - \dfrac{1}{\csc t} = \csc t - \sin t$

7.3: Trigonometric Equations

1. $\theta = 120°$ is in quadrant II, $\theta_R = 180° - 120° = 60°$.

3. $\theta = 85°$ is in quadrant I, $\theta_R = 85°$.

5. $\theta = -65°$ is coterminal with $295°$ in quadrant IV, $\theta_R = 360° - 295° = 65°$.

7. $\theta = \dfrac{5\pi}{6}$ is in quadrant II, $\theta_R = \pi - \dfrac{5\pi}{6} = \dfrac{\pi}{6}$.

9. $\theta = -\dfrac{2\pi}{3}$ is coterminal with $\dfrac{4\pi}{3}$ in quadrant III, $\theta_R = \dfrac{4\pi}{3} - \pi = \dfrac{\pi}{3}$.

11. $\theta = \dfrac{5\pi}{4}$ is in quadrant III, $\theta_R = \dfrac{5\pi}{3} - \pi = \dfrac{\pi}{4}$.

13. (a) Start by solving $\sin \theta_R = 1$: $\theta_R = \sin^{-1} 1 = 90°$. This angle is quadrantal, thus $\theta = 90°$ or $\theta = \dfrac{\pi}{2}$.

   (b) Start by solving $\sin \theta_R = 1$: $\theta_R = \sin^{-1} 1 = 90°$. This angle is quadrantal, thus $\theta = 270°$ or $\theta = \dfrac{3\pi}{2}$.

15. (a) Start by solving $\tan \theta_R = \sqrt{3}$: $\theta_R = \tan^{-1} \sqrt{3} = 60°$. The tangent function is positive in quadrants I and III.

   Angles in these quadrants with a 60° reference angle are 60° and 240° or $\dfrac{\pi}{3}$ and $\dfrac{4\pi}{3}$.

   (b) Start by solving $\tan \theta_R = \sqrt{3}$: $\theta_R = \tan^{-1} \sqrt{3} = 60°$. The tangent function is negative in quadrants II and IV.

   Angles in these quadrants with a 60° reference angle are 120° and 300° or $\dfrac{2\pi}{3}$ and $\dfrac{5\pi}{3}$.

17. First note that $\sec \theta = 2$ when $\cos \theta = \dfrac{1}{2}$ and $\sec \theta = -2$ when $\cos \theta = -\dfrac{1}{2}$.

   (a) Start by solving $\cos \theta_R = \dfrac{1}{2}$: $\theta_R = \cos^{-1} \dfrac{1}{2} = 60°$. The secant function is positive in quadrants I and IV.

   Angles in these quadrants with a 60° reference angle are 60° and 300° or $\dfrac{\pi}{3}$ and $\dfrac{5\pi}{3}$.

   (b) Start by solving $\cos \theta_R = \dfrac{1}{2}$: $\theta_R = \cos^{-1} \dfrac{1}{2} = 60°$. The secant function is negative in quadrants II and III.

   Angles in these quadrants with a 60° reference angle are 120° and 240° or $\dfrac{2\pi}{3}$ and $\dfrac{4\pi}{3}$.

19. (a) No solution since $\sin \theta \neq 3$ for any θ.

   (b) No solution since $\sin \theta \neq -3$ for any θ.

21. (a) Since $t_R = \sin^{-1}\left(\dfrac{1}{2}\right) = \dfrac{\pi}{6}$ and sine is positive in quadrants I and II, $t = \dfrac{\pi}{6} + 2\pi n$ and $\dfrac{5\pi}{6} + 2\pi n$.

   (b) Since $t_R = \sin^{-1}\left(\dfrac{1}{2}\right) = \dfrac{\pi}{6}$ and sine is negative in quadrants III and IV, $t = \dfrac{7\pi}{6} + 2\pi n$ and $\dfrac{11\pi}{6} + 2\pi n$.

23. (a) Since $t_R = \tan^{-1} (1) = \dfrac{\pi}{4}$ and tangent is positive in quadrants I and III, $t = \dfrac{\pi}{4} + \pi n$.

   (b) Since $t_R = \tan^{-1} (1) = \dfrac{\pi}{4}$ and tangent is negative in quadrants II and IV, $t = \dfrac{3\pi}{4} + \pi n$.

25. First note that $\cos t = \dfrac{1}{2}$ when $\sec t = 2$ and $\cos t = -\dfrac{1}{2}$ when $\sec t = -2$.

   (a) Since $t_R = \cos^{-1}\left(\dfrac{1}{2}\right) = \dfrac{\pi}{3}$ and the secant function is positive in quadrants I and IV,

   $t = \dfrac{\pi}{3} + 2\pi n$ and $\dfrac{5\pi}{3} + 2\pi n$.

   (b) Since $t_R = \cos^{-1}\left(\dfrac{1}{2}\right) = \dfrac{\pi}{3}$ and the secant function is negative in quadrants II and III,

   $t = \dfrac{2\pi}{3} + 2\pi n$ and $\dfrac{4\pi}{3} + 2\pi n$.

27. At the intersection points $t = \dfrac{\pi}{4}$ and $t = \dfrac{5\pi}{4}$ therefore, $\sin t = \cos t$ for $t = \dfrac{\pi}{4}, \dfrac{5\pi}{4}$.

  Symbolically: $\sin t = \cos t \Rightarrow \dfrac{\sin t}{\cos t} = 1 \Rightarrow \tan t = 1 \Rightarrow t = \dfrac{\pi}{4}, \dfrac{5\pi}{4}$.

29. At the intersection points $t = \dfrac{\pi}{3}$ and $t = \dfrac{5\pi}{3}$ therefore, $3 \cot t = 2 \sin t$ for $t = \dfrac{\pi}{3}, \dfrac{5\pi}{3}$.

  Symbolically: $3 \cot t = 2 \sin t \Rightarrow 3 \dfrac{\cos t}{\sin t} = 2 \sin t \Rightarrow 3 \cos t = 2 \sin^2 t \Rightarrow 3 \cos t = 2(1 - \cos^2 t) \Rightarrow$

  $2 \cos^2 t + 3 \cos t - 2 = 0 \Rightarrow (2 \cos t - 1)(\cos t + 2) = 0 \Rightarrow \cos t = \dfrac{1}{2}$ or $\cos t = -2$.

  If $\cos t = \dfrac{1}{2}$, then $t = \dfrac{\pi}{3}, \dfrac{5\pi}{3}$. If $\cos t = -2$, then t is undefined.

31. (a) $2x - 1 = 0 \Rightarrow 2x = 1 \Rightarrow x = \dfrac{1}{2}$

  (b) $2 \sin \theta - 1 = 0 \Rightarrow 2 \sin \theta = 1 \Rightarrow \sin \theta = \dfrac{1}{2}$

  Since $\theta_R = \sin^{-1} \dfrac{1}{2} = 30°$ and sine is positive in quadrants I and II, $\theta = 30°, 150°$.

33. (a) $x^2 = x \Rightarrow x^2 - x = 0 \Rightarrow x(x - 1) = 0 \Rightarrow x = 0, 1$

  (b) $\sin^2 \theta = \sin \theta \Rightarrow \sin^2 \theta - \sin \theta = 0 \Rightarrow \sin \theta (\sin \theta - 1) = 0 \Rightarrow \sin \theta = 0, 1$

  Since $\theta_R = \sin^{-1} 0 = 0°$ or $\theta_R = \sin^{-1} 1 = 90°$ and sine is positive in quadrants I and II, $\theta = 0°, 90°, 180°$.

35. (a) $x^2 + 1 = 2 \Rightarrow x^2 = 1 \Rightarrow x = \pm 1$

  (b) $\tan^2 t + 1 = 2 \Rightarrow \tan^2 t = 1 \Rightarrow \tan t = \pm 1$

  Since $t_R = \tan^{-1} 1 = \dfrac{\pi}{4}$ and tangent is positive or negative in all quadrants, $t = \dfrac{\pi}{4}, \dfrac{3\pi}{4}, \dfrac{5\pi}{4}, \dfrac{7\pi}{4}$.

37. (a) $x^2 + x = 2 \Rightarrow x^2 + x - 2 = 0 \Rightarrow (x + 2)(x - 1) = 0 \Rightarrow x = -2, 1$

  (b) $\cos^2 t + \cos t = 2 \Rightarrow \cos^2 t + \cos t - 2 = 0 \Rightarrow (\cos t + 2)(\cos t - 1) = 0 \Rightarrow \cos t = -2, 1$

  Since $t_R = \cos^{-1} 1 = 0$ ($\cos t = -2$ is not possible) and cosine is positive in quadrants I and IV, $t = 0$.

39. $\tan^2 t - 3 = 0 \Rightarrow \tan^2 t = 3 \Rightarrow \tan t = \pm \sqrt{3}$

  Since $t_R = \tan^{-1} \sqrt{3} = \dfrac{\pi}{3}$ and tangent is positive or negative in all quadrants, $t = \dfrac{\pi}{3}, \dfrac{2\pi}{3}, \dfrac{4\pi}{3}, \dfrac{5\pi}{3}$.

41. $3 \cos t + 4 = 0 \Rightarrow 3 \cos t = -4 \Rightarrow \cos t = -\dfrac{4}{3}$

  $\cos t = -\dfrac{4}{3}$ is not possible. No solution.

43. $\sin t \cos t = \cos t \Rightarrow \sin t \cos t - \cos t = 0 \Rightarrow \cos t (\sin t - 1) = 0 \Rightarrow \cos t = 0$ or $\sin t = 1$

  Since $t_R = \cos^{-1} 0 = \dfrac{\pi}{2}$ or $t_R = \sin^{-1} 1 = \dfrac{\pi}{2}$ and the angle is quadrantal, $t = \dfrac{\pi}{2}, \dfrac{3\pi}{2}$.

45. $\csc^2 t = 2 \cot t \Rightarrow 1 + \cot^2 t = 2 \cot t \Rightarrow \cot^2 t - 2 \cot t + 1 = 0 \Rightarrow (\cot t - 1)^2 = 0 \Rightarrow \cot t = 1$

  Since $t_R = \cot^{-1} 1 = \tan^{-1} 1 = \dfrac{\pi}{4}$ and tangent is positive in quadrants I and III,

  $t = \dfrac{\pi}{4}, \dfrac{5\pi}{4}$. $t = \dfrac{\pi}{4} + \pi n$ for $n = 0, \pm 1, \pm 2, \pm 3, \ldots$.

47. $\sin^2 t = \dfrac{1}{4} \Rightarrow \sin t = \pm\dfrac{1}{2}$

    Since $t_R = \sin^{-1}\dfrac{1}{2} = \dfrac{\pi}{6}$ and sine is positive or negative in all quadrants, $t = \dfrac{\pi}{6}, \dfrac{5\pi}{6}, \dfrac{7\pi}{6}, \dfrac{11\pi}{6}$.

49. $\sin t \cos t = 0 \Rightarrow \sin t = 0$ or $\cos t = 0$. Since $t_R = \sin^{-1} 0 = 0$ or $t_R = \cos^{-1} 0 = \dfrac{\pi}{2}, t = 0, \dfrac{\pi}{2}, \pi, \dfrac{3\pi}{2}$.

51. $2\sec t = \tan^2 t + 1 \Rightarrow 2\sec t = \sec^2 t \Rightarrow \sec^2 t - 2\sec t = 0 \Rightarrow \sec t(\sec t - 2) = 0 \Rightarrow$

    $\sec t = 0, \sec t - 2 = 0$. For $\sec t = 0, t$ is undefined. For $\sec t - 2 = 0 \Rightarrow \sec t = 2 \Rightarrow t = \dfrac{\pi}{3}, \dfrac{5\pi}{3}$.

53. $\tan t + \sec t = 1 \Rightarrow \tan t = 1 - \sec t \Rightarrow \tan^2 t = (1 - \sec t)^2 \Rightarrow \sec^2 t - 1 = 1 - 2\sec t + \sec^2 t \Rightarrow$

    $-2 = -2\sec t \Rightarrow \sec t = 1 \Rightarrow t = 0$.

55. $2\tan t - 1 = 1 \Rightarrow 2\tan t = 2 \Rightarrow \tan t = 1 \Rightarrow t = \dfrac{\pi}{4}$ and tangent is positive in quadrants I and III,

    $t = \dfrac{\pi}{4}, \dfrac{5\pi}{4}, \dfrac{9\pi}{4}, \ldots . t = \dfrac{\pi}{4} + \pi n$ for $n = 0, \pm 1, \pm 2, \pm 3, \ldots$.

57. $2\sin t + 2 = 3 \Rightarrow 2\sin t = 1 \Rightarrow \sin t = \dfrac{1}{2} \Rightarrow t = \dfrac{\pi}{6}, \dfrac{5\pi}{6} \Rightarrow t = \dfrac{\pi}{6} + 2\pi n,$

    $t = \dfrac{5\pi}{6} + 2\pi n$ for $n = 0, \pm 1, \pm 2, \pm 3, \ldots$.

59. $2\sin^2 t - 3\sin t = -1 \Rightarrow 2\sin^2 t - 3\sin t + 1 = 0 \Rightarrow (2\sin t - 1)(\sin t - 1) = 0 \Rightarrow$

    $(2\sin t - 1) = 0, (\sin t - 1) = 0. \ 2\sin t - 1 = 0 \Rightarrow \sin t = \dfrac{1}{2} \Rightarrow t = \dfrac{\pi}{6}, \dfrac{5\pi}{6}$.

    $\sin t - 1 = 0 \Rightarrow \sin t = 1 \Rightarrow t = \dfrac{\pi}{2}. \ t = \dfrac{\pi}{6} + 2\pi n, \dfrac{5\pi}{6} + 2\pi n, \dfrac{\pi}{2} + 2\pi n$ for $n = 0, \pm 1, \pm 2, \pm 3, \ldots$.

61. $\sec^2 t + 3\sec t + 2 = 0 \Rightarrow (\sec t + 2)(\sec t + 1) = 0 \Rightarrow \sec t + 2 = 0, (\sec t + 1) = 0$.

    $\sec t + 2 = 0 \Rightarrow \sec t = -2 \Rightarrow t = \dfrac{2\pi}{3}, \dfrac{4\pi}{3}. \ \sec t + 1 = 0 \Rightarrow \sec t = -1 \Rightarrow t = \pi$.

    $t = \dfrac{2\pi}{3} + 2\pi n, \dfrac{4\pi}{3} + 2\pi n, \pi + 2\pi n$ for $n = 0, \pm 1, \pm 2, \pm 3, \ldots$.

63. $\tan^2 t - 1 = 0 \Rightarrow \tan^2 t = 1 \Rightarrow \tan t = \pm 1$

    Since $t_R = \tan^{-1} 1 = \dfrac{\pi}{4}$ and tangent is positive or negative in all quadrants, $t = \pm\dfrac{\pi}{4}, \pm\dfrac{3\pi}{4}, \pm\dfrac{5\pi}{4}, \ldots$.

    $t = \dfrac{\pi}{4} + \dfrac{\pi}{2}n$ for $n = 0, \pm 1, \pm 2, \pm 3, \ldots$.

65. $\sin^2 t + \sin t - 20 = 0 \Rightarrow (\sin t + 5)(\sin t - 4) = 0 \Rightarrow \sin t = -5, 4$

    $\sin t = -5$ and $\sin t = 4$ are both not possible. No solution.

67. $\cos t \sin t = \sin t \Rightarrow \cos t \sin t - \sin t = 0 \Rightarrow \sin t (\cos t - 1) = 0 \Rightarrow \sin t = 0$ or $\cos t = 1$

    Since $t_R = \sin^{-1} 0 = 0$ or $t_R = \cos^{-1} 1 = 0$ and the angle is quadrantal, $t = 0, \pm\pi, \pm 2\pi, \ldots$.

    $t = \pi n$ for $n = 0, \pm 1, \pm 2, \pm 3, \ldots$.

69. $\sec^2 t = 2 \tan t \Rightarrow 1 + \tan^2 t = 2 \tan t \Rightarrow \tan^2 t - 2 \tan t + 1 = 0 \Rightarrow (\tan t - 1)^2 = 0 \Rightarrow \tan t = 1$

    Since $t_R = \tan^{-1} 1$ and tangent is positive in quadrants I and III, $t = \dfrac{\pi}{4}, \dfrac{5\pi}{4}$.

    $t = \dfrac{\pi}{4} + \pi n$ for $n = 0, \pm 1, \pm 2, \pm 3, \ldots$.

71. $\sin^2 t \cos^2 t = 0 \Rightarrow \sin^2 t = 0$ or $\cos^2 t = 0 \Rightarrow \sin t = 0$ or $\cos t = 0$

    Since $t_R = \sin^{-1} 0 = 0$ or $t_R = \cos^{-1} 0 = \dfrac{\pi}{2}$ and the angle is quadrantal, $t = 0, \pm\dfrac{\pi}{2}, \pm\pi, \pm\dfrac{3\pi}{2}, \ldots$.

    $t = \dfrac{\pi}{2} n$ for $n = 0, \pm 1, \pm 2, \pm 3, \ldots$.

73. $\sin t + \cos t = 1 \Rightarrow \sin^2 t + 2 \sin t \cos t + \cos^2 t = 1 \Rightarrow 2 \sin t \cos t = 0 \Rightarrow \sin t = 0$ or $\cos t = 0$

    Since $t_R = \sin^{-1} 0 = 0$ or $t_R = \cos^{-1} 0 = \dfrac{\pi}{2}$ and the angle is quadrantal, $t = \ldots, -\dfrac{3\pi}{2}, 0, \dfrac{\pi}{2}, 2\pi, \ldots$.

    $t = 2\pi n, \dfrac{\pi}{2} + 2\pi n$ for $n = 0, \pm 1, \pm 2, \pm 3, \ldots$.

75. $\sin 3t = \dfrac{1}{2}$, let $\theta = 3t$ so then the equation becomes $\sin \theta = \dfrac{1}{2}$ and in radian measure these solutions are

    $\theta = \dfrac{\pi}{6}$ and $\dfrac{5\pi}{6}$. For all real number solutions $\theta = \dfrac{\pi}{6} + 2\pi n$ or $\theta = \dfrac{5\pi}{6} + 2\pi n$. Because $\theta = 3t$, we can

    determine t by substituting $3t$ for $\theta \Rightarrow 3t = \dfrac{\pi}{6} + 2\pi n$ or $3t = \dfrac{5\pi}{6} + 2\pi n$. Therefore $t = \dfrac{\pi}{18} + \dfrac{2\pi}{3} n$ or

    $t = \dfrac{5\pi}{18} + \dfrac{2\pi}{3} n$.

77. $\cos 4t = -\dfrac{\sqrt{3}}{2}$, let $\theta = 4t$ so then the equation becomes $\cos \theta = -\dfrac{\sqrt{3}}{2}$ and in radian measure these solutions

    are $\theta = \dfrac{5\pi}{6}$ and $\dfrac{7\pi}{6}$. For all real number solutions $\theta = \dfrac{5\pi}{6} + 2\pi n$ or $\theta = \dfrac{7\pi}{6} + 2\pi n$. Because $\theta = 4t$, we can

    determine t by substituting $4t$ for $\theta \Rightarrow 4t = \dfrac{5\pi}{6} + 2\pi n$ or $4t = \dfrac{7\pi}{6} + 2\pi n$. Therefore $t = \dfrac{5\pi}{24} + \dfrac{\pi}{2} n$ or

    $t = \dfrac{7\pi}{24} + \dfrac{\pi}{2} n$.

79. $\tan 5t = 1$, let $\theta = 5t$ so then the equation becomes $\tan \theta = 1$ and in radian measure these solutions are

    $\theta = \dfrac{\pi}{4}$ and $\dfrac{5\pi}{4}$. For all real number solutions $\theta = \dfrac{\pi}{4} + \pi n$ or $\theta = \dfrac{5\pi}{4} + \pi n$. Since these representations are

    equivalent, we use $\theta = \dfrac{\pi}{4} + \pi n$. Because $\theta = 5t$, we can determine t by substituting

    $5t$ for $\theta \Rightarrow 5t = \dfrac{\pi}{4} + \pi n$. Therefore $t = \dfrac{\pi}{20} + \dfrac{\pi}{5} n$.

81. $2 \sin 4t = -1 \Rightarrow \sin 4t = -\dfrac{1}{2}$, let $\theta = 4t$ so then the equation becomes $\sin \theta = -\dfrac{1}{2}$ and in radian measure

    these solutions are $\theta = \dfrac{7\pi}{6}$ and $\dfrac{11\pi}{6}$. For all real number solutions $\theta = \dfrac{7\pi}{6} + 2\pi n$ or $\theta = \dfrac{11\pi}{6} + 2\pi n$.

    Because $\theta = 4t$, we can determine t by substituting $4t$ for $\theta \Rightarrow 4t = \dfrac{7\pi}{6} + 2\pi n$ or $4t = \dfrac{11\pi}{6} + 2\pi n$.

    Therefore $t = \dfrac{7\pi}{24} + \dfrac{\pi}{2} n$ or $t = \dfrac{11\pi}{24} + \dfrac{\pi}{2} n$.

83. $-\sec 4t = \sqrt{2} \Rightarrow \sec 4t = -\sqrt{2}$, let $\theta = 4t$ so then the equation becomes $\sec \theta = -\sqrt{2}$ or $\cos \theta = -\dfrac{1}{\sqrt{2}}$

and in radian measure these solutions are $\theta = \dfrac{3\pi}{4}$ and $\dfrac{5\pi}{4}$. For all real number solutions $\theta = \dfrac{3\pi}{4} + 2\pi n$ or

$\theta = \dfrac{5\pi}{4} + 2\pi n$. Because $\theta = 4t$, we can determine t by substituting $4t$ for $\theta \Rightarrow 4t = \dfrac{3\pi}{4} + 2\pi n$ or

$4t = \dfrac{5\pi}{4} + 2\pi n$. Therefore $t = \dfrac{3\pi}{16} + \dfrac{\pi}{2}n$ or $t = \dfrac{5\pi}{16} + \dfrac{\pi}{2}n$.

85. $2 \sin 8t - 3 = -1 \Rightarrow 2 \sin 8t = 2 \Rightarrow \sin 8t = 1$, let $\theta = 8t$ so then the equation becomes $\sin \theta = 1$ and in

radian measure this solution is $\theta = \dfrac{\pi}{2}$. For all real number solutions $\theta = \dfrac{\pi}{2} + 2\pi n$. Because $\theta = 8t$, we can

determine t by substituting $8t$ for $\theta \Rightarrow 8t = \dfrac{\pi}{2} + 2\pi n$. Therefore $t = \dfrac{\pi}{16} + \dfrac{\pi}{4}n$.

87. $\cot 4t + 5 = 6 \Rightarrow \cot 4t = 1$, let $\theta = 4t$ so then the equation becomes $\cot \theta = 1$ or $\tan \theta = 1$ and in radian

measure this solution is $\theta = \dfrac{\pi}{4}$. For all real number solutions $\theta = \dfrac{\pi}{4} + \pi n$. Because $\theta = 4t$, we can determine

t by substituting $4t$ for $\theta \Rightarrow 4t = \dfrac{\pi}{4} + \pi n$. Therefore $t = \dfrac{\pi}{16} + \dfrac{\pi}{4}n$.

89. $5 \cos 3t = 1 \Rightarrow \cos 3t = \dfrac{1}{5}$, let $\theta = 3t$ so then the equation becomes $\cos \theta = \dfrac{1}{5}$ and in radian measure these

solutions are $\theta \approx 1.369$ and 4.914. For all real number solutions $\theta \approx 1.369 + 2\pi n$ or $\theta \approx 4.914 + 2\pi n$.

Because $\theta = 3t$ we can determine t by substituting $3t$ for $\theta \Rightarrow 3t \approx 1.369 + 2\pi n$ or $3t \approx 4.914 + 2\pi n$.

Therefore $t \approx 0.456 + \dfrac{2}{3}\pi n$ or $t \approx 1.638 + \dfrac{2}{3}\pi n$.

91. $\sin 2t = \dfrac{1}{3}$, let $\theta = 2t$ so then the equation becomes $\sin \theta = \dfrac{1}{3}$ and in radian measure these solutions are

$\theta \approx 0.340$ and 2.802. For all real number solutions $\theta \approx 0.340 + 2\pi n$ or $\theta \approx 2.802 + 2\pi n$. Because

$\theta = 2t$ we can determine t by substituting $2t$ for $\theta \Rightarrow 2t \approx 0.340 + 2\pi n$ or $2t \approx 2.802 + 2\pi n$. Therefore

$t \approx 0.170 + \pi n$ or $t \approx 1.401 + \pi n$.

93. $2.1 \sec t - 4.5 = 0 \Rightarrow 2.1 \sec t = 4.5 \Rightarrow \sec t = \dfrac{4.5}{2.1} \Rightarrow \cos t = \dfrac{2.1}{4.5} \Rightarrow \cos^{-1}\left(\dfrac{2.1}{4.5}\right) = t \Rightarrow t \approx 1.085$.

The cosine function is positive in quadrants I and IV. For t in $[0, 2\pi)$, the two solutions are $t \approx 1.085$ and

$2\pi - \cos^{-1}\left(\dfrac{2.1}{4.5}\right) \approx 5.198$.

95. $5.8 \sin t - 3.7 = 0.2 \Rightarrow 5.8 \sin t = 3.9 \Rightarrow \sin t = \dfrac{3.9}{5.8} \Rightarrow \sin^{-1}\left(\dfrac{3.9}{5.8}\right) = t \Rightarrow t \approx 0.737$.

The sine function is positive in quadrants I and II. For t in $[0, 2\pi)$, the two solutions are $t \approx 0.737$ and

$\pi - \sin^{-1}\left(\dfrac{3.9}{5.8}\right) \approx 2.404$.

97. $5\tan^2 t - 3 = 0 \Rightarrow 5\tan^2 t = 3 \Rightarrow \tan^2 t = \frac{3}{5} \Rightarrow \tan t = \pm\sqrt{\frac{3}{5}} \Rightarrow \tan^{-1}\left(\pm\sqrt{\frac{3}{5}}\right) = t$. The reference number $t_R = \tan^{-1}\left(\sqrt{\frac{3}{5}}\right) \approx 0.659$. The tangent function is both positive and negative in all four quadrants. For t in $[0, 2\pi)$, the solutions are $t \approx 0.659$, $\pi - \tan^{-1}\left(\sqrt{\frac{3}{5}}\right) \approx 2.483$, $\pi + \tan^{-1}\left(\sqrt{\frac{3}{5}}\right) \approx 3.801$ and $2\pi - \tan^{-1}\left(\sqrt{\frac{3}{5}}\right) \approx 5.624$.

99. Graph $Y_1 = \tan(X)$ and $Y_2 = X$ in dot mode in $[-\pi/6, 2\pi, \pi/2]$ by $[-2, 6, 1]$.
The intersections of the two graphs are shown in Figure 97a and Figure 97b.
Table $Y_1 = \tan(X)$ and $Y_2 = X$ starting at $x = 0$, incrementing by 0.9. See Figure 99c.
The solutions are $x = 0$, $x \approx 4.49$.

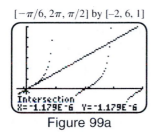

$[-\pi/6, 2\pi, \pi/2]$ by $[-2, 6, 1]$

Figure 99a

$[-\pi/6, 2\pi, \pi/2]$ by $[-2, 6, 1]$

Figure 99b

X	Y1	Y2
0	0	0
.9	1.2602	.9
1.8	-4.286	1.8
2.7	-.4727	2.7
3.6	.49347	3.6
4.5	4.6373	4.5
5.4	-1.218	5.4

Y1■tan(X)

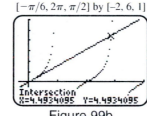

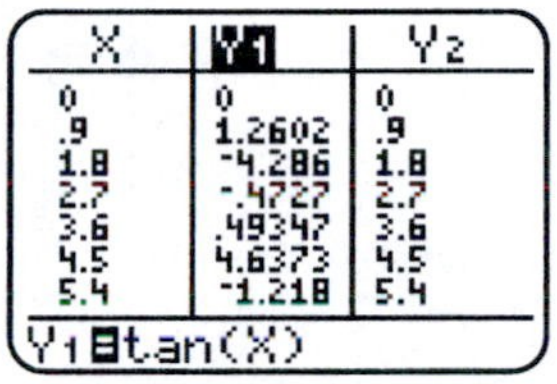

Figure 99c

101. Graph $Y_1 = \sin(X)$ and $Y_2 = (X - 1)$^2 in $[-\pi/6, 2\pi, \pi/2]$ by $[-3, 7, 1]$.
The intersections of the two graphs are shown in Figure 101a and Figure 101b.
Table $Y_1 = \sin(X)$ and $Y_2 = (X - 1)$^2 starting at $x = 0$, incrementing by 0.4. See Figure 101c.
The solutions are $x \approx 0.39$, $x \approx 1.96$.

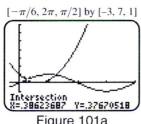

$[-\pi/6, 2\pi, \pi/2]$ by $[-3, 7, 1]$

Figure 101a

$[-\pi/6, 2\pi, \pi/2]$ by $[-3, 7, 1]$

Figure 101b

X	Y1	Y2
0	0	1
.4	.38942	.36
.8	.71736	.04
1.2	.93204	.04
1.6	.99957	.36
2	.9093	1
2.4	.67546	1.96

Y1■sin(X)

Figure 101c

$[-\pi/6, 2\pi, \pi/2]$ by $[-5, 5, 1]$

Figure 103a

X	Y1	Y2
3	-3.922	-.836
3.1	-3.564	-.841
3.2	-3.138	-.8405
3.3	-2.645	-.8346
3.4	-2.09	-.8231
3.5	-1.476	-.8055
3.6	-.8075	-.7813

X=3.6

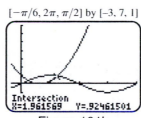

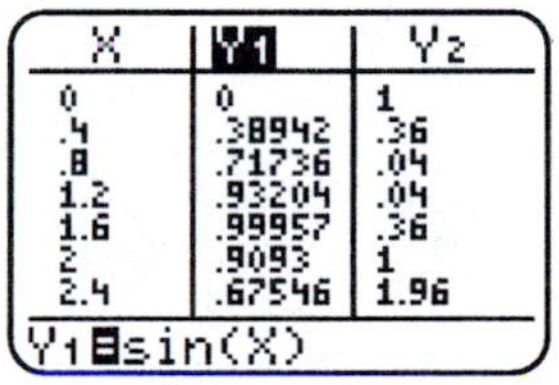

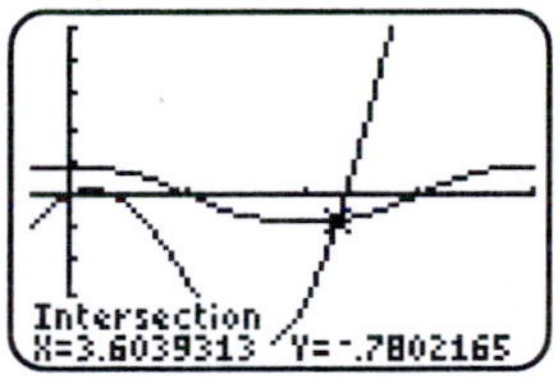

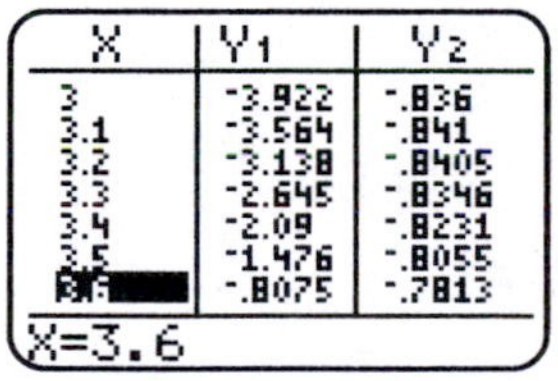

Figure 103b

103. Graph $Y_1 = 2X\cos(X + 1)$ and $Y_2 = \sin(\cos(X))$ in $[-\pi/6, 2\pi, \pi/2]$ by $[-5, 5, 1]$.
The intersection of the two graphs is shown in Figure 103a.
Table $Y_1 = 2X\cos(X + 1)$ and $Y_2 = \sin(\cos(X))$ starting at $x = 3.0$, incrementing by 0.1.
See Figure 103b. The solution is $x \approx 3.60$.

105. $Y_1 = 0$ for $x = 30°, 210°$

We can write all solutions to this equation as $30° + 180°n$, where $n = 0, \pm1, \pm2, \ldots$.

107. $\sin^{-1} x = \dfrac{\pi}{2} \Rightarrow \sin(\sin^{-1} x) = \sin\dfrac{\pi}{2} \Rightarrow x = 1$

109. $2\cos^{-1} x = \dfrac{5\pi}{3} \Rightarrow \cos^{-1} x = \dfrac{5\pi}{6} \Rightarrow \cos(\cos^{-1} x) = \cos\dfrac{5\pi}{6} \Rightarrow x = -\dfrac{\sqrt{3}}{2} \approx -0.866$

111. $\pi + \tan^{-1} x = \dfrac{3\pi}{4} \Rightarrow \tan^{-1} x = -\dfrac{\pi}{4} \Rightarrow \tan(\tan^{-1} x) = \tan\left(-\dfrac{\pi}{4}\right) \Rightarrow x = -1$

113. $\tan^{-1}(3x+1) = \dfrac{\pi}{4} \Rightarrow \tan(\tan^{-1}(3x+1)) = \tan\left(\dfrac{\pi}{4}\right) \Rightarrow 3x+1 = 1 \Rightarrow 3x = 0 \Rightarrow x = 0$

115. $\cos^{-1} x + 3\cos^{-1} x = \pi \Rightarrow 4\cos^{-1} x = \pi \Rightarrow \cos^{-1} x = \dfrac{\pi}{4} \Rightarrow \cos(\cos^{-1} x) = \cos\dfrac{\pi}{4} \Rightarrow$

   $x = \dfrac{1}{\sqrt{2}} \approx 0.707$

117. $0.25 = \dfrac{1}{2}(1 - \cos\theta) \Rightarrow 0.5 = 1 - \cos\theta \Rightarrow \cos\theta = 0.5.$ $\theta_R = \cos^{-1}(0.5) = 60°.$ The cosine function is

   positive in quadrants I and IV. Angles in these quadrants with $\theta_R = 60°$ are $60°$ and $300°$. Since the cosine

   function has period $360°$, all solutions can be written as $60° + 360° \cdot n$ or $300° + 360° \cdot n$ where n is an integer.

119. $200 = \dfrac{44^2}{4.5 + 32.2\tan\theta} \Rightarrow 4.5 + 32.2\tan\theta = \dfrac{44^2}{200} \Rightarrow 32.2\tan\theta = \dfrac{44^2}{200} - 4.5 \Rightarrow \tan\theta = \dfrac{\frac{44^2}{200} - 4.5}{32.2} \Rightarrow$

   $\tan\theta \approx 0.1609.$ $\theta = \tan^{-1}(0.1609) \approx 9.1°.$ The superelevation should be about $9.1°$.

121. Graph $Y_1 = 6.5\sin(\pi/6(X - 3.65)) + 12.4$ and $Y_2 = 9$ in $[0, 13, 1]$ by $[0, 24, 2]$.

   The intersections of the two graphs are shown in Figure 119a and Figure 119b.

   Table $Y_1 = 6.5\sin(\pi/6(X - 3.65)) + 12.4$ starting at $x = 2.6$, incrementing by 1.35. See Figure 119c.

   The solutions are $x \approx 2.6$ and $x \approx 10.7$; near February 17 and October 22.

[0, 13, 1] by [0, 24, 2]   [0, 13, 1] by [0, 24, 2]

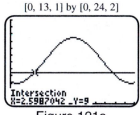   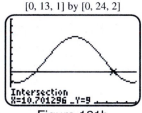   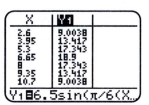

Figure 121a   Figure 121b   Figure 121c

123. $6.5\sin\left(\dfrac{\pi}{6}(x - 3.65)\right) + 12.4 = 9 \Rightarrow 6.5\sin\left(\dfrac{\pi}{6}(x - 3.65)\right) = -3.4 \Rightarrow$

   $\sin\left(\dfrac{\pi}{6}(x - 3.65)\right) = -0.5231$

   $\Rightarrow \dfrac{\pi}{6}(x - 3.65) = \sin^{-1}(-0.5231)$

   $\Rightarrow \dfrac{\pi}{6}(x - 3.65) \approx -0.5505$ or $\dfrac{\pi}{6}(x - 3.65) \approx 3.6920$

   $\Rightarrow x - 3.65 \approx -1.0513$ or $x - 3.65 \approx 7.0513$

   $\Rightarrow x \approx 2.6$ or $x \approx 10.7$

125. (a) Using the sine regression function on the calculator we get $f(x) \approx 122.3 \sin (0.524x - 1.7) + 367$.

Answers may vary.

(b) Graph $Y_1 = 122.3 \sin (0.524X - 1.7) + 367$ and $Y_2 = 350$ together in $[0, 13, 1]$ by $[200, 600, 50]$.

The intersections of the two graphs are shown in Figure 125a and Figure 125b.

We see that $f(x) \geq 350$ for $k_1 \leq x \leq k_2$ where $k_1 \approx 2.98$ and $k_2 \approx 9.51$. At 50° N latitude the

maximum monthly hours of sunshine is greater than or equal to 350 hours from roughly March through

September.

$[0, 13, 1]$ by $[200, 600, 50]$ $[0, 13, 1]$ by $[200, 600, 50]$ $[0, 0.5, 0.1]$ by $[-0.2, 0.2, 0.1]$

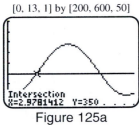
Figure 125a

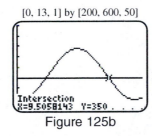
Figure 125b

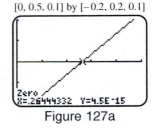
Figure 127a

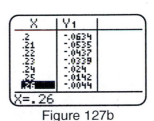
Figure 127b

127. Use the intercept method to solve the equation.

Graph $Y_1 = X - 0.26 - 0.017 \sin (X)$ in $[0, 0.5, 0.1]$ by $[-0.2, 0.2, 0.1]$. See Figure 127a.

Table $Y_1 = X - 0.26 - 0.017 \sin (X)$ starting at $x = 0.20$, incrementing by 0.01. See Figure 127b.

The solution is approximately 0.26.

129. (a) Let $Y_1 = 0.003 \sin (880\pi X - 0.7)$, $Y_2 = 0.002 \sin (880\pi X + 0.6)$.

Graph Y_1, Y_2 and $Y_3 = Y_1 + Y_2$ seperately in $[0, 0.01, 0.005]$ by $[-0.005, 0.005, 0.001]$.

Graph Y_1 is shown in Figure 129a. Graph Y_2 is shown in Figure 129b. Graph Y_3 is shown in Figure 129c.

(b) The maximum pressure is $P \approx 0.004$. This can be found at any peak on the graph in Figure 129c.

(c) No. The maximum of P_1 is 0.003 and the maximum of P_2 is 0.002.

$[0, 0.01, 0.005]$ by $[-0.005, 0.005, 0.001]$ $[0, 0.01, 0.005]$ by $[-0.005, 0.005, 0.001]$ $[0, 0.01, 0.005]$ by $[-0.005, 0.005, 0.001]$

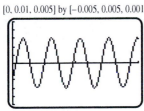
Figure 129a

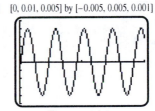
Figure 129b

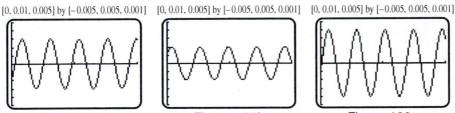
Figure 129c

7.4: Sum and Difference Identities

1. $\sin 15° = \sin (45° - 30°) = \sin 45° \cos 30° - \cos 45° \sin 30° = \left(\dfrac{\sqrt{2}}{2}\right)\left(\dfrac{\sqrt{3}}{2}\right) - \left(\dfrac{\sqrt{2}}{2}\right)\left(\dfrac{1}{2}\right) =$

    $\dfrac{\sqrt{6} - \sqrt{2}}{4}$

3. $\tan 15° = \tan (60° - 45°) = \dfrac{\tan 60° - \tan 45°}{1 + \tan 60° \tan 45°} = \dfrac{\sqrt{3} - 1}{1 + \sqrt{3}} = \dfrac{\sqrt{3} - 1}{\sqrt{3} + 1} \cdot \dfrac{\sqrt{3} - 1}{\sqrt{3} - 1} =$

    $\dfrac{3 - 2\sqrt{3} + 1}{3 - 1} = 2 - \sqrt{3}$

5. $\cos 75° = \cos (135° - 60°) = \cos 135° \cos 60° + \sin 135° \sin 60° = \left(-\dfrac{\sqrt{2}}{2}\right)\left(\dfrac{1}{2}\right) + \left(\dfrac{\sqrt{2}}{2}\right)\left(\dfrac{\sqrt{3}}{2}\right) =$

$\dfrac{-\sqrt{2} + \sqrt{6}}{4} = \dfrac{\sqrt{6} - \sqrt{2}}{4}$

7. $\sin \dfrac{\pi}{12} = \sin \left(\dfrac{\pi}{3} - \dfrac{\pi}{4}\right) = \sin \dfrac{\pi}{3} \cos \dfrac{\pi}{4} - \cos \dfrac{\pi}{3} \sin \dfrac{\pi}{4} = \left(\dfrac{\sqrt{3}}{2}\right)\left(\dfrac{\sqrt{2}}{2}\right) - \left(\dfrac{1}{2}\right)\left(\dfrac{\sqrt{2}}{2}\right) = \dfrac{\sqrt{6} - \sqrt{2}}{4}$

9. $\sin \dfrac{5\pi}{12} = \sin \left(\dfrac{2\pi}{3} - \dfrac{\pi}{4}\right) = \sin \dfrac{2\pi}{3} \cos \dfrac{\pi}{4} - \cos \dfrac{2\pi}{3} \sin \dfrac{\pi}{4} = \left(\dfrac{\sqrt{3}}{2}\right)\left(\dfrac{\sqrt{2}}{2}\right) - \left(-\dfrac{1}{2}\right)\left(\dfrac{\sqrt{2}}{2}\right) =$

$\dfrac{\sqrt{6} + \sqrt{2}}{4}$

11. (a) Graphical: Graph $Y_1 = \sin (X + \pi/2)$ and $Y_2 = \cos (X)$ seperately in $[-2\pi, 2\pi, \pi/2]$ by $[-4, 4, 1]$.

    Graph Y_1 is shown in Figure 11a. Graph Y_2 is shown in Figure 11b. The graphs are the same.

    (b) $\sin \left(t + \dfrac{\pi}{2}\right) = \sin t \cos \dfrac{\pi}{2} + \cos t \sin \dfrac{\pi}{2} = \sin t \,(0) + \cos t \,(1) = \cos t$

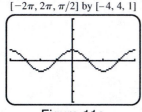

$[-2\pi, 2\pi, \pi/2]$ by $[-4, 4, 1]$ $[-2\pi, 2\pi, \pi/2]$ by $[-4, 4, 1]$

Figure 11a   Figure 11b

13. (a) Graphical: Graph $Y_1 = \cos (X + \pi)$ and $Y_2 = - \cos (X)$ seperately in $[-2\pi, 2\pi, \pi/2]$ by $[-4, 4, 1]$.

    Graph Y_1 is shown in Figure 13a. Graph Y_2 is shown in Figure 13b. The graphs are the same.

    (b) $\cos (t + \pi) = \cos t \cos \pi - \sin t \sin \pi = \cos t \,(-1) - \sin t \,(0) = -\cos t$

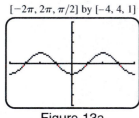

$[-2\pi, 2\pi, \pi/2]$ by $[-4, 4, 1]$ $[-2\pi, 2\pi, \pi/2]$ by $[-4, 4, 1]$

Figure 13a   Figure 13b

15. (a) Graphical: Graph $Y_1 = 1/\cos (X - \pi/2)$ and $Y_2 = 1/\sin (X)$ seperately in $[-2\pi, 2\pi, \pi/2]$ by $[-4, 4, 1]$.

    Graph Y_1 is shown in Figure 15a. Graph Y_2 is shown in Figure 15b. The graphs are the same.

    (b) $\sec \left(t - \dfrac{\pi}{2}\right) = \dfrac{1}{\cos \left(t - \frac{\pi}{2}\right)} = \dfrac{1}{\cos t \cos \frac{\pi}{2} + \sin t \sin \frac{\pi}{2}} = \dfrac{1}{\cos t \,(0) + \sin t \,(1)} = \dfrac{1}{\sin t} = \csc t$

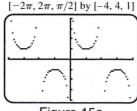 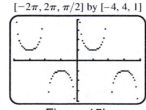

$[-2\pi, 2\pi, \pi/2]$ by $[-4, 4, 1]$ $[-2\pi, 2\pi, \pi/2]$ by $[-4, 4, 1]$

Figure 15a   Figure 15b

17. $\sin\left(\dfrac{\pi}{2} - t\right) = \sin\dfrac{\pi}{2}\cos t - \cos\dfrac{\pi}{2}\sin t = (1)\cos t - (0)\sin t = \cos t$

19. $\sec\left(\dfrac{\pi}{2} - t\right) = \dfrac{1}{\cos\left(\frac{\pi}{2} - t\right)} = \dfrac{1}{\cos\frac{\pi}{2}\cos t + \sin\frac{\pi}{2}\sin t} = \dfrac{1}{(0)\cos t + (1)\sin t} = \dfrac{1}{\sin t} = \csc t$

21. First sketch possible angles for α and β. See Figures 21a & 21b. Note: $\cos\alpha = \dfrac{4}{5}$ and $\cos\beta = \dfrac{12}{13}$.

(a) $\sin(\alpha + \beta) = \sin\alpha\cos\beta + \cos\alpha\sin\beta = \left(\dfrac{3}{5}\right)\left(\dfrac{12}{13}\right) + \left(\dfrac{4}{5}\right)\left(\dfrac{5}{13}\right) = \dfrac{36}{65} + \dfrac{20}{65} = \dfrac{56}{65}$

(b) $\cos(\alpha + \beta) = \cos\alpha\cos\beta - \sin\alpha\sin\beta = \left(\dfrac{4}{5}\right)\left(\dfrac{12}{13}\right) - \left(\dfrac{3}{5}\right)\left(\dfrac{5}{13}\right) = \dfrac{48}{65} - \dfrac{15}{65} = \dfrac{33}{65}$

(c) $\tan(\alpha + \beta) = \dfrac{\sin(\alpha + \beta)}{\cos(\alpha + \beta)} = \dfrac{\frac{56}{65}}{\frac{33}{65}} = \dfrac{56}{65}\cdot\dfrac{65}{33} = \dfrac{56}{33}$

(d) Since both $\sin(\alpha + \beta)$ and $\cos(\alpha + \beta)$ are positive, $\alpha + \beta$ is in quadrant I.

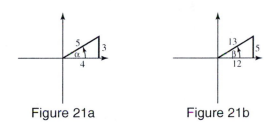

Figure 21a   Figure 21b

23. First sketch possible angles for α and β. See Figures 23a & 23b. Note: $\cos\alpha = -\dfrac{15}{17}$ and $\sin\beta = \dfrac{60}{61}$.

(a) $\sin(\alpha + \beta) = \sin\alpha\cos\beta + \cos\alpha\sin\beta = \left(-\dfrac{8}{17}\right)\left(\dfrac{11}{61}\right) + \left(-\dfrac{15}{17}\right)\left(\dfrac{60}{61}\right) = -\dfrac{88}{1037} - \dfrac{900}{1037} =$

$-\dfrac{988}{1037}$

(b) $\cos(\alpha + \beta) = \cos\alpha\cos\beta - \sin\alpha\sin\beta = \left(-\dfrac{15}{17}\right)\left(\dfrac{11}{61}\right) - \left(-\dfrac{8}{17}\right)\left(\dfrac{60}{61}\right) = -\dfrac{165}{1037} + \dfrac{480}{1037} = \dfrac{315}{1037}$

(c) $\tan(\alpha + \beta) = \dfrac{\sin(\alpha + \beta)}{\cos(\alpha + \beta)} = \dfrac{-\frac{988}{1037}}{\frac{315}{1037}} = -\dfrac{988}{1037}\cdot\dfrac{1037}{315} = -\dfrac{988}{315}$

(d) Since $\sin(\alpha + \beta)$ is negative and $\cos(\alpha + \beta)$ is positive, $\alpha + \beta$ is in quadrant IV.

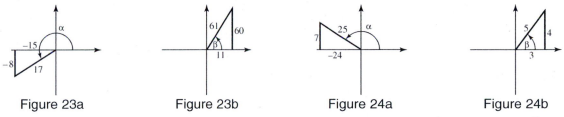

Figure 23a   Figure 23b   Figure 24a   Figure 24b

25. First sketch possible angles for α and β. See Figures 25a & 25b. Note: $\sin\alpha = -\dfrac{4}{5}$ and $\sin\beta = -\dfrac{5}{13}$.

(a) $\sin(\alpha + \beta) = \sin\alpha\cos\beta + \cos\alpha\sin\beta = \left(-\dfrac{4}{5}\right)\left(\dfrac{12}{13}\right) + \left(-\dfrac{3}{5}\right)\left(-\dfrac{5}{13}\right) = -\dfrac{48}{65} + \dfrac{15}{65} = -\dfrac{33}{65}$

(b) $\cos{(\alpha + \beta)} = \cos{\alpha} \cos{\beta} - \sin{\alpha} \sin{\beta} = \left(-\dfrac{3}{5}\right)\left(\dfrac{12}{13}\right) - \left(-\dfrac{4}{5}\right)\left(-\dfrac{5}{13}\right) = -\dfrac{36}{65} - \dfrac{20}{65} = -\dfrac{56}{65}$

(c) $\tan{(\alpha + \beta)} = \dfrac{\sin{(\alpha + \beta)}}{\cos{(\alpha + \beta)}} = \dfrac{-\frac{33}{65}}{-\frac{56}{65}} = -\dfrac{33}{65} \cdot \left(-\dfrac{65}{56}\right) = \dfrac{33}{56}$

(d) Since both $\sin{(\alpha + \beta)}$ and $\cos{(\alpha + \beta)}$ are negative, $\alpha + \beta$ is in quadrant III.

    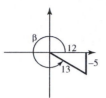

Figure 25a   Figure 25b

27. First sketch possible angles for α and β. See Figures 27a & 27b.

Note: $\sin{\alpha} = \dfrac{5}{13}$, $\cos{\alpha} = -\dfrac{12}{13}$, $\sin{\beta} = \dfrac{60}{61}$, $\cos{\beta} = -\dfrac{11}{60}$

(a) $\sin{(\alpha + \beta)} = \sin{\alpha} \cos{\beta} + \cos{\alpha} \sin{\beta} = \left(\dfrac{5}{13}\right)\left(-\dfrac{11}{61}\right) + \left(-\dfrac{12}{13}\right)\left(\dfrac{60}{61}\right) =$

$\left(-\dfrac{55}{793}\right) + \left(-\dfrac{720}{793}\right) = -\dfrac{775}{793}$

(b) $\cos{(\alpha + \beta)} = \cos{\alpha} \cos{\beta} - \sin{\alpha} \sin{\beta} = \left(-\dfrac{12}{13}\right)\left(-\dfrac{11}{61}\right) - \left(\dfrac{5}{13}\right)\left(\dfrac{60}{61}\right) = \dfrac{132}{793} - \dfrac{300}{793} = -\dfrac{168}{793}$

(c) $\tan{(\alpha + \beta)} = \dfrac{\sin{(\alpha + \beta)}}{\cos{(\alpha + \beta)}} = \dfrac{-\frac{775}{793}}{-\frac{168}{793}} = -\dfrac{775}{793} \cdot \left(-\dfrac{793}{168}\right) = \dfrac{775}{168}$

(d) Since both $\sin{(\alpha + \beta)}$ and $\cos{(\alpha + \beta)}$ are negative, $\alpha + \beta$ is in quadrant III.

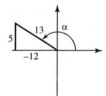   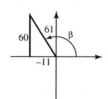

Figure 27a   Figure 27b

29. $\cos{\left(t - \dfrac{\pi}{4}\right)} = \cos{t} \cos{\dfrac{\pi}{4}} + \sin{t} \sin{\dfrac{\pi}{4}} = \cos{t}\left(\dfrac{\sqrt{2}}{2}\right) + \sin{t}\left(\dfrac{\sqrt{2}}{2}\right) = \dfrac{\sqrt{2}}{2}(\cos{t} + \sin{t})$

31. $\tan{\left(t + \dfrac{\pi}{4}\right)} = \dfrac{\tan{t} + \tan{\frac{\pi}{4}}}{1 - \tan{t} \tan{\frac{\pi}{4}}} = \dfrac{\tan{t} + 1}{1 - \tan{t}\,(1)} = \dfrac{1 + \tan{t}}{1 - \tan{t}}$

33. $\dfrac{\cos{(x - y)}}{\cos{(x + y)}} = \dfrac{\cos{x} \cos{y} + \sin{x} \sin{y}}{\cos{x} \cos{y} - \sin{x} \sin{y}} = \dfrac{\cos{x} \cos{y} + \sin{x} \sin{y}}{\cos{x} \cos{y} - \sin{x} \sin{y}} \cdot \dfrac{\sec{x} \sec{y}}{\sec{x} \sec{y}} = \dfrac{1 + \tan{x} \tan{y}}{1 - \tan{x} \tan{y}}$

35. $\dfrac{\cos{(\alpha - \beta)}}{\cos{\alpha} \sin{\beta}} = \dfrac{\cos{\alpha} \cos{\beta} + \sin{\alpha} \sin{\beta}}{\cos{\alpha} \sin{\beta}} = \dfrac{\cos{\alpha} \cos{\beta}}{\cos{\alpha} \sin{\beta}} + \dfrac{\sin{\alpha} \sin{\beta}}{\cos{\alpha} \sin{\beta}} = \cot{\beta} + \tan{\alpha} = \tan{\alpha} + \cot{\beta}$

37. $\sin{2t} = \sin{(t + t)} = \sin{t} \cos{t} + \cos{t} \sin{t} = 2 \sin{t} \cos{t}$

39. $\sin(\alpha + \beta) + \sin(\alpha - \beta) = (\sin\alpha\cos\beta + \cos\alpha\sin\beta) + (\sin\alpha\cos\beta - \cos\alpha\sin\beta) = 2\sin\alpha\cos\beta$

41. $\tan(\pi - \theta) = \dfrac{\tan\pi - \tan\theta}{1 + \tan\pi\tan\theta} = \dfrac{0 - \tan\theta}{1 + (0)\tan\theta} = -\tan\theta$

43. $\tan(x - y) - \tan(y - x) = \dfrac{\tan x - \tan y}{1 + \tan x\tan y} - \dfrac{\tan y - \tan x}{1 + \tan y\tan x} = \dfrac{\tan x - \tan y - \tan y + \tan x}{1 + \tan x\tan y} =$

$\dfrac{2\tan x - 2\tan y}{1 + \tan x\tan y} = \dfrac{2(\tan x - \tan y)}{1 + \tan x\tan y}$

45. $\dfrac{\sin(x + y)}{\cos x\cos y} = \dfrac{\sin x\cos y + \cos x\sin y}{\cos x\cos y} = \dfrac{\sin x\cos y}{\cos x\cos y} + \dfrac{\cos x\sin y}{\cos x\cos y} = \dfrac{\sin x}{\cos x} + \dfrac{\sin y}{\cos y} = \tan x + \tan y$

47. $\tan\gamma = \tan(\alpha - \beta) = \dfrac{\tan\alpha - \tan\beta}{1 + \tan\alpha\tan\beta} = \dfrac{\left(\frac{6}{7}\right) - \left(\frac{5}{7}\right)}{1 + \left(\frac{6}{7}\right)\left(\frac{5}{7}\right)} = \dfrac{\frac{1}{7}}{1 + \frac{30}{49}} = \dfrac{\frac{1}{7}}{\frac{79}{49}} = \dfrac{1}{7}\cdot\dfrac{49}{79} = \dfrac{7}{79}$

49. $\tan\theta = \tan(\beta - \alpha) = \dfrac{\tan\beta - \tan\alpha}{1 + \tan\beta\tan\alpha} = \dfrac{m_2 - m_1}{1 + m_2 m_1} = \dfrac{m_2 - m_1}{1 + m_1 m_2}$

51. For $y = 2x - 3$, $m_2 = 2$ and for $y = \dfrac{3}{5}x + 1$, $m_1 = \dfrac{3}{5}$.

$\tan\theta = \dfrac{2 - \frac{3}{5}}{1 + 2\left(\frac{3}{5}\right)} = \dfrac{\frac{7}{5}}{\frac{11}{5}} = \dfrac{7}{5}\cdot\dfrac{5}{11} = \dfrac{7}{11} \Rightarrow \theta = \tan^{-1}\dfrac{7}{11} \approx 32.5°$

53. (a) $F = 2.89W\cos\theta \Rightarrow F = 2.89(200)\cos\dfrac{\pi}{4} \approx 409$ lb$s$

(b) $2.89(200)\cos\theta = 400 \Rightarrow \cos\theta = \dfrac{400}{2.89(200)} \Rightarrow \theta = \cos^{-1}\left(\dfrac{2}{2.89}\right) \approx 0.81$ or about $46.2°$.

55. (a) Graph $Y_1 = 4\cos(220\pi X) + 3\sin(220\pi X)$ in $[0, 0.02, 0.001]$ by $[-6, 6, 1]$. See Figure 55.

(b) A maximum occurs at approximately $(0.00093114, 5)$. Thus $a = 5$. Since $\sin\theta$ is maximum when $\theta = \dfrac{\pi}{2}$,

we let $220\pi(0.00093114) + k = \dfrac{\pi}{2}$. Thus $k = \dfrac{\pi}{2} - 220\pi(0.00093114) \approx 0.9272$.

(c) $a\sin(220\pi t + k) \approx 5\sin(220\pi t + 0.9272) =$

$5[\sin(220\pi t)\cos 0.9272 + \cos(220\pi t)\sin 0.9272] \approx$

$5[\sin(220\pi t)(0.6) + \cos(220\pi t)(0.8)] = 3\sin(220\pi t) + 4\cos(220\pi t)$

$[0, 0.02, 0.001]$ by $[-6, 6, 1]$

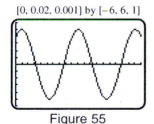

Figure 55

Extended and Discovery Exercises for Section 7.4

1. (a) Graph $Y_1 = 0.006 \sin(880\pi X) + 0.004 \sin(886\pi X)$ as shown in Figure 1a. There are 3 beats per second.

    (b) Graph $Y_1 = 0.006 \sin(440\pi X) + 0.004 \sin(448\pi X)$ as shown in Figure 1b. There are 4 beats per second.

    (c) When the frequencies are F_1 and F_2, the rate of beats per second is given by $|F_2 - F_1|$.

[0.15, 1.15, 0.05] by [−0.01, 0.01, 0.001] [0.15, 1.15, 0.05] by [−0.01, 0.01, 0.001] [0.2, 1.2, 0.05] by [−0.01, 0.01, 0.001]

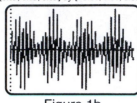 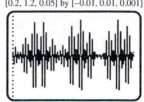

Figure 1a   Figure 1b   Figure 3

3. Graph $Y_1 = 0.004 \cos(830\pi X) + 0.005 \sin(836\pi X)$ as shown in Figure 3. There are 3 beats per second.

Checking Basic Concepts for Sections 7.3 and 7.4

1. (a) $225°$ is in quadrant III, $\theta_R = 225° - 180° = 45°$

    (b) $\dfrac{5\pi}{6}$ is in quadrant II, $\theta_R = \pi - \dfrac{5\pi}{6} = \dfrac{\pi}{6}$

3. (a) $\sin t = -\cos t \Rightarrow \dfrac{\sin t}{\cos t} = -1 \Rightarrow \tan t = -1 \Rightarrow t = \tan^{-1}(-1) = \dfrac{3\pi}{4} + \pi n$

    (b) $2 \sin^2 t = 1 - \cos t \Rightarrow 2(1 - \cos^2 t) = 1 - \cos t \Rightarrow 2 - 2\cos^2 t = 1 - \cos t \Rightarrow$

    $2 \cos^2 t - \cos t - 1 = 0 \Rightarrow (2 \cos t + 1)(\cos t - 1) = 0$

    $\Rightarrow \cos t = -\dfrac{1}{2} \qquad$ or $\qquad \cos t = 1$

    $\Rightarrow t = \cos^{-1}\left(-\dfrac{1}{2}\right) \qquad$ or $\qquad t = \cos^{-1} 1$

    $\Rightarrow t = \dfrac{2\pi}{3} + 2\pi n, \dfrac{4\pi}{3} + 2\pi n, 2\pi n$

5. $\sin(t - \pi) = \sin t \cos \pi - \cos t \sin \pi = \sin t \,(-1) - \cos t \,(0) = -\sin t$

    Graph $Y_1 = \sin(X - \pi)$ and $Y_2 = -\sin(X)$ in $[-2\pi, 2\pi, \pi/2]$ by $[-2, 2, 1]$.

    Graph Y_1 is shown in Figure 5a. Graph Y_2 is shown in Figure 5b. The graphs are the same.

    Table $Y_1 = \sin(X - \pi)$ and $Y_2 = -\sin(X)$ starting at $x = 0$, incrementing by $\dfrac{\pi}{3}$. See Figure 5c.

    The tables are the same.

$[-2\pi, 2\pi, \pi/2]$ by $[-2, 2, 1]$ $[-2\pi, 2\pi, \pi/2]$ by $[-2, 2, 1]$

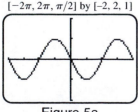 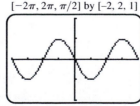 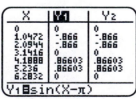

Figure 5a   Figure 5b   Figure 5c

7.5: Multiple-Angle Identities

1. $\sin^2 10t \Rightarrow \theta = 10t \Rightarrow \sin^2 10t = \dfrac{1 - \cos(2(10t))}{2} = \dfrac{1 - \cos(20t)}{2}$

3. $\tan^2 5t \Rightarrow \theta = 5t \Rightarrow \tan^2 5t = \dfrac{1 - \cos(2(5t))}{1 + \cos(2(5t))} = \dfrac{1 - \cos(10t)}{1 + \cos(10t)}$

5. $\sin 20x \Rightarrow \theta = 10x \Rightarrow \sin 20x = 2 \sin 10x \cos 10x$

7. $\tan 5x \Rightarrow \theta = 10x \Rightarrow \tan \dfrac{\theta}{2} = \dfrac{1 - \cos(10x)}{\sin(10x)}$

9. (a) $\sin 30° + \sin 30° = \dfrac{1}{2} + \dfrac{1}{2} = 1$

    (b) $\sin 60° = \dfrac{\sqrt{3}}{2}$

    The results are not the same.

11. (a) $\cos 60° + \cos 60° = \dfrac{1}{2} + \dfrac{1}{2} = 1$

    (b) $\cos 120° = -\dfrac{1}{2}$

    The results are not the same.

13. (a) $\tan 45° + \tan 45° = 1 + 1 = 2$

    (b) $\tan 90°$ is undefined

    The results are not the same.

15. Graphical: Graph $Y_1 = \tan(2X)$ and $Y_2 = 2 \tan(X)$ in dot mode in $[-2\pi, 2\pi, \pi/2]$ by $[-4, 4, 1]$.

    Graph Y_1 is shown in Figure 15a. Graph Y_2 is shown in Figure 15b. Note that the graphs are different.

    Symbolic: $\tan 2\theta = \dfrac{2 \tan \theta}{1 - \tan^2 \theta} \neq 2 \tan \theta$, unless $\tan \theta = 0$

$[-2\pi, 2\pi, \pi/2]$ by $[-4, 4, 1]$ $[-2\pi, 2\pi, \pi/2]$ by $[-4, 4, 1]$

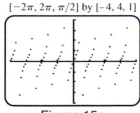

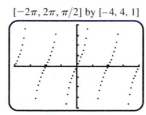

Figure 15a   Figure 15b

17. (a) $\sin 2\theta = 2 \sin \theta \cos \theta = 2\left(\dfrac{3}{5}\right)\left(\dfrac{4}{5}\right) = \dfrac{24}{25}$

    $\cos 2\theta = \cos^2 \theta - \sin^2 \theta = \left(\dfrac{4}{5}\right)^2 - \left(\dfrac{3}{5}\right)^2 = \dfrac{16 - 9}{25} = \dfrac{7}{25}$

    $\tan 2\theta = \dfrac{\sin 2\theta}{\cos 2\theta} = \dfrac{\frac{24}{25}}{\frac{7}{25}} = \dfrac{24}{25} \cdot \dfrac{25}{7} = \dfrac{24}{7}$

(b) Since θ is in quadrant I, $\theta = \cos^{-1}\dfrac{4}{5}$. Numerical support is shown in Figures 17a & 17b.

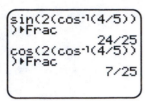

| Figure 17a | Figure 17b |

19. Since $\sin\theta$ is negative and $\cos\theta$ is positive, θ is in quadrant IV. One possibility is shown in Figure 19a.

We see that $\cos\theta = \dfrac{7}{25}$.

(a) $\sin 2\theta = 2\sin\theta\cos\theta = 2\left(-\dfrac{24}{25}\right)\left(\dfrac{7}{25}\right) = -\dfrac{336}{625}$

$\cos 2\theta = \cos^2\theta - \sin^2\theta = \left(\dfrac{7}{25}\right)^2 - \left(-\dfrac{24}{25}\right)^2 = \dfrac{49 - 576}{625} = -\dfrac{527}{625}$

$\tan 2\theta = \dfrac{\sin 2\theta}{\cos 2\theta} = \dfrac{-\frac{336}{625}}{-\frac{527}{625}} = -\dfrac{336}{625}\cdot\left(-\dfrac{625}{527}\right) = \dfrac{336}{527}$

(b) Since θ is in quadrant IV, $\theta = \sin^{-1}\left(-\dfrac{24}{25}\right)$. Numerical support is shown in Figures 19b & 19c.

 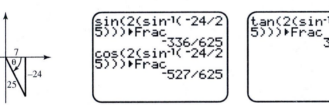 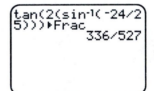

| Figure 19a | Figure 19b | Figure 19c |

21. Since $\sin\theta$ is negative and $\sec\theta$ is positive, θ is in quadrant IV. One possibility is shown in Figure 21a.

We see that $\cos\theta = \dfrac{60}{61}$.

(a) $\sin 2\theta = 2\sin\theta\cos\theta = 2\left(-\dfrac{11}{61}\right)\left(\dfrac{60}{61}\right) = -\dfrac{1320}{3721}$

$\cos 2\theta = \cos^2\theta - \sin^2\theta = \left(\dfrac{60}{61}\right)^2 - \left(-\dfrac{11}{61}\right)^2 = \dfrac{3600 - 121}{3721} = \dfrac{3479}{3721}$

$\tan 2\theta = \dfrac{\sin 2\theta}{\cos 2\theta} = \dfrac{-\frac{1320}{3721}}{\frac{3479}{3721}} = -\dfrac{1320}{3721}\cdot\dfrac{3721}{3479} = -\dfrac{1320}{3479}$

(b) Since θ is in quadrant IV, $\theta = \sin^{-1}\left(-\dfrac{11}{61}\right)$. Numerical support is shown in Figures 21b & 21c.

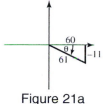

Figure 21a

Figure 21b

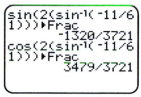

Figure 21c

23. Since $\tan \theta$ is positive and $\cos \theta$ is negative, θ is in quadrant III. One possibility is shown in Figure 23a.

We see that $\sin \theta = -\dfrac{7}{25}$ and $\cos \theta = -\dfrac{24}{25}$.

(a) $\sin 2\theta = 2 \sin \theta \cos \theta = 2\left(-\dfrac{7}{25}\right)\left(-\dfrac{24}{25}\right) = \dfrac{336}{625}$

$\cos 2\theta = \cos^2 \theta - \sin^2 \theta = \left(-\dfrac{24}{25}\right)^2 - \left(-\dfrac{7}{25}\right)^2 = \dfrac{576 - 49}{625} = \dfrac{527}{625}$

$\tan 2\theta = \dfrac{\sin 2\theta}{\cos 2\theta} = \dfrac{\frac{336}{625}}{\frac{527}{625}} = \dfrac{336}{625} \cdot \dfrac{625}{527} = \dfrac{336}{527}$

(b) Since θ is in quadrant III, $\theta = 360° - \cos^{-1}\left(-\dfrac{24}{25}\right)$. Numerical support is shown in Figures 23b & 23c.

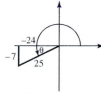

Figure 23a

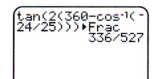

Figure 23b

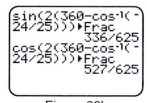

Figure 23c

25. Let $\theta = \cos^{-1} 1$. It follows that $\cos \theta = 1$ and $\sin \theta = 0$ (since $\theta = 0°$).

$\sin \left(2 \cos^{-1} 1\right) = \sin (2\theta) = 2 \sin \theta \cos \theta = 2(0)(1) = 0$

27. Let $\theta = \sin^{-1} \dfrac{7}{25}$. It follows that $\sin \theta = \dfrac{7}{25}$.

$\cos \left(2 \sin^{-1} \dfrac{7}{25}\right) = \cos (2\theta) = 1 - 2\sin^2 \theta = 1 - 2\left(\dfrac{7}{25}\right)^2 = 1 - \dfrac{98}{625} = \dfrac{527}{625}$

Figure 28

29. Let $\theta = \sin^{-1} \dfrac{5}{13}$. It follows that $\sin \theta = \dfrac{5}{13}$ and $\cos \theta = \dfrac{12}{13}$. See Figure 29.

$\cos \left(3 \sin^{-1} \dfrac{5}{13}\right) = \cos (3\theta) = 4\cos^3 \theta - 3\cos \theta = 4\left(\dfrac{12}{13}\right)^3 - 3\left(\dfrac{12}{13}\right) = \dfrac{828}{2197}$

31. Let $\theta = \tan^{-1} x$, where $x < 0$. It follows that $\sin \theta = \dfrac{x}{\sqrt{x^2 + 1}}$ and $\cos \theta = \dfrac{1}{\sqrt{x^2 + 1}}$. See Figure 31.

$$\sin\left(2\tan^{-1} x\right) = \sin(2\theta) = 2\sin\theta\cos\theta = 2\left(\dfrac{x}{\sqrt{x^2 + 1}}\right)\left(\dfrac{1}{\sqrt{x^2 + 1}}\right) = \dfrac{2x}{x^2 + 1}$$

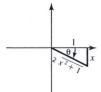

Figure 31

33. Let $\alpha = \sin^{-1}\dfrac{3}{5}$. It follows that $\sin\alpha = \dfrac{3}{5}$ and $\cos\alpha = \dfrac{4}{5}$. See Figure 33a.

Let $\beta = \sin^{-1}\dfrac{4}{5}$. It follows that $\sin\beta = \dfrac{4}{5}$ and $\cos\beta = \dfrac{3}{5}$. See Figure 33b.

$$\cos\left(\sin^{-1}\dfrac{3}{5} - \sin^{-1}\dfrac{4}{5}\right) = \cos(\alpha - \beta) = \cos\alpha\cos\beta + \sin\alpha\sin\beta = \left(\dfrac{4}{5}\right)\left(\dfrac{3}{5}\right) + \left(\dfrac{3}{5}\right)\left(\dfrac{4}{5}\right) = \dfrac{24}{25}$$

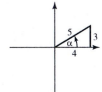   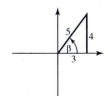

Figure 33a   Figure 33b

35. $2\cos\theta\sin\theta = 2\sin\theta\cos\theta = \sin 2\theta$

37. $\sin\theta\cos\theta = \dfrac{1}{2}(2\sin\theta\cos\theta) = \dfrac{1}{2}\sin 2\theta$

39. $2\cos^2 2\theta - 1 = \cos 4\theta$ \qquad (let $t = 2\theta$, then $2\cos^2 t - 1 = \cos 2t$)

41. $\sin^2 3\theta + \cos^2 3\theta = 1$ \qquad (let $t = 3\theta$, then $\sin^2 t + \cos^2 t = 1$)

43. $\csc^2 5x - 1 = \cot^2 5x$ \qquad (let $t = 5x$, then $\csc^2 t - 1 = \cot^2 t$)

45. $\cos^2(22.5°) = \dfrac{1 + \cos(2\cdot 22.5°)}{2} = \dfrac{1 + \cos 45°}{2} = \dfrac{1 + \frac{\sqrt{2}}{2}}{2} = \dfrac{\frac{2 + \sqrt{2}}{2}}{2} = \dfrac{2 + \sqrt{2}}{4}$

Numerical support is shown in Figure 45.

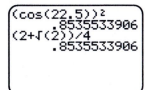

```
(cos(22.5))²
    .8535533906
(2+√(2))/4
    .8535533906
```

```
(tan(75))²
    13.92820323
(2+√(3))/(2-√(3))
)
    13.92820323
```

Figure 45   Figure 47

47. $\tan^2(75°) = \dfrac{1 - \cos(2\cdot 75°)}{1 + \cos(2\cdot 75°)} = \dfrac{1 - \cos 150°}{1 + \cos 150°} = \dfrac{1 - \left(-\frac{\sqrt{3}}{2}\right)}{1 + \left(-\frac{\sqrt{3}}{2}\right)} = \dfrac{\frac{2 + \sqrt{3}}{2}}{\frac{2 - \sqrt{3}}{2}} = \dfrac{2 + \sqrt{3}}{2 - \sqrt{3}}$

Numerical support is shown in Figure 47.

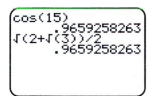

Figure 49a

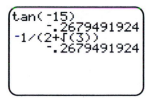

Figure 49b

49. (a) Since the angle is in quadrant I, we use the positive value of the half-angle formula.

$$\cos 15° = \sqrt{\frac{1 + \cos 30°}{2}} = \sqrt{\frac{1 + \frac{\sqrt{3}}{2}}{2}} = \sqrt{\frac{\frac{2 + \sqrt{3}}{2}}{2}} = \sqrt{\frac{2 + \sqrt{3}}{4}} = \frac{\sqrt{2 + \sqrt{3}}}{2}$$

(b) Since the angle is in quadrant IV, we use the negative value of the half-angle formula.

$$\tan(-15°) = -\sqrt{\frac{1 - \cos(-30°)}{1 + \cos(-30°)}} = -\sqrt{\frac{1 - \frac{\sqrt{3}}{2}}{1 + \frac{\sqrt{3}}{2}}} = -\sqrt{\frac{\frac{2 - \sqrt{3}}{2}}{\frac{2 + \sqrt{3}}{2}}} = -\sqrt{\frac{2 - \sqrt{3}}{2 + \sqrt{3}}} \text{ or}$$

$$-\frac{1}{2 + \sqrt{3}} \text{ or } \sqrt{3} - 2. \text{ Numerical support is shown in Figure 49a \& 49b.}$$

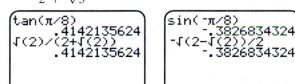

Figure 51a

Figure 51b

51. (a) Since the angle is in quadrant I, we use the positive value of the half-angle formula.

$$\tan\left(\frac{\pi}{8}\right) = \sqrt{\frac{1 - \cos\frac{\pi}{4}}{1 + \cos\frac{\pi}{4}}} = \sqrt{\frac{1 - \frac{\sqrt{2}}{2}}{1 + \frac{\sqrt{2}}{2}}} = \sqrt{\frac{\frac{2 - \sqrt{2}}{2}}{\frac{2 + \sqrt{2}}{2}}} = \sqrt{\frac{2 - \sqrt{2}}{2 + \sqrt{2}}} \text{ or } \frac{\sqrt{2}}{2 + \sqrt{2}} \text{ or } \sqrt{2} - 1$$

(b) Since the angle is in quadrant IV, we use the negative value of the half-angle formula.

$$\sin\left(-\frac{\pi}{8}\right) = -\sqrt{\frac{1 - \cos\left(-\frac{\pi}{4}\right)}{2}} = -\sqrt{\frac{1 - \left(\frac{\sqrt{2}}{2}\right)}{2}} = -\sqrt{\frac{\frac{2 - \sqrt{2}}{2}}{2}} = -\sqrt{\frac{2 - \sqrt{2}}{4}} = -\frac{\sqrt{2 - \sqrt{2}}}{2}$$

Numerical support is shown in Figure 51a & 51b.

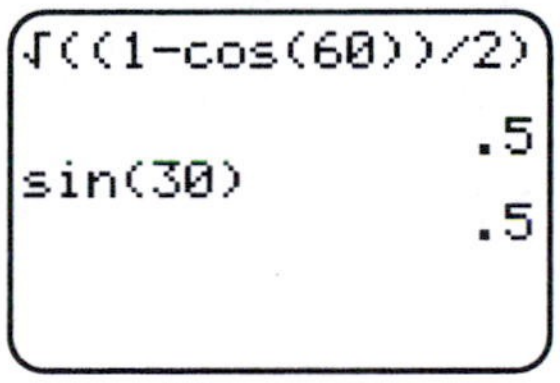

[TI calculator screen: √((1-cos(60))/2) = .5, sin(30) = .5]

Figure 53

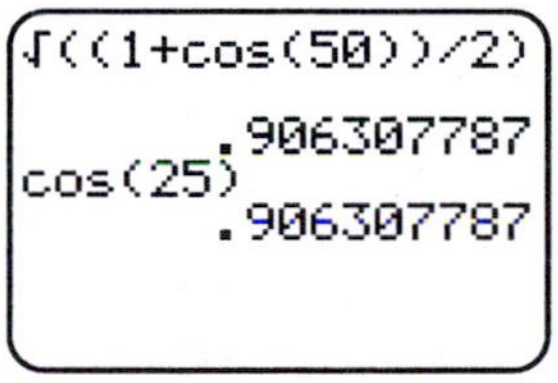

[TI calculator screen: √((1+cos(50))/2) = .906307787, cos(25) = .906307787]

Figure 55

53. $\sqrt{\dfrac{1 - \cos 60°}{2}} = \sin\left(\dfrac{60°}{2}\right) = \sin 30°.$ Numerical support is shown in Figure 53.

55. $\sqrt{\dfrac{1 + \cos 50°}{2}} = \cos\left(\dfrac{50°}{2}\right) = \cos 25°.$ Numerical support is shown in Figure 55.

57. $\sqrt{\dfrac{1 - \cos 40°}{1 + \cos 40°}} = \tan\left(\dfrac{40°}{2}\right) = \tan 20°$. Numerical support is shown in Figure 57.

59. Since $0° < \dfrac{\theta}{2} < 45°$, we know that $\dfrac{\theta}{2}$ is in quadrant I.

$$\sin\dfrac{\theta}{2} = \sqrt{\dfrac{1 - \cos\theta}{2}} = \sqrt{\dfrac{1 - \frac{4}{5}}{2}} = \sqrt{\dfrac{\frac{5-4}{5}}{2}} = \sqrt{\dfrac{1}{10}} = \dfrac{1}{\sqrt{10}}$$

$$\cos\dfrac{\theta}{2} = \sqrt{\dfrac{1 + \cos\theta}{2}} = \sqrt{\dfrac{1 + \frac{4}{5}}{2}} = \sqrt{\dfrac{\frac{5+4}{5}}{2}} = \sqrt{\dfrac{9}{10}} = \dfrac{3}{\sqrt{10}}$$

$$\tan\dfrac{\theta}{2} = \sqrt{\dfrac{1 - \cos\theta}{1 + \cos\theta}} = \sqrt{\dfrac{1 - \frac{4}{5}}{1 + \frac{4}{5}}} = \sqrt{\dfrac{\frac{5-4}{5}}{\frac{5+4}{5}}} = \sqrt{\dfrac{\frac{1}{5}}{\frac{9}{5}}} = \sqrt{\dfrac{1}{9}} = \dfrac{1}{3}$$

61. Since $-45° < \dfrac{\theta}{2} < 0°$, we know that $\dfrac{\theta}{2}$ is in quadrant IV.

By sketching an appropriate triangle in quadrant IV, we see that $\cos\theta = \dfrac{12}{13}$.

$$\sin\dfrac{\theta}{2} = -\sqrt{\dfrac{1 - \cos\theta}{2}} = -\sqrt{\dfrac{1 - \frac{12}{13}}{2}} = -\sqrt{\dfrac{\frac{13-12}{13}}{2}} = -\sqrt{\dfrac{1}{26}} = -\dfrac{1}{\sqrt{26}}$$

$$\cos\dfrac{\theta}{2} = \sqrt{\dfrac{1 + \cos\theta}{2}} = \sqrt{\dfrac{1 + \frac{12}{13}}{2}} = \sqrt{\dfrac{\frac{13+12}{13}}{2}} = \sqrt{\dfrac{25}{26}} = \dfrac{5}{\sqrt{26}}$$

$$\tan\dfrac{\theta}{2} = -\sqrt{\dfrac{1 - \cos\theta}{1 + \cos\theta}} = -\sqrt{\dfrac{1 - \frac{12}{13}}{1 + \frac{12}{13}}} = -\sqrt{\dfrac{\frac{13-12}{13}}{\frac{13+12}{13}}} = -\sqrt{\dfrac{\frac{1}{13}}{\frac{25}{13}}} = -\sqrt{\dfrac{1}{25}} = -\dfrac{1}{5}$$

63. Since $45° < \dfrac{\theta}{2} < 90°$, we know that $\dfrac{\theta}{2}$ is in quadrant I.

By sketching an appropriate triangle in quadrant II, we see that $\cos\theta = -\dfrac{7}{25}$.

$$\sin\dfrac{\theta}{2} = \sqrt{\dfrac{1 - \cos\theta}{2}} = \sqrt{\dfrac{1 - \left(-\frac{7}{25}\right)}{2}} = \sqrt{\dfrac{\frac{25+7}{25}}{2}} = \sqrt{\dfrac{32}{50}} = \sqrt{\dfrac{16}{25}} = \dfrac{4}{5}$$

$$\cos\dfrac{\theta}{2} = \sqrt{\dfrac{1 + \cos\theta}{2}} = \sqrt{\dfrac{1 + \left(-\frac{7}{25}\right)}{2}} = \sqrt{\dfrac{\frac{25-7}{25}}{2}} = \sqrt{\dfrac{18}{50}} = \sqrt{\dfrac{9}{25}} = \dfrac{3}{5}$$

$$\tan\dfrac{\theta}{2} = \sqrt{\dfrac{1 - \cos\theta}{1 + \cos\theta}} = \sqrt{\dfrac{1 - \left(-\frac{7}{25}\right)}{1 + \left(-\frac{7}{25}\right)}} = \sqrt{\dfrac{\frac{25+7}{25}}{\frac{25-7}{25}}} = \sqrt{\dfrac{\frac{32}{25}}{\frac{18}{25}}} = \sqrt{\dfrac{32}{18}} = \sqrt{\dfrac{16}{9}} = \dfrac{4}{3}$$

65. $4\sin 2x = 4(2\sin x\cos x) = 8\sin x\cos x$

Graph $Y_1 = 4\sin(2X)$ and $Y_2 = 8\sin(X)\cos(X)$ in $[-2\pi, 2\pi, \pi/2]$ by $[-6, 6, 1]$.

Graph Y_1 is shown in Figure 65a. Graph Y_2 is shown in Figure 65b. The graphs are the same.

Table $Y_1 = 4\sin(2X)$ and $Y_2 = 8\sin(X)\cos(X)$ starting at $x = 0$, incrementing by $\dfrac{\pi}{6}$. See Figure 65c.

The tables are the same.

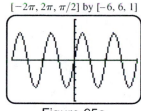

$[-2\pi, 2\pi, \pi/2]$ by $[-6, 6, 1]$

Figure 65a

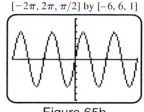

$[-2\pi, 2\pi, \pi/2]$ by $[-6, 6, 1]$

Figure 65b

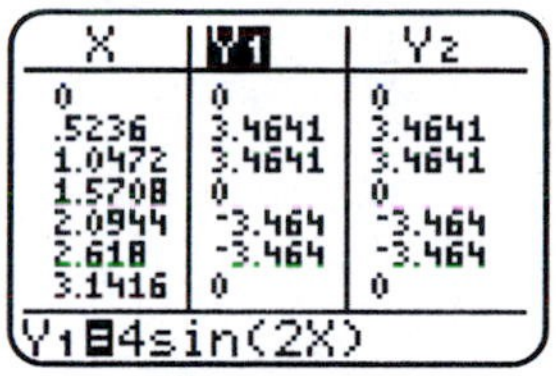

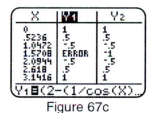

Figure 65c

67. $\dfrac{2 - \sec^2 x}{\sec^2 x} = \dfrac{2}{\sec^2 x} - \dfrac{\sec^2 x}{\sec^2 x} = 2\cos^2 x - 1 = \cos 2x$

Graph $Y_1 = (2 - 1/(\cos (X))^\wedge 2)/(1/(\cos (X))^\wedge 2)$ and $Y_2 = \cos (2X)$ in $[-2\pi, 2\pi, \pi/2]$ by $[-2, 2, 1]$.

Graph Y_1 is shown in Figure 67a. Graph Y_2 is shown in Figure 67b. The graphs are the same.

Table $Y_1 = (2 - 1/(\cos (X))^\wedge 2)/(1/(\cos (X))^\wedge 2)$ and $Y_2 = \cos (2X)$ starting at $x = 0$, incrementing by $\dfrac{\pi}{6}$.

See Figure 67c. The tables are the same.

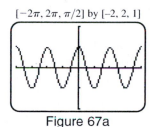

$[-2\pi, 2\pi, \pi/2]$ by $[-2, 2, 1]$

Figure 67a

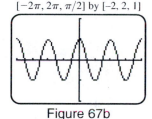

$[-2\pi, 2\pi, \pi/2]$ by $[-2, 2, 1]$

Figure 67b

Figure 67c

69. $\sec 2x = \dfrac{1}{\cos 2x} = \dfrac{1}{1 - 2\sin^2 x}$

Graph $Y_1 = 1/\cos (2X)$ and $Y_2 = 1/(1 - 2 (\sin (X))^\wedge 2)$ in $[-2\pi, 2\pi, \pi/2]$ by $[-4, 4, 1]$.

Graph Y_1 is shown in Figure 69a. Graph Y_2 is shown in Figure 69b. The graphs are the same.

Table $Y_1 = 1/\cos (2X)$ and $Y_2 = 1/(1 - 2 (\sin (X))^\wedge 2)$ starting at $x = 0$, incrementing by $\dfrac{\pi}{6}$.

See Figure 69c. The tables are the same.

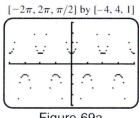

$[-2\pi, 2\pi, \pi/2]$ by $[-4, 4, 1]$

Figure 69a

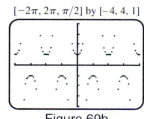

$[-2\pi, 2\pi, \pi/2]$ by $[-4, 4, 1]$

Figure 69b

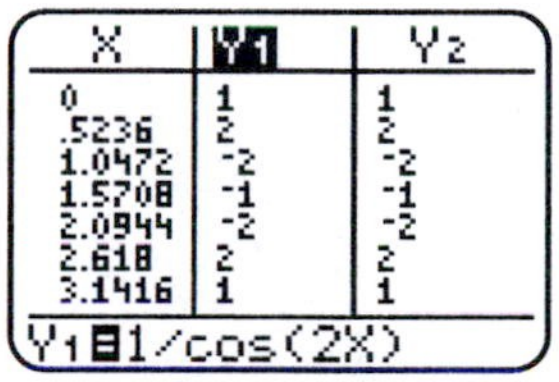

Figure 69c

71. $\sin 3\theta = \sin (2\theta + \theta) = \sin 2\theta \cos \theta + \cos 2\theta \sin \theta = (2 \sin \theta \cos \theta) \cos \theta + (1 - 2\sin^2 \theta) \sin \theta =$

$2 \sin \theta \cos^2 \theta + \sin \theta - 2 \sin^3 \theta = 2 \sin \theta (1 - \sin^2 \theta) + \sin \theta - 2 \sin^3 \theta =$

$2 \sin \theta - 2 \sin^3 \theta + \sin \theta - 2 \sin^3 \theta = 3 \sin \theta - 4 \sin^3 \theta$

Graph $Y_1 = \sin (3X)$ and $Y_2 = 3 \sin (X) - 4 (\sin (X))^\wedge 3$ in $[-2\pi, 2\pi, \pi/2]$ by $[-2, 2, 1]$.

Graph Y_1 is shown in Figure 71a. Graph Y_2 is shown in Figure 71b. The graphs are the same.

Table $Y_1 = \sin(3X)$ and $Y_2 = 3 \sin(X) - 4 (\sin(X))^{\wedge}3$ starting at $x = 0$, incrementing by $\dfrac{\pi}{6}$.

See Figure 71c. The tables are the same.

$[-2\pi, 2\pi, \pi/2]$ by $[-2, 2, 1]$ $[-2\pi, 2\pi, \pi/2]$ by $[-2, 2, 1]$

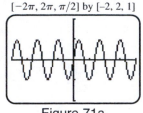   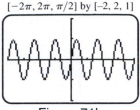   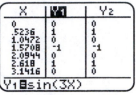

     Figure 71a   Figure 71b   Figure 71c

73. $\sin 4\theta = \sin(2 \cdot 2\theta) = 2 \sin 2\theta \cos 2\theta = 2(2 \sin \theta \cos \theta) \cos 2\theta = 4 \sin \theta \cos \theta \cos 2\theta$

    Graph $Y_1 = \sin(4X)$ and $Y_2 = 4 \sin(X) \cos(X) \cos(2X)$ in $[-2\pi, 2\pi, \pi/2]$ by $[-2, 2, 1]$.

    Graph Y_1 is shown in Figure 73a. Graph Y_2 is shown in Figure 73b. The graphs are the same.

    Table $Y_1 = \sin(4X)$ and $Y_2 = 4 \sin(X) \cos(X) \cos(2X)$ starting at $x = 0$, incrementing by $\dfrac{\pi}{6}$.

    See Figure 73c. The tables are the same.

$[-2\pi, 2\pi, \pi/2]$ by $[-2, 2, 1]$ $[-2\pi, 2\pi, \pi/2]$ by $[-2, 2, 1]$

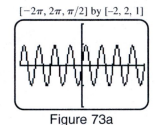   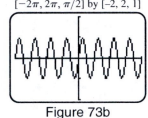

     Figure 73a   Figure 73b   Figure 73c

75. $\dfrac{\sin 2\theta}{\sin \theta} = \dfrac{2 \sin \theta \cos \theta}{\sin \theta} = 2 \cos \theta$

77. $2 \cos^2 \dfrac{\theta}{2} = 2\left(\dfrac{1 + \cos \theta}{2}\right) = 1 + \cos \theta$

79. $\cos^4 \theta - \sin^4 \theta = (\cos^2 \theta - \sin^2 \theta)(\cos^2 \theta + \sin^2 \theta) = (\cos 2\theta)(1) = \cos 2\theta$

81. $\csc 2t = \dfrac{1}{\sin 2t} = \dfrac{1}{2 \sin t \cos t} = \dfrac{\csc t}{2 \cos t}$

83. $\tan \dfrac{x}{2} = \dfrac{\sin \frac{x}{2}}{\cos \frac{x}{2}} = \dfrac{2 \sin \frac{x}{2}}{2 \cos \frac{x}{2}} = \dfrac{2 \sin \frac{x}{2} \cos \frac{x}{2}}{2 \cos^2 \frac{x}{2}} = \dfrac{\sin x}{1 + (2 \cos^2 \frac{x}{2} - 1)} = \dfrac{\sin x}{1 + \cos x}$

85. (a) $\cos 50° \sin 20° = \dfrac{1}{2}(\sin(50° + 20°) - \sin(50° - 20°)) = \dfrac{1}{2}(\sin 70° - \sin 30°)$

    (b) $\cos 2x \cos x = \dfrac{1}{2}(\cos(2x + x) + \cos(2x - x)) = \dfrac{1}{2}(\cos(3x) + \cos(x))$

87. (a) $\sin 7\theta \cos 3\theta = \dfrac{1}{2}(\sin(7\theta + 3\theta) + \sin(7\theta - 3\theta)) = \dfrac{1}{2}(\sin 10\theta + \sin 4\theta)$

    (b) $\sin 8x \sin 4x = \dfrac{1}{2}(\cos(8x - 4x) - \cos(8x + 4x)) = \dfrac{1}{2}(\cos 4x - \cos 12x)$

89. (a) $\sin 40° + \sin 30° = 2 \sin\left(\dfrac{40° + 30°}{2}\right) \cos\left(\dfrac{40° - 30°}{2}\right) = 2 \sin 35° \cos 5°$

    (b) $\cos 45° + \cos 35° = 2 \cos\left(\dfrac{45° + 35°}{2}\right) \cos\left(\dfrac{45° - 35°}{2}\right) = 2 \cos 40° \cos 5°$

91. (a) $\cos 6\theta + \cos 4\theta = 2 \cos \left(\dfrac{6\theta + 4\theta}{2}\right) \cos \left(\dfrac{6\theta - 4\theta}{2}\right) = 2 \cos 5\theta \cos \theta$

(b) $\sin 7x + \sin 4x = 2 \sin \left(\dfrac{7x + 4x}{2}\right) \cos \left(\dfrac{7x - 4x}{2}\right) = 2 \sin \dfrac{11x}{2} \cos \dfrac{3x}{2}$

93. (a) Symbolic: $\cos 2\theta = 1 \Rightarrow 2\theta = \cos^{-1} 1 \Rightarrow 2\theta = 0°, 360° \Rightarrow \theta = 0°, 180°$

(b) Graphical: Graph $Y_1 = \cos(2X)$ and $Y_2 = 1$ in degree mode in $[0°, 360°, 30°]$ by $[-2, 2, 1]$.

　　In $[0°, 360°)$, the x-coordinates of the intersection points are $x = 0°, 180°$. See Figure 93b.

(c) Numerical: Table $Y_1 = \cos(2X)$ and $Y_2 = 1$ in degree mode starting at $x = 0$, incrementing by 60.

　　In the interval $[0°, 360°)$, the tables match when $x = 0°, 180°$. See Figure 93c.

$[0°, 360°, 30°]$ by $[-2, 2, 1]$　　　　　　　　　　　　$[0°, 360°, 30°]$ by $[-2, 2, 1]$

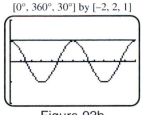
Figure 93b

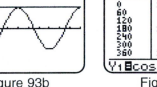
Figure 93c

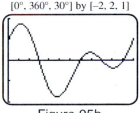
Figure 95b

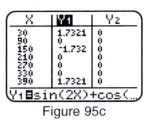
Figure 95c

95. (a) Symbolic: $\sin 2\theta + \cos \theta = 0 \Rightarrow 2\sin \theta \cos \theta + \cos \theta = 0 \Rightarrow \cos \theta(2\sin \theta + 1) = 0 \Rightarrow \cos \theta = 0,$

$2 \sin \theta + 1 = 0.$ $\cos \theta = 0 \Rightarrow \theta = 90°, 270°.$ $2 \sin \theta + 1 = 0 \Rightarrow \sin \theta = -\dfrac{1}{2} \Rightarrow \theta = 210°, 330°.$

(b) Graphical: Graph $Y_1 = \sin(2X) + \cos(X)$ and $Y_2 = 0$ in degree mode in $[0°, 360°, 30°]$ by $[-2, 2, 1]$.

　　In $[0°, 360°)$, the x-coordinates of the intersection points are $x = 90°, 210°, 270°, 330°$. See Figure 95b.

(c) Numerical: Table $Y_1 = \sin(2X) + \cos(X)$ and $Y_2 = 0$ in degree mode starting at $x = 30$, incrementing

　　by 60. In the interval $[0°, 360°)$, the tables match when $x = 90°, 210°, 270°, 330°$. See Figure 95c.

97. (a) Symbolic: $\sin \dfrac{\theta}{2} = 1 \Rightarrow \dfrac{\theta}{2} = \sin^{-1} 1 \Rightarrow \dfrac{\theta}{2} = 90° \Rightarrow \theta = \pi$

(b) Graphical: Graph $Y_1 = \sin(X/2)$ and $Y_2 = 1$ in radian mode in $\left[0, 2\pi, \dfrac{\pi}{6}\right)$ by $[-2, 2, 1]$.

　　In $[0, 2\pi)$, the x-coordinate of the intersection point is $x = 180°$. See Figure 97b.

(c) Numerical: Table $Y_1 = \sin(X/2)$ and $Y_2 = 1$ in degree mode starting at $x = 0$, incrementing by 60.

　　In the interval $[0°, 360°)$, the tables match when $x = \pi$. See Figure 97c.

$[0, 2\pi, \pi/6]$ by $[-2, 2, 1]$　　　　　　　　　　　　$[0°, 2\pi, \pi/6)$ by $[-2, 2, 1]$

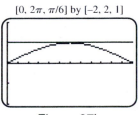
Figure 97b

Figure 97c

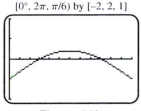
Figure 99b

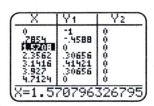
Figure 99c

99. (a) Symbolic: $\sqrt{2} \sin \frac{\theta}{2} - 1 = 0 \Rightarrow \sqrt{2} \sin \frac{\theta}{2} = 1 \Rightarrow \sin \frac{\theta}{2} = \frac{\sqrt{2}}{2} \Rightarrow \frac{\theta}{2} = \sin^{-1} \frac{\sqrt{2}}{2} \Rightarrow$

$\frac{\theta}{2} = \frac{\pi}{4}, \frac{3\pi}{4} \Rightarrow \theta = \frac{\pi}{2}, \frac{3\pi}{2}$

(b) Graphical: graph $Y_1 = \sqrt{2} \sin(X/2) - 1$ and $Y_2 = 0$ in radian mode in $\left[0, 2\pi, \frac{\pi}{6}\right)$ by $[-2, 2, 1]$. In

$[0, 2\pi)$, the x-coordinates of the intersection points are $x = \frac{\pi}{2}, \frac{3\pi}{2}$. See Figure 99b.

(c) Numeric: table $Y_1 = \sqrt{2} \sin(X/2) - 1$ and $Y_2 = 0$ in radian mode starting at $x = 0$, incrementing by $\frac{\pi}{4}$.

In the interval $[0, 2\pi)$, the tables match when $x = \frac{\pi}{2}, \frac{3\pi}{2}$. See Figure 99c.

101. $2 \cos 2t = \sqrt{3} \Rightarrow \cos 2t = \frac{\sqrt{3}}{2}. \cos 2t = \frac{\sqrt{3}}{2} \Rightarrow 2t = \frac{\pi}{6} + 2\pi n, \frac{11\pi}{6} + 2\pi n \Rightarrow$

$t = \frac{\pi}{12} + \pi n, \frac{11\pi}{12} + \pi n.$

103. $\sin 2t + \sin t = 0 \Rightarrow 2 \sin t \cos t + \sin t = 0 \Rightarrow \sin t (2 \cos t + 1) = 0$

$\Rightarrow \sin t = 0 \qquad\qquad \text{or} \qquad 2 \cos t + 1 = 0$

$\Rightarrow \sin t = 0 \qquad\qquad \text{or} \qquad \cos t = -\frac{1}{2}$

$\Rightarrow t = \sin^{-1} 0 \qquad\qquad \text{or} \qquad t = \cos^{-1} \left(-\frac{1}{2}\right)$

$\Rightarrow t = \pi n, \frac{2\pi}{3} + 2\pi n, \frac{4\pi}{3} + 2\pi n$

105. $2 \sin \frac{t}{2} - 1 = 0 \Rightarrow \sin \frac{t}{2} = \frac{1}{2} \Rightarrow \frac{t}{2} = \sin^{-1} \frac{1}{2}$

$\Rightarrow \frac{t}{2} = \frac{\pi}{6} + 2\pi n, \frac{5\pi}{6} + 2\pi n$

$\Rightarrow t = \frac{\pi}{3} + 4\pi n, \frac{5\pi}{3} + 4\pi n$

107. $\cos 2t = \sin t \Rightarrow 1 - 2 \sin^2 t = \sin t \Rightarrow 2 \sin^2 t + \sin t - 1 = 0 \Rightarrow (2 \sin t - 1)(\sin t + 1) = 0$

$\Rightarrow 2 \sin t - 1 = 0 \qquad\qquad \text{or} \qquad \sin t + 1 = 0$

$\Rightarrow \sin t = \frac{1}{2} \qquad\qquad \text{or} \qquad \sin t = -1$

$\Rightarrow t = \sin^{-1} \frac{1}{2} \qquad\qquad \text{or} \qquad t = \sin^{-1}(-1)$

$\Rightarrow t = \frac{\pi}{6} + 2\pi n, \frac{5\pi}{6} + 2\pi n, \frac{3\pi}{2} + 2\pi n$

109. $\tan 2t = 1 \Rightarrow 2t = \tan^{-1} 1 \Rightarrow 2t = \frac{\pi}{4} + 2\pi n$ or $2t = \frac{5\pi}{4} + 2\pi n \Rightarrow t = \frac{\pi}{8} + \pi n$ or $t = \frac{5\pi}{8} + \pi n$

or equivalently,

111. $2 \cos \frac{t}{2} = 1 \Rightarrow \cos \frac{t}{2} = \frac{1}{2} \Rightarrow \frac{t}{2} = \cos^{-1} \frac{1}{2} \Rightarrow \frac{t}{2} = \frac{\pi}{3} + 2\pi n$ or $\frac{t}{2} = \frac{5\pi}{3} + 2\pi n \Rightarrow$

$t = \frac{2\pi}{3} + 4\pi n$ or $t = \frac{10\pi}{3} + 4\pi n$

113. $\cos 2t = 2 \sin t \cos t$, let $2 \sin t \cos t = \sin 2t$ and let $\theta = 2t$. So then the equation becomes $\cos \theta = \sin \theta$,

divide both sides by $\cos \theta \Rightarrow \dfrac{\cos \theta}{\cos \theta} = \dfrac{\sin \theta}{\cos \theta} \Rightarrow 1 = \tan \theta \Rightarrow \theta = \dfrac{\pi}{4}$ and $\dfrac{5\pi}{4}$. For all real number

solutions $\theta = \dfrac{\pi}{4} + 2\pi n$ or $\theta = \dfrac{5\pi}{4} + 2\pi n$. Because $\theta = 2t$, we can determine t by substituting $2t$ for $\theta \Rightarrow$

$2t = \dfrac{\pi}{4} + 2\pi n$ or $2t = \dfrac{5\pi}{4} + 2\pi n$. Therefore $t = \dfrac{\pi}{8} + \pi n$ or $t = \dfrac{5\pi}{8} + \pi n$ or equivalently, $\dfrac{\pi}{8} + \dfrac{\pi n}{2}$.

115. $2 \sin^2 2t + \sin 2t - 1 = 0 \Rightarrow (2 \sin 2t - 1)(\sin 2t + 1) = 0$

$\Rightarrow 2 \sin 2t - 1 = 0 \quad$ or $\quad \sin 2t + 1 = 0$

$\Rightarrow \sin 2t = \dfrac{1}{2} \qquad$ or $\quad \sin 2t = -1$

Let $\theta = 2t$, then

$\Rightarrow \sin \theta = \dfrac{1}{2} \qquad$ or $\quad \sin \theta = -1$

$\Rightarrow \theta = \sin^{-1}\left(\dfrac{1}{2}\right) \quad$ or $\quad \theta = \sin^{-1}(-1)$

$\Rightarrow \theta = \dfrac{\pi}{6}$ or $\dfrac{5\pi}{6} \quad$ or $\quad \theta = \dfrac{3\pi}{2}$

For all real number solutions $\theta = \dfrac{\pi}{6} + 2\pi n$ or $\theta = \dfrac{5\pi}{6} + 2\pi n$ or $\theta = \dfrac{3\pi}{2} + 2\pi n$. Because $\theta = 2t$, we can

determine t by substituting $2t$ for $\theta \Rightarrow 2t = \dfrac{\pi}{6} + 2\pi n$ or $2t = \dfrac{5\pi}{6} + 2\pi n$ or $2t = \dfrac{3\pi}{2} + 2\pi n$.

Therefore $t = \dfrac{\pi}{12} + \pi n$ or $t = \dfrac{5\pi}{12} + \pi n$ or $t = \dfrac{3\pi}{4} + \pi n$.

117. $\sin t + \sin 2t = \cos t$. Graph $Y_1 = \sin(X) + \sin(2X)$ and $Y_2 = \cos(X)$ in radian mode in $[0, 2\pi, \pi/2]$ by

$[-3, 3, 1]$. See Figure 117. The x-coordinates of the intersection points are approximately 0.333 and 4.379.

$[0, 2\pi, \pi/2]$ by $[-3, 3, 1]$ $[0, 0.04, 0.01]$ by $[-500, 2500, 500]$

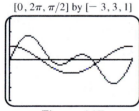
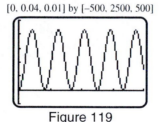

Figure 117   Figure 119

119. (a) Graph $Y_1 = (310 \sin(120\pi X))(7 \sin(120\pi X))$ in $[0, 0.04, 0.01]$ by $[-500, 2500, 500]$.

See Figure 119.

(b) $W = (310 \sin(120\pi x))(7 \sin(120\pi x)) = 2170 \sin^2(120\pi x) = 2170 \cdot \dfrac{1 - \cos(240\pi x)}{2} =$

$1085(1 - \cos(240\pi x)) = 1085 - 1085 \cos(240\pi x) = -1085 \cos(240\pi x) + 1085 \Rightarrow$

$a = -1085, k = 240, d = 1085$

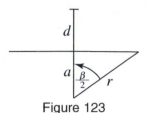

Figure 123

121. We must solve the equation $V(t) = 160$ for t: $320 \sin(120\pi t) = 160 \Rightarrow \sin(120\pi t) = \dfrac{1}{2}$.

Let $\theta = 120\pi t$ then $\sin \theta = \dfrac{1}{2} \Rightarrow \theta = \dfrac{\pi}{6} + 2\pi n$ or $\theta = \dfrac{5\pi}{6} + 2\pi n$. Substituting $120\pi t$ for θ yields

$\Rightarrow 120\pi t = \dfrac{\pi}{6} + 2\pi n \qquad$ or $\qquad 120\pi t = \dfrac{5\pi}{6} + 2\pi n$

$\Rightarrow t = \dfrac{1}{720} + \dfrac{n}{60} \qquad$ or $\qquad t = \dfrac{5}{720} + \dfrac{n}{60}$ sec

123. (a) $d = 600\left(1 - \cos\dfrac{80°}{2}\right) = 600(1 - \cos 40°) \approx 140.4$ ft

(b) By bisecting β, we create a diagram as shown in Figure 123.

Here $r = d + a$ and from right triangle trigonometry, $\cos\dfrac{\beta}{2} = \dfrac{a}{r} \Rightarrow a = r\cos\dfrac{\beta}{2}$.

Thus, $r = d + r\cos\dfrac{\beta}{2} \Rightarrow d = r - r\cos\dfrac{\beta}{2} \Rightarrow d = r\left(1 - \cos\dfrac{\beta}{2}\right)$

(c) No, $\cos\dfrac{\beta}{2} = \pm\sqrt{\dfrac{1 + \cos\beta}{2}} \neq \dfrac{1}{2}\cos\beta$

125. Let $f(t)$ model the tone for number 3 and let $g(t)$ model the tone for number 4.

(a) $f(t) = \cos(2(697)\pi t) + \cos(2(1477)\pi t) = \cos(1394\pi t) + \cos(2954\pi t)$

$g(t) = \cos(2(770)\pi t) + \cos(2(1209)\pi t) = \cos(1540\pi t) + \cos(2418\pi t)$

(b) $f(t) = 2\cos\left(\dfrac{1394\pi t + 2954\pi t}{2}\right)\cos\left(\dfrac{1394\pi t - 2954\pi t}{2}\right) = 2\cos(2174\pi t)\cos(-780\pi t) =$

$2\cos(2174\pi t)\cos(780\pi t)$

$g(t) = 2\cos\left(\dfrac{1540\pi t + 2418\pi t}{2}\right)\cos\left(\dfrac{1540\pi t - 2418\pi t}{2}\right) = 2\cos(1979\pi t)\cos(-439\pi t) =$

$2\cos(1979\pi t)\cos(439\pi t)$

Checking Basic Concepts for Section 17.5

1. $\sin 2\theta = 2\sin\theta\cos\theta = 2\left(\dfrac{24}{25}\right)\left(-\dfrac{7}{25}\right) = -\dfrac{336}{625}$

$\cos 2\theta = \cos^2\theta - \sin^2\theta = \left(-\dfrac{7}{25}\right)^2 - \left(\dfrac{24}{25}\right)^2 = -\dfrac{527}{625}$

3. When θ is acute and $\sin \theta = \dfrac{4}{5}$, then $\cos \theta = \dfrac{3}{5}$.

$$\sin \frac{\theta}{2} = \sqrt{\frac{1 - \cos \theta}{2}} = \sqrt{\frac{1 - \frac{3}{5}}{2}} = \sqrt{\frac{\frac{5-3}{5}}{2}} = \sqrt{\frac{2}{10}} = \sqrt{\frac{1}{5}} = \frac{1}{\sqrt{5}}$$

$$\cos \frac{\theta}{2} = \sqrt{\frac{1 + \cos \theta}{2}} = \sqrt{\frac{1 + \frac{3}{5}}{2}} = \sqrt{\frac{\frac{5+3}{5}}{2}} = \sqrt{\frac{8}{10}} = \sqrt{\frac{4}{5}} = \frac{2}{\sqrt{5}}$$

5. $\sin 2\theta = 2 \cos \theta \Rightarrow 2 \sin \theta \cos \theta = 2 \cos \theta \Rightarrow 2 \sin \theta \cos \theta - 2 \cos \theta = 0 \Rightarrow 2 \cos \theta (\sin \theta - 1) = 0$

$\Rightarrow \cos \theta = 0 \qquad$ or $\qquad \sin \theta - 1 = 0$

$\Rightarrow \cos \theta = 0 \qquad$ or $\qquad \sin \theta = 1$

$\Rightarrow \theta = \cos^{-1} 0 \qquad$ or $\qquad \theta = \sin^{-1} 1$

$\Rightarrow \theta = \dfrac{\pi}{2}, \dfrac{3\pi}{2}$

Chapter 7 Review Exercises

1. If $\sec \theta < 0$ and $\sin \theta > 0$ then any point (x, y) on the terminal side of θ must satisfy

$y > 0$ and $\dfrac{r}{x} < 0 \Rightarrow x < 0$. Thus, θ is contained in quadrant II.

3. $\tan \theta = \dfrac{\sin \theta}{\cos \theta} = \dfrac{\frac{3}{5}}{-\frac{4}{5}} = -\dfrac{3}{4} \qquad\qquad \csc \theta = \dfrac{1}{\sin \theta} = \dfrac{1}{\frac{3}{5}} = \dfrac{5}{3}$

$\sec \theta = \dfrac{1}{\cos \theta} = \dfrac{1}{-\frac{4}{5}} = -\dfrac{5}{4} \qquad\qquad \cot \theta = \dfrac{1}{\tan \theta} = \dfrac{1}{-\frac{3}{4}} = -\dfrac{4}{3}$

5. $\sin \theta = \cos \theta \tan \theta = \dfrac{24}{25}\left(-\dfrac{7}{24}\right) = -\dfrac{7}{25} \qquad \csc \theta = \dfrac{1}{\sin \theta} = \dfrac{1}{-\frac{7}{25}} = -\dfrac{25}{7}$

$\sec \theta = \dfrac{1}{\cos \theta} = \dfrac{1}{\frac{24}{25}} = \dfrac{25}{24} \qquad\qquad \cot \theta = \dfrac{1}{\tan \theta} = \dfrac{1}{-\frac{7}{24}} = -\dfrac{24}{7}$

7. Since the sine function is an odd function $\sin(-13°) = -\sin 13°$.

9. Since the secant function is an even function $\sec\left(-\dfrac{3\pi}{7}\right) = \sec \dfrac{3\pi}{7}$.

11. $\sec \theta \cot \theta \sin \theta = \dfrac{1}{\cos \theta} \cdot \dfrac{\cos \theta}{\sin \theta} \cdot \dfrac{\sin \theta}{1} = 1$

13. $(\sec^2 t - 1)(\csc^2 t - 1) = (\tan^2 t)(\cot^2 t) = 1$

15. $\dfrac{\csc \theta \sin \theta}{\sec \theta} = \dfrac{1}{\sec \theta} = \cos \theta$

17. $\tan \theta = 1.2367 \Rightarrow \theta = \tan^{-1} 1.2367 \qquad\qquad \sin \theta = \sin(\tan^{-1} 1.2367) \approx 0.7776$

$$\cos \theta = \cos (\tan^{-1} 1.2367) \approx 0.6288 \qquad \csc \theta = \frac{1}{\sin (\tan^{-1} 1.2367)} \approx 1.2860$$

$$\sec \theta = \frac{1}{\cos (\tan^{-1} 1.2367)} \approx 1.5904 \qquad \cot \theta = \frac{1}{\tan \theta} \approx 0.8086$$

19. $\cos \theta = -0.4544 \Rightarrow \theta = \cos^{-1} (-0.4544)$

$$\sin \theta = \sin (\cos^{-1} (-0.4544)) \approx 0.8908 \qquad \tan \theta = \tan (\cos^{-1} (-0.4544)) \approx -1.9604$$

$$\csc \theta = \frac{1}{\sin \theta} \approx 1.1226 \qquad \sec \theta = \frac{1}{\cos \theta} \approx -2.2007 \qquad \cot \theta = \frac{1}{\tan \theta} \approx -0.5101$$

21. $\sin^2 \theta + 2 \sin \theta + 1 = (\sin \theta + 1)(\sin \theta + 1)$

23. $\tan^2 \theta - 9 = (\tan \theta + 3)(\tan \theta - 3)$

25. $(\sec \theta - 1)(\sec \theta + 1) = \sec^2 \theta - 1 = \tan^2 \theta$

27. $(1 + \tan t)^2 = 1 + 2 \tan t + \tan^2 t = \sec^2 t + 2 \tan t$

29. $\sin (x - \pi) = \sin x \cos \pi - \cos x \sin \pi = \sin x (-1) - \cos x (0) = -\sin x$

31. $\sin 8x = \sin (2 \cdot 4x) = 2 \sin 4x \cos 4x$

33. $\sec 2x = \dfrac{1}{\cos 2x} = \dfrac{1}{2 \cos^2 x - 1}$

35. $\cos^4 x \sin^3 x = \cos^4 x (\sin^2 x) \sin x = \cos^4 x (1 - \cos^2 x) \sin x = (\cos^4 x - \cos^6 x) \sin x$

37. $\sec^4 \theta - \tan^4 \theta = (\sec^2 \theta - \tan^2 \theta)(\sec^2 \theta + \tan^2 \theta) = (1)((1 + \tan^2 \theta) + \tan^2 \theta) = 1 + 2 \tan^2 \theta$

39. $\theta = 240°$ is in quadrant III, $\theta_R = 240° - 180° = 60°$

41. $\theta = \dfrac{9\pi}{7}$ is in quadrant III, $\theta_R = \dfrac{9\pi}{7} - \pi = \dfrac{2\pi}{7}$

43. Estimate: $\theta = \dfrac{\pi}{2}, \dfrac{3\pi}{2}$

   Symbolic: $\cos^2 \theta - 2 \cos \theta = 0 \Rightarrow \cos \theta (\cos \theta - 2) = 0$

   $\Rightarrow \cos \theta = 0 \qquad$ or $\qquad \cos \theta - 2 = 0$

   $\Rightarrow \cos \theta = 0 \qquad$ or $\qquad \cos \theta = 2$ (not possible)

   $\Rightarrow \theta = \cos^{-1} 0 \Rightarrow \theta = \dfrac{\pi}{2}, \dfrac{3\pi}{2}$

45. (a) $\tan \theta = \sqrt{3} \Rightarrow \theta = \tan^{-1} \sqrt{3} = 60°, 240°$

   (b) $\cot \theta = -\sqrt{3} \Rightarrow \tan \theta = -\dfrac{1}{\sqrt{3}} \Rightarrow \theta = \tan^{-1} \left(-\dfrac{1}{\sqrt{3}} \right) = 150°, 330°$

47. $2 \cos \theta - 1 = 0 \Rightarrow \cos \theta = \dfrac{1}{2} \Rightarrow \theta = \cos^{-1} \dfrac{1}{2} = \dfrac{\pi}{6}, \dfrac{5\pi}{6}$

49. $2 \sin^2 \theta + \sin \theta - 3 = 0 \Rightarrow (\sin \theta - 1)(2 \sin \theta + 3)$

   $\Rightarrow \sin \theta - 1 = 0 \qquad$ or $\qquad 2 \sin \theta + 3 = 0$

   $\Rightarrow \sin \theta = 1 \qquad$ or $\qquad \sin \theta = -\dfrac{3}{2}$ (not possible)

   $\Rightarrow \theta = \sin^{-1} 1 = \dfrac{\pi}{2}$

51. $\tan^2 t - 2 \tan t + 1 = 0 \Rightarrow (\tan t - 1)(\tan t - 1) = 0 \Rightarrow \tan t = 1 \Rightarrow t = \tan^{-1} 1 \Rightarrow t = \dfrac{\pi}{4}, \dfrac{5\pi}{4}$

53. $3 \tan^2 t - 1 = 0 \Rightarrow \tan^2 t = \dfrac{1}{3} \Rightarrow \tan t = \pm \dfrac{1}{\sqrt{3}} \Rightarrow t = \tan^{-1} \left(\pm \dfrac{1}{\sqrt{3}} \right)$

   $t = \dfrac{\pi}{6} + \pi n, \, -\dfrac{\pi}{6} + \pi n$ or $t = 30° + 180°n, \, -30° + 180°n$

55. $\sin 2t + 3 \cos t = 0 \Rightarrow 2 \sin t \cos t + 3 \cos t = 0 \Rightarrow \cos t \, (2 \sin t + 3) = 0$

   $\Rightarrow \cos t = 0 \qquad \text{or} \qquad 2 \sin t + 3 = 0$

   $\Rightarrow \cos t = 0 \qquad \text{or} \qquad \sin t = -\dfrac{3}{2} \text{ (not possible)}$

   $\Rightarrow t = \cos^{-1} 0$

   $t = \dfrac{\pi}{2} + \pi n$ or $t = 90° + 180°n$

57. $\cos 105° = \cos \dfrac{210°}{2} = -\sqrt{\dfrac{1 + \cos 210°}{2}} = -\sqrt{\dfrac{1 - \left(\dfrac{\sqrt{3}}{2}\right)}{2}} = -\sqrt{\dfrac{\dfrac{2 - \sqrt{3}}{2}}{2}} = -\sqrt{\dfrac{2 - \sqrt{3}}{4}} =$

   $-\sqrt{\dfrac{2 - \sqrt{3}}{2}}$. See Figure 57.

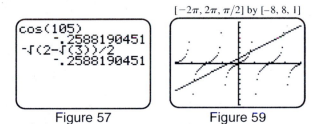

$[-2\pi, 2\pi, \pi/2]$ by $[-8, 8, 1]$

Figure 57   Figure 59

59. Graph $Y_1 = \tan(X)$ and $Y_2 = X + 1$ using dot mode in $[-2\pi, 2\pi, \pi/2]$ by $[-8, 8, 1]$.

    See Figure 59. The x-coordinates of the intersection points are approximately -4.43, 1.13 and 4.53.

61. If $\cos \alpha = \dfrac{3}{5}$ and α is in quadrant I, then $\sin \alpha = \dfrac{4}{5}$. If $\cos \beta = \dfrac{12}{13}$ and β is in quadrant I, then $\sin \beta = \dfrac{5}{13}$.

    (a) $\sin(\alpha + \beta) = \sin \alpha \cos \beta + \cos \alpha \sin \beta = \left(\dfrac{4}{5}\right)\left(\dfrac{12}{13}\right) + \left(\dfrac{3}{5}\right)\left(\dfrac{5}{13}\right) = \dfrac{48 + 15}{65} = \dfrac{63}{65}$

    (b) $\cos(\alpha + \beta) = \cos \alpha \cos \beta - \sin \alpha \sin \beta = \left(\dfrac{3}{5}\right)\left(\dfrac{12}{13}\right) - \left(\dfrac{4}{5}\right)\left(\dfrac{5}{13}\right) = \dfrac{36 - 20}{65} = \dfrac{16}{65}$

    (c) $\tan(\alpha + \beta) = \dfrac{\sin(\alpha + \beta)}{\cos(\alpha + \beta)} = \dfrac{\frac{63}{65}}{\frac{16}{65}} = \dfrac{63}{65} \cdot \dfrac{65}{16} = \dfrac{63}{16}$

    (d) Since $\sin(\alpha + \beta)$ and $\cos(\alpha + \beta)$ are positive, $\alpha + \beta$ is in quadrant I.

63. Since $\tan \theta = \dfrac{\sin \theta}{\cos \theta}$, $\cos \theta = \dfrac{\sin \theta}{\tan \theta} = \dfrac{\frac{4}{5}}{-\frac{4}{3}} = \dfrac{4}{5} \cdot \left(-\dfrac{3}{4}\right) = -\dfrac{3}{5}$

    (a) $\sin 2\theta = 2 \sin \theta \cos \theta = 2 \left(\dfrac{4}{5}\right)\left(-\dfrac{3}{5}\right) = -\dfrac{24}{25}$

$$\cos 2\theta = 1 - 2\sin^2\theta = 1 - 2\left(\frac{4}{5}\right)^2 = 1 - \frac{32}{25} = -\frac{7}{25}$$

$$\tan 2\theta = \frac{\sin 2\theta}{\cos 2\theta} = \frac{-\frac{24}{25}}{-\frac{7}{25}} = -\frac{24}{25}\cdot\left(-\frac{25}{7}\right) = \frac{24}{7}$$

(b) Since θ is in quadrant II, $\theta = \cos^{-1}\left(-\frac{3}{5}\right)$. Numerical support is shown in Figures 63a & 63b.

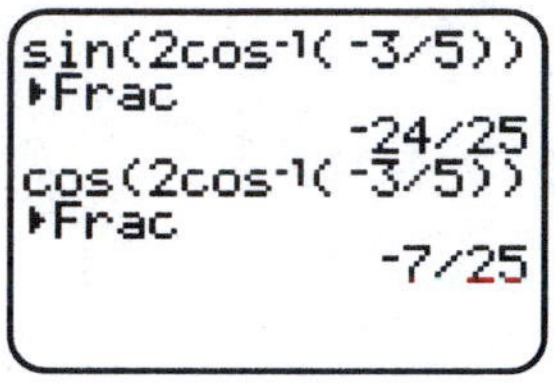
sin(2cos⁻¹(-3/5))
▸Frac
-24/25
cos(2cos⁻¹(-3/5))
▸Frac
-7/25
Figure 63a

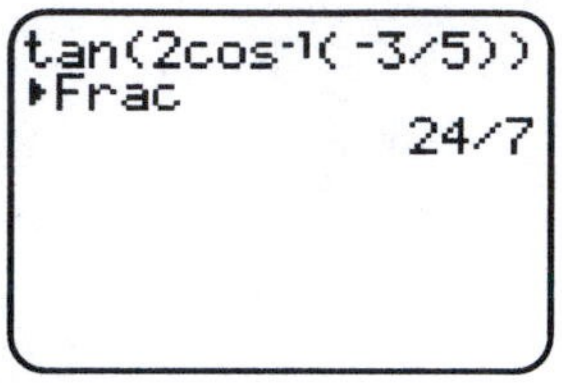
tan(2cos⁻¹(-3/5))
▸Frac
24/7
Figure 63b

65. Since $\cos\theta = \frac{1}{4}$ and θ is in quadrant I, $\sin\theta = \sqrt{1 - \cos^2\theta} = \sqrt{1 - \left(\frac{1}{4}\right)^2} = \sqrt{\frac{15}{16}} = \frac{\sqrt{15}}{4}$.

(a) $\sin\frac{\theta}{2} = \sqrt{\frac{1 - \cos\theta}{2}} = \sqrt{\frac{1 - \frac{1}{4}}{2}} = \sqrt{\frac{\frac{4-1}{4}}{2}} = \sqrt{\frac{3}{8}}$

$$\cos\frac{\theta}{2} = \sqrt{\frac{1 + \cos\theta}{2}} = \sqrt{\frac{1 + \frac{1}{4}}{2}} = \sqrt{\frac{\frac{4+1}{4}}{2}} = \sqrt{\frac{5}{8}}$$

$$\tan\frac{\theta}{2} = \frac{\sin\frac{\theta}{2}}{\cos\frac{\theta}{2}} = \frac{\sqrt{\frac{3}{8}}}{\sqrt{\frac{5}{8}}} = \sqrt{\frac{3}{8}}\cdot\sqrt{\frac{8}{5}} = \sqrt{\frac{3}{5}}$$

(b) Since θ is in quadrant I, $\theta = \cos^{-1}\frac{1}{4}$. Numerical support is shown in Figures 65a, 65b and 65c.

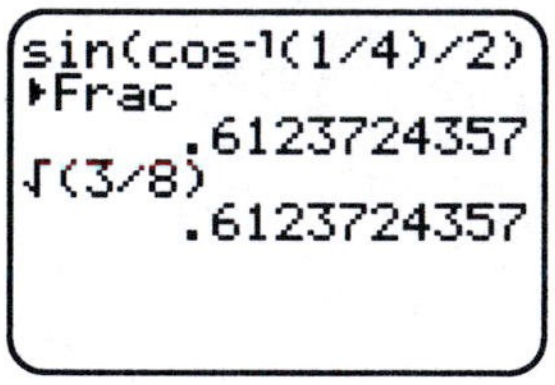
sin(cos⁻¹(1/4)/2)
▸Frac
.6123724357
√(3/8)
.6123724357
Figure 65a

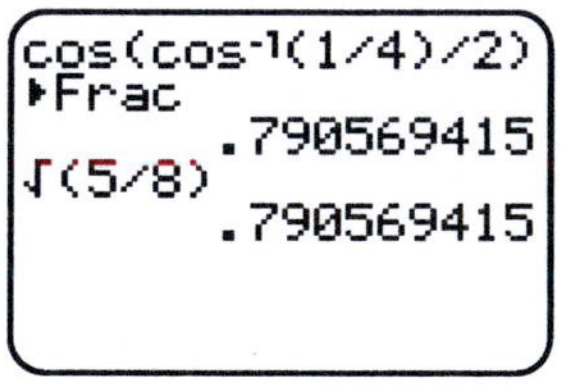
cos(cos⁻¹(1/4)/2)
▸Frac
.790569415
√(5/8)
.790569415
Figure 65b

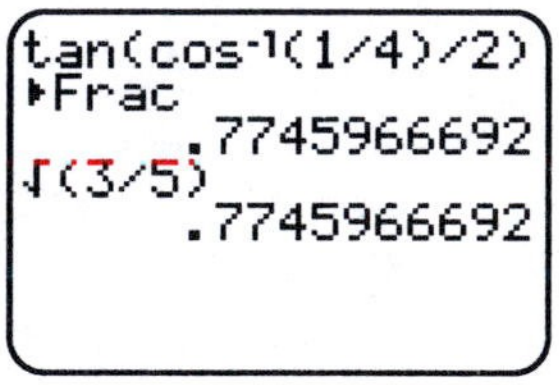
tan(cos⁻¹(1/4)/2)
▸Frac
.7745966692
√(3/5)
.7745966692
Figure 65c

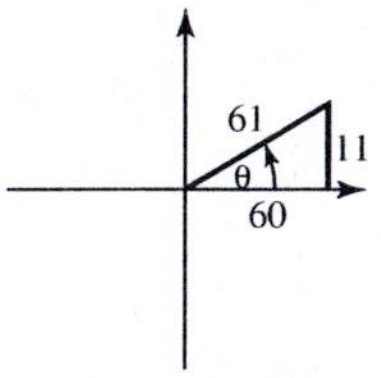
61
11
θ
60
Figure 67

67. Let $\theta = \tan^{-1}\frac{11}{60}$. It follows that $\sin\theta = \frac{11}{61}$. See Figure 67.

$$\cos\left(2\tan^{-1}\frac{11}{60}\right) = \cos(2\theta) = 1 - 2\sin^2\theta = 1 - 2\left(\frac{11}{61}\right)^2 = 1 - \frac{242}{3721} = \frac{3479}{3721}$$

69. Graph $Y_1 = 1.2 \cos(\pi/6(X - 0.7)) + 12.1$ and $Y_2 = 11.5$ in $[0, 13, 1]$ by $[10, 15, 1]$.

See Figures 69a and 69b. The x-coordinates of the intersection points are approximately 4.7 and 8.7. The number of daylight hours equals 11.5 on about April 21 and again on about August 22.

[0, 13, 1] by [10, 15, 1]   [0, 13, 1] by [10, 15, 1]   [0, 0.06, 0.01] by [−0.012, 0.012, 0.002]

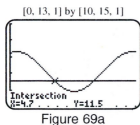
Figure 69a

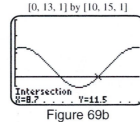
Figure 69b

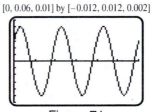
Figure 71

71. (a) Graph $Y_1 = 0.006 \cos(100\pi X) + 0.008 \sin(100\pi X)$ in $[0, 0.06, 0.01]$ by $[-0.012, 0.012, 0.002]$.

See Figure 71.

(b) The peaks of the graph are located at $y = 0.01$, thus $a = 0.01$. The x-value at the first peak is

$x \approx 0.00295089$. $100\pi(0.00295089) + k = \dfrac{\pi}{2} \Rightarrow k \approx 0.6435$.

(c) $P = 0.01 \sin(100\pi t + 0.6435) = 0.01[\sin(100\pi t)\cos(0.6435) + \cos(100\pi t)\sin(0.6435)] \approx$

$0.01[\sin(100\pi t)(0.8) + \cos(100\pi t)(0.6)] \approx 0.008 \sin(100\pi t) + 0.006 \cos(100\pi t)$

73. $W(t) = 7\cos^2(240\pi t) \Rightarrow W(t) = 7(1 - \sin^2(240\pi t)) \Rightarrow W(t) = 7 - 7\sin^2(240\pi t)$

When V is maximum or minimum, $W = 0$.

75. Let $F = 250 \Rightarrow 250 = 289\cos\theta \Rightarrow \cos\theta = \dfrac{250}{289}$ or $\theta = \cos^{-1}\left(\dfrac{250}{289}\right) \Rightarrow \theta \approx 30.11°$ or 0.53 radians.

Extended and Discovery Exercises for Chapter 7

1. (a) The graphs of functions i, ii, iii, iv and v are shown in Figures 1a, 1b, 1c, 1d and 1e respectively.

(b) The graph approximates a saw-tooth shape.

(c) The maximum pressure of P is approximately 0.00317. See Figure 1f.

(d) The pure tone is modeled by a smooth graph whereas the piano tone is modeled by a saw-tooth shape.

[0, 0.01, 0.002] by [−0.005, 0.005, 0.001]   [0, 0.01, 0.002] by [−0.005, 0.005, 0.001]   [0, 0.01, 0.002] by [−0.005, 0.005, 0.001]   [0, 0.01, 0.002] by [−0.005, 0.005, 0.001]

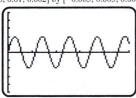
Figure 1a

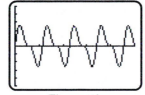
Figure 1b

Figure 1c

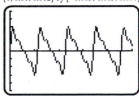
Figure 1d

[0, 0.01, 0.002] by [−0.005, 0.005, 0.001]   [0, 0.01, 0.002] by [−0.005, 0.005, 0.001]

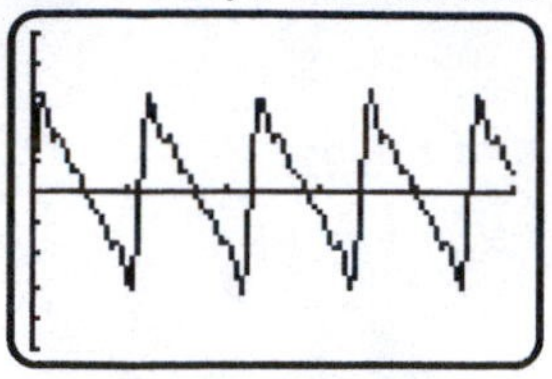

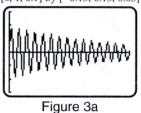

Figure 1e

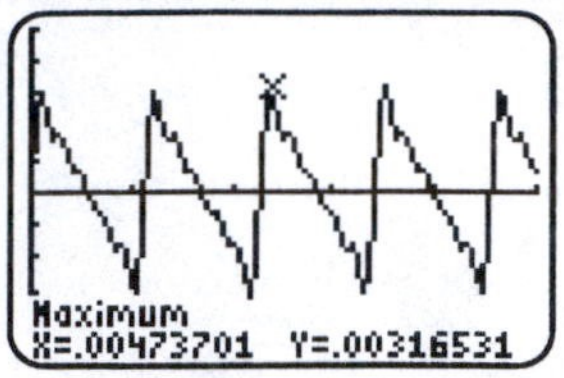

Figure 1f

3.   (a) Graph $Y_3 = 0.1e^{(-1.2X)} \sin(2\pi * 15X)$ in $[0, 1, 0.1]$ by $[-0.15, 0.15, 0.05]$. See Figure 3a.

(b) Graph Y_3 along with $Y_1 = -0.1e^{(-1.2X)}$ and $Y_2 = 0.1e^{(-1.2X)}$ in the same window. See Figure 3b.

These graphs relate to the amplitude of A.

(c) The initial amplitude of A is 0.1. Graph Y_3 along with $Y_4 = 0.05$ in the same window. See Figure 3c.

Using the trace feature of the calculator, we see that the amplitude is nearly 0.05 when $t \approx 0.55$ seconds.

[0, 1, 0.1] by [−0.15, 0.15, 0.05]   [0, 1, 0.1] by [−0.15, 0.15, 0.05]   [0, 1, 0.1] by [−0.15, 0.15, 0.05]

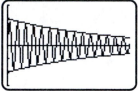

Figure 3a

Figure 3b

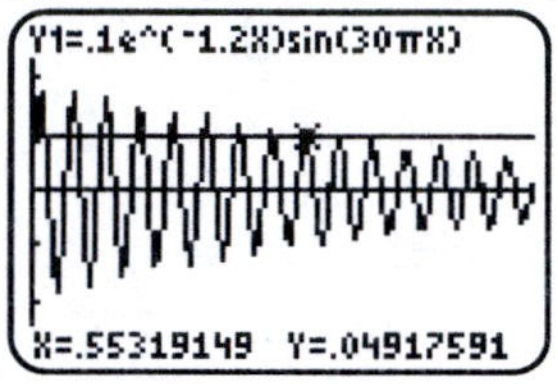

Figure 3c

Chapter 8: Further Topics in Trigonometry

8.1: Law of Sines

1. This triangle is of the form AAS. There is only one solution.

$$\beta = 180° - 45° - 75° = 60°$$

$$\frac{a}{\sin \alpha} = \frac{c}{\sin \gamma} \Rightarrow a = \frac{c \sin \alpha}{\sin \gamma} = \frac{4 \sin 75°}{\sin 45°} \approx 5.46 \qquad \frac{b}{\sin \beta} = \frac{c}{\sin \gamma} \Rightarrow b = \frac{c \sin \beta}{\sin \gamma} = \frac{4 \sin 60°}{\sin 45°} \approx 4.90$$

$$\beta = 60°, \; a \approx 5.5, \; b \approx 4.9$$

3. This triangle is of the form ASA. There is only one solution.

$$\beta = 180° - 120° - 25° = 35°$$

$$\frac{a}{\sin \alpha} = \frac{b}{\sin \beta} \Rightarrow a = \frac{b \sin \alpha}{\sin \beta} = \frac{10 \sin 120°}{\sin 35°} \approx 15.1 \qquad \frac{c}{\sin \gamma} = \frac{b}{\sin \beta} \Rightarrow c = \frac{b \sin \gamma}{\sin \beta} = \frac{10 \sin 25°}{\sin 35°} \approx 7.37$$

$$\beta = 35°, \; a \approx 15.1, \; c \approx 7.4$$

5. The triangle is of the form ASA. There is only one solution. $\gamma = 180° - 60° - 40° = 80°$

$$\frac{a}{\sin \alpha} = \frac{c}{\sin \gamma} \Rightarrow a = \frac{c \sin \alpha}{\sin \gamma} = \frac{10 \sin 40°}{\sin 80°} \approx 6.5 \qquad \frac{b}{\sin \beta} = \frac{c}{\sin \gamma} \Rightarrow b = \frac{c \sin \beta}{\sin \gamma} = \frac{10 \sin 60°}{\sin 80°} \approx 8.8$$

$$\gamma = 80°, \; a = 6.5, \; b = 8.8$$

7. The triangle is of the form AAS. There is only one solution. $\beta = 180° - 40° - 25° = 115°$

$$\frac{a}{\sin \alpha} = \frac{c}{\sin \gamma} \Rightarrow c = \frac{c \sin \gamma}{\sin \alpha} = \frac{9.7 \sin 40°}{\sin 25°} \approx 14.8 \qquad \frac{\alpha}{\sin \alpha} = \frac{b}{\sin \beta} \Rightarrow b = \frac{a \sin \beta}{\sin \alpha} = \frac{9.7 \sin 115°}{\sin 25°} \approx 20.8,$$

$$\beta = 115°, \; b = 20.8, \; c = 14.8$$

9. The triangle is of the form ASA. There is only one solution. $\alpha = 180° - 40.2° - 60.7° = 79.1°$

$$\frac{a}{\sin \alpha} = \frac{c}{\sin \gamma} \Rightarrow c = \frac{a \sin \gamma}{\sin \alpha} = \frac{5.5 \sin 60.7°}{\sin 79.1°} \approx 4.9 \qquad \frac{b}{\sin \beta} = \frac{a}{\sin \alpha} \Rightarrow b = \frac{a \sin \beta}{\sin \alpha} = \frac{5.5 \sin 40.2°}{\sin 79.1°} \approx 3.6,$$

$$\alpha = 79.1°, \; b = 3.6, \; c = 4.9$$

11. The triangle is of the form ASA. There is only one solution. $\beta = 180° - 27° - 49° = 104°$

$$\frac{a}{\sin \alpha} = \frac{b}{\sin \beta} \Rightarrow a = \frac{b \sin \alpha}{\sin \beta} = \frac{67 \sin 27°}{\sin 104°} \approx 31.3 \qquad \frac{b}{\sin \beta} = \frac{c}{\sin \gamma} \Rightarrow c = \frac{b \sin \gamma}{\sin \beta} = \frac{67 \sin 49°}{\sin 104°} \approx 52.1,$$

$$\beta = 104°, \; a = 31.3, \; c = 52.1$$

13. No. This situation is AAS.

15. No. This situation is SSS.

17. Yes. This situation is SSA.

19. Yes. This situation is SSA.

21. This triangle is of the form SSA. It is ambiguous.

$$\frac{\sin \beta}{b} = \frac{\sin \alpha}{a} \Rightarrow \sin \beta = \frac{b \sin \alpha}{a} = \frac{7 \sin 46°}{6} \Rightarrow \beta_R = \sin^{-1}\left(\frac{7 \sin 46°}{6}\right) \approx 57.1°$$

Thus, $\beta \approx 57.1°$ or $\beta \approx 180° - 57.1° \approx 122.9°$.

Solution 1: Let $\beta \approx 57.1°$. Then $\gamma \approx 180° - 46° - 57.1° \approx 76.9°$.

$\dfrac{c}{\sin \gamma} = \dfrac{a}{\sin \alpha} \Rightarrow c = \dfrac{a \sin \gamma}{\sin \alpha} = \dfrac{6 \sin 76.9°}{\sin 46°} \approx 8.12 \qquad \beta \approx 57.1°, \ \gamma \approx 76.9°, \ c \approx 8.1$

Solution 2: Let $\beta \approx 122.9°$. Then $\gamma \approx 180° - 46° - 122.9° \approx 11.1°$.

$\dfrac{c}{\sin \gamma} = \dfrac{a}{\sin \alpha} \Rightarrow c = \dfrac{a \sin \gamma}{\sin \alpha} = \dfrac{6 \sin 11.1°}{\sin 46°} \approx 1.61 \qquad \beta \approx 122.9°, \ \gamma \approx 11.1°, \ c \approx 1.6$

23. This triangle is of the form SSA. It is ambiguous.

$\dfrac{\sin \alpha}{a} = \dfrac{\sin \beta}{b} \Rightarrow \sin \alpha = \dfrac{a \sin \beta}{b} = \dfrac{5 \sin 52°}{3} \approx 1.31 \qquad$ This is not possible. No solution.

25. This triangle is of the form SSA. It is ambiguous.

$\dfrac{\sin \beta}{b} = \dfrac{\sin \alpha}{a} \Rightarrow \sin \beta = \dfrac{b \sin \alpha}{a} = \dfrac{12 \sin 50.2°}{10} \approx 0.9219 \Rightarrow$

$\beta_R \approx \sin^{-1}(0.9219) \approx 67.2°$ or $\beta_1 \approx 67° \ 13'$

Then $\gamma_1 = 180° - 50.2° - 67.2° \approx 62.6°$ or $\gamma_1 \approx 62° \ 35'$

$\dfrac{c_1}{\sin \gamma} = \dfrac{a}{\sin \alpha} \Rightarrow c_1 = \dfrac{a \sin \gamma}{\sin \alpha} = \dfrac{10 \sin 62.6°}{\sin 50.2°} \approx 11.6$ or $c_1 \approx 11.6 \qquad \beta_1 \approx 67°13', \ \gamma_1 \approx 62°35', \ c_1 \approx 11.6$

However if $\beta_2 = 180° - 67.2° = 112.8°$ or $112° \ 48'$ then $\gamma_2 = 180° - 50.2° - 112.8° = 17°$ or $\gamma_2 \approx 17° \ 0'$

$\dfrac{c_2}{\sin \gamma_2} = \dfrac{a}{\sin \alpha} \Rightarrow c_2 = \dfrac{a \sin \gamma_2}{\sin \alpha} = \dfrac{10 \sin 17°}{\sin 50.2°} \approx 3.8$ or $c_2 \approx 3.8 \qquad \beta_2 \approx 112°48', \ \gamma_2 \approx 17°0', \ c_2 \approx 3.8$

27. This triangle is of the form AAS. There is only one solution.

$\gamma = 180° - 32° - 55° = 93°$

$\dfrac{a}{\sin \alpha} = \dfrac{b}{\sin \beta} \Rightarrow a = \dfrac{b \sin \alpha}{\sin \beta} = \dfrac{12 \sin 32°}{\sin 55°} \approx 7.76 \qquad \dfrac{c}{\sin \gamma} = \dfrac{b}{\sin \beta} \Rightarrow c = \dfrac{b \sin \gamma}{\sin \beta} = \dfrac{12 \sin 93°}{\sin 55°} \approx 14.6$

$\gamma = 93°, \ a \approx 7.8, \ c \approx 14.6$

29. This triangle is of the form SSA. It is ambiguous.

$\dfrac{\sin \beta}{b} = \dfrac{\sin \alpha}{a} \Rightarrow \sin \beta = \dfrac{b \sin \alpha}{a} = \dfrac{9 \sin 20°}{7} \Rightarrow \beta_R = \sin^{-1}\left(\dfrac{9 \sin 20°}{7}\right) \approx 26.1°$

Thus, $\beta \approx 26.1°$ or $\beta \approx 180° - 26.1° \approx 153.9°$.

Solution 1: Let $\beta \approx 26.1°$. Then $\gamma \approx 180° - 20° - 26.1° \approx 133.9°$.

$\dfrac{c}{\sin \gamma} = \dfrac{a}{\sin \alpha} \Rightarrow c = \dfrac{a \sin \gamma}{\sin \alpha} = \dfrac{7 \sin 133.9°}{\sin 20°} \approx 14.7 \qquad \beta \approx 26.1°, \ \gamma \approx 133.9°, \ c \approx 14.7$

Solution 2: Let $\beta \approx 153.9°$. Then $\gamma \approx 180° - 20° - 153.9° \approx 6.1°$.

$\dfrac{c}{\sin \gamma} = \dfrac{a}{\sin \alpha} \Rightarrow c = \dfrac{a \sin \gamma}{\sin \alpha} = \dfrac{7 \sin 6.1°}{\sin 20°} \approx 2.17 \qquad \beta \approx 153.9°, \ \gamma \approx 6.1°, \ c \approx 2.2$

31. This triangle is of the form SSA. It is ambiguous.

$\dfrac{\sin \gamma}{c} = \dfrac{\sin \beta}{b} \Rightarrow \sin \gamma = \dfrac{c \sin \beta}{b} = \dfrac{20 \sin 30°}{10} = 1 \Rightarrow \gamma_R = \sin^{-1} 1 = 90°$

Thus, $\gamma = 90°$. Then $\alpha = 180° - 30° - 90° = 60°$.

$\dfrac{a}{\sin \alpha} = \dfrac{b}{\sin \beta} \Rightarrow a = \dfrac{b \sin \alpha}{\sin \beta} = \dfrac{10 \sin 60°}{\sin 30°} = 10\sqrt{3} \approx 17.3 \qquad \gamma = 90°, \ \alpha = 60°, \ a = 10\sqrt{3} \approx 17.3$

33. This triangle is of the form SSA. It is ambiguous.

$\dfrac{\sin \alpha}{a} = \dfrac{\sin \gamma}{c} \Rightarrow \sin \alpha = \dfrac{a \sin \gamma}{c} = \dfrac{42.1 \sin 102°}{51.6} \Rightarrow \alpha_R = \sin^{-1}\left(\dfrac{42.1 \sin 102°}{51.6}\right) \approx 52.9°$

Thus, $\alpha \approx 52.9°$ or $\alpha \approx 180° - 52.9° \approx 127.1°$ (which is not possible).

So $\alpha \approx 52.9°$. Then $\beta \approx 180° - 102° - 52.9° \approx 25.1°$.

$$\frac{b}{\sin \beta} = \frac{c}{\sin \gamma} \Rightarrow b = \frac{c \sin \beta}{\sin \gamma} = \frac{51.6 \sin 25.1°}{\sin 102°} \approx 22.4 \qquad \alpha \approx 52.9°, \ \beta \approx 25.1°, \ b \approx 22.4$$

35. This triangle is of the form ASA. There is only one solution.

$$\beta = 180° - 55.2° - 114.8° = 10°$$

$$\frac{a}{\sin \alpha} = \frac{b}{\sin \beta} \Rightarrow a = \frac{b \sin \alpha}{\sin \beta} = \frac{19.5 \sin 55.2°}{\sin 10°} \approx 92.2$$

$$\frac{c}{\sin \gamma} = \frac{b}{\sin \beta} \Rightarrow c = \frac{b \sin \gamma}{\sin \beta} = \frac{19.5 \sin 114.8°}{\sin 10°} \approx 101.9$$

$$\beta = 10°, \ a \approx 92.2, \ c \approx 101.9$$

37. This triangle is of the form SSA. It is ambiguous.

$$\frac{\sin \gamma}{c} = \frac{\sin \beta}{b} \Rightarrow \sin \gamma = \frac{c \sin \beta}{b} = \frac{7.4 \sin 73°}{6.2} \approx 1.14 \quad \text{This is not possible. No solution.}$$

39. This triangle is of the form SSA. It is ambiguous.

$$\frac{\sin \beta}{b} = \frac{\sin \alpha}{a} \Rightarrow \sin \beta = \frac{b \sin \alpha}{a} = \frac{12 \sin 35° 15'}{5} \approx 1.39 \quad \text{This is not possible. No solution.}$$

41. This triangle is of the form SSA. It is ambiguous.

$$\frac{\sin \alpha}{a} = \frac{\sin \beta}{b} \Rightarrow \sin \alpha = \frac{a \sin \beta}{b} = \frac{6 \sin 46° 45'}{5} \approx 0.8740 \Rightarrow \alpha_R \approx \sin^{-1}(0.8740) \approx 60.93°$$

or $\alpha_1 \approx 60° 56'$ then $\gamma_1 \approx 180° - 46.75° - 60.93° \approx 72.32°$ or $\gamma_1 \approx 72° 19'$

$$\frac{c_1}{\sin \gamma_1} = \frac{b}{\sin \beta} \Rightarrow c_1 = \frac{b \sin \gamma_1}{\sin \beta} = \frac{5 \sin 72.32°}{\sin 46.75°} \approx 6.54 \text{ or } c_1 \approx 6.5$$

However, if $\alpha_2 \approx 180° - 60.93° \approx 119.07°$ or $119° 4'$ then $\gamma_2 \approx 180° - 46.75° - 119.07° \approx 14.18°$ or

$\gamma_2 = 14° 11'.$ $\quad \dfrac{c_2}{\sin \gamma_2} = \dfrac{b}{\sin \beta} \Rightarrow c_2 = \dfrac{b \sin \gamma_2}{\sin \beta} = \dfrac{5 \sin 14.18°}{\sin 46.75°} \approx 1.68$ or $c_2 \approx 1.7$

43. This triangle is of the form ASA. There is only one solution. Note that $\alpha = 56° 30'$ or $56.5°$ and

$\beta = 23° 45'$ or $23.75°$. $\gamma = 180° - 56.5° - 23.75° = 99.75°$ or $99° 45'$

$$\frac{a}{\sin \alpha} = \frac{c}{\sin \gamma} \Rightarrow a = \frac{c \sin \alpha}{\sin \gamma} = \frac{100 \sin 56.5°}{\sin 99.75°} \approx 84.6$$

$$\frac{b}{\sin \beta} = \frac{c}{\sin \gamma} \Rightarrow b = \frac{c \sin \beta}{\sin \gamma} = \frac{100 \sin 23.75°}{\sin 99.75°} \approx 40.9$$

$$\gamma = 99° 45', \ a \approx 84.6, \ b \approx 40.9$$

45. Using standard labels we have the following: $\beta = 52° - 7° = 45°$ and $\alpha = 180° - 70° - 45° = 65°$.

$$\frac{c}{\sin 70°} = \frac{3500}{\sin 65°} \Rightarrow c = \frac{3500 \sin 70°}{\sin 65°} \approx 3629 \text{ ft. The ground distance is about 3629 feet.}$$

47. $\gamma = \theta - \alpha = 52.7430° - 52.6901° = 0.0529°$

$$\frac{a}{\sin \alpha} = \frac{c}{\sin \gamma} \Rightarrow a = \frac{c \sin \alpha}{\sin \gamma} = \frac{398.02 \sin 52.6901°}{\sin 0.0529°} \approx 342,878.7$$

The calculated distance to the moon changes to about 343,000 km, a difference of about 76,000 km. A small error in measuring lunar angle could result in large errors when calculating the distance to the moon.

49. Using standard labels we have the following:

$\alpha = 90° - 54.3° = 35.7°$, $\beta = 325.2° - 270° = 55.2°$ and $\gamma = 180° - 35.7° - 55.2° = 89.1°$.

$$\frac{a}{\sin \alpha} = \frac{c}{\sin \gamma} \Rightarrow a = \frac{c \sin \alpha}{\sin \gamma} = \frac{15 \sin 35.7°}{\sin 89.1°} \approx 8.75$$

Then the perpendicular distance d from the ship to the shore can be found as follows:

$\sin 55.2° = \dfrac{d}{8.75} \Rightarrow d = 8.75 \sin 55.2° \approx 7.2$ mi. The ship is about 7.2 miles from the shore.

51. Angle DAB has measure $180° - 118° - 28° = 34°$. $\dfrac{BD}{\sin 34°} = \dfrac{24.2}{\sin 28°} \Rightarrow BD = \dfrac{24.2 \sin 34°}{\sin 28°} \approx 28.8$ ft.

53. Note that $\beta = 112° \, 10'$ or $112.17°$ and $\gamma = 15° \, 20'$ or $15.33°$.

Therefore $\alpha = 180° - 112.17° - 15.33° = 52.5°$ or $52° \, 30'$

$$\frac{c}{\sin \gamma} = \frac{a}{\sin \alpha} \Rightarrow c = \frac{a \sin \gamma}{\sin \alpha} = \frac{354 \sin 15.33°}{\sin 52.5°} \approx 118 \text{ meters}$$ Therefore c ≈ 118 meters from point A to B.

55. Let A, B, and C be the positions of the observer on the left, the helicopter, and observer on the right, respectively. Triangle ABC has the following measurements: $A = 20.5°$, $C = 180° - 27.8° = 152.2°$, $B = 180° - 20.5° - 152.2° = 7.3°$, and $b = 3$. Using the law of sines,

$$\frac{a}{\sin 20.5°} = \frac{3}{\sin 7.3°} \Rightarrow a = \frac{3 \sin 20.5°}{\sin 7.3°} \Rightarrow a \approx 8.27 \text{ miles.}$$ If the perpendicular line segment joining B to the ground intersects at D, we can use right triangle CBD to find the distance d to the helipcopter:

$\sin 27.8° = \dfrac{d}{8.27} \Rightarrow d = 8.27 \sin 27.8° \approx 3.86$ miles.

57. Let A, B, and C be the positions of the town on the left, the balloon, and town on the right, respectively. Triangle ABC has the following measurements: $A = 35°$, $B = 180° - 35° - 31° = 114°$, $C = 31°$, and $b = 1.5$. Using the law of sines, $\dfrac{c}{\sin 31°} = \dfrac{1.5}{\sin 114°} \Rightarrow c = \dfrac{1.5 \sin 31°}{\sin 114°} \Rightarrow c \approx 0.85$ mile. If the perpendicular line segment joining B to the ground intersects at D, we can use right triangle ABD to find the distance d to the balloon: $\sin 35° = \dfrac{d}{0.85} \Rightarrow d = 0.85 \sin 35° \approx 0.49$ mile.

59. Using standard labels we have the distance $AB = c$, $\gamma = 55.1°$, $\alpha = 75.7°$ and $b = 97.3$.

$\beta = 180° - 75.7° - 55.1° = 49.2°$ $\dfrac{c}{\sin \gamma} = \dfrac{b}{\sin \beta} \Rightarrow c = \dfrac{b \sin \gamma}{\sin \beta} = \dfrac{97.3 \sin 55.1°}{\sin 49.2°} \approx 105.4$

The distance AB is approximately 105.4 feet.

61. (a) Using standard labels we calculate the distance as follows:

$\gamma = 180° - 28° - 60° = 92°$ $\dfrac{b}{\sin \beta} = \dfrac{c}{\sin \gamma} \Rightarrow b = \dfrac{c \sin \beta}{\sin \gamma} = \dfrac{4.12 \sin 60°}{\sin 92°} \approx 3.57$ mi.

(b) Let D represent the point $(4, 0)$ and let θ represent the angle DAB.

$$\tan \theta = \frac{1}{4} \Rightarrow \theta = \tan^{-1} \frac{1}{4} \approx 14.0° \quad \text{Thus, the bearing is } 90° - 28° - 14.0° \approx 48°.$$

63. Represent the top of the arch with point C and use standard labels.

Note that $\alpha = 180° - 64.91° = 115.09°$. Then $\gamma = 180° - 115.09° - 60.81° = 4.1°$

First we find side $b = AC$: $\dfrac{b}{\sin \beta} = \dfrac{c}{\sin \gamma} \Rightarrow b = \dfrac{c \sin \beta}{\sin \gamma} = \dfrac{57 \sin 60.81°}{\sin 4.1°} \approx 696$ ft.

Label a point D on the ground directly below the top of the arch and let h represent the height of the arch.

From triangle ADC: $\sin 64.91° \approx \dfrac{h}{696} \Rightarrow h \approx 696 \sin 64.91° \approx 630$ ft.

8.2: Law of Cosines

1. (a) This triangle is SAS.

   (b) Law of cosines should be used.

3. (a) This triangle is SSA.

   (b) This case is ambiguous. Law of sines should be used.

5. (a) This triangle is ASA.

   (b) Law of sines should be used.

7. (a) This triangle is ASA.

   (b) Law of sines should be used.

9. $c^2 = 3^2 + 8^2 - 2(3)(8) \cos 60° \Rightarrow c^2 = 73 - (48)(0.5) \Rightarrow c^2 = 73 - 24 \Rightarrow c^2 = 49 \Rightarrow c = 7$

11. $12^2 = 11^2 + 14^2 - 2(11)(14) \cos \theta \Rightarrow 144 = 317 - 308 \cos \theta \Rightarrow \cos \theta = \dfrac{173}{308} \Rightarrow \theta \approx 55.8$

13. $a^2 = b^2 + c^2 - 2bc \cos \alpha = 4^2 + 6^2 - 2(4)(6) \cos 61° \Rightarrow a = \sqrt{52 - 48 \cos 61°} \approx 5.35996$

   $c^2 = a^2 + b^2 - 2ab \cos \gamma \Rightarrow \cos \gamma = \dfrac{c^2 - a^2 - b^2}{-2ab} \Rightarrow \gamma = \cos^{-1}\left(\dfrac{6^2 - 5.35996^2 - 4^2}{-2(5.35996)(4)}\right) \approx 78.25399°$

   $\beta = 180° - 61° - 78.25399° \approx 40.74601° \quad a \approx 5.4,\ \beta \approx 40.7°,\ \gamma \approx 78.3°$

15. $a^2 = b^2 + c^2 - 2bc \cos \alpha \Rightarrow \cos \alpha = \dfrac{a^2 - b^2 - c^2}{-2bc} \Rightarrow \alpha = \cos^{-1}\left(\dfrac{4^2 - 10^2 - 8^2}{-2(10)(8)}\right) \approx 22.33165°$

   $b^2 = a^2 + c^2 - 2ac \cos \beta \Rightarrow \cos \beta = \dfrac{b^2 - a^2 - c^2}{-2ac} \Rightarrow \beta = \cos^{-1}\left(\dfrac{10^2 - 4^2 - 8^2}{-2(4)(8)}\right) \approx 108.20996°$

   $\gamma = 180° - 22.33165° - 108.20996° \approx 49.45839° \quad \alpha \approx 22.3°,\ \beta \approx 108.2°,\ \gamma \approx 49.5°$

17. $a^2 = b^2 + c^2 - 2bc \cos \alpha \Rightarrow \cos \alpha = \dfrac{a^2 - b^2 - c^2}{-2bc} \Rightarrow \alpha = \cos^{-1}\left(\dfrac{5^2 - 7^2 - 9^2}{-2(7)(9)}\right) \approx 33.55731°$

   $b^2 = a^2 + c^2 - 2ac \cos \beta \Rightarrow \cos \beta = \dfrac{b^2 - a^2 - c^2}{-2ac} \Rightarrow \beta = \cos^{-1}\left(\dfrac{7^2 - 5^2 - 9^2}{-2(5)(9)}\right) \approx 50.70352°$

   $\gamma = 180° - 33.55731° - 50.70352° \approx 95.73917° \quad \alpha \approx 33.6°,\ \beta \approx 50.7°,\ \gamma \approx 95.7°$

19. $c^2 = a^2 + b^2 - 2ab \cos \gamma = 45^2 + 24^2 - 2(45)(24) \cos 35° \implies c = \sqrt{2601 - 2160 \cos 35°} \approx 28.83802$

$a^2 = b^2 + c^2 - 2bc \cos \alpha \implies \cos \alpha = \dfrac{a^2 - b^2 - c^2}{-2bc} \implies$

$\alpha = \cos^{-1}\left(\dfrac{45^2 - 24^2 - 28.83802^2}{-2(24)(28.83802)}\right) \approx 116.48753°$

$\beta = 180° - 35° - 116.48753° \approx 28.51247°$ $\qquad c \approx 28.8,\ \alpha \approx 116.5°,\ \beta \approx 28.5°$

21. $a^2 = b^2 + c^2 - 2bc \cos \alpha \implies \cos \alpha = \dfrac{a^2 - b^2 - c^2}{-2bc} \implies \alpha = \cos^{-1}\left(\dfrac{2.4^2 - 1.7^2 - 1.4^2}{-2(1.7)(1.4)}\right) \approx 101.02145°$

$b^2 = a^2 + c^2 - 2ac \cos \beta \implies \cos \beta = \dfrac{b^2 - a^2 - c^2}{-2ac} \implies \beta = \cos^{-1}\left(\dfrac{1.7^2 - 2.4^2 - 1.4^2}{-2(2.4)(1.4)}\right) \approx 44.04863°$

$\gamma = 180° - 101.02145° - 44.04863° \approx 34.92992°$ $\qquad \alpha \approx 101.0°,\ \beta \approx 44.0°,\ \gamma \approx 34.9°$

(The angles do not sum to 180° due to rounding)

23. $a^2 = b^2 + c^2 - 2bc \cos \alpha = 24.1^2 + 15.8^2 - 2(24.1)(15.8) \cos 10°30' \implies$

$a = \sqrt{830.45 - 761.56 \cos 10°30'} \approx 9.03562$

$b^2 = a^2 + c^2 - 2ac \cos \beta \implies \cos \beta = \dfrac{b^2 - a^2 - c^2}{-2ac} \implies$

$\beta = \cos^{-1}\left(\dfrac{24.1^2 - 9.03562^2 - 15.8^2}{-2(9.03562)(15.8)}\right) \approx 150.91785°$

$\gamma = 180° - 10.5° - 150.91785° \approx 18.58215°$ $\qquad a \approx 9.0,\ \beta \approx 150.9°,\ \gamma \approx 18.6°$

25. $a^2 = b^2 + c^2 - 2bc \cos \alpha \implies \cos \alpha = \dfrac{a^2 - b^2 - c^2}{-2bc} \implies$

$\alpha = \cos^{-1}\left(\dfrac{10.6^2 - 25.8^2 - 20.6^2}{-2(25.8)(20.6)}\right) \approx 23.11284°$

$b^2 = a^2 + c^2 - 2ac \cos \beta \implies \cos \beta = \dfrac{b^2 - a^2 - c^2}{-2ac} \implies \beta = \cos^{-1}\left(\dfrac{25.8^2 - 10.6^2 - 20.6^2}{-2(10.6)(20.6)}\right) \approx$

$107.16957°$

$\gamma = 180° - 23.11284° - 107.16957° \approx 49.71759°$ $\qquad \alpha \approx 23.1°,\ \beta \approx 107.2°,\ \gamma \approx 49.7°$

27. $b^2 = a^2 + c^2 - 2ac \cos \beta = 20^2 + 15^2 - 2(20)(15) \cos 122°\,10' \implies$

$b = \sqrt{625 - 600 \cos 122°\,10'} \approx 30.73207$

$a^2 = b^2 + c^2 - 2bc \cos \alpha \implies \cos \alpha = \dfrac{a^2 - b^2 - c^2}{-2bc} \implies$

$\alpha = \cos^{-1}\left(\dfrac{20^2 - 30.73207^2 - 15^2}{-2(30.73207)(15)}\right) \approx 33.42685°$ or $33°\,26'$

$\gamma \approx 180° - 122°\,10' - 33°\,26' \approx 24°\,24'$ $\qquad b \approx 30.7,\ \alpha \approx 33°\,26',\ \gamma \approx 24°\,24'$

29. $a^2 = b^2 + c^2 - 2bc \cos \alpha \Rightarrow \cos \alpha = \dfrac{a^2 - b^2 - c^2}{-2bc} \Rightarrow \alpha = \cos^{-1}\left(\dfrac{5.3^2 - 6.7^2 - 7.1^2}{-2(6.7)(7.1)}\right) \approx 45.05460°$

$b^2 = a^2 + c^2 - 2ac \cos \beta \Rightarrow \cos \beta = \dfrac{b^2 - a^2 - c^2}{-2ac} \Rightarrow \beta = \cos^{-1}\left(\dfrac{6.7^2 - 5.3^2 - 7.1^2}{-2(5.3)(7.1)}\right) \approx 63.47520°$

$\gamma = 180° - 45.05460° - 63.47520° \approx 71.4702°$ $\alpha \approx 45.1,\ \beta = 63.5°,\ \gamma \approx 71.5°$

(The angles do not sum to 180° due to rounding)

31. No, since $a + b < c$.

33. No, since $89° + 112° > 180°$.

35. Yes, since we are given ASA and $\alpha + \gamma < 180°$.

37. Area $= K = \dfrac{1}{2}(18)(15) \sin 40° \approx 86.8$

39. $s = \dfrac{1}{2}(3 + 4 + 6) = 6.5 \Rightarrow$ Area $= K = \sqrt{6.5(6.5 - 3)(6.5 - 4)(6.5 - 6)} \approx 5.3$

41. Area $= K = \dfrac{1}{2}(10)(12) \sin 58° \approx 50.9$

43. Area $= K = \dfrac{1}{2}(5.5)(6.8) \sin 78° \approx 18.3$

45. Note that $\gamma = 180° - 31° - 54° = 95°$. Then $\dfrac{b}{\sin \beta} = \dfrac{a}{\sin \alpha} \Rightarrow b = \dfrac{a \sin \beta}{\sin \alpha} = \dfrac{2.6 \sin 31°}{\sin 54°} \approx 1.65522$.

Area $= K \approx \dfrac{1}{2}(1.65522)(2.6) \sin 95° \approx 2.1$

47. $s = \dfrac{1}{2}(5.5 + 6.7 + 9.2) = 10.7 \Rightarrow$ Area $= K = \sqrt{10.7(10.7 - 5.5)(10.7 - 6.7)(10.7 - 9.2)} \approx 18.3$

49. $s = \dfrac{1}{2}(11 + 13 + 20) = 22 \Rightarrow$ Area $= K = \sqrt{22(22 - 11)(22 - 13)(22 - 20)} = 66$

51. To apply the formula

$K = \dfrac{1}{2}ac \sin \beta$, we need to find β, or to use the formula $K = \dfrac{1}{2}bc \sin \alpha$, we need to find b.

This triangle is in the form of SSA. It is ambiguous.

$\dfrac{\sin \gamma}{c} = \dfrac{\sin \alpha}{a} \Rightarrow \sin \gamma = \dfrac{c \sin \alpha}{a} = \dfrac{16 \sin 42°}{21} \approx 0.50981 \Rightarrow \gamma_R = \sin^{-1}(0.50981) \approx 30.65°$

$\gamma_1 = 30.65°$ then $\beta = 180° - 42° - 30.65° = 107.35°$

Since $\beta = 107.35°$, then Area $= K = \dfrac{1}{2}(21)(16) \sin 107.35° \approx 160.4°$

53. Using standard labels, $a = BC, b = AC, c = AB$ and $\gamma =$ angle ACB.

$c^2 = a^2 + b^2 - 2ab \cos \gamma \Rightarrow c = \sqrt{123^2 + 143^2 - 2(123)(143) \cos 78° \, 35'} \approx 169$ ft.

55. After 1.5 hours, the first ship has gone 30 miles and the second ship has gone 21 miles. See Figure 55.

$d^2 = 30^2 + 21^2 - 2(30)(21) \cos 69° \Rightarrow d = \sqrt{1341 - 1260 \cos 69°} \approx 29.82$

After 1.5 hours, the ships are about 29.8 miles apart.

Figure 55

57. The shorter diagonal and the two given sides form a triangle of the form SAS. Using the law of cosines we will be able to find the length of the shorter diagonal. We will use $\alpha = 56°$, $b = 5.2$ and $c = 3.5 \Rightarrow$

$a^2 = b^2 + c^2 - 2bc \cos \alpha = 5.2^2 + 3.5^2 - 2(5.2)(3.5) \cos 56° \Rightarrow$

$a = \sqrt{39.29 - 36.4 \cos 56°} \approx 4.35148$ or $a = 4.4$ feet. The longer diagonal and the two given sides form a triangle of the form SAS. Since adjacent angles of a parallelogram are supplementary, $\alpha = 180° - 56° = 124°$. Using the law of cosines we will be able to find the length of the longer diagonal. We will use

$\alpha = 124°$, $b = 5.2$ and $c = 3.5 \Rightarrow a^2 = b^2 + c^2 - 2bc \cos \alpha = 5.2^2 + 3.5^2 - 2(5.2)(3.5) \cos 124° \Rightarrow$

$a = \sqrt{39.29 - 36.4 \cos 124°} \approx 7.72299$ or $a \approx 7.7$ feet.

59. Using standard labels, $a = 500$, $b = 410$, $c = 400$, $\alpha = $ angle CAB, $\beta = $ angle ABC and $\gamma = $ angle ACB.

To find the bearing when traveling from A to B we first need to find α.

$a^2 = b^2 + c^2 - 2bc \cos \alpha \Rightarrow \cos \alpha = \dfrac{a^2 - b^2 - c^2}{-2bc} \Rightarrow \alpha = \cos^{-1}\left(\dfrac{500^2 - 410^2 - 400^2}{-2(410)(400)}\right) \approx 76.2°$

Thus, the bearing is approximately 76.2°.

To find the bearing when traveling from B to C we first need to find β.

$b^2 = a^2 + c^2 - 2ac \cos \beta \Rightarrow \cos \beta = \dfrac{b^2 - a^2 - c^2}{-2ac} \Rightarrow \beta = \cos^{-1}\left(\dfrac{410^2 - 500^2 - 400^2}{-2(500)(400)}\right) \approx 52.8°$

Thus, the bearing is approximately $180° + 76.2° + 52.8° \approx 309°$.

61. (a) $\cos \alpha = \dfrac{a^2 - b^2 - c^2}{-2bc} \Rightarrow \alpha = \cos^{-1}\left(\dfrac{145.2^2 - 136.8^2 - 95.3^2}{-2(136.8)(95.3)}\right) \approx 75.08001°$

$\cos \beta = \dfrac{b^2 - a^2 - c^2}{-2ac} \Rightarrow \beta = \cos^{-1}\left(\dfrac{136.8^2 - 145.2^2 - 95.3^2}{-2(145.2)(95.3)}\right) \approx 65.55867°$

$\gamma = 180° - 75.08001° - 65.55867° \approx 39.36132° \qquad \alpha \approx 75.1°, \ \beta \approx 65.6°, \ \gamma \approx 39.4°$

(The angles do not sum to 180° due to rounding)

(b) $s = \dfrac{1}{2}(145.2 + 136.8 + 95.3) = 188.65 \Rightarrow$

Area $= K = \sqrt{188.65(188.65 - 145.2)(188.65 - 136.8)(188.65 - 95.3)} \approx 6298.76 \approx 6299$ ft^2.

63. $13^2 = 16^2 + 20^2 - 2(16)(20) \cos \theta \Rightarrow \theta = \cos^{-1}\left(\dfrac{13^2 - 16^2 - 20^2}{-2(16)(20)}\right) \approx 40.5°$

65. Refer to Figure 65. Mark point D directly south of C on the segment joining A and B. Then triangle CDB is a right triangle. Since the bearing from C to B is 125°, angle $DCB = 180° - 125° = 55°$.

Thus $\beta = 90° - 55° = 35°$. We will use $a = 357$, $c = 515$ and $\beta = 35°$.

$b^2 = a^2 + c^2 - 2ac \cos \beta = 357^2 + 515^2 - 2(357)(515) \cos 35° \Rightarrow b = \sqrt{392{,}674 - 367{,}710 \cos 35°} \approx$

302 miles. Thus, the distance between A and C is approximately 302 miles.

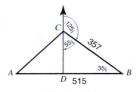

Figure 65

67. We will use the labels A = Airplane, B = Battleship, and C = Submarine, with corresponding angles

α, β, and γ, respectively. Since the angle of elevation from B to A equals the angle of depression from A to B,

we have $\beta = 17° \ 30'$. Also $\alpha = 24° \ 10' - 17° \ 30' = 6° \ 40'$ and $\gamma = 180° - 17° \ 30' - 6° \ 40' = 155° \ 50'$.

Using the law of sines $\dfrac{a}{\sin 6° \ 40'} = \dfrac{5120}{\sin 155° \ 50'} \Rightarrow a = \dfrac{5120 \sin 6° \ 40'}{\sin 155° \ 50'} \approx 1451.9$.

The distance between the battleship and the submarine is about 1452 ft.

69. Use the law of cosines to find the distance d. $d^2 = 402^2 + 402^2 - 2(402)(402) \cos 135° \ 40' \Rightarrow$

$d = \sqrt{323{,}208 - 323{,}208 \cos 135° \ 40'} \Rightarrow d \approx 745$ mi.

71. The painter's triangular region is in the form of SAS where we will use $c = 25$, $b = 15$ and $\alpha = 128°$.

If $K = \dfrac{1}{2} bc \sin \alpha$, then $K = \dfrac{1}{2}(15)(25) \sin 128° \approx 147.8$. Thus, the area of the region is 147.8 ft^2.

73. (a) $L = 6$ and $n = 3 \Rightarrow A = \dfrac{3(6)^2}{4} \cot \dfrac{\pi}{3} = 27\left(\dfrac{1}{\sqrt{3}}\right) = 9\sqrt{3} \approx 15.6$ in^2.

   (b) $s = \dfrac{1}{2}(6 + 6 + 6) = 9 \Rightarrow \text{Area} = K = \sqrt{9(9 - 6)(9 - 6)(9 - 6)} \approx 15.6$ in^2.

   The results are equal.

75. Area $= K = \dfrac{1}{2}(500)(600) \sin 85° \approx 149{,}429$ ft^2.

77. In triangle ADE,

$s = \dfrac{1}{2}(105 + 190 + 125) = 210 \Rightarrow K_1 = \sqrt{210(210 - 105)(210 - 190)(210 - 125)} \approx 6122.5$ ft^2.

In triangle ABD, $s = \dfrac{1}{2}(190 + 110 + 195) = 247.5 \Rightarrow$

$K_2 = \sqrt{247.5(247.5 - 190)(247.5 - 110)(247.5 - 195)} \approx 10{,}135.7$ ft^2.

In triangle BCD,

$s = \dfrac{1}{2}(195 + 75 + 150) = 210 \Rightarrow K_3 = \sqrt{210(210 - 195)(210 - 75)(210 - 150)} \approx 5051.2$ ft^2.

The total area is $K_1 + K_2 + K_3 = 6122.5 + 10{,}135.7 + 5051.2 \approx 21{,}309$ ft^2.

Extended and Discovery Exercises for Section 8.2

1. Since the satellite makes a complete orbit in 120 minutes, it forms a central angle of $\frac{3}{120} \cdot 360 = 9°$ in 3 minutes. Let d represent the distance between the satellite and the tracking station at 12:03 P.M. The other two sides of the triangle shown in the diagram have lengths 6400 km and $6400 + 1600 = 8000$ km. Using the law of cosines $d^2 = 6400^2 + 8000^2 - 2(6400)(8000)\cos 9° \Rightarrow$ $d = \sqrt{6400^2 + 8000^2 - 2(6400)(8000)\cos 9°} \approx 1954.7$. The distance is about 2000 km.

Checking Basic Concepts for Sections 8.1 and 8.2

1. This triangle is of the form AAS. There is only one solution.

$$\beta = 180° - 44° - 62° = 74°$$

$$\frac{b}{\sin\beta} = \frac{a}{\sin\alpha} \Rightarrow b = \frac{a\sin\beta}{\sin\alpha} = \frac{12\sin 74°}{\sin 44°} \approx 16.6$$

$$\frac{c}{\sin\gamma} = \frac{a}{\sin\alpha} \Rightarrow c = \frac{a\sin\gamma}{\sin\alpha} = \frac{12\sin 62°}{\sin 44°} \approx 15.3 \quad \beta = 74°,\ b \approx 16.6,\ c \approx 15.3$$

3. (a) $b^2 = a^2 + c^2 - 2ac\cos\beta = 8.1^2 + 8.3^2 - 2(8.1)(8.3)\cos 51° \Rightarrow$

$$b = \sqrt{134.5 - 134.46\cos 51°} \approx 7.06269$$

$$c^2 = a^2 + b^2 - 2ab\cos\gamma \Rightarrow \cos\gamma = \frac{c^2 - a^2 - b^2}{-2ab} \Rightarrow$$

$$\gamma = \cos^{-1}\left(\frac{8.3^2 - 8.1^2 - 7.06269^2}{-2(8.1)(7.06269)}\right) \approx 65.96459°$$

$$\alpha = 180° - 51° - 65.96459° \approx 63.03541° \quad b \approx 7.1,\ \alpha \approx 63.0°,\ \gamma \approx 66.0°$$

(b) $a^2 = b^2 + c^2 - 2bc\cos\alpha \Rightarrow \cos\alpha = \frac{a^2 - b^2 - c^2}{-2bc} \Rightarrow \alpha = \cos^{-1}\left(\frac{14^2 - 9^2 - 8^2}{-2(9)(8)}\right) \approx 110.74238°$

$$b^2 = a^2 + c^2 - 2ac\cos\beta \Rightarrow \cos\beta = \frac{b^2 - a^2 - c^2}{-2ac} \Rightarrow \beta = \cos^{-1}\left(\frac{9^2 - 14^2 - 8^2}{-2(14)(8)}\right) \approx 36.95507°$$

$$\gamma = 180° - 110.74238° - 36.95507° \approx 32.30255° \quad \alpha \approx 110.7°,\ \beta \approx 37.0°,\ \gamma \approx 32.3°$$

8.3: Vectors

1. (a) $a_1 \approx 3,\ a_2 \approx 4$

   (b) $\|\mathbf{v}\| = \sqrt{3^2 + 4^2} = \sqrt{9 + 16} = \sqrt{25} = 5$

3. (a) $a_1 \approx -5,\ a_2 \approx -12$

   (b) $\|\mathbf{v}\| = \sqrt{(-5)^2 + (-12)^2} = \sqrt{25 + 144} = \sqrt{169} = 13$

5. (a) See Figure 5.

(b) $\mathbf{v} = \langle 0, -20 \rangle$

(c) $2\mathbf{v} = 2\langle 0, -20 \rangle = \langle 0, -40 \rangle$; this represents a 40 mph north wind.

$-\dfrac{1}{2}\mathbf{v} = -\dfrac{1}{2}\langle 0, -20 \rangle = \langle 0, 10 \rangle$; this represents a 10 mph south wind.

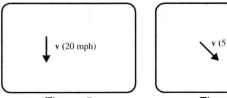

Figure 5   Figure 7

7. (a) See Figure 7.

(b) We may think of this vector as a hypotenuse of an isosceles right triangle. Since the legs l of such a triangle have the same length, we have the following:

$$l^2 + l^2 = 5^2 \Rightarrow 2l^2 = 25 \Rightarrow l^2 = \frac{25}{2} \Rightarrow l = \sqrt{\frac{25}{2}} \Rightarrow l = \frac{5}{\sqrt{2}}.$$

The vector can be represented by $\mathbf{v} = \left\langle \dfrac{5}{\sqrt{2}}, -\dfrac{5}{\sqrt{2}} \right\rangle$ or $\left\langle \dfrac{5}{2}\sqrt{2}, -\dfrac{5}{2}\sqrt{2} \right\rangle$.

(c) $2\mathbf{v} = 2\left\langle \dfrac{5}{\sqrt{2}}, -\dfrac{5}{\sqrt{2}} \right\rangle = \left\langle \dfrac{10}{\sqrt{2}}, -\dfrac{10}{\sqrt{2}} \right\rangle$ or $\langle 5\sqrt{2}, -5\sqrt{2} \rangle$; this represents a 10 mph northwest wind.

$-\dfrac{1}{2}\mathbf{v} = -\dfrac{1}{2}\left\langle \dfrac{5}{2}\sqrt{2}, -\dfrac{5}{2}\sqrt{2} \right\rangle = \left\langle -\dfrac{5}{4}\sqrt{2}, \dfrac{5}{4}\sqrt{2} \right\rangle$; this represents a 2.5 mph southeast wind.

9. (a) See Figure 9.

(b) $\mathbf{v} = \langle 0, 30 \rangle$

(c) $2\mathbf{v} = 2\langle 0, 30 \rangle = \langle 0, 60 \rangle$; this represents a 60 lb force upward.

$-\dfrac{1}{2}\mathbf{v} = -\dfrac{1}{2}\langle 0, 30 \rangle = \langle 0, -15 \rangle$; this represents a 15 lb force downward.

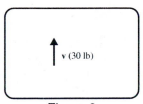

Figure 9

11. (a) Horizontal $= 1$, vertical $= 1$

(b) $\|\mathbf{v}\| = \sqrt{1^2 + 1^2} = \sqrt{1 + 1} = \sqrt{2}$; $\mathbf{v}$ is not a unit vector.

(c) $\|\mathbf{v}\|$ represents the length of $\mathbf{v}$. See Figure 11.

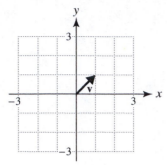

Figure 11

13. (a) Horizontal $= 3$, vertical $= -4$

    (b) $\|\mathbf{v}\| = \sqrt{3^2 + (-4)^2} = \sqrt{9 + 16} = \sqrt{25} = 5$; $\mathbf{v}$ is not a unit vector.

    (c) $\|\mathbf{v}\|$ represents the length of $\mathbf{v}$. See Figure 13.

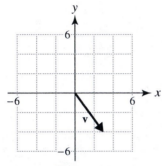

Figure 13

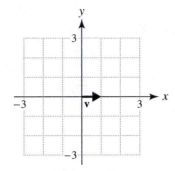

Figure 15

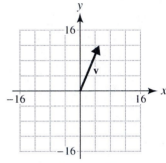

Figure 17

15. (a) Horizontal $= 1$, vertical $= 0$

    (b) $\|\mathbf{v}\| = \sqrt{1^2 + 0^2} = \sqrt{1 + 0} = \sqrt{1} = 1$; $\mathbf{v}$ is a unit vector.

    (c) $\|\mathbf{v}\|$ represents the length of $\mathbf{v}$. See Figure 15.

17. (a) Horizontal $= 5$, vertical $= 12$

    (b) $\|\mathbf{v}\| = \sqrt{5^2 + 12^2} = \sqrt{25 + 144} = \sqrt{169} = 13$; $\mathbf{v}$ is not a unit vector.

    (c) $\|\mathbf{v}\|$ represents the length of $\mathbf{v}$. See Figure 17.

19. $\|\mathbf{v}\| = \sqrt{3^2 + 0^2} = 3$; $\mathbf{v} = \langle 3, 0 \rangle$ lies on the positive x-axis and $\tan\theta = \dfrac{0}{3} \Rightarrow \theta = 0$.

    Therefore, the direction angle is $0°$.

21. $\|\mathbf{v}\| = \sqrt{(-1)^2 + 1^2} = \sqrt{2}$; $\mathbf{v} = \langle -1, 1 \rangle$ is in quadrant II and $\tan\theta = -1 \Rightarrow \theta = 135°$.

    Therefore, the direction angle is $135°$.

23. $\|\mathbf{v}\| = \sqrt{(\sqrt{3})^2 + (-1)^2} = 2$; $\mathbf{v} = \langle \sqrt{3}, -1 \rangle$ is in quadrant IV and $\tan\theta = \dfrac{-1}{\sqrt{3}} \Rightarrow \theta = 330°$.

    Therefore, the direction angle is $330°$.

25. $\|\mathbf{v}\| = \sqrt{(-5)^2 + (-12)^2} = 13$; $\mathbf{v} = \langle -5, -12 \rangle$ is in quadrant III and $\tan\theta = \dfrac{-12}{-5} \Rightarrow \theta = 247.4°$.

    Therefore, the direction angle is $247.4°$.

27. $\|\mathbf{v}\| = \sqrt{13^2 + (-84)^2} = 85$; $\mathbf{v} = \langle 13, -84 \rangle$ is in quadrant IV and $\tan \theta = \dfrac{-84}{13} \Rightarrow \theta = 278.8°$.

Therefore, the direction angle is $278.8°$.

29. $\|\mathbf{v}\| = \sqrt{(-20)^2 + (21)^2} = 29$; $\mathbf{v} = \langle -20, 21 \rangle$ is in quadrant II and $\tan \theta = \dfrac{21}{-20} \Rightarrow \theta = 133.6°$.

Therefore, the direction angle is $133.6°$.

31. $a_1 = 4 \cos 180° = -4$; $a_2 = 4 \sin 180° = 0$. Horizontal component $= -4$, vertical component $= 0$.

33. $a_1 = \sqrt{2} \cos 135° = -1$; $a_2 = \sqrt{2} \sin 135° = 1$. Horizontal component $= -1$, vertical component $= 1$.

35. $a_1 = 23 \cos 54° = 13.5$; $a_2 = 23 \sin 54° = 18.6$. Horizontal component $= 13.5$, vertical component $= 18.6$.

37. $a_1 = 34 \cos 312° = 22.8$; $a_2 = 34 \sin 312° = -25.3$. Horizontal component $= 22.8$,

vertical component $= -25.3$.

39. $\mathbf{v} = \langle 6 \cos 30°, 6 \sin 30° \rangle = \langle 3\sqrt{3}, 3 \rangle$

41. $\mathbf{v} = \langle 9 \cos 225°, 9 \sin 225° \rangle = \left\langle -\dfrac{9\sqrt{2}}{2}, -\dfrac{9\sqrt{2}}{2} \right\rangle$

43. horizontal: $\cos 40° = \dfrac{v_1}{4} \Rightarrow v_1 = 4 \cos 40° \approx 3.06$

vertical: $\sin 40° = \dfrac{v_2}{4} \Rightarrow v_2 = 4 \sin 40° \approx 2.57$

45. horizontal: $\cos (-35°) = \dfrac{v_1}{5} \Rightarrow v_1 = 5 \cos (-35°) \approx 4.10$

vertical: $\sin (-35°) = \dfrac{v_2}{5} \Rightarrow v_2 = 5 \sin (-35°) \approx -2.87$

47. (a) See Figure 47.

(b) $\mathbf{v} = \langle -1 - 0, 2 - 0 \rangle = \langle -1, 2 \rangle$

(c) $\|\overrightarrow{PQ}\| = \sqrt{(-1)^2 + (2)^2} = \sqrt{5}$

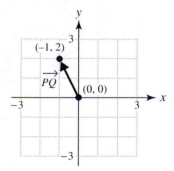

Figure 47

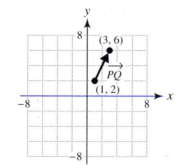

Figure 49

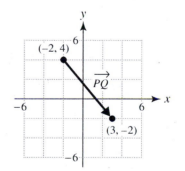

Figure 51

49. (a) See Figure 49.

(b) $\mathbf{v} = \langle 3 - 1, 6 - 2 \rangle = \langle 2, 4 \rangle$

(c) $\|\overrightarrow{PQ}\| = \sqrt{(2)^2 + (4)^2} = \sqrt{20}$

51. (a) See Figure 51.

(b) $\mathbf{v} = \langle 3 - (-2), -2 - 4 \rangle = \langle 5, -6 \rangle$

(c) $\|\overrightarrow{PQ}\| = \sqrt{(5)^2 + (-6)^2} = \sqrt{61}$

53. Complete the parallelogram and find the unknown angle: $180° - 40° = 140°$.

    Find the magnitude of the resultant force by using the law of cosines to find the length of the diagonal d.

    $d^2 = 40^2 + 60^2 - 2(40)(60)\cos 140° \Rightarrow d = \sqrt{5200 - 4800\cos 140°} \approx 94.2$ lb

55. Complete the parallelogram and find the unknown angle: $180° - 110° = 70°$.

    Find the magnitude of the resultant force by using the law of cosines to find the length of the diagonal d.

    $d^2 = 15^2 + 25^2 - 2(15)(25)\cos 70° \Rightarrow d = \sqrt{850 - 750\cos 70°} \approx 24.4$ lb

57. $F_1 + F_2 = (3\mathbf{i} - 4\mathbf{j}) + (-8\mathbf{i} + 16\mathbf{j}) = (3 - 8)\mathbf{i} + (-4 + 16)\mathbf{j} = -5\mathbf{i} + 12\mathbf{j}$,

    $\|\mathbf{F}\| = \sqrt{(-5)^2 + 12^2} = \sqrt{169} = 13$

59. $F_1 + F_2 = \langle 8, 9 \rangle + \langle 1, -31 \rangle = \langle 8 + 1, 9 + (-31) \rangle = \langle 9, -22 \rangle$, $\|\mathbf{F}\| = \sqrt{9^2 + (-22)^2} = \sqrt{565}$

61. $F_1 + F_2 = (0.5\mathbf{i} + 0.7\mathbf{j}) + (-1.5\mathbf{i} - 5.7\mathbf{j}) = (0.5 - 1.5)\mathbf{i} + (0.7 - 5.7)\mathbf{j} = -\mathbf{i} - 5\mathbf{j}$,

    $\|\mathbf{F}\| = \sqrt{(-1)^2 + (-5)^2} = \sqrt{26}$

63. (a) $\mathbf{a} + \mathbf{b} = \langle 0, 2 \rangle + \langle 3, 0 \rangle = \langle 0 + 3, 2 + 0 \rangle = \langle 3, 2 \rangle$

    (b) $\mathbf{a} - \mathbf{b} = \langle 0, 2 \rangle - \langle 3, 0 \rangle = \langle 0 - 3, 2 - 0 \rangle = \langle -3, 2 \rangle$

65. (a) $\mathbf{a} + \mathbf{b} = (2\mathbf{i} + \mathbf{j}) + (\mathbf{i} - 2\mathbf{j}) = 3\mathbf{i} - \mathbf{j}$

    (b) $\mathbf{a} - \mathbf{b} = (2\mathbf{i} + \mathbf{j}) - (\mathbf{i} - 2\mathbf{j}) = \mathbf{i} + 3\mathbf{j}$

67. (a) $\mathbf{a} + \mathbf{b} = \left\langle -\sqrt{2}, \dfrac{1}{2} \right\rangle + \left\langle \sqrt{2}, -\dfrac{3}{4} \right\rangle = \left\langle -\sqrt{2} + \sqrt{2}, \dfrac{1}{2} + \left(-\dfrac{3}{4} \right) \right\rangle = \left\langle 0, -\dfrac{1}{4} \right\rangle$

    (b) $\mathbf{a} - \mathbf{b} = \left\langle -\sqrt{2}, \dfrac{1}{2} \right\rangle - \left\langle \sqrt{2}, -\dfrac{3}{4} \right\rangle = \left\langle -\sqrt{2} - \sqrt{2}, \dfrac{1}{2} - \left(-\dfrac{3}{4} \right) \right\rangle = \left\langle -2\sqrt{2}, \dfrac{5}{4} \right\rangle$

69. (a) $\mathbf{a} + \mathbf{b} = \left[\left(\cos \dfrac{\pi}{4} \right)\mathbf{i} + \left(\sin \dfrac{\pi}{4} \right)\mathbf{j} \right] + \left[\left(\cos \dfrac{\pi}{2} \right)\mathbf{i} + \left(\sin \dfrac{\pi}{2} \right)\mathbf{j} \right] =$

    $\left[\left(\cos \dfrac{\pi}{4} \right)\mathbf{i} + \left(\cos \dfrac{\pi}{2} \right)\mathbf{i} + \left(\sin \dfrac{\pi}{4} \right)\mathbf{j} + \left(\sin \dfrac{\pi}{2} \right)\mathbf{j} \right] = \left(\dfrac{\sqrt{2}}{2}\mathbf{i} + 0\mathbf{i} \right) + \left(\dfrac{\sqrt{2}}{2}\mathbf{j} + 1\mathbf{j} \right) =$

    $\dfrac{\sqrt{2}}{2}\mathbf{i} + \dfrac{\sqrt{2} + 2}{2}\mathbf{j}$

    (b) $\mathbf{a} - \mathbf{b} = \left[\left(\cos \dfrac{\pi}{4} \right)\mathbf{i} + \left(\sin \dfrac{\pi}{4} \right)\mathbf{j} \right] - \left[\left(\cos \dfrac{\pi}{2} \right)\mathbf{i} + \left(\sin \dfrac{\pi}{2} \right)\mathbf{j} \right] =$

    $\left[\left(\cos \dfrac{\pi}{4} \right)\mathbf{i} - \left(\cos \dfrac{\pi}{2} \right)\mathbf{i} + \left(\sin \dfrac{\pi}{4} \right)\mathbf{j} - \left(\sin \dfrac{\pi}{2} \right)\mathbf{j} \right] = \left(\dfrac{\sqrt{2}}{2}\mathbf{i} - 0\mathbf{i} \right) + \left(\dfrac{\sqrt{2}}{2}\mathbf{j} - 1\mathbf{j} \right) =$

    $\dfrac{\sqrt{2}}{2}\mathbf{i} + \dfrac{\sqrt{2} - 2}{2}\mathbf{j}$

71. (a) $\mathbf{a} + \mathbf{b} = \langle -8, 8 \rangle + \langle 4, 8 \rangle = \langle -8 + 4, 8 + 8 \rangle = \langle -4, 16 \rangle$

    (b) $\mathbf{a} - \mathbf{b} = \langle -8, 8 \rangle - \langle 4, 8 \rangle = \langle -8 - 4, 8 - 8 \rangle = \langle -12, 0 \rangle$

    (c) $-\mathbf{a} = -\langle -8, 8 \rangle = \langle 8, -8 \rangle$

73. (a) $\mathbf{a} + \mathbf{b} = \langle 4, 8 \rangle + \langle 4, -8 \rangle = \langle 4 + 4, 8 + (-8) \rangle = \langle 8, 0 \rangle$

(b) $\mathbf{a} - \mathbf{b} = \langle 4, 8 \rangle - \langle 4, -8 \rangle = \langle 4 - 4, 8 - (-8) \rangle = \langle 0, 16 \rangle$

(c) $-\mathbf{a} = -\langle 4, 8 \rangle = \langle -4, -8 \rangle$

75. (a) $\mathbf{a} + \mathbf{b} = \langle -8, 4 \rangle + \langle 8, 8 \rangle = \langle -8 + 8, 4 + 8 \rangle = \langle 0, 12 \rangle$

(b) $\mathbf{a} - \mathbf{b} = \langle -8, 4 \rangle - \langle 8, 8 \rangle = \langle -8 - 8, 4 - 8 \rangle = \langle -16, -4 \rangle$

(c) $-\mathbf{a} = -\langle -8, 4 \rangle = \langle 8, -4 \rangle$

77. (a) Graphical: See Figure 77a. The length appears to be about 2.

    Symbolic: $\|\mathbf{a}\| = \|2\mathbf{i}\| = \sqrt{2^2} = \sqrt{4} = 2$

(b) Graphical: $2\mathbf{a} = 4\mathbf{i}$. See Figure 77b.

    Symbolic: $2\mathbf{a} = 2(2\mathbf{i}) = 4\mathbf{i}$

(c) Graphical: $2\mathbf{a} + 3\mathbf{b} = 7\mathbf{i} + 3\mathbf{j}$. See Figure 77c.

    Symbolic: $2\mathbf{a} + 3\mathbf{b} = 2(2\mathbf{i}) + 3(\mathbf{i} + \mathbf{j}) = 4\mathbf{i} + 3\mathbf{i} + 3\mathbf{j} = 7\mathbf{i} + 3\mathbf{j}$

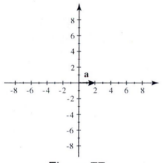

Figure 77a

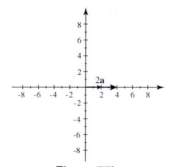

Figure 77b

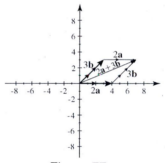

Figure 77c

79. (a) Graphical: Graphical: See Figure 79a. The length appears to be about 2.25.

    Symbolic: $\|\mathbf{a}\| = \sqrt{(-1)^2 + 2^2} = \sqrt{5}$

(b) Graphical: $2\mathbf{a} = \langle -2, 4 \rangle$. See Figure 79b.

    Symbolic: $2\mathbf{a} = 2\langle -1, 2 \rangle = \langle -2, 4 \rangle$

(c) Graphical: $2\mathbf{a} + 3\mathbf{b} = \langle 7, 4 \rangle$. See Figure 79c.

    Symbolic: $2\mathbf{a} + 3\mathbf{b} = 2\langle -1, 2 \rangle + 3\langle 3, 0 \rangle = \langle -2, 4 \rangle + \langle 9, 0 \rangle = \langle -2 + 9, 4 + 0 \rangle = \langle 7, 4 \rangle$

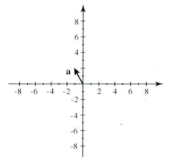

Figure 79a

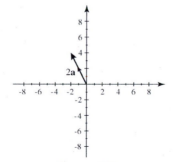

Figure 79b

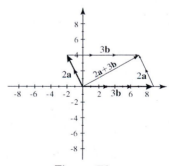

Figure 79c

81. (a) $-a + 4b = -(\mathbf{i} - 2\mathbf{j}) + 4(-5\mathbf{i} + 2\mathbf{j}) = (-\mathbf{i} + 2\mathbf{j}) + (-20\mathbf{i} + 8\mathbf{j}) = (-1 - 20)\mathbf{i} + (2 + 8)\mathbf{j} =$
$-21\mathbf{i} + 10\mathbf{j}$

    (b) $2a - 3b = 2(\mathbf{i} - 2\mathbf{j}) - 3(-5\mathbf{i} + 2\mathbf{j}) = (2\mathbf{i} - 4\mathbf{j}) + (15\mathbf{i} - 6\mathbf{j}) = (2 + 15)\mathbf{i} + (-4 - 6)\mathbf{j} =$
$17\mathbf{i} - 10\mathbf{j}$

83. (a) $-a + 4b = -\langle 1, -4 \rangle + 4\langle -3, 5 \rangle = \langle -1, 4 \rangle + \langle -12, 20 \rangle = \langle -1 - 12, 4 + 20 \rangle = \langle -13, 24 \rangle =$
$-13\mathbf{i} + 24\mathbf{j}$

    (b) $2a - 3b = 2\langle 1, -4 \rangle - 3\langle -3, 5 \rangle = \langle 2, -8 \rangle + \langle 9, -15 \rangle = \langle 2 + 9, -8 - 15 \rangle = \langle 11, -23 \rangle = 11\mathbf{i} - 23\mathbf{j}$

85. (a) $-a + 4b = -\langle -7, -2 \rangle + 4\langle 9, -1 \rangle = \langle 7, 2 \rangle + \langle 36, -4 \rangle = \langle 7 + 36, 2 - 4 \rangle = \langle 43, -2 \rangle = 43\mathbf{i} - 2\mathbf{j}$

    (b) $2a - 3b = 2\langle -7, -2 \rangle - 3\langle 9, -1 \rangle = \langle -14, -4 \rangle + \langle -27, 3 \rangle = \langle -14 - 27, -4 + 3 \rangle = \langle -41, -1 \rangle =$
$-41\mathbf{i} - \mathbf{j}$

87. (a) $\mathbf{a} \cdot \mathbf{b} = (1)(3) + (-2)(1) = 3 - 2 = 1$

    (b) $\|\mathbf{a}\| = \sqrt{1^2 + (-2)^2} = \sqrt{1 + 4} = \sqrt{5},\ \|\mathbf{b}\| = \sqrt{3^2 + 1^2} = \sqrt{9 + 1} = \sqrt{10}$

$$\theta = \cos^{-1}\left(\frac{\mathbf{a} \cdot \mathbf{b}}{\|\mathbf{a}\|\|\mathbf{b}\|}\right) = \cos^{-1}\left(\frac{1}{\sqrt{5}\,\sqrt{10}}\right) = \cos^{-1}\left(\frac{1}{\sqrt{50}}\right) \approx 81.9°$$

    (c) The vectors are neither parallel nor perpendicular.

89. (a) $\mathbf{a} \cdot \mathbf{b} = (6)(-4) + (8)(3) = -24 + 24 = 0$

    (b) $\|\mathbf{a}\| = \sqrt{6^2 + 8^2} = \sqrt{36 + 64} = \sqrt{100} = 10,\ \|\mathbf{b}\| = \sqrt{(-4)^2 + 3^2} = \sqrt{16 + 9} = \sqrt{25} = 5$

$$\theta = \cos^{-1}\left(\frac{\mathbf{a} \cdot \mathbf{b}}{\|\mathbf{a}\|\|\mathbf{b}\|}\right) = \cos^{-1}\left(\frac{0}{10 \cdot 5}\right) = \cos^{-1} 0 = 90°$$

    (c) The vectors are perpendicular.

91. (a) $\mathbf{a} \cdot \mathbf{b} = (5)(10) + (6)(12) = 50 + 72 = 122$

    (b) $\|\mathbf{a}\| = \sqrt{5^2 + 6^2} = \sqrt{25 + 36} = \sqrt{61},\ \|\mathbf{b}\| = \sqrt{10^2 + 12^2} = \sqrt{100 + 144} = \sqrt{244}$

$$\theta = \cos^{-1}\left(\frac{\mathbf{a} \cdot \mathbf{b}}{\|\mathbf{a}\|\|\mathbf{b}\|}\right) = \cos^{-1}\left(\frac{122}{\sqrt{61}\,\sqrt{244}}\right) = \cos^{-1}\left(\frac{122}{\sqrt{14{,}884}}\right) = \cos^{-1}\left(\frac{122}{122}\right) = \cos^{-1} 1 = 0°$$

    (c) The vectors are parallel and they point in the same direction.

93. (a) $\mathbf{a} \cdot \mathbf{b} = (1)(0.5) + (3)(-1.5) = 0.5 - 4.5 = -4$

    (b) $\|\mathbf{a}\| = \sqrt{1^2 + 3^2} = \sqrt{1 + 9} = \sqrt{10},\ \|\mathbf{b}\| = \sqrt{0.5^2 + (-1.5)^2} = \sqrt{0.25 + 2.25} = \sqrt{2.5}$

$$\theta = \cos^{-1}\left(\frac{\mathbf{a} \cdot \mathbf{b}}{\|\mathbf{a}\|\|\mathbf{b}\|}\right) = \cos^{-1}\left(\frac{-4}{\sqrt{10}\,\sqrt{2.5}}\right) = \cos^{-1}\left(\frac{-4}{\sqrt{25}}\right) = \cos^{-1}\left(\frac{-4}{5}\right) \approx 143.1°$$

    (c) The vectors are neither parallel nor perpendicular.

95. $W = 30 \cdot 5 = 150$ ft-lb

97. $W = 100 \cdot 1000 = 100{,}000$ ft-lb

99. $W = \mathbf{F} \cdot \mathbf{D} = \langle 10, 20 \rangle \cdot \langle 15, 22 \rangle = (10)(15) + (20)(22) = 590$ ft-lb

    $\|\mathbf{F}\| = \sqrt{10^2 + 20^2} = \sqrt{100 + 400} = \sqrt{500} = 10\sqrt{5} \approx 22.4$ lb

101. $W = \mathbf{F} \cdot \mathbf{D} = (5\mathbf{i} - 3\mathbf{j}) \cdot (3\mathbf{i} - 4\mathbf{j}) = (5)(3) + (-3)(-4) = 27$ ft-lb

$\|\mathbf{F}\| = \sqrt{5^2 + (-3)^2} = \sqrt{25 + 9} = \sqrt{34} \approx 5.8$ lb

103. $\overrightarrow{PQ} = \langle 1 - (-2), 6 - 3 \rangle = \langle 3, 3 \rangle = 3\mathbf{i} + 3\mathbf{j}$

$W = \mathbf{F} \cdot \mathbf{D} = (5\mathbf{i} + 3\mathbf{j}) \cdot (3\mathbf{i} + 3\mathbf{j}) = (5)(3) + (3)(3) = 24$

105. $\overrightarrow{PQ} = \langle 4 - 2, -5 - (-3) \rangle = \langle 2, -2 \rangle = 2\mathbf{i} - 2\mathbf{j}$

$W = \mathbf{F} \cdot \mathbf{D} = (5\mathbf{i} + 3\mathbf{j}) \cdot (2\mathbf{i} - 2\mathbf{j}) = (5)(2) + (3)(-2) = 4$

107. The swimmer's velocity is modeled by $\mathbf{a} = \langle 0, 3 \rangle$. The velocity of the current is modeled by $\mathbf{b} = \langle 2, 0 \rangle$.

The resultant velocity of the swimmer is modeled by $\mathbf{a} + \mathbf{b} = \langle 0, 3 \rangle + \langle 2, 0 \rangle = \langle 0 + 2, 3 + 0 \rangle = \langle 2, 3 \rangle$.

The speed of the swimmer is $\|\mathbf{a} + \mathbf{b}\| = \sqrt{2^2 + 3^2} = \sqrt{4 + 9} = \sqrt{13} \approx 3.6$ mph.

109. The velocity of the airplane is modeled by $\mathbf{a} = \langle -400, 0 \rangle$. Refer to the referenced example when finding

the appropriate vector to represent the wind. The velocity of the wind can be modeled by

$\mathbf{b} = \left\langle \dfrac{50}{\sqrt{2}}, -\dfrac{50}{\sqrt{2}} \right\rangle$. The resultant velocity of the airplane can be found by vector addition and is modeled by

$\mathbf{a} + \mathbf{b} = \langle -400, 0 \rangle + \left\langle \dfrac{50}{\sqrt{2}}, -\dfrac{50}{\sqrt{2}} \right\rangle = \left\langle -400 + \dfrac{50}{\sqrt{2}}, 0 + \left(-\dfrac{50}{\sqrt{2}} \right) \right\rangle \approx \langle -364.6, -35.4 \rangle$.

The ground speed of the airplane is $\|\mathbf{a} + \mathbf{b}\| = \sqrt{(-364.6)^2 + (-35.4)^2} \approx 366.3$ mph.

The bearing of the airplane is $270° - \tan^{-1}\left(\dfrac{-35.4}{-364.6} \right) \approx 264.5°$.

111. Since the bearing is $160°$, the plane is headed for a location which makes a $-70°$ angle with the horizontal.

Thus the vector location is modeled by $\langle 450 \cos(-70°), 450 \sin(-70°) \rangle \approx \langle 153.9, -422.9 \rangle$. The wind can

be modeled by the vector $\langle 0, 20 \rangle$. The vector $\mathbf{a}$ that the plane should take can be found as follows:

$\mathbf{a} + \langle 0, 20 \rangle \approx \langle 153.9, -422.9 \rangle \Rightarrow \mathbf{a} \approx \langle 153.9, -402.9 \rangle$. The airplane's speed is the magnitude of this vec-

tor: $\|\mathbf{a}\| = \sqrt{153.9^2 + (-402.9)^2} \approx 431.3$ mph = groundspeed. The angle from the horizontal of vector $\mathbf{a}$

is $\tan^{-1}\left(\dfrac{-402.9}{153.9} \right) \approx -69.1°$. Thus the final bearing should be about $90° + 69.1° = 159.1°$.

113. Since the bearing is $75°$, the plane is headed for a location which makes a $15°$ angle with the horizontal. The

airspeed is found by the use of tangent $\Rightarrow \tan 15° = \dfrac{40}{x} \Rightarrow 0.2679x = 40 \Rightarrow x = 149.3$ mph. The

groundspeed is found by the use of sine $\Rightarrow \sin 15° = \dfrac{40}{x} \Rightarrow 0.2588x = 40 \Rightarrow x = 154.6$ mph.

115. (a) $\|\mathbf{R}\| = \sqrt{1^2 + (-2)^2} = \sqrt{1 + 4} = \sqrt{5} \approx 2.2$, $\|\mathbf{A}\| = \sqrt{0.5^2 + 1^2} = \sqrt{0.25 + 1} = \sqrt{1.25} \approx 1.1$

About 2.2 inches of rain fell. The area of the opening of the rain gauge was about 1.1 in^2.

(b) $V = |\mathbf{R} \cdot \mathbf{A}| = |(1)(0.5) + (-2)(1)| = |-1.5| = 1.5$. The volume of the rain collected was 1.5 in^3.

(c) $\mathbf{R}$ and $\mathbf{A}$ should be parallel and point in opposite directions.

117. (a) $\mathbf{c} = \mathbf{a} + \mathbf{b} = \langle 3, 2 \rangle + \langle -2, 2 \rangle = \langle 1, 4 \rangle$

(b) $\|\mathbf{c}\| = \sqrt{1^2 + 4^2} = \sqrt{1 + 16} = \sqrt{17} \approx 4.1$ ft.

(c) $3\mathbf{a} + 0.5\mathbf{b} = 3\langle 3, 2 \rangle + 0.5\langle -2, 2 \rangle = \langle 9, 6 \rangle + \langle -1, 1 \rangle = \langle 8, 7 \rangle$

119. (a) Represent the point $(-2, 4)$ by the vector $\mathbf{a} = \langle -2, 4 \rangle$. The new location is given by

$$\mathbf{b} = \mathbf{a} + \mathbf{v} = \langle -2, 4 \rangle + \langle 4, -2 \rangle = \langle -2 + 4, 4 + (-2) \rangle = \langle 2, 2 \rangle \Rightarrow (2, 2)$$

(b) See Figure 119.

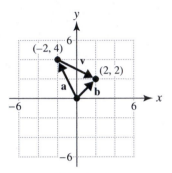

Figure 119

121. Note that 1.5 miles = 7920 feet. The x-component of the displacement vector is $d_1 = 7920 \cos 15°$.

The y-component of the displacement vector is $d_2 = 7920 \sin 15°$.

$$W = \mathbf{F} \cdot \mathbf{D} = \langle 0, 145 \rangle \cdot \langle 7920 \cos 15°, 7920 \sin 15° \rangle =$$

$$(0)(7920 \cos 15°) + (145)(7920 \sin 15°) \approx 297{,}228 \text{ ft-lb}$$

123. Since the bearing is 135°, the city is located on a vector which makes a $-45°$ angle with the horizontal. Thus the vector to the city is modeled by $\langle 200 \cos (-45°), 200 \sin (-45°) \rangle \approx \langle 141.4, -141.4 \rangle$. The wind can be modeled by the vector $\langle 0, -30 \rangle$. The vector $\mathbf{a}$ that the pilot should take can be found as follows:

$\mathbf{a} + \langle 0, -30 \rangle \approx \langle 141.4, -141.4 \rangle \Rightarrow \mathbf{a} \approx \langle 141.4, -111.4 \rangle$. The airplane's speed is the magnitude of this vector: $\|\mathbf{a}\| = \sqrt{141.4^2 + (-111.4)^2} \approx 180$ mph. The angle from horizontal of vector $\mathbf{a}$ is

$\tan^{-1}\left(\dfrac{-111.4}{141.4}\right) \approx -38.2°$. Thus, the bearing should be about $90° + 38.2° = 128.2°$. The pilot should fly

approximately 180 mph on a bearing of about 128.2°.

Extended and Discovery Exercises for Section 8.3

1. First, find vector $\overrightarrow{PS}$.

$$\overrightarrow{PS} = \langle -2 - (-1), -5 - 2 \rangle = \langle -1, -7 \rangle \text{ or } -\mathbf{i} - 7\mathbf{j}$$

Then compute the dot product $\overrightarrow{PS} \cdot \overrightarrow{PR}$.

$$\overrightarrow{PS} \cdot \overrightarrow{PR} = (-\mathbf{i} - 7\mathbf{j}) \cdot (5\mathbf{i} - \mathbf{j}) = 2 > 0$$

Thus, θ is acute and the point $(-2, -5)$ is in the blue region.

3. When $\overrightarrow{PQ} = 5\mathbf{i} - \mathbf{j}$ has initial point $P = (2, 1)$, the terminal point $Q = (2 + 5, 1 - 1) = (7, 0)$.

By sketching the appropriate vectors (not shown) we see that the pixel located at $S = (100, -10)$ is blue and

the pixel located at $S = (-500, 50)$ is white.

(a) Blue

(b) White.

8.4: Parametric Equations

1. (a)

t	0	1	2	3
x	-1	0	1	2
y	0	2	4	6

(b) See Figure 1.

(c) This curve is a line segment.

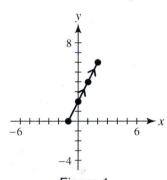

Figure 1   Figure 3

3. (a)

t	0	1	2	3
x	2	3	4	5
y	4	1	0	1

(b) See Figure 3.

(c) This curve is the lower portion of a parabola.

5. (a)

t	0	1	2	3
x	3	$\sqrt{8}$	$\sqrt{5}$	0
y	0	1	2	3

(b) See Figure 5.

(c) This curve is a portion of a circle with radius 3.

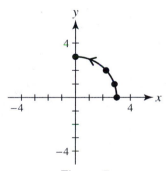

Figure 5

7. (a)

t	0	1	2	3
x	0	1	2	3
y	3	$\sqrt{8}$	$\sqrt{5}$	0

(b) See Figure 7.

(c) This curve is a portion of a circle with radius 3.

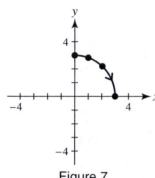

Figure 7

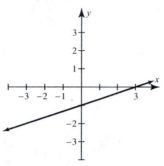

Figure 9

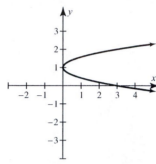

Figure 11

9. $x = 3t \Rightarrow t = \dfrac{1}{3}x$, substituting for t in $y = t - 1$ results in $y = \dfrac{1}{3}x - 1$. This curve is a line segment.
See Figure 9.

11 $y = t + 1 \Rightarrow t = y - 1$, substituting for t in $x = 3t^2$ results in $x = 3(y - 1)^2$. This curve is a parabola.
See Figure 11.

13. Since $y = t$, substituting y for t in $x = \sqrt{1 - t^2}$ yields $x = \sqrt{1 - y^2}$. This curve is a portion of a circle with
radius 1. See Figure 13.

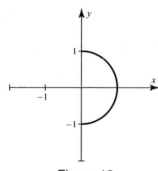

Figure 13

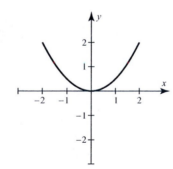

Figure 15

15. $x = t \Rightarrow y = \dfrac{1}{2}x^2$. This curve is a portion of a parabola. See Figure 15.

17. $x = t - 2 \Rightarrow x + 2 = t \Rightarrow y = (x + 2)^2 + 1 \Rightarrow y = x^2 + 4x + 5$

This curve is a portion of a parabola. See Figure 17.

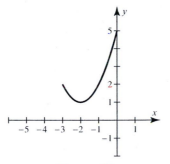

Figure 17

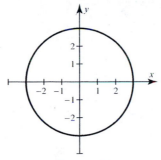

Figure 19

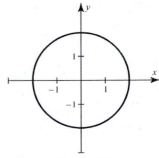

Figure 21

19. $x^2 + y^2 = (3 \sin t)^2 + (3 \cos t)^2 = 9 \sin^2 t + 9 \cos^2 t = 9(\sin^2 t + \cos^2 t) = 9 \Rightarrow x^2 + y^2 = 9$

This curve is a circle with radius 3. See Figure 19.

21. $x^2 + y^2 = (2 \sin t)^2 + (-2 \cos t)^2 = 4 \sin^2 t + 4 \cos^2 t = 4(\sin^2 t + \cos^2 t) = 4 \Rightarrow x^2 + y^2 = 4$

This curve is a circle with radius 2. See Figure 21.

23. $x^2 + y^2 = (3 \cos 2t)^2 + (3 \sin 2t)^2 = 9 \cos^2 2t + 9 \sin^2 2t = 9(\cos^2 2t + \sin^2 2t) = 9 \Rightarrow x^2 + y^2 = 9$

This curve is a circle with radius 3. See Figure 23.

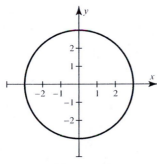

Figure 23

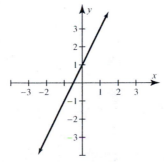

Figure 25

25. See Figure 25.

27. See Figure 27.

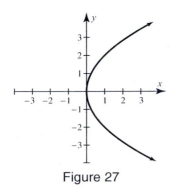

Figure 27

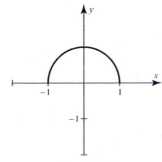

Figure 29

29. See Figure 29.

31. See Figure 31.

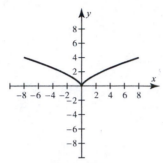

Figure 31

33. See Figure 33.

$[0, 6, 1]$ by $[-2, 2, 1]$

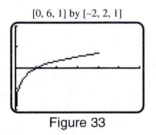

Figure 33

$[-2, 20, 2]$ by $[-1, 3, 1]$

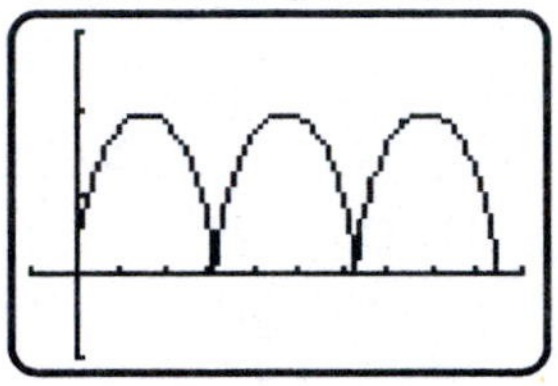

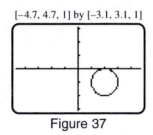

Figure 35

35. See Figure 35.

37. See Figure 37.

$[-4.7, 4.7, 1]$ by $[-3.1, 3.1, 1]$

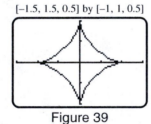

Figure 37

$[-1.5, 1.5, 0.5]$ by $[-1, 1, 0.5]$

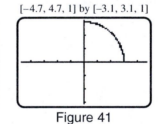

Figure 39

$[-4.7, 4.7, 1]$ by $[-3.1, 3.1, 1]$

Figure 41

39. See Figure 39.

41. See Figure 41.

43. Let $x = t$. Then $2x + y = 4 \Rightarrow 2t + y = 4 \Rightarrow y = 4 - 2t$.

One representation in parametric form is $x = t, y = 4 - 2t$. *Answers may vary.*

45. Let $x = t$. Then $y = 4 - x^2 \Rightarrow y = 4 - t^2$.

One representation in parametric form is $x = t, y = 4 - t^2$. *Answers may vary.*

47. Let $y = t$. Then $x = y^2 + y - 3 \Rightarrow x = t^2 + t - 3$.

One representation in parametric form is $x = t^2 + t - 3, y = t$. *Answers may vary.*

49. Since $\cos^2 t + \sin^2 t = 1$ for all t, $4\cos^2 t + 4\sin^2 t = 4 \Rightarrow (2\cos t)^2 + (2\sin t)^2 = 4$.

One representation in parametric form is $x = 2\cos t, y = 2\sin t, 0 \le t \le 2\pi$. *Answers may vary.*

51. Let $x = t$. Then $\ln y = 0.1x^2 \Rightarrow \ln y = 0.1t^2 \Rightarrow y = e^{0.1t^2}$.

One representation in parametric form is $x = t$, $y = e^{0.1t^2}$. *Answers may vary.*

53. Let $y = t$. Then $x = y^2 - 2y + 1 \Rightarrow x = t^2 - 2t + 1$. One representation in parametric form is $x = t^2 - 2t + 1$, $y = t$. *Answers may vary.*

55. (a) The curve traces a circle of radius 3 once counterclockwise, starting at $(3, 0)$. See Figure 55a.

    (b) The curve traces a circle of radius 3 twice counterclockwise, starting at $(3, 0)$. See Figure 55b.

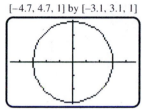 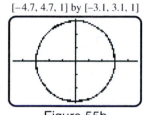

[-4.7, 4.7, 1] by [-3.1, 3.1, 1]　　　[-4.7, 4.7, 1] by [-3.1, 3.1, 1]　　　　　　　]

　　　　　Figure 55a　　　　　　　　Figure 55b

57. (a) The curve traces a circle of radius 3 once counterclockwise, starting at $(3, 0)$. See Figure 57a.

    (b) The curve traces a circle of radius 3 once clockwise, starting at $(0, 3)$. See Figure 57b.

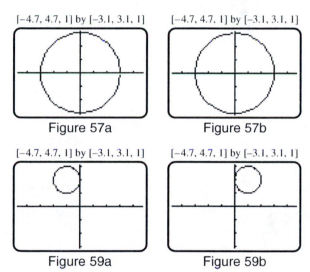

[-4.7, 4.7, 1] by [-3.1, 3.1, 1]　　　[-4.7, 4.7, 1] by [-3.1, 3.1, 1]

　　　　　Figure 57a　　　　　　　　Figure 57b

[-4.7, 4.7, 1] by [-3.1, 3.1, 1]　　　[-4.7, 4.7, 1] by [-3.1, 3.1, 1]

　　　　　Figure 59a　　　　　　　　Figure 59b

59. (a) The curve traces a circle of radius 1 centered at $(-1, 2)$. See Figure 59a.

    (b) The curve traces a circle of radius 1 centered at $(1, 2)$. See Figure 59b.

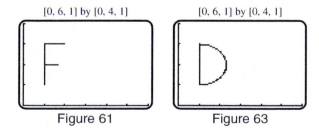

[0, 6, 1] by [0, 4, 1]　　　　　　[0, 6, 1] by [0, 4, 1]

　　　　　Figure 61　　　　　　　　　Figure 63

61. This is the letter F. See Figure 61.

63. This is the letter D. See Figure 63.

[−3, 3, 1] by [−2, 2, 1]

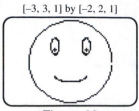

Figure 69

65. One possible solution is $x_1 = 0$, $y_1 = 2t$; $x_2 = t$, $y_2 = 0$ for $0 \le t \le 1$. *Answers may vary.*

67. One possible solution is $x_1 = \sin t$, $y_1 = \cos t$; $x_2 = 0$, $y_2 = t - 2$ for $0 \le t \le \pi$. *Answers may vary.*

69. One possible solution is $x_1 = 2 \cos t$, $y_1 = 2 \sin t$; $x_2 = 0.75 + 0.2 \cos t$, $y_2 = 0.75 + 0.4 \sin t$;

    $x_3 = -0.75 + 0.2 \cos t$, $y_3 = 0.75 + 0.4 \sin t$; $x_4 = 0.75 \cos 0.5t$, $y_4 = -0.7 - 0.3 \sin 0.5t$ for $0 \le t \le 2\pi$.

    Add pupils by plotting the points $(0.75, 0.6)$ and $(-0.75, 0.6)$. See Figure 69. *Answers may vary.*

71. The parametric equations which model the flight of the golf ball hit at a 35° angle are given by

    $x_1 = 66 \cos 35°(t)$, $y_1 = 66 \sin 35°(t) - 16t^2$ for $0 \le t \le 4$. This ball travels about 128 feet. See Figure 71a.

    The parametric equations which model the flight of the golf ball hit at a 50° angle are given by

    $x_2 = 66 \cos 50°(t)$, $y_2 = 66 \sin 50°(t) - 16t^2$ for $0 \le t \le 4$. This ball travels about 134 feet. See Figure 71b.

[0, 150, 10] by [−10, 50, 10]   [0, 150, 10] by [−10, 50, 10]

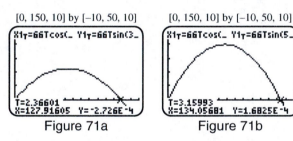

Figure 71a   Figure 71b

73. The parametric equations $x_1 = 88 \cos 45°(t)$, $y_1 = 88 \sin 45°(t) - 16t^2$ for $0 \le t \le 5$ model the flight of this

    ball. The fence can be modeled by the parametric equations $x_2 = 200$, $y_2 = 2t$ for $0 \le t \le 5$. The ball will

    go over the fence. See Figure 73.

[0, 250, 25] by [0, 65, 5]

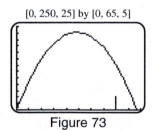

Figure 73

75. Let $v = 88$, $\theta = 45°$, and $h = 50$. The parametric equations are initially

    $x = (88 \cos 45°)t$ and $y = (88 \sin 45°)t - 16t^2 + 50$.

    Since $\cos 45° = \sin 45° = \dfrac{1}{\sqrt{2}}$, these equations become $x = \left(\dfrac{88}{\sqrt{2}}\right)t$ and $y = \left(\dfrac{88}{\sqrt{2}}\right)t - 16t^2 + 50$.

    Since the ball hits the ground when $y = 0$, we can determine how long the ball was in flight by solving the following equation: $-16t^2 + \left(\dfrac{88}{\sqrt{2}}\right)t + 50 = 0$. By the quadratic formula, the solution is $t \approx 4.573$ seconds.

After 4.573 seconds the ball has traveled horizontally $x = \left(\dfrac{88}{\sqrt{2}}\right)(4.573) \approx 285$ feet.

77. See Figure 77.

[−6, 6, 1] by [−4, 4, 1]   [−6, 6, 1] by [−4, 4, 1]

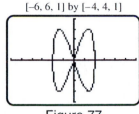   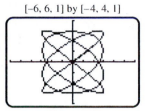

Figure 77   Figure 79

79. See Figure 79.

Extended and Discovery Exercises for Section 8.4

1. $\dfrac{F_1}{F_2} = \dfrac{3}{2} \Rightarrow F_2 = \dfrac{2}{3}F_1 = \dfrac{2}{3}(150) = 100$

3. $\dfrac{F_1}{F_2} = \dfrac{4}{3} \Rightarrow F_2 = \dfrac{3}{4}F_1 = \dfrac{3}{4}(400) = 300$

Checking Basic Concepts for Sections 8.3 and 8.4

1. (a) $\mathbf{v} = \overrightarrow{QP} = \langle 3 - (-1), 7 - 3 \rangle = \langle 4, 4 \rangle$

   (b) $\|\mathbf{v}\| = \sqrt{4^2 + 4^2} = \sqrt{16 + 16} = \sqrt{32} = 4\sqrt{2}$

   (c) $\overrightarrow{PQ} = \langle -1 - 3, 3 - 7 \rangle = \langle -4, -4 \rangle \Rightarrow \overrightarrow{PQ} + \overrightarrow{QP} = \langle 4, 4 \rangle + \langle -4, -4 \rangle =$
   $\langle 4 + (-4), 4 + (-4) \rangle = \langle 0, 0 \rangle$

3. (a) $\mathbf{a} \cdot \mathbf{b} = \langle 3, -2 \rangle \cdot \langle -1, 3 \rangle = (3)(-1) + (-2)(3) = -3 - 6 = -9$

   (b) $\|\mathbf{a}\| = \sqrt{3^2 + (-2)^2} = \sqrt{9 + 4} = \sqrt{13}, \|\mathbf{b}\| = \sqrt{(-1)^2 + 3^2} = \sqrt{1 + 9} = \sqrt{10}$

   $\theta = \cos^{-1}\left(\dfrac{\mathbf{a} \cdot \mathbf{b}}{\|\mathbf{a}\|\|\mathbf{b}\|}\right) = \cos^{-1}\left(\dfrac{-9}{\sqrt{13}\sqrt{10}}\right) = \cos^{-1}\left(\dfrac{-9}{\sqrt{130}}\right) \approx 142.1°$

8.5: Polar Equations

1.   See Figure 1.

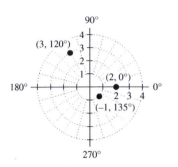

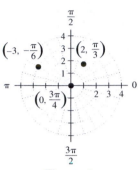

Figure 1   Figure 3

3.   See Figure 3.

5.   Yes. The angles $180°$ and $-180°$ are coterminal.

7.   No. The angles $45°$ and $-45°$ are not coterminal.

9.   Yes. The points are both located at the pole.

11. $x = r \cos \theta = 3 \cos 45° = \dfrac{3}{\sqrt{2}}, y = r \sin \theta = 3 \sin 45° = \dfrac{3}{\sqrt{2}}; \left(\dfrac{3}{\sqrt{2}}, \dfrac{3}{\sqrt{2}} \right)$ or $\left(\dfrac{3\sqrt{2}}{2}, \dfrac{3\sqrt{2}}{2} \right)$

13. $x = r \cos \theta = 10 \cos 90° = 0, y = r \sin \theta = 10 \sin 90° = 10; \ (0, 10)$

15. $x = r \cos \theta = 5 \cos 2\pi = 5, y = r \sin \theta = 5 \sin 2\pi = 0; \ (5, 0)$

17. $x = r \cos \theta = -3 \cos 60° = -3 \left(\dfrac{1}{2} \right) = -\dfrac{3}{2}, y = r \sin \theta = -3 \sin 60° = -3 \left(\dfrac{\sqrt{3}}{2} \right) = -\dfrac{3\sqrt{3}}{2};$

   $\left(-\dfrac{3}{2}, -\dfrac{3\sqrt{3}}{2} \right)$

19. $x = r \cos \theta = -2 \cos \left(-\dfrac{3\pi}{2} \right) = -2(0) = 0, y = r \sin \theta = -2 \sin \left(-\dfrac{3\pi}{2} \right) = -2(1) = -2; \ (0, -2)$

21.   (a) $r^2 = x^2 + y^2 \Rightarrow r^2 = (0)^2 + (3)^2 \Rightarrow r^2 = 9 \Rightarrow r = \pm 3 \Rightarrow r = 3 \ (\text{since } r > 0)$

   $\tan \theta = \dfrac{y}{x} \Rightarrow \tan \theta = \dfrac{3}{0} \Rightarrow \tan \theta$ is undefined $\Rightarrow \theta = 90°$ or $270°$ (since $0° \leq \theta < 360°$)

   Since $(0, 3)$ is located on the positive y-axis, $\theta = 90°$. The polar coordinates are $(3, 90°)$.

   (b) $r^2 = x^2 + y^2 \Rightarrow r^2 = (0)^2 + (3)^2 \Rightarrow r^2 = 9 \Rightarrow r = \pm 3 \Rightarrow r = -3 \ (\text{since } r < 0)$

   $\tan \theta = \dfrac{y}{x} \Rightarrow \tan \theta = \dfrac{3}{0} \Rightarrow \tan \theta$ is undefined $\Rightarrow \theta = -90°$ or $-270°$ (since $-180° < \theta \leq 180°$)

   Since $(0, 3)$ is located on the positive y-axis, $\theta = -90°$. The polar coordinates are $(-3, -90°)$.

23.   (a) $r^2 = x^2 + y^2 \Rightarrow r^2 = (-1)^2 + (-\sqrt{3})^2 \Rightarrow r^2 = 4 \Rightarrow r = \pm 2 \Rightarrow r = 2 \ (\text{since } r > 0)$

   $\tan \theta = \dfrac{y}{x} \Rightarrow \tan \theta = \dfrac{-\sqrt{3}}{-1} \Rightarrow \theta = \tan^{-1} \sqrt{3} \Rightarrow \theta = 60°$ or $240°$ (since $0° \leq \theta < 360°$)

   Since $(-1, -\sqrt{3})$ is located in quadrant III, $\theta = 240°$. The polar coordinates are $(2, 240°)$.

(b) $r^2 = x^2 + y^2 \Rightarrow r^2 = (-1)^2 + (-\sqrt{3})^2 \Rightarrow r^2 = 4 \Rightarrow r = \pm 2 \Rightarrow r = -2$ (since $r < 0$)

$\tan\theta = \dfrac{y}{x} \Rightarrow \tan\theta = \dfrac{-\sqrt{3}}{-1} \Rightarrow \theta = \tan^{-1}\sqrt{3} \Rightarrow \theta = -120°$ or $60°$ (since $-180° < \theta \le 180°$)

Since $(-1, -\sqrt{3}\,)$ is located in quadrant III, $\theta = 60°$. The polar coordinates are $(-2, 60°)$.

25. (a) $r^2 = x^2 + y^2 \Rightarrow r^2 = (3)^2 + (-3)^2 \Rightarrow r^2 = 18 \Rightarrow r = \pm\sqrt{18} \Rightarrow r = \sqrt{18}$ (since $r > 0$)

$\tan\theta = \dfrac{y}{x} \Rightarrow \tan\theta = \dfrac{-3}{3} \Rightarrow \theta = \tan^{-1}(-1) \Rightarrow \theta = 135°$ or $315°$ (since $0° \le \theta < 360°$)

Since $(3, -3)$ is located in quadrant IV, $\theta = 315°$. The polar coordinates are $(\sqrt{18}, 315°)$.

(b) $r^2 = x^2 + y^2 \Rightarrow r^2 = (3)^2 + (-3)^2 \Rightarrow r^2 = 18 \Rightarrow r = \pm\sqrt{18} \Rightarrow r = -\sqrt{18}$ (since $r < 0$)

$\tan\theta = \dfrac{y}{x} \Rightarrow \tan\theta = \dfrac{-3}{3} \Rightarrow \theta = \tan^{-1}(-1) \Rightarrow \theta = -45°$ or $135°$ (since $-180° < \theta \le 180°$)

Since $(3, -3)$ is located in quadrant IV, $\theta = 135°$. The polar coordinates are $(-\sqrt{18}, 135°)$.

27. $r^2 = x^2 + y^2 \Rightarrow r^2 = (7)^2 + (24)^2 \Rightarrow r^2 = 625 \Rightarrow r = \pm 25 \Rightarrow r = 25$ (since $r > 0$)

$\tan\theta = \dfrac{y}{x} \Rightarrow \tan\theta = \dfrac{24}{7} \Rightarrow \theta = \tan^{-1}\dfrac{24}{7} \Rightarrow \theta \approx 1.29$ or $\pi + 1.29 \approx 4.43$ (since $0 \le \theta < 2\pi$)

Since $(7, 24)$ is located in quadrant I, $\theta = 1.29$. The polar coordinates are $(25, 1.29)$.

29. $r^2 = x^2 + y^2 \Rightarrow r^2 = (-5)^2 + (12)^2 \Rightarrow r^2 = 169 \Rightarrow r = \pm 13 \Rightarrow r = 13$ (since $r > 0$)

$\tan\theta = \dfrac{y}{x} \Rightarrow \tan\theta = \dfrac{12}{-5} \Rightarrow \theta = \tan^{-1}\left(-\dfrac{12}{5}\right) \Rightarrow \theta \approx 1.97$ or $\pi + 1.97 \approx 5.11$ (since $0 \le \theta < 2\pi$)

Since $(-5, 12)$ is located in quadrant II, $\theta = 1.97$. The polar coordinates are $(13, 1.97)$.

31. See Figure 31.

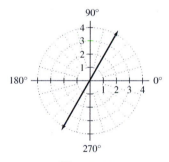

Figure 31

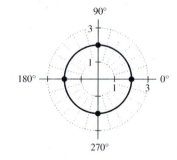

Figure 33

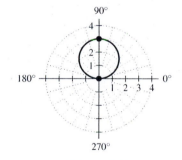

Figure 35

33. (a)

θ	0°	90°	180°	270°
r	2	2	2	2

(b) See Figure 33.

35. (a)

θ	0°	90°	180°	270°
r	0	3	0	-3

(b) See Figure 35.

37. (a)

θ	0°	90°	180°	270°
r	2	4	2	0

(b) See Figure 37.

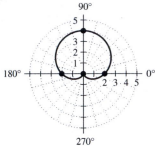

Figure 37

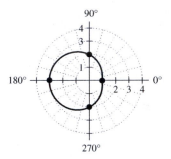

Figure 39

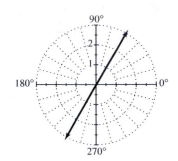

Figure 41

39. (a)

θ	0°	90°	180°	270°
r	1	2	3	2

(b) See Figure 39.

41. See Figure 41.

43. See Figure 43.

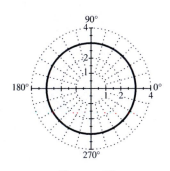

Figure 43

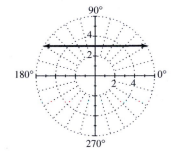

Figure 45

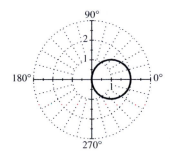

Figure 47

45. See Figure 45.

47. See Figure 47.

49. See Figure 49.

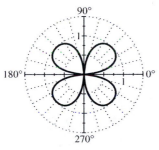

Figure 49

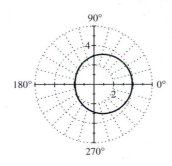

Figure 51

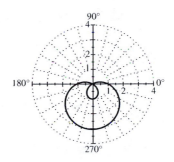

Figure 53

51. See Figure 51.

53. See Figure 53.

55. See Figure 55.

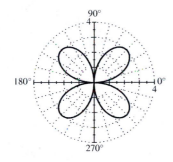

Figure 55

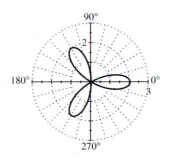

Figure 57

57. See Figure 57.

59. $y = 3 \Rightarrow r \sin\theta = 3$ or $r = 3\csc\theta$

61. $y = x \Rightarrow r\sin\theta = r\cos\theta \Rightarrow \dfrac{r\sin\theta}{r\cos\theta} = 1 \Rightarrow \tan\theta = 1$ or $\theta = \dfrac{\pi}{4}$

63. $x^2 + y^2 = 9 \Rightarrow (r\cos\theta)^2 + (r\sin\theta)^2 = 9 \Rightarrow r^2(\cos^2\theta + \sin^2\theta) = 9 \Rightarrow r^2 = 9$ or $r = 3$

65. $x^2 + y^2 = 2x \Rightarrow (r\cos\theta)^2 + (r\sin\theta)^2 = 2(r\cos\theta) \Rightarrow r^2(\cos^2\theta + \sin^2\theta) = 2(r\cos\theta) \Rightarrow$
    $r^2 = 2(r\cos\theta) \Rightarrow r = 2\cos\theta$

67. $r = 3 \Rightarrow r^2 = 9 \Rightarrow x^2 + y^2 = 9$

69. $r = 2\sec\theta \Rightarrow r = \dfrac{2}{\cos\theta} \Rightarrow r\cos\theta = 2 \Rightarrow x = 2$

71. $r = \dfrac{3}{2\cos\theta + 4\sin\theta} \Rightarrow 2r\cos\theta + 4r\sin\theta = 3 \Rightarrow 2x + 4y = 3$

73. $r = \cos\theta \Rightarrow r^2 = r\cos\theta \Rightarrow x^2 + y^2 = x$

75. Graph $r_1 = 3 + 3\cos(\theta)$ for $0° \le \theta \le 360°$ in $[-9.4, 9.4, 1]$ by $[-6.2, 6.2, 1]$. See Figure 75.

[–9.4, 9.4, 1] by [–6.2, 6.2, 1]   [–9.4, 9.4, 1] by [–6.2, 6.2, 1]

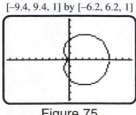 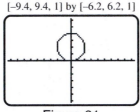

Figure 75   Figure 77

77. Graph $r_1 = 3 - 2 \sin(\theta)$ for $0° \le \theta \le 360°$ in $[-9.4, 9.4, 1]$ by $[-6.2, 6.2, 1]$. See Figure 77.

79. Graph $r_1 = 2 - 4 \cos(\theta)$ for $0° \le \theta \le 360°$ in $[-9.4, 9.4, 1]$ by $[-6.2, 6.2, 1]$. See Figure 79.

[–9.4, 9.4, 1] by [–6.2, 6.2, 1]   [–9.4, 9.4, 1] by [–6.2, 6.2, 1]

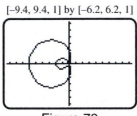 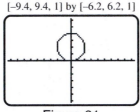

Figure 79   Figure 81

81. Graph $r_1 = 4 \sin(\theta)$ for $0° \le \theta \le 180°$ in $[-9.4, 9.4, 1]$ by $[-6.2, 6.2, 1]$. See Figure 81.

83. Graph $r_1 = 2 \cos(5\theta)$ for $0° \le \theta \le 180°$ in $[-4.7, 4.7, 1]$ by $[-3.1, 3.1, 1]$. See Figure 83.

[–4.7, 4.7, 1] by [–3.1, 3.1, 1]   [–9.4, 9.4, 1] by [–6.2, 6.2, 1]   [–3, 3, 1] by [–2, 2, 1]

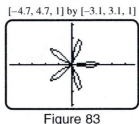 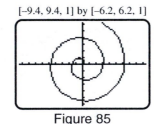 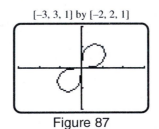

Figure 83   Figure 85   Figure 87

85. Graph $r_1 = \theta/2$ for $0 \le \theta \le \dfrac{9\pi}{2}$ in $[-9.4, 9.4, 1]$ by $[-6.2, 6.2, 1]$. See Figure 85.

87. Let $r_1 = \sqrt{2 \sin 2\theta}$ and $r_2 = -\sqrt{2 \sin 2\theta}$

   Graph r_1 and r_2 for $0° \le \theta \le 180°$ in $[-3, 3, 1]$ by $[-2, 2, 1]$. See Figure 87.

89. $r_1 = r_2 \Rightarrow 3 = 2 + 2 \sin \theta \Rightarrow 1 = 2 \sin \theta \Rightarrow \sin \theta = \dfrac{1}{2} \Rightarrow \theta_R = \sin^{-1} \dfrac{1}{2} = 30° \Rightarrow \theta = 30°, 150°$

91. (a) $r_1 = r_2 \Rightarrow 3 = 2 - 2 \sin \theta \Rightarrow 2 \sin \theta = -1 \Rightarrow \sin \theta = -\dfrac{1}{2} \Rightarrow \theta_R = \sin^{-1} \dfrac{1}{2} = 30° \Rightarrow$

   $\theta = 210°, 330°$

   (b) Graph $r_1 = 3$ and $r_2 = 2 - 2 \sin(\theta)$ for $0° \le \theta \le 360°$ in $[-9.4, 9.4, 1]$ by $[-6.2, 6.2, 1]$.

   Use the calculator's trace feature to find the solutions: $\theta = 210°, 330°$. See Figures 91a and 91b.

   (c) Table $r_1 = 3$ and $r_2 = 2 - 2 \sin(\theta)$ starting at $\theta = 180$, incrementing by 30.

   The solutions are $\theta = 210°, 330°$. See Figure 91c.

[−9.4, 9.4, 1] by [−6.2, 6.2, 1]   [−9.4, 9.4, 1] by [−6.2, 6.2, 1]

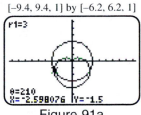

Figure 91a   Figure 91b   Figure 91c

93. (a) $r_1 = r_2 \Rightarrow 1 = 2 \sin \theta \Rightarrow \sin \theta = \dfrac{1}{2} \Rightarrow \theta_R = \sin^{-1} \dfrac{1}{2} = 30° \Rightarrow \theta = 30°, 150°$

(b) Graph $r_1 = 1$ and $r_2 = 2 \sin (\theta)$ for $0° \le \theta \le 360°$ in $[-4.7, 4.7, 1]$ by $[-3.1, 3.1, 1]$.

Use the calculator's trace feature to find the solutions: $\theta = 30°, 150°$. See Figures 93a and 93b.

(c) Table $r_1 = 1$ and $r_2 = 2 \sin (\theta)$ starting at $\theta = 0$, incrementing by 30.

The solutions are $\theta = 30°, 150°$. See Figure 93c.

[−4.7, 4.7, 1] by [−3.1, 3.1, 1]   [−4.7, 4.7, 1] by [−3.1, 3.1, 1]

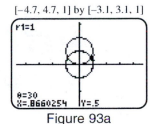
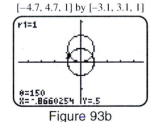
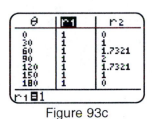

Figure 93a   Figure 93b   Figure 93c

95. For Saturn $r_1 = \dfrac{9.54(1 - 0.056^2)}{1 + 0.056 \cos \theta} = (9.54(1 - 0.056^2))/(1 + 0.056 \cos (\theta))$

For Uranus $r_2 = \dfrac{19.2(1 - 0.047^2)}{1 + 0.047 \cos \theta} = (19.2(1 - 0.047^2))/(1 + 0.047 \cos (\theta))$

Graph r_1 and r_2 for $0° \le \theta \le 360°$ in $[-30, 30, 10]$ by $[-20, 20, 10]$. See Figure 95.

[−30, 30, 10] by [−20, 20, 10]   [−3, 3, 1] by [−2, 2, 1]

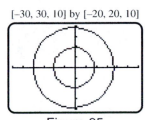
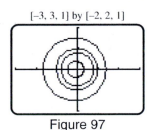

Figure 95   Figure 97

97. For Mercury $r_1 = \dfrac{0.39(1 - 0.206^2)}{1 + 0.206 \cos \theta} = (0.39(1 - 0.206^2))/(1 + 0.206 \cos (\theta))$

For Venus $r_2 = \dfrac{0.78(1 - 0.007^2)}{1 + 0.007 \cos \theta} = (0.78(1 - 0.007^2))/(1 + 0.007 \cos (\theta))$

For Earth $r_3 = \dfrac{1.00(1 - 0.017^2)}{1 + 0.017 \cos \theta} = (1.00(1 - 0.017^2))/(1 + 0.017 \cos (\theta))$

For Mars $r_4 = \dfrac{1.52(1 - 0.093^2)}{1 + 0.093 \cos \theta} = (1.52(1 - 0.093^2))/(1 + 0.093 \cos (\theta))$

Graph r_1, r_2, r_3 and r_4 for $0° \le \theta \le 360°$ in $[-3, 3, 1]$ by $[-2, 2, 1]$. See Figure 97.

99. Let $r_1 = \sqrt{40{,}000 \cos 2\theta}$ and $r_2 = -\sqrt{40{,}000 \cos 2\theta}$

Graph r_1 and r_2 for $0° \le \theta \le 180°$ in $[-300, 300, 100]$ by $[-200, 200, 100]$. See Figure 99.

The radio signal is received inside the "figure eight". This region is generally in an east-west direction from the two towers with a maximum distance of 200 miles.

$[-300, 300, 100]$ by $[-200, 200, 100]$

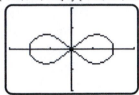

Figure 99

Extended and Discovery Exercise for Section 8.5

1. $y = x \tan (\ln (x^2 + y^2)) \Rightarrow \dfrac{y}{x} = \tan (\ln r^2) \Rightarrow \tan \theta = \tan (\ln r^2) \Rightarrow \theta = \ln r^2 \Rightarrow e^\theta = r^2 \Rightarrow r = e^{\theta/2}$

8.6: Trigonometric Form and Roots of Complex Numbers

1. See Figure 1.

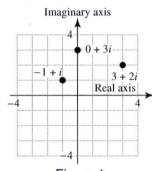

Figure 1

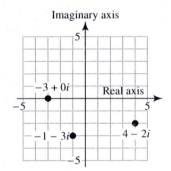

Figure 3

3. See Figure 3.

5. $|1 + i| = \sqrt{1^2 + 1^2} = \sqrt{1 + 1} = \sqrt{2}$

7. $|12 - 5i| = \sqrt{12^2 + (-5)^2} = \sqrt{144 + 25} = \sqrt{169} = 13$

9. $|-6 + 0i| = \sqrt{(-6)^2 + 0^2} = \sqrt{36} = 6$

11. $|2 - 3i| = \sqrt{2^2 + (-3)^2} = \sqrt{4 + 9} = \sqrt{13}$

13. $r = \sqrt{1^2 + 1^2} = \sqrt{1 + 1} = \sqrt{2}$, $\tan \theta = \dfrac{1}{1} \Rightarrow \theta_R = \tan^{-1} 1 = 45°$

Since $-1 + i$ is located in quadrant III of the complex plane, $\theta = 225°$; $\sqrt{2}(\cos 225° + i \sin 225°)$

15. $r = \sqrt{5^2 + 0^2} = \sqrt{25 + 0} = \sqrt{25} = 5$, $\tan \theta = \dfrac{0}{5} \Rightarrow \theta_R = \tan^{-1} 0 = 0°$

Since 5 is positive and quadrantal in the complex plane, $\theta = 0°$; $5(\cos 0° + i \sin 0°)$

17. $r = \sqrt{0^2 + 4^2} = \sqrt{0 + 16} = \sqrt{16} = 4$, $\tan \theta = \dfrac{4}{0}$ is undefined $\Rightarrow \theta_R = 90°$

Since $4i$ is quadrantal in the complex plane, $\theta = 90°$; $4(\cos 90° + i \sin 90°)$

19. $r = \sqrt{(-1)^2 + (\sqrt{3})^2} = \sqrt{1 + 3} = \sqrt{4} = 2$, $\tan \theta = \dfrac{\sqrt{3}}{-1} \Rightarrow \theta_R = \tan^{-1} \sqrt{3} = 60°$

Since $-1 + i\sqrt{3}$ is located in quadrant II of the complex plane, $\theta = 120°$; $2(\cos 120° + i \sin 120°)$

21. $r = \sqrt{(\sqrt{3})^2 + 1^2} = \sqrt{3 + 1} = \sqrt{4} = 2$, $\tan \theta = \dfrac{1}{\sqrt{3}} \Rightarrow \theta_R = \tan^{-1} \dfrac{1}{\sqrt{3}} = 30°$

Since $\sqrt{3} + i$ is located in quadrant I of the complex plane, $\theta = 30°$; $2(\cos 30° + i \sin 30°)$

23. $r = \sqrt{(-2)^2 + 0^2} = \sqrt{4 + 0} = \sqrt{4} = 2$, $\tan \theta = \dfrac{0}{-2} \Rightarrow \theta_R = \tan^{-1} 0 = 0°$

Since -2 is negative and quadrantal in the complex plane, $\theta = \pi$; $2(\cos \pi + i \sin \pi)$

25. $r = \sqrt{(-2)^2 + 2^2} = \sqrt{4 + 4} = \sqrt{8}$, $\tan \theta = \dfrac{2}{-2} \Rightarrow \theta_R = \tan^{-1} 1 = \dfrac{\pi}{4}$

Since $-2 + 2i$ is located in quadrant II of the complex plane, $\theta = \dfrac{3\pi}{4}$; $\sqrt{8}\left(\cos \dfrac{3\pi}{4} + i \sin \dfrac{3\pi}{4}\right)$

27. $5(\cos 180° + i \sin 180°) = 5(-1 + i(0)) = -5$

29. $2(\cos 45° + i \sin 45°) = 2\left(\dfrac{1}{\sqrt{2}} + i\left(\dfrac{1}{\sqrt{2}}\right)\right) = \sqrt{2} + i\sqrt{2}$

31. $2\left(\cos \dfrac{\pi}{6} + i \sin \dfrac{\pi}{6}\right) = 2\left(\dfrac{\sqrt{3}}{2} + i\left(\dfrac{1}{2}\right)\right) = \sqrt{3} + i$

33. $3(\cos 2\pi + i \sin 2\pi) = 3(1 + i(0)) = 3$

35. $z_1 z_2 = (9 \cdot 3)(\cos (45° + 15°) + i \sin (45° + 15°)) = 27(\cos 60° + i \sin 60°) =$

$27\left(\dfrac{1}{2} + i\left(\dfrac{\sqrt{3}}{2}\right)\right) = \dfrac{27}{2} + \dfrac{27\sqrt{3}}{2}i$

$\dfrac{z_1}{z_2} = \left(\dfrac{9}{3}\right)(\cos (45° - 15°) + i \sin (45° - 15°)) = 3(\cos 30° + i \sin 30°) =$

$3\left(\dfrac{\sqrt{3}}{2} + i\left(\dfrac{1}{2}\right)\right) = \dfrac{3\sqrt{3}}{2} + \dfrac{3}{2}i$

37. $z_1 z_2 = (6 \cdot 1)\left(\cos \left(\dfrac{3\pi}{4} + \dfrac{\pi}{4}\right) + i \sin \left(\dfrac{3\pi}{4} + \dfrac{\pi}{4}\right)\right) = 6(\cos \pi + i \sin \pi) = 6(-1 + i(0)) = -6$

$\dfrac{z_1}{z_2} = \left(\dfrac{6}{1}\right)\left(\cos \left(\dfrac{3\pi}{4} - \dfrac{\pi}{4}\right) + i \sin \left(\dfrac{3\pi}{4} - \dfrac{\pi}{4}\right)\right) = 6\left(\cos \dfrac{\pi}{2} + i \sin \dfrac{\pi}{2}\right) = 6(0 + i(1)) = 6i$

39. $z_1 z_2 = (1 \cdot 1)(\cos (15° + (-45°)) + i \sin (15° + (-45°))) = 1(\cos (-30°) + i \sin (-30°)) = \dfrac{\sqrt{3}}{2} - \dfrac{1}{2}i$

$\dfrac{z_1}{z_2} = \left(\dfrac{1}{1}\right)(\cos (15° - (-45°)) + i \sin (15° - (-45°))) = 1(\cos 60° + i \sin 60°) = \dfrac{1}{2} + \dfrac{\sqrt{3}}{2}i$

41. $(2(\cos 30° + i \sin 30°))^3 = 2^3(\cos (3 \cdot 30°) + i \sin (3 \cdot 30°)) = 8(\cos 90° + i \sin 90°) = 8(0 + i(1)) = 8i$

43. $(\cos 10° + i \sin 10°)^{36} = 1^{36}(\cos (36 \cdot 10°) + i \sin (36 \cdot 10°)) = 1(\cos 360° + i \sin 360°) = 1 + i(0) = 1$

45. $(5(\cos 60° + i \sin 60°))^2 = 5^2(\cos (2 \cdot 60°) + i \sin (2 \cdot 60°)) = 25(\cos 120° + i \sin 120°) =$

$-\dfrac{25}{2} + \dfrac{25\sqrt{3}}{2}i$

47. $(1 + i)^3 = (\sqrt{2}(\cos 45° + i \sin 45°))^3 = (\sqrt{2})^3(\cos (3 \cdot 45°) + i \sin (3 \cdot 45°)) =$

    $2\sqrt{2}(\cos 135° + i \sin 135°) = -2 + 2i$ This result is verified with a calculator in Figure 47.

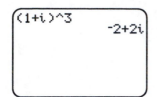

| Figure 47 | Figure 49 | Figure 51 |

49. $(\sqrt{3} + i)^5 = (2(\cos 30° + i \sin 30°))^5 = 2^5(\cos (5 \cdot 30°) + i \sin (5 \cdot 30°)) =$

    $32(\cos 150° + i \sin 150°) = -16\sqrt{3} + 16i$ This result is verified with a calculator in Figure 49.

51. $w_0 = \sqrt{4}\left(\cos \dfrac{120° + 360° \cdot 0}{2} + i \sin \dfrac{120° + 360° \cdot 0}{2}\right) = 2(\cos 60° + i \sin 60°) = 1 + i\sqrt{3}$

    $w_1 = \sqrt{4}\left(\cos \dfrac{120° + 360° \cdot 1}{2} + i \sin \dfrac{120° + 360° \cdot 1}{2}\right) = 2(\cos 240° + i \sin 240°) = -1 - i\sqrt{3}$

    This result is verified with a calculator in Figure 51.

53. $w_0 = \sqrt[3]{1}\left(\cos \dfrac{180° + 360° \cdot 0}{3} + i \sin \dfrac{180° + 360° \cdot 0}{3}\right) = 1(\cos 60° + i \sin 60°) = \dfrac{1}{2} + \dfrac{\sqrt{3}}{2}i$

    $w_1 = \sqrt[3]{1}\left(\cos \dfrac{180° + 360° \cdot 1}{3} + i \sin \dfrac{180° + 360° \cdot 1}{3}\right) = 1(\cos 180° + i \sin 180°) = -1$

    $w_2 = \sqrt[3]{1}\left(\cos \dfrac{180° + 360° \cdot 2}{3} + i \sin \dfrac{180° + 360° \cdot 2}{3}\right) = 1(\cos 300° + i \sin 300°) = \dfrac{1}{2} - \dfrac{\sqrt{3}}{2}i$

    These results are verified with a calculator in Figure 53a and 53b.

```
cos(180)+isin(18
0)
                -1
(1/2+√(3)/2i)^3
                -1
```

```
(-1)^3
                -1
(1/2-√(3)/2i)^3
                -1
```

| Figure 53a | Figure 53b |

55. $i = \cos 90° + i \sin 90°$

    $w_0 = \sqrt{1}\left(\cos \dfrac{90° + 360° \cdot 0}{2} + i \sin \dfrac{90° + 360° \cdot 0}{2}\right) = 1(\cos 45° + i \sin 45°) = \dfrac{\sqrt{2}}{2} + \dfrac{\sqrt{2}}{2}i$

    $w_1 = \sqrt{1}\left(\cos \dfrac{90° + 360° \cdot 1}{2} + i \sin \dfrac{90° + 360° \cdot 1}{2}\right) = 1(\cos 225° + i \sin 225°) = -\dfrac{\sqrt{2}}{2} - \dfrac{\sqrt{2}}{2}i$

    These results are verified with a calculator in Figure 55.

```
(√(2)/2+√(2)/2i)
²
                i
(-√(2)/2-√(2)/2i
)²
                i
```

Figure 55

57. $-8 = 8(\cos 180° + i \sin 180°)$

$$w_0 = \sqrt[3]{8}\left(\cos \frac{180° + 360° \cdot 0}{3} + i \sin \frac{180° + 360° \cdot 0}{3}\right) = 2(\cos 60° + i \sin 60°) = 1 + i\sqrt{3}$$

$$w_1 = \sqrt[3]{8}\left(\cos \frac{180° + 360° \cdot 1}{3} + i \sin \frac{180° + 360° \cdot 1}{3}\right) = 2(\cos 180° + i \sin 180°) = -2$$

$$w_2 = \sqrt[3]{8}\left(\cos \frac{180° + 360° \cdot 2}{3} + i \sin \frac{180° + 360° \cdot 2}{3}\right) = 2(\cos 300° + i \sin 300°) = 1 - i\sqrt{3}$$

These results are verified with a calculator in Figure 57.

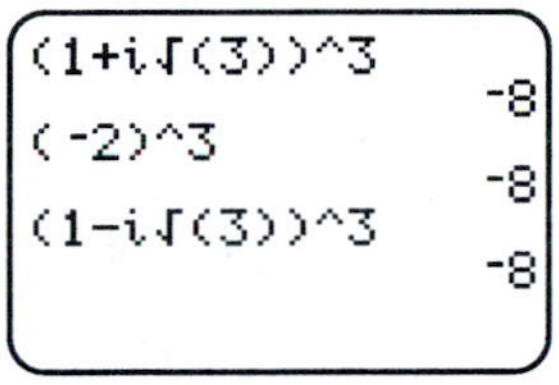

```
(1+i√(3))^3
            -8
(-2)^3
            -8
(1-i√(3))^3
            -8
```

Figure 57

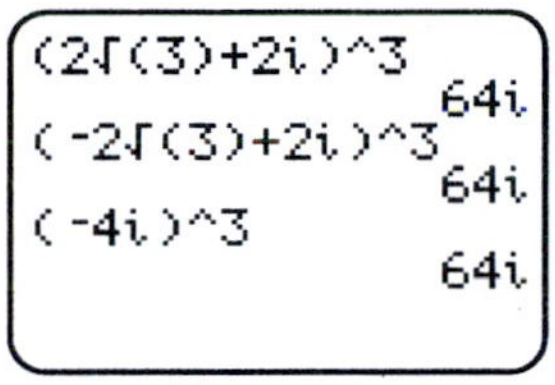

```
(2√(3)+2i)^3
            64i
(-2√(3)+2i)^3
            64i
(-4i)^3
            64i
```

Figure 59

59. $64i = 64(\cos 90° + i \sin 90°)$

$$w_0 = \sqrt[3]{64}\left(\cos \frac{90° + 360° \cdot 0}{3} + i \sin \frac{90° + 360° \cdot 0}{3}\right) = 4(\cos 30° + i \sin 30°) = 2\sqrt{3} + 2i$$

$$w_1 = \sqrt[3]{64}\left(\cos \frac{90° + 360° \cdot 1}{3} + i \sin \frac{90° + 360° \cdot 1}{3}\right) = 4(\cos 150° + i \sin 150°) = -2\sqrt{3} + 2i$$

$$w_2 = \sqrt[3]{64}\left(\cos \frac{90° + 360° \cdot 2}{3} + i \sin \frac{90° + 360° \cdot 2}{3}\right) = 4(\cos 270° + i \sin 270°) = -4i$$

These results are verified with a calculator in Figure 59.

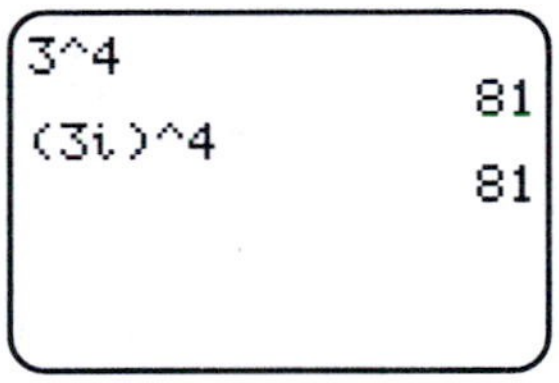

```
3^4
            81
(3i)^4
            81
```

Figure 61a

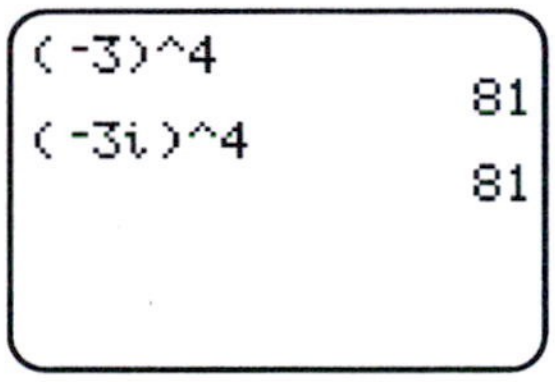

```
(-3)^4
            81
(-3i)^4
            81
```

Figure 61b

61. $81 = 81(\cos 0° + i \sin 0°)$

$$w_0 = \sqrt[4]{81}\left(\cos \frac{0° + 360° \cdot 0}{4} + i \sin \frac{0° + 360° \cdot 0}{4}\right) = 3(\cos 0° + i \sin 0°) = 3$$

$$w_1 = \sqrt[4]{81}\left(\cos \frac{0° + 360° \cdot 1}{4} + i \sin \frac{0° + 360° \cdot 1}{4}\right) = 3(\cos 90° + i \sin 90°) = 3i$$

$$w_2 = \sqrt[4]{81}\left(\cos \frac{0° + 360° \cdot 2}{4} + i \sin \frac{0° + 360° \cdot 2}{4}\right) = 3(\cos 180° + i \sin 180°) = -3$$

$$w_3 = \sqrt[4]{81}\left(\cos \frac{0° + 360° \cdot 3}{4} + i \sin \frac{0° + 360° \cdot 3}{4}\right) = 3(\cos 270° + i \sin 270°) = -3i$$

These results are verified with a calculator in Figure 61a and 61b.

63. $z_0 = -0.4i$

   $z_1 = (-0.4i)^2 + (-0.4i) = -0.16 - 0.4i \Rightarrow |z_1| \approx 0.4308$

   $z_2 = (-0.16 - 0.4i)^2 + (-0.4i) = -0.1344 - 0.272i \Rightarrow |z_2| \approx 0.3034$

   $z_3 = (-0.1344 - 0.272i)^2 + (-0.4i) \approx -0.0559 - 0.3269i \Rightarrow |z_3| \approx 0.3316$

   The modulus of each consecutive z_k never exceeds 2. Thus, $-0.4i$ is in the Mandelbrot set.

65. $z_0 = 1 + i$

   $z_1 = (1 + i)^2 + (1 + i) = 1 + 3i \Rightarrow |z_1| = \sqrt{10} \approx 3.162 > 2$

   The modulus of z_1 exceeds 2. Thus, $1 + i$ is not in the Mandelbrot set.

67. (a) $Z = 50 + 60 + 15i + 17i = 110 + 32i$

   (b) $|110 + 32i| = \sqrt{110^2 + 32^2} = \sqrt{12,100 + 1024} = \sqrt{13,124} \approx 114.6$ ohms

Extended and Discovery Exercise for Section 8.6

1. (a) Given $\langle \sqrt{3}, 1 \rangle$, then $\|\mathbf{v}\| = \sqrt{(\sqrt{3})^2 + (1)^2} = 2$. $\tan \theta = \dfrac{1}{\sqrt{3}} \Rightarrow \theta = 30°$. $\langle 2 \cos 30°, 2 \sin 30° \rangle$

   (b) Given $(\sqrt{3}, 1)$, then $r = \sqrt{(\sqrt{3})^2 + (1)^2} = 2$. $\tan \theta = \dfrac{1}{\sqrt{3}} \Rightarrow \theta = 30°$. $(r, \theta) = (2, 30°)$

   (c) Given that $\sqrt{3} + i$, then $r = \sqrt{(\sqrt{3})^2 + (1)^2} = 2$ and $\tan \theta = \dfrac{1}{\sqrt{3}} \Rightarrow \theta = 30°$. $2(\cos 30° + i \sin 30°)$

3. (a) Given $\langle 4, -3 \rangle$, then $\|\mathbf{v}\| = \sqrt{4^2 + (-3)^2} = 5$. $\tan \theta = \dfrac{-3}{4} \Rightarrow \theta = 323.1°$. $\langle 5 \cos 323.1°, 5 \sin 323.1° \rangle$

   (b) Given $(4, -3)$, then $r = \sqrt{4^2 + (-3)^2} = 5$. $\tan \theta = \dfrac{-3}{4} \Rightarrow \theta = 323.1°$. $(r, \theta) = (5, 323.1°)$

   (c) Given that $4 + 3i$, then $r = \sqrt{4^2 + (-3)^2} = 5$ and $\tan \theta = \dfrac{-3}{4} \Rightarrow \theta = 323.1°$.

   $5(\cos 323.1° + i \sin 323.1°)$

Checking Basic Concepts for Sections 8.5 and 8.6

1. See Figure 1.

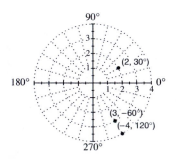

Figure 1

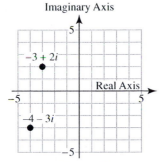

Figure 3

3. See Figure 3.

5. $i = 1(\cos 90° + i \sin 90°)$

$$w_0 = \sqrt[3]{1}\left(\cos \frac{90° + 360° \cdot 0}{3} + i \sin \frac{90° + 360° \cdot 0}{3}\right) = 1(\cos 30° + i \sin 30°) = \frac{\sqrt{3}}{2} + \frac{1}{2}i$$

$$w_1 = \sqrt[3]{1}\left(\cos \frac{90° + 360° \cdot 1}{3} + i \sin \frac{90° + 360° \cdot 1}{3}\right) = 1(\cos 150° + i \sin 150°) = -\frac{\sqrt{3}}{2} + \frac{1}{2}i$$

$$w_2 = \sqrt[3]{1}\left(\cos \frac{90° + 360° \cdot 2}{3} + i \sin \frac{90° + 360° \cdot 2}{3}\right) = 1(\cos 270° + i \sin 270°) = -i$$

Chapter 8 Review Exercises

1. This triangle is of the form ASA. There is only one solution.

$\gamma = 180° - 70° - 40° = 70°$

$\dfrac{a}{\sin \alpha} = \dfrac{c}{\sin \gamma} \Rightarrow a = \dfrac{c \sin \alpha}{\sin \gamma} = \dfrac{10.1 \sin 70°}{\sin 70°} = 10.1$

$\dfrac{b}{\sin \beta} = \dfrac{c}{\sin \gamma} \Rightarrow b = \dfrac{c \sin \beta}{\sin \gamma} = \dfrac{10.1 \sin 40°}{\sin 70°} \approx 6.91$

$\gamma = 70°, a = 10.1, b \approx 6.9$

3. This triangle is of the form SAS. There is only one solution.

$a^2 = b^2 + c^2 - 2bc \cos \alpha = 7^2 + 8^2 - 2(7)(8) \cos 42° \Rightarrow a = \sqrt{113 - 112 \cos 42°} \approx 5.45599$

$b^2 = a^2 + c^2 - 2ac \cos \beta \Rightarrow \cos \beta = \dfrac{b^2 - a^2 - c^2}{-2ac} \Rightarrow \beta = \cos^{-1}\left(\dfrac{7^2 - 5.45599^2 - 8^2}{-2(5.45599)(8)}\right) \approx 59.14756°$

$\gamma = 180° - 42° - 59.14756° \approx 78.85244°$ $a \approx 5.5, \beta \approx 59.1°, \gamma \approx 78.9°$

5. This triangle is of the form AAS. There is only one solution.

$\gamma = 180° - 19° - 46° = 115°$

$\dfrac{a}{\sin \alpha} = \dfrac{b}{\sin \beta} \Rightarrow a = \dfrac{b \sin \alpha}{\sin \beta} = \dfrac{13 \sin 19°}{\sin 46°} \approx 5.9$ $\dfrac{c}{\sin \gamma} = \dfrac{b}{\sin \beta} \Rightarrow c = \dfrac{b \sin \gamma}{\sin \beta} = \dfrac{13 \sin 115°}{\sin 46°} \approx 16.4$

$\gamma = 115°, a \approx 5.9, c \approx 16.4$

7. This triangle is of the form SSA. It is ambiguous.

$$\frac{\sin \beta}{b} = \frac{\sin \gamma}{c} \Rightarrow \sin \beta = \frac{b \sin \gamma}{c} = \frac{8 \sin 20°}{11} \Rightarrow \beta_R = \sin^{-1}\left(\frac{8 \sin 20°}{11}\right) \approx 14.4°$$

Thus, $\beta \approx 14.4°$ or $\beta \approx 180° - 14.4° \approx 165.6°$ (which is not possible).

So $\beta \approx 14.4°$. Then $\alpha \approx 180° - 20° - 14.4° \approx 145.6°$.

$$\frac{a}{\sin \alpha} = \frac{c}{\sin \gamma} \Rightarrow a = \frac{c \sin \alpha}{\sin \gamma} = \frac{11 \sin 145.6°}{\sin 20°} \approx 18.2 \qquad \alpha \approx 145.6°, \beta \approx 14.4°, a \approx 18.2$$

9. This triangle is of the form SAS. There is only one solution.

$$c^2 = a^2 + b^2 - 2ab \cos \gamma = 18^2 + 23^2 - 2(18)(23) \cos 35° \Rightarrow c = \sqrt{853 - 828 \cos 35°} \approx 13.21901$$

$$a^2 = b^2 + c^2 - 2bc \cos \alpha \Rightarrow \cos \alpha = \frac{a^2 - b^2 - c^2}{-2bc} \Rightarrow \alpha = \cos^{-1}\left(\frac{18^2 - 23^2 - 13.21901^2}{-2(23)(13.21901)}\right) \approx$$

$51.35453°$

$$\beta = 180° - 35° - 51.35453° \approx 93.64547° \qquad c \approx 13.2, \alpha \approx 51.4°, \beta \approx 93.6°$$

11. Area $= K = \dfrac{1}{2}(12.3)(13.7) \sin 39° \approx 53.0$

13. $s = \dfrac{1}{2}(34 + 67 + 53) = 77 \Rightarrow$ Area $= K = \sqrt{77(77 - 34)(77 - 67)(77 - 53)} \approx 891.4$

15. (a) Horizontal $= 3$, vertical $= 4$

    (b) $\|\mathbf{v}\| = \sqrt{3^2 + 4^2} = \sqrt{9 + 16} = \sqrt{25} = 5$

    (c) $\|\mathbf{v}\|$ represents the length of $\mathbf{v}$. See Figure 15.

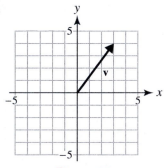

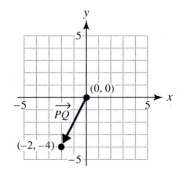

Figure 15   Figure 17

17. (a) See Figure 17.

    (b) $\mathbf{v} = \langle -2 - 0, -4 - 0 \rangle = \langle -2, -4 \rangle = -2\mathbf{i} - 4\mathbf{j}$

    (c) $\|\overrightarrow{PQ}\| = \sqrt{(-2)^2 + (-4)^2} = \sqrt{4 + 16} = \sqrt{20}$

19. (a) $2\mathbf{a} = 2\langle 3, -2 \rangle = \langle 6, -4 \rangle$

    (b) $\mathbf{a} - 3\mathbf{b} = \langle 3, -2 \rangle - 3\langle 1, 1 \rangle = \langle 3, -2 \rangle - \langle 3, 3 \rangle = \langle 3 - 3, -2 - 3 \rangle = \langle 0, -5 \rangle$

    (c) $\mathbf{a} \cdot \mathbf{b} = \langle 3, -2 \rangle \cdot \langle 1, 1 \rangle = (3)(1) + (-2)(1) = 3 - 2 = 1$

    (d) $\|\mathbf{a}\| = \sqrt{3^2 + (-2)^2} = \sqrt{9 + 4} = \sqrt{13}$, $\|\mathbf{b}\| = \sqrt{1^2 + 1^2} = \sqrt{1 + 1} = \sqrt{2}$

    $$\theta = \cos^{-1}\left(\frac{\mathbf{a} \cdot \mathbf{b}}{\|\mathbf{a}\|\|\mathbf{b}\|}\right) = \cos^{-1}\left(\frac{1}{\sqrt{13}\sqrt{2}}\right) = \cos^{-1}\left(\frac{1}{\sqrt{26}}\right) \approx 78.7°$$

21. (a) $2\mathbf{a} = 2(2\mathbf{i} + 2\mathbf{j}) = 4\mathbf{i} + 4\mathbf{j}$

    (b) $\mathbf{a} - 3\mathbf{b} = (2\mathbf{i} + 2\mathbf{j}) - 3(\mathbf{i} + \mathbf{j}) = 2\mathbf{i} + 2\mathbf{j} - 3\mathbf{i} - 3\mathbf{j} = -\mathbf{i} - \mathbf{j}$

(c) $\mathbf{a} \cdot \mathbf{b} = (2\mathbf{i} + 2\mathbf{j}) \cdot (\mathbf{i} + \mathbf{j}) = (2)(1) + (2)(1) = 2 + 2 = 4$

(d) $\|\mathbf{a}\| = \sqrt{2^2 + 2^2} = \sqrt{4 + 4} = \sqrt{8}$, $\|\mathbf{b}\| = \sqrt{1^2 + 1^2} = \sqrt{1 + 1} = \sqrt{2}$

$$\theta = \cos^{-1}\left(\frac{\mathbf{a} \cdot \mathbf{b}}{\|\mathbf{a}\|\|\mathbf{b}\|}\right) = \cos^{-1}\left(\frac{4}{\sqrt{8}\sqrt{2}}\right) = \cos^{-1}\left(\frac{4}{\sqrt{16}}\right) = \cos^{-1} 1 = 0°$$

23. Complete the parallelogram and find the unknown angle: $180° - 52° = 128°$.

Find the magnitude of the resultant force by using the law of cosines to find the length of the diagonal d.

$$d^2 = 100^2 + 130^2 - 2(100)(130)\cos 128° \Rightarrow d = \sqrt{26{,}900 - 26{,}000\cos 128°} \approx 207.1 \text{ lb}$$

25. See Figure 25.

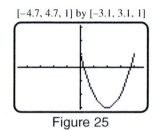
[−4.7, 4.7, 1] by [−3.1, 3.1, 1]

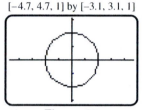
[−4.7, 4.7, 1] by [−3.1, 3.1, 1]

Figure 25   Figure 27

27. See Figure 27.

29. See Figure 29.

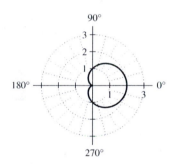

Figure 29

31. The graph of $r = 3 \sin(3\theta)$ for $0 \le \theta \le \pi$ is shown in Figure 31.

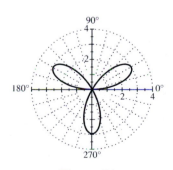

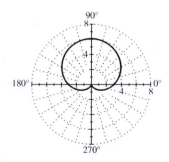

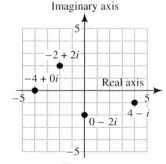

Figure 31   Figure 33   Figure 35

33. The graph of $r = 3 + 3 \sin(\theta)$ for $0 \le \theta \le 2\pi$ is shown in Figure 33.

35. See Figure 35.

37. $z_1 z_2 = (4 \cdot 2)(\cos(150° + 30°) + i \sin(150° + 30°)) = 8(\cos 180° + i \sin 180°) = 8(-1 + i(0)) = -8$

$$\frac{z_1}{z_2} = \left(\frac{4}{2}\right)(\cos(150° - 30°) + i\sin(150° - 30°)) = 2(\cos 120° + i\sin 120°) =$$

$$2\left(-\frac{1}{2} + i\left(\frac{\sqrt{3}}{2}\right)\right) = -1 + i\sqrt{3}$$

39. $w_0 = \sqrt{4}\left(\cos\frac{60° + 360°\cdot 0}{2} + i\sin\frac{60° + 360°\cdot 0}{2}\right) = 2(\cos 30° + i\sin 30°) = \sqrt{3} + i$

    $w_1 = \sqrt{4}\left(\cos\frac{60° + 360°\cdot 1}{2} + i\sin\frac{60° + 360°\cdot 1}{2}\right) = 2(\cos 210° + i\sin 210°) = -\sqrt{3} - i$

    This result is verified with a calculator in Figure 39.

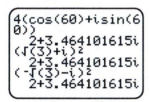

Figure 39

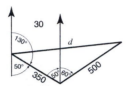

Figure 41

41. Refer to Figure 41.

    $d^2 = 350^2 + 500^2 - 2(350)(500)\cos 110° \Rightarrow d = \sqrt{372{,}500 - 350{,}000\cos 110°} \approx 701.6$

    The plane is about 701.6 miles from its takeoff point.

43. Let the airplane be located at position C and use standard labels.

    Note that $\beta = 180° - 57° = 123°$. Then $\gamma = 180° - 123° - 52° = 5°$.

    First we find side $a = BC$: $\dfrac{a}{\sin\alpha} = \dfrac{c}{\sin\gamma} \Rightarrow a = \dfrac{c\sin\alpha}{\sin\gamma} = \dfrac{950\sin 52°}{\sin 5°} \approx 8589.34$ ft.

    Label a point D on the ground directly below the airplane and let h represent the height of the airplane.

    From triangle BDC: $\sin 57° = \dfrac{h}{8589.34} \Rightarrow h = 8589.34\sin 57° \approx 7204$ ft.

45. In triangle ABC, $s = \dfrac{1}{2}(150 + 140 + 200) = 245 \Rightarrow$

    $K_1 = \sqrt{245(245 - 150)(245 - 140)(245 - 200)} \approx 10{,}486.9$ ft^2.

    In triangle ACD, $s = \dfrac{1}{2}(100 + 125 + 200) = 212.5 \Rightarrow$

    $K_2 = \sqrt{212.5(212.5 - 100)(212.5 - 125)(212.5 - 200)} \approx 5113.5$ ft^2.

    The total area is $K_1 + K_2 = 10{,}486.9 + 5113.5 \approx 15{,}600$ ft^2.

47. (a) $\mathbf{c} = \mathbf{a} + \mathbf{b} = (40\mathbf{i} - 20\mathbf{j}) + (20\mathbf{i} + 30\mathbf{j}) = 60\mathbf{i} + 10\mathbf{j}$

    (b) $\|\mathbf{c}\| = \sqrt{60^2 + 10^2} = \sqrt{3600 + 100} = \sqrt{3700} \approx 60.8$ cm

    (c) $\mathbf{a} + 2\mathbf{b} = (40\mathbf{i} - 20\mathbf{j}) + 2(20\mathbf{i} + 30\mathbf{j}) = (40\mathbf{i} - 20\mathbf{j}) + (40\mathbf{i} + 60\mathbf{j}) = 80\mathbf{i} + 40\mathbf{j}$

49. The parametric equations which model the flight of the ball are given by:

    $x_1 = 50\cos 45°(t)$, $y_1 = 50\sin 45°(t) - 16t^2$ for $0 \le t \le 3$. The ball travels about 78.1 feet. See Figure 49.

[0, 90, 10] by [−10, 30, 10]

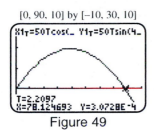

Figure 49

Extended and Discovery Exercises for Chapter 8

1. (a) Convert $0° \, 0' \, 10.34'' = 0.0028722222°$ to radians: $0.0028722222° \left(\dfrac{\pi}{180°} \right) \approx 5.013 \times 10^{-5}$ radians.

Then $\| \mathbf{v}_t \| \approx s = 3.5 \times 10^{13}(5.013 \times 10^{-5}) \approx 1{,}754{,}550{,}000$ miles per year.

That is $\dfrac{1{,}754{,}550{,}000 \text{ mi}}{1 \text{ year}} \cdot \dfrac{1 \text{ year}}{365 \text{ days}} \cdot \dfrac{1 \text{ day}}{24 \text{ hr}} \cdot \dfrac{1 \text{ hr}}{60 \text{ min}} \cdot \dfrac{1 \text{ min}}{60 \text{ sec}} \approx 56$ miles per second.

(b) Since $\mathbf{v} = \mathbf{v}_r + \mathbf{v}_t = \langle 0, -67 \rangle + \langle 56, 0 \rangle = \langle 56, -67 \rangle$, $\| \mathbf{v} \| = \sqrt{56^2 + (-67)^2} \approx 87$ miles per second.

3. (a) $d = \sqrt{(2.1 - 0.9)^2 + (-2.4 - 3.5)^2} \approx 6.02$ in.

(b) $X_H = \dfrac{7400(0.9)}{6 \sec 4.1° - 3.5 \sin 4.1°} \approx 1155.2$, $\ Y_H = \dfrac{7400(3.5) \cos 4.1°}{6 \sec 4.1° - 3.5 \sin 4.1°} \approx 4481.0$

$X_F = \dfrac{7400(2.1)}{6 \sec 4.1° - (-2.4) \sin 4.1°} \approx 2511.7$, $\ Y_F = \dfrac{7400(-2.4) \cos 4.1°}{6 \sec 4.1° - (-2.4) \sin 4.1°} \approx -2863.2$

$d = \sqrt{(2511.7 - 1155.2)^2 + (-2863.2 - 4481.0)^2} \approx 7468.4$ ft.

The ground distance between the house and the fire is approximately 7470 feet.

5. $\| \mathbf{a} - \mathbf{b} \|^2 = \| \mathbf{a} \|^2 + \| \mathbf{b} \|^2 - 2 \| \mathbf{a} \| \| \mathbf{b} \| \cos \theta \Rightarrow \| \mathbf{a} - \mathbf{b} \|^2 - \| \mathbf{a} \|^2 - \| \mathbf{b} \|^2 = -2 \| \mathbf{a} \| \| \mathbf{b} \| \cos \theta$

$\Rightarrow \left(\sqrt{(a_1 - b_1)^2 + (a_2 - b_2)^2} \right)^2 - \left(\sqrt{a_1^2 + a_2^2} \right)^2 - \left(\sqrt{b_1^2 + b_2^2} \right)^2 = -2 \| \mathbf{a} \| \| \mathbf{b} \| \cos \theta$

$\Rightarrow (a_1 - b_1)^2 + (a_2 - b_2)^2 - (a_1^2 + a_2^2) - (b_1^2 + b_2^2) = -2 \| \mathbf{a} \| \| \mathbf{b} \| \cos \theta$

$\Rightarrow a_1^2 - 2a_1 b_1 + b_1^2 + a_2^2 - 2a_2 b_2 + b_2^2 - a_1^2 - a_2^2 - b_1^2 - b_2^2 = -2 \| \mathbf{a} \| \| \mathbf{b} \| \cos \theta$

$\Rightarrow -2a_1 b_1 - 2a_2 b_2 = -2 \| \mathbf{a} \| \| \mathbf{b} \| \cos \theta$

$\Rightarrow a_1 b_1 + a_2 b_2 = \| \mathbf{a} \| \| \mathbf{b} \| \cos \theta$

$\Rightarrow \mathbf{a} \cdot \mathbf{b} = \| \mathbf{a} \| \| \mathbf{b} \| \cos \theta$

Chapters 1-8 Cumulative Review Exercises

1. $D = \sqrt{[7-3]^2 + [(-9)-(-2)]^2} = \sqrt{16+49} = \sqrt{65}$

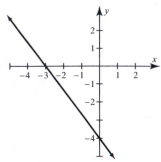

Figure 7

3   The domain of f includes all real numbers less than or equal to 4

$\Rightarrow D = \{x \mid x \le 4\}; \; f(-5) = \sqrt{4-(-5)} = \sqrt{9} = 3.$

5. $f(x) = 4x^2; \; f(x+h) = 4(x+h)^2 = 4x^2 + 8xh + 4h^2$, the difference quotient $= \dfrac{f(x+h) - f(x)}{h} \Rightarrow$

$\dfrac{4x^2 + 8xh + 4h^2 - 4x^2}{h} = \dfrac{8xh + 4h^2}{h} = 8x + 4h$

7. For $4x + 3y = -12$: x-intercept, then $y = 0 \Rightarrow 4x + 3(0) = -12 \Rightarrow 4x = -12 \Rightarrow x = -3,$

y-intercept, then $x = 0 \Rightarrow 4(0) + 3y = -12 \Rightarrow 3y = -12 \Rightarrow y = -4$. See Figure 7.

9. (a) $|2x - 5| = 6$ then $2x - 5 = 6$ or $2x - 5 = -6$. When $2x - 5 = -6 \Rightarrow 2x = -1 \Rightarrow x = -\dfrac{1}{2}.$

   When $2x - 5 = 6 \Rightarrow 2x = 11 \Rightarrow x = \dfrac{11}{2}$. Thus $x = -\dfrac{1}{2}$ or $\dfrac{11}{2}.$

   (b) $6x^2 + 22x = 8 \Rightarrow 6x^2 + 22x - 8 = 0 \Rightarrow 2(3x^2 + 11x - 4) = 0 \Rightarrow 2(3x - 1)(x + 4) = 0 \Rightarrow$

   $3x - 1 = 0 \Rightarrow x = \dfrac{1}{3}$ or $x + 4 = 0 \Rightarrow x = -4$. Thus $x = \dfrac{1}{3}$ or -4

   (c) $x^3 = x \Rightarrow x^3 - x = 0 \Rightarrow x(x^2 - 1) = 0 \Rightarrow x(x + 1)(x - 1) = 0 \Rightarrow x = 0, -1$ or 1

   $x = -1$ or $x - 1 = 0 \Rightarrow x = 1 \Rightarrow x = 0, -1$ or 1

   (d) $x^4 - 2x^2 - 3 = 0 \Rightarrow (x^2 - 3)(x^2 + 1) = 0 \Rightarrow x^2 - 3 = 0 \Rightarrow x^2 = 3 \Rightarrow$

   $x = \pm\sqrt{3}$ or $x^2 + 1 = 0 \Rightarrow x^2 = -1 \Rightarrow x = \pm\sqrt{-1}$. Thus $x = \pm\sqrt{3}$ or $\pm i.$

   (e) $2e^{3x} - 1 = 50 \Rightarrow 2e^{3x} = 51 \Rightarrow e^{3x} = \dfrac{51}{2} \Rightarrow 3x = \ln\dfrac{51}{2} \Rightarrow x = \dfrac{\ln\left(\frac{51}{2}\right)}{3} \approx 1.08$

   (f) $3x^{2/3} = 12 \Rightarrow x^{2/3} = 4 \Rightarrow (x^{2/3})^3 = 4^3 \Rightarrow x^2 = 64 \Rightarrow x = \pm 8$

   (g) $\sin t = \dfrac{1}{2}, \; t = \sin^{-1}\left(\dfrac{1}{2}\right) \Rightarrow t = \dfrac{\pi}{6}$ or $\dfrac{5\pi}{6}$

   (h) $\tan 2t = -\sqrt{3} \Rightarrow$ let $\theta = 2t$ so then the equation becomes $\tan\theta = -\sqrt{3}$ and in radian measure these

   solutions are $\theta = \dfrac{2\pi}{3}$ and $\dfrac{5\pi}{3}$. For all real number solutions $\theta = \dfrac{2\pi}{3} + \pi n.$

   Because $\theta = 2t$, we can determine t by substituting $2t$ for $\theta \Rightarrow 2t = \dfrac{2\pi}{3} + \pi n.$

   Therefore $t = \dfrac{\pi}{3} + \dfrac{\pi}{2}n.$

(i) $2\cos^2 t + \cos t = 1 \Rightarrow 2\cos^2 + \cos t - 1 = 0 \Rightarrow (2\cos t - 1)(\cos t + 1) = 0$. Either

$2\cos t - 1 = 0 \Rightarrow \cos t = \dfrac{1}{2} \Rightarrow t = \dfrac{\pi}{3} + 2\pi n$ or $t = \dfrac{5\pi}{3} + 2\pi n$

or $\cos t + 1 = 0 \Rightarrow \cos t = -1 \Rightarrow t = \pi + 2\pi n$

11. (a) $4(x - 3) > 1 - x \Rightarrow 4x - 12 > 1 - x \Rightarrow 5x > 13 \Rightarrow x > \dfrac{13}{5} \Rightarrow \left(\dfrac{13}{5}, \infty\right)$

(b) The solutions to $|2x - 1| \le 3$ are $c \le x \le d$ where c and d are the solutions to $|2x - 1| = 3$.

$|2x - 1| = 3 \Rightarrow 2x - 1 = 3 \Rightarrow 2x = 4 \Rightarrow x = 2$ or $2x - 1 = -3 \Rightarrow 2x = -2 \Rightarrow x = -1$

Here $c = -1$ and $d = 2$. The solution interval is $[-1, 2]$.

(c) First solve $x^2 - 2x - 3 = 0 \Rightarrow (x + 1)(x - 3) = 0 \Rightarrow x = -1$ or $x = 3$.

Since the graph of $y = x^2 - 2x - 3$ is a parabola opening upward, the graph is above the x-axis for

$x < -1$ or $x > 3$. The solution interval is $(-\infty, -1) \cup (3, \infty)$.

(d) First solve $x^3 - 4x = 0 \Rightarrow x(x^2 - 4) = 0 \Rightarrow x(x + 2)(x - 2) = 0 \Rightarrow x = -2, 0,$ or 2.

The graph of $y = x^3 - 4x$ is above the x-axis for $-2 < x < 0$ or $x > 2$.

The solution interval is $(-2, 0) \cup (2, \infty)$.

(e) First solve $\dfrac{x}{x - 1} = 0 \Rightarrow x = 0$. The graph of $y = \dfrac{x}{x - 1}$ intersects or is on or below the x-axis for

$0 \le x < 1$. The solution interval is $[0, 1)$.

(f) $-4 \le 4 - 3x \le 12 \Rightarrow -8 \le -3x \le 8 \Rightarrow \dfrac{8}{3} \ge x \ge -\dfrac{8}{3} \Rightarrow -\dfrac{8}{3} \le x \le \dfrac{8}{3}$.

The solution interval is $\left[-\dfrac{8}{3}, \dfrac{8}{3}\right]$.

13. (a)

$$
\begin{array}{r}
3x^2 - 6x + 11 \\
x + 2 \overline{)3x^3 + 0x^2 - x + 2} \\
\underline{3x^3 + 6x^2} \\
-6x^2 - x + 2 \\
\underline{-6x^2 - 12x} \\
11x + 2 \\
\underline{11x + 22} \\
-20
\end{array}
$$

The solution is $3x^2 - 6x + 11 + \dfrac{-20}{x + 2}$.

(b)

$$
\begin{array}{r}
x^2 - x \\
2x - 1 \overline{)2x^3 - 3x^2 + x - 1} \\
\underline{2x^3 - x^2} \\
-2x^2 + x - 1 \\
\underline{-2x^2 + x} \\
-1
\end{array}
$$

The solution is $x^2 - x + \dfrac{-1}{2x - 1}$.

15. The denominator cannot equal zero $\Rightarrow 2 - 3x \ne 0 \Rightarrow x \ne \dfrac{2}{3}$. Then $D = \left\{x \,|\, x \ne \dfrac{2}{3}\right\}$. The vertical

asymptotes are values when the function is undefined. The vertical asymptote is $x = \dfrac{2}{3}$. The horizontal

asymptote is $y = \dfrac{3x}{-3x} = -1 \Rightarrow y = -1$.

17. If $f(x) = 3x - 2$ then $y = 3x - 2 \Rightarrow y + 2 = 3x \Rightarrow \dfrac{y + 2}{3} = x \Rightarrow f^{-1}(x) = \dfrac{x + 2}{3}$

19. If the initial bacteria count is 5000 then $C = 5000$. Since it doubles every 1.5 hours, $10,000 = 5000a^{3/2} \Rightarrow 2 = a^{3/2} \Rightarrow a = 2^{2/3}$.

21. (a) $\log_3 \dfrac{1}{27} = \log_3 \dfrac{1}{3^3} = \log_3 3^{-3} = -3$

    (b) $\ln \dfrac{1}{e^3} = \ln e^{-3} = -3$

    (c) $\log \sqrt[3]{10} = \log 10^{1/3} = \dfrac{1}{3}$

    (d) $\log_4 32 - \log_4 \dfrac{1}{2} = \log_4 \dfrac{32}{\frac{1}{2}} = \log_4 64 = \log_4 4^3 = 3$

23. $225° \left(\dfrac{\pi}{180} \right) = \dfrac{5\pi}{4}$

25. $c = \sqrt{5^2 + 12^2} = \sqrt{169} = 13$

    $\sin \theta = \dfrac{5}{13}$ $\qquad\qquad \cos \theta = \dfrac{12}{13}$ $\qquad\qquad \tan \theta = \dfrac{5}{12}$

    $\csc \theta = \dfrac{13}{5}$ $\qquad\qquad \sec \theta = \dfrac{13}{12}$ $\qquad\qquad \cot \theta = \dfrac{12}{5}$

27. Since $\sec \theta < 0$, $\cos \theta = -\sqrt{1 - \sin^2 \theta} = -\sqrt{1 - \left(-\frac{7}{25}\right)^2} = -\sqrt{1 - \frac{49}{625}} = -\sqrt{\frac{576}{625}} = -\dfrac{24}{25}$

    $\sin \theta = -\dfrac{7}{25}$ $\qquad\qquad \cos \theta = -\dfrac{24}{25}$ $\qquad\qquad \tan \theta = -\dfrac{7}{24}$

    $\csc \theta = -\dfrac{25}{7}$ $\qquad\qquad \sec \theta = -\dfrac{25}{24}$ $\qquad\qquad \cot \theta = \dfrac{24}{7}$

29. $c = \sqrt{5^2 + 12^2} = \sqrt{169} = 13$. $\sin \theta = \dfrac{5}{13} \Rightarrow \sin^{-1}\left(\dfrac{5}{13}\right) \approx 22.6° \Rightarrow \theta \approx 22.6°$

    $180° - 90° - 22.6° = 67.4° \Rightarrow \beta = 67.4°$

31. $\cot^2 \theta - 2 \cot \theta + 1 = (\cot \theta - 1)(\cot \theta - 1) = (\cot \theta - 1)^2$

33. (a) This triangle is in the form ASA. There is only one solution. $\beta = 180° - 31° - 53° = 96°$

    $\dfrac{a}{\sin \alpha} = \dfrac{b}{\sin \beta} \Rightarrow a = \dfrac{b \sin \alpha}{\sin \beta} = \dfrac{15 \sin 31°}{\sin 96°} \approx 7.8$

    $\dfrac{c}{\sin \gamma} = \dfrac{b}{\sin \beta} \Rightarrow c = \dfrac{b \sin \gamma}{\sin \beta} = \dfrac{15 \sin 53°}{\sin 96°} \approx 12.0$ $\quad \beta = 96°, a \approx 7.8, c \approx 12.0$

    (b) This triangle is in the form SSA. It is ambiguous.

    $\dfrac{\sin \beta}{b} = \dfrac{\sin \alpha}{a} \Rightarrow \sin \beta = \dfrac{b \sin \alpha}{a} = \dfrac{5 \sin 31°}{6} = 0.42920 \Rightarrow \beta_R \approx \sin^{-1}(0.42920) \approx 25.4°$

    Thus $\beta = 25.4°$. Then $\gamma = 180° - 31° - 25.4° \approx 123.6°$

    $\dfrac{c}{\sin \gamma} = \dfrac{a}{\sin \alpha} \Rightarrow c = \dfrac{a \sin \gamma}{\sin \alpha} = \dfrac{6 \sin 123.6°}{\sin 31°} \approx 9.7$ $\quad \beta \approx 25.4°, \gamma \approx 123.6°, c \approx 9.7$

(c) $b^2 = a^2 + c^2 - 2ac \cos \beta = 6^2 + 8^2 - 2(6)(8) \cos 56° \Rightarrow b = \sqrt{100 - 96 \cos 56°} \approx 6.8$

$a^2 = b^2 + c^2 - 2bc \cos \alpha \Rightarrow \cos \alpha = \dfrac{a^2 - b^2 - c^2}{-2bc} \Rightarrow \alpha = \cos^{-1}\left(\dfrac{6^2 - (6.8)^2 - 8^2}{-2(6.8)(8)}\right) \approx 47.0°$

$\gamma = 180° - 56° - 47.0° \approx 77.0°$ $b \approx 6.8,\ \alpha \approx 47.0°,\ \gamma \approx 77.0°$

(d) $a^2 = b^2 + c^2 - 2bc \cos \alpha \Rightarrow \cos \alpha = \dfrac{a^2 - b^2 - c^2}{-2bc} \Rightarrow \cos^{-1}\left(\dfrac{6^2 - 7^2 - 8^2}{-2(7)(8)}\right) \approx 46.6°$

$b^2 = a^2 + c^2 - 2ac \cos \beta \Rightarrow \cos \beta = \dfrac{b^2 - a^2 - c^2}{-2ac} \Rightarrow \beta = \cos^{-1}\left(\dfrac{7^2 - 6^2 - 8^2}{-2(6)(8)}\right) \approx 57.9°$

$\gamma = 180° - 46.6° - 57.9° \approx 75.5°$ $\alpha \approx 46.6°,\ \beta \approx 57.9°,\ \gamma \approx 75.5°$

35. (a) $\|\mathbf{b}\| = \sqrt{7^2 + (-24)^2} = \sqrt{49 + 576} = \sqrt{625} = 25$

(b) $2\mathbf{a} - 3\mathbf{b} = 2\langle -5, 12 \rangle - 3\langle 7, -24 \rangle = \langle -10, 24 \rangle - \langle 21, -72 \rangle \Rightarrow \langle -10 - 21, 24 - (-72) \rangle \Rightarrow$

$\langle -31, 96 \rangle$

(c) $\mathbf{a} \cdot \mathbf{b} = (-5)(7) + (12)(-24) = -35 + (-288) = -323$

(d) $\|\mathbf{a}\| = \sqrt{(-5)^2 + 12^2} = \sqrt{25 + 144} = \sqrt{169} = 13,$

$\|\mathbf{b}\| = \sqrt{7^2 + (-24)^2} = \sqrt{49 + 576} = \sqrt{625} = 25$

$\theta = \cos^{-1}\left(\dfrac{\mathbf{a} \cdot \mathbf{b}}{\|\mathbf{a}\|\|\mathbf{b}\|}\right) = \cos^{-1}\left(\dfrac{-323}{(13)(25)}\right) = \cos^{-1}\left(\dfrac{-323}{325}\right) \Rightarrow \theta \approx 173.6°$

37. See Figure 37.

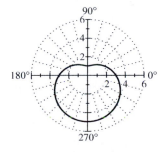

Figure 37

39. Let x = width and $x + 4$ = length. The new width of the box is $x - 6$, because two square corners, with sides of 3 inches have been removed. Similarly, the length of the bottom of the box is $x - 2$ inches. Because height times the width times the length must equal the volume, or 351 cubic inches, it follows that:

$3(x - 6)(x - 2) = 351$ or $(x - 6)(x - 2) = 117 \Rightarrow x^2 - 2x - 6x + 12 = 117 \Rightarrow$

$x^2 - 8x + 12 = 117 \Rightarrow x^2 - 8x - 105 = 0 \Rightarrow (x + 7)(x - 15) = 0 \Rightarrow x = -7 \text{ or } x = 15$. Since the dimensions cannot be negative, the width of the cardboard is 15 inches and the length is 4 inches more or 19 inches.

41. Let x = length, $\dfrac{1}{2}x$ = width and h = height. Then the surface area is given by

$A = 2\left(\dfrac{1}{2}x^2\right) + 2(xh) + 2\left(\dfrac{1}{2}xh\right) = x^2 + 2xh + xh = x^2 + 3xh$. Since the surface area is 288, we have

$x^2 + 3xh = 288$. Also, the volume is given by $V = x\left(\dfrac{1}{2}x\right)h = \dfrac{1}{2}x^2 h$. Since the volume is 288, we have

$\dfrac{1}{2}x^2 h = 288 \Rightarrow x^2 h = 576 \Rightarrow h = \dfrac{576}{x^2}$. Substitute this value for h in the surface area equation.

$x^2 + 3x\left(\dfrac{576}{x^2}\right) = 288 \Rightarrow x^2 + \dfrac{1728}{x} = 288 \Rightarrow x^3 + 1728 = 288x \Rightarrow x^3 - 288x + 1728 = 0$

The solutions can be found by graphing the left side of this equation in $[0, 20, 1]$ by $[-250, 250, 50]$. (Not shown.)

$x \approx 7.4 \Rightarrow$ length ≈ 7.4 in., width $\approx \dfrac{1}{2}(7.4) = 3.7$ in., and height $\approx \dfrac{576}{7.4^2} = 10.5$ in.

$x = 12 \Rightarrow$ length $= 12$ in., width $= \dfrac{1}{2}(12) = 6$ in., and height $\approx \dfrac{576}{12^2} = 4$ in.

43. The length of the shadow is found by the use of tangent. $\tan 63° = \dfrac{5}{x} \Rightarrow 5 = x \tan 63° \Rightarrow x = \dfrac{5}{\tan 63°} \Rightarrow$

$x \approx 2.5$ feet.

45. The angle of elevation can be found by the use of tangent. $\tan^{-1}\dfrac{85}{57} = x \Rightarrow x = 56.2°$.

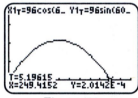

Figure 48

47. (a) $c^2 = a^2 + b^2 - 2ab \cos \gamma = 242^2 + 165^2 - 2(242)(165) \cos 72° \Rightarrow$

$c = \sqrt{85,789 - 79,860 \cos 72°} \approx 247°$

(b) Area $= K = \dfrac{1}{2}(242)(165) \sin 72° \approx 18,988$ ft^2.

Chapter 9: Systems of Equations and Inequalities

9.1: Functions and Systems of Equations in Two Variables

1. $A(5, 8) = \frac{1}{2}(5)(8) = 20$. The area of a triangle with a base of 5 and height of 8 is 20 square units.

3. $f(2, -3)$ if $f(x, y) = x^2 + y^2 \Rightarrow f(2, -3) = 2^2 + (-3)^2 = 4 + 9 = 13$

5. $f(-2, 3)$ if $f(x, y) = 3x - 4y \Rightarrow f(-2, 3) = 3(-2) - 4(3) = -6 - 12 = -18$

7. $f\left(\frac{1}{2}, -\frac{7}{4}\right)$ if $f(x, y) = \frac{2x}{y + 3} \Rightarrow f\left(\frac{1}{2}, -\frac{7}{4}\right) = \frac{2\left(\frac{1}{2}\right)}{\left(-\frac{7}{4}\right) + 3} = \frac{1}{\frac{5}{4}} = \frac{4}{5}$

9. The sum of y and twice x is computed by $f(x, y) = y + 2x$.

11. The product of x and y divided by $1 + x$ is computed by $f(x, y) = \frac{xy}{1 + x}$.

13. $3x - 4y = 7 \Rightarrow 3x = 4y + 7 \Rightarrow x = \frac{4y + 7}{3}$; $3x - 4y = 7 \Rightarrow -4y = -3x + 7 \Rightarrow y = \frac{3x - 7}{4}$

15. $x - y^2 = 5 \Rightarrow x = y^2 + 5$; $x - y^2 = 5 \Rightarrow -y^2 = -x + 5 \Rightarrow y^2 = x - 5 \Rightarrow y = \pm\sqrt{x - 5}$

17. $\frac{2x - y}{3y} = 1 \Rightarrow 2x - y = 3y \Rightarrow 2x = 4y \Rightarrow x = 2y$; $\frac{2x - y}{3y} = 1 \Rightarrow 2x - y = 3y \Rightarrow 2x = 4y \Rightarrow$

$y = \frac{x}{2}$

19. The only ordered pair that satisfies both equations is $(2, 1)$. The system is linear.

$2(2) + 1 = 5\bigstar$	$2(-2) + 1 = -3$	$2(1) + 0 = 2$
$2 + 1 = 3\bigstar$	$-2 + 1 = -1$	$1 + 0 = 1$

21. The only ordered pair that satisfies both equations is $(4, -3)$. The system is non-linear.

$4^2 + (-3)^2 = 25\bigstar$	$0^2 + 5^2 = 25\bigstar$	$4^2 + 3^2 = 25\bigstar$
$2(4) + 3(-3) = -1\bigstar$	$2(0) + 3(5) = 15$	$2(4) + 3(3) = 17$

23. From the graph the solution is $(2, 2)$. The solution satisfies both $x - y = 0$ and $x + y = 4$.

25. From the graph the solution is $\left(\frac{1}{2}, -2\right)$. The solution satisfies both $6x + 4y = -5$ and $2x - 3y = 7$.

27. Since the lines intersect, the system is consistent with a unique solution at $(2, 2)$.

$x + y = 4 \Rightarrow y = 4 - x$; We substitute $4 - x$ for y in the other equation.

$2x - (4 - x) = 2 \Rightarrow 3x = 6 \Rightarrow x = 2$, then $2 + y = 4 \Rightarrow y = 2$. The solution is $(2, 2)$.

29. Since the lines are parallel, the system is inconsistent with no solutions.

31. Parallel lines $\Rightarrow$ inconsistent system. See Figure 31.

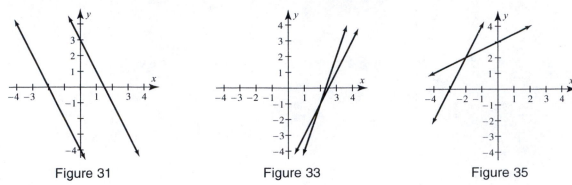

Figure 31   Figure 33   Figure 35

33. The lines intersect at $(2, -1)$. The system is consistent and independent. See Figure 33.

35. The lines intersect at $(-2, 2)$. The system is consistent and independent. See Figure 35.

37. The system has an infinite number of solutions. $\{(x, y)|2x - y = -4\}$.   The system is consistent and dependent. See Figure 37.

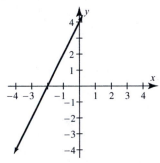

Figure 37

39. Solve the first equation for x: $x + 2y = 0 \Rightarrow x = -2y$. Substitute this into the second equation.

    $3x + 7y = 1 \Rightarrow 3(-2y) + 7y = 1 \Rightarrow -6y + 7y = 1 \Rightarrow y = 1$.

    If $y = 1$, then $x = -2(1) = -2$. The solution is $(-2, 1)$.

    Check: $x + 2y = 0 \Rightarrow (-2) + 2(1) = 0 \Rightarrow 0 = 0$; $x = -2y \Rightarrow -2 = -2(1) \Rightarrow -2 = -2$

41. Solve the first equation for x: $2x - 9y = -17 \Rightarrow 2x = 9y - 17 \Rightarrow x = \dfrac{9y - 17}{2}$. Substitute this into the

    second equation. $8x + 5y = 14 \Rightarrow 8\left(\dfrac{9y - 17}{2}\right) + 5y = 14 \Rightarrow 4(9y - 17) + 5y = 14 \Rightarrow$

    $36y - 68 + 5y = 14 \Rightarrow 41y = 82 \Rightarrow y = 2$. If $y = 2$, then $x = \dfrac{9(2) - 17}{2} = \dfrac{1}{2}$. The solution is $(0.5, 2)$.

    Check: $2x - 9y = -17 \Rightarrow 2(0.5) - 9(2) = -17 \Rightarrow 1 - 18 = -17 \Rightarrow -17 = -17$;

    $8x + 5y = 14 \Rightarrow 8(0.5) + 5(2) = 14 \Rightarrow 4 + 10 = 14 \Rightarrow 14 = 14$

43. Solve the second equation for x: $x + \frac{1}{2}y = 10 \Rightarrow x = -\frac{1}{2}y + 10$. Substitute this into the first equation.

$$\frac{1}{2}x - y = -5 \Rightarrow \frac{1}{2}\left(-\frac{1}{2}y + 10\right) - y = -5 \Rightarrow -\frac{1}{4}y + 5 - y = -5 \Rightarrow -\frac{5}{4}y = -10 \Rightarrow y = 8.$$

If $y = 8$, then $x = -\frac{1}{2}(8) + 10 = 6$. The solution is $(6, 8)$.

Check: $x + \frac{1}{2}y = 10 \Rightarrow 6 + \frac{1}{2}(8) = 10 \Rightarrow 6 + 4 = 10 \Rightarrow 10 = 10$

$\frac{1}{2}x - y = -5 \Rightarrow \frac{1}{2}(6) - 8 = -5 \Rightarrow 3 - 8 = -5 \Rightarrow -5 = -5$

45. Solve the first equation for y: $3x - 2y = 5 \Rightarrow -2y = -3x + 5 \Rightarrow y = \frac{3}{2}x - \frac{5}{2}$. Substitute this into the

second equation. $-6x + 4\left(\frac{3}{2}x - \frac{5}{2}\right) = -10 \Rightarrow -6x + 6x - 10 = -10 \Rightarrow -10 = -10 \Rightarrow$ there are

infinitely many solutions. $\{(x, y) \mid 3x - 2y = 5\}$.

47. Solve the first equation for x: $2x - 7y = 8 \Rightarrow 2x = 7y + 8 \Rightarrow x = \frac{7}{2}y + 4$. Substitute this into the second

equation. $-3\left(\frac{7}{2}y + 4\right) + \frac{21}{2}y = 5 \Rightarrow -\frac{21}{2}y - 12 + \frac{21}{2}y = 5 \Rightarrow -12 = 5 \Rightarrow$ there are no real

solutions.

49. $0.2x - 0.1y = 0.5 \Rightarrow 2x - y = 5$ and $0.4x + 0.3y = 2.5 \Rightarrow 4x + 3y = 25$

Solve the first equation for y: $2x - y = 5 \Rightarrow y = 2x - 5$. Substitute this into the second equation.

$4x + 3y = 25 \Rightarrow 4x + 3(2x - 5) = 25 \Rightarrow 4x + 6x - 15 = 25 \Rightarrow 10x = 40 \Rightarrow x = 4$.

If $x = 4$, then $y = 2(4) - 5 \Rightarrow y = 3$. The solution is $(4, 3)$.

51. Solve the second equation for y: $2x + y = 0 \Rightarrow y = -2x$. Substitute this into the first equation.

$x^2 - y = 0 \Rightarrow x^2 - (-2x) = 0 \Rightarrow x^2 + 2x = 0 \Rightarrow x(x + 2) = 0 \Rightarrow x = 0$ or $x = -2$.

When $x = 0$, $y = -2(0) = 0$ and when $x = -2$, $y = -2(-2) = 4$. The solutions are $(0, 0)$ and $(-2, 4)$.

53. Solve the second equation for y: $x + y = 6 \Rightarrow y = 6 - x$. Substitute this into the first equation.

$xy = 8 \Rightarrow x(6 - x) = 8 \Rightarrow 6x - x^2 = 8 \Rightarrow x^2 - 6x + 8 = 0 \Rightarrow (x - 4)(x - 2) = 0 \Rightarrow$

$x = 4$ or $x = 2$.

When $x = 4$, $y = 6 - 4 = 2$ and when $x = 2$, $y = 6 - 2 = 4$. The solutions are $(4, 2)$ and $(2, 4)$.

55. Substitute the second equation, $y = 2x$, into the first equation.

$x^2 + y^2 = 20 \Rightarrow x^2 + (2x)^2 = 20 \Rightarrow x^2 + 4x^2 = 20 \Rightarrow 5x^2 = 20 \Rightarrow x^2 = 4 \Rightarrow x = \pm 2$.

When $x = -2$, $y = 2(-2) = -4$ and when $x = 2$, $y = 2(2) = 4$. The solutions are $(-2, -4)$ and $(2, 4)$.

57. Solve the second equation for x: $x - y = -2 \Rightarrow x = y - 2$. Substitute this into the first equation.

$\sqrt{y - 2} - 2y = 0 \Rightarrow \sqrt{y - 2} = 2y \Rightarrow y - 2 = 4y^2 \Rightarrow 4y^2 - y + 2 = 0$. Using the quadratic formula

to solve we get: $\dfrac{1 \pm \sqrt{1 - 4(4)(2)}}{2(4)} = \dfrac{1 \pm \sqrt{-31}}{8} \Rightarrow$ no real solutions.

59. Solve the first equation for y: $2x^2 - y = 5 \Rightarrow -y = -2x^2 + 5 \Rightarrow y = 2x^2 - 5$. Substitute this into the second equation. $-4x^2 + 2(2x^2 - 5) = -10 \Rightarrow -4x^2 + 4x^2 - 10 = -10 \Rightarrow -10 = -10 \Rightarrow$ there are infinitely many solutions, $\{(x, y) | 2x^2 - y = 5\}$.

61. Solve the second equation for y: $x^2 + y = 4 \Rightarrow y = 4 - x^2$. Substitute this into the first equation.
    $x^2 - y = 4 \Rightarrow x^2 - (4 - x^2) = 4 \Rightarrow 2x^2 = 8 \Rightarrow x^2 = 4 \Rightarrow x = \pm 2$.
    When $x = -2$, $y = 4 - (-2)^2 = 0$ and when $x = 2$, $y = 4 - (2)^2 = 0$.
    The solutions are $(-2, 0)$ and $(2, 0)$.

63. Solve the second equation for y: $x - y = 0 \Rightarrow y = x$. Substitute this into the first equation.
    $x^3 - x = 3y \Rightarrow x^3 - x = 3x \Rightarrow x^3 - 4x = 0 \Rightarrow x(x + 2)(x - 2) = 0 \Rightarrow x = 0, x = -2,$ or $x = 2$.
    When $x = 0$, $y = 0$, when $x = -2$, $y = -2$, and when $x = 2$, $y = 2$.
    The solutions are $(-2, -2)$, $(0, 0)$, and $(2, 2)$.

65. $x - y = 2$ and $2x + 2y = 38$. Multiply the second equation by $\dfrac{1}{2}$ and add to eliminate the y-variable.

$$\begin{aligned} x - y &= \ 2 \\ \underline{x + y} &= \underline{19} \\ 2x &= 21 \end{aligned} \Rightarrow x = 10.5$$

Since, $x - y = 2$, $y = 8.5$. The solution is $(10.5, 8.5)$.

67. $x + y = 75$ and $4x + 7y = 456$. Multiply the first equation by 7 and subtract to eliminate the y-variable.

$$\begin{aligned} 7x + 7y &= 525 \\ \underline{4x + 7y} &= \underline{456} \\ 3x &= \ 69 \end{aligned} \Rightarrow x = 23$$

Since, $x + y = 75$, $y = 52$. The solution is $(23, 52)$.

69. The given equations result in the following nonlinear system of equations.
    $A(l, w) = 35 \Rightarrow lw = 35$ and $P(l, w) = 24 \Rightarrow 2l + 2w = 24$
    Begin solving the second equation for l. $2l + 2w = 24 \Rightarrow l + w = 12 \Rightarrow l = 12 - w$. Substitute this into the first equation. $lw = 35 \Rightarrow (12 - w)w = 35 \Rightarrow 12w - w^2 = 35 \Rightarrow w^2 - 12w + 35 = 0$. This is a quadratic equation that can be solved by factoring, graphing, or the quadratic formula. The solutions to this quadratic are found using factoring. $w^2 - 12w + 35 = 0 \Rightarrow (w - 5)(w - 7) = 0 \Rightarrow w = 5$ or 7. Since $l = 12 - w$, if $w = 5$, then $l = 7$, and if $w = 7$, then $l = 5$. If the length is greater than the width, the solution is $l = 7$ and $w = 5$. A rectangle with length 7 and width 5 has an area of 35 and a perimeter of 24.

71. Add the two equations together to eliminate the y-variable.

$$\begin{aligned} x + y &= 20 \\ \underline{x - y} &= \ \underline{8} \\ 2x &= 28 \end{aligned} \Rightarrow x = 14$$

Since $x + y = 20$, it follows that $y = 6$. The unique solution is $(14, 6)$. The system is consistent and independent. Graphical and numerical support are shown in Figure 71a & 71b, where
$Y_1 = 20 - X$ and $Y_2 = X - 8$.

[0, 24, 4] by [0, 16, 4]

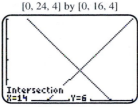

X	Y1	Y2
11	9	3
12	8	4
13	7	5
14	6	6
15	5	7
16	4	8
17	3	9

Intersection
X=14   Y=6

X=14

Figure 71a   Figure 71b

73. Subtract the two equations to eliminate the x-variable.

$$\begin{array}{rcl} x + 3y &=& 10 \\ x - 2y &=& -5 \\ \hline 5y &=& 15 \end{array} \Rightarrow y = 3$$

Since $x + 3y = 10$, it follows that $x = 1$. The solution is $(1, 3)$. The system is consistent and independent.

75. Multiply the second equation by -1 and subtract to eliminate both variables.

$$\begin{array}{rcl} x + y &=& 500 \\ x + y &=& 500 \\ \hline 0 &=& 0 \end{array} \Rightarrow \text{infinite number of solutions}$$

The solution is $\{(x, y) \mid x + y = 500\}$. The system is consistent but dependent.

77. Multiply the second equation by 2 and add. This eliminates both variables.

$$\begin{array}{rcl} 2x + 4y &=& 7 \\ -2x - 4y &=& 10 \\ \hline 0 &=& 17 \end{array} \Rightarrow \text{no solution}$$

Since $0 \neq 17$, the system is inconsistent.

79. Multiply the second equation by 2 and subtract to eliminate the x-variable.

$$\begin{array}{rcl} 2x + 3y &=& 2 \\ 2x - 4y &=& -10 \\ \hline 7y &=& 12 \end{array} \Rightarrow y = \frac{12}{7}$$

Since $2x + 3y = 2$, it follows that $x = \dfrac{2 - 3y}{2} \Rightarrow x = -\dfrac{11}{7}$. The solution is $\left(-\dfrac{11}{7}, \dfrac{12}{7}\right)$. The

system is consistent and independent.

81. Multiply the second equation by $-\dfrac{1}{2}$ and add to eliminate the x-variable.

$$\begin{array}{rcl} \dfrac{1}{2}x - y &=& 5 \\ -\dfrac{1}{2}x + \dfrac{1}{4}y &=& -2 \\ \hline -\dfrac{3}{4}y &=& 3 \end{array} \Rightarrow y = -4$$

Then, $x - \dfrac{1}{2}y = 4 \Rightarrow x = 4 + \dfrac{1}{2}y \Rightarrow x = 2$. The solution is $(2, -4)$.

83. Multiply the first equation by 3 and add to eliminate both variables.

$$
\begin{aligned}
21x - 9y &= -51 \\
-21x + 9y &= 51 \\
\hline
0 &= 0
\end{aligned}
\Rightarrow \text{infinite number of solutions}
$$

There are infinitely many solutions of the form $\{(x, y) \mid 7x - 3y = -17\}$.

85. Multiply the first equation by 3 and add to eliminate both variables.

$$
\begin{aligned}
2x + 4y &= 1 \\
-2x - 4y &= 5 \\
\hline
0 &= 6
\end{aligned}
\Rightarrow \text{no solutions}
$$

87. Clear decimals: $0.2x + 0.3y = 8 \Rightarrow 2x + 3y = 80$ and $-0.4x + 0.2y = 0 \Rightarrow -4x + 2y = 0$.

    Multiply the first equation by 2 and add to eliminate the x-variable.

$$
\begin{aligned}
4x + 6y &= 160 \\
-4x + 2y &= 0 \\
\hline
8y &= 160
\end{aligned}
\Rightarrow y = 20
$$

Then, $-4x + 2y = 0 \Rightarrow 4x = 2y \Rightarrow x = \dfrac{1}{2}y \Rightarrow x = \dfrac{1}{2}(20) = 10$.   The solution is $(10, 20)$.

89. Multiply the first equation by 3 and the second equation by 2. Add to eliminate the x-variable.

$$
\begin{aligned}
6x + 9y &= 21 \\
-6x + 4y &= -8 \\
\hline
13y &= 13
\end{aligned}
\Rightarrow y = 1
$$

Then, $2x + 3y = 7 \Rightarrow 2x + 3 = 7 \Rightarrow 2x = 4 \Rightarrow x = 2$. The solution is $(2, 1)$.

91. Multiply the first equation by 3, and the second equation by 5. Add to elimate the y-variable.

$$
\begin{aligned}
21x - 15y &= -45 \\
-10x + 15y &= -10 \\
\hline
11x &= -55
\end{aligned}
\Rightarrow x = -5
$$

Then, $-10x + 15y = -10 \Rightarrow 50 + 15y = -10 \Rightarrow 15y = -60 \Rightarrow y = -4$. The solution is $(-5, -4)$.

93. Add the two equations:

$$
\begin{aligned}
x^2 + y &= 12 \\
x^2 - y &= 6 \\
\hline
2x^2 &= 18 \Rightarrow x^2 = 9 \Rightarrow x = \pm 3 .
\end{aligned}
$$

If $x = 3$, then $3^2 + y = 12 \Rightarrow y = 3$, and if $x = -3$ then $(-3)^2 + y = 12 \Rightarrow y = 3$. Therefore the solutions are: $(3, 3)$ and $(-3, 3)$.

95. Subtract the two equations:

$$
\begin{aligned}
x^2 + y^2 &= 25 \\
x^2 + 7y &= 37 \\
\hline
y^2 - 7y &= -12 \Rightarrow y^2 - 7y + 12 = 0 \Rightarrow (y - 3)(y - 4) = 0 \Rightarrow y = 3, 4.
\end{aligned}
$$

If $y = 3$, then $x^2 + 7(3) = 37x^2 = 16 \Rightarrow x = \pm 4$, and if $y = 4$ then $x^2 + 7(4) = 37 \Rightarrow x^2 = 9 \Rightarrow x = \pm 3$. Therefore the solutions are: $(4, 3)$, $(-4, 3)$, $(3, 4)$, and $(-3, 4)$.

97. Subtract the two equations:

$$\begin{array}{rcl} x^2 + y^2 &=& 4 \\ 2x^2 + y^2 &=& 8 \\ \hline -x^2 &=& -4 \end{array} \Rightarrow x^2 = 4 \Rightarrow x = \pm 2.$$

If $x = -2$, then $(-2)^2 + y^2 = 4 \Rightarrow 4 + y^2 = 4 \Rightarrow y^2 = 0 \Rightarrow y = 0$,

and if $x = 2$ then $(2)^2 + y^2 = 4 \Rightarrow y^2 = 0 \Rightarrow y = 0$. Therefore the solutions are: $(-2, 0)$ and $(2, 0)$.

99. $x^2 + y^2 = 16 \Rightarrow y = \pm\sqrt{16 - x^2}$ and $x - y = 0 \Rightarrow y = x$

   Graph $Y_1 = \sqrt{(16 - X^2)}$, $Y_2 = -\sqrt{(16 - X^2)}$, and $Y_3 = X$. Their graphs intersect near the points

   $(-2.828, -2.828)$ and $(2.828, 2.828)$. See Figures 99a & 99b.

   Substituting $y = x$ into the first equation gives $x^2 + x^2 = 16 \Rightarrow 2x^2 = 16 \Rightarrow x^2 = 8 \Rightarrow x = \pm\sqrt{8}$

   Since $y = x$, the solutions are $(-\sqrt{8}, -\sqrt{8})$ and $(\sqrt{8}, \sqrt{8})$

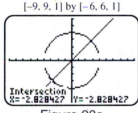
$[-9, 9, 1]$ by $[-6, 6, 1]$

Figure 99a

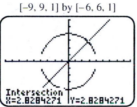
$[-9, 9, 1]$ by $[-6, 6, 1]$

Figure 99b

101. $xy = 12 \Rightarrow y = \dfrac{12}{x}$ and $x - y = 4 \Rightarrow y = x - 4$. Graph $Y_1 = 12/X$, $Y_2 = X - 4$.

   Their graphs intersect near the points $(6, 2)$ and $(-2, -6)$. See Figures 101a & 101b.

   Substituting $y = x - 4$ into the first equation gives $x(x - 4) = 12 \Rightarrow x^2 - 4x - 12 = 0 \Rightarrow$

   $(x + 2)(x - 6) = 0 \Rightarrow x = -2$ or 6. Since $y = x - 4$, the solutions are $(-2, -6)$ and $(6, 2)$.

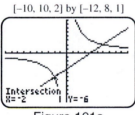
$[-10, 10, 2]$ by $[-12, 8, 1]$

Figure 101a

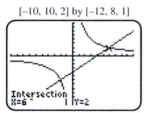
$[-10, 10, 2]$ by $[-12, 8, 1]$

Figure 101b

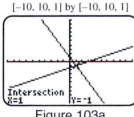
$[-10, 10, 1]$ by $[-10, 10, 1]$

Figure 103a

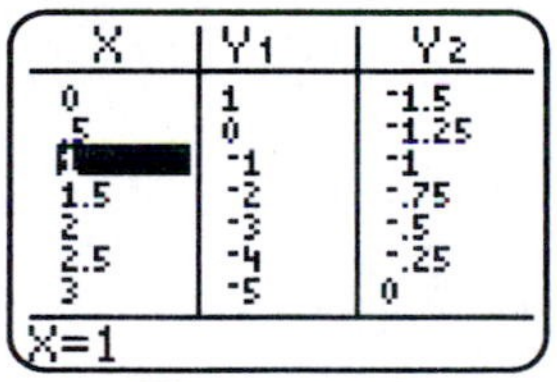

Figure 103b

103. (a) $2x + y = 1 \Rightarrow y = 1 - 2x$ and $x - 2y = 3 \Rightarrow y = \dfrac{1}{2}(x - 3)$

   Graph $Y_1 = 1 - 2X$ and $Y_2 = 0.5(X - 3)$. Their graphs intersect at the point $(1, -1)$, which is the

   solution. See Figure 103a.

(b) Table $Y_1 = 1 - 2X$ and $Y_2 = 0.5(X - 3)$ starting at 0 and incrementing by 0.5. See Figure 103b.

Here $Y_1 = Y_2 = -1$ when $x = 1$. The solution is $(1, -1)$.

(c) Substituting $y = 1 - 2x$ into the equation $x - 2y = 3$ gives $x - 2(1 - 2x) = 3 \Rightarrow$

$x - 2 + 4x = 3 \Rightarrow 5x = 5 \Rightarrow x = 1$. If $x = 1$, then $y = 1 - 2(1) = -1$. The solution is $(1, -1)$.

105. (a) $-2x + y = 0 \Rightarrow y = 2x$ and $7x - 2y = 3 \Rightarrow 7x - 3 = 2y \Rightarrow y = \dfrac{7x - 3}{2}$

Graph $Y_1 = 2X$ and $Y_2 = (7X - 3)/2$. Their graphs intersect at the point $(1, 2)$, which is the

solution. See Figure 105a.

(b) Table $Y_1 = 2X$ and $Y_2 = (7X - 3)/2$ starting at 0 and incrementing by 0.5. See Figure 105b.

Here $Y_1 = Y_2 = 2$ when $x = 1$. The solution is $(1, 2)$.

(c) Substituting $y = 2x$ into the second equation gives $7x - 2(2x) = 3 \Rightarrow 3x = 3 \Rightarrow x = 1$.

If $x = 1$, then $y = 2(1) = 2$. The solution is $(1, 2)$.

[−10, 10, 1] by [−10, 10, 1]

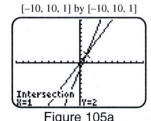

Figure 105a   Figure 105b

107. $x^3 - 3x + y = 1 \Rightarrow y = 1 + 3x - x^3$ and $x^2 + 2y = 3 \Rightarrow y = \dfrac{3 - x^2}{2}$. Graph $Y_1 = 1 + 3X - X^3$

and $Y_2 = (3 - X^2)/2$. See Figure 107. There are three points of intersection. The coordinates of these

points are near $(-1.588, 0.239)$, $(0.164, 1.487)$, and $(1.924, -0.351)$.

[−4, 4, 1] by [−4, 4, 1]

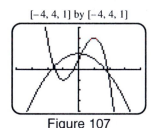

Figure 107

109. $2x^3 - x^2 = 5y \Rightarrow y = \dfrac{2x^3 - x^2}{5}$ and $2^{-x} - y = 0 \Rightarrow y = 2^{-x}$. Graph $Y_1 = (2X^3 - X^2)/5$

and $Y_2 = 2^{\wedge}(-X)$. See Figure 109. There is one point of intersection. The coordinates of this point are near

$(1.220, 0.429)$.

[-5, 5, 1] by [-3, 3, 1]

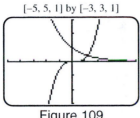

Figure 109

[-3, 3, 1] by [-5, 5, 1]

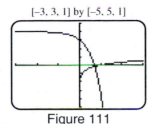

Figure 111

111. $e^{2x} + y = 4 \Rightarrow y = 4 - e^{2x}$ and $\ln x - 2y = 0 \Rightarrow y = \dfrac{\ln x}{2}$. Graph $Y_1 = 4 - e^{\wedge}(2X)$

and $Y_2 = \ln (X)/2$. See Figure 111. There is one point of intersection. The coordinates of this point are

near $(0.714, -0.169)$.

113. (a) Let $x =$ population of Minneapolis and $y =$ population of St. Paul. Then $x + y = 670$ and $x - y = 96$.

(b) $\begin{array}{r} x + y = 670 \\ x - y = 98 \\ \hline 2x = 768 \end{array} \Rightarrow x = 384.$

Then, $x - y = 78 \Rightarrow 384 - y = 98 \Rightarrow y = 286$. The solution is $(384, 286)$.

(c) The system is consistent and independent.

115. $W_1 + \sqrt{2}W_2 = 300 \Rightarrow W_2 = \dfrac{300 - W_1}{\sqrt{2}}$ and $\sqrt{3}W_1 - \sqrt{2}W_2 = 0 \Rightarrow W_2 = \dfrac{\sqrt{3}W_1}{\sqrt{2}}$

Graph $Y_1 = (300 - X)/\sqrt{(2)}$ and $Y_2 = \sqrt{(3)}X/\sqrt{(2)}$. Their graphs intersect near the point

$(109.81, 134.49)$ as shown in Figure 115a. The forces on the rafters are approximately 110 and 134 pounds.

To find the solution numerically, table Y_1 and Y_2 starting at 107 and incrementing by 1. We find that

$Y_1 = Y_2 \approx 134$ when $x \approx 110$. See Figure 115b.

We can find the solution symbolically by using the substitution method. Since $W_1 + \sqrt{2}W_2 = 300 \Rightarrow$

$W_1 = 300 - \sqrt{2}W_2$, we will substitute into the other equation. $\sqrt{3}(300 - \sqrt{2}W_2) - \sqrt{2}W_2 = 0 \Rightarrow$

$300\sqrt{3} - \sqrt{6}W_2 - \sqrt{2}W_2 = 0 \Rightarrow 300\sqrt{3} = (\sqrt{6} + \sqrt{2})W_2 \Rightarrow W_2 = \dfrac{300\sqrt{3}}{\sqrt{6} + \sqrt{2}}$ and

$W_1 = 300 - \sqrt{2}\left[\dfrac{300\sqrt{3}}{\sqrt{6} + \sqrt{2}}\right] \Rightarrow W_1 = 300 - \dfrac{300\sqrt{3}}{1 + \sqrt{3}} \Rightarrow W_1 = \dfrac{300}{1 + \sqrt{3}}$

$W_1 = \dfrac{300}{1 + \sqrt{3}} \approx 109.8$ lbs, $W_2 = \dfrac{300\sqrt{3}}{\sqrt{6} + \sqrt{2}} \approx 134.5$ lbs

[0, 200, 50] by [0, 200, 50]

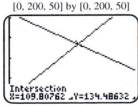

Figure 115a

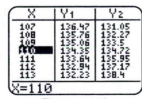

Figure 115b

[0, 4, 1] by [0, 20, 1]

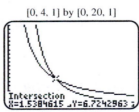

Figure 117

117. We must solve the system of nonlinear equations: $\pi r^2 h = 50$ and $2\pi r h = 65$.

Solving each equation for h results in the following: $\pi r^2 h = 50 \Rightarrow h = \dfrac{50}{\pi r^2}$ and $2\pi r h = 65 \Rightarrow h = \dfrac{65}{2\pi r}$.

Graph $Y_1 = 50/(\pi X^2)$ and $Y_2 = 65/(2\pi X)$ Their graphs intersect near $(1.538, 6.724)$. See Figure 117.

A cylinder with approximate measurements of $r \approx 1.538$ inches and $h \approx 6.724$ inches has a volume of 50

cubic inches and a lateral surface area of 65 square inches.

119. Let x = the length of each side of the base and let y = the height of the box. Since the volume = 576 in^3,

$x^2 y = 576$, and so $y = \dfrac{576}{x^2}$. Since the surface are is 336 in^2, $x^2 + 4xy = 336$. Substituting for y in this

equation yields $x^2 + 4x \cdot \dfrac{576}{x^2} = 336$. Simplifying we get: $x^3 - 336x + 2304 = 0$. By graphing the left side

of the equation, the x-intercepts are approximately 9.1 and 12. See Figure 119. When $x \approx 9.1$, the

dimensions are: 9.1 by 9.1 by $\dfrac{576}{(9.1)^2} \approx 6.96$ inches. When

$x = 12$, the dimensions are: 12 by 12 by $\dfrac{576}{(12)^2} = 4$ inches.

[0, 20, 2] by [−500, 1000, 100]   [0, 800,000, 100,000] by [0, 800,000, 100,000]

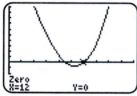
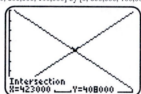

Figure 119   Figure 121

121. (a) Let x represent the number of bank thefts in 2009 and let y represent the number of bank thefts in 2010.

Then $x + y = 11,693$ and $x - y = 437$.

(b) $x + y = 11,693$

$x - y = 437$

$2x \phantom{{}={}} = 12,130 \Rightarrow x = 6065.$

Then, $x - y = 437 \Rightarrow 6065 - y = 437 \Rightarrow y = 5628$. The solution is $(6065, 5628)$.

(c) Solve each equation for y and graph $Y_1 = 11,693 - X$ and $Y_2 = X - 437$ as shown in Figure 121.

The solution is the intersection point $(6065, 5628)$.

123. (a) To solve this problem start by letting x represent the amount of the 8% loan and y the 10% loan. Since the

total of both loans is \$3000, the equation $x + y = 3000$ must be satisfied. The annual interest rate for the

8% loan is given by $0.08x$, while the annual interest rate for the 10% loan is expressed by $0.10y$. the total

interest for both loans is their sum $0.08x + 0.10y = 264$.

(b) Thus, to determine a solution, the following linear system of equations could be solved.

$x + y = 3000$ and $0.08x + 0.10y = 264$. Start by solving each for y.

$x + y = 3000 \Rightarrow y = 3000 - x$

and $0.08x + 0.10y = 264 \Rightarrow 0.10y = 264 - 0.08x \Rightarrow y = \dfrac{264 - 0.08x}{0.10} \Rightarrow y = 2640 - 0.8x$. The two equations can be solved using the intersection-of-graphs method. Let $Y_1 = 3000 - X$ and $Y_2 = 2640 - 0.8X$. Since the system of equations is linear, each graph is a line. The lines are not parallel and intersect at the point $(1800, 1200)$, as shown in Figure 123. Thus, the 8% loan is for $\$1800$ and the 10% loan is for $\$1200$.

[0, 3000, 1000] by [0, 3000, 1000] [0, 3000, 1000] by [0, 3000, 1000]

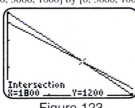

Figure 123   Figure 125

125. With these conditions the system of equations becomes $x + y = 3000$ and $0.10x + 0.10y = 264$. Solving each equation for y provides the following results, $y = 3000 - x$ and $y = 2640 - x$. Graphs of $Y_1 = 3000 - X$ and $Y_2 = 2640 - X$ are shown in Figure 125. Notice that their graphs are parallel lines with slope -1 that do not intersect. There is no solution. This means that there is no way to have two loans totaling $\$3000$, both with an interest rate of 10%, and only pay $\$264$ in interest each year. The interest must be 10% of $\$3000$ or $\$300$. This system of equations is inconsistent - there is no solution. Graphs of inconsistent systems in two variables consist of parallel lines.

127. Let x represent the air speed of the plane and y the wind speed. Traveling with the wind, the average ground speed of the plane is $\dfrac{1680}{3} = 560$ mph, while its ground speed against the wind was $\dfrac{1680}{3.5} = 480$ mph. Thus,

$$\begin{array}{r} x + y = 560 \\ x - y = 480 \\ \hline 2x = 1040 \end{array} \Rightarrow x = 520$$

Thus, $y = 560 - x = 40$. The air speed of the plane is 520 mph and the wind speed is 40 mph.

129. (a) The perimeter is 40, thus $2l + 2w = 40 \Rightarrow l + w = 20 \Rightarrow l = 20 - w$. Since the area is 91, we have that $lw = 91$. Use the substitution method to solve this system.

$(20 - w)w = 91 \Rightarrow 20w - w^2 = 91 \Rightarrow w^2 - 20w + 91 = 0 \Rightarrow (w - 13)(w - 7) = 0 \Rightarrow$

$w = 7$ or $w = 13$. When $w = 7$, $l = 13$ and when $w = 13$, $l = 7$. Since length is longer than width, the solution is $l = 13$ feet and $w = 7$ feet.

(b) $P = 2l + 2w = 40 \Rightarrow l + w = 20 \Rightarrow l = 20 - w$. Then, $A = lw = (20 - w)w = 20w - w^2$. Graph $Y_1 = 20X - X^2$ in $[0, 25, 5]$ by $[0, 150, 25]$.

(c) The area can be any positive number less than or equal to 100 square feet. The maximum area of 100 square feet occurs when $w = 10$ and $l = 10$. See Figure 129. Since all sides are equal to 10, the shape of the rectangle is a square. That is, a square pen will provide the largest area.

[0, 25, 5] by [0, 150, 25]

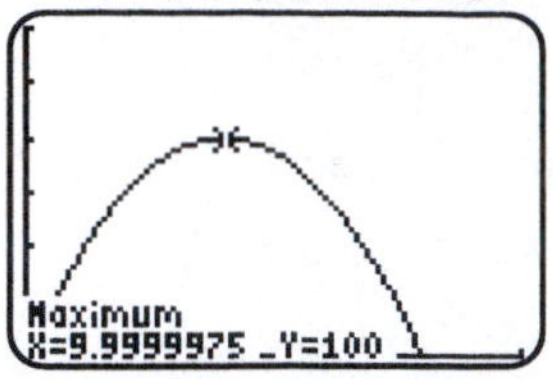

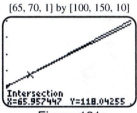

Figure 129

131. (a) A 6' 11" person is 83 inches tall.

$$w = 7.46(83) - 374 = 245.18 \approx 245 \text{ lbs}; \quad w = 7.93(83) - 405 = 253.19 \approx 253 \text{ lbs}$$

(b) Graph $Y_1 = 7.46X - 374$ and $Y_2 = 7.93X - 405$. Their graphs intersect near $(65.96, 118.04)$.

See Figure 131. The models agree when $h \approx 65.96$ inches and $w \approx 118$ pounds.

(c) The first model's coefficient for h is 7.46. Thus, for each increase in height of 1 inch, the weight increases

by 7.46 lbs. Similarly, for the second equation the increase is 7.93 lbs.

[65, 70, 1] by [100, 150, 10]

Figure 131

133. Substitute 165.1 for h and 70 for w: $S(70, 165.1) = 0.007184(70)^{0.425}(165.1)^{0.725} \approx 1.77 \text{m}^2$

135. $w = 132 \text{ lb} \approx \dfrac{132}{2.2} \text{ kg} = 60 \text{ kg}; \quad h = 62 \text{ inches} \approx 62 \cdot 2.54 \text{ cm} = 157.48 \text{ cm}$

$$S(w, h) = 0.007184(w^{0.425})(h^{0.725}) \implies S(60, 157.48) = 0.007184(60^{0.425})(157.48^{0.725}) \approx 1.6 \text{ m}^2$$

137. Since $z = kx^2y^3$ and $z = 31.9$ when $x = 2$ and $y = 2.5$.

$$31.9 = k(2)^2(2.5)^3 \implies 31.9 = 62.5k \implies k = \frac{31.9}{62.5} \approx 0.51$$

139. $z = k\sqrt{x} \cdot \sqrt[3]{y} \implies 10.8 = k\sqrt{4} \cdot \sqrt[3]{8} \implies 10.8 = k \cdot 2 \cdot 2 \implies 10.8 = 4k \implies k = 2.7$

Therefore: $z = 2.7\sqrt{x} \cdot \sqrt[3]{y}$ and now $z = 2.7\sqrt{16} \cdot \sqrt[3]{27} \implies z = 2.7(4)(3) \implies z = 32.4$.

141. Let d represent the diameter of the blades, v represent the wind velocity, and w represent the watts of power

generated by the windmill. Then $w = kd^2v^3$ and $w = 2405$ when $d = 8$ and $v = 10$.

$$2405 = k(8)^2(10)^3 \implies 2405 = 64{,}000k \implies k = \frac{481}{12{,}800}. \text{ The variation equation becomes}$$

$w = \dfrac{481}{12{,}800}d^2v^3$. Thus, when $d = 6$ and $v = 20$; $w = \dfrac{481}{12{,}800}(6)^2(20)^3 = 10{,}822.5$.

With six-foot blades and a 20 mile-per-hour wind, the windmill will produce about 10,823 watts.

143. From the example $V = 0.00132h^{1.12}d^{1.98}$. When $h = 105$ and $d = 38$ we have

$V = 0.00132(105)^{1.12}(38)^{1.98} \approx 325.295$A tree which is 105 feet tall with a diameter of 38 inches contains

approximately 325.295 cubic feet of wood. To find the number of cords, divide this result by 128:

$\dfrac{325.295}{128} \approx 2.54$ cords.

145. From exercise 133, $S(w, h) = 0.007184(w^{0.425})(h^{0.725})$ where w is weight in kilograms and h is height in centimeters. Let $S = 1.77$, $w = 154$, and $h = 65$. Then, $1.77 = k(154^{0.425})(65^{0.725})$, which results in $k \approx 0.0101$. Thus, $S(w, h) = 0.0101(w^{0.425})(h^{0.725})$, where w is in pounds, h is in inches, and S is in square meters.

9.2: Systems of Inequalities in Two Variables

1. Graph the boundary line $y = x$. Choose $(1, 0)$ as a test point. Since $1 \geq 0$, the region containing $(1, 0)$ is part of the solution set. The inequality $x \geq y$ includes this region and the line $y = x$. See Figure 1.

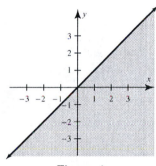

Figure 1

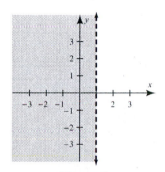

Figure 3

3. Graph the boundary line $x = 1$. Choose $(0, 0)$ as a test point. Since $0 < 1$, the region containing $(0, 0)$ is the solution set. The boundary line $x = 1$ is not part of the solution set. See Figure 3.

5. Graph the boundary line $x + y = 2 \Rightarrow y = 2 - x$. Choose $(0, 0)$ as a test point. Since $0 + 0 \leq 2$, the region containing $(0, 0)$ is part of the solution set. The boundary line $y = 2 - x$ is also part of the solution set. See Figure 5.

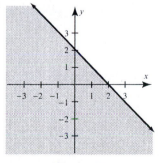

Figure 5

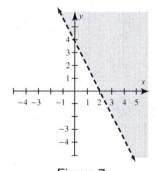

Figure 7

7. Graph the boundary line $2x + y = 4 \Rightarrow y = 4 - 2x$. Choose $(3, 0)$ as a test point. Since $0 > 4 - 2(3)$, the region containing $(3, 0)$ is the solution set. The boundary line $y = 4 - 2x$ is not part of the solution set. See Figure 7.

9. Graph the circle determined by $x^2 + y^2 = 4$. Choose $(0, 0)$ as a test point. Since $0 > 4$, the region inside the circle is not part of the solution $\Rightarrow$ shade the region outside the circle. Note, the circle is not part of the solution. See Figure 9.

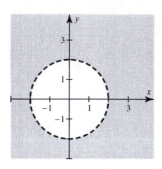

Figure 9

11. Graph the boundary parabola determined by $x^2 + y = 2 \Rightarrow y = -x^2 + 2$. Choose $(0, 0)$ as a test point. Since $0 \le 2$, the region inside the parabola is part of the solution $\Rightarrow$ shade the region inside of the parabola. Note, the parabola is part of the solution. See Figure 11.

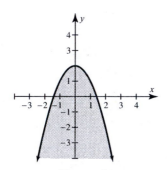

Figure 11

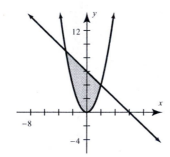

Figure 17

13. $x + y \ge 2 \Rightarrow y \ge 2 - x$ is above the line $y = 2 - x$. It includes the line.

$x - y \le 1 \Rightarrow y \ge x - 1$ is above the line $y = x - 1$. It includes the line.

The solution is above two lines, which matches Figure c. One solution is $(2, 3)$. *Answers may vary.*

15. $\dfrac{1}{2}x^3 - y > 0 \Rightarrow y < \dfrac{1}{2}x^3$ is below the curve $y = \dfrac{1}{2}x^3$. It does not include the curve.

$2x - y \le 1 \Rightarrow y \ge 2x - 1$ is above the line $y = 2x - 1$. It includes the line.

The solution is below a dotted curve and above a solid line, which matches Figure d. One solution is $(-1, -1)$.

Answers may vary.

17. The solution region is above the parabola $y = x^2$ and below the line $y = 6 - x$. It includes the boundary. See Figure 17. One solution is $(0, 2)$. *Answers may vary.*

19. The solution region lies between the parallel lines $y = -\dfrac{1}{2}x - 1$ and $y = -\dfrac{1}{2}x + \dfrac{5}{2}$. It does not include the

    boundary. See Figure 19. One solution is $(0, 0)$. *Answers may vary.*

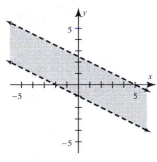

Figure 19

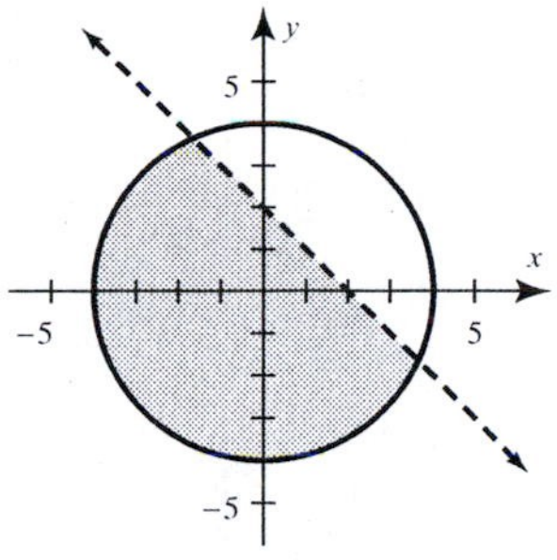

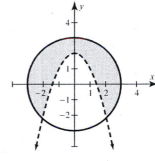

Figure 21

Figure 23

21. The solution region lies inside the circle centered at the origin with radius 4 and below the line $y = -x + 2$. It

    does not include the boundary determined by the line. See Figure 21. One solution is $(-1, 1)$.

    Answers may vary.

23. The solution region lies inside the circle centered at the origin with radius 3 and above the parabola

    $y = 2 - x^2$. It does not include the parabola as a boundary line. See Figure 23. One solution is $(2, 1)$. **Answers**

    may vary.

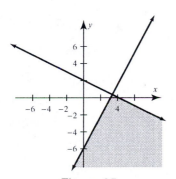

Figure 25

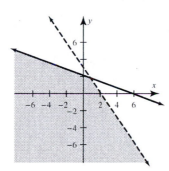

Figure 27

25. $x + 2y \le 4 \Rightarrow y \le -\dfrac{1}{2}x + 2$ and $2x - y \ge 6 \Rightarrow y \le 2x - 6$.   Graph the boundary lines

    $y = -\dfrac{1}{2}x + 2$ and $y = 2x - 6$. The region satisfying the system is below the line $x + 2y = 4$ and below the

    line $2x - y = 6$. Because equality is included, the boundaries are part of the region. See Figure 25.

27. $3x + 2y < 6 \Rightarrow y < -\dfrac{3}{2}x + 3$ and $x + 3y \le 6 \Rightarrow y \le -\dfrac{1}{3}x + 2$.   Graph the boundary lines

    $y = -\dfrac{3}{2}x + 3$ and $y = -\dfrac{1}{3}x + 2$. The region satisfying the system is below the line $3x + 2y = 6$, not

    including the boundary and below the line $x + 3y = 6$, including the boundary. See Figure 27.

29. $x - 2y \geq 0 \Rightarrow y \leq \frac{1}{2}x$ and $x - 3y \leq 3 \Rightarrow y \geq \frac{1}{3}x - 1$. Graph the boundary lines $y = \frac{1}{2}x$ and

$y = \frac{1}{3}x - 1$. The region satisfying the system is below the line $y = \frac{1}{2}x$ and above the line $y = \frac{1}{3}x - 1$.

Because equality is included, the boundaries are part of the region. See Figure 29.

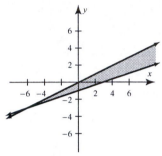

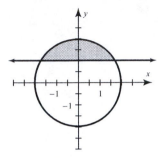

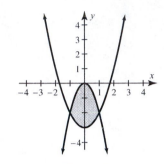

Figure 29   Figure 31   Figure 33

31. $x^2 + y^2 \leq 4$ and $y \geq 1$. Graph the boundary circle $x^2 + y^2 = 4$ and the boundary line $y = 1$. The region

satisfying the system is inside the circle $x^2 + y^2 = 4$ and above the line $y = 1$. Because equality is included,

the boundaries are part of the region. See Figure 31.

33. $2x^2 + y \leq 0 \Rightarrow y \leq -2x^2$ and $x^2 - y \leq 3 \Rightarrow y \geq x^2 - 3$.   Graph the boundary parabolas $y = -2x^2$ and

$y = x^2 - 3$. Because equality is included, the boundaries are part of the region. See Figure 33.

35. $x^2 + 2y \leq 2 \Rightarrow y \leq 1 - \frac{x^2}{2}$ and $x^2 + y^2 \leq 4$.   Graph the boundary parabola $y = 1 - \frac{x^2}{2}$ and the boundary

circle $x^2 + y^2 = 4$. Because equality is included, the boundaries are part of the region. See Figure 35.

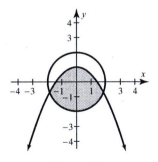

Figure 35

37. The person will probably experience jitters.

39. The selected person could weigh 180 pounds or more.

41. The total number of vehicles entering intersection A is $500 + 150 = 650$ vehicles per hour. The expression

$x + y$ represents the number of vehicles leaving intersection A each hour. Therefore, we have $x + y = 650$.

The total number of vehicles leaving intersection B is $50 + 400 = 450$. There are 100 vehicles entering

intersection B from the south and y vehicles entering intersection B from the west. Thus, $y + 100 = 450$. We must solve the system;

$$\begin{array}{rcl} x + y & = & 650 \\ y + 100 & = & 450 \\ \hline x \quad\quad - 100 & = & 200 \end{array} \implies x = 300$$

Thus, $y = 350$ and $x = 300$. At intersection A, a stoplight should allow for 300 vehicles per hour to travel south and 350 vehicles per hour to continue traveling east.

43. This region corresponds to weights that are less and heights that are greater than recommended. This individual has weight that is less than recommended for his or her height.

45. The upper left boundary is given by $25h - 7w = 800$ or $h = \dfrac{7w + 800}{25}$. The region is below and includes this line, which is described by $h \leq \dfrac{7w + 800}{25}$ or $25h - 7w \leq 800$. The lower right boundary of this region is given by $5h - w = 170$. The region is above and includes this line, which is described by $5h - w \geq 170$. Thus, the region can be described by the system of inequalities: $25h - 7w \leq 800,\ 5h - w \geq 170$.

47. See Figure 47.

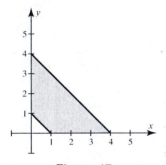

Figure 47

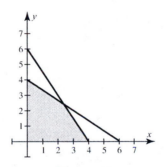

Figure 49

49. See Figure 49.

51. To find the maximum and minimum values of P in the region, we must evaluate P at each of the vertices. These values are shown in the table. See Figure 51. From the table the maximum is 65 and the minimum is 8.

Vertex	$P = 3x + 5y$
(1, 1)	$3(1) + 5(1) = 8$
(6, 3)	$3(6) + 5(3) = 33$
(5, 10)	$3(5) + 5(10) = 65$
(2, 7)	$3(2) + 5(7) = 41$

Figure 51

Vertex	$C = 3x + 5y$
(1, 0)	$3(1) + 5(0) = 3$
(7, 6)	$3(7) + 5(6) = 51$
(7, 9)	$3(7) + 5(9) = 66$
(1, 10)	$3(1) + 5(10) = 53$

Figure 53

53. To find the maximum and minimum values of C in the region, we must evaluate C at each of the vertices. These values are shown in the table. See Figure 53. From the table the maximum is 66 and the minimum is 3.

55. To find the maximum and minimum values of C in the region, we must evaluate C at each of the vertices.

These values are shown in the table. See Figure 55. From the table the maximum is 100 and the minimum is 0.

Vertex	$C = 10y$
$(1, 0)$	$10(0) = 0$
$(7, 6)$	$10(6) = 60$
$(7, 9)$	$10(9) = 90$
$(1, 10)$	$10(10) = 100$

Figure 55

57. The line that goes through the points $(0, 4)$ and $(4, 0)$ has the equation $x + y = 4$. The shaded region is also

bounded by the line $x = 0$ and the line $y = 0$. Thus, the shaded region is described by the system:

$x + y \leq 4$, $x \geq 0$, and $y \geq 0$.

59. The region of feasible solutions is shown in Figure 59a. The vertices of this region are $(3, 0)$, $(6, 0)$, $(0, 4)$, and

$(0, 3)$. To find the minimum value of C in the region, we must evaluate C at each of the vertices. These values

are shown in Figure 59b. From the table the minimum is 6 at the point $(0, 3)$.

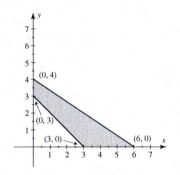

Figure 59a

Vertex	$C = 4x + 2y$
$(3, 0)$	$4(3) + 2(0) = 12$
$(6, 0)$	$4(6) + 2(0) = 24$
$(0, 4)$	$4(0) + 2(4) = 8$
$(0, 3)$	$4(0) + 2(3) = 6$

Figure 59b

61a. The region of the feasible solutions is shown in Figure 61a. The vertices of this region are $(0, 4)$, $(0, 8)$, $(4, 0)$,

and $(8, 0)$. To find the maximum and minimum values of $z = 7x + 6y$ in the region, we must evaluate z at

each of the vertices. These values are shown in Figure 61b. From the table the maximum is 56 and the

minimum is 24.

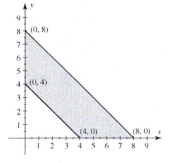

Figure 61a

Vertex	$z = 7x + 6y$
$(0, 4)$	$7(0) + 6(4) = 24$
$(0, 8)$	$7(0) + 6(8) = 48$
$(4, 0)$	$7(4) + 6(0) = 28$
$(8, 0)$	$7(8) + 6(0) = 56$

Figure 61b

63. The new objective equation is $P = 20x + 15y$. To find the maximum profit we must evaluate P at each of the vertices. These values are shown in Figure 63. From the table the maximum is 950 at the vertex $(25, 30)$. The maximum profit will be $\$950$ when 25 radios and 30 CD players are manufactured.

Vertex	$P = 20x + 15y$
$(5, 5)$	$20(5) + 15(5) = 175$
$(25, 25)$	$20(25) + 15(25) = 875$
$(25, 30)$	$20(25) + 15(30) = 950$
$(5, 30)$	$20(5) + 15(30) = 550$

Figure 63

65. Make a table to list the information given. See Figure 65a. Using the table, we can write the linear programming problem as follows;

Cost: $C = 80x + 50y$, Protein: $15x + 20y \geq 60$, Fat: $10x + 5y \geq 30$, $x \geq 0$, and $y \geq 0$.

Brand	Units	Protein	Fat	Cost
A	x	15	10	80¢
B	y	20	5	50¢
Minimum		60	30	

Figure 65a

The region of feasible solutions is shown in Figure 65b. The vertices of this region are $(0, 6)$, $(2.4, 1.2)$, and $(4, 0)$. To find the minimum value of C in the region, we must evaluate C at each of the vertices. These values are shown in Figure 65c. The minimum cost occurs when 2.4 units of Brand A and 1.2 units of Brand B are mixed, to give a cost of $\$2.52$ per serving.

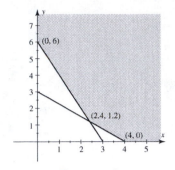

Figure 65b

Vertex	$C = 80x + 50y$
$(0, 6)$	$80(0) + 50(6) = 300$
$(2.4, 1.2)$	$80(2.4) + 50(1.2) = 252$
$(4, 0)$	$80(4) + 50(0) = 320$

Figure 65c

67. Let x and y represent the number of hamsters and mice respectively. Since the total number of animals cannot exceed 50, $x + y \leq 50$. Because no more than 20 hamsters can be raised, $x \leq 20$. Here the revenue function is $R = 15x + 10y$. From the graph of the region of feasible solutions (not shown), the vertices are $(0, 0)$, $(0, 50)$, $(20, 30)$, and $(20, 0)$. To find $(20, 30)$ solve the equations $x + y = 50$ and $x = 20$. The maximum value of R occurs at one of the vertices. For $(0, 50)$, $R = 15(0) + 10(50) = 500$.

For $(20, 30)$, $R = 15(20) + 10(30) = 600$. For $(20, 0)$, $R = 15(20) + 10(0) = 300$.

The maximum revenue is $600.

69. The number of hours on machine A needed to manufacture x units of part X is $4x$ while the number of hours on machine A needed to manufacture y units of part Y is $1y$. Since machine A is only available for 40 hours each week we have the constraint $4x + y \leq 40$. Similarly, the number of hours on machine B needed to manufacture x units of part X is $2x$ while the number of hours on machine B to make y units of part Y is $3y$. Since machine B is only available for 30 hours each week we have the constraint $2x + 3y \leq 30$. It should be noted that the number of parts of each type cannot be negative. This gives the constraint $x \geq 0$ and $y \geq 0$. The profit earned on x units of part X is $500x$ while the profit earned on y units of part Y is $600y$. Thus, the total weekly profit is $P = 500x + 600y$. This is our objective function. Graph the constraints and shade the region. The region of feasible solutions is shown in Figure 69a. The vertices of this region are $(0, 0)$, $(10, 0)$, $(9, 4)$, and $(0, 10)$. To find the maximum value of P in the region, we must evaluate P at each of the vertices. These values are shown in Figure 69b. From the table the maximum is 6900 at the point $(9, 4)$. The maximum profit is $6900 when there are 9 parts of type X and 4 parts of type Y manufactured.

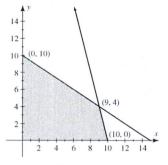

Vertex	$P = 500x + 600y$
$(0, 0)$	$500(0) + 600(0) = 0$
$(10, 0)$	$500(10) + 600(0) = 5000$
$(9, 4)$	$500(9) + 600(4) = 6900$
$(0, 10)$	$500(0) + 600(10) = 6000$

Figure 69a   Figure 69b

Checking Basic Concepts for Sections 9.1 and 9.2

1. $d(13, 18) = \sqrt{(13 - 1)^2 + (18 - 2)^2} = \sqrt{12^2 + 16^2} = \sqrt{400} = 20$

3. $z = x^2 + y^2 \Rightarrow y^2 = z - x^2 \Rightarrow y = \pm\sqrt{z - x^2}$

5. Graph the boundary line $3x - 2y = 6 \Rightarrow y = \dfrac{3}{2}x - 3$. Choose $(0, 0)$ as a test point. Since $0 \le 6$,

    the region containing $(0, 0)$ is the solution set. The boundary line $y = \dfrac{3}{2}x - 3$ is part of the solution set.

    See Figure 5.

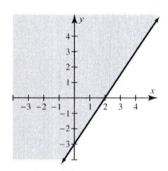

Figure 5

7. (a) Let x represent the number of Plasma TV's sold and y represent the number of HD TV's

    sold. Then, $x + y = 220$ and $9x = 2y \Rightarrow 4.5x - y = 0$.

    (b) $x + y = 220$

    $\dfrac{4.5\,x - y = \ \ 0}{5.5x \qquad = 220} \Rightarrow x = 40$

    Then, $x + y = 220 \Rightarrow 40 + y = 220 \Rightarrow y = 180$. The solution is 40 plasma televisions were sold and 180

    HD TV's were sold.

9.3 Systems of Linear Equations in Three Variables

1. No, systems of linear equations can have zero, one, or infinitely many solutions.

3. 2, the same as the number of variables.

5. A system with an infinite number of solutions means that the equations are dependent.

7. Testing $(0, 2, -2)$: $0 + 2 - (-2) = 4$, true; $-0 + 2 + (-2) = 2$, false; $0 + 2 + (-2) = 0$, true; since the second equation is false, $(0, 2, -2)$ is not a solution for the system.

    Testing $(-1, 3, -2)$: $-1 + 3 - (-2) = 4$, true; $-(-1) + 3 + (-2) = 2$, true; $-1 + 3 + (-2) = 0$, true; since all equation are true, $(-1, 3, -2)$ is a solution for the system.

9. Testing $\left(-\dfrac{5}{11}, \dfrac{20}{11}, -2\right)$: $-\dfrac{5}{11} + 3\left(\dfrac{20}{11}\right) - 2(-2) = 9 \Rightarrow -\dfrac{5}{11} + \dfrac{60}{11} + \dfrac{44}{11} = 9 \Rightarrow \dfrac{99}{11} = 9$, true;

    $-3\left(-\dfrac{5}{11}\right) + 2\left(\dfrac{20}{11}\right) + 4(-2) = -3 \Rightarrow \dfrac{15}{11} + \dfrac{40}{11} - \dfrac{88}{11} = -3 \Rightarrow -\dfrac{33}{11} = -3$, true;

    $-2\left(-\dfrac{5}{11}\right) + 5\left(\dfrac{20}{11}\right) + 2(-2) = 6 \Rightarrow \dfrac{10}{11} + \dfrac{100}{11} - \dfrac{44}{11} = 6 \Rightarrow \dfrac{66}{11} = 6$, true; since all of the equations

    are true, $\left(-\dfrac{5}{11}, \dfrac{20}{11}, -2\right)$ is a solution for the system.

    Testing $(1, 2, -1)$: $1 + 3(2) - 2(-1) = 9 \Rightarrow 1 + 6 + 2 = 9 \Rightarrow 9 = 9$, true;

    $-3(1) + 2(2) + 4(-1) = -3 \Rightarrow -3 + 4 + (-4) = -3 \Rightarrow -3 = -3$, true;

    $-2(1) + 5(2) + 2(-1) = 6 \Rightarrow (-2) + 10 + (-2) = 6 \Rightarrow 6 = 6$, true; since all equations are true,

    $(1, 2, -1)$ is a solution for the system.   Therefore they are both solutions.

11. Add The first two equations:

    $$
    \begin{array}{r}
    x + y + z = 6 \\
    -x + 2y + z = 6 \\
    \hline
    3y + 2z = 12.
    \end{array}
    $$

    Subtract 2 times the third equation from this equation:

    $$
    \begin{array}{r}
    3y + 2z = 12 \\
    2y + 2z = 10 \\
    \hline
    y \qquad = 2.
    \end{array}
    $$

    Substitute $y = 2$ into the equation: $3y + 2z = 12 \Rightarrow 3(2) + 2z = 12 \Rightarrow 6 + 2z = 12 \Rightarrow 2z = 6 \Rightarrow$

    $z = 3$. Finally, substitute $y = 2$ and $z = 3$ into one of the original equations: $x + 2 + 3 = 6 \Rightarrow x = 1$.

    The solution is $(1, 2, 3)$.

13. Multiply the first equation by 2 and subtract the second equation:

$$2x + 4y + 6z = 8$$
$$\underline{2x + y + 3z = 5}$$
$$3y + 3z = 3.$$

Subtract the third equation from the first:

$$x + 2y + 3z = 4$$
$$\underline{x - y + z = 2}$$
$$3y + 2z = 2.$$

Now subtract these two equations:

$$3y + 3z = 3$$
$$\underline{3y + 2z = 2}$$
$$z = 1.$$

Substitute $z = 1$ into $3y + 2z = 2$: $3y + 2(1) = 2 \Rightarrow 3y = 0 \Rightarrow y = 0$.

Finally substitute $y = 0$ and $z = 1$ into the original equation $x - y + z = 2$: $x - 0 + 1 = 2 \Rightarrow x = 1$.

The solution is: $(1, 0, 1)$.

15. Add the second and third equations:

$$4x + 2y + z = 1$$
$$\underline{2x - 2y - z = 2}$$
$$6x = 3 \Rightarrow x = \frac{1}{2}.$$

Subtract the first two equations:

$$3x + y + z = 0$$
$$\underline{4x + 2y + z = 1}$$
$$-x - y = -1.$$

Substitute $x = \frac{1}{2}$ into $-x - y = -1$: $-\frac{1}{2} - y = -1 \Rightarrow -y = -\frac{1}{2} \Rightarrow y = \frac{1}{2}$.

Substitute $x = \frac{1}{2}$ and $y = \frac{1}{2}$ into $3x + y + z = 0$: $3\left(\frac{1}{2}\right) + \frac{1}{2} + z = 0 \Rightarrow \frac{3}{2} + \frac{1}{2} + z = 0 \Rightarrow$

$2 + z = 0 \Rightarrow z = -2$.

The solution is: $\left(\frac{1}{2}, \frac{1}{2}, -2\right)$.

17. Subtract the third equation from the first:

$$x + 3y + z = 6$$
$$\underline{x - y - z = 0}$$
$$4y + 2z = 6.$$

Multiply the third equation by 3 and subtract it from the second equation:

$$3x + y - z = 6$$
$$\underline{3x - 3y - 3z = 0}$$
$$4y + 2z = 6.$$

Subtracting these two equations we get:

$$
\begin{aligned}
4y + 2z &= 6 \\
\underline{4y + 2z} &= \underline{6} \\
0 &= 0.
\end{aligned}
$$

Therefore infinitely many solutions and $4y + 2z = 6 \Rightarrow 4y = -2z + 6 \Rightarrow y = \dfrac{-z + 3}{2}$.

Adding the last two original equations we get:

$$
\begin{aligned}
3x + y - z &= 6 \\
\underline{x - y - z} &= \underline{0} \\
4x \qquad - 2z &= 6 \Rightarrow 4x = 2z + 6 \Rightarrow x = \dfrac{z + 3}{2}.
\end{aligned}
$$

We have infinitely many solutions: $\left(\dfrac{z + 3}{2}, \dfrac{-z + 3}{2}, z \right)$.

19. Add the first two equations:

$$
\begin{aligned}
x - 4y + 2z &= -2 \\
\underline{x + 2y - 2z} &= \underline{-3} \\
2x - 2y \qquad &= -5.
\end{aligned}
$$

Multiply the third equation by 2 and subtract from $2x - 2y = -5$:

$$
\begin{aligned}
2x - 2y &= -5 \\
\underline{2x - 2y} &= \underline{\;\;8} \\
0 &= -13.
\end{aligned}
$$

Therefore we have no solutions.

21. Add the last two equations:

$$
\begin{aligned}
2a + b - c &= -11 \\
\underline{2a - 2b + c} &= \underline{\;\;3} \\
4a - b \qquad &= -8.
\end{aligned}
$$

Multiply the second equation by 2 and add to the first equation:

$$
\begin{aligned}
4a - b + 2c &= \;\;\;0 \\
\underline{4a + 2b - 2c} &= \underline{-22} \\
8a + b \qquad &= -22.
\end{aligned}
$$

Now add the equation $4a - b = -8$ to $8a + b = -8$:

$$
\begin{aligned}
4a - b &= \;\;-8 \\
\underline{8a + b} &= \underline{-22} \\
12a \qquad &= -30 \Rightarrow a = \dfrac{-30}{12} \Rightarrow a = \dfrac{-5}{2}.
\end{aligned}
$$

Substitute $a = \dfrac{-5}{2}$: $8\left(\dfrac{-5}{2}\right) + b = -22 \Rightarrow -20 + b = -22 \Rightarrow b = -2$.

Substitute $a = \dfrac{-5}{2}$ and $b = -2$ into $4\left(\dfrac{-5}{2}\right) - (-2) + 2c = 0 \Rightarrow -10 + 2 + 2c = 0 \Rightarrow 2c = 8 \Rightarrow$

$c = 4$. The solution is: $\left(\dfrac{-5}{2}, -2, 4 \right)$.

23. Subtract the first and third equations:

$$\begin{array}{r} a + b + c = 0 \\ \underline{a + 3b + 3c = 5} \\ -2b - 2c = -5. \end{array}$$

Subtract the second and third equations:

$$\begin{array}{r} a - b - c = 3 \\ \underline{a + 3b + 3c = 5} \\ -4b - 4c = -2. \end{array}$$

Multiply $-2b - 2c = -5$ by 2 and subtract $-4b - 4c = -2$:

$$\begin{array}{r} -4b - 4c = -10 \\ \underline{-4b - 4c = -2} \\ 0 = -8. \end{array}$$

Therefore, we have no solution.

25. Add the first two equations:

$$\begin{array}{r} 3x + 2y + z = -1 \\ \underline{3x + 4y - z = 1} \\ 6x + 6y = 0. \end{array}$$

Add the last two equations:

$$\begin{array}{r} 3x + 4y - z = 1 \\ \underline{x + 2y + z = 0} \\ 4x + 6y = 1. \end{array}$$

Now, subtract $6x + 6y = 0$ and $4x + 6y = 1$:

$$\begin{array}{r} 6x + 6y = 0 \\ \underline{4x + 6y = 1} \\ 2x = -1 \Rightarrow x = -\frac{1}{2}. \end{array}$$

Substitute $x = -\frac{1}{2}$ into $6x + 6y = 0$: $6\left(-\frac{1}{2}\right) + 6y = 0 \Rightarrow 6y = 3 \Rightarrow y = \frac{1}{2}$.

Substitute $x = -\frac{1}{2}$ and $y = \frac{1}{2}$ into $x + 2y + x = 0 \Rightarrow -\frac{1}{2} + 2\left(\frac{1}{2}\right) + z = 0 \Rightarrow z = -\frac{1}{2}$.

The solution is: $\left(-\frac{1}{2}, \frac{1}{2}, -\frac{1}{2}\right)$.

27. Multiply the first equation by 2 and add the second equation:

$$-2x + 6y + 2z = 6$$
$$\underline{2x + 7y + 4z = 13}$$
$$13y + 6z = 19.$$

Multiply the second equation by 2 and subtract the third equation:

$$4x + 14y + 8z = 26$$
$$\underline{4x + y + 2z = 7}$$
$$13y + 6z = 19.$$

Subtracting the two new equations we get:

$$13y + 6z = 19$$
$$\underline{13y + 6z = 19}$$
$$0 = 0.$$

Therefore, we have infinitely many solutions.

$$13y + 6z = 19 \Rightarrow 13y = -6z + 19 \Rightarrow y = \frac{-6z + 19}{13}.$$

Multiply the third equation by 3 and subtract from the first equation:

$$-x + 3y + z = 3$$
$$\underline{12x + 3y + 6z = 21}$$
$$-13x - 5z = -18 \Rightarrow x = \frac{-5z + 18}{13}.$$

We have infinitely many solutions: $\left(\dfrac{-5z + 18}{13}, \dfrac{-6z + 19}{13}, z \right)$.

29. Subtract the second and third equations:

$$y + 4z = -13$$
$$\underline{3x + y = 13}$$
$$-3x + 4z = -26.$$

Multiply the first equation by 3 and subtract $-3x + 4z = -26$ from it

$$-3x + 6z = -27$$
$$\underline{-3x + 4z = -26}$$
$$2z = -1 \Rightarrow z = -\frac{1}{2}.$$

Substitute $z = -\dfrac{1}{2}$ into $y + 4z = -13$: $y + 4\left(-\dfrac{1}{2} \right) = -13 \Rightarrow y = -11.$

Substitute $z = -\dfrac{1}{2}$ into $-x + 2z = -9 \Rightarrow -x + 2\left(-\dfrac{1}{2} \right) = -9 \Rightarrow -x = -8 \Rightarrow x = 8.$

The solution is: $\left(8, -11, -\dfrac{1}{2} \right)$.

31. Multiply the first equation by 2 and subtract the second equation:

$$\begin{array}{rcrcrcr}
x & - & 2y & + & z & = & -8 \\
x & + & 2y & - & 3z & = & 20 \\
\hline
 & & -4y & + & 4z & = & -28.
\end{array}$$

Add the first and third equations:

$$\begin{array}{rcrcrcr}
\frac{1}{2}x & - & y & + & \frac{1}{2}z & = & -4 \\
-\frac{1}{2}x & + & 3y & + & 2z & = & 0 \\
\hline
 & & 2y & + & \frac{5}{2}z & = & -4.
\end{array}$$

Multiply $2y + \dfrac{5}{2}z = -4$ by 2 and add $-4y + 4z = -28$:

$$\begin{array}{rcrcr}
4y & + & 5z & = & -8 \\
-4y & + & 4z & = & -28 \\
\hline
 & & 9z & = & -36 \Rightarrow z = -4.
\end{array}$$

Substitute $z = -4$ into $-4y + 4z = -28$: $-4y + 4(-4) = -28 \Rightarrow -4y - 16 = -28 \Rightarrow$
$-4y = -12 \Rightarrow y = 3$.

Substitute $y = 3$ and $z = -4$ into $x + 2y - 3z = 20$: $x + 2(3) - 3(-4) = 20 \Rightarrow x + 18 = 20 \Rightarrow x = 2$

The solution is: $(2, 3, -4)$.

33. Let x = children tickets sold, y = student tickets sold, and z = adult tickets sold. Then: $x + y + z = 500$,

$5x + 7y + 10z = 3560$, and $y = z + 180$ or $y - z = 180$.

Multiply the first equation by 5 and subtract the second equation:

$$\begin{array}{rcrcrcr}
5x & + & 5y & + & 5z & = & 2500 \\
5x & + & 7y & + & 10z & = & 3560 \\
\hline
 & & -2y & - & 5z & = & -1060.
\end{array}$$

Now, multiply $y - z = 180$ by 2 and add $-2y - 5z = -1060$:

$$\begin{array}{rcrcr}
2y & - & 2z & = & 360 \\
-2y & - & 5z & = & -1060 \\
\hline
 & & -7z & = & -700 \Rightarrow z = 100.
\end{array}$$

Substitute $z = 100$ into $y - z = 180$: $y - 100 = 180 \Rightarrow y = 280$.

Substitute $y = 280$ and $z = 100$ into $x + y + z = 500$: $x + 280 + 100 = 500 \Rightarrow x = 120$.

There were 120 children tickets, 280 student tickets, and 100 adult tickets sold.

35. Let x = cost of a hamburger, y = cost of fries, and z = cost of a soda. Then $2x + 2y + z = 9$;
$x + y + z = 5$; and $x + y = 5$.

Multiply the equation $x + y + z = 5$ by 2 and subtract from the equation $2x + 2y + z = 9$:

$$\begin{array}{rcrcrcr}
2x & + & 2y & + & z & = & 9 \\
2x & + & 2y & + & 2z & = & 10 \\
\hline
 & & & & -z & = & -1 \Rightarrow z = 1.
\end{array}$$

Subtract $x + y + z = 5$ and $x + y = 5$:

$$
\begin{aligned}
x + y + z &= 5 \\
\underline{x + y \qquad = 5} \\
z &= 0.
\end{aligned}
$$

z cannot equal both 0 and 1 therefore there is no solution, at least one student was charged incorrectly.

37. (a) $x + y + z = 180$; $x = z + 25 \Rightarrow x - z = 25$; and $y + z = x + 30 \Rightarrow -x + y + z = 30$.

   (b) Add $x - z = 25$ to $-x + y + z = 30$:

$$
\begin{aligned}
x \qquad - z &= 25 \\
\underline{-x + y + z = 30} \\
y \qquad = 55.
\end{aligned}
$$

   Add $x + y + z = 180$ to $x - z = 25$:

$$
\begin{aligned}
x + y + z &= 180 \\
\underline{x \qquad - z = \ 25} \\
2x + y \qquad = 205.
\end{aligned}
$$

   Substitute $y = 55$ into $2x + y = 205$: $2x + 55 = 205 \Rightarrow 2x = 150 \Rightarrow x = 75$.

   Substitute $x = 75$ and $y = 55$ into $x + y + z = 180$: $75 + 55 + z = 180 \Rightarrow z = 50$.

   The angles are: $75°$, $55°$, and $50°$.

   Check: $75 + 55 + 50 = 180 \Rightarrow 180 = 180$; $75 - 50 = 25 \Rightarrow 25 = 25$; $-75 + 55 + 50 = 30 \Rightarrow$

   $30 = 30$.

39. Let x, y, and z equal the amounts invested in the three mutual funds.

   Then $x + y + z = 20{,}000$; $0.05x + 0.07y + 0.10z = 1650$; and $4x = z \Rightarrow 4x - z = 0$.

   Multiply $x + y + z = 20{,}000$ by 0.07 and subtract $0.05x + 0.07y + 0.10z = 1650$:

$$
\begin{aligned}
0.07x + 0.07y + 0.07z &= \ \ 1400 \\
\underline{0.05x + 0.07y + 0.10z = \ \ 1650} \\
0.02x \qquad\qquad - 0.03z &= -250.
\end{aligned}
$$

   Now multiply $0.02x - 0.03z = -250$ by 200 and subtract $4x - z = 0$:

$$
\begin{aligned}
4x - 6z &= -50{,}000 \\
\underline{4x - \ z = \qquad 0} \\
-5z &= -50{,}000 \Rightarrow z = 10{,}000.
\end{aligned}
$$

   Substitute $z = 10{,}000$ into $4x - z = 0$: $4x - 10{,}000 = 0 \Rightarrow 4x = 10{,}000 \Rightarrow x = 2500$.

   Substitute $x = 2500$ and $z = 10{,}000$ into $x + y + z = 20{,}000$: $2500 + y + 10{,}000 = 20{,}000 \Rightarrow y = 7500$.

   The fund amounts are: $2500 at 5%, $7500 at 7%, and $10,000 at 10%.

41. (a)
$$
\begin{aligned}
N + P + K &= 80 \\
N + P - K &= \ 8 \\
9P - K &= \ 0
\end{aligned}
$$

   (b) Using technology to solve the system, the solution is $(40, 4, 36)$.

   The sample contains 40 pounds of nitrogen, 4 pounds of phosphorus and 36 pounds of potassium.

9.4 Solutions to Linear Systems Using Matrices

1. Since there are three rows and one column, its dimension is 3×1.

3. Since there are two rows and two columns, its dimension is 2×2.

5. 3×2

7. This system can be written using a 2×3 matrix:

$$\begin{bmatrix} 5 & -2 & | & 3 \\ -1 & 3 & | & -1 \end{bmatrix}$$

9. This system can be written using a 3×4 matrix:

$$\begin{bmatrix} -3 & 2 & 1 & | & -4 \\ 5 & 0 & -1 & | & 9 \\ 1 & -3 & -6 & | & -9 \end{bmatrix}$$

11. $3x + 2y = 4$ and $y = 5$

13. $3x + y + 4z = 0, 5y + 8z = -1$, and $-7z = 1$

15. (a) Yes

    (b) No, since $a_{22} = -1$ and $a_{32} \neq 0$. The diagonal is not all 1's and there are not all 0's below the diagonal.

    (c) Yes

17. The system can be written as $x + 2y = 3$ and $y = -1$. Substituting $y = -1$ into the first equation gives

    $x + 2(-1) = 3 \Rightarrow x = 5$. The solution is $(5, -1)$.

19. The system can be written as $x - y = 2$ and $y = 0$. Substituting $y = 0$ into the first equation gives

    $x - 0 = 2 \Rightarrow x = 2$. The solution is $(2, 0)$.

21. The system can be written as $x + y - z = 4, y - z = 2$, and $z = 1$. Substituting $z = 1$ into the second

    equation gives $y - (1) = 2 \Rightarrow y = 3$. Substituting $y = 3$ and $z = 1$ into the first equation gives

    $x + (3) - (1) = 4 \Rightarrow x = 2$. The solution is $(2, 3, 1)$.

23. The system can be written as $x + 2y - z = 5, y - 2z = 1$, and $0 = 0$. Since $0 = 0$, there are an infinite

    number of solutions. The second equation gives $y = 1 + 2z$. Substituting this into the first equation gives

    $x + 2(1 + 2z) - z = 5 \Rightarrow x = 3 - 3z$. The solution can be written as

    $\{(3 - 3z, 1 + 2z, z) | z \text{ is a real number}\}$.

25. The system can be written as $x + 2y + z = -3, y - 3z = \dfrac{1}{2}$, and $0 = 4$. Since $0 = 4$ is false, there are no

    solutions.

27.
$$\begin{array}{c} (1/2)R_1 \rightarrow \\ \\ (1/4)R_3 \rightarrow \end{array} \begin{bmatrix} 1 & -2 & 3 & | & 5 \\ -3 & 5 & 3 & | & 2 \\ 1 & 2 & 1 & | & -2 \end{bmatrix}$$

29.
$$\begin{array}{c} \\ R_2 + R_1 \rightarrow \\ R_3 - R_1 \rightarrow \end{array} \begin{bmatrix} 1 & -1 & 1 & | & 2 \\ 0 & 1 & -1 & | & 2 \\ 0 & 8 & -1 & | & 3 \end{bmatrix}$$

31. The system can be written as follows:

$$\begin{bmatrix} 1 & 2 & | & 3 \\ -1 & -1 & | & 7 \end{bmatrix} \begin{array}{c} \\ R_2 + R_1 \to \end{array} \begin{bmatrix} 1 & 2 & | & 3 \\ 0 & 1 & | & 10 \end{bmatrix}$$

The solution is $y = 10$ and $x + 2y = 3 \Rightarrow x + 2(10) = 3 \Rightarrow x = -17$. The solution is $(-17, 10)$.

33. The system can be written as follows:

$$\begin{bmatrix} 1 & 2 & 1 & | & 3 \\ 1 & 1 & -1 & | & 3 \\ -1 & -2 & 1 & | & -5 \end{bmatrix} \begin{array}{c} \\ R_2 - R_1 \to \\ R_3 + R_1 \to \end{array} \begin{bmatrix} 1 & 2 & 1 & | & 3 \\ 0 & -1 & -2 & | & 0 \\ 0 & 0 & 2 & | & -2 \end{bmatrix} \begin{array}{c} \\ (-1)R_2 \to \\ (1/2)R_3 \to \end{array} \begin{bmatrix} 1 & 2 & 1 & | & 3 \\ 0 & 1 & 2 & | & 0 \\ 0 & 0 & 1 & | & -1 \end{bmatrix}$$

Back substitution produces $z = -1$; $y + 2z = 0 \Rightarrow y = 2$; $x + 2y + z = 3 \Rightarrow x = 0$.

The solution is $(0, 2, -1)$.

35. The system can be written as follows:

$$\begin{bmatrix} 1 & 2 & -1 & | & -1 \\ 2 & -1 & 1 & | & 0 \\ -1 & -1 & 2 & | & 7 \end{bmatrix} \begin{array}{c} \\ -2R_1 + R_2 \to \\ R_1 + R_3 \to \end{array} \begin{bmatrix} 1 & 2 & -1 & | & -1 \\ 0 & -5 & 3 & | & 2 \\ 0 & 1 & 1 & | & 6 \end{bmatrix} \begin{array}{c} \\ R_3 \Leftrightarrow R_2 \to \end{array} \begin{bmatrix} 1 & 2 & -1 & | & -1 \\ 0 & 1 & 1 & | & 6 \\ 0 & -5 & 3 & | & 2 \end{bmatrix}$$

$$5R_2 + R_3 \to \begin{bmatrix} 1 & 2 & -1 & | & -1 \\ 0 & 1 & 1 & | & 6 \\ 0 & 0 & 8 & | & 32 \end{bmatrix} \begin{array}{c} \\ \frac{1}{8}R_3 \to \end{array} \begin{bmatrix} 1 & 2 & -1 & | & -1 \\ 0 & 1 & 1 & | & 6 \\ 0 & 0 & 1 & | & 4 \end{bmatrix}$$

Back substitution produces $z = 4$; $y + z = 6 \Rightarrow y = 2$; $x + 2y - z = -1 \Rightarrow x = -1$.

The solution is $(-1, 2, 4)$.

37. The system can be written as follows:

$$\begin{bmatrix} 3 & 1 & 3 & | & 14 \\ 1 & 1 & 1 & | & 6 \\ -2 & -2 & 3 & | & -7 \end{bmatrix} \begin{array}{c} R_2 \to \\ R_1 \to \end{array} \begin{bmatrix} 1 & 1 & 1 & | & 6 \\ 3 & 1 & 3 & | & 14 \\ -2 & -2 & 3 & | & -7 \end{bmatrix} \begin{array}{c} \\ R_2 - 3R_1 \to \\ R_3 + 2R_1 \to \end{array} \begin{bmatrix} 1 & 1 & 1 & | & 6 \\ 0 & -2 & 0 & | & -4 \\ 0 & 0 & 5 & | & 5 \end{bmatrix}$$

$$\begin{array}{c} (-1/2)R_2 \to \\ (1/5)R_3 \to \end{array} \begin{bmatrix} 1 & 1 & 1 & | & 6 \\ 0 & 1 & 0 & | & 2 \\ 0 & 0 & 1 & | & 1 \end{bmatrix}$$

Back substitution produces $z = 1$; $y = 2$; $x + y + z = 6 \Rightarrow x = 3$. The solution is $(3, 2, 1)$.

39. The system can be written as follows:

$$\begin{bmatrix} 1 & 2 & -1 & | & 2 \\ 2 & 5 & 1 & | & 8 \\ 3 & 7 & 0 & | & 5 \end{bmatrix} \begin{array}{c} \\ R_2 - 2R_1 \to \\ R_3 - 3R_1 \to \end{array} \begin{bmatrix} 1 & 2 & -1 & | & 2 \\ 0 & 1 & 3 & | & 4 \\ 0 & 1 & 3 & | & -1 \end{bmatrix} \begin{array}{c} \\ R_3 - R_2 \to \end{array} \begin{bmatrix} 1 & 2 & -1 & | & 2 \\ 0 & 1 & 3 & | & 4 \\ 0 & 0 & 0 & | & -5 \end{bmatrix}$$

The last equation indicates that $0 = -5$, which is false. Therefore, there are no solutions.

41. The system can be written as follows:

$$\begin{bmatrix} -1 & 2 & 4 & | & 10 \\ 3 & -2 & -2 & | & -12 \\ 1 & 2 & 6 & | & 8 \end{bmatrix} \begin{array}{c} (-1)R_1 \to \\ R_2 + 3R_1 \to \\ R_1 + R_3 \to \end{array} \begin{bmatrix} 1 & -2 & -4 & | & -10 \\ 0 & 4 & 10 & | & 18 \\ 0 & 4 & 10 & | & 18 \end{bmatrix} \begin{array}{c} R_1 + (1/2)R_2 \to \\ (1/4)R_2 \to \\ R_2 - R_3 \to \end{array} \begin{bmatrix} 1 & 0 & 1 & | & -1 \\ 0 & 1 & \frac{5}{2} & | & \frac{9}{2} \\ 0 & 0 & 0 & | & 0 \end{bmatrix}$$

The last equation indicates that $0 = 0$, which is true. Therefore, there is an infinite number of solutions.

The second equation gives: $y + \dfrac{5}{2}z = \dfrac{9}{2} \Rightarrow y = \dfrac{-5z + 9}{2}$. The first equation gives: $x + z = -1 \Rightarrow$

$x = -1 - z$. Therefore there are infinitely many solutions which can be written as $\left(-1 - z, \dfrac{-5z + 9}{2}, z\right)$.

43. $\begin{bmatrix} 1 & -1 & 1 & | & 1 \\ 1 & 2 & -1 & | & 2 \\ 0 & 1 & -1 & | & 0 \end{bmatrix} \begin{array}{c} \\ R_2 - R_1 \rightarrow \\ \\ \end{array} \begin{bmatrix} 1 & -1 & 1 & | & 1 \\ 0 & 3 & -2 & | & 1 \\ 0 & 1 & -1 & | & 0 \end{bmatrix} (1/3)R_2 \rightarrow \begin{bmatrix} 1 & -1 & 1 & | & 1 \\ 0 & 1 & -\frac{2}{3} & | & \frac{1}{3} \\ 0 & 1 & -1 & | & 0 \end{bmatrix}$

$\begin{array}{c} \\ \\ R_3 - R_2 \rightarrow \end{array} \begin{bmatrix} 1 & -1 & 1 & | & 1 \\ 0 & 1 & -\frac{2}{3} & | & \frac{1}{3} \\ 0 & 0 & -\frac{1}{3} & | & -\frac{1}{3} \end{bmatrix} -3R_3 \rightarrow \begin{bmatrix} 1 & -1 & 1 & | & 1 \\ 0 & 1 & -\frac{2}{3} & | & \frac{1}{3} \\ 0 & 0 & 1 & | & 1 \end{bmatrix}$

The matrix is now in row-echelon form. We see that $z = 1$. Thus, $y - \dfrac{2}{3}z = \dfrac{1}{3} \Rightarrow$

$y - \dfrac{2}{3}(1) = \dfrac{1}{3} \Rightarrow y = 1$ and $x - y + z = 1 \Rightarrow x - 1 + 1 = 1 \Rightarrow x = 1$. The solution is $(1, 1, 1)$.

45. $\begin{bmatrix} 2 & -4 & 2 & | & 11 \\ 1 & 3 & -2 & | & -9 \\ 4 & -2 & 1 & | & 7 \end{bmatrix} (1/2)R_1 \rightarrow \begin{bmatrix} 1 & -2 & 1 & | & \frac{11}{2} \\ 1 & 3 & -2 & | & -9 \\ 4 & -2 & 1 & | & 7 \end{bmatrix} \begin{array}{c} \\ R_2 - R_1 \rightarrow \\ R_3 - 4R_1 \rightarrow \end{array} \begin{bmatrix} 1 & -2 & 1 & | & \frac{11}{2} \\ 0 & 5 & -3 & | & -\frac{29}{2} \\ 0 & 6 & -3 & | & -15 \end{bmatrix}$

$(1/5)R_2 \rightarrow \begin{bmatrix} 1 & -2 & 1 & | & \frac{11}{2} \\ 0 & 1 & -\frac{3}{5} & | & -\frac{29}{10} \\ 0 & 6 & -3 & | & -15 \end{bmatrix} \begin{array}{c} \\ \\ R_3 - 6R_2 \rightarrow \end{array} \begin{bmatrix} 1 & -2 & 1 & | & \frac{11}{2} \\ 0 & 1 & -\frac{3}{5} & | & -\frac{29}{10} \\ 0 & 0 & \frac{3}{5} & | & \frac{24}{10} \end{bmatrix} (5/3)R_3 \rightarrow \begin{bmatrix} 1 & -2 & 1 & | & \frac{11}{2} \\ 0 & 1 & -\frac{3}{5} & | & -\frac{29}{10} \\ 0 & 0 & 1 & | & 4 \end{bmatrix}$

The matrix is now in row-echelon form. We see that $z = 4$. Thus, $y - \dfrac{3}{5}z = -\dfrac{29}{10} \Rightarrow$

$y - \dfrac{3}{5}(4) = -\dfrac{29}{10} \Rightarrow y = -\dfrac{1}{2}$ and $x - 2y + z = \dfrac{11}{2} \Rightarrow x - 2\left(-\dfrac{1}{2}\right) + 4 = \dfrac{11}{2} \Rightarrow x = \dfrac{1}{2}$.

The solution is $\left(\dfrac{1}{2}, -\dfrac{1}{2}, 4\right)$.

47. $\begin{bmatrix} 3 & -2 & 2 & | & -18 \\ -1 & 2 & -4 & | & 16 \\ 4 & -3 & -2 & | & -21 \end{bmatrix} (1/3)R_1 \rightarrow \begin{bmatrix} 1 & -\frac{2}{3} & \frac{2}{3} & | & -6 \\ -1 & 2 & -4 & | & 16 \\ 4 & -3 & -2 & | & -21 \end{bmatrix} \begin{array}{c} \\ R_2 + R_1 \rightarrow \\ R_3 - 4R_1 \rightarrow \end{array} \begin{bmatrix} 1 & -\frac{2}{3} & \frac{2}{3} & | & -6 \\ 0 & \frac{4}{3} & -\frac{10}{3} & | & 10 \\ 0 & -\frac{1}{3} & -\frac{14}{3} & | & 3 \end{bmatrix}$

$\begin{array}{c} (3/4)R_2 \rightarrow \\ 4R_3 + R_2 \rightarrow \end{array} \begin{bmatrix} 1 & -\frac{2}{3} & \frac{2}{3} & | & -6 \\ 0 & 1 & -\frac{5}{2} & | & \frac{15}{2} \\ 0 & 0 & -22 & | & 22 \end{bmatrix} (-1/22)R_3 \rightarrow \begin{bmatrix} 1 & -\frac{2}{3} & \frac{2}{3} & | & -6 \\ 0 & 1 & -\frac{5}{2} & | & \frac{15}{2} \\ 0 & 0 & 1 & | & -1 \end{bmatrix}$

The matrix is now in row-echelon form. We see that $z = -1$. Thus, $y - \dfrac{5}{2}z = \dfrac{15}{2} \Rightarrow$

$y - \dfrac{5}{2}(-1) = \dfrac{15}{2} \Rightarrow y = 5$ and $x - \dfrac{2}{3}y + \dfrac{2}{3}z = -6 \Rightarrow x - \dfrac{2}{3}(5) + \dfrac{2}{3}(-1) = -6 \Rightarrow x = -2$.

The solution is $(-2, 5, -1)$.

49. $\begin{bmatrix} 1 & -4 & 3 & | & 26 \\ -1 & 3 & -2 & | & -19 \\ 0 & -1 & 1 & | & 10 \end{bmatrix} R_2 + R_1 \rightarrow \begin{bmatrix} 1 & -4 & 3 & | & 26 \\ 0 & -1 & 1 & | & 7 \\ 0 & -1 & 1 & | & 10 \end{bmatrix} \begin{array}{c} -R_2 \rightarrow \\ R_3 - R_2 \rightarrow \end{array} \begin{bmatrix} 1 & -4 & 3 & | & 26 \\ 0 & 1 & -1 & | & -7 \\ 0 & 0 & 0 & | & 3 \end{bmatrix}$

The last equation indicates that $0 = 3$, which is always false. Therefore, the system has no solutions.

51. $\begin{bmatrix} 5 & 0 & 4 & | & 7 \\ 2 & -4 & 0 & | & 6 \\ 0 & 3 & 3 & | & 3 \end{bmatrix}$ $\begin{matrix} (1/5)R_1 \rightarrow \\ 2R_1 - 5R_2 \rightarrow \\ (1/3)R_3 \rightarrow \end{matrix}$ $\begin{bmatrix} 1 & 0 & \frac{4}{5} & | & \frac{7}{5} \\ 0 & 20 & 8 & | & -16 \\ 0 & 1 & 1 & | & 1 \end{bmatrix}$ $\begin{matrix} (1/20)R_2 \rightarrow \\ \\ R_2 - 20R_3 \rightarrow \end{matrix}$ $\begin{bmatrix} 1 & 0 & \frac{4}{5} & | & \frac{7}{5} \\ 0 & 1 & \frac{2}{5} & | & -\frac{4}{5} \\ 0 & 0 & -12 & | & -36 \end{bmatrix}$

$(-1/12)R_3 \rightarrow$ $\begin{bmatrix} 1 & 0 & \frac{4}{5} & | & \frac{7}{5} \\ 0 & 1 & \frac{2}{5} & | & -\frac{4}{5} \\ 0 & 0 & 1 & | & 3 \end{bmatrix}$ $\begin{matrix} (-4/5)R_3 + R_1 \rightarrow \\ (-2/5)R_3 + R_2 \rightarrow \\ \end{matrix}$ $\begin{bmatrix} 1 & 0 & 0 & | & -1 \\ 0 & 1 & 0 & | & -2 \\ 0 & 0 & 1 & | & 3 \end{bmatrix}$ The solution is $(-1, -2, 3)$.

53. $\begin{bmatrix} 5 & -2 & 1 & | & 5 \\ 1 & 1 & -2 & | & -2 \\ 4 & -3 & 3 & | & 7 \end{bmatrix}$ $\begin{matrix} R_2 \rightarrow \\ 5R_2 - R_1 \rightarrow \\ 4R_2 - R_3 \rightarrow \end{matrix}$ $\begin{bmatrix} 1 & 1 & -2 & | & -2 \\ 0 & 7 & -11 & | & -15 \\ 0 & 7 & -11 & | & -15 \end{bmatrix}$ $\begin{matrix} (1/7)R_2 \rightarrow \\ \\ R_2 - R_3 \rightarrow \end{matrix}$ $\begin{bmatrix} 1 & 1 & -2 & | & -2 \\ 0 & 1 & -\frac{11}{7} & | & -\frac{15}{7} \\ 0 & 0 & 0 & | & 0 \end{bmatrix}$

$(-1)R_2 + R_1 \rightarrow$ $\begin{bmatrix} 1 & 0 & -\frac{3}{7} & | & \frac{1}{7} \\ 0 & 1 & -\frac{11}{7} & | & -\frac{15}{7} \\ 0 & 0 & 0 & | & 0 \end{bmatrix}$

The last equation indicates that $0 = 0$, which is true. Therefore, there is an infinite number of solutions.

The second equation gives: $y - \dfrac{11}{7}z = -\dfrac{15}{7} \Rightarrow y = \dfrac{11z - 15}{7}$. The first equation gives: $x - \dfrac{3}{7}z = \dfrac{1}{7} \Rightarrow$

$x = \dfrac{3z + 1}{7}$. Therefore, there are infinitely many solutions which can be written as: $\left(\dfrac{3z + 1}{7}, \dfrac{11z - 15}{7}, z \right)$.

55. The equations are: $x = 12$ and $y = 3 \Rightarrow (12, 3)$.

57. The equations are: $x = -2$, $y = 4$, and $z = \dfrac{1}{2} \Rightarrow \left(-2, 4, \dfrac{1}{2} \right)$.

59. The last equation indicates that $0 = 0$, which is true. Therefore, there is an infinite number of solutions.

The first equation gives: $x + 2z = 4 \Rightarrow x = -2z + 4$. The second equation gives: $y - z = -3 \Rightarrow$

$y = z - 3$. The system has infinitely many solutions which can be written as: $(-2z + 4, z - 3, z)$.

61. The last equation indicates that $0 = \dfrac{2}{3}$, which is always false. Therefore, there are no solutions.

63. $\begin{bmatrix} 1 & -1 & | & 1 \\ 1 & 1 & | & 5 \end{bmatrix}$ $R_2 - R_1 \rightarrow$ $\begin{bmatrix} 1 & -1 & | & 1 \\ 0 & 2 & | & 4 \end{bmatrix}$ $(1/2)R_2 \rightarrow$ $\begin{bmatrix} 1 & -1 & | & 1 \\ 0 & 1 & | & 2 \end{bmatrix}$ $R_1 + R_2 \rightarrow$ $\begin{bmatrix} 1 & 0 & | & 3 \\ 0 & 1 & | & 2 \end{bmatrix}$

The solution is $(3, 2)$

65. $\begin{bmatrix} 1 & 2 & 1 & | & 3 \\ 0 & 1 & -1 & | & -2 \\ -1 & -2 & 2 & | & 6 \end{bmatrix}$ $R_1 + R_3 \rightarrow$ $\begin{bmatrix} 1 & 2 & 1 & | & 3 \\ 0 & 1 & -1 & | & -2 \\ 0 & 0 & 3 & | & 9 \end{bmatrix}$ $\begin{matrix} R_1 - 2R_2 \rightarrow \\ \\ (1/3)R_3 \rightarrow \end{matrix}$ $\begin{bmatrix} 1 & 0 & 3 & | & 7 \\ 0 & 1 & -1 & | & -2 \\ 0 & 0 & 1 & | & 3 \end{bmatrix}$

$\begin{matrix} R_1 - 3R_3 \rightarrow \\ R_2 + R_3 \rightarrow \\ \end{matrix}$ $\begin{bmatrix} 1 & 0 & 0 & | & -2 \\ 0 & 1 & 0 & | & 1 \\ 0 & 0 & 1 & | & 3 \end{bmatrix}$ The solution is $(-2, 1, 3)$.

67. $\begin{bmatrix} 1 & -1 & 2 & | & 7 \\ 2 & 1 & -4 & | & -27 \\ -1 & 1 & -1 & | & 0 \end{bmatrix} \begin{matrix} \\ R_2 - 2R_1 \rightarrow \\ R_1 + R_3 \rightarrow \end{matrix} \begin{bmatrix} 1 & -1 & 2 & | & 7 \\ 0 & 3 & -8 & | & -41 \\ 0 & 0 & 1 & | & 7 \end{bmatrix} (1/3)R_2 \rightarrow \begin{bmatrix} 1 & -1 & 2 & | & 7 \\ 0 & 1 & -\frac{8}{3} & | & -\frac{41}{3} \\ 0 & 0 & 1 & | & 7 \end{bmatrix}$

$R_2 + R_1 \rightarrow \begin{bmatrix} 1 & 0 & -\frac{2}{3} & | & -\frac{20}{3} \\ 0 & 1 & -\frac{8}{3} & | & -\frac{41}{3} \\ 0 & 0 & 1 & | & 7 \end{bmatrix} \begin{matrix} (2/3)R_3 + R_1 \rightarrow \\ (8/3)R_3 + R_2 \rightarrow \end{matrix} \begin{bmatrix} 1 & 0 & 0 & | & -2 \\ 0 & 1 & 0 & | & 5 \\ 0 & 0 & 1 & | & 7 \end{bmatrix}$ The solution is $(-2, 5, 7)$.

69. $\begin{bmatrix} 2 & 1 & -1 & | & 2 \\ 1 & -2 & 1 & | & 0 \\ 1 & 3 & -2 & | & 4 \end{bmatrix} \begin{matrix} R_2 \rightarrow \\ 2R_2 - R_1 \rightarrow \\ R_2 - R_3 \rightarrow \end{matrix} \begin{bmatrix} 1 & -2 & 1 & | & 0 \\ 0 & -5 & 3 & | & -2 \\ 0 & -5 & 3 & | & -4 \end{bmatrix} \begin{matrix} \\ (-1/5)R_2 \rightarrow \\ R_2 - R_3 \rightarrow \end{matrix} \begin{bmatrix} 1 & -2 & 1 & | & 0 \\ 0 & 1 & -\frac{3}{5} & | & \frac{2}{5} \\ 0 & 0 & 0 & | & 2 \end{bmatrix}$

The last equation indicates that $0 = 2$, which is always false. Therefore, there are no solutions.

71. Enter the coefficients of the linear system into a 3×4 matrix, as shown in Figure 71a. Reduce the matrix to reduced row-echelon form, as shown in Figure 71b. The solution is the ordered triple $(-9.266, -9.167, 2.440)$.

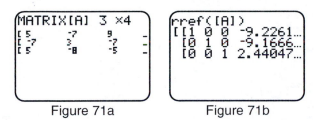

Figure 71a   Figure 71b

73. Enter the coefficients of the linear system into a 3×4 matrix, as shown in Figure 73a. Reduce the matrix to reduced row-echelon form, as shown in Figure 73b. The solution is the ordered triple $(5.211, 3.739, -4.655)$.

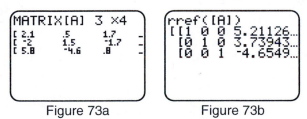

Figure 73a   Figure 73b

75. Enter the coefficients of the linear system into a 3×4 matrix, as shown in Figure 75a. Reduce the matrix to reduced row-echelon form, as shown in Figure 75b. The solution is the ordered triple $(7.993, 1.609, -0.401)$.

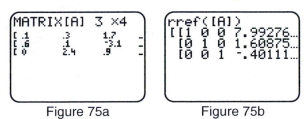

Figure 75a   Figure 75b

77. (a) The constants a, b, and c should satisfy the following three equations:

$1300 = a(1800) + b(5000) + c, 5300 = a(3200) + b(12,000) + c,$

and $6500 = a(4500) + b(13,000) + c$

This can be written:

$1800a + 5000b + c = 1300, 3200a + 12,000b + c = 5300,$ and $4500a + 13,000b + c = 6500$

The associated augmented matrix is: $\begin{bmatrix} 1800 & 5000 & 1 & \vline & 1300 \\ 3200 & 12{,}000 & 1 & \vline & 5300 \\ 4500 & 13{,}000 & 1 & \vline & 6500 \end{bmatrix}$

Using technology, the solution is $a \approx 0.5714$, $b \approx 0.4571$, and $c \approx -2014$. See Figure 73. Thus, the equation modeling the data can be expressed (approximately) as $F = 0.5714N + 0.4571R - 2014$.

(b) To predict the food costs for a shelter that serves 3500 people and receives charitable receipts of $\$12{,}500$, let $N = 3500$ and $R = 12{,}500$ and evaluate the equation.

$F = 0.5714(3500) + 0.4571(12{,}500) - 2014 = 5699.65$. This model predicts monthly food costs of approximately $\$5700$.

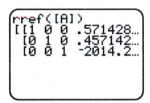

Figure 77

79. Let x represent the fraction of the pool that the first pump can empty each hour, and y the fraction of the pool that the second pump can empty each hour, and z the fraction for the third pump. Since the first pump is twice as fast we have $x = 2y$ or $x - 2y = 0$. Since the first two pumps can empty the pool in 8 hours, it follows that they can empty $\dfrac{1}{8}$ of the pool in an hour. Thus, $x + y = \dfrac{1}{8}$. Similarly, all three pumps can empty the pool in 6 hours, so $x + y + z = \dfrac{1}{6}$. Putting these equations in an augmented matrix results in the following:

$$\begin{bmatrix} 1 & -2 & 0 & \vline & 0 \\ 1 & 1 & 0 & \vline & \frac{1}{8} \\ 1 & 1 & 1 & \vline & \frac{1}{6} \end{bmatrix} \begin{bmatrix} 1 & -2 & 0 & \vline & 0 \\ 0 & 3 & 0 & \vline & \frac{1}{8} \\ 0 & 3 & 1 & \vline & \frac{1}{6} \end{bmatrix} \begin{bmatrix} 1 & -2 & 0 & \vline & 0 \\ 0 & 1 & 0 & \vline & \frac{1}{24} \\ 0 & 0 & 1 & \vline & \frac{1}{24} \end{bmatrix} \begin{bmatrix} 1 & 0 & 0 & \vline & \frac{1}{12} \\ 0 & 1 & 0 & \vline & \frac{1}{24} \\ 0 & 0 & 1 & \vline & \frac{1}{24} \end{bmatrix}$$

Thus, $x = \dfrac{1}{12}$, $y = z = \dfrac{1}{24}$. The first pump could empty $\dfrac{1}{12}$ of the pool in one hour or the entire pool in 12 hours, while the second and third pumps individually could empty the pool in 24 hours. Using technology, this solution can be obtained as shown in Figure 79.

Figure 79

81. $I_1 = I_2 + I_3 \Rightarrow I_1 - I_2 - I_3 = 0$, $15 + 4I_3 = 14I_2 \Rightarrow -14I_2 + 4I_3 = -15$, and

$10 + 4I_3 = 5I_1 \Rightarrow -5I_1 + 4I_3 = -10$

Therefore the matrix is:

$$\begin{bmatrix} 1 & -1 & -1 & | & 0 \\ 0 & -14 & 4 & | & -15 \\ -5 & 0 & 4 & | & -10 \end{bmatrix} \begin{matrix} \\ (-1/14)R_2 \to \\ 5R_1 + R_3 \to \end{matrix} \begin{bmatrix} 1 & -1 & -1 & | & 0 \\ 0 & 1 & -\frac{2}{7} & | & \frac{15}{14} \\ 0 & -5 & -1 & | & -10 \end{bmatrix} 5R_2 + R_3 \to \begin{bmatrix} 1 & -1 & -1 & | & 0 \\ 0 & 1 & -\frac{2}{7} & | & \frac{15}{14} \\ 0 & 0 & -\frac{17}{7} & | & -\frac{65}{14} \end{bmatrix}$$

$$(-7/17)R_3 \to \begin{bmatrix} 1 & -1 & -1 & | & 0 \\ 0 & 1 & -\frac{2}{7} & | & \frac{15}{14} \\ 0 & 0 & 1 & | & \frac{65}{34} \end{bmatrix} \begin{matrix} R_1 + R_3 \to \\ (2/7)R_3 + R_2 \to \end{matrix} \begin{bmatrix} 1 & -1 & 0 & | & \frac{65}{34} \\ 0 & 1 & 0 & | & \frac{385}{238} \\ 0 & 0 & 1 & | & \frac{65}{34} \end{bmatrix} R_2 + R_1 \to \begin{bmatrix} 1 & 0 & 0 & | & \frac{840}{238} \\ 0 & 1 & 0 & | & \frac{385}{238} \\ 0 & 0 & 1 & | & \frac{65}{34} \end{bmatrix}$$

The solution is: $(3.53, 1.62, 1.91)$.

83. Let x equal the amount invested at 8%, y equal the amount invested at 11% and z equal the amount invested at 14%.

(a) $x + y + z = 5000$, $x + y - z = 0$, and $0.08x + 0.11y + 0.14z = 595$

(b) Enter the coefficients of the following 3×4 augmented matrix into a calculator.

$$\begin{bmatrix} 1 & 1 & 1 & | & 5000 \\ 1 & 1 & -1 & | & 0 \\ 0.08 & 0.11 & 0.14 & | & 595 \end{bmatrix}$$

Using technology, the solution is $x = 1000$, $y = 1500$, and $z = 2500$. See Figure 83. Thus, $1000 needs

to be invested at 8%, $1500 at 11%, and $2500 at 14% in order to earn the total amount interest of $595.

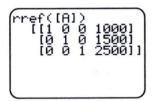

Figure 83

85. (a) At intersection A incoming traffic is equal to $x + 5$. The outgoing traffic is given by $y + 7$. Therefore,

$x + 5 = y + 7$, which is the first equation. The incoming traffic at intersection B is $z + 6$ and the

outgoing traffic is $x + 3$, so $z + 6 = x + 3$. Finally at intersection C, the incoming flow is $y + 3$ and the

outgoing flow is $z + 4$, so $y + 3 = z + 4$.

(b) These three equations can be written as: $x - y = 2$, $x - z = 3$, and $y - z = 1$.

The system of linear equations can be represented by the following augmented matrix:

$$\begin{bmatrix} 1 & -1 & 0 & | & 2 \\ 1 & 0 & -1 & | & 3 \\ 0 & 1 & -1 & | & 1 \end{bmatrix}$$

Begin by subtracting the first row from the second, followed by subtracting the second row from the third.

Gaussian elimination results in the following augmented matrix:

$$\begin{bmatrix} 1 & -1 & 0 & | & 2 \\ 0 & 1 & -1 & | & 1 \\ 0 & 0 & 0 & | & 0 \end{bmatrix}$$

The last row of zeros indicates that the linear system is dependent and has an infinite number of solutions.

Back solving produces $y - z = 1 \Rightarrow y = z + 1$. Substituting into the first equation gives

$x - (z + 1) = 2 \Rightarrow x = z + 3$. Thus, the solution can be written

$\{(z + 3, z + 1, z) | z$ is any nonnegative real number$\}$.

(c) There are an infinite number of solutions to the system. However, solutions such as $z = 1000, x = 1003$,

and $y = 1001$ are likely, unless a large number of people are simply driving around the block. In reality there is

an average traffic flow rate for z that could be measured. From this, values for both x and y could be determined.

87. (a) The three equations can be written as: $3 = 1^2 a + 1b + c, 29 = 5^2 a + 5b + c$, and

$40 = 6^2 a + 6b + c \Rightarrow 36a + 6b + c = 40, 25a + 5b + c = 29, a + b + c = 3$.

Therefore, the matrix is:

$$\begin{bmatrix} 36 & 6 & 1 & | & 40 \\ 25 & 5 & 1 & | & 29 \\ 1 & 1 & 1 & | & 3 \end{bmatrix}$$

(b) Using technology, $a = \dfrac{9}{10}, b = \dfrac{11}{10}, c = 1 \Rightarrow f(x) = \dfrac{9}{10}x^2 + \dfrac{11}{10}x + 1$

(c) See Figure 87.

(d) For example, 6 quarters after its release the sales was $f(6) = \dfrac{9}{10}(6)^2 + \dfrac{11}{10}(6) + 1 = 40$

$[-0.5, 5, 1]$ by $[-10, 175, 25]$

$[1985, 2035, 5]$ by $[5, 12, 1]$

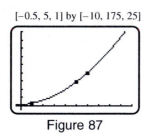

Figure 87

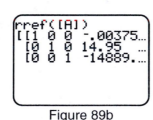

Figure 89b

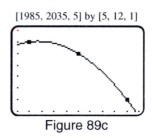

Figure 89c

89. (a) Let $f(x) = ax^2 + bx + c$.   The constants a, b, and c must satisfy the following equations:

$f(1990) = a(1990)^2 + b(1990) + c = 11$, $f(2010) = a(2010)^2 + b(2010) + c = 10$, and

$f(2030) = a(2030)^2 + b(2030) + c = 6$. This system of equations can be represented by the following

augmented matrix:

$$\begin{bmatrix} 1990^2 & 1990 & 1 & | & 11 \\ 2010^2 & 2010 & 1 & | & 10 \\ 2030^2 & 2030 & 1 & | & 6 \end{bmatrix}$$

(b) Using technology to solve the system results in the reduced row-echelon form shown in Figure 89b. The solution is $a = -0.00375$, $b = 14.95$, and $c = -14{,}889.125$. Therefore, the symbolic representation of f is $f(x) = -0.00375x^2 + 14.95x - 14{,}889.125$.

(c) The data and f are graphed in Figure 89c. Notice that the graph of f passes through each data point.

(d) *Answers may vary.* For example, in 2015 the ratio could be $f(2015) \approx 9.3$.

Extended and Discovery Exercise for Section 9.4

1. Using technology: $w = 1$, $x = -1$, $y = 2$, $z = 0$. The solution is $(1, -1, 2, 0)$.

Checking Basic Concepts for Sections 9.3 and 9.4

1. (a) $\begin{bmatrix} 1 & -2 & 1 & -2 \\ 1 & 1 & 2 & 3 \\ 2 & -1 & -1 & 5 \end{bmatrix}$ $\begin{matrix} \\ R_1 - R_2 \rightarrow \\ 2R_2 - R_3 \rightarrow \end{matrix}$ $\begin{bmatrix} 1 & -2 & 1 & -2 \\ 0 & -3 & -1 & -5 \\ 0 & 3 & 5 & 1 \end{bmatrix}$ $\begin{matrix} (-1/3)R_2 \rightarrow \\ \\ R_2 + R_3 \rightarrow \end{matrix}$ $\begin{bmatrix} 1 & -2 & 1 & -2 \\ 0 & 1 & \frac{1}{3} & \frac{5}{3} \\ 0 & 0 & 4 & -4 \end{bmatrix}$

$\begin{matrix} \\ \\ (1/4)R_3 \rightarrow \end{matrix}$ $\begin{bmatrix} 1 & -2 & 1 & -2 \\ 0 & 1 & \frac{1}{3} & \frac{5}{3} \\ 0 & 0 & 1 & -1 \end{bmatrix}$ $\begin{matrix} (-1)R_3 + R_1 \rightarrow \\ (-1/3)R_3 + R_2 \rightarrow \\ \end{matrix}$ $\begin{bmatrix} 1 & -2 & 0 & -1 \\ 0 & 1 & 0 & 2 \\ 0 & 0 & 1 & -1 \end{bmatrix}$

$2R_2 + R_1 \rightarrow \begin{bmatrix} 1 & 0 & 0 & 3 \\ 0 & 1 & 0 & 2 \\ 0 & 0 & 1 & -1 \end{bmatrix}$

The solution is: $(3, 2, -1)$.

(b) $\begin{bmatrix} 1 & -2 & 1 & -2 \\ 1 & 1 & 2 & 3 \\ 2 & -1 & 3 & 1 \end{bmatrix}$ $\begin{matrix} \\ R_1 - R_2 \rightarrow \\ 2R_2 - R_3 \rightarrow \end{matrix}$ $\begin{bmatrix} 1 & -2 & 1 & -2 \\ 0 & -3 & -1 & -5 \\ 0 & 3 & 1 & 5 \end{bmatrix}$ $\begin{matrix} (-1/3)R_2 \rightarrow \\ \\ R_2 + R_3 \rightarrow \end{matrix}$ $\begin{bmatrix} 1 & -2 & 1 & -2 \\ 0 & 1 & \frac{1}{3} & \frac{5}{3} \\ 0 & 0 & 0 & 0 \end{bmatrix}$

$2R_2 + R_1 \rightarrow \begin{bmatrix} 1 & 0 & \frac{5}{3} & \frac{4}{3} \\ 0 & 1 & \frac{1}{3} & \frac{5}{3} \\ 0 & 0 & 0 & 0 \end{bmatrix}$

The last equation indicates that $0 = 0$, which is true. Therefore, there is an infinite number of solutions. The second equation gives: $y + \frac{1}{3}z = \frac{5}{3} \Rightarrow y = \frac{5 - z}{3}$. The first equation gives: $x + \frac{5}{3}z = \frac{4}{3} \Rightarrow$ $x = \frac{4 - 5z}{3}$. Therefore, there are infinitely many solutions which can be written as: $\left(\frac{4 - 5z}{3}, \frac{-z + 5}{3}, z \right)$.

(c) $\begin{bmatrix} 1 & -2 & 1 & -2 \\ 1 & 1 & 2 & 3 \\ 2 & -1 & 3 & 5 \end{bmatrix}$ $\begin{matrix} \\ R_1 - R_2 \rightarrow \\ 2R_2 - R_3 \rightarrow \end{matrix}$ $\begin{bmatrix} 1 & -2 & 1 & -2 \\ 0 & -3 & -1 & -5 \\ 0 & 3 & 1 & 1 \end{bmatrix}$ $\begin{matrix} (-1/3)R_2 \rightarrow \\ \\ R_2 + R_3 \rightarrow \end{matrix}$ $\begin{bmatrix} 1 & -2 & 1 & -2 \\ 0 & 1 & \frac{1}{3} & \frac{5}{3} \\ 0 & 0 & 0 & -4 \end{bmatrix}$

The last equation gives: $0 = -4$, which is false. Therefore there is no solution.

3. $\begin{bmatrix} 1 & 0 & 1 & | & 2 \\ 1 & 1 & -1 & | & 1 \\ -1 & -2 & -1 & | & 0 \end{bmatrix} \begin{matrix} \\ R_2 - R_1 \rightarrow \\ R_3 + R_1 \rightarrow \end{matrix} \begin{bmatrix} 1 & 0 & 1 & | & 2 \\ 0 & 1 & -2 & | & -1 \\ 0 & -2 & 0 & | & 2 \end{bmatrix} \begin{matrix} \\ \\ (-1/2)R_3 \rightarrow \end{matrix} \begin{bmatrix} 1 & 0 & 1 & | & 2 \\ 0 & 1 & -2 & | & -1 \\ 0 & 1 & 0 & | & -1 \end{bmatrix}$

$\begin{matrix} \\ \\ R_3 - R_2 \rightarrow \end{matrix} \begin{bmatrix} 1 & 0 & 0 & | & -2 \\ 0 & 1 & -2 & | & -1 \\ 0 & 0 & 2 & | & 0 \end{bmatrix}$

Backward substitution produces $z = 0$, $y - 2z = -1 \Rightarrow y - 0 = -1 \Rightarrow y = -1$. Substituting $z = 0$ and $y = -1$ into the first equation results in $x + 0(-1) + (0) = 2 \Rightarrow x = 2$. The solution is $(2, -1, 0)$.

9.5 Properties and Applications of Matrices

1. (a) $a_{12} = 3, b_{32} = 1, b_{22} = 0$

    (b) $a_{11} = 1, b_{11} = 1 \Rightarrow a_{11}b_{11} = 1$ $a_{12} = 3, b_{21} = 3 \Rightarrow a_{12}b_{21} = 9$,

    $a_{13} = -4, b_{31} = 3 \Rightarrow a_{13}b_{31} = -12 \Rightarrow a_{11}b_{11} + a_{12}b_{21} + a_{13}b_{31} = -2$

    (c) If $A = B$, then all elements of A must be equal to all elements of B. Then $x = 3$.

3. $x = 1$ and $y = 1$

5. The matrices are not the same size, therefore not possible.

7. (a) $\begin{bmatrix} 4 & -1 \\ -1 & 4 \end{bmatrix} + \begin{bmatrix} -1 & 4 \\ 4 & -1 \end{bmatrix} = \begin{bmatrix} 3 & 3 \\ 3 & 3 \end{bmatrix}$

    (b) $\begin{bmatrix} -1 & 4 \\ 4 & -1 \end{bmatrix} + \begin{bmatrix} 4 & -1 \\ -1 & 4 \end{bmatrix} = \begin{bmatrix} 3 & 3 \\ 3 & 3 \end{bmatrix}$

    (c) $\begin{bmatrix} 4 & -1 \\ -1 & 4 \end{bmatrix} - \begin{bmatrix} -1 & 4 \\ 4 & -1 \end{bmatrix} = \begin{bmatrix} 5 & -5 \\ -5 & 5 \end{bmatrix}$

9. (a) $\begin{bmatrix} 3 & 4 & -1 \\ 0 & -3 & 2 \\ -2 & 5 & 10 \end{bmatrix} + \begin{bmatrix} 11 & 5 & -2 \\ 4 & -7 & 12 \\ 6 & 6 & 6 \end{bmatrix} = \begin{bmatrix} 14 & 9 & -3 \\ 4 & -10 & 14 \\ 4 & 11 & 16 \end{bmatrix}$

    (b) $\begin{bmatrix} 11 & 5 & -2 \\ 4 & -7 & 12 \\ 6 & 6 & 6 \end{bmatrix} + \begin{bmatrix} 3 & 4 & -1 \\ 0 & -3 & 2 \\ -2 & 5 & 10 \end{bmatrix} = \begin{bmatrix} 14 & 9 & -3 \\ 4 & -10 & 14 \\ 4 & 11 & 16 \end{bmatrix}$

    (c) $\begin{bmatrix} 3 & 4 & -1 \\ 0 & -3 & 2 \\ -2 & 5 & 10 \end{bmatrix} - \begin{bmatrix} 11 & 5 & -2 \\ 4 & -7 & 12 \\ 6 & 6 & 6 \end{bmatrix} = \begin{bmatrix} -8 & -1 & 1 \\ -4 & 4 & -10 \\ -8 & -1 & 4 \end{bmatrix}$

11. (a) $A + B = \begin{bmatrix} 2 & -6 \\ 3 & 1 \end{bmatrix} + \begin{bmatrix} -1 & 0 \\ -2 & 3 \end{bmatrix} = \begin{bmatrix} 1 & -6 \\ 1 & 4 \end{bmatrix}$

   (b) $3A = 3\begin{bmatrix} 2 & -6 \\ 3 & 1 \end{bmatrix} = \begin{bmatrix} 6 & -18 \\ 9 & 3 \end{bmatrix}$

   (c) $2A - 3B = 2\begin{bmatrix} 2 & -6 \\ 3 & 1 \end{bmatrix} - 3\begin{bmatrix} -1 & 0 \\ -2 & 3 \end{bmatrix} = \begin{bmatrix} 7 & -12 \\ 12 & -7 \end{bmatrix}$

13. (a) $A + B$ is undefined since A is 3×3 and B is 2×3. They do not have the same dimension.

   (b) $3A = 3\begin{bmatrix} 1 & -1 & 0 \\ 1 & 5 & 9 \\ -4 & 8 & -5 \end{bmatrix} = \begin{bmatrix} 3 & -3 & 0 \\ 3 & 15 & 27 \\ -12 & 24 & -15 \end{bmatrix}$

   (c) $2A - 3B$ is undefined since A is 3×3 and B is 2×3. They do not have the same dimension.

15. (a) $A + B = \begin{bmatrix} -2 & -1 \\ -5 & 1 \\ 2 & -3 \end{bmatrix} + \begin{bmatrix} 2 & -1 \\ 3 & 1 \\ 7 & -5 \end{bmatrix} = \begin{bmatrix} 0 & -2 \\ -2 & 2 \\ 9 & -8 \end{bmatrix}$

   (b) $3A = 3\begin{bmatrix} -2 & -1 \\ -5 & 1 \\ 2 & -3 \end{bmatrix} = \begin{bmatrix} -6 & -3 \\ -15 & 3 \\ 6 & -9 \end{bmatrix}$

   (c) $2A - 3B = 2\begin{bmatrix} -2 & -1 \\ -5 & 1 \\ 2 & -3 \end{bmatrix} - 3\begin{bmatrix} 2 & -1 \\ 3 & 1 \\ 7 & -5 \end{bmatrix} = \begin{bmatrix} -10 & 1 \\ -19 & -1 \\ -17 & 9 \end{bmatrix}$

17. $2\begin{bmatrix} 2 & -1 \\ 5 & 1 \\ 0 & 3 \end{bmatrix} + \begin{bmatrix} 5 & 0 \\ 7 & -3 \\ 1 & 1 \end{bmatrix} - \begin{bmatrix} 9 & -4 \\ 4 & 4 \\ 1 & 6 \end{bmatrix} = \begin{bmatrix} 0 & 2 \\ 13 & -5 \\ 0 & 1 \end{bmatrix}$

19. $\begin{bmatrix} 4 & 6 \\ 3 & -7 \end{bmatrix} - 2\begin{bmatrix} 1 & 0 \\ -4 & 1 \end{bmatrix} = \begin{bmatrix} 2 & 6 \\ 11 & -9 \end{bmatrix}$

21. $2\begin{bmatrix} 2 & -1 & -1 \\ -1 & 2 & -1 \\ -1 & -1 & 2 \end{bmatrix} + 3\begin{bmatrix} 1 & 2 & 3 \\ 2 & 1 & 3 \\ 2 & 3 & 1 \end{bmatrix} = \begin{bmatrix} 7 & 4 & 7 \\ 4 & 7 & 7 \\ 4 & 7 & 7 \end{bmatrix}$

23. The "1" is dark gray and the background is light gray.

   $A = \begin{bmatrix} 1 & 2 & 1 \\ 1 & 2 & 1 \\ 1 & 2 & 1 \end{bmatrix}$

25. To enhance the contrast, change light gray to white and change dark gray to black. This could be accomplished by adding the 3×3 matrix B to A.

   $B = \begin{bmatrix} -1 & 1 & -1 \\ -1 & 1 & -1 \\ -1 & 1 & -1 \end{bmatrix}$; $A + B = \begin{bmatrix} 1 & 2 & 1 \\ 1 & 2 & 1 \\ 1 & 2 & 1 \end{bmatrix} + \begin{bmatrix} -1 & 1 & -1 \\ -1 & 1 & -1 \\ -1 & 1 & -1 \end{bmatrix} = \begin{bmatrix} 0 & 3 & 0 \\ 0 & 3 & 0 \\ 0 & 3 & 0 \end{bmatrix}$

27. A and B are both 2×2 so AB and BA are also both 2×2.

$$AB = \begin{bmatrix} 1 & -1 \\ 2 & 0 \end{bmatrix}\begin{bmatrix} -2 & 3 \\ 1 & 2 \end{bmatrix} = \begin{bmatrix} -3 & 1 \\ -4 & 6 \end{bmatrix}; \; BA = \begin{bmatrix} -2 & 3 \\ 1 & 2 \end{bmatrix}\begin{bmatrix} 1 & -1 \\ 2 & 0 \end{bmatrix} = \begin{bmatrix} 4 & 2 \\ 5 & -1 \end{bmatrix}$$

29. Since both A and B are 2×3, the number of rows in B is not equal to the number of columns in A, so AB is undefined. Also, the number of rows in A is not equal to the number of columns in B so BA is undefined.

31. $AB = \begin{bmatrix} 3 & -1 \\ 1 & 0 \\ -2 & -4 \end{bmatrix}\begin{bmatrix} -2 & 5 & -3 \\ 9 & -7 & 0 \end{bmatrix} =$

$$\begin{bmatrix} 3(-2) + (-1)(9) & 3(5) + (-1)(-7) & 3(-3) + (-1)(0) \\ 1(-2) + 0(9) & 1(5) + 0(-7) & 1(-3) + 0(0) \\ -2(-2) + (-4)(9) & -2(5) + (-4)(-7) & -2(-3) + (-4)(0) \end{bmatrix} \Rightarrow AB = \begin{bmatrix} -15 & 22 & -9 \\ -2 & 5 & -3 \\ -32 & 18 & 6 \end{bmatrix}$$

$$BA = \begin{bmatrix} -2 & 5 & -3 \\ 9 & -7 & 0 \end{bmatrix}\begin{bmatrix} 3 & -1 \\ 1 & 0 \\ -2 & -4 \end{bmatrix} =$$

$$\begin{bmatrix} -2(3) + 5(1) + (-3)(-2) & -2(-1) + (5)(0) + (-3)(-4) \\ 9(3) + (-7)(1) + 0(-2) & 9(-1) + (-7)(0) + 0(-4) \end{bmatrix} \Rightarrow BA = \begin{bmatrix} 5 & 14 \\ 20 & -9 \end{bmatrix}$$

33. AB is undefined, we cannot multiply a 3×3 by a 2×3.

$$BA = \begin{bmatrix} -1 & 3 & -1 \\ 7 & -7 & 1 \end{bmatrix}\begin{bmatrix} 1 & -1 & 0 \\ 2 & -1 & 5 \\ 6 & 1 & -4 \end{bmatrix} =$$

$$\begin{bmatrix} -1(1) + 3(2) + (-1)(6) & -1(-1) + 3(-1) + (-1)(1) & -1(0) + 3(5) + (-1)(-4) \\ 7(1) + (-7)(2) + 1(6) & 7(-1) + (-7)(-1) + 1(1) & 7(0) + (-7)(5) + 1(-4) \end{bmatrix} \Rightarrow$$

$$BA = \begin{bmatrix} -1 & -3 & 19 \\ -1 & 1 & -39 \end{bmatrix}$$

35. Since A is 2×2 and B is 3×1, the number of rows in B is not equal to the number of columns in A, so AB is undefined. Also, the number of rows in A is not equal to the number of columns in B so BA is undefined.

37. A and B are both 3×3 so AB and BA are also both 3×3.

$$AB = \begin{bmatrix} 2 & -1 & 3 \\ 0 & 1 & 0 \\ 2 & -2 & 3 \end{bmatrix}\begin{bmatrix} 1 & 5 & -1 \\ 0 & 1 & 3 \\ -1 & 2 & 1 \end{bmatrix} = \begin{bmatrix} -1 & 15 & -2 \\ 0 & 1 & 3 \\ -1 & 14 & -5 \end{bmatrix};$$

$$BA = \begin{bmatrix} 1 & 5 & -1 \\ 0 & 1 & 3 \\ -1 & 2 & 1 \end{bmatrix}\begin{bmatrix} 2 & -1 & 3 \\ 0 & 1 & 0 \\ 2 & -2 & 3 \end{bmatrix} = \begin{bmatrix} 0 & 6 & 0 \\ 6 & -5 & 9 \\ 0 & 1 & 0 \end{bmatrix}$$

39. A is 2×2 and B is 2×1 so AB is 2×1. However BA is undefined since the number of rows in A is not equal to the number of columns in B.

$$AB = \begin{bmatrix} 2 & -1 \\ 3 & 1 \end{bmatrix}\begin{bmatrix} 1 \\ 3 \end{bmatrix} = \begin{bmatrix} -1 \\ 6 \end{bmatrix}$$

41. *A* is 2 × 2 and *B* is 2 × 3 so *AB* is 2 × 3. However *BA* is undefined since the number of rows in *A* is not equal to the number of columns in *B*.

$$AB = \begin{bmatrix} -3 & 1 \\ 2 & -4 \end{bmatrix}\begin{bmatrix} 1 & 0 & -2 \\ -4 & 8 & 1 \end{bmatrix} = \begin{bmatrix} -7 & 8 & 7 \\ 18 & -32 & -8 \end{bmatrix}$$

43. *A* is 3 × 3 and *B* is 3 × 1 so *AB* is 3 × 1. However *BA* is undefined since the number of rows in *A* is not equal to the number of columns in *B*.

$$AB = \begin{bmatrix} 1 & 0 & -2 \\ 3 & -4 & 1 \\ 2 & 0 & 5 \end{bmatrix}\begin{bmatrix} 1 \\ -1 \\ 3 \end{bmatrix} = \begin{bmatrix} -5 \\ 10 \\ 17 \end{bmatrix}$$

45. Using technology; $BA = \begin{bmatrix} 3 & -2 & 4 \\ 5 & 2 & 3 \\ 7 & 5 & 4 \end{bmatrix}\begin{bmatrix} 1 & 1 & -5 \\ -1 & 0 & -7 \\ -6 & 4 & 3 \end{bmatrix} = \begin{bmatrix} -19 & 19 & 11 \\ 21 & -7 & -48 \\ -22 & 23 & -58 \end{bmatrix}$

47. Using technology; $3A^2 + 2B = 3\begin{bmatrix} 3 & -2 & 4 \\ 5 & 2 & 3 \\ 7 & 5 & 4 \end{bmatrix}^2 + 2\begin{bmatrix} 1 & 1 & -5 \\ -1 & 0 & -7 \\ -6 & 4 & 3 \end{bmatrix} = \begin{bmatrix} 83 & 32 & 92 \\ 10 & -63 & -8 \\ 210 & 56 & 93 \end{bmatrix}$

49. (a) $B + C = \begin{bmatrix} 6 & 2 & 7 \\ 3 & -4 & -5 \\ 7 & 1 & 0 \end{bmatrix} + \begin{bmatrix} 1 & 4 & -3 \\ 8 & 1 & -1 \\ 4 & 6 & -2 \end{bmatrix} = \begin{bmatrix} 7 & 6 & 4 \\ 11 & -3 & -6 \\ 11 & 7 & -2 \end{bmatrix}$

$$A(B + C) = \begin{bmatrix} 2 & -1 & 3 \\ 1 & 3 & -5 \\ 0 & -2 & 1 \end{bmatrix}\begin{bmatrix} 7 & 6 & 4 \\ 11 & -3 & -6 \\ 11 & 7 & -2 \end{bmatrix} = \begin{bmatrix} 36 & 36 & 8 \\ -15 & -38 & -4 \\ -11 & 13 & 10 \end{bmatrix}$$

(b) $AB = \begin{bmatrix} 2 & -1 & 3 \\ 1 & 3 & -5 \\ 0 & -2 & 1 \end{bmatrix}\begin{bmatrix} 6 & 2 & 7 \\ 3 & -4 & -5 \\ 7 & 1 & 0 \end{bmatrix} = \begin{bmatrix} 30 & 11 & 19 \\ -20 & -15 & -8 \\ 1 & 9 & 10 \end{bmatrix}$

$AC = \begin{bmatrix} 2 & -1 & 3 \\ 1 & 3 & -5 \\ 0 & -2 & 1 \end{bmatrix}\begin{bmatrix} 1 & 4 & -3 \\ 8 & 1 & -1 \\ 4 & 6 & -2 \end{bmatrix} = \begin{bmatrix} 6 & 25 & -11 \\ 5 & -23 & 4 \\ -12 & 4 & 0 \end{bmatrix}$; $AB + AC = \begin{bmatrix} 36 & 36 & 8 \\ -15 & -38 & -4 \\ -11 & 13 & 10 \end{bmatrix}$

$A(B + C) = AB + AC$, which indicates that the distributive property holds for matrices.

51. (a) $(A - B)^2 = \begin{bmatrix} -4 & -3 & -4 \\ -2 & 7 & 0 \\ -7 & -3 & 1 \end{bmatrix}\begin{bmatrix} -4 & -3 & -4 \\ -2 & 7 & 0 \\ -7 & -3 & 1 \end{bmatrix} = \begin{bmatrix} 50 & 3 & 12 \\ -6 & 55 & 8 \\ 27 & -3 & 29 \end{bmatrix}$

(b) $A^2 - AB - BA + B^2 =$

$$\begin{bmatrix} 3 & -11 & 14 \\ 5 & 18 & -17 \\ -2 & -8 & 11 \end{bmatrix} - \begin{bmatrix} 30 & 11 & 19 \\ -20 & -15 & -8 \\ 1 & 9 & 10 \end{bmatrix} - \begin{bmatrix} 14 & -14 & 15 \\ 2 & -5 & 24 \\ 15 & -4 & 16 \end{bmatrix} + \begin{bmatrix} 91 & 11 & 32 \\ -29 & 17 & 41 \\ 45 & 10 & 44 \end{bmatrix} =$$

$$\begin{bmatrix} 50 & 3 & 12 \\ -6 & 55 & 8 \\ 27 & -3 & 29 \end{bmatrix}$$

$(A - B)^2 = A^2 - AB - BA + B^2$, which indicates that matrices seem to conform to common rules of algebra except for the commutative property since $AB \neq BA$, in general.

53. Because person 1, likes person 4, we put a 1 in row 1 column 4. Similarly, Person 4 likes person 2, so we put a 1 in row 4 column 2. When no arrow exists to indicate that one person likes another, we place a 0 in the appropriate row and column matrix. Using this process results in the following matrix.

$$\begin{bmatrix} 0 & 0 & 1 & 1 \\ 1 & 0 & 0 & 0 \\ 1 & 0 & 0 & 1 \\ 1 & 1 & 1 & 0 \end{bmatrix}$$

55. Since there is only one 1 in the second column then Person 2 is the least liked person in the network.

57. Because person 1, likes person 3, there is a 1 in row 1 column 3. Similarly, Person 3 likes person 2, so there is a 1 in row 3 column 2. When no arrow exists to indicate that one person likes another, there is a 0 in the appropriate row and column matrix. Using this process results in the following diagram.

59. No one likes Person 4.

61. To make a negative image, subtract the matrix A from a completely black image matrix B.

$$B = \begin{bmatrix} 3 & 3 & 3 \\ 3 & 3 & 3 \\ 3 & 3 & 3 \end{bmatrix}; \; B - A = \begin{bmatrix} 3 & 3 & 3 \\ 3 & 3 & 3 \\ 3 & 3 & 3 \end{bmatrix} - \begin{bmatrix} 0 & 3 & 0 \\ 0 & 3 & 0 \\ 0 & 3 & 0 \end{bmatrix} = \begin{bmatrix} 3 & 0 & 3 \\ 3 & 0 & 3 \\ 3 & 0 & 3 \end{bmatrix}$$

63. $A = \begin{bmatrix} 3 & 3 & 3 & 3 \\ 3 & 0 & 0 & 0 \\ 3 & 3 & 3 & 0 \\ 3 & 0 & 0 & 0 \\ 3 & 0 & 0 & 0 \end{bmatrix}$

65. (a) One possible solution for a "Z" is $A = \begin{bmatrix} 3 & 3 & 3 & 3 \\ 0 & 0 & 3 & 0 \\ 0 & 3 & 0 & 0 \\ 3 & 3 & 3 & 3 \end{bmatrix}$

(b) If A is the matrix in part (a) then

$$B = \begin{bmatrix} 3 & 3 & 3 & 3 \\ 3 & 3 & 3 & 3 \\ 3 & 3 & 3 & 3 \\ 3 & 3 & 3 & 3 \end{bmatrix} \text{ and } B - A = \begin{bmatrix} 3 & 3 & 3 & 3 \\ 3 & 3 & 3 & 3 \\ 3 & 3 & 3 & 3 \\ 3 & 3 & 3 & 3 \end{bmatrix} - \begin{bmatrix} 3 & 3 & 3 & 3 \\ 0 & 0 & 3 & 0 \\ 0 & 3 & 0 & 0 \\ 3 & 3 & 3 & 3 \end{bmatrix} = \begin{bmatrix} 0 & 0 & 0 & 0 \\ 3 & 3 & 0 & 3 \\ 3 & 0 & 3 & 3 \\ 0 & 0 & 0 & 0 \end{bmatrix}$$

67. (a) One possible solution for a "L" is $A = \begin{bmatrix} 3 & 0 & 0 & 0 \\ 3 & 0 & 0 & 0 \\ 3 & 0 & 0 & 0 \\ 3 & 3 & 3 & 3 \end{bmatrix}$

(b) If A is the matrix in part (a) then

$$B = \begin{bmatrix} 3 & 3 & 3 & 3 \\ 3 & 3 & 3 & 3 \\ 3 & 3 & 3 & 3 \\ 3 & 3 & 3 & 3 \end{bmatrix} \text{ and } B - A = \begin{bmatrix} 3 & 3 & 3 & 3 \\ 3 & 3 & 3 & 3 \\ 3 & 3 & 3 & 3 \\ 3 & 3 & 3 & 3 \end{bmatrix} - \begin{bmatrix} 3 & 0 & 0 & 0 \\ 3 & 0 & 0 & 0 \\ 3 & 0 & 0 & 0 \\ 3 & 3 & 3 & 3 \end{bmatrix} = \begin{bmatrix} 0 & 3 & 3 & 3 \\ 0 & 3 & 3 & 3 \\ 0 & 3 & 3 & 3 \\ 0 & 0 & 0 & 0 \end{bmatrix}$$

69. (a) These tables can be represented by the matrices A and B where $A = \begin{bmatrix} 12 & 4 \\ 8 & 7 \end{bmatrix}$ and $B = \begin{bmatrix} 55 \\ 70 \end{bmatrix}$.

(b) The product AB of these matrices calculates tuition cost for each student.

$$AB = \begin{bmatrix} 12 & 4 \\ 8 & 7 \end{bmatrix}\begin{bmatrix} 55 \\ 70 \end{bmatrix} = \begin{bmatrix} 12(55) + 4(70) \\ 8(55) + 7(70) \end{bmatrix} = \begin{bmatrix} 940 \\ 930 \end{bmatrix}$$

Student 1 is taking 12 credits at $55 each and 4 credits at $70 each. The total tuition for student 1 is

$12(\$55) + 4(\$70) = \$940$. Similarly, the tuition for student 2 is $930.

71. (a) These tables can be represented by the matrices A and B where $A = \begin{bmatrix} 10 & 5 \\ 9 & 8 \\ 11 & 3 \end{bmatrix}$ and $B = \begin{bmatrix} 60 \\ 70 \end{bmatrix}$.

(b) The product AB of these matrices calculates tuition cost for each student.

$$AB = \begin{bmatrix} 10 & 5 \\ 9 & 8 \\ 11 & 3 \end{bmatrix}\begin{bmatrix} 60 \\ 70 \end{bmatrix} = \begin{bmatrix} 10(60) + 5(70) \\ 9(60) + 8(70) \\ 11(60) + 3(70) \end{bmatrix} = \begin{bmatrix} 950 \\ 1100 \\ 870 \end{bmatrix}$$

The total tuition for student 1 is $10(\$60) + 5(\$70) = \$950$. Similarly, the tuition for student 2 is $1100

and the tuition for student 3 is $870.

73. $AB = \begin{bmatrix} 3 & 4 & 8 \\ 5 & 6 & 2 \end{bmatrix}\begin{bmatrix} 10 \\ 20 \\ 30 \end{bmatrix} = \begin{bmatrix} 350 \\ 230 \end{bmatrix}$; The total cost of Order 1 is $350, and the total cost of Order 2 is $230.

75. Because there is a link from Page 1 to Page 3 we put a 1 in row 1 column 3. Similarly, there is a link from Page

2 to Page 4, so we put a 1 in row 2 column 4. When no link exists from one web page to another, we place a 0

in the appropriate row and column. Using this process results in the following matrix.

$$\begin{bmatrix} 0 & 0 & 1 & 1 \\ 1 & 0 & 0 & 1 \\ 0 & 0 & 0 & 1 \\ 0 & 0 & 1 & 0 \end{bmatrix}$$

77. $\begin{bmatrix} 0 & 0 & 1 & 1 \\ 1 & 0 & 0 & 1 \\ 0 & 0 & 0 & 1 \\ 0 & 0 & 1 & 0 \end{bmatrix} \cdot \begin{bmatrix} 0 & 0 & 1 & 1 \\ 1 & 0 & 0 & 1 \\ 0 & 0 & 0 & 1 \\ 0 & 0 & 1 & 0 \end{bmatrix} = \begin{bmatrix} 0 & 0 & 1 & 1 \\ 0 & 0 & 2 & 1 \\ 0 & 0 & 1 & 0 \\ 0 & 0 & 0 & 1 \end{bmatrix}$

79. There are two different 2-click paths from page 2 to page 3.

Extended and Discovery Exercises for Section 9.5

1. $\begin{bmatrix} C \\ M \\ Y \end{bmatrix} = \begin{bmatrix} 1 \\ 1 \\ 1 \end{bmatrix} - \begin{bmatrix} 0.631 \\ 1 \\ 0.933 \end{bmatrix} = \begin{bmatrix} 0.369 \\ 0 \\ 0.067 \end{bmatrix}$

Aquamarine is represented by (0.369, 0, 0.067) in *CMY*.

3. $\begin{bmatrix} R \\ G \\ B \end{bmatrix} = \begin{bmatrix} 1 \\ 1 \\ 1 \end{bmatrix} - \begin{bmatrix} C \\ M \\ Y \end{bmatrix}$

9.6 Inverses of Matrices

1. *B* is the inverse of *A*.

$AB = \begin{bmatrix} 4 & 3 \\ 5 & 4 \end{bmatrix}\begin{bmatrix} 4 & -3 \\ -5 & 4 \end{bmatrix} = \begin{bmatrix} 1 & 0 \\ 0 & 1 \end{bmatrix}$ and $BA = \begin{bmatrix} 4 & -3 \\ -5 & 4 \end{bmatrix}\begin{bmatrix} 4 & 3 \\ 5 & 4 \end{bmatrix} = \begin{bmatrix} 1 & 0 \\ 0 & 1 \end{bmatrix}$

3. *B* is the inverse of *A*.

$AB = \begin{bmatrix} 1 & -1 & 2 \\ 0 & 1 & -1 \\ 1 & 0 & 2 \end{bmatrix}\begin{bmatrix} 2 & 2 & -1 \\ -1 & 0 & 1 \\ -1 & -1 & 1 \end{bmatrix} = \begin{bmatrix} 1 & 0 & 0 \\ 0 & 1 & 0 \\ 0 & 0 & 1 \end{bmatrix};$

$BA = \begin{bmatrix} 2 & 2 & -1 \\ -1 & 0 & 1 \\ -1 & -1 & 1 \end{bmatrix}\begin{bmatrix} 1 & -1 & 2 \\ 0 & 1 & -1 \\ 1 & 0 & 2 \end{bmatrix} = \begin{bmatrix} 1 & 0 & 0 \\ 0 & 1 & 0 \\ 0 & 0 & 1 \end{bmatrix}$

5. *B* is not the inverse of *A*.

$AB = \begin{bmatrix} 2 & 1 & -1 \\ 3 & 0 & 2 \\ -1 & 0 & 1 \end{bmatrix}\begin{bmatrix} 0 & 1 & -2 \\ 1 & -3 & 7 \\ 0 & -1 & 3 \end{bmatrix} = \begin{bmatrix} 1 & 0 & 0 \\ 0 & 1 & 0 \\ 0 & -2 & 5 \end{bmatrix};$

$BA = \begin{bmatrix} 0 & 1 & -2 \\ 1 & -3 & 7 \\ 0 & -1 & 3 \end{bmatrix}\begin{bmatrix} 2 & 1 & -1 \\ 3 & 0 & 2 \\ -1 & 0 & 1 \end{bmatrix} = \begin{bmatrix} 5 & 0 & 0 \\ -14 & 1 & 0 \\ -6 & 0 & 1 \end{bmatrix}$

7. $AA^{-1} = \begin{bmatrix} 1 & 1 \\ 1 & 2 \end{bmatrix}\begin{bmatrix} 2 & -1 \\ -1 & k \end{bmatrix} = \begin{bmatrix} 1 & -1 + k \\ 0 & 2k - 1 \end{bmatrix}$

We must have $-1 + k = 0$ and $2k - 1 = 1$. The solution to both equations is $k = 1$.

9. $AA^{-1} = \begin{bmatrix} 1 & 3 \\ -1 & -5 \end{bmatrix}\begin{bmatrix} k & 1.5 \\ -0.5 & -0.5 \end{bmatrix} = \begin{bmatrix} k - 1.5 & 0 \\ -k + 2.5 & 1 \end{bmatrix}$

We must have $k - 1.5 = 1$ and $-k + 2.5 = 0$. The solution to both equations is $k = 2.5$.

11. I_2 multiplied by any 2×2 matrix *A* is equal to *A*, that is, $I_2A = AI_2 = A$.

13. I_3 multiplied by any 3×3 matrix *A* is equal to *A*, that is, $I_3A = AI_3 = A$.

15. $A|I_2 = \begin{bmatrix} 1 & 2 & | & 1 & 0 \\ 1 & 3 & | & 0 & 1 \end{bmatrix} \begin{matrix} \\ R_2 - R_1 \rightarrow \end{matrix} \begin{bmatrix} 1 & 2 & | & 1 & 0 \\ 0 & 1 & | & -1 & 1 \end{bmatrix} \begin{matrix} R_1 - 2R_2 \rightarrow \\ \end{matrix} \begin{bmatrix} 1 & 0 & | & 3 & -2 \\ 0 & 1 & | & -1 & 1 \end{bmatrix}; A^{-1} = \begin{bmatrix} 3 & -2 \\ -1 & 1 \end{bmatrix}$

17. $A|I_2 = \begin{bmatrix} -1 & 2 & | & 1 & 0 \\ 3 & -5 & | & 0 & 1 \end{bmatrix} \xrightarrow{-1R_1} \begin{bmatrix} 1 & -2 & | & -1 & 0 \\ 3 & -5 & | & 0 & 1 \end{bmatrix} \xrightarrow{R_2 - 3R_1} \begin{bmatrix} 1 & -2 & | & -1 & 0 \\ 0 & 1 & | & 3 & 1 \end{bmatrix}$

$R_1 + 2R_2 \rightarrow \begin{bmatrix} 1 & 0 & | & 5 & 2 \\ 0 & 1 & | & 3 & 1 \end{bmatrix}$; $A^{-1} = \begin{bmatrix} 5 & 2 \\ 3 & 1 \end{bmatrix}$

19. $A|I_2 = \begin{bmatrix} 8 & 5 & | & 1 & 0 \\ 2 & 1 & | & 0 & 1 \end{bmatrix} \begin{array}{c} (1/8)R_1 \rightarrow \\ R_1 - 4R_2 \rightarrow \end{array} \begin{bmatrix} 1 & \frac{5}{8} & | & \frac{1}{8} & 0 \\ 0 & 1 & | & 1 & -4 \end{bmatrix} \xrightarrow{(-5/8)R_2 + R_1} \begin{bmatrix} 1 & 0 & | & -\frac{1}{2} & \frac{5}{2} \\ 0 & 1 & | & 1 & -4 \end{bmatrix}$;

$A^{-1} = \begin{bmatrix} -\frac{1}{2} & \frac{5}{2} \\ 1 & -4 \end{bmatrix}$

21. $A|I_3 = \begin{bmatrix} 0 & 0 & 1 & | & 1 & 0 & 0 \\ 1 & 0 & 0 & | & 0 & 1 & 0 \\ 0 & 1 & 0 & | & 0 & 0 & 1 \end{bmatrix} \begin{array}{c} R_2 \rightarrow \\ R_3 \rightarrow \\ R_1 \rightarrow \end{array} \begin{bmatrix} 1 & 0 & 0 & | & 0 & 1 & 0 \\ 0 & 1 & 0 & | & 0 & 0 & 1 \\ 0 & 0 & 1 & | & 1 & 0 & 0 \end{bmatrix}$; $A^{-1} = \begin{bmatrix} 0 & 1 & 0 \\ 0 & 0 & 1 \\ 1 & 0 & 0 \end{bmatrix}$

23. $A|I_3 = \begin{bmatrix} 1 & 0 & 1 & | & 1 & 0 & 0 \\ 2 & 1 & 3 & | & 0 & 1 & 0 \\ -1 & 1 & 1 & | & 0 & 0 & 1 \end{bmatrix} \begin{array}{c} R_2 - 2R_1 \rightarrow \\ R_3 + R_1 \rightarrow \end{array} \begin{bmatrix} 1 & 0 & 1 & | & 1 & 0 & 0 \\ 0 & 1 & 1 & | & -2 & 1 & 0 \\ 0 & 1 & 2 & | & 1 & 0 & 1 \end{bmatrix} R_3 - R_2 \rightarrow$

$\begin{bmatrix} 1 & 0 & 1 & | & 1 & 0 & 0 \\ 0 & 1 & 1 & | & -2 & 1 & 0 \\ 0 & 0 & 1 & | & 3 & -1 & 1 \end{bmatrix} \begin{array}{c} R_1 - R_3 \rightarrow \\ R_2 - R_3 \rightarrow \end{array} \begin{bmatrix} 1 & 0 & 0 & | & -2 & 1 & -1 \\ 0 & 1 & 0 & | & -5 & 2 & -1 \\ 0 & 0 & 1 & | & 3 & -1 & 1 \end{bmatrix}$; $A^{-1} = \begin{bmatrix} -2 & 1 & -1 \\ -5 & 2 & -1 \\ 3 & -1 & 1 \end{bmatrix}$

25. $\begin{bmatrix} 1 & 2 & -1 & | & 1 & 0 & 0 \\ 2 & 5 & 0 & | & 0 & 1 & 0 \\ -1 & -1 & 2 & | & 0 & 0 & 1 \end{bmatrix} \begin{array}{c} 2R_1 - R_2 \rightarrow \\ R_3 + R_1 \rightarrow \end{array} \begin{bmatrix} 1 & 2 & -1 & | & 1 & 0 & 0 \\ 0 & -1 & -2 & | & 2 & -1 & 0 \\ 0 & 1 & 1 & | & 1 & 0 & 1 \end{bmatrix}$

$\begin{array}{c} (-1)R_2 \rightarrow \\ R_2 + R_3 \rightarrow \end{array} \begin{bmatrix} 1 & 2 & -1 & | & 1 & 0 & 0 \\ 0 & 1 & 2 & | & -2 & 1 & 0 \\ 0 & 0 & -1 & | & 3 & -1 & 1 \end{bmatrix} \begin{array}{c} R_1 - R_3 \rightarrow \\ 2R_3 + R_2 \rightarrow \\ (-1)R_3 \rightarrow \end{array} \begin{bmatrix} 1 & 2 & 0 & | & -2 & 1 & -1 \\ 0 & 1 & 0 & | & 4 & -1 & 2 \\ 0 & 0 & 1 & | & -3 & 1 & -1 \end{bmatrix}$

$R_1 - 2R_2 \rightarrow \begin{bmatrix} 1 & 0 & 0 & | & -10 & 3 & -5 \\ 0 & 1 & 0 & | & 4 & -1 & 2 \\ 0 & 0 & 1 & | & -3 & 1 & -1 \end{bmatrix}$; $A^{-1} = \begin{bmatrix} -10 & 3 & -5 \\ 4 & -1 & 2 \\ -3 & 1 & -1 \end{bmatrix}$

27. $\begin{bmatrix} -2 & 1 & -3 & | & 1 & 0 & 0 \\ 0 & 1 & 2 & | & 0 & 1 & 0 \\ 1 & -2 & 1 & | & 0 & 0 & 1 \end{bmatrix} \begin{array}{c} (-1/2)R_1 \rightarrow \\ 2R_3 + R_1 \rightarrow \end{array} \begin{bmatrix} 1 & -\frac{1}{2} & \frac{3}{2} & | & -\frac{1}{2} & 0 & 0 \\ 0 & 1 & 2 & | & 0 & 1 & 0 \\ 0 & -3 & -1 & | & 1 & 0 & 2 \end{bmatrix}$

$3R_2 + R_3 \rightarrow \begin{bmatrix} 1 & -\frac{1}{2} & \frac{3}{2} & | & -\frac{1}{2} & 0 & 0 \\ 0 & 1 & 2 & | & 0 & 1 & 0 \\ 0 & 0 & 5 & | & 1 & 3 & 2 \end{bmatrix} (1/5)R_3 \rightarrow \begin{bmatrix} 1 & -\frac{1}{2} & \frac{3}{2} & | & -\frac{1}{2} & 0 & 0 \\ 0 & 1 & 2 & | & 0 & 1 & 0 \\ 0 & 0 & 1 & | & \frac{1}{5} & \frac{3}{5} & \frac{2}{5} \end{bmatrix}$

$\begin{array}{c} R_1 - (3/2)R_3 \rightarrow \\ (-2)R_3 + R_2 \rightarrow \end{array} \begin{bmatrix} 1 & -\frac{1}{2} & 0 & | & -\frac{4}{5} & -\frac{9}{10} & -\frac{3}{5} \\ 0 & 1 & 0 & | & -\frac{2}{5} & -\frac{1}{5} & -\frac{4}{5} \\ 0 & 0 & 1 & | & \frac{1}{5} & \frac{3}{5} & \frac{2}{5} \end{bmatrix} R_1 + (1/2)R_2 \rightarrow \begin{bmatrix} 1 & 0 & 0 & | & -1 & -1 & -1 \\ 0 & 1 & 0 & | & -\frac{2}{5} & -\frac{1}{5} & -\frac{4}{5} \\ 0 & 0 & 1 & | & \frac{1}{5} & \frac{3}{5} & \frac{2}{5} \end{bmatrix}$;

$A^{-1} = \begin{bmatrix} -1 & -1 & -1 \\ -\frac{2}{5} & -\frac{1}{5} & -\frac{4}{5} \\ \frac{1}{5} & \frac{3}{5} & \frac{2}{5} \end{bmatrix}$

29. $A = \begin{bmatrix} 0.5 & -1.5 \\ 0.2 & -0.5 \end{bmatrix} \Rightarrow A^{-1} = \begin{bmatrix} -10 & 30 \\ -4 & 10 \end{bmatrix}$ as shown in Figure 29.

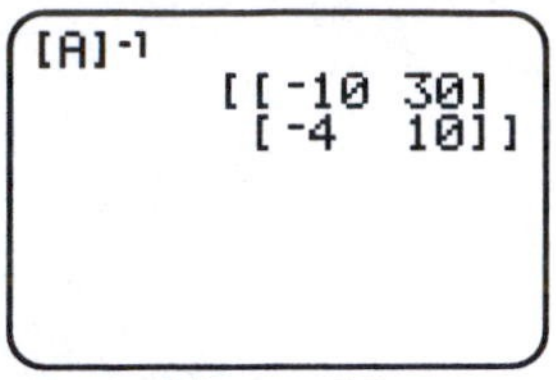
[A]-1
[[-10 30]
[-4 10]]

Figure 29

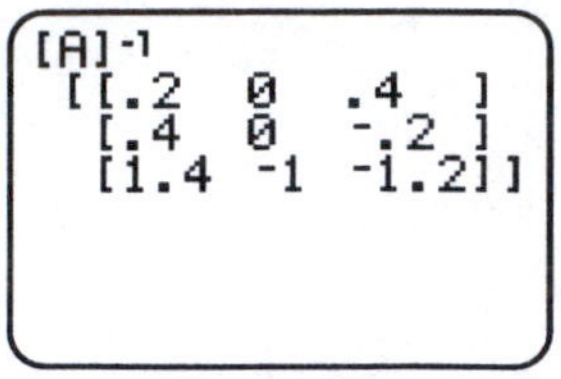
[A]-1
[[.2 0 .4]
[.4 0 -.2]
[1.4 -1 -1.2]]

Figure 31

31. $A = \begin{bmatrix} 1 & 2 & 0 \\ -1 & 4 & -1 \\ 2 & -1 & 0 \end{bmatrix} \Rightarrow A^{-1} = \begin{bmatrix} 0.2 & 0 & 0.4 \\ 0.4 & 0 & -0.2 \\ 1.4 & -1 & -1.2 \end{bmatrix}$ as shown in Figure 31.

33. $A = \begin{bmatrix} 2 & -2 & 1 \\ 0 & 5 & 8 \\ 0 & 0 & -1 \end{bmatrix} \Rightarrow A^{-1} = \begin{bmatrix} 0.5 & 0.2 & 2.1 \\ 0 & 0.2 & 1.6 \\ 0 & 0 & -1 \end{bmatrix}$ as shown in Figure 33.

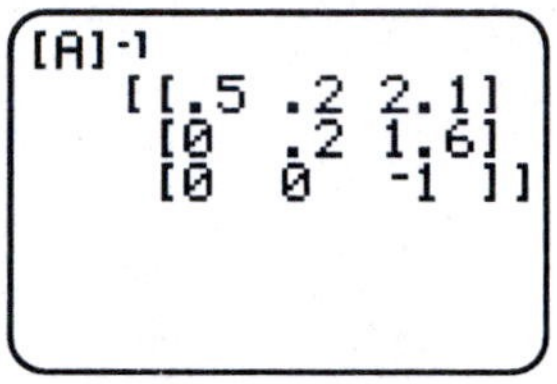
[A]-1
[[.5 .2 2.1]
[0 .2 1.6]
[0 0 -1]]

Figure 33

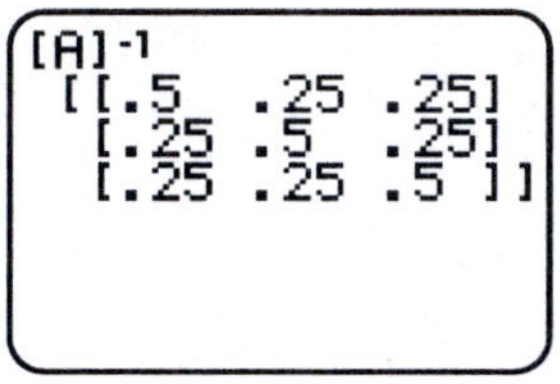
[A]-1
[[.5 .25 .25]
[.25 .5 .25]
[.25 .25 .5]]

Figure 35

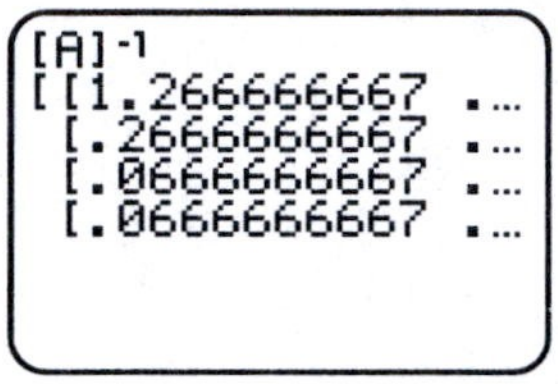
[A]-1
[[1.266666667 ...
[.2666666667 ...
[.0666666667 ...
[.0666666667 ...

Figure 37

35. $A = \begin{bmatrix} 3 & -1 & -1 \\ -1 & 3 & -1 \\ -1 & -1 & 3 \end{bmatrix} \Rightarrow A^{-1} = \begin{bmatrix} 0.5 & 0.25 & 0.25 \\ 0.25 & 0.5 & 0.25 \\ 0.25 & 0.25 & 0.5 \end{bmatrix}$ as shown in Figure 35.

37. $A = \begin{bmatrix} 1 & -1 & 0 & 0 \\ -1 & 5 & -1 & 0 \\ 0 & -1 & 5 & -1 \\ 0 & 0 & -1 & 1 \end{bmatrix} \Rightarrow A^{-1} = \begin{bmatrix} 1.2\overline{6} & 0.2\overline{6} & 0.0\overline{6} & 0.0\overline{6} \\ 0.2\overline{6} & 0.2\overline{6} & 0.0\overline{6} & 0.0\overline{6} \\ 0.0\overline{6} & 0.0\overline{6} & 0.2\overline{6} & 0.2\overline{6} \\ 0.0\overline{6} & 0.0\overline{6} & 0.2\overline{6} & 1.2\overline{6} \end{bmatrix}$ as shown in Figure 37.

39. $\begin{aligned} 2x - 3y &= 7 \\ -3x - 4y &= 9 \end{aligned} \Rightarrow AX = \begin{bmatrix} 2 & -3 \\ -3 & -4 \end{bmatrix}\begin{bmatrix} x \\ y \end{bmatrix} = \begin{bmatrix} 7 \\ 9 \end{bmatrix} = B$

41. $\begin{aligned} \tfrac{1}{2}x - \tfrac{3}{2}y &= \tfrac{1}{4} \\ -x + 2y &= 5 \end{aligned} \Rightarrow AX = \begin{bmatrix} \frac{1}{2} & -\frac{3}{2} \\ -1 & 2 \end{bmatrix}\begin{bmatrix} x \\ y \end{bmatrix} = \begin{bmatrix} \frac{1}{4} \\ 5 \end{bmatrix} = B$

43. $\begin{aligned} x - 2y + z &= 5 \\ 3y - z &= 6 \\ 5x - 4y - 7z &= 0 \end{aligned} \Rightarrow AX = \begin{bmatrix} 1 & -2 & 1 \\ 0 & 3 & -1 \\ 5 & -4 & -7 \end{bmatrix}\begin{bmatrix} x \\ y \\ z \end{bmatrix} = \begin{bmatrix} 5 \\ 6 \\ 0 \end{bmatrix} = B$

45. $\begin{aligned} 4x - y + 3z &= -2 \\ x + 2y + 5z &= 11 \\ 2x - 3y &= -1 \end{aligned} \Rightarrow AX = \begin{bmatrix} 4 & -1 & 3 \\ 1 & 2 & 5 \\ 2 & -3 & 0 \end{bmatrix}\begin{bmatrix} x \\ y \\ z \end{bmatrix} = \begin{bmatrix} -2 \\ 11 \\ -1 \end{bmatrix} = B$

47. (a) $\begin{aligned} x + 2y &= 3 \\ x + 3y &= 6 \end{aligned} \Rightarrow AX = \begin{bmatrix} 1 & 2 \\ 1 & 3 \end{bmatrix}\begin{bmatrix} x \\ y \end{bmatrix} = \begin{bmatrix} 3 \\ 6 \end{bmatrix} = B$

(b) If $AX = B \Rightarrow \begin{bmatrix} 1 & 2 \\ 1 & 3 \end{bmatrix}\begin{bmatrix} x \\ y \end{bmatrix} = \begin{bmatrix} 3 \\ 6 \end{bmatrix} \Rightarrow X = A^{-1}B \Rightarrow \begin{bmatrix} x \\ y \end{bmatrix} = \begin{bmatrix} 3 & -2 \\ -1 & 1 \end{bmatrix}\begin{bmatrix} 3 \\ 6 \end{bmatrix} = \begin{bmatrix} -3 \\ 3 \end{bmatrix}$

The solution to the system is $(-3, 3)$.

49. (a) $\begin{aligned} -x + 2y &= 5 \\ 3x - 5y &= -2 \end{aligned} \Rightarrow AX = \begin{bmatrix} -1 & 2 \\ 3 & -5 \end{bmatrix}\begin{bmatrix} x \\ y \end{bmatrix} = \begin{bmatrix} 5 \\ -2 \end{bmatrix} = B$

(b) If $AX = B \Rightarrow \begin{bmatrix} -1 & 2 \\ 3 & -5 \end{bmatrix}\begin{bmatrix} x \\ y \end{bmatrix} = \begin{bmatrix} 5 \\ -2 \end{bmatrix} \Rightarrow X = A^{-1}B \Rightarrow \begin{bmatrix} x \\ y \end{bmatrix} = \begin{bmatrix} 5 & 2 \\ 3 & 1 \end{bmatrix}\begin{bmatrix} 5 \\ -2 \end{bmatrix} = \begin{bmatrix} 21 \\ 13 \end{bmatrix}$

The solution to the system is $(21, 13)$.

51. (a) $\begin{aligned} x \quad\;\; + z &= -7 \\ 2x + y + 3z &= -13 \\ -x + y + z &= -4 \end{aligned} \Rightarrow AX = \begin{bmatrix} 1 & 0 & 1 \\ 2 & 1 & 3 \\ -1 & 1 & 1 \end{bmatrix}\begin{bmatrix} x \\ y \\ z \end{bmatrix} = \begin{bmatrix} -7 \\ -13 \\ -4 \end{bmatrix} = B$

(b) If $AX = B \Rightarrow \begin{bmatrix} 1 & 0 & 1 \\ 2 & 1 & 3 \\ -1 & 1 & 1 \end{bmatrix}\begin{bmatrix} x \\ y \\ z \end{bmatrix} = \begin{bmatrix} -7 \\ -13 \\ -4 \end{bmatrix} \Rightarrow X = A^{-1}B = \begin{bmatrix} x \\ y \\ z \end{bmatrix} = \begin{bmatrix} -2 & 1 & -1 \\ -5 & 2 & -1 \\ 3 & -1 & 1 \end{bmatrix}\begin{bmatrix} -7 \\ -13 \\ -4 \end{bmatrix} =$

$\begin{bmatrix} 5 \\ 13 \\ -12 \end{bmatrix}$ The solution to the system is $(5, 13, -12)$.

53. (a) $\begin{aligned} x + 2y - z &= 2 \\ 2x + 5y \quad\;\; &= -1 \\ -x - y + 2z &= 0 \end{aligned} \Rightarrow AX = \begin{bmatrix} 1 & 2 & -1 \\ 2 & 5 & 0 \\ -1 & -1 & 2 \end{bmatrix}\begin{bmatrix} x \\ y \\ z \end{bmatrix} = \begin{bmatrix} 2 \\ -1 \\ 0 \end{bmatrix} = B$

(b) If $AX = B \Rightarrow \begin{bmatrix} 1 & 2 & -1 \\ 2 & 5 & 0 \\ -1 & -1 & 2 \end{bmatrix}\begin{bmatrix} x \\ y \\ z \end{bmatrix} = \begin{bmatrix} 2 \\ -1 \\ 0 \end{bmatrix} \Rightarrow X = A^{-1}B = \begin{bmatrix} x \\ y \\ z \end{bmatrix} = \begin{bmatrix} -10 & 3 & -5 \\ 4 & -1 & 2 \\ -3 & 1 & -1 \end{bmatrix}\begin{bmatrix} 2 \\ -1 \\ 0 \end{bmatrix} =$

$\begin{bmatrix} -23 \\ 9 \\ -7 \end{bmatrix}$ The solution to the system is $(-23, 9, -7)$.

55. (a) $AX = B \Rightarrow \begin{bmatrix} 1.5 & 3.7 \\ -0.4 & -2.1 \end{bmatrix}\begin{bmatrix} x \\ y \end{bmatrix} = \begin{bmatrix} 0.32 \\ 0.36 \end{bmatrix}$

(b) See Figure 55. $X = A^{-1}B \Rightarrow X = \begin{bmatrix} 1.2 \\ -0.4 \end{bmatrix}$

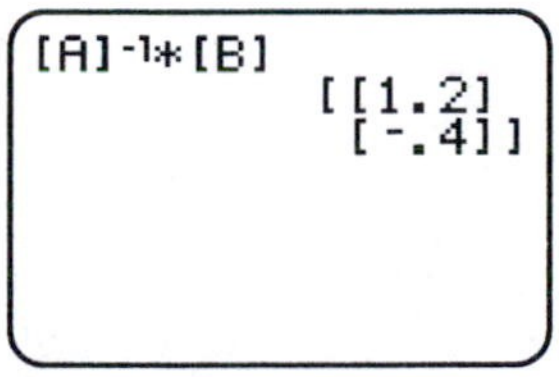

```
[A]⁻¹*[B]
            [[1.2]
             [-.4]]
```

```
[A]⁻¹*[B]
                [[.7]
                 [.8]]
```

Figure 55 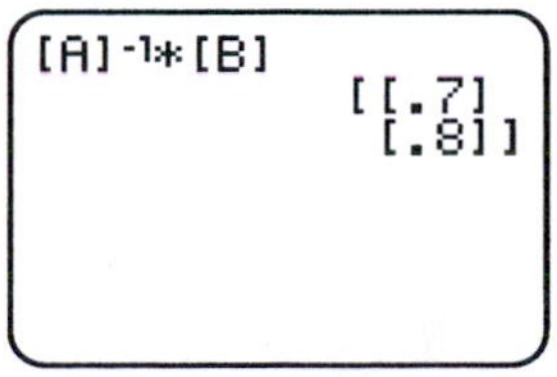   Figure 57

57. (a) $AX = B \Rightarrow \begin{bmatrix} 0.08 & -0.7 \\ 1.1 & -0.05 \end{bmatrix} \begin{bmatrix} x \\ y \end{bmatrix} = \begin{bmatrix} -0.504 \\ 0.73 \end{bmatrix}$

    (b) See Figure 57. $X = A^{-1}B \Rightarrow X = \begin{bmatrix} 0.7 \\ 0.8 \end{bmatrix}$

59. (a) $AX = B \Rightarrow \begin{bmatrix} 3.1 & 1.9 & -1 \\ 6.3 & 0 & -9.9 \\ -1 & 1.5 & 7 \end{bmatrix} \begin{bmatrix} x \\ y \\ z \end{bmatrix} = \begin{bmatrix} 1.99 \\ -3.78 \\ 5.3 \end{bmatrix}$

    (b) See Figure 59. $X = A^{-1}B \Rightarrow X = \begin{bmatrix} 0.5 \\ 0.6 \\ 0.7 \end{bmatrix}$

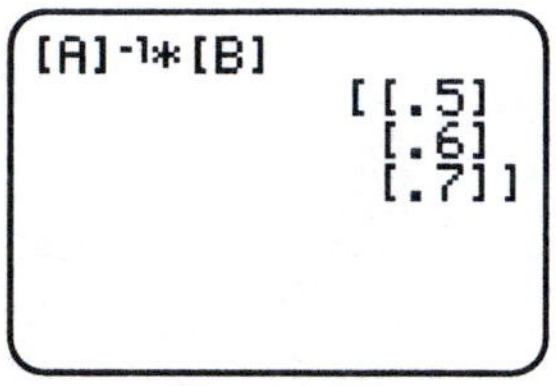

```
[A]⁻¹*[B]
        [[.5]
         [.6]
         [.7]]
```

```
[A]⁻¹*[B]
  [[9.262253521]
   [27.39098592]
   [4.504225352]]
```

Figure 59 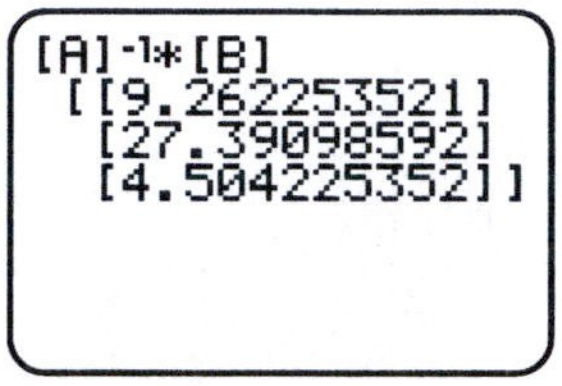   **Figure 61**

61. (a) $AX = B \Rightarrow \begin{bmatrix} 3 & -1 & 1 \\ 5.8 & -2.1 & 0 \\ -1 & 0 & 2.9 \end{bmatrix} \begin{bmatrix} x \\ y \\ z \end{bmatrix} = \begin{bmatrix} 4.9 \\ -3.8 \\ 3.8 \end{bmatrix}$

    (b) See Figure 61. $X = A^{-1}B \Rightarrow X \approx \begin{bmatrix} 9.26 \\ 27.39 \\ 4.50 \end{bmatrix}$

63. (a) Since $h = 2$ and $k = 3$, the matrix A will translate the point $(0, 1)$ to the right 2 units and up 3 units. Its new location will be $(0 + 2, 1 + 3) = (2, 4)$. This is verified by the following computation:

$$AX = \begin{bmatrix} 1 & 0 & 2 \\ 0 & 1 & 3 \\ 0 & 0 & 1 \end{bmatrix} \begin{bmatrix} 0 \\ 1 \\ 1 \end{bmatrix} = \begin{bmatrix} 2 \\ 4 \\ 1 \end{bmatrix}$$

    (b) $A^{-1}Y = X$. That is, A^{-1} will translate $(2, 4)$ back to $(0, 1)$ by moving it left 2 units and down 3 units. Thus $h = -2$ and $k = -3$ in A^{-1}.

$$A^{-1}Y = \begin{bmatrix} 1 & 0 & -2 \\ 0 & 1 & -3 \\ 0 & 0 & 1 \end{bmatrix} \begin{bmatrix} 2 \\ 4 \\ 1 \end{bmatrix} = \begin{bmatrix} 0 \\ 1 \\ 1 \end{bmatrix}$$

    (c) The product $AA^{-1} = I_3 = A^{-1}A$, since they are 3×3 inverse matrices.

65. Three units left implies that $h = -3$ and five units down implies that $k = -5$.

$$A = \begin{bmatrix} 1 & 0 & -3 \\ 0 & 1 & -5 \\ 0 & 0 & 1 \end{bmatrix} \text{ and } A^{-1} = \begin{bmatrix} 1 & 0 & 3 \\ 0 & 1 & 5 \\ 0 & 0 & 1 \end{bmatrix}$$

A^{-1} will translate a point 3 units to the right and 5 units up.

67. (a) $BX = \begin{bmatrix} \dfrac{1}{\sqrt{2}} & \dfrac{1}{\sqrt{2}} & 0 \\[2mm] -\dfrac{1}{\sqrt{2}} & \dfrac{1}{\sqrt{2}} & 0 \\[2mm] 0 & 0 & 1 \end{bmatrix} \begin{bmatrix} -\sqrt{2} \\ -\sqrt{2} \\ 1 \end{bmatrix} = \begin{bmatrix} -2 \\ 0 \\ 1 \end{bmatrix} = Y$

(b) $B^{-1}Y = \begin{bmatrix} \dfrac{1}{\sqrt{2}} & -\dfrac{1}{\sqrt{2}} & 0 \\[2mm] \dfrac{1}{\sqrt{2}} & \dfrac{1}{\sqrt{2}} & 0 \\[2mm] 0 & 0 & 1 \end{bmatrix} \begin{bmatrix} -2 \\ 0 \\ 1 \end{bmatrix} = \begin{bmatrix} -\sqrt{2} \\ -\sqrt{2} \\ 1 \end{bmatrix} = X$

B^{-1} rotates the point represented by Y counterclockwise 45° about the origin.

69. (a) In the computation ABX, B translates $(1, 1)$ left 3 units and up 3 units to $(-2, 4)$. Then, A translates $(-2, 4)$ right 4 units and down 2 units to $(2, 2)$.

$ABX = \begin{bmatrix} 1 & 0 & 4 \\ 0 & 1 & -2 \\ 0 & 0 & 1 \end{bmatrix} \begin{bmatrix} 1 & 0 & -3 \\ 0 & 1 & 3 \\ 0 & 0 & 1 \end{bmatrix} \begin{bmatrix} 1 \\ 1 \\ 1 \end{bmatrix} = \begin{bmatrix} 2 \\ 2 \\ 1 \end{bmatrix} = Y$; This represents the point $(2, 2)$ as expected.

(b) The net result of A and B is to translate a point 1 unit right and 1 unit up. Therefore, it is reasonable to expect that $h = 1$ and $k = 1$ in the matrix of the form AB.

$AB = \begin{bmatrix} 1 & 0 & 4 \\ 0 & 1 & -2 \\ 0 & 0 & 1 \end{bmatrix} \begin{bmatrix} 1 & 0 & -3 \\ 0 & 1 & 3 \\ 0 & 0 & 1 \end{bmatrix} = \begin{bmatrix} 1 & 0 & 1 \\ 0 & 1 & 1 \\ 0 & 0 & 1 \end{bmatrix}$

(c) Yes. If a point is translated left 3 units and up 3 units followed by right 4 units and down 2 units, the final result will be the same as the translation obtained when the point is first translated right 4 units and down 2 units followed by left 3 units and up 3. Therefore, we might expect that $AB = BA$.

$BA = \begin{bmatrix} 1 & 0 & -3 \\ 0 & 1 & 3 \\ 0 & 0 & 1 \end{bmatrix} \begin{bmatrix} 1 & 0 & 4 \\ 0 & 1 & -2 \\ 0 & 0 & 1 \end{bmatrix} = \begin{bmatrix} 1 & 0 & 1 \\ 0 & 1 & 1 \\ 0 & 0 & 1 \end{bmatrix} = AB$

(d) Since AB translates a point 1 unit right and 1 unit up, the inverse of AB would translate a point 1 unit left and 1 unit down. So $h = -1$ and $k = -1$ in $(AB)^{-1}$.

$(AB)^{-1} = \begin{bmatrix} 1 & 0 & -1 \\ 0 & 1 & -1 \\ 0 & 0 & 1 \end{bmatrix}$; Notice that $(AB)(AB)^{-1} = I_3$ as expected.

71. Let x be the number of CDs of type A purchased, let y be the CDs of type B and let z be the CDs of type C. The first row in the table implies that $2x + 3y + 4z = 120.91$. The other rows can be interpreted similarly. The system in matrix form is shown below.

$AX = B \implies \begin{bmatrix} 2 & 3 & 4 \\ 1 & 4 & 0 \\ 2 & 1 & 3 \end{bmatrix} \begin{bmatrix} x \\ y \\ z \end{bmatrix} = \begin{bmatrix} 120.91 \\ 62.95 \\ 79.94 \end{bmatrix}$

Use a graphing calculator to find the solution as shown in Figure 71. Type A CDs cost $10.99, type B cost $12.99, and type C cost $14.99.

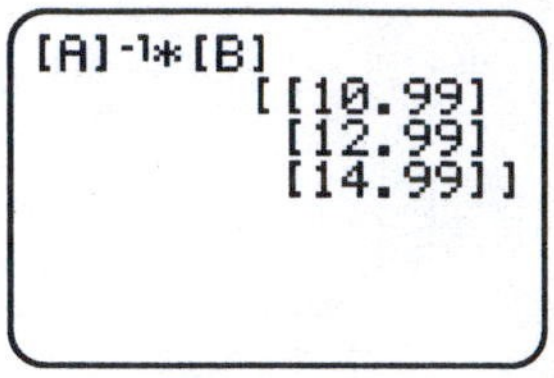

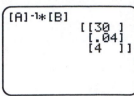

Figure 71

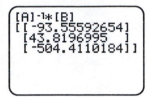

Figure 73

[A]⁻¹*[B]
[[-93.55592654]
[43.8196995]
[-504.4110184]]

Figure 76

73. (a) The following three equations must be solved, using the equation $P = a + bS + cC$.

$$a + b(1500) + c(\ 8) = 122$$
$$a + b(2000) + c(\ 5) = 130$$
$$a + b(2200) + c(10) = 158$$

These equations can be written in matrix form as follows:

$$AX = B \Rightarrow \begin{bmatrix} 1 & 1500 & 8 \\ 1 & 2000 & 5 \\ 1 & 2200 & 10 \end{bmatrix} \begin{bmatrix} a \\ b \\ c \end{bmatrix} = \begin{bmatrix} 122 \\ 130 \\ 158 \end{bmatrix}$$

The solution is given by $X = A^{-1}B$ as shown in Figure 73. It is $a = 30$, $b = 0.04$, and $c = 4$.

That is, P is given by $P = 30 + 0.04S + 4C$

(b) $P = 30 + 0.04(1800) + 4(7) = 130$, or $130,000

75. (a) Although this system of equations can be solved using matrices, we will use elimination to avoid performing row operations on variables. Subtracting the second equation from the third equations gives

$$\begin{array}{r} 0.4S + 0.4E + 0.1T = T \\ -\ 0.4S + 0.2E + 0.1T = E \\ \hline 0.2E \qquad\quad = T - E \Rightarrow E = \dfrac{5}{6}T. \end{array}$$

Substituting $E = \dfrac{5}{6}T$ in the first equation gives $0.2S + 0.4\left(\dfrac{5}{6}T\right) + 0.8T = S \Rightarrow$

$\dfrac{17}{15}T = \dfrac{4}{5}S \Rightarrow S = \dfrac{17}{12}T.$ The solution is $\left(\dfrac{17}{12}T, \dfrac{5}{6}T, T\right)$.

(b) When $T = 60$ units, $S = \dfrac{17}{12}(60) = 85$ units and $E = \dfrac{5}{6}(60) = 50$ units. That is, service should produce 85 units and electrical should produce 50 units.

Checking Basic Concepts for Sections 9.5 and 9.6

1. (a) $A + B = \begin{bmatrix} 1 & 0 & 1 \\ -1 & 1 & 2 \\ 1 & 3 & 0 \end{bmatrix} + \begin{bmatrix} -1 & 1 & 2 \\ 0 & 4 & 1 \\ 1 & -2 & 0 \end{bmatrix} = \begin{bmatrix} 0 & 1 & 3 \\ -1 & 5 & 3 \\ 2 & 1 & 0 \end{bmatrix}$

(b) $2A - B = 2\begin{bmatrix} 1 & 0 & 1 \\ -1 & 1 & 2 \\ 1 & 3 & 0 \end{bmatrix} - \begin{bmatrix} -1 & 1 & 2 \\ 0 & 4 & 1 \\ 1 & -2 & 0 \end{bmatrix} = \begin{bmatrix} 3 & -1 & 0 \\ -2 & -2 & 3 \\ 1 & 8 & 0 \end{bmatrix}$

(c) $AB = \begin{bmatrix} 1 & 0 & 1 \\ -1 & 1 & 2 \\ 1 & 3 & 0 \end{bmatrix}\begin{bmatrix} -1 & 1 & 2 \\ 0 & 4 & 1 \\ 1 & -2 & 0 \end{bmatrix} = \begin{bmatrix} 0 & -1 & 2 \\ 3 & -1 & -1 \\ -1 & 13 & 5 \end{bmatrix}$

3. (a) If $\begin{matrix} x - 2y = 13 \\ 2x + 3y = 5 \end{matrix} \Rightarrow AX = \begin{bmatrix} 1 & -2 \\ 2 & 3 \end{bmatrix}\begin{bmatrix} x \\ y \end{bmatrix} = \begin{bmatrix} 13 \\ 5 \end{bmatrix}$

$A^{-1} = \begin{bmatrix} 1 & -2 & | & 1 & 0 \\ 2 & 3 & | & 0 & 1 \end{bmatrix} 2R_1 - R_2 \rightarrow \begin{bmatrix} 1 & -2 & | & 1 & 0 \\ 0 & -7 & | & 2 & -1 \end{bmatrix} (-1/7)R_2 \rightarrow \begin{bmatrix} 1 & -2 & | & 1 & 0 \\ 0 & 1 & | & -\frac{2}{7} & \frac{1}{7} \end{bmatrix}$

$2R_2 + R_1 \rightarrow \begin{bmatrix} 1 & 0 & | & \frac{3}{7} & \frac{2}{7} \\ 0 & 1 & | & -\frac{2}{7} & \frac{1}{7} \end{bmatrix} \Rightarrow A^{-1} = \begin{bmatrix} \frac{3}{7} & \frac{2}{7} \\ -\frac{2}{7} & \frac{1}{7} \end{bmatrix} \Rightarrow X = A^{-1}B \Rightarrow \begin{bmatrix} x \\ y \end{bmatrix} = \begin{bmatrix} \frac{3}{7} & \frac{2}{7} \\ -\frac{2}{7} & \frac{1}{7} \end{bmatrix}\begin{bmatrix} 13 \\ 5 \end{bmatrix} = \begin{bmatrix} 7 \\ -3 \end{bmatrix}$

The solution to the system is $(7, -3)$. See Figure 3a.

(b) If $\begin{matrix} x - y + z = 2 \\ -x + y + z = 4 \\ y - z = -1 \end{matrix} \Rightarrow AX = \begin{bmatrix} 1 & -1 & 1 \\ -1 & 1 & 1 \\ 0 & 1 & -1 \end{bmatrix}\begin{bmatrix} x \\ y \\ z \end{bmatrix} = \begin{bmatrix} 2 \\ 4 \\ -1 \end{bmatrix} \Rightarrow$

$A^{-1} = \begin{bmatrix} 1 & -1 & 1 & | & 1 & 0 & 0 \\ -1 & 1 & 1 & | & 0 & 1 & 0 \\ 0 & 1 & -1 & | & 0 & 0 & 1 \end{bmatrix}$

$R_1 + R_2 \rightarrow \begin{bmatrix} 1 & -1 & 1 & | & 1 & 0 & 0 \\ 0 & 0 & 2 & | & 1 & 1 & 0 \\ 0 & 1 & -1 & | & 0 & 0 & 1 \end{bmatrix} \begin{matrix} R_3 \rightarrow \\ R_2 \rightarrow \end{matrix} \begin{bmatrix} 1 & -1 & 1 & | & 1 & 0 & 0 \\ 0 & 1 & -1 & | & 0 & 0 & 1 \\ 0 & 0 & 2 & | & 1 & 1 & 0 \end{bmatrix}$

$\begin{matrix} R_2 + R_3 \rightarrow \\ (1/2)R_3 \rightarrow \end{matrix} \begin{bmatrix} 1 & -1 & 1 & | & 1 & 0 & 0 \\ 0 & 1 & 1 & | & 1 & 1 & 1 \\ 0 & 0 & 1 & | & \frac{1}{2} & \frac{1}{2} & 0 \end{bmatrix} R_1 - R_3 \rightarrow \begin{bmatrix} 1 & -1 & 0 & | & \frac{1}{2} & -\frac{1}{2} & 0 \\ 0 & 1 & 1 & | & 1 & 1 & 1 \\ 0 & 0 & 1 & | & \frac{1}{2} & \frac{1}{2} & 0 \end{bmatrix}$

$R_2 - R_3 \rightarrow \begin{bmatrix} 1 & -1 & 0 & | & \frac{1}{2} & -\frac{1}{2} & 0 \\ 0 & 1 & 0 & | & \frac{1}{2} & \frac{1}{2} & 1 \\ 0 & 0 & 1 & | & \frac{1}{2} & \frac{1}{2} & 0 \end{bmatrix} R_1 + R_2 \rightarrow \begin{bmatrix} 1 & 0 & 0 & | & 1 & 0 & 1 \\ 0 & 1 & 0 & | & \frac{1}{2} & \frac{1}{2} & 1 \\ 0 & 0 & 1 & | & \frac{1}{2} & \frac{1}{2} & 0 \end{bmatrix} \Rightarrow$

$A^{-1} = \begin{bmatrix} 1 & 0 & 1 \\ \frac{1}{2} & \frac{1}{2} & 1 \\ \frac{1}{2} & \frac{1}{2} & 0 \end{bmatrix} \Rightarrow X = A^{-1}B \Rightarrow \begin{bmatrix} x \\ y \\ z \end{bmatrix} = \begin{bmatrix} 1 & 0 & 1 \\ \frac{1}{2} & \frac{1}{2} & 1 \\ \frac{1}{2} & \frac{1}{2} & 0 \end{bmatrix}\begin{bmatrix} 2 \\ 4 \\ -1 \end{bmatrix} = \begin{bmatrix} 1 \\ 2 \\ 3 \end{bmatrix}$

The solution to the system is $(1, 2, 3)$. See Figure 3b.

(c) $AX = B \Rightarrow \begin{bmatrix} 3.1 & -5.3 \\ -0.1 & 1.8 \end{bmatrix}\begin{bmatrix} x \\ y \end{bmatrix} = \begin{bmatrix} -2.682 \\ 0.787 \end{bmatrix}$. Then $X = A^{-1}B \Rightarrow X = \begin{bmatrix} -0.13 \\ 0.43 \end{bmatrix}$ as shown in Figure 3c.

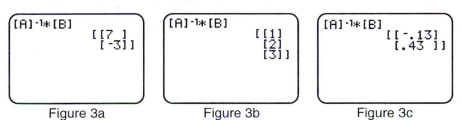

Figure 3a Figure 3b Figure 3c

9.7 Determinants

1. $\det A = \det\begin{bmatrix} 4 & 3 \\ 5 & 4 \end{bmatrix} = (4)(4) - (5)(3) = 1 \neq 0$; A is invertible.

3. $\det A = \det\begin{bmatrix} -4 & 6 \\ -8 & 12 \end{bmatrix} = (-4)(12) - (-8)(6) = 0$; A is not invertible.

5. Deleting the first row and second column gives $M_{12} = \det A = \det\begin{bmatrix} 2 & -2 \\ 0 & 5 \end{bmatrix} = (2)(5) - (0)(-2) = 10$.

    The cofactor is $A_{12} = (-1)^{1+2}M_{12} = -1(10) = -10$.

7. Deleting the second row and second column gives $M_{22} = \det A = \det\begin{bmatrix} 7 & 1 \\ 1 & -2 \end{bmatrix} = (7)(-2) - (1)(1) = -15$.

    The cofactor is $A_{22} = (-1)^{2+2}M_{22} = 1(-15) = -15$.

9. $\det A = a_{11}A_{11} + a_{21}A_{21} + a_{31}A_{31} = a_{11}M_{11} - a_{21}M_{21} + a_{31}M_{31} \Rightarrow$

    $\det A = (1)\det\begin{bmatrix} 2 & -3 \\ -1 & 3 \end{bmatrix} - (0)\det\begin{bmatrix} 4 & -7 \\ -1 & 3 \end{bmatrix} + (0)\det\begin{bmatrix} 4 & -7 \\ 2 & -3 \end{bmatrix} = (1)(6 - 3) = 3$

    Since $\det A = 3 \neq 0$, A^{-1} exists.

11. $\det A = a_{11}A_{11} + a_{21}A_{21} + a_{31}A_{31} = a_{11}M_{11} - a_{21}M_{21} + a_{31}M_{31} \Rightarrow$

    $\det A = (5)\det\begin{bmatrix} -2 & 0 \\ 4 & 0 \end{bmatrix} - (0)\det\begin{bmatrix} 1 & 6 \\ 4 & 0 \end{bmatrix} + (0)\det\begin{bmatrix} 1 & 6 \\ -2 & 0 \end{bmatrix} = (5)(0 - 0) = 0$

    Since $\det A = 0$, A^{-1} does not exist.

13. Expanding about the first column results in $\det A = 2\det\begin{bmatrix} 3 & 0 \\ 0 & 5 \end{bmatrix} - 0 + 0 = (2)(15 - 0) = 30$.

15. Expanding about the first row results in $\det A = 0$.

17. Expanding about the first column results in

    $\det A = 3\det\begin{bmatrix} 5 & 7 \\ 0 & -1 \end{bmatrix} - 0 + 1\det\begin{bmatrix} -1 & 2 \\ 5 & 7 \end{bmatrix} = (3)(-5 - 0) + (1)(-7 - 10) = -32$

19. Expanding about the first column results in

    $\det A = 1\det\begin{bmatrix} 1 & 3 \\ 4 & -2 \end{bmatrix} - (-7)\det\begin{bmatrix} -5 & 2 \\ 4 & -2 \end{bmatrix} + 0 = (1)(-2 - 12) - (-7)(10 - 8) = 0$

21. $\det A = \det\begin{bmatrix} 11 & -32 \\ 1.2 & 55 \end{bmatrix} = 643.4$

23. $\det A = \det\begin{bmatrix} 2.3 & 5.1 & 2.8 \\ 1.2 & 4.5 & 8.8 \\ -0.4 & -0.8 & -1.2 \end{bmatrix} = -4.484$

25. By Cramer's rule, the solution can be found as follows:

    $E = \det\begin{bmatrix} 5 & 2 \\ 1 & 3 \end{bmatrix} = 13; \quad F = \det\begin{bmatrix} -1 & 5 \\ 3 & 1 \end{bmatrix} = -16; \quad D = \det\begin{bmatrix} -1 & 2 \\ 3 & 3 \end{bmatrix} = -9$

    Thus $x = \dfrac{E}{D} = \dfrac{13}{-9}$ and $y = \dfrac{F}{D} = \dfrac{-16}{-9} = \dfrac{16}{9}$. The solution is $\left(-\dfrac{13}{9}, \dfrac{16}{9}\right)$.

27. By Cramer's rule, the solution can be found as follows:

    $E = \det\begin{bmatrix} 8 & 3 \\ 3 & -5 \end{bmatrix} = -49; \quad F = \det\begin{bmatrix} -2 & 8 \\ 4 & 3 \end{bmatrix} = -38; \quad D = \det\begin{bmatrix} -2 & 3 \\ 4 & -5 \end{bmatrix} = -2$

    Thus $x = \dfrac{E}{D} = \dfrac{-49}{-2} = \dfrac{49}{2}$ and $y = \dfrac{F}{D} = \dfrac{-38}{-2} = 19$. The solution is $\left(\dfrac{49}{2}, 19\right)$.

29. By Cramer's rule, the solution can be found as follows:

$$E = \det\begin{bmatrix} 23 & 4 \\ 70 & -5 \end{bmatrix} = -395; \quad F = \det\begin{bmatrix} 7 & 23 \\ 11 & 70 \end{bmatrix} = 237; \quad D = \det\begin{bmatrix} 7 & 4 \\ 11 & -5 \end{bmatrix} = -79$$

Thus $x = \dfrac{E}{D} = \dfrac{-395}{-79} = 5$ and $y = \dfrac{F}{D} = \dfrac{237}{-79} = -3$. The solution is $(5, -3)$.

31. By Cramer's rule, the solution can be found as follows:

$$E = \det\begin{bmatrix} -0.91 & -2.5 \\ 0.423 & 0.9 \end{bmatrix} = 0.2385; \quad F = \det\begin{bmatrix} 1.7 & -0.91 \\ -0.4 & 0.423 \end{bmatrix} = 0.3551; \quad D = \det\begin{bmatrix} 1.7 & -2.5 \\ -0.4 & 0.9 \end{bmatrix} = 0.53$$

Thus $x = \dfrac{E}{D} = \dfrac{0.2385}{0.53} = 0.45$ and $y = \dfrac{F}{D} = \dfrac{0.3551}{0.53} = 0.67$. The solution is $(0.45, 0.67)$.

33. Enter the vertices as columns in a counterclockwise direction.

$$D = \frac{1}{2}\det\begin{bmatrix} 0 & 4 & 1 \\ 0 & 2 & 4 \\ 1 & 1 & 1 \end{bmatrix} = 7; \quad \text{The area of the triangle is 7 square units.}$$

35. A line segment between $(1, 3)$ and $(3, 2)$ divides the quadrangle into two triangles whose areas can be found using determinants. Enter the vertices of each triangle as columns in a counterclockwise direction.

$$D = \frac{1}{2}\det\begin{bmatrix} 0 & 3 & 1 \\ 0 & 2 & 3 \\ 1 & 1 & 1 \end{bmatrix} + \frac{1}{2}\det\begin{bmatrix} 1 & 3 & 5 \\ 3 & 2 & 4 \\ 1 & 1 & 1 \end{bmatrix} = 6.5; \quad \text{The area of the quadrangle is 6.5 square units.}$$

37. If the three points form a triangle with no area ($D = 0$ using determinants), then the points must be collinear.

$$D = \frac{1}{2}\det\begin{bmatrix} 1 & -3 & 2 \\ 3 & 11 & 1 \\ 1 & 1 & 1 \end{bmatrix} = 0; \quad \text{The points are collinear.}$$

39. If the three points form a triangle with no area ($D = 0$ using determinants), then the points must be collinear.

$$D = \frac{1}{2}\det\begin{bmatrix} -2 & 4 & 2 \\ -5 & 4 & 3 \\ 1 & 1 & 1 \end{bmatrix} = 6 \neq 0; \quad \text{The points are not collinear.}$$

41. Use cofactors to expand about row 1 of $\begin{bmatrix} x & y & 1 \\ 2 & 1 & 1 \\ -1 & 4 & 1 \end{bmatrix} = 0 \Rightarrow x(-3) - y(3) + 9 = 0 \Rightarrow$

$-3x - 3y = -9 \Rightarrow x + y = 3$.

43. Use cofactors to expand about row 1 of $\begin{bmatrix} x & y & 1 \\ 6 & -7 & 1 \\ 4 & -3 & 1 \end{bmatrix} = 0 \Rightarrow x(-4) - y(2) + (10) = 0 \Rightarrow 2x + y = 5$.

Extended and Discovery Exercises for Section 9.7

1. $D = 1[(1)(3) - (1)(2)] - 2[(1)(3) - (1)(1)] + 0[(1)(2) - (1)(1)] = 1 - 4 + 0 = -3$

   $E = 6[(1)(3) - (1)(2)] - 9[(1)(3) - (1)(1)] + 9[(1)(2) - (1)(1)] = 6 - 18 + 9 = -3$

   $F = 1[(9)(3) - (9)(2)] - 2[(6)(3) - (9)(1)] + 0[(6)(2) - (9)(1)] = 9 - 18 + 0 = -9$

   $G = 1[(1)(9) - (1)(9)] - 2[(1)(9) - (1)(6)] + 0[(1)(9) - (1)(6)] = 0 - 6 + 0 = -6$

   $x = \dfrac{E}{D} = \dfrac{-3}{-3} = 1, y = \dfrac{F}{D} = \dfrac{-9}{-3} = 3, z = \dfrac{G}{D} = \dfrac{-6}{-3} = 2$. The solution is $(1, 3, 2)$.

3. $D = 1[(1)(2) - (1)(0)] - 1[(0)(2) - (1)(1)] + 0[(0)(0) - (1)(1)] = 2 + 1 + 0 = 3$

   $E = 2[(1)(2) - (1)(0)] - 0[(0)(2) - (1)(1)] + 1[(0)(0) - (1)(1)] = 4 + 0 - 1 = 3$

   $F = 1[(0)(2) - (1)(0)] - 1[(2)(2) - (1)(1)] + 0[(0)(2) - (1)(0)] = 0 - 3 + 0 = -3$

   $G = 1[(1)(1) - (1)(0)] - 1[(0)(1) - (1)(2)] + 0[(0)(0) - (1)(2)] = 1 + 2 + 0 = 3$

   $x = \dfrac{E}{D} = \dfrac{3}{3} = 1, y = \dfrac{F}{D} = \dfrac{-3}{3} = -1, z = \dfrac{G}{D} = \dfrac{3}{3} = 1$. The solution is $(1, -1, 1)$.

5. $D = 1[(1)(2) - (-1)(1)] - (-1)[(0)(2) - (-1)(2)] + 2[(0)(1) - (1)(2)] = 3 + 2 - 4 = 1$

   $E = 7[(1)(2) - (-1)(1)] - 5[(0)(2) - (-1)(2)] + 6[(0)(1) - (1)(2)] = 21 - 10 - 12 = -1$

   $F = 1[(5)(2) - (6)(1)] - (-1)[(7)(2) - (6)(2)] + 2[(7)(1) - (5)(2)] = 4 + 2 - 6 = 0$

   $G = 1[(1)(6) - (-1)(5)] - (-1)[(0)(6) - (-1)(7)] + 2[(0)(5) - (1)(7)] = 11 + 7 - 14 = 4$

   $x = \dfrac{E}{D} = \dfrac{-1}{1} = -1, y = \dfrac{F}{D} = \dfrac{0}{1} = 0, z = \dfrac{G}{D} = \dfrac{4}{1} = 4$. The solution is $(-1, 0, 4)$.

7. Use cofactors to expand about row 1 of $\begin{bmatrix} x^2 + y^2 & x & y & 1 \\ 4 & 0 & 2 & 1 \\ 4 & 2 & 0 & 1 \\ 4 & -2 & 0 & 1 \end{bmatrix} = 0 \Rightarrow$

   $(x^2 + y^2)[-8] - x(0) + y(0) - (-32) = 0 \Rightarrow x^2 + y^2 - 4 = 0.$

9. Use cofactors to expand about row 1 of $\begin{bmatrix} x^2 + y^2 & x & y & 1 \\ 1 & 0 & 1 & 1 \\ 2 & 1 & -1 & 1 \\ 8 & 2 & 2 & 1 \end{bmatrix} = 0 \Rightarrow$

   $(x^2 + y^2)(5) - x(15) + y(-5) - 0 = 0 \Rightarrow 5x^2 + 5y^2 - 15x - 5y = 0.$

Checking Basic Concepts for Section 9.7

1. $\det A = a_{11}A_{11} + a_{21}A_{21} + a_{31}A_{31} = a_{11}M_{11} - a_{21}M_{21} + a_{31}M_{31} \Rightarrow$

   $\det A = (1) \det \begin{bmatrix} 3 & 1 \\ -2 & 5 \end{bmatrix} - (2) \det \begin{bmatrix} -1 & 2 \\ -2 & 5 \end{bmatrix} + (0) \det \begin{bmatrix} -1 & 2 \\ 3 & 1 \end{bmatrix} =$

   $(1)(15 - (-2)) - (2)(-5 - (-4)) + 0 = 19$ Since $\det A = 19 \neq 0$ A is invertible.

Chapter 9 Review Exercises

1. $A(b, h) = \dfrac{1}{2}bh \Rightarrow A(3, 6) = \dfrac{1}{2}(3)(6) = 9$

3. The solution is the intersection point $(3, 1)$. The equation of the red line is $y = 2x - 5$ and the equation of the blue line is $x + y = 4$. Substituting the first equation into the second equation we have

    $x + (2x - 5) = 4 \Rightarrow 3x - 5 = 4 \Rightarrow 3x = 9 \Rightarrow x = 3$ and $3 + y = 4 \Rightarrow y = 1$.

5. (a) $3x + y = 1 \Rightarrow y = 1 - 3x$ and $2x - 3y = 8 \Rightarrow y = \dfrac{2x - 8}{3}$

    Graph $Y_1 = 1 - 3X$ and $Y_2 = (2X - 8)/3$ The graphs intersect at the point $(1, -2)$. See Figure 5.

    (b) Substituting $y = 1 - 3x$ into the second equation gives $2x - 3(1 - 3x) = 8 \Rightarrow 2x - 3 + 9x = 8 \Rightarrow$

    $11x = 11 \Rightarrow x = 1$. If $x = 1$, then $y = 1 - 3(1) = -2$. The solution is $(1, -2)$.

[−10, 10, 1] by [−10, 10, 1]

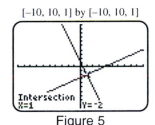

Figure 5

7. Multiply the first equation by 2 and add to eliminate the y-variable.

    $$\begin{aligned} 4x + 2y &= 14 \\ \underline{x - 2y} &= \underline{-4} \\ 5x &= 10 \end{aligned} \Rightarrow x = 2$$

    Since $2x + y = 7$, it follows that $y = 3$. The solution is $(2, 3)$. The system is consistent.

9. Multiply the first equation by 2, the second equation by 3, and add to eliminate the y-variable.

    $$\begin{aligned} 12x - 30y &= 24 \\ \underline{-12x + 30y} &= \underline{-24} \\ 0 &= 0 \end{aligned} \Rightarrow \text{infinitely many solutions of the form } \{(x, y)\,|\,2x - 5y = 4\}$$

    Since $0 = 0$, the system is consistent and there are infinitely many solutions.

11. Subtract the equations to eliminate the x-variable.

    $$\begin{aligned} x^2 - 3y &= 3 \\ x^2 + 2y^2 &= 5 \end{aligned}$$

    $-3y - 2y^2 = -2 \Rightarrow 0 = 2y^2 + 3y - 2 \Rightarrow (2y - 1)(y + 2) = 0 \Rightarrow y = -2, \dfrac{1}{2}$.

    If $y = -2$ then $x^2 - 3(-2) = 3 \Rightarrow x^2 + 6 = 3 \Rightarrow x^2 = -3$. There is no real solution to this $\Rightarrow y \neq -2$.

    If $y = \dfrac{1}{2}$ then $x^2 - 3\left(\dfrac{1}{2}\right) = 3 \Rightarrow x^2 = \dfrac{9}{2} \Rightarrow x = \pm\dfrac{3}{\sqrt{2}} \Rightarrow x = \pm\dfrac{3\sqrt{2}}{2}$.

    The solutions are $\left(\dfrac{3\sqrt{2}}{2}, \dfrac{1}{2}\right)$ and $\left(\dfrac{-3\sqrt{2}}{2}, \dfrac{1}{2}\right)$.

13. See Figure 13.

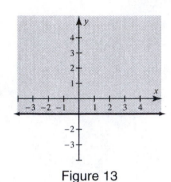

Figure 13   Figure 15

15. The solution region is inside the circle or radius 3 and above the line $y = 3 - x$. Their graphs intersect at the

    points $(0, 3)$ and $(3, 0)$. It does not include the boundary. See Figure 15. One solution to the system is $(2, 2)$.

17. First, subtract the first two equations to eliminate the x-variable.

    $$\begin{array}{rcrcrcr} x & - & y & + & z & = & -2 \\ x & + & 2y & - & z & = & 2 \\ \hline & & -3y & + & 2z & = & -4 \end{array}$$

    Multiply this equation by 2 and multiply the last equation by 3 to eliminate the y-variable.

    $$\begin{array}{rcrcr} -6y & + & 4z & = & -8 \\ 6y & + & 9z & = & 21 \\ \hline & & 13z & = & 13 \Rightarrow z = 1 \end{array}$$

    Substituting $z = 1$ into $2y + 3z = 7$ gives $2y + 3 = 7 \Rightarrow 2y = 4 \Rightarrow y = 2$. Substituting $y = 2$ and $z = 1$

    into the original first equation gives $x - 2 + 1 = -2 \Rightarrow x - 1 = -2 \Rightarrow x = -1$.

    The solution to the system is $(-1, 2, 1)$.

19. Add the first two equations to eliminate the x-variable.

    $$\begin{array}{rcrcrcr} -x & + & 2y & + & 2z & = & 9 \\ x & + & y & - & 3z & = & 6 \\ \hline & & 3y & - & z & = & 15 \end{array}$$

    Subtract this new equation from the third equation.

    $$\begin{array}{rcrcr} 3y & - & z & = & 8 \\ 3y & - & z & = & 15 \\ \hline & & 0 & = & -7 \end{array}$$

    Since $0 \neq -7$, there is no solution.

21. The system is $x + 5y = 6$ and $y = 3$.

    Substituting $y = 3$ into the first equation gives $x + 5(3) = 6 \Rightarrow x = -9$. The solution is $(-9, 3)$.

23. The augmented matrix is in Row-Echelon Form $\Rightarrow x = -2, y = 3, z = 0 \Rightarrow (-2, 3, 0)$.

25. $\begin{bmatrix} 2 & -1 & 2 & | & 10 \\ 1 & -2 & 1 & | & 8 \\ 3 & -1 & 2 & | & 11 \end{bmatrix} \begin{array}{c} (1/2)R_1 \to \\ R_1 - 2R_2 \to \\ 3R_2 - R_3 \to \end{array} \begin{bmatrix} 1 & -\frac{1}{2} & 1 & | & 5 \\ 0 & 3 & 0 & | & -6 \\ 0 & -5 & 1 & | & 13 \end{bmatrix} (1/3)R_2 \to \begin{bmatrix} 1 & -\frac{1}{2} & 1 & | & 5 \\ 0 & 1 & 0 & | & -2 \\ 0 & -5 & 1 & | & 13 \end{bmatrix}$

$5R_2 + R_3 \to \begin{bmatrix} 1 & -\frac{1}{2} & 1 & | & 5 \\ 0 & 1 & 0 & | & -2 \\ 0 & 0 & 1 & | & 3 \end{bmatrix}$

Backward substitution produces $z = 3; y = -2; x - \frac{1}{2}(-2) + 3 = 5 \Rightarrow x + 1 + 3 = 5 \Rightarrow x = 1$.

The solution is $(1, -2, 3)$.

27. (a) $a_{12} = 3$ and $a_{22} = 2 \Rightarrow a_{12} + a_{22} = 3 + 2 \Rightarrow a_{12} + a_{22} = 5$

   (b) $a_{11} = -2$ and $a_{23} = 4 \Rightarrow a_{11} - 2a_{23} = -2 - 2(4) \Rightarrow a_{11} - 2a_{23} = -10$

29. (a) $A + 2B = \begin{bmatrix} 1 & -3 \\ 2 & -1 \end{bmatrix} + 2\begin{bmatrix} 3 & 2 \\ -5 & 1 \end{bmatrix} = \begin{bmatrix} 7 & 1 \\ -8 & 1 \end{bmatrix}$

   (b) $A - B = \begin{bmatrix} 1 & -3 \\ 2 & -1 \end{bmatrix} - \begin{bmatrix} 3 & 2 \\ -5 & 1 \end{bmatrix} = \begin{bmatrix} -2 & -5 \\ 7 & -2 \end{bmatrix}$

   (c) $-4A = -4\begin{bmatrix} 1 & -3 \\ 2 & -1 \end{bmatrix} = \begin{bmatrix} -4 & 12 \\ -8 & 4 \end{bmatrix}$

31. A and B are both 2×2 so AB and BA are also both 2×2.

   $AB = \begin{bmatrix} 2 & 0 \\ -5 & 3 \end{bmatrix}\begin{bmatrix} -1 & -2 \\ 4 & 7 \end{bmatrix} = \begin{bmatrix} -2 & -4 \\ 17 & 31 \end{bmatrix}; BA = \begin{bmatrix} -1 & -2 \\ 4 & 7 \end{bmatrix}\begin{bmatrix} 2 & 0 \\ -5 & 3 \end{bmatrix} = \begin{bmatrix} 8 & -6 \\ -27 & 21 \end{bmatrix}$

33. A is 2×3 and B is 3×2 so AB is 2×2 and BA is 3×3.

   $AB = \begin{bmatrix} 2 & -1 & 3 \\ 2 & 4 & 0 \end{bmatrix}\begin{bmatrix} 1 & 0 \\ -1 & 2 \\ 0 & 3 \end{bmatrix} = \begin{bmatrix} 3 & 7 \\ -2 & 8 \end{bmatrix}; BA = \begin{bmatrix} 1 & 0 \\ -1 & 2 \\ 0 & 3 \end{bmatrix}\begin{bmatrix} 2 & -1 & 3 \\ 2 & 4 & 0 \end{bmatrix} = \begin{bmatrix} 2 & -1 & 3 \\ 2 & 9 & -3 \\ 6 & 12 & 0 \end{bmatrix}$

35. B is the inverse of A.

   $AB = \begin{bmatrix} 8 & 5 \\ 6 & 4 \end{bmatrix}\begin{bmatrix} 2 & -2.5 \\ -3 & 4 \end{bmatrix} = \begin{bmatrix} 1 & 0 \\ 0 & 1 \end{bmatrix}$ and $BA = \begin{bmatrix} 2 & -2.5 \\ -3 & 4 \end{bmatrix}\begin{bmatrix} 8 & 5 \\ 6 & 4 \end{bmatrix} = \begin{bmatrix} 1 & 0 \\ 0 & 1 \end{bmatrix}$

37. $A|I_2 = \begin{bmatrix} 1 & -2 & | & 1 & 0 \\ -1 & 1 & | & 0 & 1 \end{bmatrix} R_2 + R_1 \to \begin{bmatrix} 1 & -2 & | & 1 & 0 \\ 0 & -1 & | & 1 & 1 \end{bmatrix} (-1)R_2 \to \begin{bmatrix} 1 & -2 & | & 1 & 0 \\ 0 & 1 & | & -1 & -1 \end{bmatrix}$

   $R_1 + 2R_2 \to \begin{bmatrix} 1 & 0 & | & -1 & -2 \\ 0 & 1 & | & -1 & -1 \end{bmatrix}; A^{-1} = \begin{bmatrix} -1 & -2 \\ -1 & -1 \end{bmatrix}$

39. (a) $\begin{array}{c} x - 3y = 4 \\ 2x - y = 3 \end{array} \Rightarrow AX = \begin{bmatrix} 1 & -3 \\ 2 & -1 \end{bmatrix}\begin{bmatrix} x \\ y \end{bmatrix} = \begin{bmatrix} 4 \\ 3 \end{bmatrix} = B$

   (b) $X = A^{-1}B \Rightarrow \begin{bmatrix} x \\ y \end{bmatrix} = \begin{bmatrix} -\frac{1}{5} & \frac{3}{5} \\ -\frac{2}{5} & \frac{1}{5} \end{bmatrix}\begin{bmatrix} 4 \\ 3 \end{bmatrix} = \begin{bmatrix} 1 \\ -1 \end{bmatrix}$

41. (a) $AX = B \Rightarrow \begin{bmatrix} 12 & 7 & -3 \\ 8 & -11 & 13 \\ -23 & 0 & 9 \end{bmatrix} \begin{bmatrix} x \\ y \\ z \end{bmatrix} = \begin{bmatrix} 14.6 \\ -60.4 \\ -14.6 \end{bmatrix}$

(b) See Figure 41. $X = A^{-1}B \Rightarrow X = \begin{bmatrix} -0.5 \\ 1.7 \\ -2.9 \end{bmatrix}$

```
[A]-1*[B]
        [[-.5 ]
         [1.7 ]
         [-2.9]]
```

Figure 41

43. Expanding about the first column results in

$$\det A = 2 \det\begin{bmatrix} 3 & 4 \\ 0 & 5 \end{bmatrix} - 0 + 1 \det\begin{bmatrix} 1 & 3 \\ 3 & 4 \end{bmatrix} = (2)(15 - 0) + (1)(4 - 9) = 25.$$

45. $\det A = \det\begin{bmatrix} 13 & 22 \\ 55 & -57 \end{bmatrix} = (13)(-57) - (55)(22) = -1951 \neq 0$ A is invertible.

47. The given equations result in the following nonlinear system of equations: $A(l, w) = 77 \Rightarrow lw = 77$ and

$P(l, w) = 36 \Rightarrow 2l + 2w = 36$. Begin by solving the second equation for l.

$2l + 2w = 36 \Rightarrow 2l = 36 - 2w \Rightarrow l = 18 - w$. Substitute this into the first equation.

$lw = 77 \Rightarrow (18 - w)w = 77 \Rightarrow 18w - w^2 = 77 \Rightarrow w^2 - 18w + 77 = 0$. This is a quadratic equation

that can be solved by factoring. $w^2 - 18w + 77 = 0 \Rightarrow (w - 7)(w - 11) = 0 \Rightarrow w = 7$ or 11.

Since $l = 18 - w$, if $w = 7$, then $l = 11$ and if $w = 11$ then $l = 7$. For this rectangle $l = 11$ and $w = 7$.

[0, 2000, 100] by [0, 2000, 100]

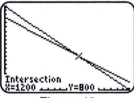

Figure 49

49. (a) Let x represent the amount of the 7% loan and let y represent the amount of the 9% loan. Then the system

of equations is $x + y = 2000$ and $0.07x + 0.09y = 156$. Multiply the first equation by 0.09 and subtract.

$$\begin{array}{r} 0.09x + 0.09y = 180 \\ \underline{0.07x + 0.09y = 156} \\ 0.02x \qquad\quad = 24 \end{array} \Rightarrow x = 1200 \text{ and } y = 2000 - 1200 = 800$$

The loan amounts are $1200 at 7% and $800 at 9%.

(b) Graph $Y_1 = 2000 - X$ and $Y_2 = (156 - 0.07X)/0.09$ The graph intersect at (1200, 800). See Figure 49.

    The loan amounts are $1200 at 7% and $800 at 9%.

51. Let x be the number of CDs of type A purchased and let y be the number of CDs of type B. Then from the table we see that $1x + 2y = 37.47$ and $2x + 3y = 61.95$.

$$AX = B \Rightarrow \begin{bmatrix} 1 & 2 \\ 2 & 3 \end{bmatrix}\begin{bmatrix} x \\ y \end{bmatrix} = \begin{bmatrix} 37.47 \\ 61.95 \end{bmatrix}$$

Using a graphing calculator to solve the system yeilds that type A CDs cost $11.49 and type B CD's cost $12.99.

53. Enter the vertices as columns in a counterclockwise direction.

$$D = \frac{1}{2}\det\begin{bmatrix} 0 & 5 & 2 \\ 0 & 2 & 5 \\ 1 & 1 & 1 \end{bmatrix} = 10.5;\ \text{ The area of the triangle is 10.5 square units.}$$

55. Since P varies jointly as the square of x and the cube of y, the variation equation $P = kx^2y^3$ must hold.

If $P = 432$ when $x = 2$ and $y = 3$, then $432 = k(2)^2(3)^3 \Rightarrow 432 = 108k \Rightarrow k = 4$.

Our variation equation becomes $P = 4x^2y^3$. Thus, when $x = 3$ and $y = 5$, $P = 4(3)^2(5)^3 = 4500$.

Extended and Discovery Exercises for Chapter 9

1. (a) $A^{\mathrm{T}} = \begin{bmatrix} 3 & 2 & 4 \\ -3 & 6 & 2 \end{bmatrix}$ (b) $A^{\mathrm{T}} = \begin{bmatrix} 0 & 2 & -4 \\ 1 & 5 & 3 \\ -2 & 4 & 9 \end{bmatrix}$ (c) $A^{\mathrm{T}} = \begin{bmatrix} 5 & 1 & 6 & -9 \\ 7 & -7 & 3 & 2 \end{bmatrix}$

3. Start by forming the following matrix equation:

$$AX = B \Rightarrow \begin{bmatrix} 0 & 1 \\ 1 & 1 \\ 2 & 1 \\ 3 & 1 \\ 4 & 1 \\ 5 & 1 \end{bmatrix}\begin{bmatrix} a \\ b \end{bmatrix} = \begin{bmatrix} 2.2 \\ 4.5 \\ 7.9 \\ 10.5 \\ 13 \\ 15 \end{bmatrix}$$

To find the least-squares solution to this system of linear equations, solve the matrix equation $A^{\mathrm{T}}AX = A^{\mathrm{T}}B$ for X. The solution is given by $X = (A^{\mathrm{T}}A)^{-1}A^{\mathrm{T}}B$. Enter the matrices A and B and compute the solution as shown in Figure 3a. Thus, $f(x) = 2.6314x + 2.2714$. The data and f are graphed in Figure 3b.

[-1, 6, 1] by [0, 18, 2]

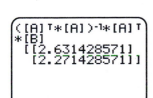
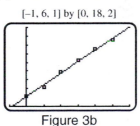

Figure 3a   Figure 3b

Chapter 10: Conic Sections

10.1: Parabolas

1. A parabola always opens toward its focus.

3. A parabola with equation $x^2 = 4py$ opens upward if p > 0 and it opens downward if p < 0.

5. A parabola with equation $x^2 = 4py$ has a vertical axis of symmetry.

7. See Figure 7.

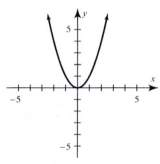

Figure 7

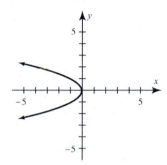

Figure 9

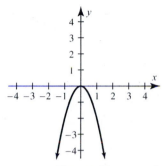

Figure 11

9. See Figure 9.

11. See Figure 11.

13. See Figure 13.

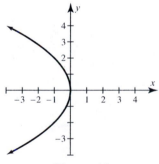

Figure 13

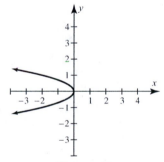

Figure 15

15. See Figure 15.

17. Opens upward; e

19. Opens to the left; a

21. Opens to the right, passes through (2, 2); d

23. The equation $16y = x^2$ is in the form $x^2 = 4py$, and the vertex is $V(0, 0)$. Thus, $16 = 4p$ or $p = 4$. The focus is $F(0, 4)$, the equation of the directrix is $y = -4$, and the parabola opens upward. The graph of $y = \dfrac{1}{16}x^2$ is shown in Figure 23.

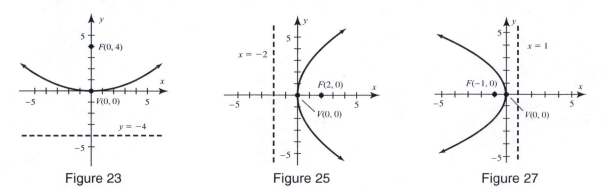

Figure 23   Figure 25   Figure 27

25. The equation $x = \dfrac{1}{8}y^2$ can be written as $y^2 = 8x$, which is in the form $y^2 = 4px$. The vertex is $V(0, 0)$. Thus, $8 = 4p$ or $p = 2$. The focus is $F(2, 0)$, the equation of the directrix is $x = -2$, and the parabola opens to the right. The graph of $x = \dfrac{1}{8}y^2$ is shown in Figure 25.

27. The equation $-4x = y^2$ can be written as $y^2 = -4x$, which is in the form $y^2 = 4px$. The vertex is $V(0, 0)$. Thus, $-4 = 4p$ or $p = -1$. The focus is $F(-1, 0)$, the equation of the directrix is $x = 1$, and the parabola opens to the left. The graph of $x = -\dfrac{1}{4}y^2$ is shown in Figure 27.

29. The equation $x^2 = -8y$ is in the form $x^2 = 4py$, and the vertex is $V(0, 0)$. Thus, $-8 = 4p$ or $p = -2$. The focus is $F(0, -2)$, the equation of the directrix is $y = 2$, and the parabola opens downward. The graph of $y = -\dfrac{1}{8}x^2$ is shown in Figure 29.

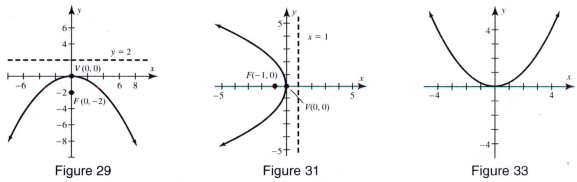

Figure 29   Figure 31   Figure 33

31. The equation $2y^2 = -8x$ can be written as $y^2 = -4x$, which is in the form $y^2 = 4px$. The vertex is $V(0, 0)$. Thus, $-4 = 4p$ or $p = -1$. The focus is $F(-1, 0)$, the equation of the directrix is $x = 1$, and the parabola opens to the left. The graph of $x = -\dfrac{1}{4}y^2$ is shown in Figure 31.

33. The focus is $F(0, 1)$ and the vertex is $V(0, 0)$. The distance between these points is 1. Since the focus is above the directrix, the parabola opens upward, so $p = 1$. Since the line passing through F and V is vertical, the parabola has a vertical axis. Its equation is given by $x^2 = 4py$ or $x^2 = 4y$. The graph of $y = \dfrac{1}{4}x^2$ is shown in Figure 33.

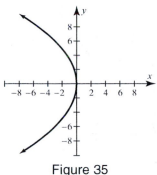

Figure 35

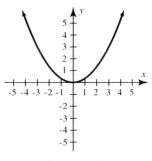

Figure 37

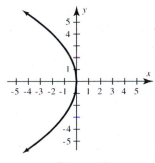

Figure 39

35. The focus is $F(-3, 0)$ and the vertex is $V(0, 0)$. The distance between these points is 3. Since the focus is left of the directrix, the parabola opens to the left, so $p = -3$. Since the line passing through F and V is horizontal, the parabola has a horizontal axis. Its equation is given by $y^2 = 4px$ or $y^2 = -12x$. The graph of $x = -\dfrac{1}{12}y^2$ is shown in Figure 35.

37. If the vertex is $V(0, 0)$ and the focus is $F\left(0, \dfrac{3}{4}\right)$, then the parabola opens upward and $p = \dfrac{3}{4}$. Thus, $x^2 = 4py \Rightarrow x^2 = 3y$. Its graph is shown in Figure 37.

39. If the vertex is $V(0, 0)$ and the directrix is $x = 2$, then the parabola opens to the left and $p = -2$. Thus, $y^2 = 4px \Rightarrow y^2 = -8x$. Its graph is shown in Figure 39.

41. If the vertex is $V(0, 0)$ and the focus is $F(1, 0)$, then the parabola opens to the right and $p = 1$. Thus, $y^2 = 4px \Rightarrow y^2 = 4x$. Its graph is shown in Figure 41.

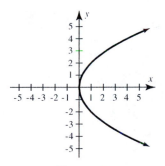

Figure 41

43. If the vertex is $V(0, 0)$ and the directrix is $x = \dfrac{1}{4}$, then the parabola opens to the left and $p = -\dfrac{1}{4}$. Thus, $y^2 = 4px \Rightarrow y^2 = -x$. Its graph is shown in Figure 43.

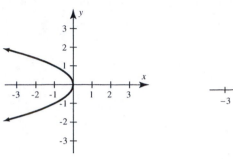

Figure 43

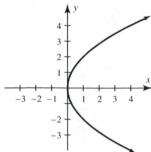

Figure 45

45. If the vertex is $V(0,0)$ and the parabola has a horizontal axis, the equation is in the form $y^2 = 4px$. Find the value of p by using the fact that the parabola passes through $(1, -2)$. Thus, $(-2)^2 = 4p(1) \Rightarrow p = 1$. The equation is $y^2 = 4x$. Its graph is shown in Figure 45.

47. If the focus is $F(0, -3)$ and the equation of the directrix is $y = 3$, the vertex is $V(0,0)$, the parabola opens downward, and $p = -3$. Thus, $x^2 = 4py \Rightarrow x^2 = -12y$.

49. If the focus is $F(-1, 0)$ and the equation of the directrix is $x = 1$, the vertex is $V(0,0)$, the parabola opens to the left, and $p = -1$. Thus, $y^2 = 4px \Rightarrow y^2 = -4x$.

51. See Figure 51.

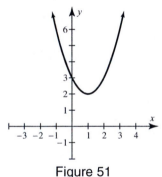

Figure 51

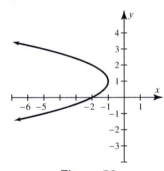

Figure 53

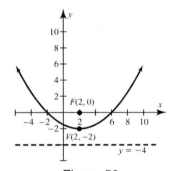

Figure 59

53. See Figure 53.

55. Vertex at $(1, 1)$; c

57. Vertex at $(0, 2)$; a

59. The equation $(x - 2)^2 = 8(y + 2)$ is in the form $(x - h)^2 = 4p(y - k)$, with $(h, k) = (2, -2)$ and $p = 2$. The parabola opens upward, with vertex at $(2, -2)$, focus at $(2, 0)$, and equation of directrix $y = -4$. The graph is shown in Figure 59.

61. The equation $x = -\dfrac{1}{4}(y + 3)^2 + 2$ can be written as $(y + 3)^2 = -4(x - 2)$, which is in the form $(y - k)^2 = 4p(x - h)$, with $(h, k) = (2, -3)$ and $p = -1$. The parabola opens to the left, with vertex at $(2, -3)$, focus at $(1, -3)$, and equation of directrix $x = 3$. The graph is shown in Figure 61.

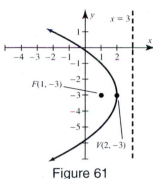

Figure 61

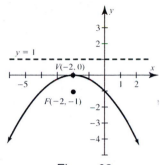

Figure 63

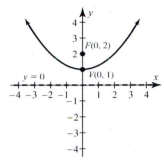

Figure 65

63. The equation $y = -\dfrac{1}{4}(x + 2)^2$ can be written as $(x + 2)^2 = -4y$, which is in the form

$(x - h)^2 = 4p(y - k)$, with $(h, k) = (-2, 0)$ and $p = -1$. The parabola opens downward, with vertex at

$(-2, 0)$, focus at $(-2, -1)$, and equation of directrix $y = 1$. The graph is shown in Figure 63.

65. If the focus is at $(0, 2)$ and the vertex at $(0, 1)$, the parabola opens upward and $p = 1$. Substituting in

$(x - h)^2 = 4p(y - k)$, we get $(x - 0)^2 = 4(1)(y - 1)$ or $x^2 = 4(y - 1)$. See Figure 65.

67. If the focus is at $(0, 0)$ and the directrix has equation $x = -2$, the vertex is at $(-1, 0)$, $p = 1$, and the parabola

opens to the right. Substituting in $(y - k)^2 = 4p(x - h)$, we get

$(y - 0)^2 = 4(1)(x - (-1))$ or $y^2 = 4(x + 1)$. See Figure 67.

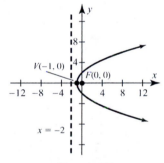

Figure 67

69. If the focus is at $(-1, 3)$ and the directrix has equation $y = 7$, the vertex is at $(-1, 5)$, $p = -2$, and the

parabola opens downward. Substituting in $(x - h)^2 = 4p(y - k)$, we get $(x + 1)^2 = 4(-2)(y - 5)$ or

$(x + 1)^2 = -8(y - 5)$.

71. Since the parabola has a horizontal axis, the equation is in the form $(y - k)^2 = a(x - h)$. Find the value of a

by using the fact that the parabola passes through $(-4, 0)$ and the vertex is $V(-2, 3)$.

Substituting $x = -4$, $y = 0$, $h = -2$ and $k = 3$ yields $(0 - 3)^2 = a(-4 - (-2)) \Rightarrow a = -\dfrac{9}{2}$.

The equation is $(y - 3)^2 = -\dfrac{9}{2}(x + 2)$.

73. $-2x = y^2 + 6x + 10 \Rightarrow y^2 = -8x - 10 \Rightarrow (y - 0)^2 = -8\left(x + \dfrac{5}{4}\right)$

75. $x = 2y^2 + 4y - 1 \Rightarrow 2y^2 + 4y = x + 1 \Rightarrow y^2 + 2y = \frac{1}{2}(x + 1) \Rightarrow$

$y^2 + 2y + 1 = \frac{1}{2}(x + 1) + 1 \Rightarrow (y + 1)^2 = \frac{1}{2}(x + 1 + 2) \Rightarrow (y + 1)^2 = \frac{1}{2}(x + 3)$

77. $x^2 - 3x + 4 = 2y \Rightarrow x^2 - 3x = 2y - 4 \Rightarrow x^2 - 3x + \frac{9}{4} = 2y - 4 + \frac{9}{4} \Rightarrow$

$\left(x - \frac{3}{2}\right)^2 = 2y - \frac{7}{4} \Rightarrow \left(x - \frac{3}{2}\right)^2 = 2\left(y - \frac{7}{8}\right)$

79. $4y^2 + 4y - 5 = 5x \Rightarrow 4y^2 + 4y = 5x + 5 \Rightarrow y^2 + y = \frac{5}{4}(x + 1) \Rightarrow$

$y^2 + y + \frac{1}{4} = \frac{5}{4}(x + 1) + \frac{1}{4} \Rightarrow \left(y + \frac{1}{2}\right)^2 = \frac{5}{4}\left(x + 1 + \frac{1}{5}\right) \Rightarrow \left(y + \frac{1}{2}\right)^2 = \frac{5}{4}\left(x + \frac{6}{5}\right)$

81. $y = -0.75 \pm \sqrt{-3x}$; See Figure 81.

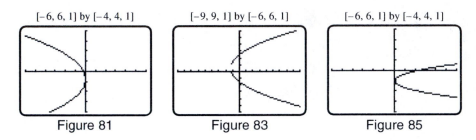

$[-6, 6, 1]$ by $[-4, 4, 1]$ $[-9, 9, 1]$ by $[-6, 6, 1]$ $[-6, 6, 1]$ by $[-4, 4, 1]$

Figure 81   Figure 83   Figure 85

83. $y = 0.5 \pm \sqrt{3.1(x + 1.3)}$; See Figure 83. Note: If a break in the graph appears near the vertex, it should not be there. It is a result of the low resolution of the graphing calculator screen.

85. $y = -1 \pm \sqrt{\frac{x}{2.3}}$; See Figure 85.

87. $x^2 = 2y$ and $x^2 = y + 1 \Rightarrow 2y = y + 1 \Rightarrow y = 1$; $x^2 = 2y$ when $y = 1 \Rightarrow x^2 = 2 \Rightarrow x = \pm\sqrt{2}$; the solution is $(\pm\sqrt{2}, 1)$.

89. $\frac{1}{3}y^2 = -3x$ and $y^2 = x + 1 \Rightarrow -9x = x + 1 \Rightarrow x = -\frac{1}{10}$; $y^2 = x + 1$ when $x = -\frac{1}{10} \Rightarrow$

$y^2 = -\frac{1}{10} + 1 \Rightarrow y = \pm\sqrt{0.9}$; the solution is $\left(-\frac{1}{10}, \pm\sqrt{0.9}\right)$.

91. $(y - 1)^2 = x + 1$ and $(y + 2)^2 = -x + 4 \Rightarrow (y - 1)^2 + (y + 2)^2 = 5 \Rightarrow$

$y^2 - 2y + 1 + y^2 + 4y + 4 = 5 \Rightarrow 2y^2 + 2y = 0 \Rightarrow 2y(y + 1) = 0 \Rightarrow y = 0$ or

$y = -1$; $(y - 1)^2 = x + 1$ when $y = 0 \Rightarrow 1 = x + 1 \Rightarrow x = 0$, when $y = -1 \Rightarrow 4 = x + 1 \Rightarrow$

$x = 3$, the solution is $(0, 0)$, $(3, -1)$.

93. Substitute the point $(3, 0.75)$ into $x^2 = 4py$ and solve for p; $9 = 4p(0.75) \Rightarrow 9 = 3p \Rightarrow p = 3$. The receiver should be 3 feet from the vertex.

95. (a) Substitute the point $(105, 32)$ into $y = ax^2$ and solve for a; $32 = a(105)^2 \Rightarrow a = \frac{32}{11,025}$.

The equation is $y = \frac{32}{11,025}x^2$.

(b) Rewriting the answer in (a) we have $x^2 = \dfrac{11,025}{32}y$, so $4p = \dfrac{11,025}{32}$ and $p = \dfrac{11,025}{128} \approx 86.1$.

The receiver should be located about 86.1 feet from the vertex.

97. (a) Since $y^2 = 100x$, $4p = 100$ and $p = 25$. Thus, the coordinates of the sun are $(25, 0)$.

(b) The minimum distance occurs when the comet is at the vertex of the parabola, so the minimum distance is

25 million miles.

99. The pipe should be at the focus, so $p = 18$, $k = 4p = 72$ inches or 6 ft.

10.2: Ellipses

1. The endpoints of the major axis of an ellipse are called the vertices of the ellipse.

3. An ellipse with equation $\dfrac{x^2}{a^2} + \dfrac{y^2}{b^2} = 1 (a > b > 0)$ has a horizontal major axis.

5. $\dfrac{x^2}{4} + \dfrac{y^2}{9} = 1 \Rightarrow a = 3$ and $b = 2$. $a^2 - b^2 = 3^2 - 2^2 = 5 = c^2 \Rightarrow c = \sqrt{5}$. The foci are $(0, \pm\sqrt{5})$, the

endpoints of the major axis (vertices) are $(0, \pm 3)$, while the endpoints of the minor axis are $(\pm 2, 0)$. The

ellipse is graphed in Figure 5.

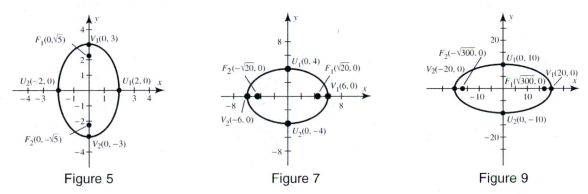

Figure 5   Figure 7   Figure 9

7. $\dfrac{x^2}{36} + \dfrac{y^2}{16} = 1 \Rightarrow a = 6$ and $b = 4$. $a^2 - b^2 = 6^2 - 4^2 = 20 = c^2 \Rightarrow c = \sqrt{20}$. The foci are $(\pm\sqrt{20}, 0)$,

the endpoints of the major axis (vertices) are $(\pm 6, 0)$, while the endpoints of the minor axis are $(0, \pm 4)$. The

ellipse is graphed in Figure 7.

9. $x^2 + 4y^2 = 400 \Rightarrow \dfrac{x^2}{400} + \dfrac{y^2}{100} = 1 \Rightarrow a = 20$ and $b = 10$.

$a^2 - b^2 = 400 - 100 = 300 = c^2 \Rightarrow c = \sqrt{300}$. The foci are $(\pm\sqrt{300}, 0)$, the endpoints of the major

axis (vertices) are $(\pm 20, 0)$, while the endpoints of the minor axis are $(0, \pm 10)$. The ellipse is graphed in

Figure 9.

11. $25x^2 + 9y^2 = 225 \Rightarrow \dfrac{x^2}{9} + \dfrac{y^2}{25} = 1 \Rightarrow a = 5$ and $b = 3$. $a^2 - b^2 = 25 - 9 = 16 = c^2 \Rightarrow c = 4$. The foci are $(0, \pm 4)$, the endpoints of the major axis (vertices) are $(0, \pm 5)$, while the endpoints of the minor axis are $(\pm 3, 0)$. The ellipse is graphed in Figure 11.

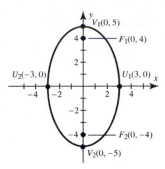

Figure 11

13. Vertices $(0, \pm 6)$; b

15. Vertices $(\pm 4, 0)$; c

17. The ellipse is centered at $(0, 0)$ and has a horizontal major axis. Its standard equation has the form $\dfrac{x^2}{a^2} + \dfrac{y^2}{b^2} = 1$. The endpoints of the major axis are $(\pm 6, 0)$ and the endpoints of the minor axis are $(0, \pm 4)$. It follows that $a = 6$ and $b = 4$, and the standard equation is $\dfrac{x^2}{36} + \dfrac{y^2}{16} = 1$. The foci lie on the horizontal major axis and can be determined as follows. $c^2 = a^2 - b^2 = 36 - 16 = 20$. Thus, $c = \sqrt{20}$, and the coordinates of the foci are $(\pm \sqrt{20}, 0)$.

19. The ellipse is centered at $(0, 0)$ and has a vertical major axis. Its standard equation has the form $\dfrac{x^2}{b^2} + \dfrac{y^2}{a^2} = 1$. The endpoints of the major axis are $(0, \pm 4)$ and the endpoints of the minor axis are $(\pm 2, 0)$. It follows that $a = 4$ and $b = 2$, and the standard equation is $\dfrac{x^2}{4} + \dfrac{y^2}{16} = 1$. The foci lie on the vertical major axis and can be determined as follows. $c^2 = a^2 - b^2 = 16 - 4 = 12$. Thus, $c = \sqrt{12}$, and the coordinates of the foci are $(0, \pm \sqrt{12})$.

21. To sketch a graph of an ellipse centered at the origin, it is helpful to plot the vertices and the endpoints of the minor axis. The vertices are $V(\pm 5, 0)$ so $a = 5$, the foci are $F(\pm 4, 0)$ so $c = 4$, and the endpoints of the minor axis are $U(0, \pm 3)$ so $b = 3$. A graph of the ellipse is shown in Figure 21. Its equation is $\dfrac{x^2}{25} + \dfrac{y^2}{9} = 1$.

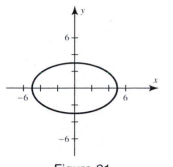

Figure 21

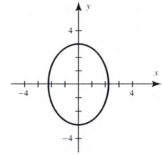

Figure 23

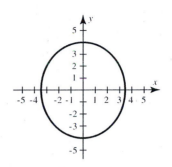

Figure 25

23. The vertices are $V(0, \pm 3)$ so $a = 3$, the foci are $F(0, \pm 2)$ so $c = 2$, and the endpoints of the minor axis are

 $U(\pm\sqrt{5}, 0)$ so $b = \sqrt{5}$. A graph of the ellipse is shown in Figure 23. Its equation is $\dfrac{x^2}{5} + \dfrac{y^2}{9} = 1$.

25. Foci of $F(0, \pm 2) \Rightarrow c = 2$ and $V(0, \pm 4) \Rightarrow a = 4$. The major axis lies on the y-axis. The value of b is as

 follows: $a^2 - b^2 = c^2 \Rightarrow a^2 - c^2 = b^2 \Rightarrow 4^2 - 2^2 = b^2 \Rightarrow b^2 = 12 \Rightarrow b = \sqrt{12}$. The equation of the

 ellipse is $\dfrac{x^2}{12} + \dfrac{y^2}{16} = 1$. Its graph is shown in Figure 25.

27. Foci of $F(\pm 5, 0) \Rightarrow c = 5$ and $V(\pm 6, 0) \Rightarrow a = 6$. The major axis lies on the x-axis. The value of b is as

 follows: $a^2 - b^2 = c^2 \Rightarrow a^2 - c^2 = b^2 \Rightarrow 6^2 - 5^2 = b^2 \Rightarrow b^2 = 11 \Rightarrow b = \sqrt{11}$. The equation of the

 ellipse is $\dfrac{x^2}{36} + \dfrac{y^2}{11} = 1$. Its graph is shown in Figure 27.

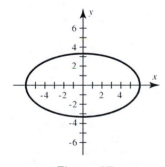

Figure 27

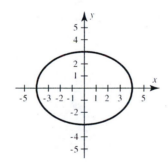

Figure 29

29. Horizontal major axis of length $8 \Rightarrow a = 4$. Minor axis of length $6 \Rightarrow b = 3$. The major axis lies on the

 x-axis. The equation of the ellipse is $\dfrac{x^2}{16} + \dfrac{y^2}{9} = 1$. Its graph is shown in Figure 29.

31. $e = \dfrac{c}{a} = \dfrac{2}{3} \Rightarrow 3c = 2a \Rightarrow c = \dfrac{2}{3}a$. Since the major axis is length $6 \Rightarrow a = 3$. Thus, $c = \dfrac{2}{3}(3) = 2$.

 Then the value of b is given by the following: $a^2 - b^2 = c^2 \Rightarrow a^2 - c^2 = b^2 \Rightarrow 3^2 - 2^2 = b^2 \Rightarrow$

 $b^2 = 5 \Rightarrow b = \sqrt{5}$. The equation of the ellipse is $\dfrac{x^2}{9} + \dfrac{y^2}{5} = 1$. Its graph is shown in Figure 31.

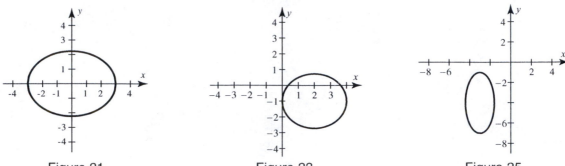

Figure 31   Figure 33   Figure 35

33. To translate the center from $(0, 0)$ to $(2, -1)$ replace x with $(x - 2)$ and y with $(y + 1)$. This new equation is $\dfrac{(x - 2)^2}{4} + \dfrac{(y + 1)^2}{3} = 1$. See Figure 33.

35. To translate the center from $(0, 0)$ to $(-3, -4)$ replace x with $(x + 3)$ and y with $(y + 4)$. This new equation is $\dfrac{(x + 3)^2}{2} + \dfrac{(y + 4)^2}{9} = 1$. See Figure 35.

37. The ellipse is centered at $(2, 1)$. The major axis has length $2a = 6$ and the length of the minor axis is $2b = 4$. The major axis is parallel to the y-axis. The graph is shown in Figure 37.

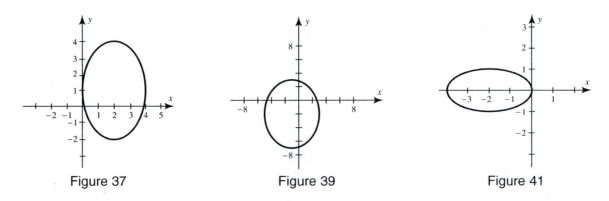

Figure 37   Figure 39   Figure 41

39. The ellipse is centered at $(-1, -2)$. The major axis has length $2a = 10$ and the length of the minor axis is $2b = 8$. The major axis is parallel to the y-axis. The graph is shown in Figure 39.

41. The ellipse is centered at $(-2, 0)$. The horizontal major axis has length $2a = 4$ and the vertical minor axis has length $2b = 2$. The graph is shown in Figure 41.

43. Center at $(2, -4)$; d

45. Center at $(-1, 1)$; c

47. Center at $(1, 1)$, $a = 5$, $b = 3$, major axis vertical. $c^2 = a^2 - b^2 = 25 - 9 = 16 \Rightarrow c = 4$.

Foci: $(1, 1 \pm 4)$; veritices: $(1, 1 \pm 5)$; the graph is shown in Figure 47.

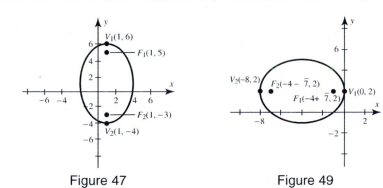

Figure 47   Figure 49

49. Center at $(-4, 2)$, $a = 4$, $b = 3$, major axis horizontal. $c^2 = a^2 - b^2 = 16 - 9 = 7 \Rightarrow c = \sqrt{7}$.

Foci: $(-4 \pm \sqrt{7}, 2)$; veritices: $(-4 \pm 4, 2)$; the graph is shown in Figure 49.

51. Since the center is $(2, 1)$ and a focus is $(2, 3)$, $c = 2$. Since the vertex is $(2, 4)$, $a = 3$;

$b^2 = a^2 - c^2 = 9 - 4 = 5$; the major axis is vertical. The equation is $\dfrac{(x - 2)^2}{5} + \dfrac{(y - 1)^2}{9} = 1$.

53. The center is halfway between the vertices at $(0, 2)$; $a = 3$ and $c = 2$; $b^2 = a^2 - c^2 = 9 - 4 = 5$; the

major axis is horizontal. The equation is $\dfrac{x^2}{9} + \dfrac{(y - 2)^2}{5} = 1$.

55. Center at $(2, 4)$, $a = 4$ and $b = 2$, major axis parallel to the x-axis; the equation is

$\dfrac{(x - 2)^2}{16} + \dfrac{(y - 4)^2}{4} = 1$.

57. $9x^2 + 18x + 4y^2 - 8y - 23 = 0 \Rightarrow 9(x^2 + 2x) + 4(y^2 - 2y) = 23 \Rightarrow$

$9(x^2 + 2x + 1) + 4(y^2 - 2y + 1) = 23 + 9 + 4 \Rightarrow 9(x + 1)^2 + 4(y - 1)^2 = 36 \Rightarrow$

$\dfrac{(x + 1)^2}{4} + \dfrac{(y - 1)^2}{9} = 1$; The center is $(-1, 1)$. The vertices are

$(-1, 1 - 3), (-1, 1 + 3)$ or $(-1, -2), (-1, 4)$.

59. $4x^2 + 8x + y^2 + 2y + 1 = 0 \Rightarrow 4(x^2 + 2x) + (y^2 + 2y) = -1 \Rightarrow$

$4(x^2 + 2x + 1) + (y^2 + 2y + 1) = -1 + 4 + 1 \Rightarrow 4(x + 1)^2 + (y + 1)^2 = 4 \Rightarrow$

$\dfrac{(x + 1)^2}{1} + \dfrac{(y + 1)^2}{4} = 1$; The center is $(-1, -1)$. The vertices are

$(-1, -1 - 2), (-1, -1 + 2)$ or $(-1, -3), (-1, 1)$.

61. $4x^2 + 16x + 5y^2 - 10y + 1 = 0 \Rightarrow 4(x^2 + 4x) + 5(y^2 - 2y) = -1 \Rightarrow$

$4(x^2 + 4x + 4) + 5(y^2 - 2y + 1) = -1 + 16 + 5 \Rightarrow 4(x + 2)^2 + 5(y - 1)^2 = 20 \Rightarrow$

$\dfrac{(x + 2)^2}{5} + \dfrac{(y - 1)^2}{4} = 1$; The center is $(-2, 1)$. The vertices are $(-2 - \sqrt{5}, 1), (-2 + \sqrt{5}, 1)$.

63. $16x^2 - 16x + 4y^2 + 12y = 51 \Rightarrow 16(x^2 - x) + 4(y^2 + 3y) = 51 \Rightarrow$

$16\left(x^2 - x + \dfrac{1}{4}\right) + 4\left(y^2 + 3y + \dfrac{9}{4}\right) = 51 + 4 + 9 \Rightarrow 16\left(x - \dfrac{1}{2}\right)^2 + 4\left(y + \dfrac{3}{2}\right)^2 = 64 \Rightarrow$

$$\frac{(x - \frac{1}{2})^2}{4} + \frac{(y + \frac{3}{2})^2}{16} = 1$$

The center is $\left(\frac{1}{2}, -\frac{3}{2}\right)$. The vertices are $\left(\frac{1}{2}, -\frac{3}{2} - 4\right), \left(\frac{1}{2}, -\frac{3}{2} + 4\right)$ or $\left(\frac{1}{2}, -\frac{11}{2}\right), \left(\frac{1}{2}, \frac{5}{2}\right)$.

65. $y = \pm\sqrt{10\left(1 - \frac{x^2}{15}\right)}$; See Figure 65.

67. $y = \pm\sqrt{\frac{25 - 4.1x^2}{6.3}}$; See Figure 67.

[−6, 6, 1] by [−4, 4, 1]   [−4.7, 4.7, 1] by [−3.1, 3.1, 1]

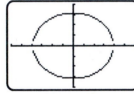

Figure 65

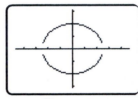

Figure 67

69. $\frac{x^2}{4} + \frac{y^2}{9} = 1 \Rightarrow 9x^2 + 4y^2 = 36 \Rightarrow 9x^2 + 4(3 - x)^2 = 36 \Rightarrow 9x^2 + 4(9 - 6x + x^2) = 36 \Rightarrow$

$13x^2 - 24x = 0$. Then $x(13x - 24) = 0 \Rightarrow x = 0$ or $x = \frac{24}{13}$. Since $y = 3 - x$, the corresponding y values

are $3 - 0 = 3$ and $3 - \frac{24}{13} = \frac{15}{13}$. The solutions are $(0, 3)$ and $\left(\frac{24}{13}, \frac{15}{13}\right)$. The system is graphed in Figure 69.

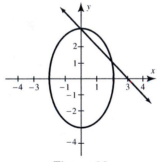

Figure 69

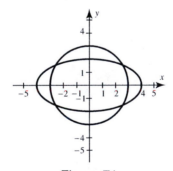

Figure 71

71. $x^2 + y^2 = 9 \Rightarrow x^2 = 9 - y^2$; then $4(9 - y^2) + 16y^2 = 64 \Rightarrow 36 - 4y^2 + 16y^2 = 64 \Rightarrow$

$y^2 = \frac{7}{3} \Rightarrow y = \pm\sqrt{\frac{7}{3}}$. Substituting in

$x^2 + y^2 = 9$ we find $x^2 + \frac{7}{3} = 9$, $x^2 = \frac{27}{3} - \frac{7}{3} = \frac{20}{3}$, so $x = \pm\sqrt{\frac{20}{3}}$.

There are four solutions: $\left(\pm\sqrt{\frac{20}{3}}, \pm\sqrt{\frac{7}{3}}\right)$. The system is graphed in Figure 71.

73. $x^2 + y^2 = 9 \Rightarrow y^2 = 9 - x^2$; then $2x^2 + 3(9 - x^2) = 18 \Rightarrow 2x^2 + 27 - 3x^2 = 18 \Rightarrow$

$x^2 = 9 \Rightarrow x = \pm 3$;

Substituting in $x^2 + y^2 = 9$ we find $9 + y^2 = 9$, so $y = 0$. There are two solutions: $(\pm 3, 0)$. The system is

graphed in Figure 73.

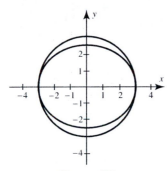

Figure 73

75. $\dfrac{x^2}{2} + \dfrac{y^2}{4} = 1 \Rightarrow 2x^2 + y^2 = 4 \Rightarrow 2(2y - 4) + y^2 = 4 \Rightarrow 4y - 8 + y^2 = 4 \Rightarrow y^2 + 4y - 12 = 0$

Then $(y + 6)(y - 2) = 0 \Rightarrow y = -6$ or $y = 2$. Since $x = \pm\sqrt{2y - 4}$, the corresponding x values are

$\pm\sqrt{2(-6) - 4}$, which is undefined, and $\pm\sqrt{2(2) - 4} = 0$. The solution is $(0, 2)$.

77. From the first equation $\dfrac{x^2}{2} + \dfrac{y^2}{4} = 1 \Rightarrow 2x^2 + y^2 = 4 \Rightarrow y^2 = 4 - 2x^2$.

From the second equation $\dfrac{x^2}{4} + \dfrac{y^2}{2} = 1 \Rightarrow x^2 + 2y^2 = 4 \Rightarrow y^2 = 2 - \dfrac{1}{2}x^2$. That is $4 - 2x^2 = 2 - \dfrac{1}{2}x^2$.

$4 - 2x^2 = 2 - \dfrac{1}{2}x^2 \Rightarrow \dfrac{3}{2}x^2 = 2 \Rightarrow x^2 = \dfrac{4}{3} \Rightarrow x = \pm\dfrac{2}{\sqrt{3}} = \pm\dfrac{2\sqrt{3}}{3}$

Since $y = \pm\sqrt{4 - 2x^2}$, the y values are $\pm\sqrt{4 - 2\left(\dfrac{2}{\sqrt{3}}\right)^2} = \pm\sqrt{4 - \dfrac{8}{3}} = \pm\sqrt{\dfrac{4}{3}}$.

There are four solutions: $\left(\pm\sqrt{\dfrac{4}{3}}, \pm\sqrt{\dfrac{4}{3}}\right)$.

79. Subtracting the second equation from the first equation yields $(x - 2)^2 - x^2 = 0$.

$(x - 2)^2 - x^2 = 0 \Rightarrow x^2 - 4x + 4 - x^2 = 0 \Rightarrow -4x + 4 = 0 \Rightarrow -4x = -4 \Rightarrow x = 1$

Since $y = \pm\sqrt{9 - x^2}$, the y values are $\pm\sqrt{9 - (1)^2} = \pm\sqrt{8}$.

The solutions are $(1, -\sqrt{8})$, $(1, \sqrt{8})$.

81. The system is $(x - 1)^2 + (y + 1)^2 < 4$ and $(x + 1)^2 + y^2 > 1$. See Figure 81.

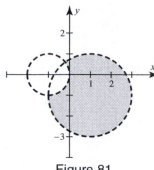

Figure 81

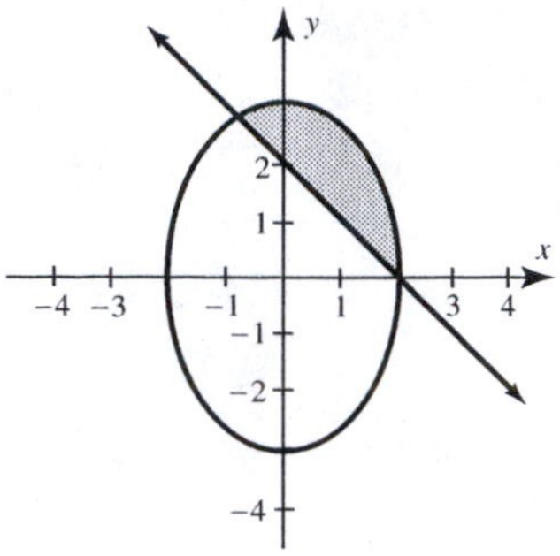

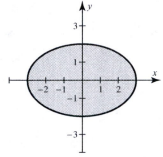

Figure 83

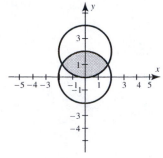

Figure 85

83. The system is $\dfrac{x^2}{4} + \dfrac{y^2}{9} \leq 1$ and $x + y \geq 2$. See Figure 83.

85. The system is $x^2 + y^2 \leq 4$ and $x^2 + (y - 2)^2 \leq 4$. See Figure 85.

87. The system is $x^2 + y^2 \leq 4$ and $(x + 1)^2 - y \leq 0$. See Figure 87.

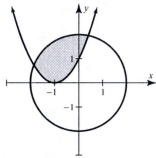

Figure 87

Figure 89

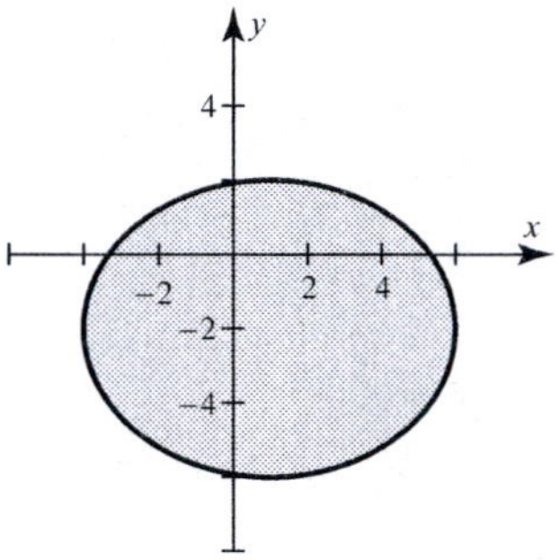

Figure 91

89. The inequality can be written $\dfrac{x^2}{9} + \dfrac{y^2}{4} \leq 1$. The shaded region is shown in Figure 89.

Here $a = 3$ and $b = 2$. The area is $A = \pi ab = \pi(3)(2) = 6\pi \approx 18.85$ ft^2.

91. The shaded region is shown in Figure 91.

Here $a = 5$ and $b = 4$. The area is $A = \pi ab = \pi(5)(4) = 20\pi \approx 62.83$ ft^2.

93. $e = \dfrac{c}{a} = 0.206 \Rightarrow c = 0.206a$. Since $a = 0.387$, it follows that $c = 0.206(0.387) \approx 0.0797$. Then, the value

of b is given by the following: $a^2 - b^2 = c^2 \Rightarrow a^2 - c^2 = b^2 \Rightarrow 0.387^2 - 0.0797^2 = b^2 \Rightarrow$

$b^2 \approx 0.1434 \Rightarrow b \approx 0.379$. The major axis could be located on either the x- or y-axis. We will choose the

x-axis. Thus, the equation of the orbit is $\dfrac{x^2}{0.387^2} + \dfrac{y^2}{0.379^2} = 1$. The sun can be located on either of the foci.

We will locate the sun at $(0.0797, 0)$. To graph the orbit with a graphing calculator, solve the equation for the

ellipse for the variable y: $\dfrac{x^2}{0.387^2} + \dfrac{y^2}{0.379^2} = 1 \Rightarrow y = \pm 0.379\sqrt{1 - \dfrac{x^2}{0.387^2}}$. Graph each of the equations

and plot the sun at $(0.0797, 0)$. $Y_1 = 0.379\sqrt{((1 - X^2)/0.387^2)}$, $Y_2 = -Y_1$. See Figure 93.

[−0.6, 0.6, 0.1] by [−0.4, 0.4, 0.1]

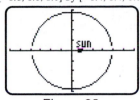

Figure 93

95. The source and the stone are at the two foci of the ellipse, so the distance between them is $2c$.

$c^2 = a^2 - b^2 = 4^2 - (2.5)^2 = 16 - 6.25 = 9.75 \Rightarrow c = \sqrt{9.75}$. Thus, $2c = 2\sqrt{9.75} \approx 6.245$. The stone should be 6.245 inches from the source.

97. $c = 50$ and $b = 40$ for the elliptical floor. $a = \sqrt{50^2 + 40^2} = 10 \cdot \sqrt{41} \approx 64.03$. The area of the ellipse is πab, so the area of the floor is $\pi(64.03)(40) \approx 8046.25$ square feet.

99. The ellipse with a mjor axis of 620 feet and minor axis of 513 feet inplies vertices of $(\pm 310, 0)$ and $(0, \pm 256.5)$. $c^2 = a^2 - b^2 \Rightarrow c^2 = (310)^2 - (256.5)^2 \Rightarrow c \approx 174.1$. The distance between the foci is given as $2c = 2(174.1) = 348.2$ feet.

101. The equation of the ellipse is $\dfrac{x^2}{30^2} + \dfrac{y^2}{25^2} = 1$. Solving for y we get $y = 25\sqrt{1 - \dfrac{x^2}{900}}$. When $x = 15$,

$y = 25\sqrt{1 - \dfrac{225}{900}} \approx 21.65$ feet.

103. The minimum height is $4464 - (3960 + 164) = 340$ miles; the maximum height is $4464 - (3960 - 164) = 668$ miles.

Extended and Discovery Exercises for Section 10.2

1. The slope of the line through $(-2, 6)$ and $(4, -3)$ is $-\dfrac{3}{2}$ and has equation $y - 6 = -\dfrac{3}{2}(x + 2)$.

   $0 - 6 = -\dfrac{3}{2}(x + 2) \Rightarrow x = 2$, x-intercept is $(2, 0)$. $y - 6 = -\dfrac{3}{2}(0 + 2) \Rightarrow y = 3$, y-intercept is $(0, 3)$.

   The equation of the line in intercept form is $\dfrac{x}{2} + \dfrac{y}{3} = 1$.

3. The equation of the line through $(3, -1)$ with slope of -2 is $y + 1 = -2(x - 3)$.

   $0 + 1 = -2(x - 3) \Rightarrow x = 2.5$, x-intercept is $(2.5, 0)$. $y + 1 = -2(0 - 3) \Rightarrow y = 5$, y-intercept is $(0, 5)$.

   The equation of the line in intercept form is $\dfrac{x}{2.5} + \dfrac{y}{5} = 1$.

5. The ellipse $\dfrac{x^2}{25} + \dfrac{y^2}{9} = 1 \Rightarrow a = 5$ and $b = 3$. Therefore, the x-intercept is $(\pm 5, 0)$ and y-intercept is $(0, \pm 3)$.

Checking Basic Concepts for Sections 10.1 and 10.2

1. The equation $x = \frac{1}{2}y^2$ can be written as $y^2 = 2x$, which is in the form $y^2 = 4px$. The vertex is $V(0, 0)$. Thus,

    $2 = 4p$ or $p = \frac{1}{2}$. The focus is $F\left(\frac{1}{2}, 0\right)$, the equation of the directrix is $x = -\frac{1}{2}$, and the parabola opens to

    the right. The graph of $x = \frac{1}{2}y^2$ is shown in Figure 1.

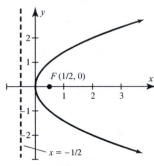

Figure 1

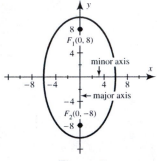

Figure 3

3. $\frac{x^2}{36} + \frac{y^2}{100} = 1 \Rightarrow a = 10$ and $b = 6$. $a^2 - b^2 = 10^2 - 6^2 = 64 = c^2 \Rightarrow c = 8$. The foci are $(0, \pm 8)$, the

    endpoints of the major axis (vertical) are $(0, \pm 10)$, while the endpoints of the minor axis are $(\pm 6, 0)$. The

    ellipse is graphed in Figure 3.

5. For a parabola with vertex at the origin and passing through $(2, 1)$, the equation $x^2 = 4py$ becomes $4 = 4p$, so

    $p = 1$. The filament should be located 1 foot from the vertex of the reflector.

7. $x^2 - 4x + 4y^2 + 8y - 8 = 0 \Rightarrow (x^2 - 4x) + 4(y^2 + 2y) = 8 \Rightarrow$

    $(x^2 - 4x + 4) + 4(y^2 + 2y + 1) = 8 + 4 + 4 \Rightarrow (x - 2)^2 + 4(y + 1)^2 = 16 \Rightarrow$

    $\frac{(x - 2)^2}{16} + \frac{(y + 1)^2}{4} = 1$; The center is $(2, -1)$. The vertices are

    $(2 - 4, -1), (2 + 4, -1)$ or $(-2, -1), (6, -1)$.

Section 10.3: Hyperbolas

1. The vertices are the endpoints of the transverse axis of a hyperbola.

3. A hyperbola with an equation of the form $\frac{x^2}{a^2} - \frac{y^2}{b^2} = 1$ has a horizontal transverse axis.

5. The transverse axis is horizontal with $a = 3$ and $b = 7$. The vertices are $(\pm 3, 0)$. The asymptotes are

    $y = \pm \frac{7}{3}x$. See Figure 5.

    Since $c^2 = a^2 + b^2 \Rightarrow c = \pm \sqrt{9 + 49} \Rightarrow \sqrt{58}$, the foci are $(\pm \sqrt{58}, 0)$.

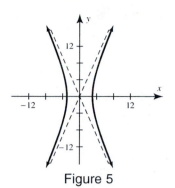

Figure 5

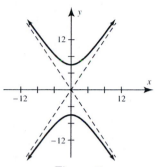

Figure 7

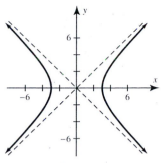

Figure 9

7. The transverse axis is vertical with $a = 6$ and $b = 4$. The vertices are $(0, \pm 6)$. The asymptotes are

$y = \pm\dfrac{3}{2}x$. See Figure 7.

Since $c^2 = a^2 + b^2 \Rightarrow c = \pm\sqrt{36 + 16} \Rightarrow \sqrt{52}$, the foci are $(0, \pm\sqrt{52})$.

9. $x^2 - y^2 = 9 \Rightarrow \dfrac{x^2}{9} - \dfrac{y^2}{9} = 1$. The transverse axis is horizontal with $a = 3$ and $b = 3$. The vertices are

$(\pm 3, 0)$. The asymptotes are $y = \pm x$. See Figure 9.

Since $c^2 = a^2 + b^2 \Rightarrow c = \pm\sqrt{9 + 9} \Rightarrow \sqrt{18}$, the foci are $(\pm\sqrt{18}, 0)$.

11. $9y^2 - 16x^2 = 144 \Rightarrow \dfrac{y^2}{16} - \dfrac{x^2}{9} = 1$. The transverse axis is vertical with $a = 4$ and $b = 3$. The vertices are

$(0, \pm 4)$. The asymptotes are $y = \pm\dfrac{4}{3}x$. See Figure 11.

Since $c^2 = a^2 + b^2 \Rightarrow c = \pm\sqrt{16 + 9} \Rightarrow \pm 5$, the foci are $(0, \pm 5)$.

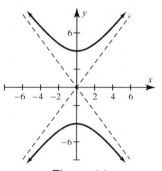

Figure 11

13. Horizontal transverse axis, vertices $(\pm 2, 0)$; d

15. Vertical transverse axis, vertices $(0, \pm 3)$; a

17. Since the foci and the vertices lie on the *x*-axis, the hyperbola has a horizontal transverse axis with an equation

of the form $\dfrac{x^2}{a^2} - \dfrac{y^2}{b^2} = 1$. $F(\pm 5, 0) \Rightarrow c = 5$ and $V(\pm 4, 0) \Rightarrow a = 4$. $c^2 = a^2 + b^2 \Rightarrow$

$b^2 = c^2 - a^2 = 25 - 16 = 9 \Rightarrow b = 3$. The equation of the hyperbola is $\dfrac{x^2}{16} - \dfrac{y^2}{9} = 1$, and the asymptotes

have the equation $y = \pm\dfrac{b}{a}x$ or $y = \pm\dfrac{3}{4}x$. The hyperbola is graphed in Figure 17.

19. Since the foci and the vertices lie on the y-axis, the hyperbola has a vertical transverse axis with an equation of the form $\dfrac{y^2}{a^2} - \dfrac{x^2}{b^2} = 1$. $F(0, \pm 10) \Rightarrow c = 10$ and $V(0, \pm 6) \Rightarrow a = 6$. $c^2 = a^2 + b^2 \Rightarrow$ $b^2 = c^2 - a^2 = 100 - 36 = 64 \Rightarrow b = 8$. The equation of the hyperbola is $\dfrac{y^2}{36} - \dfrac{x^2}{64} = 1$, and the asymptotes have the equation $y = \pm \dfrac{a}{b}x$ or $y = \pm \dfrac{3}{4}x$. The hyperbola is graphed in Figure 19.

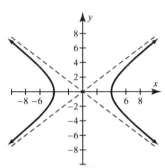

Figure 17

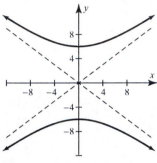

Figure 19

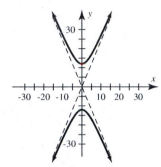

Figure 21

21. Since the foci and the vertices lie on the y-axis, the hyperbola has a vertical transverse axis with an equation of the form $\dfrac{y^2}{a^2} - \dfrac{x^2}{b^2} = 1$. $F(0, \pm 13) \Rightarrow c = 13$ and $V(0, \pm 12) \Rightarrow a = 12$. $c^2 = a^2 + b^2 \Rightarrow$ $b^2 = c^2 - a^2 = 169 - 144 = 25 \Rightarrow b = 5$. The equation of the hyperbola is $\dfrac{y^2}{144} - \dfrac{x^2}{25} = 1$, and the asymptotes have the equation $y = \pm \dfrac{a}{b}x$ or $y = \pm \dfrac{12}{5}x$. The hyperbola is graphed in Figure 21.

23. Since the foci are $(0, \pm 5)$, $c = 5$. Since the transverse axis is vertical of length 4, $a = 2$. The equation has the form $\dfrac{y^2}{a^2} - \dfrac{x^2}{b^2} = 1$. $c^2 = a^2 + b^2 \Rightarrow b^2 = c^2 - a^2 = 25 - 4 = 21 \Rightarrow b = \sqrt{21} \approx 4.58$. The equation of the hyperbola is $\dfrac{y^2}{4} - \dfrac{x^2}{21} = 1$, and the asymptotes have the equation $y = \pm \dfrac{a}{b}x$ or $y = \pm \dfrac{2}{\sqrt{21}}x$. The hyperbola is graphed in Figure 23.

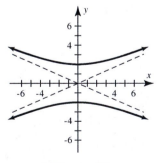

Figure 23

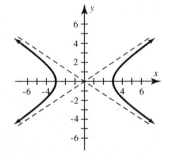

Figure 25

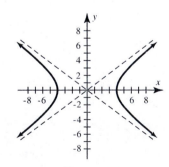

Figure 27

25. Since the vertices lie on the x-axis and are $(\pm 3, 0)$, $a = 3$. The equation has the form $\dfrac{x^2}{a^2} - \dfrac{y^2}{b^2} = 1$. Since $y = \pm \dfrac{b}{a}x = \pm \dfrac{2}{3}x$ and $a = 3$, it follows that $b = 2$. The equation of the hyperbola is $\dfrac{x^2}{9} - \dfrac{y^2}{4} = 1$. The hyperbola is graphed in Figure 25.

27. Since the endpoints of the conjugate axis are $(0, \pm 3)$, $b = 3$. The vertices $(\pm 4, 0)$ lie on the x-axis so $a = 4$

    and the equation of the hyperbola is $\dfrac{x^2}{a^2} - \dfrac{y^2}{b^2} = 1$ or $\dfrac{x^2}{16} - \dfrac{y^2}{9} = 1$. The asymptotes have the equation

    $y = \pm \dfrac{b}{a} x$ or $y = \pm \dfrac{3}{4} x$. The hyperbola is graphed in Figure 27.

29. Since the vertices lie on the x-axis and are $(\pm \sqrt{10}, 0)$, $a^2 = 10$. The equation has the form $\dfrac{x^2}{a^2} - \dfrac{y^2}{b^2} = 1$.

    The value of b^2 can be found by substituting $a^2 = 10$, $x = 10$, and $y = 9$ in this equation.

    $\dfrac{(10)^2}{10} - \dfrac{(9)^2}{b^2} = 1 \Rightarrow \dfrac{100}{10} - \dfrac{81}{b^2} = 1 \Rightarrow 10 - \dfrac{81}{b^2} = 1 \Rightarrow 9 = \dfrac{81}{b^2} \Rightarrow b^2 = \dfrac{81}{9} \Rightarrow b^2 = 9$

    The equation of the hyperbola is $\dfrac{x^2}{10} - \dfrac{y^2}{9} = 1$. The asymptotes have the equation $y = \pm \dfrac{b}{a} x$ or $y = \pm \dfrac{3}{\sqrt{10}} x$.

    The hyperbola is graphed in Figure 29.

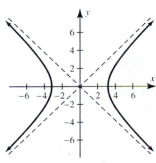

Figure 29

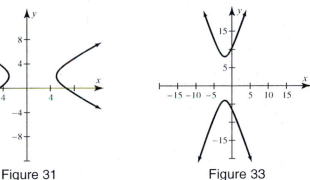

Figure 31   Figure 33

31. See Figure 31. The hyperbola has a horizontal transverse axis and its center is $(1, 2)$. Since

    $a^2 = 16$ and $b^2 = 4$, it follows that $c^2 = a^2 + b^2 = 16 + 4 = 20$. Thus, $a = 4$, $b = 2$, and $c = \sqrt{20}$. The

    vertices are located 4 units to the left and right of center and the foci are located $\sqrt{20}$ units to the left and right

    of center. That is, the vertices are $(1 \pm 4, 2)$ and the foci are $(1 \pm \sqrt{20}, 2)$. The asymptotes are given by

    $y = \pm \dfrac{b}{a}(x - h) + k \Rightarrow y = \pm \dfrac{1}{2}(x - 1) + 2.$

33. See Figure 33. The hyperbola has a vertical transverse axis and its center is $(-2, 2)$. Since

    $a^2 = 36$ and $b^2 = 4$, it follows that $c^2 = a^2 + b^2 = 36 + 4 = 40$. Thus, $a = 6$, $b = 2$, and $c = \sqrt{40}$. The

    vertices are located 6 units above and below center and the foci are located $\sqrt{40}$ units above and below center.

    That is, the vertices are $(-2, 2 \pm 6)$ and the foci are $(-2, 2 \pm \sqrt{40})$. The asymptotes are given by

    $y = \pm \dfrac{b}{a}(x - h) + k \Rightarrow y = \pm 3(x + 2) + 2.$

35. See Figure 35. The hyperbola has a horizontal transverse axis and its center is $(0, 1)$. Since $a^2 = 4$ and $b^2 = 1$,

    it follows that $c^2 = a^2 + b^2 = 4 + 1 = 5$. Thus, $a = 2$, $b = 1$, and $c = \sqrt{5}$. The vertices are located 2 units

    to the left and right of center and the foci are located $\sqrt{5}$ units to the left and right of center. That is, the ver-

    tices are $(\pm 2, 1)$ and the foci are $(\pm \sqrt{5}, 1)$. The asymptotes are given by

    $y = \pm \dfrac{b}{a}(x - h) + k \Rightarrow y = \pm \dfrac{1}{2} x + 1.$

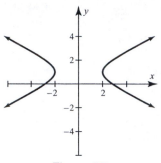

Figure 35

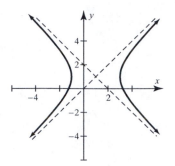

Figure 43

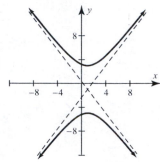

Figure 45

37. Center at $(2, -4)$; b

39. Center at $(2, -1)$; c

41. The hyperbola is centered at $(4, -4)$. The vertical transverse axis has length 8 so $a = 4$. The conjugate axis has length 4, so $b = 2$. Its equation is $\dfrac{(y + 4)^2}{16} - \dfrac{(x - 4)^2}{4} = 1$ with vertices $(4, -4 \pm 4)$, foci $(4, -4 \pm \sqrt{20})$. The asymptotes are given by $y = \pm\dfrac{b}{a}(x - h) + k \Rightarrow y = \pm2(x - 4) - 4$.

43. The hyperbola has a horizontal transverse axis, and its center is $(1, 1)$. Since $a^2 = 4$, $b^2 = 4$, and $c^2 = a^2 + b^2 = 8$, $a = 2$, $b = 2$, and $c = \sqrt{8}$. Thus, the vertices are $(1 \pm 2, 1)$ and the foci are $(1 \pm \sqrt{8}, 1)$. The asymptotes are the lines $y = \pm(x - 1) + 1$. See Figure 43.

45. The hyperbola has a vertical transverse axis, and its center is $(1, -1)$. Since $a^2 = 16$, $b^2 = 9$, and $c^2 = a^2 + b^2 = 25$, $a = 4$, $b = 3$, and $c = 5$. Thus, the vertices are $(1, -1 \pm 4)$ and the foci are $(1, -1 \pm 5)$. The asymptotes are the lines $y = \pm\dfrac{4}{3}(x - 1) - 1$. See Figure 45.

47. Center $(2, -2) \Rightarrow h = 2$ and $k = -2$. Rewrite the coordinates for the given vertex: $(3, -2) \Rightarrow (1 + 2, -2)$. Rewrite the coordinates for the given focus: $(4, -2) \Rightarrow (2 + 2, -2)$. Thus, the transverse axis is horizontal with $a = 1$ and $c = 2$. $b^2 = c^2 - a^2 = 3$. Using the standard equation form, we get the equation: $\dfrac{(x - h)^2}{a^2} - \dfrac{(y - k)^2}{b^2} = 1 \Rightarrow (x - 2)^2 - \dfrac{(y + 2)^2}{3} = 1$.

49. Since the vertices $(-1, \pm1)$ and the foci $(-1, \pm3)$ have the same x-coordinate, $h = -1$ and the transverse axis is vertical. Rewrite the vertices: $(-1, \pm1) \Rightarrow (-1, 0 \pm 1)$, so $k = 0$ and $a = 1 \Rightarrow a^2 = 1$. Rewrite the foci: $(-1, \pm3) \Rightarrow (-1, 0 \pm 3)$, so $k = 0$ and $c = 3$. So, $b^2 = c^2 - a^2 = 8$. Using the standard equation form, we get the equation: $\dfrac{(y - k)^2}{a^2} - \dfrac{(x - h)^2}{b^2} = 1 \Rightarrow y^2 - \dfrac{(x + 1)^2}{8} = 1$.

51. $x^2 - 2x - y^2 + 2y = 4 \Rightarrow (x^2 - 2x + 1) - (y^2 - 2y + 1) = 4 + 1 - 1 \Rightarrow$ $(x - 1)^2 - (y - 1)^2 = 4 \Rightarrow \dfrac{(x - 1)^2}{4} - \dfrac{(y - 1)^2}{4} = 1$. The center is $(1, 1)$. The vertices are $(1 - 2, 1), (1 + 2, 1)$ or $(-1, 1), (3, 1)$.

53. $3y^2 + 24y - 2x^2 + 12x + 24 = 0 \Rightarrow 3(y^2 + 8y) - 2(x^2 - 6x) = -24 \Rightarrow$

$3(y^2 + 8y + 16) - 2(x^2 - 6x + 9) = -24 + 48 - 18 \Rightarrow 3(y + 4)^2 - 2(x - 3)^2 = 6 \Rightarrow$

$\dfrac{(y + 4)^2}{2} - \dfrac{(x - 3)^2}{3} = 1$. The center is $(3, -4)$. The vertices are $(3, -4 - \sqrt{2}\,), (3, -4 + \sqrt{2}\,)$.

55. $x^2 - 6x - 2y^2 + 7 = 0 \Rightarrow (x^2 - 6x + 9) - 2y^2 = -7 + 9 \Rightarrow (x - 3)^2 - 2(y - 0)^2 = 2 \Rightarrow$

$\dfrac{(x - 3)^2}{2} - \dfrac{(y - 0)^2}{1} = 1$. The center is $(3, 0)$. The vertices are $(3 - \sqrt{2}, 0), (3 + \sqrt{2}, 0)$.

57. $4y^2 + 32y - 5x^2 - 10x + 39 = 0 \Rightarrow 4(y^2 + 8y) - 5(x^2 + 2x) = -39 \Rightarrow$

$4(y^2 + 8y + 16) - 5(x^2 + 2x + 1) = -39 + 64 - 5 \Rightarrow 4(y + 4)^2 - 5(x + 1)^2 = 20 \Rightarrow$

$\dfrac{(y + 4)^2}{5} - \dfrac{(x + 1)^2}{4} = 1$. The center is $(-1, -4)$. The vertices are $(-1, -4 - \sqrt{5}), (-1, -4 + \sqrt{5})$.

59. Solve for y: $\dfrac{(y - 1)^2}{11} - \dfrac{x^2}{5.9} = 1 \Rightarrow \dfrac{(y - 1)^2}{11} = 1 + \dfrac{x^2}{5.9} \Rightarrow (y - 1)^2 = 11\left(1 + \dfrac{x^2}{5.9}\right) \Rightarrow$

$y - 1 = \pm\sqrt{11\left(1 + \dfrac{x^2}{5.9}\right)} \Rightarrow y = 1 \pm \sqrt{11\left(1 + \dfrac{x^2}{5.9}\right)}$. Graph $Y_1 = 1 + \sqrt{(11 \cdot (1 + (X\text{\textasciicircum}2/5.9)))}$ and

$Y_2 = 1 - \sqrt{(11 \cdot (1 + (X\text{\textasciicircum}2/5.9)))}$. See Figure 59.

[-15, 15, 5] by [-10, 10, 5]　　　　[-9, 9, 1] by [-6, 6, 1]

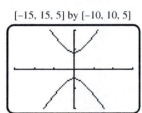　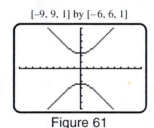

Figure 59　　　　　　　Figure 61

61. Solve for y: $3y^2 - 4x^2 = 15 \Rightarrow 3y^2 = 4x^2 + 15 \Rightarrow y^2 = \dfrac{4x^2 + 15}{3} \Rightarrow y = \pm\sqrt{\dfrac{4x^2 + 15}{3}}$.

Graph $Y_1 = \sqrt{((4X\text{\textasciicircum}2 + 15)/3)}$ and $Y_2 = -\sqrt{((4X\text{\textasciicircum}2 + 15)/3)}$. See Figure 61.

63. Add both equations together to eliminate the y^2-term:

$$\begin{array}{r} x^2 - y^2 = 4 \\ \underline{x^2 + y^2 = 9} \\ 2x^2 \quad\;\; = 13 \end{array} \Rightarrow x^2 = \dfrac{13}{2} \Rightarrow x = \pm\sqrt{\dfrac{13}{2}}$$

Substitute $x^2 = \dfrac{13}{2}$ into the first equation and solve for y: $x^2 - y^2 = 4 \Rightarrow \dfrac{13}{2} - y^2 = 4 \Rightarrow y^2 = \dfrac{5}{2} \Rightarrow$

$y = \pm\sqrt{\dfrac{5}{2}}$ There are four solutions to the system:

$\left(\sqrt{\dfrac{13}{2}}, \sqrt{\dfrac{5}{2}}\right), \left(-\sqrt{\dfrac{13}{2}}, \sqrt{\dfrac{5}{2}}\right), \left(\sqrt{\dfrac{13}{2}}, -\sqrt{\dfrac{5}{2}}\right)$, and $\left(-\sqrt{\dfrac{13}{2}}, -\sqrt{\dfrac{5}{2}}\right)$. See Figure 63.

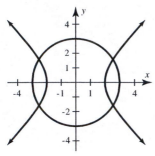

Figure 63

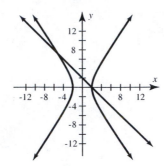

Figure 65

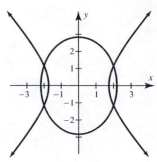

Figure 67

65. Solve the second equation for y: $x + y = 2 \Rightarrow y = 2 - x$. Substitute this result into the first equation and

solve for x: $\dfrac{x^2}{4} - \dfrac{y^2}{9} = 1 \Rightarrow \dfrac{x^2}{4} - \dfrac{(2-x)^2}{9} = 1 \Rightarrow 9x^2 - 4(4 - 4x + x^2) = 36 \Rightarrow$

$9x^2 - 16 + 16x - 4x^2 = 36 \Rightarrow 5x^2 + 16x - 52 = 0 \Rightarrow (5x + 26)(x - 2) = 0 \Rightarrow$

$x = -\dfrac{26}{5} = -5.2$ or $x = 2$. Substitute for x to find y: $y = 2 - x \Rightarrow y = 2 - (-5.2) = 7.2$ and

$y = 2 - x \Rightarrow y = 2 - 2 = 0$. There are two solutions to the system: $(2, 0)$ and $(-5.2, 7.2)$. See Figure 65.

67. Multiply the second equation by 2 and add the equations to eliminate the y^2-term:

$$8x^2 - 6y^2 = 24$$
$$10x^2 + 6y^2 = 48$$
$$\overline{\rule{4cm}{0.4pt}}$$
$$18x^2 \qquad = 72 \Rightarrow x^2 = 4 \Rightarrow x = \pm 2$$

Substitute $x^2 = 4$ into the first equation and solve for y:

$8x^2 - 6y^2 = 24 \Rightarrow 8(4) - 6y^2 = 24 \Rightarrow 6y^2 = 8 \Rightarrow y^2 = \dfrac{4}{3} \Rightarrow y = \pm\sqrt{\dfrac{4}{3}} \Rightarrow y = \pm\dfrac{2}{\sqrt{3}}$.

There are four solutions to the system: $\left(-2, -\dfrac{2}{\sqrt{3}}\right), \left(-2, \dfrac{2}{\sqrt{3}}\right), \left(2, -\dfrac{2}{\sqrt{3}}\right), \left(2, \dfrac{2}{\sqrt{3}}\right)$. See Figure 67.

69. Solve the second equation for y: $3x - y = 0 \Rightarrow y = 3x$. Substitute this result into the first equation and

solve for x: $\dfrac{y^2}{3} - \dfrac{x^2}{4} = 1 \Rightarrow \dfrac{(3x)^2}{3} - \dfrac{x^2}{4} = 1 \Rightarrow \dfrac{9x^2}{3} - \dfrac{x^2}{4} = 1 \Rightarrow 3x^2 - \dfrac{x^2}{4} = 1 \Rightarrow 12x^2 - x^2 = 4 \Rightarrow$

$11x^2 = 4 \Rightarrow x^2 = \dfrac{4}{11} \Rightarrow x = \pm\dfrac{2}{\sqrt{11}}$.

Substitute for x to find y: When $x = \dfrac{2}{\sqrt{11}}, y = 3x \Rightarrow y = 3\left(\dfrac{2}{\sqrt{11}}\right) \Rightarrow y = \dfrac{6}{\sqrt{11}}$.

When $-\dfrac{2}{\sqrt{11}}, y = 3x \Rightarrow y = 3\left(-\dfrac{2}{\sqrt{11}}\right) \Rightarrow y = -\dfrac{6}{\sqrt{11}}$.

The two solutions are: $\left(\dfrac{2}{\sqrt{11}}, \dfrac{6}{\sqrt{11}}\right), \left(-\dfrac{2}{\sqrt{11}}, -\dfrac{6}{\sqrt{11}}\right)$. See Figure 69.

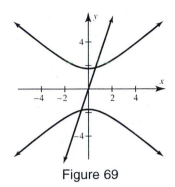

Figure 69

71. (a) $k = 2.82 \times 10^7$ and $D = 42.5 \times 10^6 \Rightarrow \dfrac{k}{\sqrt{D}} = \dfrac{2.82 \times 10^7}{\sqrt{42.5 \times 10^6}} \Rightarrow \dfrac{k}{\sqrt{D}} \approx 4325.68$. Since

$V = 2090$ and $V < \dfrac{k}{\sqrt{D}}$, the trajectory is elliptic.

(b) For $V > \dfrac{k}{\sqrt{D}}$, $V > 4326$. the speed of Explorer IV should be 4326 meters per second or greater so its

trajectory is hyperbolic.

(c) If D is larger, then $\dfrac{k}{\sqrt{D}}$ is smaller, so smaller values for V satisfy $V > \dfrac{k}{\sqrt{D}}$.

Extended and Discovery Exercises for Section 10.3

1. (a) Find a and b in the equation $\dfrac{x^2}{a^2} - \dfrac{y^2}{b^2} = 1$. Because the equations of the asymptotes of a hyperbola with

horizontal transverse axis are $y = \pm\dfrac{b}{a}x$, and the given asymptotes are $y = \pm x$, it follows that

$\dfrac{b}{a} = 1$ or $a = b$. Since the line $y = x$ intersects the x-axis at a $45°$ angle, the triangle shown in the third

quadrant is a $45°$-$45°$-$90°$ right triangle and both legs must have length d. Then by the Pythagorean

theorem, $c^2 = d^2 + d^2 = 2d^2$. That is $c = d\sqrt{2}$. Also, for a hyperbola $c^2 = a^2 + b^2$, and since $a = b$,

$c^2 = a^2 + a^2 = 2a^2$. That is $c = a\sqrt{2}$. From these two equations, $a\sqrt{2} = d\sqrt{2}$ and so $a = d$. That is,

$a = b = d = 5 \times 10^{-14}$. Thus the equation of the trajectory of A, where $x > 0$, is given by

$\dfrac{x^2}{(5 \times 10^{-14})^2} - \dfrac{y^2}{(5 \times 10^{-14})^2} = 1$. Solving for x yields

$x^2 - y^2 = (5 \times 10^{-14})^2 \Rightarrow x^2 = y^2 + 2.5 \times 10^{-27} \Rightarrow x = \sqrt{y^2 + 2.5 \times 10^{-27}}$. This equation

represents the right half of the hyperbola, as shown in the textbook.

(b) Since $a = 5 \times 10^{-14}$, the distance from the origin to the vertex is 5×10^{-14}. The distance from N to the

origin can be found using the Pythagorean theorem. Let h represent this distance, then $h^2 = d^2 + d^2$.

That is, $h^2 = (5 \times 10^{-14})^2 + (5 \times 10^{-14})^2 \Rightarrow h^2 = 5 \times 10^{-27} \Rightarrow h \approx 7 \times 10^{-14}$. The minimum distance between the centers of the alpha partical and the gold nucleus is

$5 \times 10^{-14} + 7 \times 10^{-14} \approx 1.2 \times 10^{-13}$ m.

Checking Basic Concepts for Section 10.3

1. The center is (0, 0). Since the vertices lie on the *x*-axis and are $(\pm 4, 0)$, $a = 4$. The equation of one asymptote

    is $y = \dfrac{3}{4}x$ because the line goes through the points (0, 0) and (8, 6). Thus, $a = 4$ and $b = 3$. Using the

    standard equation form, we get the equation: $\dfrac{x^2}{a^2} - \dfrac{y^2}{b^2} = 1 \Rightarrow \dfrac{x^2}{16} - \dfrac{y^2}{9} = 1$.

3. $h = 1$ and $k = 3$. Since the horizontal transverse axis has length 6, $2a = 6 \Rightarrow a = 3$. Since the conjugate

    axis has length 4, $2b = 4 \Rightarrow b = 2$. $c^2 = a^2 + b^2 = 9 + 4 = 13 \Rightarrow c = \sqrt{13}$. Thus the foci are

    $(1 \pm \sqrt{13}, 3)$. Using the standard equation form, we get the equation:

    $\dfrac{(x - h)^2}{a^2} - \dfrac{(y - k)^2}{b^2} = 1 \Rightarrow \dfrac{(x - 1)^2}{9} - \dfrac{(y - 3)^2}{4} = 1$.

Chapter 10 Review Exercises

1. The equation is $-x^2 = y$. See Figure 1.

3. The equation is $\dfrac{x^2}{25} + \dfrac{y^2}{49} = 1$. See Figure 3.

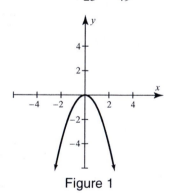

Figure 1

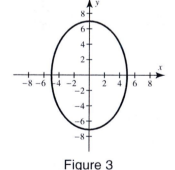

Figure 3

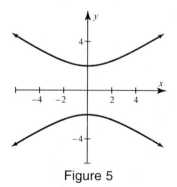

Figure 5

5. The equation is $\dfrac{y^2}{4} - \dfrac{x^2}{9} = 1$. See Figure 5.

7. A parabola opening upwards; d

9. A circle; a

11. A hyperbola with horizontal transverse axis; e

13. The parabola opens to the right and $p = 2$. $y^2 = 4px$, so $y^2 = 8x$. See Figure 13.

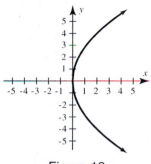

Figure 13

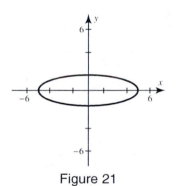

Figure 15

15. The major axis is horizontal, $a = 5$ and $c = 4$; $b = \sqrt{a^2 - c^2} = \sqrt{25 - 16} = 3$.

The equation is $\dfrac{x^2}{25} + \dfrac{y^2}{9} = 1$. See Figure 15.

17. The transverse axis is vertical, $c = 10$ and $b = 6$; $a = \sqrt{100 - 36} = 8$.

The equation is $\dfrac{y^2}{64} - \dfrac{x^2}{36} = 1$. See Figure 17.

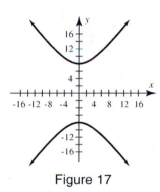

Figure 17

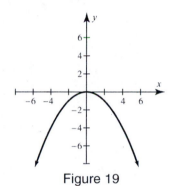

Figure 19

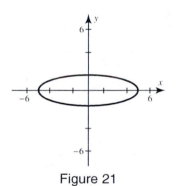

Figure 21

19. $x^2 = 4py \Rightarrow 4p = -4$, so $p = -1$. The vertex is at the origin and the focus is $(0, -1)$. See Figure 19.

21. The major axis is horizontal, the center is at the origin, $a = 5$ and $b = 2$. $c^2 = 25 - 4 = 21$, so $c = \sqrt{21}$; the foci are $(\pm\sqrt{21}, 0)$. See Figure 21.

23. The center is the origin and the transverse axis is horizontal. $c^2 = a^2 + b^2 = 16 + 9 = 25$, so $c = 5$. The foci are $(\pm 5, 0)$. See Figure 23.

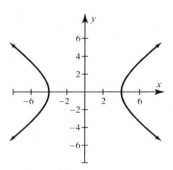

Figure 23

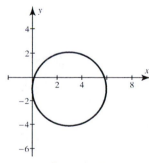

Figure 25

25. The equation represents a circle of radius 3, centered at $(3, -1)$. If we think of the circle as an ellipse, both foci are at $(3, -1)$. See Figure 25.

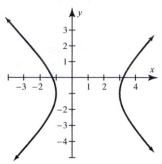

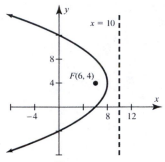

Figure 27   **Figure 29**

27. The equation $\dfrac{(x - 1)^2}{4} - \dfrac{(y + 1)^2}{4} = 1$ represents a hyperbola with horizontal transverse axis and center $(1, -1)$. See Figure 27.

29. The equation $(y - 4)^2 = -8(x - 8)$ has the form $(y - k)^2 = 4p(x - h)$ with vertex (h, k) at $(8, 4)$, and $p = -2$.

    The focus is $(6, 4)$ and the directrix is $x = 10$. See Figure 29.

31. $y = \pm\sqrt{\dfrac{1}{8.2}(60 - 7.1x^2)}$; See Figure 31.

[-5, 5, 1] by [-5, 5, 1]

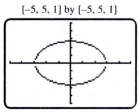

Figure 31

33. $-2x = y^2 + 8x + 14 \Rightarrow y^2 = -10x - 14 \Rightarrow (y - 0)^2 = -10\left(x + \dfrac{7}{5}\right)$

35. $4x^2 + 8x + 25y^2 - 250y = -529 \Rightarrow 4(x^2 + 2x) + 25(y^2 - 10y) = -529 \Rightarrow$

    $4(x^2 + 2x + 1) + 25(y^2 - 10y + 25) = -529 + 4 + 625 \Rightarrow 4(x + 1)^2 + 25(y - 5)^2 = 100 \Rightarrow$

    $\dfrac{(x + 1)^2}{25} + \dfrac{(y - 5)^2}{4} = 1$; The center is $(-1, 5)$. The vertices are

    $(-1 - 5, 5), (-1 + 5, 5)$ or $(-6, 5), (4, 5)$.

37. $x^2 + 4x - 4y^2 + 24y = 36 \Rightarrow (x^2 + 4x + 4) - 4(y^2 - 6y + 9) = 36 + 4 - 36 \Rightarrow$

    $(x + 2)^2 - 4(y - 3)^2 = 4 \Rightarrow \dfrac{(x + 2)^2}{4} - \dfrac{(y - 3)^2}{1} = 1.$

    The center is $(-2, 3)$. The vertices are $(-2 - 2, 3), (-2 + 2, 3)$ or $(-4, 3), (0, 3)$.

39. Clear fractions: $x^2 + y^2 = 4 \Rightarrow y^2 = 4 - x^2$. Substituting in $x^2 + 4y^2 = 8$ gives $x^2 + 4(4 - x^2) = 8$ or

$3x^2 = 8$, so $x = \pm\sqrt{\dfrac{8}{3}}$. Substituting in $x^2 + y^2 = 4$ we find $y^2 = 4 - \dfrac{8}{3} = \dfrac{4}{3}$, so $y = \pm\sqrt{\dfrac{4}{3}}$. There are four

solutions: $\left(\pm\sqrt{\dfrac{8}{3}}, \pm\sqrt{\dfrac{4}{3}}\right)$.

41. The system is $\dfrac{x^2}{9} + \dfrac{y^2}{4} \leq 1$ and $x + y \leq 3$. See Figure 41.

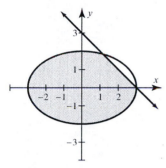

Figure 41

43. (a) $c = \sqrt{500^2 - 70^2} \approx 495.08$, and $a = 500$. The minimum distance is $a - c \approx 4.92$ million miles. The

maximum distance is $a + c \approx 995.08$ million miles.

(b) $2\pi\sqrt{\dfrac{500^2 + 70^2}{2}} \approx 2243$ million miles or 2.243 billion miles.

45. The equation of the ellipse is $\dfrac{x^2}{40^2} + \dfrac{y^2}{30^2} = 1$. Solving for y gives $y = 30\sqrt{1 - \dfrac{x^2}{40^2}}$. When $x = 10$,

$y = 30\sqrt{1 - \left(\dfrac{10}{40}\right)^2} \approx 29.05$ feet.

Extended and Discovery Exercises for Chapter 10

1. Neptune: $(0.009)(30.10) \approx 0.271$; Pluto: $(0.249)(39.44) \approx 9.82$

3. For the nearly circular orbit of Neptune, a and b are both approximately 30.10, so the equation is

$\dfrac{(x - 0.271)^2}{30.10^2} + \dfrac{y^2}{30.10^2} = 1$. For Pluto, $b^2 = a^2 - c^2 = 39.44^2 - 9.82^2 \approx 1459.08$, so $b \approx 38.20$. The

equation is $\dfrac{(x - 9.82)^2}{39.44^2} + \dfrac{y^2}{38.20^2} = 1$.

5. No. Because Pluto's orbit is so eccentric, there is a period of time when Pluto is not the farthest from the sun.

However, its average distance a from the sun is greater than any of the other planets.

Chapter 11: Further Topics in Algebra

11.1: Sequences

1. $a_1 = 2(1) + 1 = 3$; $a_2 = 2(2) + 1 = 5$; $a_3 = 2(3) + 1 = 7$; $a_4 = 2(4) + 1 = 9$.

   The first four terms are 3, 5, 7, and 9.

3. $a_1 = 4(-2)^{1-1} = 4$; $a_2 = 4(-2)^{2-1} = -8$; $a_3 = 4(-2)^{3-1} = 16$; $a_4 = 4(-2)^{4-1} = -32$.

   The first four terms are $4, -8, 16$, and -32.

5. $a_1 = \dfrac{1}{1^2 + 1} = \dfrac{1}{2}$; $a_2 = \dfrac{2}{2^2 + 1} = \dfrac{2}{5}$; $a_3 = \dfrac{3}{3^2 + 1} = \dfrac{3}{10}$; $a_4 = \dfrac{4}{4^2 + 1} = \dfrac{4}{17}$.

   The first four terms are $\dfrac{1}{2}, \dfrac{2}{5}, \dfrac{3}{10}$, and $\dfrac{4}{17}$.

7. $a_1 = (-1)^1 \left(\dfrac{1}{2}\right)^1 = -\dfrac{1}{2}$; $a_2 = (-1)^2 \left(\dfrac{1}{2}\right)^2 = \dfrac{1}{4}$; $a_3 = (-1)^3 \left(\dfrac{1}{2}\right)^3 = -\dfrac{1}{8}$; $a_4 = (-1)^4 \left(\dfrac{1}{2}\right)^4 = \dfrac{1}{16}$.

   The first four terms are $-\dfrac{1}{2}, \dfrac{1}{4}, -\dfrac{1}{8}$, and $\dfrac{1}{16}$.

9. $a_1 = (-1)^0 \left(\dfrac{2}{1 + 2}\right) = \dfrac{2}{3}$; $a_2 = (-1)^1 \left(\dfrac{4}{1 + 4}\right) = -\dfrac{4}{5}$; $a_3 = (-1)^2 \left(\dfrac{8}{1 + 8}\right) = \dfrac{8}{9}$;

   $a_4 = (-1)^3 \left(\dfrac{16}{1 + 16}\right) = -\dfrac{16}{17}$. The first four terms are $\dfrac{2}{3}, -\dfrac{4}{5}, \dfrac{8}{9}$, and $-\dfrac{16}{17}$.

11. $a_1 = 2 + 1^2 = 3$; $a_2 = 4 + 2^2 = 8$; $a_3 = 8 + 3^2 = 17$; $a_4 = 16 + 4^2 = 32$.

    The first four terms are 3, 8, 17, and 32.

13. The points $(1, 2), (2, 4), (3, 3), (4, 5), (5, 3), (6, 6), (7, 4)$ lie on the graph. Therefore, the terms of the sequence
    are 2, 4, 3, 5, 3, 6, 4.

15. (a) $a_1 = 1$; $a_2 = 2a_1 = 2 \cdot 1 = 2$; $a_3 = 2a_2 = 2 \cdot 2 = 4$; $a_4 = 2a_3 = 2 \cdot 4 = 8$.

    The first four terms are 1, 2, 4, and 8.

    (b) The graph of the points $(1, 1), (2, 2), (3, 4)$, and $(4, 8)$ is shown in Figure 15.

    [0, 5, 1] by [0, 9, 1]   [0, 5, 1] by [−4, 7, 1]

    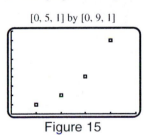   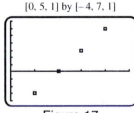

    Figure 15   Figure 17

17. (a) $a_1 = -3$; $a_2 = a_1 + 3 = -3 + 3 = 0$; $a_3 = a_2 + 3 = 0 + 3 = 3$; $a_4 = a_3 + 3 = 3 + 3 = 6$.

    The first four terms are $-3, 0, 3$, and 6.

    (b) See Figure 17.

19. (a) $a_1 = 2$; $a_2 = 3a_1 - 1 = 3(2) - 1 = 5$; $a_3 = 3a_2 - 1 = 3(5) - 1 = 14$;

    $a_4 = 3a_3 - 1 = 3(14) - 1 = 41$. The first four terms are 2, 5, 14, and 41.

    (b) See Figure 19.

[0, 5, 1] by [0, 45, 5]   [0, 5, 1] by [–3, 6, 1]

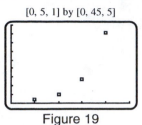   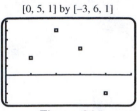

Figure 19   Figure 21

21. (a) $a_1 = 2$; $a_2 = 5$; $a_3 = a_2 - a_1 = 5 - 2 = 3$; $a_4 = a_3 - a_2 = 3 - 5 = -2$.

   The first four terms are 2, 5, 3, and –2.

   (b) The graph of the points (1, 2), (2, 5), (3, 3), and (4, –2) is shown in Figure 21.

23. (a) $a_1 = 2$; $a_2 = a_1^2 = 2^2 = 4$; $a_3 = a_2^2 = 4^2 = 16$; $a_4 = a_3^2 = 16^2 = 256$.

   The first four terms are 2, 4, 16, and 256.

   (b) The graph of the points (1, 2), (2, 4), (3, 16), and (4, 256) is shown in Figure 23.

[0, 5, 1] by [0, 300, 50]   [0, 5, 1] by [0, 12, 1]

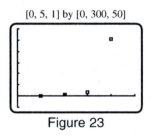   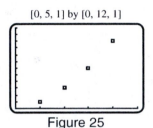

Figure 23   Figure 25

25. (a) $a_1 = 1$; $a_2 = a_1 + 2 = 1 + 2 = 3$; $a_3 = a_2 + 3 = 3 + 3 = 6$; $a_4 = a_3 + 4 = 6 + 4 = 10$.

   The first four terms are 1, 3, 6, and 10.

   (b) The graph of the points (1, 1), (2, 3), (3, 6), and (4, 10) is shown in Figure 25.

27. (a) $a_1 = 2$; $a_2 = 3$; $a_3 = a_2 \cdot a_1 = 2 \cdot 3 = 6$; $a_4 = a_3 \cdot a_2 = 6 \cdot 3 = 18$

   The first four terms are 2, 3, 6, and 18.

   (b) The graph of the points (1, 2), (2, 3), (3, 6), and (4, 18) is shown in Figure 27.

[0, 5, 1] by [0, 20, 2]

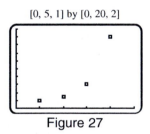

Figure 27

29. (a) Each term can be found by adding 2 to the previous term. A numerical representation for the first eight

   terms is shown in Figure 29a.

[0, 10, 1] by [0, 16, 1]

n	1	2	3	4	5	6	7	8
a_n	1	3	5	7	9	11	13	15

Figure 29a

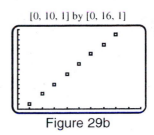

Figure 29b

(b) A graphical representation is shown in Figure 29b and includes the points in Figure 29a. Notice that the points lie on a line with slope 2 because the sequence is arithmetic.

(c) To find the symbolic representation, we will use the formula $a_n = a_1 + (n - 1)d$, where $a_n = f(n)$. The common difference of this sequence is $d = 2$ and the first term is $a_1 = 1$. Therefore, a symbolic representation of the sequence is given by $a_n = 1 + (n - 1)2$ or $a_n = 2n - 1$.

31. (a) Each term can be found by subtracting 1.5 to the previous term. A numerical representation for the first eight terms is shown in Figure 31a.

[0, 12, 1] by [-4, 8, 1]

n	1	2	3	4	5	6	7	8
a_n	7.5	6	4.5	3	1.5	0	-1.5	-3

Figure 31a

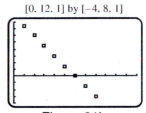

Figure 31b

(b) A graphical representation is shown in Figure 31b and includes the points in Figure 31a. Notice that the points lie on a line with slope –1.5 because the sequence is arithmetic.

(c) To find the symbolic representation, we will use the formula $a_n = a_1 + (n - 1)d$, where $a_n = f(n)$. The common difference of this sequence is $d = -1.5$ and the first term is $a_1 = 7.5$. Therefore, a symbolic representation of the sequence is given by $a_n = 7.5 + (n - 1)(-1.5)$ or $a_n = -1.5n + 9$.

33. (a) Each term can be found by adding $\frac{3}{2}$ to the previous term. A numerical representation for the first eight terms is shown in Figure 33a.

[0, 9, 1] by [0, 12, 1]

n	1	2	3	4	5	6	7	8
a_n	$\frac{1}{2}$	2	$\frac{7}{2}$	5	$\frac{13}{2}$	8	$\frac{19}{2}$	11

Figure 33a

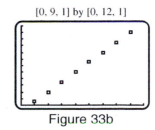

Figure 33b

(b) A graphical representation is shown in Figure 33b and includes the points in Figure 33a. Notice that the points lie on a line with slope $\frac{3}{2}$ because the sequence is arithmetic.

(c) To find the symbolic representation, we will use the formula $a_n = a_1 + (n-1)d$, where $a_n = f(n)$. The common difference of this sequence is $d = \frac{3}{2}$ and the first term is $a_1 = \frac{1}{2}$. Therefore, a symbolic representation of the sequence is given by $a_n = \frac{1}{2} + (n-1)\left(\frac{3}{2}\right)$ or $a_n = \frac{3}{2}n - 1$.

35. (a) Each term can be found by multiplying the previous term by $\frac{1}{2}$. A numerical representation for the first eight terms is shown in Figure 35a.

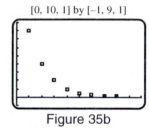

[0, 10, 1] by [−1, 9, 1]

n	1	2	3	4	5	6	7	8
a_n	8	4	2	1	$\frac{1}{2}$	$\frac{1}{4}$	$\frac{1}{8}$	$\frac{1}{16}$

Figure 35a

Figure 35b

(b) A graphical representation is shown in Figure 35b and includes the points in Figure 35a. Notice that the points lie on a curve that is decaying exponentially, because the sequence is geometric and the common ratio is less than one in absolute value.

(c) To find the symbolic representation, we will use the formula $a_n = a_1 r^{n-1}$, where $a_n = f(n)$. The common ratio of this sequence is $r = \frac{1}{2}$ and the first term is $a_1 = 8$. Therefore, a symbolic representation of the sequence is given by $a_n = 8\left(\frac{1}{2}\right)^{n-1}$.

37. (a) Each term can be found by multiplying the previous term by 2. A numerical representation for the first eight terms is shown in Figure 37a.

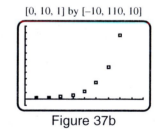

[0, 10, 1] by [−10, 110, 10]

n	1	2	3	4	5	6	7	8
a_n	$\frac{3}{4}$	$\frac{3}{2}$	3	6	12	24	48	96

Figure 37a

Figure 37b

(b) A graphical representation is shown in Figure 37b and includes the points in Figure 37a. Notice that the points lie on a curve that is increasing exponentially, because the sequence is geometric and the common ratio is greater than one in absolute value.

(c) To find the symbolic representation, we will use the formula $a_n = a_1 r^{n-1}$, where $a_n = f(n)$. The common ratio of this sequence is $r = 2$ and the first term is $a_1 = \frac{3}{4}$. Therefore, a symbolic representation of the sequence is given by $a_n = \frac{3}{4}(2)^{n-1}$.

39. (a) Each term can be found by multiplying the previous term by 2. A numerical representation for the first eight terms is shown in Figure 39a.

n	1	2	3	4	5	6	7	8
a_n	$-\frac{1}{4}$	$-\frac{1}{2}$	-1	-2	-4	-8	-16	-32

Figure 39a

[0, 9, 1] by [−36, 4, 4]

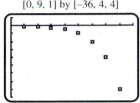

Figure 39b

(b) A graphical representation is shown in Figure 39b and includes the points in Figure 39a. Notice that the points lie on a curve that is increasing exponentially, because the sequence is geometric and the common ratio is greater than one in absolute value.

(c) To find the symbolic representation, we will use the formula $a_n = a_1 r^{n-1}$, where $a_n = f(n)$. The common ratio of this sequence is $r = 2$ and the first term is $a_1 = -\frac{1}{4}$. Therefore, a symbolic representation of the sequence is given by $a_n = -\frac{1}{4}(2)^{n-1}$.

41. Let $a_n = f(n)$, where $f(n) = dn + c$. Then, $f(n) = -2n + c$. Since $f(1) = -2(1) + c = 5 \Rightarrow c = 7$.

Thus, $a_n = f(n) = -2n + 7$. An alternate solution is to use the formula:

$a_n = a_1 + (n-1)d = 5 + (n-1)(-2) \Rightarrow a_n = -2n + 7$.

43. Let $a_n = f(n)$, where $f(n) = dn + c$. Then, $f(n) = 3n + c$. Since $f(3) = 3(3) + c = 1 \Rightarrow c = -8$.

Thus, $a_n = f(n) = 3n - 8$.

45. Let $a_n = f(n)$, where $f(n) = dn + c$. $a_2 = 5 \Rightarrow f(2) = 5$ and $a_6 = 13 \Rightarrow f(6) = 13$.

Thus, $d = \dfrac{f(6) - f(2)}{6 - 2} = \dfrac{13 - 5}{4} = 2$. Then, $f(n) = 2n + c$. $f(2) = 2(2) + c = 5 \Rightarrow c = 1$.

Thus, $a_n = f(n) = 2n + 1$.

47. Let $a_n = f(n)$, where $f(n) = dn + c$. $a_1 = 8 \Rightarrow f(1) = 8$ and $a_4 = 17 \Rightarrow f(4) = 17$.

Thus, $d = \dfrac{f(4) - f(1)}{4 - 1} = \dfrac{17 - 8}{3} = 3$. Then, $f(n) = 3n + c$. $f(1) = 3(1) + c = 8 \Rightarrow c = 5$.

Thus, $a_n = f(n) = 3n + 5$.

49. Let $a_n = f(n)$, where $f(n) = dn + c$. $a_5 = -4 \Rightarrow f(5) = -4$ and $a_8 = -2.5 \Rightarrow f(8) = -2.5$.

Thus, $d = \dfrac{f(8) - f(5)}{8 - 5} = \dfrac{-2.5 - (-4)}{3} = 0.5$.

Then, $f(n) = 0.5n + c$. $f(5) = 0.5(5) + c = -4 \Rightarrow c = -6.5$. Thus, $a_n = f(n) = 0.5n - 6.5$.

51. Let $a_n = f(n)$, where $f(n) = a_1 r^{n-1}$. Then, $a_n = f(n) = 2\left(\dfrac{1}{2}\right)^{n-1}$.

53. Let $a_n = f(n)$, where $f(n) = a_1 r^{n-1}$. Then, $a_n = f(n) = a_1\left(-\dfrac{1}{4}\right)^{n-1}$.

Since $f(3) = a_1\left(-\dfrac{1}{4}\right)^{3-1} = \dfrac{1}{32} \Rightarrow a_1 = \dfrac{16}{32} = \dfrac{1}{2}$. Thus, $a_n = f(n) = \dfrac{1}{2}\left(-\dfrac{1}{4}\right)^{n-1}$.

55. Let $a_n = f(n)$, where $f(n) = a_1 r^{n-1}$. $a_3 = 2$ and $a_6 = \dfrac{1}{4} \Rightarrow \dfrac{1}{8} = \dfrac{\frac{1}{4}}{2} = \dfrac{a_6}{a_3} = \dfrac{a_1 r^{6-1}}{a_1 r^{3-1}} = \dfrac{r^5}{r^2} = r^3 \Rightarrow$

$r = \dfrac{1}{2}$. Then, $f(n) = a_1 \left(\dfrac{1}{2}\right)^{n-1}$. $f(3) = a_1 \left(\dfrac{1}{2}\right)^{3-1} = 2 \Rightarrow a_1 = 8$. Thus, $a_n = f(n) = 8\left(\dfrac{1}{2}\right)^{n-1}$.

57. Let $a_n = f(n)$, where $f(n) = a_1 r^{n-1}$. $a_1 = -5$ and $a_3 = -125 \Rightarrow$

$25 = \dfrac{-125}{-5} = \dfrac{a_3}{a_1} = \dfrac{a_1 r^{3-1}}{a_1 r^{1-1}} = \dfrac{r^2}{r^0} = r^2 \Rightarrow r = -5$. Thus, $a_n = f(n) = -5(-5)^{n-1}$.

59. Let $a_n = f(n)$, where $f(n) = a_1 r^{n-1}$. $a_2 = -1$ and $a_7 = -32 \Rightarrow$

$32 = \dfrac{-32}{-1} = \dfrac{a_7}{a_2} = \dfrac{a_1 r^{7-1}}{a_1 r^{2-1}} = \dfrac{r^6}{r^1} = r^5 \Rightarrow r = 2$.

Then, $f(n) = a_1 (2)^{n-1}$. $f(2) = a_1 (2)^{2-1} = -1 \Rightarrow a_1 = -\dfrac{1}{2}$. Thus, $a_n = f(n) = -\dfrac{1}{2}(2)^{n-1}$.

61. Since $f(n) = 4 - 3n^3$ is not a linear function, it does not represent an arithmetic sequence.

63. Since $f(n) = 4n - (3 - n) = 5n - 3$ is a linear function, it represents an arithmetic sequence.

65. Since the plotted points appear to be collinear, the graph represents an arithmetic sequence.

67. Since the common difference is -2, the table represents an arithmetic sequence.

69. Since $f(n) = 4(2)^{n-1}$ is written in the form $f(n) = cr^{n-1}$, it represents a geometric sequence.

71. Since $f(n) = -3(n)^2$ cannot be written in the form $f(n) = cr^{n-1}$, it does not represents a geometric sequence.

73. Since the plotted points appear to be collinear, the graph does not represent a geometric sequence.

75. Since there is no common ratio, the table does not represent a geometric sequence.

77. These terms represent an arithmetic sequence. Each term can be obtained by adding 7 to the previous term.

79. These terms represent a geometric sequence. Each term can be obtained by multiplying the previous term by 4.

81. These terms represent neither an arithmetic nor a geometric sequence. There is no common ratio or difference.

83. This sequence is arithmetic since the points lie on a line (and are evenly spaced). Since the sequence is decreasing the common difference must be negative. The slope of the line passing through these points is -1, so the common difference is $d = -1$.

85. The sequence is either geometric or arithmetic. This sequence must be geometric, since the points do not lie on a line. Since the terms alternate sign, the common ratio r is negative. The absolute value of the terms are dampening to a value of 0. Therefore, $|r| < 1$.

87. The insect population increases rapidly and then levels off at 5000 per acre.

89. (a) The initial density is 500. The population density each successive year is 0.8 of the previous year. Therefore $a_1 = 500$ and $a_n = 0.8a_{n-1}$.

    (b) $a_1 = 500$, $a_2 = 0.8(500) = 400$, $a_3 = 0.8(400) = 320$, $a_4 = 0.8(320) = 256$, $a_5 = 0.8(256) = 204.8$ and $a_6 = 0.8(204.8) = 163.84$. The population density is decreasing each year by 20%

    (c) The terms of the sequence 500, 400, 320, 256, ... are a geometric sequence with $a_1 = 500$ and $r = 0.8$. Therefore, the nth term is given by $a_n = 500(0.8)^{n-1}$.

91. (a) $a_1 = 8$, $a_2 = 2.9a_1 - 0.2a_1^2 = 2.9(8) - 0.2(8)^2 = 10.4$,

$a_3 = 2.9a_2 - 0.2a_2^2 = 2.9(10.4) - 0.2(10.4)^2 = 8.528$.

(b) Figures 91a & 91b show how to enter the sequence and a graph of the first twenty terms. The population density oscillates above and below 9.5 (approximately).

[0, 21, 1] by [0, 14, 1]

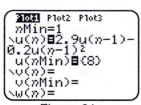

Figure 91a   Figure 91b

93. Since 100 million cell phones are thrown away per year we have the following:

$a_1 = 100$, $a_2 = 200$, $a_3 = 300$, ... , $a_n = 100n$, where n is the number of years.

$a_5 = 100(5) = 500$ After 5 years, 500 million cell phones have been thrown out.

95. (a) The first term is $a_1 = 100$ and each successive term can be found by adding $d = 94$ to the previous result.

100, $100 + 94 = 194$, $194 + 94 = 288$, $288 + 94 = 382$, $382 + 94 = 476$

(b) See Figure 95.

(c) In an arithmetic sequence with first term a_1 and common difference d, the nth term a_n is given by

$a_n = a_1 + (n - 1)d$. From part (a) we know that $a_1 = 100$ and $d = 94$ so $a_n = 100 + (n - 1)94$.

[

Figure 95

97. (a) The terms in this sequence can be found by adding the previous two terms. $a_1 = 1$, $a_2 = 1$, $a_3 = 2$,

$a_4 = 3$, $a_5 = 5$, $a_6 = 8$, $a_7 = 13$, $a_8 = 21$, $a_9 = 34$, $a_{10} = 55$, $a_{11} = 89$, and $a_{12} = 144$.

(b) $\dfrac{a_2}{a_1} = \dfrac{1}{1} = 1$, $\dfrac{a_3}{a_2} = \dfrac{2}{1} = 2$, $\dfrac{a_4}{a_3} = \dfrac{3}{2} = 1.5$, $\dfrac{a_5}{a_4} = \dfrac{5}{3} \approx 1.6667$, $\dfrac{a_6}{a_5} = \dfrac{8}{5} = 1.6$, $\dfrac{a_7}{a_6} = \dfrac{13}{8} = 1.625$,

$\dfrac{a_8}{a_7} = \dfrac{21}{13} \approx 1.6154$, $\dfrac{a_9}{a_8} = \dfrac{34}{21} = 1.6190$, $\dfrac{a_{10}}{a_9} = \dfrac{55}{34} \approx 1.6176$, $\dfrac{a_{11}}{a_{10}} = \dfrac{89}{55} = 1.6182$, and

$\dfrac{a_{12}}{a_{11}} = \dfrac{144}{89} \approx 1.6180$. These ratios seem to be approaching a number near 1.618. This number is called the golden ratio.

(c) $n = 2$: $a_1 \cdot a_3 - a_2^2 = (1)(2) - (1)^2 = 1 = (-1)^2$;

$n = 3$: $a_2 \cdot a_4 - a_3^2 = (1)(3) - (2)^2 = -1 = (-1)^3$;

$n = 4$: $a_3 \cdot a_5 - a_4^2 = (2)(5) - (3)^2 = 1 = (-1)^4$

[0, 30, 10] by [0, 150,000, 50,000]

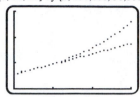

Figure 99

99. (a) The salary of the first employee at the beginning of the nth year is given by:

$$a_n = 30{,}000 + (n - 1)(2000) = 2000n + 28{,}000.$$

This is a arithmetic sequence with $a_1 = 30{,}000$ and $d = 2000$.

(b) The salary of the second employee at the beginning of the nth year is given by: $b_n = 30{,}000(1.05)^{n-1}$.

This is a geometric sequence with $b_1 = 30{,}000$ and $d = 1.05$.

(c) At the beginning of the 10th year each salary is:

$$a_{10} = 2000(10) + 28{,}000 = \$48{,}000, \ b_{10} = 30{,}000(1.05)^{10-1} \approx \$46{,}540.$$

At the beginning of the 20th year each salary is:

$$a_{20} = 2000(20) + 28{,}000 = \$68{,}000, \ b_{20} = 30{,}000(1.05)^{20-1} \approx \$75{,}809.$$

(d) The graph of each sequence is shown in Figure 99. With time the geometric sequence overtakes the arithmetic sequence since $r > 1$.

101. Let $a_1 = 2$. Then, $a_2 = \dfrac{1}{2}\left(a_1 + \dfrac{k}{a_1}\right) = \dfrac{1}{2}\left(2 + \dfrac{2}{2}\right) = 1.5$, $a_3 = \dfrac{1}{2}\left(a_2 + \dfrac{k}{a_2}\right) = \dfrac{1}{2}\left(1.5 + \dfrac{2}{1.5}\right) = 1.41\overline{6}$

In a similar manner, $a_4 \approx 1.414215686$, $a_5 \approx 1.414213562$, and $a_6 \approx 1.414213562$. Since $\sqrt{2} \approx 1.414213562$ this is a very accurate approximation.

103. Let $a_1 = 21$. Then, $a_2 = \dfrac{1}{2}\left(a_1 + \dfrac{k}{a_1}\right) = \dfrac{1}{2}\left(21 + \dfrac{21}{21}\right) = 11$, $a_3 = \dfrac{1}{2}\left(a_2 + \dfrac{k}{a_2}\right) = \dfrac{1}{2}\left(11 + \dfrac{21}{11}\right) = 6.\overline{45}$

In a similar manner, $a_4 \approx 4.854033291$, $a_5 \approx 4.59016621$, and $a_6 \approx 4.582581971$. Since $\sqrt{21} \approx 4.582575695$ this is an accurate approximation.

105. By definition $a_n = a_1 + (n - 1)d_1$ and $b_n = b_1 + (n - 1)d_2$. Then

$$c_n = a_n + b_n = [a_1 + (n - 1)d_1] + [b_1 + (n - 1)d_2] = (a_1 + b_1) + [(n - 1)d_1 + (n - 1)d_2] =$$
$$(a_1 + b_1) + (n - 1)(d_1 + d_2) = c_1 + (n - 1)d \text{ where } c_1 = a_1 + b_1 \text{ and } d = d_1 + d_2.$$

11.2: Series

1. (a) The counting numbers from 1 to 5 are 1, 2, 3, 4, 5.

(b) The series is $1 + 2 + 3 + 4 + 5$.

(c) $1 + 2 + 3 + 4 + 5 = 15$

3. (a) The integers counting down from 1 to -3 are $1, 0, -1, -2, -3$.

   (b) The series is $1 + 0 + (-1) + (-2) + (-3)$.

   (c) $1 + 0 + (-1) + (-2) + (-3) = -5$

5. Since A_n represents the number of AIDS deaths after 2000, the sum from 2005 to 2009 is given by

   $A_5 + A_6 + A_7 + A_8 + A_9$.

7. $S_5 = 3(1) + 3(2) + 3(3) + 3(4) + 3(5) = 3 + 6 + 9 + 12 + 15 = 45$

9. $S_5 = (2(1) - 1) + (2(2) - 1) + (2(3) - 1) + (2(4) - 1) + (2(5) - 1) = 1 + 3 + 5 + 7 + 9 = 25$

11. $S_5 = (1^2 + 1) + (2^2 + 1) + (3^2 + 1) + (4^2 + 1) + (5^2 + 1) = 2 + 5 + 10 + 17 + 26 = 60$

13. $S_5 = \dfrac{1}{1 + 1} + \dfrac{2}{2 + 1} + \dfrac{3}{3 + 1} + \dfrac{4}{4 + 1} + \dfrac{5}{5 + 1} = \dfrac{1}{2} + \dfrac{2}{3} + \dfrac{3}{4} + \dfrac{4}{5} + \dfrac{5}{6} = \dfrac{71}{20}$

15. The first term is $a_1 = 3$ and the last term is $a_8 = 17$. To find the sum use:

   $$S_n = n\left(\dfrac{a_1 + a_n}{2}\right) \Rightarrow S_8 = 8\left(\dfrac{3 + 17}{2}\right) = 80. \text{ The sum is } 80.$$

17. The first term is $a_1 = 1$ and the last term is $a_{50} = 50$. To find the sum use:

   $$S_n = n\left(\dfrac{a_1 + a_n}{2}\right) \Rightarrow S_{50} = 50\left(\dfrac{1 + 50}{2}\right) = 1275. \text{ The sum is } 1275.$$

19. The first term is $a_1 = -7$ and $d = 3$. The last term is 101, but we must determine n. The term of 101 is

   $\dfrac{101 - (-7)}{3} = 36$ terms after a_1. Thus, the last term is $a_{37} = 101$. To find the sum use the following:

   $$S_n = n\left(\dfrac{a_1 + a_n}{2}\right) \Rightarrow S_{37} = 37\left(\dfrac{-7 + 101}{2}\right) = 1739. \text{ The sum is } 1739.$$

21. The number of terms to be added is 40, so $n = 40$. The first term is $a_1 = 5(1) = 5$. The last term is

   $$a_{40} = 5(40) = 200. \; S_n = n\left(\dfrac{a_1 + a_n}{2}\right) \Rightarrow S_{40} = 40\left(\dfrac{5 + 200}{2}\right) = 4100. \text{ The sum is } 4100.$$

23. $S_{15} = 15\left(\dfrac{a_1 + a_{15}}{2}\right) \Rightarrow 255 = 15\left(\dfrac{3 + a_{15}}{2}\right) \Rightarrow 17 = \dfrac{3 + a_{15}}{2} \Rightarrow 34 = 3 + a_{15} \Rightarrow a_{15} = 31$

25. Since $a_1 = 4$ and $d = 2$, use the formula $S_n = \dfrac{n}{2}(2a_1 + (n - 1)d)$.

   $$S_{20} = \dfrac{20}{2}(2(4) + (20 - 1)2) = 10(8 + 19(2)) = 10(8 + 38) = 10(46) = 460$$

27. Since $a_1 = 10$ and $d = -\dfrac{1}{2}$, use the formula $S_n = \dfrac{n}{2}(2a_1 + (n - 1)d)$.

   $$S_{20} = \dfrac{20}{2}\left(2(10) + (20 - 1)\left(-\dfrac{1}{2}\right)\right) = 10\left(20 + 19\left(-\dfrac{1}{2}\right)\right) = 10\left(20 - \dfrac{19}{2}\right) = 10\left(\dfrac{21}{2}\right) = 105$$

29. Since $a_1 = 4$ and $a_{20} = 190.2$, use the formula $S_n = n\left(\dfrac{a_1 + a_n}{2}\right)$.

   $$S_{20} = 20\left(\dfrac{4 + 190.2}{2}\right) = 20\left(\dfrac{194.2}{2}\right) = 20(97.1) = 1942$$

31. Since $a_1 = -2$ and $a_{11} = 50$, the common difference is $d = \dfrac{50 - (-2)}{11 - 1} = \dfrac{52}{10} = 5.2$.

Use the formula $S_n = \dfrac{n}{2}(2a_1 + (n-1)d)$.

$$S_{20} = \frac{20}{2}(2(-2) + (20-1)5.2) = 10(-4 + 19(5.2)) = 10(-4 + 98.8) = 10(94.8) = 948$$

33. Since $a_2 = 6$ and $a_{12} = 31$, the common difference is $d = \dfrac{31-6}{12-2} = \dfrac{25}{10} = 2.5$ and so $a_1 = 6 - 2.5 = 3.5$.

Use the formula $S_n = \dfrac{n}{2}(2a_1 + (n-1)d)$.

$$S_{20} = \frac{20}{2}(2(3.5) + (20-1)2.5) = 10(7 + 19(2.5)) = 10(7 + 47.5) = 10(54.5) = 545$$

35. The first term is $a_1 = 1$ and the common ratio is $r = 2$. Since there are 8 terms, the sum is

$$S_n = a_1\left(\frac{1 - r^n}{1 - r}\right) \Rightarrow S_8 = 1\left(\frac{1 - 2^8}{1 - 2}\right) = 255.$$

This sum can be verified by adding $1 + 2 + 4 + 8 + 16 + 32 + 64 + 128 = 255$.

37. The first term is $a_1 = 0.5$ and the common ratio is $r = 3$. Since there are 7 terms, the sum is

$$S_n = a_1\left(\frac{1 - r^n}{1 - r}\right) \Rightarrow S_7 = 0.5\left(\frac{1 - (3)^7}{1 - 3}\right) = 546.5.$$

This sum can be verified by adding $0.5 + 1.5 + 4.5 + 13.5 + 40.5 + 121.5 + 364.5 = 546.5$.

39. The first term is $a_1 = 3(2)^0 = 3$ and the common ratio is $r = 2$. Since there are 20 terms, the sum is

$$S_n = a_1\left(\frac{1 - r^n}{1 - r}\right) \Rightarrow S_{20} = 3\left(\frac{1 - 2^{20}}{1 - 2}\right) = 3{,}145{,}725.$$

41. Using the formula $S_n = a_1\left(\dfrac{1 - r^n}{1 - r}\right)$ with $a_1 = 1$ and $r = -\dfrac{1}{2}$,

$$S_4 = 1\left(\frac{1 - (-\frac{1}{2})^4}{1 - (-\frac{1}{2})}\right) = 0.625; \quad S_7 = 1\left(\frac{1 - (-\frac{1}{2})^7}{1 - (-\frac{1}{2})}\right) = 0.671875; \quad S_{10} = 1\left(\frac{1 - (-\frac{1}{2})^{10}}{1 - (-\frac{1}{2})}\right) = 0.666015625$$

43. Using the formula $S_n = a_1\left(\dfrac{1 - r^n}{1 - r}\right)$ with $a_1 = \dfrac{1}{3}$ and $r = 2$,

$$S_4 = \frac{1}{3}\left(\frac{1 - 2^4}{1 - 2}\right) = 5; \quad S_7 = \frac{1}{3}\left(\frac{1 - 2^7}{1 - 2}\right) = 42.\overline{3}; \quad S_{10} = \frac{1}{3}\left(\frac{1 - 2^{10}}{1 - 2}\right) = 341$$

45. The first term is $a_1 = 1$ and the common ratio is $r = \dfrac{1}{3}$. The sum is $S = a_1\left(\dfrac{1}{1 - r}\right) \Rightarrow S = 1\left(\dfrac{1}{1 - \frac{1}{3}}\right) = \dfrac{3}{2}$.

47. The first term is $a_1 = 6$ and the common ratio is $r = -\dfrac{2}{3}$. The sum is

$$S = a_1\left(\frac{1}{1 - r}\right) \Rightarrow S = 6\left(\frac{1}{1 - (-\frac{2}{3})}\right) = \frac{18}{5}.$$

49. The first term is $a_1 = 1$ and the common ratio is $r = -\dfrac{1}{10}$ or -0.1. The sum is

$$S = a_1\left(\frac{1}{1 - r}\right) \Rightarrow S = 1\left(\frac{1}{1 - (-0.1)}\right) = \frac{1}{1.1} = \frac{10}{11}.$$

51. $\dfrac{2}{3} = 0.6666666\ldots = 0.6 + 0.06 + 0.006 + 0.0006 + 0.00006 + \cdots$

53. $\dfrac{9}{11} = 0.81818181\ldots = 0.81 + 0.0081 + 0.000081 + 0.00000081 + \cdots$

55. $\dfrac{1}{7} = 0.142857142857\ldots = 0.142857 + 0.000000142857 + 0.000000000000142857 + \cdots$

57. The series $0.8 + 0.08 + 0.008 + 0.0008 + \cdots$ is an infinite geometric series with $a_1 = 0.8$ and $r = 0.1$.

$S = \dfrac{a_1}{1 - r} = \dfrac{0.8}{1 - 0.1} = \dfrac{8}{9}.$

59. The series $0.45 + 0.0045 + 0.000045 + \cdots$ is an infinite geometric series with $a_1 = 0.45$ and $r = 0.01$.

$S = \dfrac{a_1}{1 - r} = \dfrac{0.45}{1 - 0.01} = \dfrac{45}{99} = \dfrac{5}{11}.$

61. $\displaystyle\sum_{k=1}^{4} (k + 1) = (1 + 1) + (2 + 1) + (3 + 1) + (4 + 1) = 2 + 3 + 4 + 5 = 14$

63. $\displaystyle\sum_{k=1}^{8} 4 = 4 + 4 + 4 + 4 + 4 + 4 + 4 + 4 = 32$

65. $\displaystyle\sum_{k=1}^{7} k^3 = 1^3 + 2^3 + 3^3 + 4^3 + 5^3 + 6^3 + 7^3 = 1 + 8 + 27 + 64 + 125 + 216 + 343 = 784$

67. $\displaystyle\sum_{k=4}^{5} (k^2 - k) = (4^2 - 4) + (5^2 - 5) = 12 + 20 = 32$

69. $1^4 + 2^4 + 3^4 + 4^4 + 5^4 + 6^4 = \displaystyle\sum_{k=1}^{6} k^4$

71. $1 + \dfrac{4}{3} + \dfrac{6}{4} + \dfrac{8}{5} + \dfrac{10}{6} + \dfrac{12}{7} + \dfrac{14}{8} = \displaystyle\sum_{k=1}^{7} \left(\dfrac{2k}{k + 1} \right)$

73. $1 + \dfrac{1}{2^2} + \dfrac{1}{3^2} + \dfrac{1}{4^2} + \dfrac{1}{5^2} + \cdots = \displaystyle\sum_{k=1}^{\infty} \left(\dfrac{1}{k^2} \right)$

75. Because $\displaystyle\sum_{k=6}^{9} k^3 = 6^3 + 7^3 + 8^3 + 9^3$, the summation has 4 terms. We will write $k = 6, 7, 8, 9$ as

$n = 1, 2, 3, 4$. It follows that $n + 5 = k$, and $\displaystyle\sum_{n=1}^{4} (n + 5)^3$.

77. Because $\displaystyle\sum_{k=9}^{32} (3k - 2) = (3(9) - 2) + (3(10) - 2) + \cdots + (3(32) - 2)$, the summation has 24 terms. We

will write $k = 9, 10, 11, \ldots 32$ as $n = 1, 2, 3, 4, \ldots 24$. It follows that $n + 8 = k$, and

$\displaystyle\sum_{n=1}^{24} (3(n + 8) - 2) \Rightarrow \sum_{n=1}^{24} (3n + 24 - 2) \Rightarrow \sum_{n=1}^{24} (3n + 22).$

79. Because $\displaystyle\sum_{k=16}^{52} (k^2 - 3k) = (16^2 - 3(16)) + (17^2 - 3(17)) + \cdots + (52^2 - 3(52))$, the summation has 37

terms. We will write $k = 16, 17, 18, \ldots 52$ as $n = 1, 2, 3, \ldots 37$. It follows that $n + 15 = k$, and

$\displaystyle\sum_{n=1}^{37} ((n + 15)^2 - 3(n + 15)) \Rightarrow \sum_{n=1}^{37} (n^2 + 30n + 225 - 3n - 45) \Rightarrow \sum_{n=1}^{37} (n^2 + 27n + 180).$

81. $\displaystyle\sum_{k=1}^{60} 9 = 60(9) = 540$

83. $\displaystyle\sum_{k=1}^{15} 5k = 5\sum_{k=1}^{15} k = 5\left[\frac{15(16)}{2}\right] = 600$

85. $\displaystyle\sum_{k=1}^{31}(3k-3) = 3\sum_{k=1}^{31}k - \sum_{k=1}^{31}3 = 3\left[\frac{31(32)}{2}\right] - 31(3) = 1488 - 93 = 1395$

87. $\displaystyle\sum_{k=1}^{25}k^2 = \frac{(25)(26)(51)}{6} = 5525$

89. $\displaystyle\sum_{k=1}^{16}(k^2 - k) = \sum_{k=1}^{16}k^2 - \sum_{k=1}^{16}k = \frac{(16)(17)(33)}{6} - \frac{(16)(17)}{2} = 1496 - 136 = 1360$

91. $\displaystyle\sum_{k=5}^{24}k = \sum_{k=1}^{24}k - \sum_{k=1}^{4}k = \frac{(24)(25)}{2} - \frac{(4)(5)}{2} = 300 - 10 = 290$

93. $\displaystyle\sum_{k=1}^{n}k = 1 + 2 + 3 + 4 + \cdots + n.$ This is an arithmetic series with $a_1 = 1$ and $a_n = n.$ The sum is given by

$$S_n = n\left(\frac{a_1 + a_n}{2}\right) = n\left(\frac{1+n}{2}\right) = \frac{n(n+1)}{2}.$$

95. (a) The arithmetic sequence describing the salary during year n is computed by $a_n = 42{,}000 + 1800(n-1).$

The 1st and 15th years salaries are: $a_1 = 42{,}000 + 1800(1-1) = 42{,}000$ and

$a_{15} = 42{,}000 + 1800(15-1) = 67{,}200.$ The total amount earned during the 15 year period is

$$S_{15} = \frac{15(42{,}000 + 67{,}200)}{2} = 819{,}000.$$

(b) Verify with a calculator, compute the sum $a_1 + a_2 + a_3 + \cdots + a_{15},$ where

$a_n = 42{,}000 + 1800(n-1).$ See Figure 95.

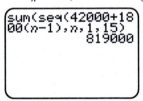

Figure 95

97. Future value is given by $S_n = A_0\left(\dfrac{(1+i)^n - 1}{i}\right),$ where $A_0 = a_1.$

$S_{20} = 2000\left(\dfrac{(1+0.08)^{20} - 1}{0.08}\right) \approx 91{,}523.93.$ If $\$2000$ is deposited in an account at the end of each year for

20 years and the account pays 8% interest, the future value of this annuity will be $\$91{,}523.93.$

99. Future value is given by $S_n = A_0\left(\dfrac{(1+i)^n - 1}{i}\right),$ where $A_0 = a_1.$

$S_5 = 10{,}000\left(\dfrac{(1+0.11)^5 - 1}{0.11}\right) \approx 62{,}278.01.$ If $\$10{,}000$ is deposited in an account at the end of each year

for 5 years and the account pays 11% interest, the future value of this annuity will be $\$62{,}278.01.$

101. The number of logs in the stack is given by $7 + 8 + 9 + 10 + 11 + 12 + 13 + 14 + 15$. This series is arithmetic with first term $a_1 = 7$ and the last term $a_9 = 15$ so its sum is given by $S_9 = 9\left(\dfrac{7 + 15}{2}\right) = 99$. So, there are 99 logs in the stack.

103. (a) Since each filter lets half of the impurities through, the series is can be written as

$0.5(1) + 0.5(0.5) + 0.5(0.25) + \cdots = \displaystyle\sum_{k=1}^{n} 0.5(0.5)^{k-1}$.

(b) The series sums to 1 only if an infinite number of filters are used.

105. The area of the largest square is 1, the area of the next largest square is $\dfrac{1}{2}$, and each successive square has an area that is $\dfrac{1}{2}$ the area of the previous square. If there are an infinite number of squares, then the area is represented by the geometric series $1 + \dfrac{1}{2} + \dfrac{1}{4} + \dfrac{1}{8} + \dfrac{1}{16} + \cdots$. This is an infinite geometric series with $a_1 = 1$ and $r = \dfrac{1}{2}$, whose sum is $S = \dfrac{1}{1 - r} = \dfrac{1}{1 - \frac{1}{2}} = 2$. The area is 2.

107. The value of e^a is approximated by $e^a = 1 + a + \dfrac{a^2}{2!} + \dfrac{a^3}{3!} + \cdots + \dfrac{a^n}{n!}$. Apply the series by letting $a = 1$, since $e^1 = e$. The first 8 terms sum to $e^1 \approx 1 + 1 + \dfrac{1}{2!} + \dfrac{1}{3!} + \dfrac{1}{4!} + \dfrac{1}{5!} + \dfrac{1}{6!} + \dfrac{1}{7!} = $

$1 + 1 + \dfrac{1}{2} + \dfrac{1}{6} + \dfrac{1}{24} + \dfrac{1}{120} + \dfrac{1}{720} + \dfrac{1}{5040} \approx 2.718254$. The actual value is $e \approx 2.718282$, so only 8 terms of the series provides an approximation for e that is accurate to 4 decimal places.

109. One can either add the terms directly or apply the formula $S_n = a_1\left(\dfrac{1 - r^n}{1 - r}\right)$. We will apply the formula. In this sequence $a_1 = 1$ and $r = \dfrac{1}{3}$. $S_2 = 1\left(\dfrac{1 - \left(\frac{1}{3}\right)^2}{1 - \left(\frac{1}{3}\right)}\right) = \dfrac{4}{3} \approx 1.3333$; $S_4 = 1\left(\dfrac{1 - \left(\frac{1}{3}\right)^4}{1 - \left(\frac{1}{3}\right)}\right) = \dfrac{40}{27} \approx 1.4815$;

$S_8 = 1\left(\dfrac{1 - \left(\frac{1}{3}\right)^8}{1 - \left(\frac{1}{3}\right)}\right) \approx 1.49977$; $S_{16} = 1\left(\dfrac{1 - \left(\frac{1}{3}\right)^{16}}{1 - \left(\frac{1}{3}\right)}\right) \approx 1.49999997$. The sum of the infinite series

$\displaystyle\sum_{k=1}^{\infty} \left(\dfrac{1}{3}\right)^{k-1}$ is given by $S = \dfrac{a_1}{1 - r} = \dfrac{1}{1 - \frac{1}{3}} = \dfrac{3}{2} = 1.5$. As n increases, the partial sums $S_2, S_4, S_8,$ and S_{16} become closer and closer to the value of 1.5.

111. One can either add the terms directly or apply the formula $S_n = a_1\left(\dfrac{1 - r^n}{1 - r}\right)$. We will apply the formula. In this sequence $a_1 = 4$ and $r = -\dfrac{1}{10}$. $S_1 = 4\left(\dfrac{1 - \left(-\frac{1}{10}\right)^1}{1 - \left(-\frac{1}{10}\right)}\right) = 4$; $S_2 = 4\left(\dfrac{1 - \left(-\frac{1}{10}\right)^2}{1 - \left(-\frac{1}{10}\right)}\right) = 3.6$;

$S_3 = 4\left(\dfrac{1 - \left(-\frac{1}{10}\right)^3}{1 - \left(-\frac{1}{10}\right)}\right) = 3.64$; $S_4 = 4\left(\dfrac{1 - \left(-\frac{1}{10}\right)^4}{1 - \left(-\frac{1}{10}\right)}\right) = 3.636$; $S_5 = 4\left(\dfrac{1 - \left(-\frac{1}{10}\right)^5}{1 - \left(-\frac{1}{10}\right)}\right) = 3.6364$;

$$S_6 = 4\left(\frac{1 - (-\frac{1}{10})^6}{1 - (-\frac{1}{10})}\right) = 3.63636;$$ The sum of the infinite series $\displaystyle\sum_{k=1}^{\infty} 4\left(-\frac{1}{10}\right)^{k-1}$ is given by

$$S = \frac{a_1}{1 - r} = \frac{4}{1 - (-0.1)} = \frac{40}{11} = 3.\overline{63}.$$ As n increases, the partial sums become closer and closer to the

value of $3.\overline{63}$.

Checking Basic Concepts for Sections 11.1 & 11.2

1. A graphical representation of $a_n = -2n + 3$ is shown in Figure 1a. A numerical representation is shown in Figure 1b. The first six terms are 1, –1, –3, –5, –7, –9.

[0, 7, 1] by [–10, 2, 1]

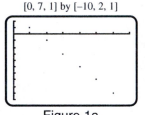

Figure 1a   Figure 1b

3. (a) The series is arithmetic since the common difference between the terms is 4. The sum of the first 10 terms shown is $10\left(\dfrac{1 + 37}{2}\right) = 190$.

    (b) The series is geometric since the common ratio is $\dfrac{1}{3}$. The sum of the first 6 terms shown is

    $$3\left(\frac{1 - (\frac{1}{3})^6}{1 - (\frac{1}{3})}\right) = 3\left(\frac{1 - (\frac{1}{729})}{1 - (\frac{1}{3})}\right) = 3\left(\frac{\frac{728}{729}}{\frac{2}{3}}\right) = 3\left(\frac{364}{243}\right) = \frac{364}{81} \approx 4.494.$$

    (c) The series is an infinite geometric series with a common ratio of $\dfrac{1}{4}$.
    The sum of the series is $\dfrac{2}{1 - \frac{1}{4}} = \dfrac{2}{0.75} = \dfrac{8}{3} \approx 2.667.$

    (d) The series is an infinite geometric series with a common ratio of 0.1.
    The sum of the series is $\dfrac{0.9}{1 - 0.1} = \dfrac{0.9}{0.9} = 1.$

5. (a) $\displaystyle\sum_{k=1}^{15} (k + 2) = \sum_{k=1}^{15} k + \sum_{k=1}^{15} 2 = \frac{15(16)}{2} + 15(2) = 150$

    (b) $\displaystyle\sum_{k=1}^{21} 2k^2 = 2\sum_{k=1}^{21} k^2 = 2\left[\frac{21(22)(43)}{6}\right] = 6622$

11.3: Counting

1. There are $2 \cdot 2 \cdot 2 \cdot 2 \cdot 2 \cdot 2 \cdot 2 \cdot 2 \cdot 2 \cdot 2 = 2^{10} = 1024$ different ways to answer the exam.

3. There are $2 \cdot 2 \cdot 2 \cdot 2 \cdot 2 \cdot 4 \cdot 4 \cdot 4 \cdot 4 \cdot 4 \cdot 4 \cdot 4 \cdot 4 \cdot 4 = 2^5 \cdot 4^{10} = 33{,}554{,}432$ different ways.

5. There are 10 digits and 26 letters. There are $10 \cdot 10 \cdot 10 \cdot 26 \cdot 26 \cdot 26 = 17{,}576{,}000$ different license plates.

7. There are 36 digits and letters. There are $26 \cdot 26 \cdot 26 \cdot 36 \cdot 36 \cdot 36 = 820{,}025{,}856$ different license plates.

9. There are 3 choices for each of the 5 letters. There are $3 \cdot 3 \cdot 3 \cdot 3 \cdot 3 = 3^5 = 243$ different strings.

11. There are 5 choices for each of the 5 letters. There are $5 \cdot 5 \cdot 5 \cdot 5 \cdot 5 = 5^5 = 3125$ different strings.

13. Since there are 2 letters that can only be used once, each string must be 2 letters long. For the first position there are 2 choices and for the second position there is only 1 choice. There are $2 \cdot 1 = 2$ possible strings.

15. There are $4 \cdot 3 \cdot 2 \cdot 1 = 24$ possible strings.

17. Each combination can vary between 000 and 999. Thus, there are 1000 possibilities for each lock. That is, there are $1000 \cdot 1000 = 1{,}000{,}000$ different combinations in all.

19. There are 2 settings (on or off) for each of the 12 switches.

    There are $2 \cdot 2 \cdot 2 \cdot 2 \cdot 2 \cdot 2 \cdot 2 \cdot 2 \cdot 2 \cdot 2 \cdot 2 \cdot 2 = 2^{12} = 4096$ different codes for the garage door opener.

21. For the first letter there are 2 possibilities, whereas there are 26 possibilities for each of the last 3 positions. There are $2 \cdot 26 \cdot 26 \cdot 26 = 35{,}152$ different call letters possible. There is no shortage of call letters.

23. There are $2 \cdot 3 \cdot 4 = 24$ different packages that can be purchased.

25. There are $8 \cdot 10 \cdot 10 \cdot 10 \cdot 10 \cdot 10 = 8{,}000{,}000$ such phone numbers.

27. $6! = 6 \cdot 5 \cdot 4 \cdot 3 \cdot 2 \cdot 1 = 720$

29. $10! = 10 \cdot 9 \cdot 8 \cdot 7 \cdot 6 \cdot 5 \cdot 4 \cdot 3 \cdot 2 \cdot 1 = 3{,}628{,}800$

31. $P(n, r) = \dfrac{n!}{(n-r)!} \Rightarrow P(5, 3) = \dfrac{5!}{(5-3)!} = \dfrac{5!}{2!} = \dfrac{120}{2} = 60$

33. $P(n, r) = \dfrac{n!}{(n-r)!} \Rightarrow P(8, 1) = \dfrac{8!}{(8-1)!} = \dfrac{8!}{7!} = \dfrac{8 \cdot 7!}{7!} = 8$

35. $P(n, r) = \dfrac{n!}{(n-r)!} \Rightarrow P(7, 3) = \dfrac{7!}{(7-3)!} = \dfrac{7!}{4!} = \dfrac{7 \cdot 6 \cdot 5 \cdot 4!}{4!} = 7 \cdot 6 \cdot 5 = 210$

37. $P(n, r) = \dfrac{n!}{(n-r)!} \Rightarrow P(25, 2) = \dfrac{25!}{(25-2)!} = \dfrac{25!}{23!} = \dfrac{25 \cdot 24 \cdot 23!}{23!} = 25 \cdot 24 = 600$

39. $P(n, r) = \dfrac{n!}{(n-r)!} \Rightarrow P(10, 4) = \dfrac{10!}{(10-4)!} = \dfrac{10!}{6!} = \dfrac{10 \cdot 9 \cdot 8 \cdot 7 \cdot 6!}{6!} = 10 \cdot 9 \cdot 8 \cdot 7 = 5040$

41. $P(4, 4) = \dfrac{4!}{(4-4)!} = \dfrac{4!}{0!} = 4! = 24$

43. There are $15 \cdot 14 \cdot 13 = 2730$ different arrangements of 3 students from a class of 15.

45. There are $7 \cdot 6 \cdot 5 = 210$ different routes.

47. The remaining 4 digits for the 7 digit number are independent events. Since the last 4 digits can be any number from 0 to 9. The total is given by $10 \cdot 10 \cdot 10 \cdot 10 = 10{,}000$. The first 3 numbers are restricted to 3 different possibilities and there will be $(3) \cdot (10{,}000)$ or 30,000 phone numbers.

49. Initially any person can sit at the table. Then the remaining 6 people can sit in $6! = 720$ different ways. Since there is no difference between the people sitting clockwise or counterclockwise around the table, we must divide this result by 2 for the final answer. There are 360 different ways to seat the 7 people.

51. $P(9, 9) = \dfrac{9!}{(9-9)!} = \dfrac{9!}{0!} = 9! = 362{,}880$

53. Counting February 29th, $P(366, 5) = \dfrac{366!}{(366 - 5)!} = \dfrac{366!}{361!} = \dfrac{366 \cdot 365 \cdot 364 \cdot 363 \cdot 362 \cdot 361!}{361!} \approx 6.39 \times 10^{12}$.

55. There are 8 choices for the first and fourth digits and 10 choices for each of the other digits.

    $8 \cdot 10 \cdot 10 \cdot 8 \cdot 10 \cdot 10 \cdot 10 \cdot 10 \cdot 10 = 6,400,000,000$

57. $C(n, r) = \dfrac{n!}{(n - r)! \, r!} \Rightarrow C(3, 1) = \dfrac{3!}{(3 - 1)! \, 1!} = \dfrac{3!}{2! \, 1!} = \dfrac{6}{2} = 3$

59. $C(n, r) = \dfrac{n!}{(n - r)! \, r!} \Rightarrow C(6, 3) = \dfrac{6!}{(6 - 3)! \, 3!} = \dfrac{6!}{3! \, 3!} = \dfrac{720}{36} = 20$

61. $C(n, r) = \dfrac{n!}{(n - r)! \, r!} \Rightarrow C(5, 0) = \dfrac{5!}{(5 - 0)! \, 0!} = \dfrac{5!}{5! \, 0!} = \dfrac{5!}{5!} = 1$

63. $\dbinom{n}{r} = \dfrac{n!}{(n - r)! \, r!} \Rightarrow \dbinom{8}{2} = \dfrac{8!}{(8 - 2)! \, 2!} = \dfrac{8!}{6! \, 2!} = \dfrac{8 \cdot 7 \cdot 6!}{6! \, 2!} = \dfrac{56}{2} = 28$

65. $\dbinom{n}{r} = \dfrac{n!}{(n - r)! \, r!} \Rightarrow \dbinom{20}{18} = \dfrac{20!}{(20 - 18)! \, 18!} = \dfrac{20!}{2! \, 18!} = \dfrac{20 \cdot 19 \cdot 18!}{2! \, 18!} = \dfrac{20 \cdot 19}{2} = 190$

67. From 39 numbers a player picks 5 numbers. Since order is unimportant, there are $C39, 52 = 575,757$ different ways of doing this.

69. Two women can be selected from 5 women in $C(5, 2)$ different ways. Two men can be selected from 3 men in $C(3, 2)$ different ways. The total number of teams is $C(5, 2) \cdot C(3, 2) = 10 \cdot 3 = 30$.

71. Three questions can be selected from 5 questions in $C(5, 3)$ different ways. Four questions can be selected from 5 questions in $C(5, 4)$ different ways. The total number of possibilities is $C(5, 3) \cdot C(5, 4) = 10 \cdot 5 = 50$.

73. Three red marbles can be drawn from 10 red marbles in $C(10, 3)$ different ways. Two blue marbles can be drawn from 12 blue marbles in $C(12, 2)$ different ways. There are $C(10, 3) \cdot C(12, 2) = 120 \cdot 66 = 7920$ ways.

75. Since order is not important, there are $C(24, 3) = 2024$ ways to do this.

77. $P(n, n - 1) = \dfrac{n!}{(n - (n - 1))!} = \dfrac{n!}{1} = n!$ and $P(n, n) = \dfrac{n!}{(n - n)!} = \dfrac{n!}{0!} = \dfrac{n!}{1} = n!$

    For example $P(7, 6) = 5040 = P(7, 7)$.

11.4: The Binomial Theorem

1. $\dbinom{5}{4} = \dfrac{5!}{1! \, 4!} = \dfrac{5 \cdot 4!}{1 \cdot 4!} = 5$

3. $\dbinom{4}{0} = \dfrac{4!}{4! \, 0!} = \dfrac{4!}{4! \cdot 1} = 1$

5. $\dbinom{6}{5} = \dfrac{6!}{1! \, 5!} = \dfrac{6 \cdot 5!}{1 \cdot 5!} = \dfrac{6}{1} = 6$

7. $\dbinom{3}{3} = \dfrac{3!}{0! \, 3!} = \dfrac{3!}{1 \cdot 3!} = 1$

9. Three a's, two b's $\Rightarrow C(5, 2) = 10$ different strings.

11. Four a's, four b's $\Rightarrow C(8, 4) = 70$ different strings.

13. Five a's, zero b's $\Rightarrow C(5, 0) = 1$ string.

15. Four a's, one $b \Rightarrow C(5, 1) = 5$ different strings.

17. $(x + y)^2 = \dbinom{2}{0}x^2y^0 + \dbinom{2}{1}x^1y^1 + \dbinom{2}{2}x^0y^2 = x^2 + 2xy + y^2$

19. $(m + 2)^3 = \dbinom{3}{0}m^3(2)^0 + \dbinom{3}{1}m^2(2)^1 + \dbinom{3}{2}m^1(2)^2 + \dbinom{3}{3}m^0(2)^3 = m^3 + 6m^2 + 12m + 8$

21. $(2x - 3)^3 = \dbinom{3}{0}(2x)^3(-3)^0 + \dbinom{3}{1}(2x)^2(-3)^1 + \dbinom{3}{2}(2x)^1(-3)^2 + \dbinom{3}{3}(2x)^0(-3)^3 =$

$8x^3 - 36x^2 + 54x - 27$

23. $(p - q)^6 = \dbinom{6}{0}p^6(-q)^0 + \dbinom{6}{1}p^5(-q)^1 + \dbinom{6}{2}p^4(-q)^2 + \dbinom{6}{3}p^3(-q)^3 + \dbinom{6}{4}p^2(-q)^4 +$

$\dbinom{6}{5}p^1(-q)^5 + \dbinom{6}{6}p^0(-q)^6 = p^6 - 6p^5q + 15p^4q^2 - 20p^3q^3 + 15p^2q^4 - 6pq^5 + q^6$

25. $(2m + 3n)^3 = \dbinom{3}{0}(2m)^3(3n)^0 + \dbinom{3}{1}(2m)^2(3n)^1 + \dbinom{3}{2}(2m)^1(3n)^2 + \dbinom{3}{3}(2m)^0(3n)^3 =$

$8m^3 + 36m^2n + 54mn^2 + 27n^3$

27. $(1 - x^2)^4 = \dbinom{4}{0}(1)^4(-x^2)^0 + \dbinom{4}{1}(1)^3(-x^2)^1 + \dbinom{4}{2}(1)^2(-x^2)^2 + \dbinom{4}{3}(1)^1(-x^2)^3 + \dbinom{4}{4}(1)^0(-x^2)^4$

$= 1 - 4x^2 + 6x^4 - 4x^6 + x^8$

29. $(2p^3 - 3)^3 = \dbinom{3}{0}(2p^3)^3(-3)^0 + \dbinom{3}{1}(2p^3)^2(-3)^1 + \dbinom{3}{2}(2p^3)^1(-3)^2 + \dbinom{3}{3}(2p^3)^0(-3)^3 =$

$8p^9 - 36p^6 + 54p^3 - 27$

31. Using Pascal's triangle, the coefficients are 1, 2, and 1.

$(x + y)^2 = 1x^2y^0 + 2x^1y^1 + 1x^0y^2 = x^2 + 2xy + y^2$

33. Using Pascal's triangle, the coefficients are 1, 4, 6, 4, and 1.

$(3x + 1)^4 = 1(3x)^4(1)^0 + 4(3x)^3(1)^1 + 6(3x)^2(1)^2 + 4(3x)^1(1)^3 + 1(3x)^0(1)^4 =$

$81x^4 + 108x^3 + 54x^2 + 12x + 1$

35. Using Pascal's triangle, the coefficients are 1, 5, 10, 10, 5, and 1.

$(2 - x)^5 = 1(2)^5(-x)^0 + 5(2)^4(-x)^1 + 10(2)^3(-x)^2 + 10(2)^2(-x)^3 + 5(2)^1(-x)^4 + 1(2)^0(-x)^5 =$

$32 - 80x + 80x^2 - 40x^3 + 10x^4 - x^5$

37. Using Pascal's triangle, the coefficients are 1, 4, 6, 4, and 1.

$(x^2 + 2)^4 = 1(x^2)^4(2)^0 + 4(x^2)^3(2)^1 + 6(x^2)^2(2)^2 + 4(x^2)^1(2)^3 + 1(x^2)^0(2)^4 =$

$x^8 + 8x^6 + 24x^4 + 32x^2 + 16$

39. Using Pascal's triangle, the coefficients are 1, 4, 6, 4, and 1.

$(4x - 3y)^4 = 1(4x)^4(-3y)^0 + 4(4x)^3(-3y)^1 + 6(4x)^2(-3y)^2 + 4(4x)^1(-3y)^3 + 1(4x)^0(-3y)^4 =$

$256x^4 - 768x^3y + 864x^2y^2 - 432xy^3 + 81y^4$

41. Using Pascal's triangle, the coefficients are 1, 6, 15, 20, 15, 6, and 1.

$(m + n)^6 = m^6 + 6m^5n + 15m^4n^2 + 20m^3n^3 + 15m^2n^4 + 6mn^5 + n^6$

43. Using Pascal's triangle, the coefficients are 1, 3, 3, and 1.

$(2x^3 - y^2)^3 = 1(2x^3)^3(-y^2)^0 + 3(2x^3)^2(-y^2)^1 + 3(2x^3)^1(-y^2)^2 + 1(2x^3)^0(-y^2)^3 =$
$8x^9 - 12x^6y^2 + 6x^3y^4 - y^6$

45. The fourth term of $(a + b)^9$ is $\binom{9}{6}a^{9-3}b^3 = \binom{9}{6}a^6b^3 = 84a^6b^3$.

47. The fifth term of $(x + y)^8$ is $\binom{8}{4}x^{8-4}y^4 = \binom{8}{4}x^4y^4 = 70x^4y^4$.

49. The fourth term of $(2x + y)^5$ is $\binom{5}{2}(2x)^{5-3}y^3 = \binom{5}{2}(2x)^2y^3 = 40x^2y^3$.

51. The sixth term of $(3x - 2y)^6$ is $\binom{6}{1}(3x)^{6-5}(-2y)^5 = \binom{6}{1}(3x)(-2y)^5 = -576xy^5$.

Checking Basic Concepts for Sections 11.3 and 11.4

1. There are $2 \cdot 2 \cdot 2 \cdot 2 \cdot 2 \cdot 2 \cdot 2 \cdot 2 = 2^8 = 256$ different ways to answer the quiz.

3. There are $26 \cdot 36 \cdot 36 \cdot 36 \cdot 36 \cdot 36 = 26 \cdot 36^5 = 1{,}572{,}120{,}576$ different license plates of this kind.

11.5: Mathematical Induction

1. $3 + 6 + 9 + \cdots + 3n = \dfrac{3n(n + 1)}{2}$

(i) Show that the statement is true for $n = 1$: $3(1) = \dfrac{3(1)(2)}{2} \Rightarrow 3 = 3$

(ii) Assume that S_k is true: $3 + 6 + 9 + \cdots + 3k = \dfrac{3k(k + 1)}{2}$

Show that S_{k+1} is true: $3 + 6 + \cdots + 3(k + 1) = \dfrac{3(k + 1)(k + 2)}{2}$

Add $3(k + 1)$ to each side of S_k: $3 + 6 + 9 + \cdots + 3k + 3(k + 1) = \dfrac{3k(k + 1)}{2} + 3(k + 1) =$

$\dfrac{3k(k + 1) + 6(k + 1)}{2} = \dfrac{(k + 1)(3k + 6)}{2} = \dfrac{3(k + 1)(k + 2)}{2}$

Since S_k implies S_{k+1}, the statement is true for every positive integer n.

3. $5 + 10 + 15 + \cdots + 5n = \dfrac{5n(n + 1)}{2}$

   (i) Show that the statement is true for $n = 1$: $5(1) = \dfrac{5(1)(2)}{2} \Rightarrow 5 = 5$

   (ii) Assume that S_k is true: $5 + 10 + 15 + \cdots + 5k = \dfrac{5k(k + 1)}{2}$

   Show that S_{k+1} is true: $5 + 10 + 5^2 + 5k + 12 = \dfrac{5k^2 + 12k + 22}{2}$

   Add $5(k + 1)$ to each side of S_k: $5 + 10 + 15 + \cdots + 5k + 5(k + 1) = \dfrac{5k(k + 1)}{2} + 5(k + 1) =$

   $\dfrac{5k(k + 1) + 10(k + 1)}{2} = \dfrac{(k + 1)(5k + 10)}{2} = \dfrac{5(k + 1)(k + 2)}{2}$

   Since S_k implies S_{k+1}, the statement is true for every positive integer n.

5. $3 + 3^2 + 3^3 + \cdots + 3^n = \dfrac{3(3^n - 1)}{2}$

   (i) Show that the statement is true for $n = 1$: $3^1 = \dfrac{3(3^1 - 1)}{2} \Rightarrow 3 = 3$

   (ii) Assume that S_k is true: $3 + 3^2 + 3^3 + \cdots + 3^k = \dfrac{3(3^k - 1)}{2}$

   Show that S_{k+1} is true: $3 + 3^2 + 3^3 + \cdots + 3^{k+1} = \dfrac{3(3^{k+1} - 1)}{2}$

   Add 3^{k+1} to each side of S_k: $3 + 3^2 + 3^3 + \cdots + 3^k + 3^{k+1} = \dfrac{3(3^k - 1)}{2} + 3^{k+1} =$

   $\dfrac{3(3^k - 1) + 2(3^{k+1})}{2} = \dfrac{3^{k+1} - 3 + 2(3^{k+1})}{2} = \dfrac{3(3^{k+1}) - 3}{2} = \dfrac{3(3^{k+1} - 1)}{2}$

   Since S_k implies S_{k+1}, the statement is true for every positive integer n.

7. $1^3 + 2^3 + 3^3 + \cdots + n^3 = \dfrac{n^2(n + 1)^2}{4}$

   (i) Show that the statement is true for $n = 1$: $1^3 = \dfrac{1^2(1 + 1)^2}{4} \Rightarrow 1 = 1$

   (ii) Assume that S_k is true: $1^3 + 2^3 + 3^3 + \cdots + k^3 = \dfrac{k^2(k + 1)^2}{4}$

   Show that S_{k+1} is true: $1^3 + 2^3 + \cdots + (k + 1)^3 = \dfrac{(k + 1)^2(k + 2)^2}{4}$

   Add $(k + 1)^3$ to each side of S_k: $1^3 + 2^3 + 3^3 + \cdots + k^3 + (k + 1)^3 = \dfrac{k^2(k + 1)^2}{4} + (k + 1)^3 =$

   $\dfrac{k^2(k + 1)^2 + 4(k + 1)^3}{4} = \dfrac{(k + 1)^2(k^2 + 4k + 4)}{4} = \dfrac{(k + 1)^2(k + 2)^2}{4}$

   Since S_k implies S_{k+1}, the statement is true for every positive integer n.

9. $\dfrac{1}{1\cdot2} + \dfrac{1}{2\cdot3} + \cdots + \dfrac{1}{n(n+1)} = \dfrac{n}{n+1}$

   (i) Show that the statement is true for $n = 1$: $\dfrac{1}{1(1+1)} = \dfrac{1}{1+1} \Rightarrow \dfrac{1}{2} = \dfrac{1}{2}$.

   (ii) Assume that S_k is true: $\dfrac{1}{1\cdot2} + \dfrac{1}{2\cdot3} + \cdots + \dfrac{1}{k(k+1)} = \dfrac{k}{k+1}$

      Show that S_{k+1} is true: $\dfrac{1}{1\cdot2} + \dfrac{1}{2\cdot3} + \cdots + \dfrac{1}{(k+1)(k+2)} = \dfrac{k+1}{k+2}$

      Add $\dfrac{1}{(k+1)(k+2)}$ to each side of S_k: $\dfrac{1}{1\cdot2} + \dfrac{1}{2\cdot3} + \cdots + \dfrac{1}{(k+1)(k+2)} =$

$$\dfrac{k}{k+1} + \dfrac{1}{(k+1)(k+2)} = \dfrac{k(k+2)+1}{(k+1)(k+2)} = \dfrac{k^2+2k+1}{(k+1)(k+2)} = \dfrac{(k+1)(k+1)}{(k+1)(k+2)} = \dfrac{k+1}{k+2}$$

   Since S_k implies S_{k+1}, the statement is true for every positive integer n.

11. $\dfrac{4}{5} + \dfrac{4}{5^2} + \dfrac{4}{5^3} + \cdots + \dfrac{4}{5^n} = 1 - \dfrac{1}{5^n}$

   (i) Show that the statement is true for $n = 1$: $\dfrac{4}{5^1} = 1 - \dfrac{1}{5^1} \Rightarrow \dfrac{4}{5} = \dfrac{4}{5}$

   (ii) Assume that S_k is true: $\dfrac{4}{5} + \dfrac{4}{5^2} + \dfrac{4}{5^3} + \cdots + \dfrac{4}{5^k} = 1 - \dfrac{1}{5^k}$

      Show that S_{k+1} is true: $\dfrac{4}{5} + \dfrac{4}{5^2} + \cdots + \dfrac{4}{5^{k+1}} = 1 - \dfrac{1}{5^{k+1}}$

      Add $\dfrac{4}{5^{k+1}}$ to each side of S_k: $\dfrac{4}{5} + \dfrac{4}{5^2} + \dfrac{4}{5^3} + \cdots + \dfrac{4}{5^k} + \dfrac{4}{5^{k+1}} = 1 - \dfrac{1}{5^k} + \dfrac{4}{5^{k+1}} =$

$$1 - \dfrac{1}{5^k}\cdot\dfrac{5}{5} + \dfrac{4}{5^{k+1}} = 1 - \dfrac{5}{5^{k+1}} + \dfrac{4}{5^{k+1}} = 1 - \dfrac{1}{5^{k+1}}$$

   Since S_k implies S_{k+1}, the statement is true for every positive integer n.

13. $\dfrac{1}{1\cdot4} + \dfrac{1}{4\cdot7} + \cdots + \dfrac{1}{(3n-2)(3n+1)} = \dfrac{n}{3n+1}$

   (i) Show that the statement is true for $n = 1$: $\dfrac{1}{1\cdot4} = \dfrac{1}{3(1)+1} \Rightarrow \dfrac{1}{4} = \dfrac{1}{4}$

   (ii) Assume that S_k is true: $\dfrac{1}{1\cdot4} + \cdots + \dfrac{1}{(3k-2)(3k+1)} = \dfrac{k}{3k+1}$

      Show that S_{k+1} is true: $\dfrac{1}{1\cdot4} + \cdots + \dfrac{1}{[3(k+1)-2][3(k+1)+1]} = \dfrac{k+1}{3(k+1)+1}$

      Add $\dfrac{1}{[3(k+1)-2][3(k+1)+1]}$ to each side of S_k: $\dfrac{1}{1\cdot4} + \cdots + \dfrac{1}{[3(k+1)-2][3(k+1)+1]}$

$$= \dfrac{k}{3k+1} + \dfrac{1}{[3(k+1)-2][3(k+1)+1]} = \dfrac{k}{3k+1} + \dfrac{1}{(3k+1)(3k+4)} = \dfrac{k(3k+4)+1}{(3k+1)(3k+4)} =$$

$$\dfrac{3k^2+4k+1}{(3k+1)(3k+4)} = \dfrac{(3k+1)(k+1)}{(3k+1)(3k+4)} = \dfrac{k+1}{3k+4} = \dfrac{k+1}{3(k+1)+1}$$

   Since S_k implies S_{k+1}, the statement is true for every positive integer n.

15. When $n = 1, 3^1 < 6(1) \Rightarrow 3 < 6$. When $n = 2, 3^2 < 6(2) \Rightarrow 9 < 12$.

    When $n = 3, 3^3 > 6(3) \Rightarrow 27 > 18$. For all $n \geq 3, 3^n > 6n$. The only values are 1 and 2.

17. When $n = 1, 2^1 > 1^2 \Rightarrow 2 > 1$. When $n = 2, 2^2 = 2^2 \Rightarrow 4 = 4$. When $n = 3, 2^3 < 3^2 \Rightarrow 8 < 9$.

    When $n = 4, 2^4 = 4^2 \Rightarrow 16 = 16$. For all $n \geq 5, 2^n > n^2$. The only values are 2, 3, and 4.

19. $(a^m)^n = a^{mn}$

    (i) Show that the statement is true for $n = 1$: $(a^m)^1 = a^{m \cdot 1} \Rightarrow a^m = a^m$

    (ii) Assume that S_k is true: $(a^m)^k = a^{mk}$

    Show that S_{k+1} is true: $(a^m)^{k+1} = a^{m(k+1)}$

    Multiply each side of S_k by a^m: $(a^m)^k \cdot (a^m)^1 = a^{mk} \cdot a^m \Rightarrow (a^m)^{k+1} = a^{mk+m} \Rightarrow (a^m)^{k+1} = a^{m(k+1)}$

    Since S_k implies S_{k+1}, the statement is true for every positive integer n.

21. $2^n > 2n$, if $n \geq 3$

    (i) Show that the statement is true for $n = 3$: $2^3 > 2(3) \Rightarrow 8 > 6$

    (ii) Assume that S_k is true: $2^k > 2k$

    Show that S_{k+1} is true: $2^{k+1} > 2(k + 1)$

    Multiply each side of S_k by 2: $2^k \cdot 2 > 2k \cdot 2 \Rightarrow 2^{k+1} > 2(k + 1)$

    Since S_k implies S_{k+1}, the statement is true for every positive integer $n \geq 3$.

23. $a^n > 1$, if $a > 1$

    (i) Show that the statement is true for $n = 1$: $a^1 > 1 \Rightarrow a > 1$, which is true by the given restriction.

    (ii) Assume that S_k is true: $a^k > 1$

    Show that S_{k+1} is true: $a^{k+1} > 1$

    Multiply each side of S_k by a: $a^k \cdot a > 1 \cdot a \Rightarrow a^{k+1} 7 a$

    Because $a > 1$, we may substitute 1 for a in the expression. That is $a^{k+1} > 1$

    Since S_k implies S_{k+1}, the statement is true for every positive integer n.

25. $a^n < a^{n-1}$, if $0 < a < 1$

    (i) Show that the statement is true for $n = 1$: $a^1 < a^0 \Rightarrow a < 1$, which is true by the given restriction.

    (ii) Assume that S_k is true: $a^k < a^{k-1}$

    Show that S_{k+1} is true: $a^{k+1} < a^k$

    Multiply each side of S_k by a: $a^k \cdot a < a^{k-1} \cdot a \Rightarrow a^{k+1} < a^k$

    Since S_k implies S_{k+1}, the statement is true for every positive integer n.

27. $n! > 2^n$, if $n \geq 4$

    (i) Show that the statement is true for $n = 4$: $4! > 2^4 \Rightarrow 24 > 16$

    (ii) Assume that S_k is true: $k! > 2^k$

    Show that S_{k+1} is true: $(k + 1)! > 2^{k+1}$

    Multiply each side of S_k by $k + 1$: $(k + 1)k! > 2^k(k + 1) \Rightarrow (k + 1)! > 2^k(k + 1)$

Because $(k + 1) > 2$ for all $k \geq 4$, we may substitute 2 for $(k + 1)$ in the expression.

That is $(k + 1)! > 2^k(2)$ or $(k + 1)! > 2^{k+1}$.

Since S_k implies S_{k+1}, the statement is true for every positive integer $n \geq 4$.

29. The number of handshakes is $\dfrac{n^2 - n}{2}$ if $n \geq 2$.

(i) Show that the statement is true for $n = 2$: The number of handshakes for 2 people is $\dfrac{2^2 - 2}{2} = \dfrac{2}{2} = 1$, which is true.

(ii) Assume that S_k is true: The number of handshakes for k people is $\dfrac{k^2 - k}{2}$.

Show that S_{k+1} is true: The number of handshakes for $k + 1$ people is

$$\frac{(k + 1)^2 - (k + 1)}{2} = \frac{k^2 + 2k + 1 - k - 1}{2} = \frac{k^2 + k}{2}.$$

When a person joins a group of k people, each person must shake hands with the new person.

Since there are a total of k people that will shake hands with the new person, the total number of handshakes

for $k + 1$ people is $\dfrac{k^2 - k}{2} + k = \dfrac{k^2 - k + 2k}{2} = \dfrac{k^2 + k}{2}$.

Since S_k implies S_{k+1}, the statement is true for every positive integer $n \geq 2$.

31. The first figure has perimeter $P = 3$. When a new figure is generated, each side if the previous figure increases

in length by a factor of $\dfrac{4}{3}$. Thus, the second figure has perimeter $P = 3\left(\dfrac{4}{3}\right)$, the third figure has perimeter

$P = 3\left(\dfrac{4}{3}\right)^2$, and so on. In general, the nth figure has perimeter $P = 3\left(\dfrac{4}{3}\right)^{n-1}$.

33. With 1 ring, 1 move is required. With 2 rings, 3 moves are required. Note that $3 = 2 + 1$. With 3 rings, 7

moves are required. Note that $7 = 2^2 + 2 + 1$.

With n rings $2^{n-1} + 2^{n-2} + \cdots + 2^1 + 1 = 2^n - 1$ moves are required.

(i) Show that the statement is true for $n = 1$: The number of moves for 1 ring is $2^1 - 1 = 1$, which is true.

(ii) Assume that S_k is true: The number of moves for k rings is $2^k - 1$.

Show that S_{k+1} is true: The number of moves for $k + 1$ rings is $2^{k+1} - 1$.

Assume $k + 1$ rings are on the first peg. Since S_k is true, the top k rings can be moved to the second peg

in $2^k - 1$ moves. Now move the bottom ring to the third peg. Since S_k is true, move the k rings from the

second peg on top of the ring on the third peg in $2^k - 1$ moves. The total number of moves is

$$(2^k - 1) + 1 + (2^k - 1) = 2 \cdot 2^k - 1 = 2^{k+1} - 1$$

Since S_k implies S_{k+1}, the statement is true for every positive integer n.

11.6: Probability

1. Yes. The number is between 0 and 1.

3. No. The number is greater 1.

5. Yes. The number is between 0 and 1, inclusive.

7. No. This number is less than 0.

9. A head when tossing a fair coin $\Rightarrow \dfrac{1}{2}$.

11. Rolling a 2 with a fair die $\Rightarrow \dfrac{1}{6}$.

13. Guessing the correct answer for a true-false question $\Rightarrow \dfrac{1}{2}$.

15. Drawing a king from a standard deck of 52 cards $\Rightarrow \dfrac{4}{52} = \dfrac{1}{13}$.

17. The access code can vary between 0000 and 9999. This is 10,000 possibilities, so the probability of guessing the ATM code at random is $\dfrac{1}{10,000}$.

19. (a) The probability that their favorite is pepperoni is 0.43, so the probability it is not pepperoni is

    $1 - 0.43 = 0.57$ or 57%.

    (b) $19\% + 14\% = 33\%$ or 0.33

21. The set $A \cup B$ is the set of elements that belong to either A or B. $A \cup B = \{10, 25, 26, 35\}$

    The set $A \cap B$ is the set of elements that belong to both A and B. $A \cap B = \{25, 26\}$

23. The set $A \cup B$ is the set of elements that belong to either A or B. $A \cup B = \{1, 3, 5, 7, 9, 11\}$

    The set $A \cap B$ is the set of elements that belong to both A and B. $A \cap B = \varnothing$

25. The set $A \cup B$ is the set of elements that belong to either A or B.

    $A \cup B = \{$Tossing a heads, Tossing a tails$\}$

    The set $A \cap B$ is the set of elements that belong to both A and B. $A \cap B = \varnothing$

27. The probability of one tail is $\dfrac{1}{2}$. The events of tossing a coin are independent so the probability of tossing two tails is $\dfrac{1}{2} \cdot \dfrac{1}{2} = \dfrac{1}{4}$.

29. The probability of rolling either a 5 or a 6 is $\dfrac{2}{6} = \dfrac{1}{3}$. The probability of obtaining a 5 or 6 on three consecutive roles is $\dfrac{1}{3} \cdot \dfrac{1}{3} \cdot \dfrac{1}{3} = \dfrac{1}{27}$.

31. To obtain a sum of 2, both die must show a 1. The probability of rolling a 1 with one die is $\dfrac{1}{6}$. Since the events are independent, the probability of obtaining a 1 on two dice is $\dfrac{1}{6} \cdot \dfrac{1}{6} = \dfrac{1}{36}$.

33. The probability of rolling a die and not obtaining a 6 is $\frac{5}{6}$. The probability of rolling a die four times and not

obtaining a 6 is $\frac{5}{6} \cdot \frac{5}{6} \cdot \frac{5}{6} \cdot \frac{5}{6} = \frac{625}{1296} \approx 0.482$.

35. The probability of drawing the first ace is $\frac{4}{52}$, the second ace $\frac{3}{51}$, the third ace $\frac{2}{50}$, and the fourth ace $\frac{1}{49}$. The

probability of drawing four aces is $\frac{4}{52} \cdot \frac{3}{51} \cdot \frac{2}{50} \cdot \frac{1}{49} = \frac{24}{6,497,400} = \frac{1}{270,725}$.

37. There are $\binom{13}{3}$ ways to draw 3 hearts, and there are $\binom{13}{2}$ ways to draw 2 diamonds. there are $\binom{52}{5}$ different

poker hands. Thus, the probability of drawing 3 hearts and 2 diamonds is

$$P(E) = \frac{n(E)}{n(S)} = \frac{\binom{13}{3} \cdot \binom{13}{2}}{\binom{52}{5}} = \frac{286 \cdot 78}{2,598,960} = 0.0086, \text{ or a } 0.86\% \text{ chance.}$$

39. There are 4 strings out of 20 that are defective. Therefore, there is a $\frac{4}{20} = 0.2$ probability or 20% chance of

drawing a defective string and rejecting the box.

41. (a) Figure 41 shows a Venn diagram of the data.

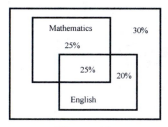

Figure 41

(b) The percentage of students needing help with mathematics, English, or both is $25 + 25 + 20 = 70\%$. The

probability of needing help in mathematics, English, or both is 0.7.

(c) If M represents the event of a student needing help with mathematics, and E represents the event that a

student needs help with English, then

$$P(M \text{ or } E) = P(M \cup E) = P(M) + P(E) - P(M \cap E) = 0.5 + 0.45 - 0.25 = 0.7.$$

43. (a) $18\% = \frac{18}{100} = \frac{9}{50}$

(b) $1 - \frac{18 + 11 + 11 + 11 + 9 + 4}{100} = \frac{36}{100} = \frac{9}{25}$

(c) $1 - \frac{18 + 11}{100} = \frac{71}{100}$

45. The probability is $\frac{634}{100,000} \approx 0.00634$.

47. (a) The probability is $\frac{223,508}{679,590} \approx 0.329$.

(b) This is the complement of the event in part (a), which is $1 - \dfrac{223{,}508}{679{,}590} = \dfrac{476{,}082}{679{,}590} \approx 0.671$.

(c) The probability is $\dfrac{65{,}947 + 64{,}573}{679{,}590} \approx 0.192$.

49. P(Pair or Sum 6) $=$ P(Pair) $+$ P(Sum 6) $-$ P(Pair and Sum 6) $= \dfrac{6}{36} + \dfrac{5}{36} - \dfrac{1}{36} = \dfrac{10}{36} = \dfrac{5}{18}$

51. There are four 2's, four 3's and four 4's in a standard deck. The probability is $\dfrac{12}{52} = \dfrac{3}{13}$.

53. (a) Since the events of rolling a 4, 5, or 6 are mutually exclusive, the probability of either a 4, 5, or 6 is

    $0.2 + 0.2 + 0.3 = 0.7$.

    (b) The events of rolling a 6 followed by a second 6 are independent events. The probability of two consecutive

    6's is $(0.3)(0.3) = 0.09$.

55. (a) The only way to obtain a sum of 12 is for both dice to show a 6. Since these events are independent, the

    probability of two 6's is $(0.3)(0.3) = 0.09$ or 9%.

    (b) There are two ways to obtain a sum of 11. The red die shows a 5 and the blue die a 6, or vice versa. The

    roll of the two dice are independent. The probability of the red die showing a 5 is 0.2 and the probability

    of the blue die showing a 6 is 0.3. The probability of a sum of 11 is $(0.2)(0.3) = 0.06$. Similarly, the

    probability of the red die showing a 6 and the blue die showing a 5 is 0.06. Since these events are mutually

    exclusive, the probability of a sum of 11 is $0.06 + 0.06 = 0.12$ or 12%.

57. There are ten possibilities for each of the three digits. By the fundamental counting principle, there are

    $10 \cdot 10 \cdot 10 = 10^3 = 1000$ different ways to pick these numbers. There is only one winning number so the

    probability is $\dfrac{1}{1000}$.

59. (a) There are a total of $22 + 18 + 10 = 50$ marbles in the jar. Since 22 of them are red, there is a probability

    of $\dfrac{22}{50} = 0.44$ of drawing a red ball.

    (b) Since there is a 0.44 probability of drawing a red ball, there is a $1 - 0.44 = 0.56$ probability of not drawing

    a red ball.

    (c) Drawing a blue or green ball is equivalent to not drawing a red ball. The probability is the same as in

    part (b), which was 0.56.

61. Since one card, a queen, has been drawn, there are 3 queens left in the set of 51 cards. So the probability of

    drawing a queen is $\dfrac{3}{51}$.

63. Since there are 4 kings in a total of 12 face cards, the probability of drawing a king is $\dfrac{4}{12} = \dfrac{1}{3}$.

65. Since 2 red out of 10 red marbles and 4 blue out of 23 blue marbles have already been drawn, 6 marbles out of 33 marbles have been removed from the jar. There are $23 - 4 = 19$ blue marbles left out of $33 - 6 = 27$ marbles in the jar. The probability of drawing a blue marble next is $\frac{19}{27}$.

67. Let E_1 denote the event that it is cloudy and E_2 denote the event that it is windy. $P(E_1) = 0.30$ and $P(E_1 \text{ and } E_2) = P(E_1 \cap E_2) = 0.12$. $P(E_1 \cap E_2) = P(E_1) \cdot P(E_2, \text{given that } E_1 \text{ has occured}) \Rightarrow 0.12 = 0.30 \cdot P(E_2, \text{given that } E_1 \text{ has occured}) \Rightarrow 0.4 = P(E_2, \text{given that } E_1 \text{ has occured})$. The probability that it will be windy given that the day is cloudy is 40%.

69. The possibilities for the result that the first die is a 2 and the sum of the two dice is 7 or more, can be represented by $\{(2, 5), (2, 6)\}$. The sample space can be represented by $\{(2, 1), (2, 2), (2, 3), (2, 4), (2, 5), \text{and } (2, 6)\}$. Out of a total of 6, 2 outcomes satisfy the conditions, so the probability is $\frac{2}{6} = \frac{1}{3}$.

71. (a) Let D represent the event that part is defective. Since 18 of the 235 parts are defective, $P(D) = \frac{18}{235}$.

   (b) Let A represent the event that part is type A. Since 7 of the 18 parts are type A, $P(A, \text{given } D) = \frac{7}{18}$.

   (c) $P(D \text{ and } A) = P(D \cap A) = P(D) \cdot P(A, \text{given } D) = \frac{18}{235} \cdot \frac{7}{18} = \frac{7}{235}$.

73. (a) Let O represent the event that the number is odd. Since 8 of the 15 numbers are odd, $P(O) = \frac{8}{15}$.

   (b) Let E represent the event that the number is even. Since 7 of the 15 numbers are even, $P(E) = \frac{7}{15}$.

   (c) Let M represent the event that the number is prime. Since 6 of the 15 numbers are prime,

   $$P(M) = \frac{6}{15} = \frac{2}{5}.$$

   (d) Since 5 of the 6 prime numbers are odd, $P(M \cap O) = \frac{5}{15} = \frac{1}{3}$.

   (e) Since 1 of the 6 prime numbers is even, $P(M \cap E) = \frac{1}{15}$.

Checking Basic Concepts for Sections 11.5 and 11.6

1. $4 + 8 + 12 + \cdots + 4n = 2n(n + 1)$

   (i) Show that the statement is true for $n = 1$: $4(1) = 2(1)(1 + 1) \Rightarrow 4 = 4$

   (ii) Assume that S_k is true: $4 + 8 + 12 + \cdots + 4k = 2k(k + 1)$

   Show that S_{k+1} is true: $4 + 8 + \cdots + 4(k + 1) = 2(k + 1)(k + 2)$

   Add $4(k + 1)$ to each side of S_k: $4 + 8 + 12 + \cdots + 4k + 4(k + 1) = 2k(k + 1) + 4(k + 1) = 2k^2 + 6k + 4 = 2(k + 1)(k + 2)$

   Since S_k implies S_{k+1}, the statement is true for every positive integer n.

3. The probability of one head is $\frac{1}{2}$. The events of tossing a coin are independent so the probability of tossing four

   heads is $\frac{1}{2} \cdot \frac{1}{2} \cdot \frac{1}{2} \cdot \frac{1}{2} = \frac{1}{16}$.

5. There are $\binom{4}{4}$ different ways to draw four aces and $\binom{4}{1}$ different ways to draw a queen. There are a total of

   $\binom{52}{5}$ different 5-card hands in a set of 52 cards. Thus, the probability of drawing four aces and a queen is

   $$\frac{\binom{4}{4} \cdot \binom{4}{1}}{\binom{52}{5}} = \frac{1 \cdot 4}{2{,}598{,}960} = 0.0000015.$$

Chapter 11 Review Exercises

1. $a_1 = -3(1) + 2 = -1; a_2 = -3(2) + 2 = -4; a_3 = -3(3) + 2 = -7; a_4 = -3(4) + 2 = -10$

   The first four terms are $-1, -4, -7,$ and -10.

3. $a_1 = 0; a_2 = 2a_1 + 1 = 2(0) + 1 = 1; a_3 = 2a_2 + 1 = 2(1) + 1 = 3; a_4 = 2a_3 + 1 = 2(3) + 1 = 7$

   The first four terms are $0, 1, 3,$ and 7.

5. The points $(1, 5), (2, 3), (3, 1), (4, 2), (5, 4),$ and $(6, 6)$ lie on the graph. Therefore, the terms of this sequence

   are $5, 3, 1, 2, 4, 6$.

7. (a) Each term can be found by adding -2 to the previous term, beginning with 3. It is an arithmetic sequence.

   A numerical representation for the first eight terms is shown in Figure 7a.

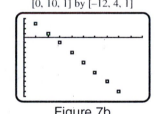

[0, 10, 1] by [−12, 4, 1]

n	1	2	3	4	5	6	7	8
a_n	3	1	−1	−3	−5	−7	−9	−11

Figure 7a

Figure 7b

   (b) A graphical representation for the first eight terms is shown in Figure 7b. Notice that the points lie on a

   line with slope -2 because the sequence is arithmetic with $d = -2$.

   (c) To find the symbolic representation, we will use the formula $a_n = a_1 + (n - 1)d$, where $a_n = f(n)$. The

   common difference of the sequence is $d = -2$ and the first term is $a_1 = 3$. Therefore a symbolic

   representation is given by $a_n = 3 + (n - 1)(-2)$ or $a_n = -2n + 5$.

9. Let $a_n = f(n)$, where $f(n) = dn + c$. Then, $f(n) = 4n + c$. Since $f(3) = 4(3) + c = -3 \Rightarrow$.

   $c = -15$. Thus $a_n = f(n) = 4n - 15$.

11. Since f is a linear function, the sequence is arithmetic.

13. Since f can be written in the form $f(n) = cr^{n-1}$, the sequence is geometric.

15. $S_5 = (4(1) + 1) + (4(2) + 1) + (4(3) + 1) + (4(4) + 1) + (4(5) + 1) =$
$5 + 9 + 13 + 17 + 21 = 65$

17. The first term is $a_1 = -2$ and the last term is $a_9 = 22$.

To find the sum use $S_n = n\left(\dfrac{a_1 + a_n}{2}\right) \Rightarrow S_9 = 9\left(\dfrac{-2 + 22}{2}\right) = 90$. The sum is 90.

19. The first term is $a_1 = 1$ and the common ratio is $r = 3$. Since there are 8 terms, the sum is

$$S_n = a_1\left(\frac{1 - r^n}{1 - r}\right) \Rightarrow S_8 = 1\left(\frac{1 - 3^8}{1 - 3}\right) = 3280.$$

21. The first term is $a_1 = 4$ and the common ratio is $r = -\dfrac{1}{3}$. The sum is

$$S_n = a_1\left(\frac{1}{1 - r}\right) \Rightarrow S = 4\left(\frac{1}{1 + \frac{1}{3}}\right) = 3.$$

23. $\displaystyle\sum_{k=1}^{5}(5k + 1) = (5 + 1) + (10 + 1) + (15 + 1) + (20 + 1) + (25 + 1) = 6 + 11 + 16 + 21 + 26$

25. $1^3 + 2^3 + 3^3 + 4^3 + 5^3 + 6^3 = \displaystyle\sum_{k=1}^{6}k^3$

27. Since $a_1 = 5$ and $d = -3$, use the formula $S_n = \dfrac{n}{2}(2a_1 + (n - 1)d)$.

$$S_{30} = \frac{30}{2}(2(5) + (30 - 1)(-3)) = 15(10 + 29(-3)) = 15(-77) = -1155$$

29. $\dfrac{2}{11} = 0.18181818\ldots = 0.18 + 0.0018 + 0.000018 + 0.00000018 + \cdots$

31. $P(n, r) = \dfrac{n!}{(n - r)!} \Rightarrow P(6, 3) = \dfrac{6!}{(6 - 3)!} = \dfrac{6!}{3!} = \dfrac{6 \cdot 5 \cdot 4 \cdot 3!}{3!} = 6 \cdot 5 \cdot 4 = 120$

33. $1 + 3 + 5 + \cdots + (2n - 1) = n^2$

   (i) Show that the statement is true for $n = 1$: $2(1) - 1 = 1^2 \Rightarrow 1 = 1$

   (ii) Assume that S_k is true: $1 + 3 + 5 + P + 2k - 12 = k^2$

   Show that S_{k+1} is true: $1 + 3 + \cdots + (2(k + 1) - 1) = (k + 1)^2$

   Add $2k + 1$ to each side of S_k: $1 + 3 + 5 + \cdots + (2k - 1) + (2k + 1) = k^2 + 2k + 1 = (k + 1)^2$

   Since S_k implies S_{k+1}, the statement is true for every positive integer n.

35. Since 1, 2, or 3 are three outcomes out of six possible outcomes, the probability is $\dfrac{3}{6} = \dfrac{1}{2}$.

37. There are $P(5, 5) = 5! = 120$ different arrangements.

39. There are $4^{20} \approx 1.1 \times 10^{12}$ different ways to answer the exam.

41. There are 50 choices for each number in the combination. So there are $50^4 = 6,250,000$ possible combinations.

43. (a) The initial height is 4 feet. On the first rebound it reaches 90% of 4 or 3.6 feet. On the second rebound it attains a height of 90% of 3.6 or 3.24 feet. Each term in this sequence is found by multiplying the previous term by 0.9. Thus, the first five terms are 4, 3.6, 3.24, 2.916, 2.6244. This is a geometric sequence.

   (b) Plot the points (1, 4), (2, 3.6), (3, 3.24), (4, 2.916), and (5, 2.6244) as shown in Figure 43.

   (c) The first term is 4 and the common ratio is 0.9. Thus, $a_n = 4(0.9)^{n-1}$.

[0, 6, 1] by [0, 6, 1]

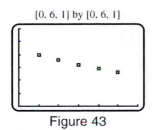

Figure 43

45. From a group of 6 people, we must select a set of 3 people. This can be done $C(6, 3) = 20$ different ways.

47. They can be selected in $C(10, 6) = 210$ different ways.

49. There are $C(16, 2)$ different ways to select 2 batteries from a pack of 16. In order to avoid testing a defective battery, we must pick 2 batteries from the 14 good ones. This can be done in $C(14, 2)$ different ways. The probability that the box is not rejected is $\dfrac{C(14, 2)}{C(16, 2)} = \dfrac{91}{120} \approx 0.758$.

51. (a) The total number of marbles in the jar is $13 + 27 + 20 = 60$. Since 27 of the marbles are blue, there is a probability of $\dfrac{27}{60} = 0.45$ of drawing a blue marble.

(b) Since there is a 0.45 probability of drawing a blue marble, there is a $1 - 0.45 = 0.55$ probability of drawing a marble that is not blue.

(c) There is a $\dfrac{13}{60} \approx 0.217$ probability of drawing a red marble.

53. The sequence is graphed in Figure 53. Initially, the population density grows slowly, then it increases rapidly. After some time, it levels off near 4000 thousand (4,000,000) per acre.

[0, 16, 1] by [0, 5000, 1000]

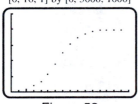

Figure 53

Extended and Discovery Exercises for Chapter 11

1. (a) $P_{kj} = \dfrac{\binom{2k}{j}\binom{4-2k}{2-j}}{\binom{4}{2}} \Rightarrow P_{00} = \dfrac{\binom{0}{0}\binom{4}{2}}{\binom{4}{2}} = 1;\ P_{01} = \dfrac{\binom{0}{1}\binom{4}{1}}{\binom{4}{2}} = 0;\ P_{02} = \dfrac{\binom{0}{2}\binom{4}{0}}{\binom{4}{2}} = 0;$

$P_{10} = \dfrac{\binom{2}{0}\binom{2}{2}}{\binom{4}{2}} = \dfrac{1}{6};\ P_{11} = \dfrac{\binom{2}{1}\binom{2}{1}}{\binom{4}{2}} = \dfrac{2}{3};\ P_{12} = \dfrac{\binom{2}{2}\binom{2}{0}}{\binom{4}{2}} = \dfrac{1}{6};\ P_{20} = \dfrac{\binom{4}{0}\binom{0}{2}}{\binom{4}{2}} = 0;$

$$P_{21} = \frac{\binom{4}{1}\binom{0}{1}}{\binom{4}{2}} = 0; \quad P_{22} = \frac{\binom{4}{2}\binom{0}{0}}{\binom{4}{2}} = 1. \text{ Thus, } P = \begin{bmatrix} P_{00} & P_{01} & P_{02} \\ P_{10} & P_{11} & P_{12} \\ P_{20} & P_{21} & P_{22} \end{bmatrix} = \begin{bmatrix} 1 & 0 & 0 \\ \frac{1}{6} & \frac{2}{3} & \frac{1}{6} \\ 0 & 0 & 1 \end{bmatrix}.$$

(b) The sum of the probabilities in each row is 1. The greatest probabilities lie along the diagonal. A mother cell is most likely to produce a daughter cell like itself. *Answers may vary.*

3. The quantity $\dfrac{a_1 + a_n}{2}$ represents not only the average of the two terms a_1 and a_n, but it also represents the average of all of the terms $a_1, a_2, a_3, \ldots, a_n$ in the series. This is true whether n is odd or even. The total sum is equal to n times the average of the terms.

Chapters 1-11 Cumulative Review Exercises

1. $34{,}500 = 3.45 \times 10^4$; $1.52 \times 10^{-4} = 0.000152$

3. $d\sqrt{(1 - (-4))^2 + (-2 - 2)^2} = \sqrt{5^2 + (-4)^2} = \sqrt{25 + 16} = \sqrt{41}$

5. (a) $f(-3) = \sqrt{1 - (-3)} = \sqrt{4} = 2$; $f(a + 1) = \sqrt{1 - (a + 1)} = \sqrt{-a}$

    (b) For f to be defined, $1 - x \geq 0$. Thus, the domain is $D = \{x \mid x \leq 1\}$.

7. $\dfrac{f(-1) - f(-2)}{-1 - (-2)} = \dfrac{((-1)^3 - 4) - ((-2)^3 - 4)}{1} = -5 - (-12) = 7$

9. A line passing through the points $(2, -4)$ and $(-3, 2)$ has slope $m = \dfrac{2 - (-4)}{-3 - 2} = \dfrac{6}{-5} = -\dfrac{6}{5}$.

    Using the point $(2, -4)$, the equation is $y = -\dfrac{6}{5}(x - 2) - 4 \Rightarrow y = -\dfrac{6}{5}x - \dfrac{8}{5}$.

11. (a) The graph passes through the points $(0, -1)$ and $(4, 2)$. The slope is $m = \dfrac{2 - (-1)}{4 - 0} = \dfrac{3}{4}$.

    The y-intercept is -1. The x-intercept can be found by noting that a 1-unit rise from the point $(0, -1)$ would require a $\dfrac{4}{3}$-unit run to return to a line with slope $\dfrac{3}{4}$. That is, $m = \dfrac{\text{rise}}{\text{run}} = \dfrac{1}{\frac{4}{3}} = \dfrac{3}{4}$. The x-intercept is $\dfrac{4}{3}$.

    (b) Since the slope is $\dfrac{3}{4}$ and the y-intercept is -1, the formula for f is $f(x) = \dfrac{3}{4}x - 1$.

    (c) $\dfrac{3}{4}x - 1 = 0 \Rightarrow \dfrac{3}{4}x = 1 \Rightarrow x = \dfrac{4}{3}$

13. (a) $4(x - 2) + 1 = 3 - \dfrac{1}{2}(2x + 3) \Rightarrow 4x - 8 + 1 = 3 - x - \dfrac{3}{2} \Rightarrow 5x = \dfrac{17}{2} \Rightarrow x = \dfrac{17}{10}$

    (b) $6x^2 = 13x + 5 \Rightarrow 6x^2 - 13x - 5 = 0 \Rightarrow (3x + 1)(2x - 5) = 0 \Rightarrow x = -\dfrac{1}{3} \text{ or } x = \dfrac{5}{2}$

    (c) By the quadratic formula, $x = \dfrac{-b \pm \sqrt{b^2 - 4ac}}{2a} \Rightarrow x = \dfrac{-(-1) \pm \sqrt{(-1)^2 - 4(1)(-3)}}{2(1)} = \dfrac{1 \pm \sqrt{13}}{2}$.

    (d) $x^3 + x^2 = 4x + 4 \Rightarrow x^3 + x^2 - 4x - 4 = 0 \Rightarrow x^2(x + 1) - 4(x + 1) = 0 \Rightarrow$

    $(x^2 - 4)(x + 1) = 0 \Rightarrow (x + 2)(x - 2)(x + 1) = 0 \Rightarrow x = -2, -1, \text{ or } 2$

(e) $x^4 - 4x^2 + 3 = 0 \Rightarrow (x^2 - 3)(x^2 - 1) = 0 \Rightarrow x^2 = 3$ or $x^2 = 1 \Rightarrow x = \pm\sqrt{3}$ or ± 1

(f) $\dfrac{1}{x - 3} = \dfrac{4}{x + 5} \Rightarrow (x - 3)(x + 5) \cdot \dfrac{1}{x - 3} = \dfrac{4}{x + 5} \cdot (x - 3)(x + 5) \Rightarrow x + 5 = 4(x - 3) \Rightarrow$

$x + 5 = 4x - 12 \Rightarrow -3x = -17 \Rightarrow x = \dfrac{17}{3}$

(g) $3e^{2x} - 5 = 23 \Rightarrow 3e^{2x} = 28 \Rightarrow e^{2x} = \dfrac{28}{3} \Rightarrow \ln e^{2x} = \ln \dfrac{28}{3} \Rightarrow 2x = \ln \dfrac{28}{3} \Rightarrow x = \dfrac{\ln\left(\frac{28}{3}\right)}{2} \approx 1.117$

(h) $2\log(x + 1) - 1 = 2 \Rightarrow 2\log(x + 1) = 3 \Rightarrow \log(x + 1) = \dfrac{3}{2} \Rightarrow 10^{\log(x+1)} = 10^{3/2} \Rightarrow$

$x + 1 = 10^{3/2} \Rightarrow x = 10^{3/2} - 1 \approx 30.623$

(i) $\sqrt{x + 3} + 4 = x + 1 \Rightarrow \sqrt{x + 3} = x - 3 \Rightarrow (\sqrt{x + 3})^2 = (x - 3)^2 \Rightarrow$

$x + 3 = x^2 - 6x + 9 \Rightarrow x^2 - 7x + 6 = 0 \Rightarrow (x - 1)(x - 6) = 0 \Rightarrow x = 1$ or $x = 6$. Note, 1 is

extraneous. The only solution is 6.

(j) $|3x - 1| = 5 \Rightarrow 3x - 1 = -5$ or $3x - 1 = 5 \Rightarrow 3x = -4$ or $3x = 6 \Rightarrow x = -\dfrac{4}{3}$ or 2

15. f is continuous. See Figure 15.

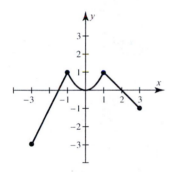

Figure 15

17. (a) The solution to $f(x) = 0$ is the x-intercepts. The solutions are $-2, -1, 1$, and 2.

(b) The graph of f is above the x-axis on the interval $(-\infty, -2)\cup(-1, 1)\cup(2, \infty)$ or

$\{x \mid x < -2$ or $-1 < x < 1$ or $x > 2\}$.

(c) The graph of f is on or below the x-axis on the interval $[-2, -1]\cup[1, 2]$ or

$\{x \mid -2 \le x \le -1$ or $1 \le x \le 2\}$.

19. The x-coordinate of the vertex is $x = -\dfrac{b}{2a} = -\dfrac{9}{2(-3)} = \dfrac{3}{2}$. The y-coordinate of the vertex is $f\left(\dfrac{3}{2}\right)$.

$y = f\left(\dfrac{3}{2}\right) = -3\left(\dfrac{3}{2}\right)^2 + 9\left(\dfrac{3}{2}\right) + 1 = -\dfrac{27}{4} + \dfrac{27}{2} + 1 = -\dfrac{27}{4} + \dfrac{54}{4} + \dfrac{4}{4} = \dfrac{31}{4}$. The vertex is $\left(\dfrac{3}{2}, \dfrac{31}{4}\right)$.

21. (a) The graph of f is increasing on $(-\infty, -2]\cup[1, \infty)$ and decreasing on $[-2, 1]$ increasing on

$\{x \mid x \le -2$ or $x \ge 1\}$ or $(-\infty, -2]\cup[1, \infty)$ decreasing on $\{x \mid -2 \le x \le 1\}$ or $[-2, 1]$.

(b) The zeros of f are the x-intercepts which are approximately $-3.3, 0$, and 1.8.

(c) The turning points are approximately $(-2, 2)$ and $(1, -0.7)$.

(d) There is a local minimum of -0.7 and a local maximum of 2.

23. (a) $\dfrac{6x^4 - 2x^2 + 1}{2x^2} = \dfrac{6x^4}{2x^2} - \dfrac{2x^2}{2x^2} + \dfrac{1}{2x^2} = 3x^2 - 1 + \dfrac{1}{2x^2}$

(b) $x + 1 \overline{\smash{\big)}\ 2x^4 - 3x^3 + 0x^2 - x + 2}$ The solution is: $2x^3 - 5x^2 + 5x - 6 + \dfrac{8}{x + 1}$

$$\begin{array}{r} 2x^3 - 5x^2 + 5x - 6 \\ \hline 2x^4 - 3x^3 + 0x^2 - x + 2 \\ \underline{2x^4 + 2x^3} \\ -5x^3 + 0x^2 - x + 2 \\ \underline{-5x^3 - 5x^2} \\ 5x^2 - x + 2 \\ \underline{5x^2 + 5x} \\ -6x + 2 \\ \underline{-6x - 6} \\ 8 \end{array}$$

25. Since $3i$ is a zero its conjugate, $-3i$, is also a zero.

The complete factored form is

$f(x) = 3(x - (-1))(x - (-3i))(x - 3i)$ or $f(x) = 3(x + 1)(x + 3i)(x - 3i)$.

To find the expanded form, first multiply $(x + 3i)(x - 3i) = x^2 + 9$. Continuing to multiply out gives

$f(x) = 3(x + 1)(x^2 + 9) = 3(x^3 + x^2 + 9x + 9)$ or $f(x) = 3x^3 + 3x^2 + 27x + 27$.

27. $x = \dfrac{-b \pm \sqrt{b^2 - 4ac}}{2a} \Rightarrow x = \dfrac{-2 \pm \sqrt{2^2 - 4(1)(5)}}{2(1)} = \dfrac{-2 \pm \sqrt{-16}}{2} = \dfrac{-2 \pm 4i}{2} = -1 \pm 2i$

29. $\sqrt[5]{(x + 1)^3} = (x + 1)^{3/5}$. When $x = 31$, $(x + 1)^{3/5} = (31 + 1)^{3/5} = 32^{3/5} = (\sqrt[5]{32})^3 = 2^3 = 8$

31. (a) $(f + g)(2) = f(2) + g(2) = 3 + 0 = 3$

(b) $(fg)(0) = f(0) \cdot g(0) = -1(1) = -1$

(c) $(g \circ f)(1) = g(f(1)) = g(0) = 1$

(d) Note, for $g^{-1}(3)$, read the graph backward. $(g^{-1} \circ f)(-2) = g^{-1}(f(-2)) = g^{-1}(3) = -4$

33. Let $y = f(x)$: $y = \dfrac{x}{x + 1}$. Interchange x and y: $x = \dfrac{y}{y + 1}$. Then solve for y.

$x = \dfrac{y}{y + 1} \Rightarrow x(y + 1) = y \Rightarrow xy + x = y \Rightarrow x = y - xy \Rightarrow x = y(1 - x) \Rightarrow \dfrac{x}{1 - x} = y$

The inverse function is $f^{-1}(x) = \dfrac{x}{1 - x}$ or $f^{-1}(x) = -\dfrac{x}{x - 1}$.

35. Since each unit increase in x results in $f(x)$ increasing by a factor of 3, the function is exponential with base 3.

Since $f(0) = 2$, the initial value is 2 and the function is $f(x) = 2(3^x)$.

37. $A = 500\left(1 + \dfrac{0.06}{4}\right)^{15 \cdot 4} \approx \$1221.61$

39. (a) A quadratic function is defined for all real inputs. The domain is $D = (-\infty, \infty)$ or $\{x \mid -\infty < x < \infty\}$.

The graph of f is a parabola opening upward with vertex $(1, 0)$. The range is $R = [0, \infty)$ or $\{x \mid x \geq 0\}$.

(b) An exponential function is defined for all real inputs. The domain is $D = (-\infty, \infty)$ or

$\{x \mid -\infty < x < \infty\}$.

The function $f(x) = 10^x$ has only positive outputs. The range is $R = (0, \infty)$ or $\{x \mid x > 0\}$.

(c) The natural logarithm function is defined for only positive real inputs. The domain is $D = (0, \infty)$ or $\{x \mid x > 0\}$.

The natural logarithm function can output any real value. The range is $R = (-\infty, \infty)$ or $\{x \mid -\infty < x < \infty\}$.

(d) The reciprocal function is defined for all real inputs except 0. The domain is $D = (-\infty, 0) \cup (0, \infty)$ or $\{x \mid x \neq 0\}$.

The reciprocal function can output any real value except 0. The range is $R = (-\infty, 0) \cup (0, \infty)$ or $\{x \mid x \neq 0\}$.

41. $2 \log x + 3 \log y - \dfrac{1}{3} \log z = \log x^2 + \log y^3 - \log z^{1/3} = \log \dfrac{x^2 y^3}{z^{1/3}} = \log \dfrac{x^2 y^3}{\sqrt[3]{z}}$

43. (a) See Figure 43a.

(b) See Figure 43b.

(c) See Figure 43c.

(d) See Figure 43d.

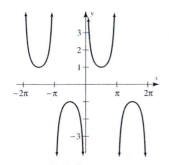

Figure 43a

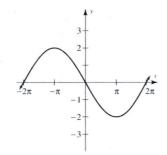

Figure 43b

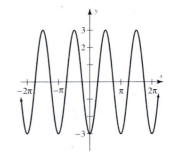

Figure 43c

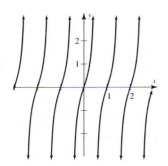

Figure 43d

45. (a) $\cos t = -\dfrac{\sqrt{3}}{2} \Rightarrow t = \cos^{-1}\left(-\dfrac{\sqrt{3}}{2}\right) = \dfrac{5\pi}{6}$ or $\dfrac{7\pi}{6}$. Therefore $t = \dfrac{5\pi}{6} + 2\pi n$ or $t = \dfrac{7\pi}{6} + 2\pi n$.

(b) $\tan^2 t - 1 = 0 \Rightarrow \tan^2 t = 1 \Rightarrow \tan t = \pm 1 \Rightarrow t_R = \tan^{-1} 1 = \dfrac{\pi}{4}$. Since tangent is positive or negative in all quadrant., $t = \dfrac{\pi}{4} + \dfrac{\pi}{2} n$.

(c) $\sin^2 t + \dfrac{1}{2}\sin t = 0 \Rightarrow 2\left(\sin^2 t + \dfrac{1}{2}\sin t\right) = 2(0) \Rightarrow 2\sin^2 t + \sin t = 0 \Rightarrow$

$\sin t\,(2\sin t + 1) = 0 \Rightarrow \sin t = 0$ or $2\sin t + 1 = 0$. When $\sin t = 0$, $t = \pi n$ and when

$2\sin t + 1 = 0 \Rightarrow 2\sin t = -1 \Rightarrow \sin t = -\dfrac{1}{2},\ t = \dfrac{7\pi}{6} + 2\pi n,\ \dfrac{11\pi}{6} + 2\pi n.$

(d) $\cos 2t = -1$, let $\theta = 2t$ so then the equation becomes $\cos\theta = -1$, in radian measure this solution is π. For

all real solutions $\theta = \pi + 2\pi n \Rightarrow$ because $\theta = 2t$, we can substitute $2t$ for $\theta \Rightarrow 2t = \pi + 2\pi n$ or

$t = \dfrac{\pi}{2} + \pi n.$

47. $5.54° = 5° + (0.54)(60) = 5° + 32.4' = 5° + 32' + (0.4)(60) = 5° + 32' + 24'' = 5°\,32'\,24''$

49. $\dfrac{5\pi}{4}\cdot\dfrac{180}{\pi} = 225°$

51. If the terminal side passes through the point $(-7, 24)$, θ lies in quadrant II.

$r^2 = x^2 + y^2 \Rightarrow r^2 = 24^2 + 7^2 \Rightarrow r^2 = 625 \Rightarrow r = 25$

$\sin\theta = \dfrac{y}{r} = \dfrac{24}{25} \qquad \cos\theta = \dfrac{x}{r} = -\dfrac{7}{25} \qquad \tan\theta = \dfrac{y}{x} = -\dfrac{24}{7}$

$\csc\theta = \dfrac{r}{y} = \dfrac{25}{24} \qquad \sec\theta = \dfrac{r}{x} = -\dfrac{25}{7} \qquad \cot\theta = \dfrac{x}{y} = -\dfrac{7}{24}$

53. If $\cos\theta < 0$ and $\csc\theta < 0$, θ lies in quadrant III, $\cos\theta = -\dfrac{11}{61}$.

$\sin^2\theta + \cos^2\theta = 1 \Rightarrow \sin^2\theta + \left(-\dfrac{11}{61}\right)^2 = 1 \Rightarrow \sin^2\theta = 1 - \dfrac{121}{3721} \Rightarrow \sin^2\theta = \dfrac{3600}{3721} \Rightarrow \sin\theta = -\dfrac{60}{61}$

$\tan\theta = \dfrac{\sin\theta}{\cos\theta} = \dfrac{-\frac{60}{61}}{-\frac{11}{61}} = \dfrac{60}{11} \qquad \csc\theta = \dfrac{1}{\sin\theta} = \dfrac{1}{-\frac{60}{61}} = -\dfrac{61}{60}$

$\sec\theta = \dfrac{1}{\cos\theta} = \dfrac{1}{-\frac{11}{61}} = -\dfrac{61}{11} \qquad \cot\theta = \dfrac{\cos\theta}{\sin\theta} = \dfrac{-\frac{11}{61}}{-\frac{60}{61}} = \dfrac{11}{60}$

55. From the pythagorean theorem $a^2 + b^2 = c^2 \Rightarrow 6^2 + b^2 = 10^2 \Rightarrow 36 + b^2 = 100 \Rightarrow b^2 = 64 \Rightarrow b = 8$

$\tan\alpha = \dfrac{6}{8} \Rightarrow \alpha = \tan^{-1}\left(\dfrac{3}{4}\right) \approx 36.9° \qquad \tan\beta = \dfrac{8}{6} \Rightarrow \beta = \tan^{-1}\left(\dfrac{4}{3}\right) \approx 53.1°.$

57. $1 - \sin^2\theta + \cot^2\theta - \sin^2\theta\cot^2\theta = 1 - \sin^2\theta - \sin^2\theta\cot^2\theta + \cot^2\theta =$

$1 - \sin^2\theta(1 + \cot^2\theta) + \cot^2\theta = 1 - \sin^2\theta(\csc^2\theta) + \cot^2\theta = 1 - \sin^2\theta\left(\dfrac{1}{\sin^2\theta}\right) + \cot^2\theta =$

$1 - 1 + \cot^2\theta = \cot^2\theta$

59. (a) This triangle is of the form of ASA. There is only one solution. $\alpha = 180° - 42° - 31° = 107°$

$\dfrac{b}{\sin\beta} = \dfrac{a}{\sin\alpha} \Rightarrow b = \dfrac{a\sin\beta}{\sin\alpha} = \dfrac{22\sin 42°}{\sin 107°} \approx 15.4$

$\dfrac{c}{\sin\gamma} = \dfrac{a}{\sin\alpha} \Rightarrow c = \dfrac{a\sin\gamma}{\sin\alpha} = \dfrac{22\sin 31°}{\sin 107°} \approx 11.8 \quad \alpha = 107°,\, b \approx 15.4,\, c \approx 11.8$

(b) This triangle is of the form of SSA. It is ambiguous.

$$\frac{\sin \beta}{b} = \frac{\sin \gamma}{c} \Rightarrow \sin \beta = \frac{b \sin \gamma}{c} = \frac{8 \sin 50°}{7} \approx 0.87548 \Rightarrow \beta_R = \sin^{-1}(0.87548) \approx 61.1°$$

Thus $\beta = 61.1°$ or $\beta = 180° - 61.1° \approx 118.9°$

Solution 1: Let $\beta = 61.1°$. Then $\alpha \approx 180° - 50° - 61.1° \approx 68.9°$.

$$\frac{a}{\sin \alpha} = \frac{c}{\sin \gamma} \Rightarrow a = \frac{c \sin \alpha}{\sin \gamma} = \frac{7 \sin 68.9°}{\sin 50°} \approx 8.5 \quad \beta \approx 61.1°, \alpha \approx 68.9°, a \approx 8.5$$

Solution 2: Let $\beta = 118.9°$. Then $\alpha \approx 180° - 50° - 118.9° \approx 11.1°$.

$$\frac{a}{\sin \alpha} = \frac{c}{\sin \gamma} \Rightarrow a = \frac{c \sin \alpha}{\sin \gamma} = \frac{7 \sin 11.1°}{\sin 50°} \approx 1.8 \quad \beta \approx 118.9°, \alpha \approx 11.1°, a \approx 1.8$$

(c) This triangle is of the form SAS.

$$a^2 = b^2 + c^2 - 2bc \cos \alpha = 7^2 + 8^2 - 2(7)(8) \cos 44° \Rightarrow a = \sqrt{113 - 112 \cos 44°} \approx 5.69508$$

$$c^2 = a^2 + b^2 - 2ab \cos \gamma \Rightarrow \cos \gamma = \frac{c^2 - a^2 - b^2}{-2ab} \Rightarrow \alpha = \cos^{-1}\left(\frac{8^2 - (5.69508)^2 - 7^2}{-2(5.69508)(7)}\right) \approx$$

$77.36971°$

$\beta = 180° - 44° - 77.36971° \approx 58.63029° \quad a \approx 5.7, \beta = 58.6°, \gamma \approx 77.4°$

(d) This Triangle is of the form SSS.

$$a^2 = b^2 + c^2 - 2bc \cos \alpha \Rightarrow \cos \alpha = \frac{a^2 - b^2 - c^2}{-2bc} \Rightarrow \alpha = \cos^{-1}\left(\frac{10^2 - 11^2 - 12^2}{-2(11)(12)}\right) \approx 51.31781°$$

$$b^2 = a^2 + c^2 - 2ac \cos \beta \Rightarrow \cos \beta = \frac{b^2 - a^2 - c^2}{-2ac} \Rightarrow \beta = \cos^{-1}\left(\frac{11^2 - 10^2 - 12^2}{-2(10)(12)}\right) \approx 59.16950°$$

$\gamma = 180° - 51.31781° - 59.16950 \approx 69.51269° \quad \alpha \approx 51.3°, \beta \approx 59.2°, \gamma \approx 69.5°$

61. (a) $\|\mathbf{a}\| = \sqrt{(3)^2 + (-4)^2} = \sqrt{25} = 5$

(b) $4\mathbf{b} - 2\mathbf{a} = 4\langle -5, 12 \rangle - 2\langle 3, -4 \rangle = \langle -20, 48 \rangle - \langle 6, -8 \rangle \Rightarrow \langle -20 - 6, 48 - (-8) \rangle = \langle -26, 56 \rangle$

(c) $\mathbf{a} \cdot \mathbf{b} = \langle 3, -4 \rangle \cdot \langle -5, 12 \rangle = (3)(-5) + (-4)(12) = -15 + (-48) = -63$

(d) $\theta = \cos^{-1}\left(\dfrac{\mathbf{a} \cdot \mathbf{b}}{\|\mathbf{a}\| \|\mathbf{b}\|}\right) \Rightarrow$ we found $\mathbf{a} \cdot \mathbf{b}$ from part (c) above to be -63.

$\|\mathbf{a}\| = \sqrt{(3)^2 + (-4)^2} = \sqrt{25} = 5$ and $\|\mathbf{b}\| = \sqrt{(-5)^2 + (12)^2} = \sqrt{169} = 13 \Rightarrow$

$\theta = \cos^{-1}\left(\dfrac{-63}{(5)(13)}\right) \approx 165.75°$

63. See Figure 63.

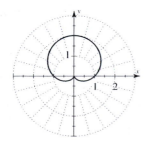

Figure 63

65. (a) Multiplying the first equation by -1 and adding the two equations will eliminate the variable x.

$$-2x - 3y = -4$$
$$\underline{2x - 5y = -12}$$
$$-8y = -16 \quad \text{Thus, } y = 2. \text{ And so } 2x + 3(2) = 4 \Rightarrow x = -1. \text{ The solution is } (-1, 2).$$

(b) Multiplying the first equation by 2 and adding the two equations will eliminate both variables.

$$-4x + y = 2$$
$$\underline{4x - y = -2}$$
$$0 = 0 \quad \text{This is an identity. The system is dependent with solutions } \{(x, y) \mid 4x - y = -2\}.$$

(c) Adding the two equations will eliminate the variable y.

$$x^2 + y^2 = 16$$
$$\underline{2x^2 - y^2 = 11}$$
$$3x^2 = 27 \quad \text{Thus, } x = \pm 3. \text{ And so } (\pm 3)^2 + y^2 = 16 \Rightarrow y^2 = 7 \Rightarrow y = \pm\sqrt{7}.$$

There are four solutions, $(\pm 3, \pm\sqrt{7})$.

(d) Multiply the first equation by 2 and subtract the second equations to eliminate the variable x.

$$2x + 2y - 4z = -12$$
$$\underline{2x - y - 3z = -18}$$
$$3y - z = 6$$

Subtract the third equation from this *new* equation to eliminate both y and z.

$$3y - z = 6$$
$$\underline{3y - z = 6}$$
$$0 = 0$$

This is an identity which means there are an infinite number of solutions to the system.

Solving the third equation for y yields $3y - z = 6 \Rightarrow 3y = z + 6 \Rightarrow y = \dfrac{z + 6}{3}$. Substituting this

result for y in the first equation yields $x + \dfrac{z + 6}{3} - 2z = -6 \Rightarrow x = 2z - \dfrac{z + 6}{3} - 6 \Rightarrow$

$x + \dfrac{z + 6}{3} - 2z = -6 \Rightarrow x = 2z - \dfrac{z + 6}{3} - 6 \Rightarrow x = \dfrac{5z - 24}{3}.$

The solutions set is $\left\{ (x, y, z) \mid x = \dfrac{5z - 24}{3}, y = \dfrac{z + 6}{3}, \text{ and } z = z \right\}.$

67. For the graphs of parts (a) and (b), see Figures 67a and 67b.

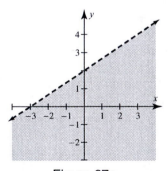

Figure 67a

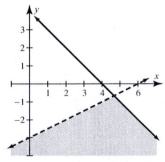

Figure 67b

69. $\begin{bmatrix} 1 & -2 & | & 1 & 0 \\ -3 & 4 & | & 0 & 1 \end{bmatrix} R_2 + 3R_1 \rightarrow \begin{bmatrix} 1 & -2 & | & 1 & 0 \\ 0 & -2 & | & 3 & 1 \end{bmatrix} R_1 - R_2 \rightarrow \begin{bmatrix} 1 & 0 & | & -2 & -1 \\ 0 & -2 & | & 3 & 1 \end{bmatrix} -\tfrac{1}{2}R_2 \rightarrow \begin{bmatrix} 1 & 0 & | & -2 & -1 \\ 0 & 1 & | & -\tfrac{3}{2} & -\tfrac{1}{2} \end{bmatrix}$

That is $A^{-1} = \begin{bmatrix} -2 & -1 \\ -1.5 & -0.5 \end{bmatrix}$.

71. $\det\left(\begin{bmatrix} -1 & 4 \\ 2 & 3 \end{bmatrix}\right) = -1(3) - 2(4) = -11$

$\det\left(\begin{bmatrix} 2 & 3 & -1 \\ 3 & -1 & 5 \\ 0 & 0 & -2 \end{bmatrix}\right) = 0(3(5) - (-1)(-1)) - 0(2(5) - 3(-1)) + -2(2(-1) - 3(3)) =$

$-2(-11) = 22$

73. If the focus is at $\left(\dfrac{3}{4}, 0\right)$ and the vertex at $(0, 0)$, the parabola opens to the right and $p = \dfrac{3}{4}$. Substituting in

$(y - h)^2 = 4p(x - k)$, we get $(y - 0)^2 = 4\left(\dfrac{3}{4}\right)(x - 0)$ or $y^2 = 3x$.

75. Since the foci and the vertices lie on the y-axis, the hyperbola has a vertical transverse axis with an equation

of the form $\dfrac{y^2}{a^2} - \dfrac{x^2}{b^2} = 1$. $F(0, \pm 13) \Rightarrow c = 13$ and $V(0, \pm 5) \Rightarrow a = 5$. $c^2 = a^2 + b^2 \Rightarrow$

$b^2 = c^2 - a^2 = 169 - 25 = 144 \Rightarrow b = 12$. The equation of the hyperbola is $\dfrac{y^2}{25} - \dfrac{x^2}{144} = 1$.

77. Let $a_n = f(n)$, where $f(n) = dn + c$. $a_1 = 4 \Rightarrow f(1) = 4$ and $a_3 = 12 \Rightarrow f(3) = 12$.

Thus, $d = \dfrac{f(3) - f(1)}{3 - 1} = \dfrac{12 - 4}{2} = 4$. Then, $f(n) = 4n + c$. $f(1) = 4(1) + c = 4 \Rightarrow c = 0$.

Thus, $a_n = f(n) = 4n$.

79. (a) The first term is $a_1 = 2$ and the last term is $a_{25} = 74$. There are 25 terms. To find the sum use:

$S_n = n\left(\dfrac{a_1 + a_n}{2}\right) \Rightarrow S_{25} = 25\left(\dfrac{2 + 74}{2}\right) = 950$. The sum is 950.

(b) The first term is $a_1 = 0.2$ and the common ratio is $r = \dfrac{1}{10}$ or 0.1. The sum is

$S = a_1\left(\dfrac{1}{1 - r}\right) \Rightarrow S = 0.2\left(\dfrac{1}{1 - (0.1)}\right) = \dfrac{0.2}{0.9} = \dfrac{2}{9}$.

81. There are $26 \cdot 26 \cdot 26 \cdot 10 \cdot 10 \cdot 10 \cdot 10 = 175{,}760{,}000$ such license plates.

83. $(2x - 1)^4 = \binom{4}{0}(2x)^4(-1)^0 + \binom{4}{1}(2x)^3(-1)^1 + \binom{4}{2}(2x)^2(-1)^2 + \binom{4}{3}(2x)^1(-1)^3 + \binom{4}{4}(2x)^0(-1)^4$

$= 16x^4 - 32x^3 + 24x^2 - 8x + 1$

85. The card must be one of the 16 cards (out of 52) that is either a heart or an ace. The probability is $\dfrac{16}{52} = \dfrac{4}{13}$.

87. There are 8 primes from 1 to 20 (2, 3, 5, 7, 11, 13, 17, 19). The probability is $\dfrac{8}{20} = \dfrac{2}{5}$.

89. From noon to 1:45 PM car A travelled $1.75(50) = 87.5$ miles and thus would be located $87.5 + 30 = 117.5$ miles north of the original location of car B. From noon to 1:45 PM car B travelled $1.75(50) = 87.5$ miles and thus would be located 87.5 miles east of its original location. The paths of the two cars form a right triangle with legs of length 117.5 and 87.5. The length of the hypotenuse is equal to the distance between the cars.

$c^2 = 117.5^2 + 87.5^2 \Rightarrow c^2 = 21,462.5 \Rightarrow c \approx 146.5$ miles

91. Let x represent the time spent jogging at 7 miles per hour and let y represent the time spent jogging at 9 miles per hour. Then the system of equations to be solved is $x + y = 1.3$ and $7x + 9y = 10.5$. Solving the first equation for y gives $y = 1.3 - x$. Substituting this expression in the second equation and solving for x yields

$7x + 9(1.3 - x) = 10.5 \Rightarrow -2x + 11.7 = 10.5 \Rightarrow -2x = -1.2 \Rightarrow x = 0.6$. From the first equation $0.6 + y = 1.3 \Rightarrow y = 0.7$. The jogger ran for 0.6 hours at 7 mph and 0.7 hours at 9 mph.

93. Let x represent the number of hours they work together. The the first person mows $\dfrac{x}{5}$ of the lawn in x hours and the second person mows $\dfrac{x}{4}$ of the lawn in x hours. Together they mow $\dfrac{x}{5} + \dfrac{x}{4}$ of the lawn in x hours. The job is complete when the fraction of the lawn mowed is 1. That is, we must solve $\dfrac{x}{5} + \dfrac{x}{4} = 1$.

The LCD is 20. The first step is to multiply each side of the equation by the LCD.

$20 \cdot \left(\dfrac{x}{5} + \dfrac{x}{4} \right) = 1 \cdot 20 \Rightarrow 4x + 5x = 20 \Rightarrow 9x = 20 \Rightarrow x = \dfrac{20}{9} \approx 2.22$ hours.

95. (a) The cost of 1 ticket is $1(405 - 1(5)) = 400$. The cost for 2 tickets is $2(405 - 2(5)) = \$790$. The cost for 3 tickets is $3(405 - 3(5)) = \$1170$. Following this pattern, the cost for x tickets is

$C(x) = x(405 - 5x)$.

(b) $x(405 - 5x) = 7000 \Rightarrow 405x - 5x^2 = 7000 \Rightarrow 5x^2 - 405x + 7000 = 0 \Rightarrow$

$x^2 - 81x + 1400 = 0 \Rightarrow (x - 25)(x - 56) = 0 \Rightarrow x = 25$ or $x = 56$. It costs \$7000 to buy either 25 or 56 tickets.

(c) The maximum for the function occurs at the vertex of its graph - a parabola. For $C(x) = -5x^2 + 405x$,

$x = -\dfrac{b}{2a} = -\dfrac{405}{2(-5)} = 40.5$. Since the number of tickets must be an integer, the maximum occurs at either 40 or 41 tickets being purchased. In either case, the maximum expenditure is

$40(405 - 5(40)) = \$8200$.

97. Substitute the point $(1.5, 0.5)$ into $x^2 = 4py$ and solve for p; $1.5^2 = 4p(0.5) \Rightarrow 2.25 = 2p \Rightarrow p = 1.125$.

The receiver should be 1.125 feet from the vertex.

99. There are 49 marbles that are not blue and there are a total of 77 marbles. The probability of drawing a marble that is not blue is $\dfrac{49}{77} = \dfrac{7}{11}$.

Chapter R: Basic Concepts from Algebra and Geometry

R.1: Formulas from Geometry

1. $A = LW = (15)(7) = 105 \text{ ft}^2$; $P = 2L + 2W = 2(15) + 2(7) = 30 + 14 = 44 \text{ ft}$

3. $A = LW = (100)(35) = 3500 \text{ m}^2$; $P = 2L + 2W = 2(100) + 2(35) = 200 + 70 = 270 \text{ m}$

5. $A = LW = (3x)(y) = 3xy$ square units; $P = 2L + 2W = 2(3x) + 2(y) = 6x + 2y$ units

7. Since the width is half of the length, $L = 2W$.

    $A = LW = (2W)(W) = 2W^2$ square units; $P = 2L + 2W = 2(2W) + 2(W) = 6W$ units

9. Since the length equals the width plus 5, $L = W + 5$.

    $A = LW = (W + 5)(W) = W(W + 5)$ square units; $P = 2L + 2W = 2(W + 5) + 2(W) = 4W + 10$ units

11. $A = \frac{1}{2}bh = \left(\frac{1}{2}\right)(8)(5) = 20 \text{ cm}^2$

13. $A = \frac{1}{2}bh = \left(\frac{1}{2}\right)(5)(8) = 20 \text{ in.}^2$

15. $A = \frac{1}{2}bh = \left(\frac{1}{2}\right)(10.1)(730) = 3686.5 \text{ m}^2$

17. $A = \frac{1}{2}bh = \left(\frac{1}{2}\right)(2x)(6x) = 6x^2$ square units

19. $A = \frac{1}{2}bh = \left(\frac{1}{2}\right)(z)(5z) = \frac{5}{2}z^2$ square units

21. $C = 2\pi r = 2\pi(4) = 8\pi \approx 25.1 \text{ m}$; $A = \pi r^2 = \pi(4)^2 = 16\pi \approx 50.3 \text{ m}^2$

23. $C = 2\pi r = 2\pi(19) = 38\pi \approx 119.4 \text{ in.}$; $A = \pi r^2 = \pi(19)^2 = 361\pi \approx 1134.1 \text{ in.}^2$

25. $C = 2\pi r = 2\pi(2x) = 4\pi x$ units; $A = \pi r^2 = \pi(2x)^2 = 4\pi x^2$ square units

27. $c^2 = a^2 + b^2 \Rightarrow c^2 = (60)^2 + (11)^2 \Rightarrow c^2 = 3721 \Rightarrow c = \sqrt{3721} \Rightarrow c = 61 \text{ ft}$

    $P = 60 + 11 + 61 = 132 \text{ ft}$

29. $c^2 = a^2 + b^2 \Rightarrow b^2 = c^2 - a^2 \Rightarrow b^2 = (13)^2 - (5)^2 \Rightarrow b^2 = 144 \Rightarrow b = \sqrt{144} \Rightarrow b = 12 \text{ cm}$

    $P = 5 + 12 + 13 = 30 \text{ cm}$

31. $c^2 = a^2 + b^2 \Rightarrow a^2 = c^2 - b^2 \Rightarrow a^2 = (10)^2 - (7)^2 \Rightarrow a^2 = 51 \Rightarrow a = \sqrt{51} \Rightarrow a \approx 7.1 \text{ mm}$

    $P \approx 7.1 + 7 + 10 = 24.1 \text{ mm}$

33. $A = \frac{1}{2}bh = \left(\frac{1}{2}\right)(6)(3) = 9 \text{ ft}^2$

35. $c^2 = a^2 + b^2 \Rightarrow b^2 = c^2 - a^2 \Rightarrow b^2 = (15)^2 - (11)^2 \Rightarrow b^2 = 104 \Rightarrow b = \sqrt{104} \Rightarrow b \approx 10.2 \text{ in.}$

    $A = \frac{1}{2}bh = \left(\frac{1}{2}\right)(11)(\sqrt{104}) \approx 56.1 \text{ in.}^2$

37. $V = LWH = (4)(3)(2) = 24 \text{ ft}^3$; $S = 2LW + 2WH + 2LH = 2(4)(3) + 2(3)(2) + 2(4)(2) = 52 \text{ ft}^2$

39. Since 1 foot is 12 inches, $V = LWH = (4.5)(4)(12) = 216 \text{ in.}^3$

    $S = 2LW + 2WH + 2LH = 2(4.5)(4) + 2(4)(12) + 2(4.5)(12) = 240 \text{ in.}^2$

41. $V = LWH = (3x)(2x)(x) = 6x^3$ cubic units

   $S = 2LW + 2WH + 2LH = 2(3x)(2x) + 2(2x)(x) + 2(3x)(x) = 22x^2$ square units

43. $V = LWH = (x)(2y)(3z) = 6xyz$ cubic units

   $S = 2LW + 2WH + 2LH = 2(x)(2y) + 2(2y)(3z) + 2(x)(3z) = 4xy + 12yz + 6xz$ square units

45. Since the length is twice W and the height is half W, $L = 2W$ and $H = 0.5W$

   $V = LWH = (2W)(W)(0.5W) = W^3$

47. $V = \frac{4}{3}\pi r^3 \Rightarrow V = \frac{4}{3}\pi(3)^3 = 36\pi \approx 113.1 \text{ ft}^3$; $S = 4\pi r^2 \Rightarrow S = 4\pi(3)^2 = 36\pi \approx 113.1 \text{ ft}^2$

49. $V = \frac{4}{3}\pi r^3 \Rightarrow V = \frac{4}{3}\pi(3.2)^3 \approx 43.7\pi \approx 137.3 \text{ m}^3$; $S = 4\pi r^2 \Rightarrow S = 4\pi(3.2)^2 \approx 41.0\pi \approx 128.8 \text{ m}^2$

51. $V = \pi r^2 h \Rightarrow V = \pi(0.5)^2(2) = 0.5\pi \approx 1.6 \text{ ft}^3$; $S_{side} = 2\pi rh \Rightarrow S_{side} = 2\pi(0.5)(2) = 2\pi \approx 6.3 \text{ ft}^2$

   $S_{total} = 2\pi rh + 2\pi r^2 \Rightarrow S_{total} = 2\pi(0.5)(2) + 2\pi(0.5)^2 = 2.5\pi \approx 7.9 \text{ ft}^2$

53. Since h is twice r, $h = 2(12) = 24$, thus $V = \pi r^2 h \Rightarrow V = \pi(12)^2(24) = 3456\pi \approx 10{,}857.3 \text{ mm}^3$

   $S_{side} = 2\pi rh \Rightarrow S_{side} = 2\pi(12)(24) = 576\pi \approx 1809.6 \text{ mm}^2$

   $S_{total} = 2\pi rh + 2\pi r^2 \Rightarrow S_{total} = 2\pi(12)(24) + 2\pi(12)^2 = 864\pi \approx 2714.3 \text{ mm}^2$

55. $V = \frac{1}{3}\pi r^2 h \Rightarrow V = \frac{1}{3}\pi(5)^2(6) \approx 157.1 \text{ cm}^3$;

   $S = \pi r \sqrt{r^2 + h^2} \Rightarrow S = \pi(5)\sqrt{(5)^2 + (6)^2} \approx 122.7 \text{ cm}^2$

57. Since 24 inches is 2 feet,

   $V = \frac{1}{3}\pi r^2 h \Rightarrow V = \frac{1}{3}\pi(2)^2(3) \approx 12.6 \text{ ft}^3$; $S = \pi r \sqrt{r^2 + h^2} \Rightarrow S = \pi(2)\sqrt{(2)^2 + (3)^2} \approx 22.7 \text{ ft}^2$

59. Since h is three times r, $h = 3(2.4) = 7.2$.

   $V = \frac{1}{3}\pi r^2 h \Rightarrow V = \frac{1}{3}\pi(2.4)^2(7.2) \approx 43.4 \text{ ft}^3$;

   $S = \pi r \sqrt{r^2 + h^2} \Rightarrow S = \pi(2.4)\sqrt{(2.4)^2 + (7.2)^2} \approx 57.2 \text{ ft}^2$

61. $\frac{x}{4} = \frac{5}{3} \Rightarrow 3x = 20 \Rightarrow x = \frac{20}{3} \approx 6.7$

63. $\frac{x}{7} = \frac{9}{6} \Rightarrow 6x = 63 \Rightarrow x = \frac{63}{6} = \frac{21}{2} = 10.5$

R.2: Integer Exponents

1. No. $2^3 = 8$ and $3^2 = 9$

3. $\frac{1}{7^n}$

5. 5^{m-n}

7. 2^{mk}

9. 5000

11. 2^3

13. 4^4

15. 3^0

17. $5^3 = 125$

19. $-2^4 = -16$

21. $5^0 = 1$

23. $\left(\dfrac{2}{3}\right)^3 = \dfrac{2^3}{3^3} = \dfrac{2 \cdot 2 \cdot 2}{3 \cdot 3 \cdot 3} = \dfrac{8}{27}$

25. $\left(-\dfrac{1}{2}\right)^4 = \left(-\dfrac{1}{2}\right) \cdot \left(-\dfrac{1}{2}\right) \cdot \left(-\dfrac{1}{2}\right) \cdot \left(-\dfrac{1}{2}\right) = \dfrac{1}{16}$

27. $4^{-3} = \dfrac{1}{4^3} = \dfrac{1}{64}$

29. $\dfrac{1}{2^{-4}} = 2^4 = 16$

31. $\left(\dfrac{3}{4}\right)^{-3} = \left(\dfrac{4}{3}\right)^3 = \left(\dfrac{4}{3}\right) \cdot \left(\dfrac{4}{3}\right) \cdot \left(\dfrac{4}{3}\right) = \dfrac{64}{27}$

33. $\dfrac{3^{-2}}{2^{-3}} = \dfrac{2^3}{3^2} = \dfrac{8}{9}$

35. $6^3 \cdot 6^{-4} = 6^{3+(-4)} = 6^{-1} = \dfrac{1}{6}$

37. $2x^2 \cdot 3x^{-3} \cdot x^4 = 6x^{2+(-3)+4} = 6x^3$

39. $10^0 \cdot 10^6 \cdot 10^2 = 10^{0+6+2} = 10^8 = 100{,}000{,}000$

41. $5^{-2} \cdot 5^3 \cdot 2^{-4} \cdot 2^3 = 5^{-2+3} \cdot 2^{-4+3} = 5^1 \cdot 2^{-1} = 5 \cdot \dfrac{1}{2} = \dfrac{5}{2}$

43. $2a^3 \cdot b^2 \cdot a^{-4} \cdot 4b^{-5} = 2a^{3+(-4)} \cdot 4b^{2+(-5)} = 8a^{-1}b^{-3} = \dfrac{8}{ab^3}$

45. $\dfrac{5^4}{5^2} = 5^{4-2} = 5^2 = 25$

47. $\dfrac{a^{-3}}{a^2 \cdot a} = a^{-3-(2+1)} = a^{-6} = \dfrac{1}{a^6}$

49. $\dfrac{24x^3}{6x} = \dfrac{24}{6}x^{3-1} = 4x^2$

51. $\dfrac{12a^2b^3}{18a^4b^2} = \dfrac{12}{18}a^{2-4} \cdot b^{3-2} = \dfrac{2}{3}a^{-2}b^1 = \dfrac{2b}{3a^2}$

53. $\dfrac{21x^{-3}y^4}{7x^4y^{-2}} = \dfrac{21}{7}x^{-3-4} \cdot y^{4-(-2)} = 3x^{-7}y^6 = \dfrac{3y^6}{x^7}$

55. $(5^{-1})^3 = 5^{-1(3)} = 5^{-3} = \dfrac{1}{5^3} = \dfrac{1}{125}$

57. $(y^4)^{-2} = y^{4(-2)} = y^{-8} = \dfrac{1}{y^8}$

59. $(4y^2)^3 = 4^3 \cdot y^{2 \cdot 3} = 64y^6$

61. $\left(\dfrac{4}{x}\right)^3 = \dfrac{4^3}{x^3} = \dfrac{64}{x^3}$

63. $\left(\dfrac{2x}{z^4}\right)^{-5} = \left(\dfrac{z^4}{2x}\right)^5 = \dfrac{(z^4)^5}{(2x)^5} = \dfrac{z^{4 \cdot 5}}{2^5 \cdot x^5} = \dfrac{z^{20}}{32x^5}$

65. $\dfrac{2}{(ab)^{-1}} = 2(ab)^1 = 2ab$

67. $\dfrac{2^{-3}}{2t^{-2}} = \dfrac{t^2}{2(2^3)} = \dfrac{t^2}{16}$

69. $\dfrac{6a^2b^{-3}}{4ab^{-2}} = \dfrac{6}{4} \cdot a^{2-1}b^{-3-(-2)} = \dfrac{3}{2} \cdot ab^{-1} = \dfrac{3a}{2b}$

71. $\dfrac{5r^2st^{-3}}{25rs^{-2}t^2} = \dfrac{5}{25} \cdot r^{2-1}s^{1-(-2)}t^{-3-2} = \dfrac{1}{5} \cdot rs^3t^{-5} = \dfrac{rs^3}{5t^5}$

73. $(3x^2y^{-3})^{-2} = \dfrac{1}{(3x^2y^{-3})^2} = \dfrac{1}{3^2(x^2)^2(y^{-3})^2} = \dfrac{1}{9x^{2\cdot2}y^{-3\cdot2}} = \dfrac{1}{9x^4y^{-6}} = \dfrac{y^6}{9x^4}$

75. $\dfrac{(d^3)^{-2}}{(d^{-2})^3} = \dfrac{d^{3\cdot(-2)}}{d^{-2\cdot3}} = \dfrac{d^{-6}}{d^{-6}} = d^{-6-(-6)} = d^0 = 1$

77. $\left(\dfrac{3t^2}{2t^{-1}}\right)^3 = \dfrac{(3t^2)^3}{(2t^{-1})^3} = \dfrac{3^3(t^2)^3}{2^3(t^{-1})^3} = \dfrac{27t^{2\cdot3}}{8t^{-1\cdot3}} = \dfrac{27t^6}{8t^{-3}} = \dfrac{27}{8} \cdot t^{6-(-3)} = \dfrac{27}{8} \cdot t^9 = \dfrac{27t^9}{8}$

79. $\dfrac{(-m^2n^{-1})^{-2}}{(mn)^{-1}} = \dfrac{(mn)^1}{(-m^2n^{-1})^2} = \dfrac{mn}{(-m^2)^2(n^{-1})^2} = \dfrac{mn}{m^{2\cdot2}n^{-1\cdot2}} = \dfrac{mn}{m^4n^{-2}} = m^{1-4}n^{1-(-2)} = m^{-3}n^3 = \dfrac{n^3}{m^3}$

81. $\left(\dfrac{2a^3}{6b}\right)^4 = \left(\dfrac{a^3}{3b}\right)^4 = \dfrac{(a^3)^4}{3^4b^4} = \dfrac{a^{3\cdot4}}{81b^4} = \dfrac{a^{12}}{81b^4}$

83. $\dfrac{8x^{-3}y^{-2}}{4x^{-2}y^{-4}} = \dfrac{8}{4}x^{-3-(-2)}y^{-2-(-4)} = 2x^{-1}y^2 = \dfrac{2y^2}{x}$

85. $\dfrac{(r^2t^2)^{-2}}{(r^3t)^{-1}} = \dfrac{r^3t}{(r^2t^2)^2} = \dfrac{r^3t}{(r^2)^2(t^2)^2} = \dfrac{r^3t}{r^{2\cdot2}t^{2\cdot2}} = \dfrac{r^3t}{r^4t^4} = r^{3-4}t^{1-4} = r^{-1}t^{-3} = \dfrac{1}{rt^3}$

87. $\dfrac{4x^{-2}y^3}{(2x^{-1}y)^2} = \dfrac{4x^{-2}y^3}{2^2(x^{-1})^2y^2} = \dfrac{4x^{-2}y^3}{4x^{-1\cdot2}y^2} = \dfrac{4x^{-2}y^3}{4x^{-2}y^2} = \left(\dfrac{4x^{-2}}{4x^{-2}}\right)y^{3-2} = 1y = y$

89. $\left(\dfrac{15r^2t}{3r^{-3}t^4}\right)^3 = \left(\dfrac{15}{3}r^{2-(-3)}t^{1-4}\right)^3 = (5r^5t^{-3})^3 = 5^3(r^5)^3(t^{-3})^3 = 125r^{5\cdot3}t^{-3\cdot3} = 125r^{15}t^{-9} = \dfrac{125r^{15}}{t^9}$

R.3: Polynomial Expressions

1. $3x^3 + 5x^3 = 8x^3$

3. $5y^7 - 8y^7 = -3y^7$

5. $5x^2 + 8x + x^2 = 6x^2 + 8x$

7. $9x^2 - x + 4x - 6x^2 = 3x^2 + 3x$

9. $x^2 + 9x - 2 + 4x^2 + 4x = 5x^2 + 13x - 2$

11. $7y + 9x^2y - 5y + x^2y = 2y + 10x^2y = 10x^2y + 2y$

13. The degree is 2. The leading coefficient is 5.

15. The degree is 3. The leading coefficient is $-\dfrac{2}{5}$.

17. The degree is 5. The leading coefficient is 1.

19. $(5x + 6) + (-2x + 6) = 3x + 12$

21. $(2x^2 - x + 7) + (-2x^2 + 4x - 9) = 3x - 2$

23. $(4x) + (1 - 4.5x) = -0.5x + 1$

25. $(x^4 - 3x^2 - 4) + \left(-8x^4 + x^2 - \frac{1}{2}\right) = -7x^4 - 2x^2 - \frac{9}{2}$

27. $(2z^3 + 5z - 6) + (z^2 - 3z + 2) = 2z^3 + z^2 + 2z - 4$

29. $-(7x^3) = -7x^3$

31. $-(19z^5 - 5z^2 + 3z) = -19z^5 + 5z^2 - 3z$

33. $-(z^4 - z^2 - 9) = -z^4 + z^2 + 9$

35. $(5x - 3) - (2x + 4) = 5x - 3 - 2x - 4 = 3x - 7$

37. $(x^2 - 3x + 1) - (-5x^2 + 2x - 4) = x^2 - 3x + 1 + 5x^2 - 2x + 4 = 6x^2 - 5x + 5$

39. $(4x^4 + 2x^2 - 9) - (x^4 - 2x^2 - 5) = 4x^4 + 2x^2 - 9 - x^4 + 2x^2 + 5 = 3x^4 + 4x^2 - 4$

41. $(x^4 - 1) - (4x^4 + 3x + 7) = x^4 - 1 - 4x^4 - 3x - 7 = -3x^4 - 3x - 8$

43. $5x(x - 5) = 5x^2 - 25x$

45. $-5(3x + 1) = -15x - 5$

47. $5(y + 2) = 5y + 10$

49. $-2(5x + 9) = -10x - 18$

51. $(y - 3)6y = 6y^2 - 18y$

53. $-4(5x - y) = -20x + 4y$

55. $(y + 5)(y - 7) = y^2 - 7y + 5y - 35 = y^2 - 2y - 35$

57. $(3 - 2x)(3 + x) = 9 + 3x - 6x - 2x^2 = -2x^2 - 3x + 9$

59. $(-2x + 3)(x - 2) = -2x^2 + 4x + 3x - 6 = -2x^2 + 7x - 6$

61. $\left(x - \frac{1}{2}\right)\left(x + \frac{1}{4}\right) = x^2 + \frac{1}{4}x - \frac{1}{2}x - \frac{1}{8} = x^2 - \frac{1}{4}x - \frac{1}{8}$

63. $(x^2 + 1)(2x^2 - 1) = 2x^4 - x^2 + 2x^2 - 1 = 2x^4 + x^2 - 1$

65. $(x + y)(x - 2y) = x^2 - 2xy + xy - 2y^2 = x^2 - xy - 2y^2$

67. $3x(2x^2 - x - 1) = 6x^3 - 3x^2 - 3x$

69. $-x(2x^4 - x^2 + 10) = -2x^5 + x^3 - 10x$

71. $(2x^2 - 4x + 1)(3x^2) = 6x^4 - 12x^3 + 3x^2$

73. $(x + 1)(x^2 + 2x - 3) = x^3 + 2x^2 - 3x + x^2 + 2x - 3 = x^3 + 3x^2 - x - 3$

75. $(2 - 3x)(5 - 2x)(x^2 - 1) \Rightarrow [(2 - 3x)(5 - 2x)](x^2 - 1) = (10 - 19x + 6x^2)(x^2 - 1) =$
    $10x^2 - 19x^3 + 6x^4 - 10 + 19x - 6x^2 = 6x^4 - 19x^3 + 4x^2 + 19x - 10$

77. $(x^2 + 2)(3x - 2) = 3x^3 - 2x^2 + 6x - 4$

79. $(x - 7)(x + 7) = x^2 - 7^2 = x^2 - 49$

81. $(3x + 4)(3x - 4) = (3x)^2 - 4^2 = 9x^2 - 16$

83. $(2x - 3y)(2x + 3y) = (2x)^2 - (3y)^2 = 4x^2 - 9y^2$

85. $(x + 4)^2 = x^2 + 2(4)x + 4^2 = x^2 + 8x + 16$

87. $(2x + 1)^2 = 4x^2 + 2(2x) + 1 = 4x^2 + 4x + 1$

89. $(x - 1)^2 = x^2 - 2(x) + 1 = x^2 - 2x + 1$

91. $(2 - 3x)^2 = 4 - 2(6x) + 9x^2 = 4 - 12x + 9x^2$

93. $3x(x + 1)(x - 1) = 3x(x^2 - 1) = 3x^3 - 3x$

95. $(2 - 5x^2)(2 + 5x^2) = 2^2 - (5x^2)^2 = 4 - 25x^4$

R.4: Factoring Polynomials

1. $10x - 15 = 5(2x - 3)$

3. $2x^3 - 5x = x(2x^2 - 5)$

5. $8x^3 - 4x^2 + 16x = 4x(2x^2 - x + 4)$

7. $5x^4 - 15x^3 + 15x^2 = 5x^2(x^2 - 3x + 3)$

9. $15x^3 + 10x^2 - 30x = 5x(3x^2 + 2x - 6)$

11. $6r^5 - 8r^4 + 12r^3 = 2r^3(3r^2 - 4r + 6)$

13. $8x^2y^2 - 24x^2y^3 = 8x^2y^2(1 - 3y)$

15. $18mn^2 - 12m^2n^3 = 6mn^2(3 - 2mn)$

17. $-4a^2 - 2ab + 6ab^2 = -2a(2a + b - 3b^2)$

19. $x^3 + 3x^2 + 2x + 6 = x^2(x + 3) + 2(x + 3) = (x + 3)(x^2 + 2)$

21. $6x^3 - 4x^2 + 9x - 6 = 2x^2(3x - 2) + 3(3x - 2) = (3x - 2)(2x^2 + 3)$

23. $z^3 - 5z^2 + z - 5 = z^2(z - 5) + 1(z - 5) = (z - 5)(z^2 + 1)$

25. $y^4 + 2y^3 - 5y^2 - 10y = y(y^3 + 2y^2 - 5y - 10) = y[y^2(y + 2) - 5(y + 2)] = y(y + 2)(y^2 - 5)$

27. $2x^3 - 3x^2 + 2x - 3 = x^2(2x - 3) + 1(2x - 3) = (x^2 + 1)(2x - 3)$

29. $2x^4 - x^3 + 4x - 2 = x^3(2x - 1) + 2(2x - 1) = (x^3 + 2)(2x - 1)$

31. $ab - 3a + 2b - 6 = a(b - 3) + 2(b - 3) = (a + 2)(b - 3)$

33. $x^2 + 7x + 10 = (x + 2)(x + 5)$

35. $x^2 + 8x + 12 = (x + 2)(x + 6)$

37. $z^2 + z - 42 = (z - 6)(z + 7)$

39. $z^2 + 11z + 24 = (z + 3)(z + 8)$

41. $24x^2 + 14x - 3 = (4x + 3)(6x - 1)$

43. $6x^2 - x - 2 = (2x + 1)(3x - 2)$

45. $1 + x - 2x^2 = (1 - x)(1 + 2x)$

47. $20 + 7x - 6x^2 = (5 - 2x)(4 + 3x)$

49. $5x^3 + x^2 - 6x = x(5x^2 + x - 6) = x(x - 1)(5x + 6)$

51. $3x^3 + 12x^2 + 9x = 3x(x^2 + 4x + 3) = 3x(x + 3)(x + 1)$

53. $2x^2 - 14x + 20 = 2(x^2 - 7x + 10) = 2(x - 5)(x - 2)$

55. $60t^4 + 230t^3 - 40t^2 = 10t^2(6t^2 + 23t - 4) = 10t^2(t + 4)(6t - 1)$

57. $4m^3 + 10m^2 - 6m = 2m(2m^2 + 5m - 3) = 2m(m + 3)(2m - 1)$

59. $x^2 - 25 = (x - 5)(x + 5)$

61. $4x^2 - 25 = (2x - 5)(2x + 5)$

63. $36x^2 - 100 = 4(9x^2 - 25) = 4(3x - 5)(3x + 5)$

65. $64z^2 - 25z^4 = z^2(64 - 25z^2) = z^2(8 - 5z)(8 + 5z)$

67. $16x^4 - y^4 = (4x^2 - y^2)(4x^2 + y^2) = (2x - y)(2x + y)(4x^2 + y^2)$

69. The sum of two squares does not factor using real numbers.

71. $4 - r^2t^2 = (2 - rt)(2 + rt)$

73. $(x - 1)^2 - 16 = ((x - 1) - 4)((x - 1) + 4) = (x - 5)(x + 3)$

75. $4 - (z + 3)^2 = (2 - (z + 3))(2 + (z + 3)) = (-1 - z)(5 + z) = -(z + 1)(z + 5)$

77. $x^2 + 2x + 1 = (x + 1)^2$

79. $4x^2 + 20x + 25 = (2x + 5)^2$

81. $x^2 - 12x + 36 = (x - 6)^2$

83. $9z^3 - 6z^2 + z = z(9z^2 - 6z + 1) = z(3z - 1)^2$

85. $9y^3 + 30y^2 + 25y = y(9y^2 + 30y + 25) = y(3y + 5)^2$

87. $4x^2 - 12xy + 9y^2 = (2x - 3y)^2$

89. $9a^3b - 12a^2b + 4ab = ab(9a^2 - 12a + 4) = ab(3a - 2)^2$

91. $x^3 - 1 = (x - 1)(x^2 + 1x + 1^2) = (x - 1)(x^2 + x + 1)$

93. $y^3 + z^3 = (y + z)(y^2 - yz + z^2)$

95. $8x^3 - 27 = (2x)^3 - 3^3 = (2x - 3)((2x)^2 + 2x(3) + 3^2) = (2x - 3)(4x^2 + 6x + 9)$

97. $x^4 + 125x = x(x^3 + 5^3) = x(x + 5)(x^2 - 5x + 5^2) = x(x + 5)(x^2 - 5x + 25)$

99. $8r^6 - t^3 = (2r^2 - t)((2r^2)^2 + 2r^2t + t^2) = (2r^2 - t)(4r^4 + 2r^2t + t^2)$

101. $10m^9 - 270n^6 = 10(m^9 - 27n^6) = 10(m^3 - 3n^2)((m^3)^2 + 3m^3n^2 + (3n^2)^2) =$

      $10(m^3 - 3n^2)(m^6 + 3m^3n^2 + 9n^4)$

103. $16x^2 - 25 = (4x)^2 - 5^2 = (4x - 5)(4x + 5)$

105. $x^3 - 64 = x^3 - 4^3 = (x - 4)(x^2 + 4x + 4^2) = (x - 4)(x^2 + 4x + 16)$

107. $x^2 + 16x + 64 = (x + 8)(x + 8) = (x + 8)^2$

109. $5x^2 - 38x - 16 = (x - 8)(5x + 2)$

111. $x^4 + 8x = x(x^3 + 8) = x(x + 2)(x^2 - 2x + 4)$

113. $64x^3 + 8y^3 = 8(8x^3 + y^3) = 8(2x + y)(4x^2 - 2xy + y^2)$

115. $3x^2 - 5x - 8 = (x + 1)(3x - 8)$

117. $7a^3 + 20a^2 - 3a = a(7a^2 + 20a - 3) = a(a + 3)(7a - 1)$

119. $2x^3 - x^2 + 6x - 3 = x^2(2x - 1) + 3(2x - 1) = (x^2 + 3)(2x - 1)$

121. $2x^4 - 5x^3 - 25x^2 = x^2(2x^2 - 5x - 25) = x^2(2x + 5)(x - 5)$

123. $2x^4 + 5x^2 + 3 = (2x^2 + 3)(x^2 + 1)$

125. $x^3 + 3x^2 + x + 3 = x^2(x + 3) + (x + 3) = (x + 3)(x^2 + 1)$

127. $5x^3 - 5x^2 + 10x - 10 = 5x^2(x - 1) + 10(x - 1) = (x - 1)(5x^2 + 10) = 5(x^2 + 2)(x - 1)$

129. $ax + bx - ay - by = x(a + b) - y(a + b) = (a + b)(x - y)$

131. $18x^2 + 12x + 2 = 2(9x^2 + 6x + 1) = 2(3x + 1)^2$

133. $-4x^3 + 24x^2 - 36x = -4x(x^2 - 6x + 9) = -4x(x - 3)^2$

135. $27x^3 - 8 = (3x - 2)(9x^2 + 6x + 4)$

137. $-x^4 - 8x = -x(x^3 + 8) = -x(x + 2)(x^2 - 2x + 4)$

139. $x^4 - 2x^3 - x + 2 = x^3(x - 2) - 1(x - 2) = (x - 2)(x^3 - 1) = (x - 2)(x - 1)(x^2 + x + 1)$

141. $r^4 - 16 = (r^2 - 4)(r^2 + 4) = (r + 2)(r - 2)(r^2 + 4)$

143. $25x^2 - 4a^2 = (5x - 2a)(5x + 2a)$

145. $(2x^4 - 2y^4) = 2(x^4 - y^4) = 2(x^2 - y^2)(x^2 + y^2) = 2(x - y)(x + y)(x^2 + y^2)$

147. $9x^3 + 6x^2 - 3x = 3x(3x^2 + 2x - 1) = 3x(3x - 1)(x + 1)$

149. $(z - 2)^2 - 9 = (z - 2 - 3)(z - 2 + 3) = (z - 5)(z + 1)$

151. $3x^5 - 27x^3 + 3x^2 - 27 = 3(x^5 - 9x^3 + x^2 - 9) = 3[x^3(x^2 - 9) + (x^2 - 9)] =$
$3(x^2 - 9)(x^3 + 1) = 3(x + 3)(x - 3)(x + 1)(x^2 - x + 1)$

153. $(x + 2)^2(x + 4)^4 + (x + 2)^3(x + 4)^3 = (x + 2)^2(x + 4)^3[x + 4 + x + 2] =$
$(x + 2)^2(x + 4)^3(2x + 6) = 2(x + 2)^2(x + 4)^3(x + 3)$

155. $(6x + 1)(8x - 3)^4 - (6x + 1)^2(8x - 3)^3 = (6x + 1)(8x - 3)^3[8x - 3 - (6x + 1)] =$
$(6x + 1)(8x - 3)^2(2x - 4) = 2(6x + 1)(8x - 3)^3(x - 2)$

157. $4x^2(5x - 1)^5 + 2x(5x - 1)^6 = 2x(5x - 1)^5[2x + 5x - 1] = 2x(5x - 1)^5(7x - 1)$

R.5: Rational Expressions

1. $\dfrac{10x^3}{5x^2} = 2x$

3. $\dfrac{(x - 5)(x + 5)}{x - 5} = x + 5$

5. $\dfrac{x^2 - 16}{x - 4} = \dfrac{(x - 4)(x + 4)}{x - 4} = x + 4$

7. $\dfrac{x + 3}{2x^2 + 5x - 3} = \dfrac{x + 3}{(2x - 1)(x + 3)} = \dfrac{1}{2x - 1}$

9. $-\dfrac{z + 2}{4z + 8} = -\dfrac{z + 2}{4(z + 2)} = -\dfrac{1}{4}$

11. $\dfrac{x^2 + 2x}{x^2 + 3x + 2} = \dfrac{x(x + 2)}{(x + 1)(x + 2)} = \dfrac{x}{x + 1}$

13. $\dfrac{a^3 + b^3}{a + b} = \dfrac{(a + b)(a^2 - ab + b^2)}{a + b} = a^2 - ab + b^2$

15. $\dfrac{5}{8} \cdot \dfrac{4}{15} = \dfrac{1}{6}$

17. $\dfrac{5}{6} \cdot \dfrac{3}{10} \cdot \dfrac{8}{3} = \dfrac{2}{3}$

19. $\dfrac{4}{7} \div \dfrac{8}{7} = \dfrac{4}{7} \cdot \dfrac{7}{8} = \dfrac{1}{2}$

21. $\dfrac{1}{2} \div \dfrac{3}{4} \div \dfrac{5}{6} = \left(\dfrac{1}{2} \cdot \dfrac{4}{3} \right) \div \dfrac{5}{6} = \dfrac{2}{3} \cdot \dfrac{6}{5} = \dfrac{4}{5}$

23. $\dfrac{3}{8} + \dfrac{5}{8} = \dfrac{8}{8} = 1$

25. $\dfrac{3}{7} - \dfrac{4}{7} = -\dfrac{1}{7}$

27. $\dfrac{2}{3} + \dfrac{5}{11} = \dfrac{22}{33} + \dfrac{15}{33} = \dfrac{37}{33}$

29. $\dfrac{4}{5} - \dfrac{1}{10} = \dfrac{8}{10} - \dfrac{1}{10} = \dfrac{7}{10}$

31. $\dfrac{1}{3} + \dfrac{3}{4} - \dfrac{3}{7} = \dfrac{28}{84} + \dfrac{63}{84} - \dfrac{36}{84} = \dfrac{55}{84}$

33. $\dfrac{1}{x^2} \cdot \dfrac{3x}{2} = \dfrac{3}{2x}$

35. $\dfrac{5x}{3} \div \dfrac{10x}{6} = \dfrac{5x}{3} \cdot \dfrac{6}{10x} = \dfrac{2}{2} = 1$

37. $\dfrac{x+1}{2x-5} \cdot \dfrac{x}{x+1} = \dfrac{x}{2x-5}$

39. $\dfrac{(x-5)(x+3)}{3x-1} \cdot \dfrac{x(3x-1)}{(x-5)} = \dfrac{x+3}{1} \cdot \dfrac{x}{1} = x(x+3)$

41. $\dfrac{x^2 - 2x - 35}{2x^3 - 3x^2} \cdot \dfrac{x^3 - x^2}{2x - 14} = \dfrac{(x-7)(x+5)}{x^2(2x-3)} \cdot \dfrac{x^2(x-1)}{2(x-7)} = \dfrac{x+5}{2x-3} \cdot \dfrac{x-1}{2} = \dfrac{(x-1)(x+5)}{2(2x-3)}$

43. $\dfrac{6b}{b+2} \div \dfrac{3b^4}{2b+4} = \dfrac{6b}{b+2} \cdot \dfrac{2(b+2)}{3b^4} = \dfrac{2}{1} \cdot \dfrac{2}{b^3} = \dfrac{4}{b^3}$

45. $\dfrac{3a+1}{a^7} \div \dfrac{a+1}{3a^8} = \dfrac{3a+1}{a^7} \cdot \dfrac{3a^8}{a+1} = \dfrac{3a+1}{1} \cdot \dfrac{3a}{a+1} = \dfrac{3a(3a+1)}{a+1}$

47. $\dfrac{x+5}{x^3 - x} \div \dfrac{x^2 - 25}{x^3} = \dfrac{x+5}{x(x^2-1)} \cdot \dfrac{x^3}{(x-5)(x+5)} = \dfrac{1}{x^2-1} \cdot \dfrac{x^2}{x-5} = \dfrac{x^2}{(x-5)(x^2-1)}$

49. $\dfrac{x-2}{x^3-x} \div \dfrac{x^2-2x}{x^2-1} = \dfrac{x-2}{x(x+1)(x-1)} \cdot \dfrac{(x+1)(x-1)}{x(x-2)} = \dfrac{1}{x^2}$

51. $\dfrac{x^2 - 3x + 2}{x^2 + 5x + 6} \div \dfrac{x^2 + x - 2}{x^2 + 2x - 3} = \dfrac{(x-2)(x-1)}{(x+3)(x+2)} \cdot \dfrac{(x+3)(x-1)}{(x+2)(x-1)} = \dfrac{(x-2)(x-1)}{(x+2)^2}$

53. $\dfrac{x^2 - 4}{x^2 + x - 2} \div \dfrac{x-1}{x-2} = \dfrac{(x+2)(x-2)}{(x+2)(x-1)} \cdot \dfrac{x-1}{x-2} = 1$

55. $\dfrac{3y}{x^2} \div \dfrac{y^2}{x} \div \dfrac{y}{5x} = \left(\dfrac{3y}{x^2} \cdot \dfrac{x}{y^2} \right) \div \dfrac{y}{5x} = \dfrac{3}{xy} \cdot \dfrac{5x}{y} = \dfrac{15}{y^2}$

57. $\dfrac{x-3}{x-1} \div \dfrac{x^2}{x-1} \div \dfrac{x-3}{x} = \left(\dfrac{x-3}{x-1} \cdot \dfrac{x-1}{x^2} \right) \div \dfrac{x-3}{x} = \dfrac{x-3}{x^2} \cdot \dfrac{x}{x-3} = \dfrac{1}{x}$

59. Since $12 = 2 \cdot 2 \cdot 3$ and $18 = 2 \cdot 3 \cdot 3$, the LCM is $2 \cdot 2 \cdot 3 \cdot 3 = 36$.

61. Since $5a^3 = 5 \cdot a \cdot a \cdot a$ and $10a = 2 \cdot 5 \cdot a$, the LCM is $2 \cdot 5 \cdot a \cdot a \cdot a = 10a^3$.

63. Since $z^2 - 4z = z(z - 4)$ and $(z - 4)^2 = (z - 4)(z - 4)$, the LCM is $z(z - 4)(z - 4) = z(z - 4)^2$.

65. Since $x^2 - 6x + 9 = (x - 3)(x - 3)$ and $x^2 - 5x + 6 = (x - 2)(x - 3)$, the LCM is
$(x - 2)(x - 3)(x - 3) = (x - 2)(x - 3)^2$.

67. The factored denominators are $x + 1$ and 7. The LCD is $7(x + 1)$.

69. The factored denominators are $x + 4$ and $(x + 4)(x - 4)$. The LCD is $(x + 4)(x - 4)$.

71. The factored denominators are 2 and $2x + 1$ and $2(x - 2)$. The LCD is $2(2x + 1)(x - 2)$.

73. $\dfrac{4}{x + 1} + \dfrac{3}{x + 1} = \dfrac{7}{x + 1}$

75. $\dfrac{2}{x^2 - 1} - \dfrac{x + 1}{x^2 - 1} = \dfrac{2 - (x + 1)}{x^2 - 1} = \dfrac{2 - x - 1}{x^2 - 1} = \dfrac{-x + 1}{(x - 1)(x + 1)} = \dfrac{-(x - 1)}{(x + 1)(x - 1)} = -\dfrac{1}{x + 1}$

77. $\dfrac{x}{x + 4} - \dfrac{x + 1}{x(x + 4)} = \dfrac{x^2}{x(x + 4)} - \dfrac{x + 1}{x(x + 4)} = \dfrac{x^2 - (x + 1)}{x(x + 4)} = \dfrac{x^2 - x - 1}{x(x + 4)}$

79. $\dfrac{2}{x^2} - \dfrac{4x - 1}{x} = \dfrac{2}{x^2} - \dfrac{x(4x - 1)}{x^2} = \dfrac{2 - (4x^2 - x)}{x^2} = \dfrac{-4x^2 + x + 2}{x^2}$

81. $\dfrac{x + 3}{x - 5} + \dfrac{5}{x - 3} = \dfrac{(x + 3)(x - 3)}{(x - 5)(x - 3)} + \dfrac{5(x - 5)}{(x - 5)(x - 3)} = \dfrac{x^2 - 9 + 5x - 25}{(x - 5)(x - 3)} = \dfrac{x^2 + 5x - 34}{(x - 5)(x - 3)}$

83. $\dfrac{3}{x - 5} - \dfrac{1}{x - 3} - \dfrac{2x}{x - 5} = \dfrac{3(x - 3)}{(x - 5)(x - 3)} - \dfrac{x - 5}{(x - 5)(x - 3)} - \dfrac{2x(x - 3)}{(x - 5)(x - 3)} = \dfrac{-2(x^2 - 4x + 2)}{(x - 5)(x - 3)}$

85. $\dfrac{x}{x^2 - 9} + \dfrac{5x}{x - 3} = \dfrac{x}{(x - 3)(x + 3)} + \dfrac{5x(x + 3)}{(x - 3)(x + 3)} = \dfrac{x + 5x^2 + 15x}{(x - 3)(x + 3)} = \dfrac{x(5x + 16)}{(x - 3)(x + 3)}$

87. $\dfrac{b}{2b - 4} - \dfrac{b - 1}{b - 2} = \dfrac{b}{2(b - 2)} - \dfrac{2(b - 1)}{2(b - 2)} = \dfrac{b - (2b - 2)}{2(b - 2)} = \dfrac{-(b - 2)}{2(b - 2)} = -\dfrac{1}{2}$

89. $\dfrac{2x}{x - 5} + \dfrac{2x - 1}{3x^2 - 16x + 5} = \dfrac{2x(3x - 1)}{(x - 5)(3x - 1)} + \dfrac{2x - 1}{(x - 5)(3x - 1)} = \dfrac{6x^2 - 2x + 2x - 1}{(x - 5)(3x - 1)} =$
$\dfrac{6x^2 - 1}{(x - 5)(3x - 1)}$

91. $\dfrac{x}{(x - 1)^2} - \dfrac{1}{(x - 1)(x + 3)} = \dfrac{x(x + 3)}{(x - 1)^2(x + 3)} - \dfrac{x - 1}{(x - 1)^2(x + 3)} = \dfrac{x^2 + 3x - x + 1}{(x - 1)^2(x + 3)} =$
$\dfrac{x^2 + 2x + 1}{(x - 1)^2(x + 3)} = \dfrac{(x + 1)^2}{(x - 1)^2(x + 3)}$

93. $\dfrac{x}{x^2 - 5x + 4} + \dfrac{2}{x^2 - 2x - 8} = \dfrac{x}{(x - 4)(x - 1)} + \dfrac{2}{(x - 4)(x + 2)} =$
$\dfrac{x(x + 2)}{(x - 4)(x + 2)(x - 1)} + \dfrac{2(x - 1)}{(x - 4)(x + 2)(x - 1)} = \dfrac{x^2 + 2x + 2x - 2}{(x - 4)(x + 2)(x - 1)} =$
$\dfrac{x^2 + 4x - 2}{(x - 4)(x + 2)(x - 1)}$

95. $\dfrac{x}{x^2-4}-\dfrac{1}{x^2+4x+4}=\dfrac{x}{(x-2)(x+2)}-\dfrac{1}{(x+2)(x+2)}=$

$\dfrac{x(x+2)}{(x+2)^2(x-2)}-\dfrac{x-2}{(x+2)^2(x-2)}=\dfrac{x^2+2x-x+2}{(x+2)^2(x-2)}=\dfrac{x^2+x+2}{(x+2)^2(x-2)}$

97. $\dfrac{3x}{x-y}-\dfrac{3y}{x^2-2xy+y^2}=\dfrac{3x}{x-y}-\dfrac{3y}{(x-y)(x-y)}=$

$\dfrac{3x(x-y)}{(x-y)^2}-\dfrac{3y}{(x-y)^2}=\dfrac{3x^2-3xy-3y}{(x-y)^2}=\dfrac{3(x^2-xy-y)}{(x-y)^2}$

99. $x+\dfrac{1}{x-1}-\dfrac{1}{x+1}=\dfrac{x(x-1)(x+1)}{(x-1)(x+1)}+\dfrac{x+1}{(x-1)(x+1)}-\dfrac{x-1}{(x-1)(x+1)}=$

$\dfrac{x^3-x+x+1-x+1}{(x-1)(x+1)}=\dfrac{x^3-x+2}{(x-1)(x+1)}$

101. $\dfrac{6}{t-1}+\dfrac{2}{t-2}+\dfrac{1}{t}=\dfrac{6t(t-2)}{t(t-1)(t-2)}+\dfrac{2t(t-1)}{t(t-1)(t-2)}+\dfrac{(t-1)(t-2)}{t(t-1)(t-2)}=$

$\dfrac{6t^2-12t+2t^2-2t+t^2-3t+2}{t(t-1)(t-2)}=\dfrac{9t^2-17t+2}{t(t-1)(t-2)}$

103. The factored denominators are x and x^2. The LCD is x^2.

$\dfrac{1}{x}+\dfrac{3}{x^2}=0\Rightarrow\dfrac{1(x^2)}{x}+\dfrac{3(x^2)}{x^2}=0(x^2)\Rightarrow x+3=0\Rightarrow x=-3$

105. The factored denominators are x and $2x-1$. The LCD is $x(2x-1)$.

$\dfrac{1}{x}+\dfrac{3x}{2x-1}=0\Rightarrow\dfrac{1(x)(2x-1)}{x}+\dfrac{3x(x)(2x-1)}{2x-1}=0(x)(2x-1)\Rightarrow 2x-1+3x^2=0\Rightarrow$

$3x^2+2x-1=0\Rightarrow(3x-1)(x+1)=0\Rightarrow x=\dfrac{1}{3}$ or $x=-1$

107. The factored denominators are $(3-x)(3+x)$ and $3-x$. The LCD is $(3-x)(3+x)$.

$\dfrac{2x}{9-x^2}+\dfrac{1}{3-x}=0\Rightarrow\dfrac{2x(3-x)(3+x)}{9-x^2}+\dfrac{1(3-x)(3+x)}{3-x}=0(3-x)(3+x)\Rightarrow$

$2x+3+x=0\Rightarrow 3x+3=0\Rightarrow x=-1$

109. The factored denominators are $2x$, $2x^2$ and x^3. The LCD is $2x^3$.

$\dfrac{1}{2x}+\dfrac{1}{2x^2}-\dfrac{1}{x^3}=0\Rightarrow\dfrac{1(2x^3)}{2x}+\dfrac{1(2x^3)}{2x^2}-\dfrac{1(2x^3)}{x^3}=0(2x^3)\Rightarrow x^2+x-2=0\Rightarrow$

$(x+2)(x-1)=0\Rightarrow x=-2$ or $x=1$

111. The factored denominators are x, $x+5$ and $x-5$. The LCD is $x(x+5)(x-5)$.

$\dfrac{1}{x}-\dfrac{2}{x+5}+\dfrac{1}{x-5}=0\Rightarrow$

$\dfrac{1(x)(x+5)(x-5)}{x}-\dfrac{2(x)(x+5)(x-5)}{x+5}+\dfrac{1(x)(x+5)(x-5)}{x-5}=0(x)(x+5)(x-5)\Rightarrow$

$(x+5)(x-5)-2x(x-5)+x(x+5)=0\Rightarrow x^2-25-2x^2+10x+x^2+5x=0\Rightarrow$

$15x-25=0\Rightarrow x=\dfrac{25}{15}=\dfrac{5}{3}$

113. $\dfrac{1 + \dfrac{1}{x}}{1 - \dfrac{1}{x}} = \dfrac{1 + \dfrac{1}{x}}{1 - \dfrac{1}{x}} \cdot \dfrac{x}{x} = \dfrac{x+1}{x-1}$

115. $\dfrac{\dfrac{1}{x-5}}{\dfrac{4}{x} - \dfrac{1}{x-5}} = \dfrac{\dfrac{1}{x-5}}{\dfrac{4}{x} - \dfrac{1}{x-5}} \cdot \dfrac{x(x-5)}{x(x-5)} = \dfrac{x}{4(x-5) - x} = \dfrac{x}{4x - 20 - x} = \dfrac{x}{3x - 20}$

117. $\dfrac{\dfrac{1}{x} + \dfrac{2-x}{x^2}}{\dfrac{3}{x^2} - \dfrac{1}{x}} = \dfrac{\dfrac{1}{x} + \dfrac{2-x}{x^2}}{\dfrac{3}{x^2} - \dfrac{1}{x}} \cdot \dfrac{x^2}{x^2} = \dfrac{x + 2 - x}{3 - x} = \dfrac{2}{3-x}$

119. $\dfrac{\dfrac{1}{x+3} + \dfrac{2}{x-3}}{2 - \dfrac{1}{x-3}} = \dfrac{\dfrac{1}{x+3} + \dfrac{2}{x-3}}{2 - \dfrac{1}{x-3}} \cdot \dfrac{(x-3)(x+3)}{(x-3)(x+3)} = \dfrac{x - 3 + 2(x+3)}{2(x-3)(x+3) - (x+3)} = \dfrac{3(x+1)}{(x+3)(2x-7)}$

121. $\dfrac{\dfrac{4}{x-5}}{\dfrac{1}{x+5} + \dfrac{1}{x}} = \dfrac{\dfrac{4}{x-5}}{\dfrac{1}{x+5} + \dfrac{1}{x}} \cdot \dfrac{x(x-5)(x+5)}{x(x-5)(x+5)} = \dfrac{4x(x+5)}{x(x-5) + (x-5)(x+5)} = \dfrac{4x(x+5)}{(x-5)(2x+5)}$

123. $\dfrac{\dfrac{1}{2a} - \dfrac{1}{2b}}{\dfrac{1}{a^2} - \dfrac{1}{b^2}} = \dfrac{\dfrac{1}{2a} - \dfrac{1}{2b}}{\dfrac{1}{a^2} - \dfrac{1}{b^2}} \cdot \dfrac{2a^2b^2}{2a^2b^2} = \dfrac{ab^2 - a^2b}{2b^2 - 2a^2} = \dfrac{ab(b-a)}{2(b+a)(b-a)} = \dfrac{ab}{2(a+b)}$

R.6: Radical Notation and Rational Exponents

1. $-\sqrt{25} = -\sqrt{5^2} = -5$ and $\sqrt{25} = \sqrt{5^2} = 5$

3. $-\sqrt{\dfrac{16}{25}} = -\sqrt{\left(\dfrac{4}{5}\right)^2} = -\dfrac{4}{5}$ and $\sqrt{\dfrac{16}{25}} = \sqrt{\left(\dfrac{4}{5}\right)^2} = \dfrac{4}{5}$

5. $-\sqrt{11} \approx -3.32$ and $\sqrt{11} \approx 3.32$

7. $\sqrt{144} = \sqrt{12^2} = 12$

9. $\sqrt{23} \approx 4.80$

11. $\sqrt{\dfrac{4}{49}} = \sqrt{\left(\dfrac{2}{7}\right)^2} = \dfrac{2}{7}$

13. Since $b < 0$, $\sqrt{b^2} = -b$.

15. $\sqrt[3]{27} = \sqrt[3]{3^3} = 3$

17. $\sqrt[3]{-8} = \sqrt[3]{(-2)^3} = -2$

19. $\sqrt[3]{\dfrac{1}{27}} = \sqrt[3]{\left(\dfrac{1}{3}\right)^3} = \dfrac{1}{3}$

21. $\sqrt[3]{b^9} = \sqrt[3]{(b^3)^3} = b^3$

23. $\sqrt{9} = 3$

25. $-\sqrt{5} \approx -2.24$

27. $\sqrt[3]{27} = 3$

29. $\sqrt[3]{-64} = -4$

31. $\sqrt[3]{5} \approx 1.71$

33. $-\sqrt[3]{x^9} = -\sqrt[3]{(x^3)^3} = -x^3$

35. $\sqrt[3]{(2x)^6} = \sqrt[3]{((2x)^2)^3} = (2x)^2 = 4x^2$

37. $\sqrt[4]{81} = 3$

39. $\sqrt[5]{-7} \approx -1.48$

41. $6^{1/2} = \sqrt{6}$

43. $(xy)^{1/2} = \sqrt{xy}$

45. $y^{-1/5} = \dfrac{1}{\sqrt[5]{y}}$

47. The expression can be written $27^{2/3} = \sqrt[3]{27^2}$ or $(\sqrt[3]{27})^2$ and evaluated as $(\sqrt[3]{27})^2 = 3^2 = 9$.

49. The expression can be written $(-1)^{4/3} = \sqrt[3]{(-1)^4}$ or $(\sqrt[3]{-1})^4$ and evaluated as $(\sqrt[3]{-1})^4 = (-1)^4 = 1$.

51. The expression can be written $8^{-1/3} = \dfrac{1}{8^{1/3}} = \dfrac{1}{\sqrt[3]{8}}$ and evaluated as $\dfrac{1}{\sqrt[3]{8}} = \dfrac{1}{2}$.

53. The expression can be written $13^{-3/5} = \dfrac{1}{13^{3/5}} = \dfrac{1}{\sqrt[5]{13^3}}$ or $\dfrac{1}{(\sqrt[5]{13})^3}$. The result is not an integer.

55. $16^{1/2} = \sqrt{16} = 4$

57. $256^{1/4} = \sqrt[4]{256} = 4$

59. $32^{1/5} = \sqrt[5]{32} = 2$

61. $(-8)^{4/3} = (\sqrt[3]{-8})^4 = (-2)^4 = 16$

63. $2^{1/2} \cdot 2^{2/3} = 2^{1/2+2/3} = 2^{7/6} \approx 2.24$

65. $\left(\dfrac{4}{9}\right)^{1/2} = \dfrac{4^{1/2}}{9^{1/2}} = \dfrac{\sqrt{4}}{\sqrt{9}} = \dfrac{2}{3}$

67. $\dfrac{4^{2/3}}{4^{1/2}} = 4^{2/3-1/2} = 4^{1/6} \approx 1.26$

69. $4^{-1/2} = \dfrac{1}{4^{1/2}} = \dfrac{1}{\sqrt{4}} = \dfrac{1}{2}$

71. $(-8)^{-1/3} = \dfrac{1}{(-8)^{1/3}} = \dfrac{1}{\sqrt[3]{-8}} = \dfrac{1}{-2} = -\dfrac{1}{2}$

73. $\left(\dfrac{1}{16}\right)^{-1/4} = 16^{1/4} = \sqrt[4]{16} = 2$

75. $(2^{1/2})^3 = 2^{1/2 \cdot 3} = 2^{3/2} \approx 2.83$

77. $(x^2)^{3/2} = x^{2 \cdot 3/2} = x^3$

79. $(x^2y^8)^{1/2} = x^{2\cdot1/2} \cdot y^{8\cdot1/2} = xy^4$

81. $\sqrt[3]{x^3y^6} = (x^3y^6)^{1/3} = x^{3\cdot1/3} \cdot y^{6\cdot1/3} = xy^2$

83. $\sqrt{\dfrac{y^4}{x^2}} = \left(\dfrac{y^4}{x^2}\right)^{1/2} = \dfrac{y^{4\cdot1/2}}{x^{2\cdot1/2}} = \dfrac{y^2}{x}$

85. $\sqrt{y^3} \cdot \sqrt[3]{y^2} = (y^3)^{1/2} \cdot (y^2)^{1/3} = y^{3\cdot1/2} \cdot y^{2\cdot1/3} = y^{3/2} \cdot y^{2/3} = y^{3/2+2/3} = y^{13/6}$

87. $\left(\dfrac{x^6}{27}\right)^{2/3} = \dfrac{x^{6\cdot2/3}}{27^{2/3}} = \dfrac{x^4}{(\sqrt[3]{27})^2} = \dfrac{x^4}{3^2} = \dfrac{x^4}{9}$

89. $\left(\dfrac{x^2}{y^6}\right)^{-1/2} = \left(\dfrac{y^6}{x^2}\right)^{1/2} = \dfrac{y^{6\cdot1/2}}{x^{2\cdot1/2}} = \dfrac{y^3}{x}$

91. $\sqrt{\sqrt{y}} = (y^{1/2})^{1/2} = y^{1/2\cdot1/2} = y^{1/4}$

93. $(a^{-1/2})^{4/3} = a^{-1/2\cdot4/3} = a^{-2/3} = \dfrac{1}{a^{2/3}}$

95. $(a^3b^6)^{1/3} = a^{3\cdot1/3} \cdot b^{6\cdot1/3} = ab^2$

97. $\dfrac{(k^{1/2})^{-3}}{(k^2)^{1/4}} = \dfrac{k^{-3/2}}{k^{1/2}} = k^{-3/2-1/2} = k^{-4/2} = k^{-2} = \dfrac{1}{k^2}$

99. $\sqrt{b} \cdot \sqrt[4]{b} = b^{1/2} \cdot b^{1/4} = b^{1/2+1/4} = b^{3/4}$

101. $\sqrt{z} \cdot \sqrt[3]{z^2} \cdot \sqrt[4]{z^3} = z^{1/2} \cdot z^{2/3} \cdot z^{3/4} = z^{1/2+2/3+3/4} = z^{23/12}$

103. $p^{1/2}(p^{3/2} + p^{1/2}) = p^{1/2+3/2} + p^{1/2+1/2} = p^2 + p$

105. $\sqrt[3]{x}(\sqrt{x} - \sqrt[3]{x^2}) = x^{1/3}(x^{1/2} - x^{2/3}) = x^{1/3+1/2} - x^{1/3+2/3} = x^{5/6} - x$

R.7: Radical Expressions

1. $\sqrt{3} \cdot \sqrt{3} = \sqrt{3\cdot3} = \sqrt{9} = 3$

3. $\sqrt{2} \cdot \sqrt{50} = \sqrt{2\cdot50} = \sqrt{100} = 10$

5. $\sqrt[3]{4} \cdot \sqrt[3]{16} = \sqrt[3]{4\cdot16} = \sqrt[3]{64} = 4$

7. $\sqrt{\dfrac{9}{25}} = \dfrac{\sqrt{9}}{\sqrt{25}} = \dfrac{3}{5}$

9. $\sqrt{\dfrac{1}{2}} \cdot \sqrt{\dfrac{1}{8}} = \sqrt{\dfrac{1\cdot1}{2\cdot8}} = \sqrt{\dfrac{1}{16}} = \dfrac{\sqrt{1}}{\sqrt{16}} = \dfrac{1}{4}$

11. $\sqrt{\dfrac{x}{2}} \cdot \sqrt{\dfrac{x}{8}} = \sqrt{\dfrac{x\cdot x}{2\cdot8}} = \sqrt{\dfrac{x^2}{16}} = \dfrac{\sqrt{x^2}}{\sqrt{16}} = \dfrac{x}{4}$

13. $\dfrac{\sqrt{45}}{\sqrt{5}} = \sqrt{\dfrac{45}{5}} = \sqrt{9} = 3$

15. $\sqrt[4]{9} \cdot \sqrt[4]{9} = \sqrt[4]{9\cdot9} = \sqrt[4]{81} = 3$

17. $\dfrac{\sqrt[5]{64}}{\sqrt[5]{-2}} = \sqrt[5]{\dfrac{64}{-2}} = \sqrt[5]{-32} = -2$

19. $\dfrac{\sqrt{a^2 b}}{\sqrt{b}} = \sqrt{\dfrac{a^2 b}{b}} = \sqrt{a^2} = a$

21. $\sqrt[3]{\dfrac{x^3}{8}} = \dfrac{\sqrt[3]{x^3}}{\sqrt[3]{8}} = \dfrac{x}{2}$

23. $\sqrt{4x^4} = \sqrt{4} \cdot \sqrt{(x^2)^2} = 2x^2$

25. $\sqrt[4]{16x^4 y} = \sqrt[4]{16} \cdot \sqrt[4]{x^4} \cdot \sqrt[4]{y} = 2x\sqrt[4]{y}$

27. $\sqrt{3x} \cdot \sqrt{12x} = \sqrt{3 \cdot 12 \cdot x \cdot x} = \sqrt{36x^2} = \sqrt{36} \cdot \sqrt{x^2} = 6x$

29. $\sqrt[3]{8x^6 y^3 z^9} = \sqrt[3]{8} \cdot \sqrt[3]{(x^2)^3} \cdot \sqrt[3]{y^3} \cdot \sqrt[3]{(z^3)^3} = 2x^2 yz^3$

31. $\sqrt[4]{\dfrac{3}{4}} \cdot \sqrt[4]{\dfrac{27}{4}} = \sqrt[4]{\dfrac{3}{4} \cdot \dfrac{27}{4}} = \sqrt[4]{\dfrac{81}{16}} = \dfrac{\sqrt[4]{81}}{\sqrt[4]{16}} = \dfrac{3}{2}$

33. $\sqrt[4]{25z} \cdot \sqrt[4]{25z} = \sqrt[4]{625z^2} = \sqrt[4]{625} \cdot \sqrt[4]{z^2} = 5\sqrt{z}$

35. $\sqrt[5]{\dfrac{7a}{b^2}} \cdot \sqrt[5]{\dfrac{b^2}{7a^6}} = \sqrt[5]{\dfrac{7ab^2}{7a^6 b^2}} \sqrt[5]{\dfrac{1}{a^5}} = \dfrac{1}{a}$

37. $\sqrt{200} = \sqrt{100 \cdot 2} = \sqrt{100} \cdot \sqrt{2} = 10\sqrt{2}$

39. $\sqrt[3]{81} = \sqrt[3]{27 \cdot 3} = \sqrt[3]{27} \cdot \sqrt[3]{3} = 3\sqrt[3]{3}$

41. $\sqrt[4]{64} = \sqrt[4]{16 \cdot 4} = \sqrt[4]{16} \cdot \sqrt[4]{4} = 2\sqrt[4]{4} = 2\sqrt[4]{2^2} = 2\sqrt{2}$

43. $\sqrt[5]{-64} = \sqrt[5]{-2^6} = \sqrt[5]{-2^5 \cdot 2} = \sqrt[5]{-2^5} \cdot \sqrt[5]{2} = -2\sqrt[5]{2}$

45. $\sqrt{8n^3} = \sqrt{(2n^2 \cdot 2n)} = \sqrt{(2n)^2} \cdot \sqrt{2n} = 2n\sqrt{2n}$

47. $\sqrt{12a^2 b^5} = \sqrt{(2ab^2)^2 \cdot 3b} = \sqrt{(2ab^2)^2} \cdot \sqrt{3b} = 2ab^2\sqrt{3b}$

49. $\sqrt[3]{-125x^4 y^5} = \sqrt[3]{(-5xy)^3 \cdot xy^2} = \sqrt[3]{(-5xy)^3} \cdot \sqrt[3]{xy^2} = -5xy\sqrt[3]{xy^2}$

51. $\sqrt[3]{5t} \cdot \sqrt[3]{125t} = \sqrt[3]{625t^2} = \sqrt[3]{5^4 t^2} = \sqrt[3]{5^3 \cdot 5t^2} = \sqrt[3]{5^3} \cdot \sqrt[3]{5t^2} = 5\sqrt[3]{5t^2}$

53. $\sqrt[4]{\dfrac{9t^5}{r^8}} \cdot \sqrt[4]{\dfrac{9r}{5t}} = \sqrt[4]{\dfrac{81rt^5}{5r^8 t}} = \sqrt[4]{\dfrac{81t^4}{5r^7}} = \dfrac{\sqrt[4]{(3t)^4}}{\sqrt[4]{r^4 \cdot 5r^3}} = \dfrac{3t}{r\sqrt[4]{5r^3}}$

55. $\sqrt{3} \cdot \sqrt[3]{3} = 3^{1/2} \cdot 3^{1/3} = 3^{1/2 + 1/3} = 3^{5/6} = \sqrt[6]{3^5}$

57. $\sqrt[4]{8} \cdot \sqrt[3]{4} = \sqrt[4]{2^3} \cdot \sqrt[3]{2^2} = 2^{3/4} \cdot 2^{2/3} = 2^{3/4 + 2/3} = 2^{17/12} = 2^{12/12 + 5/12} = 2 \cdot 2^{5/12} = 2\sqrt[12]{2^5}$

59. $\sqrt[4]{x^3} \cdot \sqrt[3]{x} = x^{3/4} \cdot x^{1/3} = x^{3/4 + 1/3} = x^{13/12} = x^{12/12} \cdot x^{1/12} = x\sqrt[12]{x}$

61. $\sqrt[4]{rt} \cdot \sqrt[3]{r^2 t} = (rt)^{1/4} \cdot (r^2 t)^{1/3} = r^{1/4} t^{1/4} \cdot r^{2/3} t^{1/3} = r^{1/4 + 2/3} t^{1/4 + 1/3} = r^{11/12} t^{7/12} = \sqrt[12]{r^{11} t^7}$

63. $2\sqrt{3} + 7\sqrt{3} = 9\sqrt{3}$

65. $\sqrt{x} + \sqrt{x} - \sqrt{y} = 2\sqrt{x} - \sqrt{y}$

67. $2\sqrt[3]{6} - 7\sqrt[3]{6} = -5\sqrt[3]{6}$

69. $3\sqrt{28} + 3\sqrt{7} = 3\sqrt{4 \cdot 7} + 3\sqrt{7} = 3 \cdot 2\sqrt{7} + 3\sqrt{7} = 9\sqrt{7}$

71. $\sqrt{44} - 4\sqrt{11} = \sqrt{4 \cdot 11} - 4\sqrt{11} = 2\sqrt{11} - 4\sqrt{11} = -2\sqrt{11}$

73. $2\sqrt[3]{16} + \sqrt[3]{2} - \sqrt{2} = 2\sqrt[3]{8 \cdot 2} + \sqrt[3]{2} - \sqrt{2} = 2 \cdot 2\sqrt[3]{2} + \sqrt[3]{2} - \sqrt{2} = 5\sqrt[3]{2} - \sqrt{2}$

75. $\sqrt[3]{xy} - 2\sqrt[3]{xy} = -\sqrt[3]{xy}$

77. $\sqrt{4x + 8} + \sqrt{x + 2} = \sqrt{4(x + 2)} + \sqrt{x + 2} = 2\sqrt{x + 2} + \sqrt{x + 2} = 3\sqrt{x + 2}$

79. $\dfrac{15\sqrt{8}}{4} - \dfrac{2\sqrt{2}}{5} = \dfrac{15 \cdot 2\sqrt{2}}{4} \cdot \dfrac{5}{5} - \dfrac{2\sqrt{2}}{5} \cdot \dfrac{4}{4} = \dfrac{150\sqrt{2}}{20} - \dfrac{8\sqrt{2}}{20} = \dfrac{150\sqrt{2} - 8\sqrt{2}}{20} = \dfrac{142\sqrt{2}}{20} = \dfrac{71\sqrt{2}}{10}$

81. $2\sqrt[4]{64} - \sqrt[4]{324} + \sqrt[4]{4} = 2\sqrt[4]{16 \cdot 4} - \sqrt[4]{81 \cdot 4} + \sqrt[4]{4} = 4\sqrt[4]{4} - 3\sqrt[4]{4} + \sqrt[4]{4} = 2\sqrt[4]{4} = 2\sqrt{2}$

83. $2\sqrt{3z} + 3\sqrt{12z} + 3\sqrt{48z} = 2\sqrt{3z} + 3\sqrt{4 \cdot 3z} + 3\sqrt{16 \cdot 3z} = 2\sqrt{3z} + 6\sqrt{3z} + 12\sqrt{3z} = 20\sqrt{3z}$

85. $\sqrt[4]{81a^5b^5} - \sqrt[4]{ab} = \sqrt[4]{(3ab)^4 \cdot ab} - \sqrt[4]{ab} = 3ab\sqrt[4]{ab} - \sqrt[4]{ab} = (3ab - 1)\sqrt[4]{ab}$

87. $5\sqrt[3]{\dfrac{n^4}{125}} - 2\sqrt[3]{n} = 5\sqrt[3]{\dfrac{n^3}{125} \cdot n} - 2\sqrt[3]{n} = 5 \cdot \dfrac{n}{5}\sqrt[3]{n} - 2\sqrt[3]{n} = n\sqrt[3]{n} - 2\sqrt[3]{n} = (n - 2)\sqrt[3]{n}$

89. $(3 + \sqrt{7})(3 - \sqrt{7}) = 3^2 - (\sqrt{7})^2 = 9 - 7 = 2$

91. $(\sqrt{x} + 8)(\sqrt{x} - 8) = (\sqrt{x})^2 - 8^2 = x - 64$

93. $(\sqrt{ab} - \sqrt{c})(\sqrt{ab} + \sqrt{c}) = (\sqrt{ab})^2 - (\sqrt{c})^2 = ab - c$

95. $(\sqrt{x} - 7)(\sqrt{x} + 8) = (\sqrt{x})^2 + 8\sqrt{x} - 7\sqrt{x} - 56 = x + \sqrt{x} - 56$

97. $\dfrac{4}{\sqrt{3}} = \dfrac{4}{\sqrt{3}} \cdot \dfrac{\sqrt{3}}{\sqrt{3}} = \dfrac{4\sqrt{3}}{3}$

99. $\dfrac{5}{3\sqrt{5}} = \dfrac{5}{3\sqrt{5}} \cdot \dfrac{\sqrt{5}}{\sqrt{5}} = \dfrac{5\sqrt{5}}{3 \cdot 5} = \dfrac{5\sqrt{5}}{15} = \dfrac{\sqrt{5}}{3}$

101. $\sqrt{\dfrac{b}{12}} = \dfrac{\sqrt{b}}{\sqrt{12}} = \dfrac{\sqrt{b}}{\sqrt{12}} \cdot \dfrac{\sqrt{12}}{\sqrt{12}} = \dfrac{\sqrt{12b}}{12} = \dfrac{\sqrt{4 \cdot 3b}}{12} = \dfrac{2\sqrt{3b}}{12} = \dfrac{\sqrt{3b}}{6}$

103. $\dfrac{1}{3 - \sqrt{2}} = \dfrac{1}{3 - \sqrt{2}} \cdot \dfrac{3 + \sqrt{2}}{3 + \sqrt{2}} = \dfrac{3 + \sqrt{2}}{9 - 2} = \dfrac{3 + \sqrt{2}}{7}$

105. $\dfrac{\sqrt{2}}{\sqrt{5} + 2} = \dfrac{\sqrt{2}}{\sqrt{5} + 2} \cdot \dfrac{\sqrt{5} - 2}{\sqrt{5} - 2} = \dfrac{\sqrt{10} - 2\sqrt{2}}{5 - 4} = \dfrac{\sqrt{10} - 2\sqrt{2}}{1} = \sqrt{10} - 2\sqrt{2}$

107. $\dfrac{1}{\sqrt{7} - \sqrt{6}} = \dfrac{1}{\sqrt{7} - \sqrt{6}} \cdot \dfrac{\sqrt{7} + \sqrt{6}}{\sqrt{7} + \sqrt{6}} = \dfrac{\sqrt{7} + \sqrt{6}}{7 - 6} = \dfrac{\sqrt{7} + \sqrt{6}}{1} = \sqrt{7} + \sqrt{6}$

109. $\dfrac{\sqrt{z}}{\sqrt{z} - 3} = \dfrac{\sqrt{z}}{\sqrt{z} - 3} \cdot \dfrac{\sqrt{z} + 3}{\sqrt{z} + 3} = \dfrac{z + 3\sqrt{z}}{z - 9}$

111. $\dfrac{\sqrt{a} + \sqrt{b}}{\sqrt{a} - \sqrt{b}} = \dfrac{\sqrt{a} + \sqrt{b}}{\sqrt{a} - \sqrt{b}} \cdot \dfrac{\sqrt{a} + \sqrt{b}}{\sqrt{a} + \sqrt{b}} = \dfrac{a + 2\sqrt{ab} + b}{a - b}$

Appendix C: Partial Fractions

1. Multiply $\dfrac{5}{3x(2x+1)} = \dfrac{A}{3x} + \dfrac{B}{2x+1}$ by $3x(2x+1) \Rightarrow 5 = A(2x+1) + B(3x)$.

   Let $x = 0 \Rightarrow 5 = A(1) \Rightarrow A = 5$. Let $x = -\dfrac{1}{2} \Rightarrow 5 = B\left(-\dfrac{3}{2}\right) \Rightarrow B = -\dfrac{10}{3}$.

   The expression can be written $\dfrac{5}{3x} + \dfrac{-10}{3(2x+1)}$.

3. Multiply $\dfrac{4x+2}{(x+2)(2x-1)} = \dfrac{A}{x+2} + \dfrac{B}{2x-1}$ by $(x+2)(2x-1) \Rightarrow 4x+2 = A(2x-1) + B(x+2)$.

   Let $x = -2 \Rightarrow -6 = A(-5) \Rightarrow A = \dfrac{6}{5}$. Let $x = \dfrac{1}{2} \Rightarrow 4 = B\left(\dfrac{5}{2}\right) \Rightarrow B = \dfrac{8}{5}$.

   The expression can be written $\dfrac{6}{5(x+2)} + \dfrac{8}{5(2x-1)}$.

5. Factoring $\dfrac{x}{x^2+4x-5}$ results in $\dfrac{x}{(x+5)(x-1)}$.

   Multiply $\dfrac{x}{(x+5)(x-1)} = \dfrac{A}{x+5} + \dfrac{B}{x-1}$ by $(x+5)(x-1) \Rightarrow x = A(x-1) + B(x+5)$.

   Let $x = -5 \Rightarrow -5 = A(-6) \Rightarrow A = \dfrac{5}{6}$. Let $x = 1 \Rightarrow 1 = B(6) \Rightarrow B = \dfrac{1}{6}$.

   The expression can be written $\dfrac{5}{6(x+5)} + \dfrac{1}{6(x-1)}$.

7. Multiply $\dfrac{2x}{(x+1)(x+2)^2} = \dfrac{A}{x+1} + \dfrac{B}{x+2} + \dfrac{C}{(x+2)^2}$ by $(x+1)(x+2)^2 \Rightarrow$

   $2x = A(x+2)^2 + B(x+1)(x+2) + C(x+1)$. Let $x = -1 \Rightarrow -2 = A(1) \Rightarrow A = -2$.

   Let $x = -2 \Rightarrow -4 = C(-1) \Rightarrow C = 4$. Let $x = 0$ with $A = -2$ and $C = 4 \Rightarrow$

   $0 = -2(4) + B(2) + 4(1) \Rightarrow 4 = 2B \Rightarrow B = 2$. The expression can be written $\dfrac{-2}{x+1} + \dfrac{2}{x+2} + \dfrac{4}{(x+2)^2}$.

9. Multiply $\dfrac{4}{x(1-x)} = \dfrac{A}{x} + \dfrac{B}{1-x}$ by $x(1-x) \Rightarrow 4 = A(1-x) + B(x)$.

   Let $x = 0 \Rightarrow 4 = A(1) \Rightarrow A = 4$. Let $x = 1 \Rightarrow 4 = B(1) \Rightarrow B = 4$.

   The expression can be written $\dfrac{4}{x} + \dfrac{4}{1-x}$.

11. Multiply $\dfrac{4x^2-x-15}{x(x+1)(x-1)} = \dfrac{A}{x} + \dfrac{B}{x+1} + \dfrac{C}{x-1}$ by $x(x+1)(x-1) \Rightarrow$

    $4x^2 - x - 15 = A(x-1)(x+1) + B(x)(x-1) + C(x)(x+1)$. Let $x = 0 \Rightarrow -15 = A(-1) \Rightarrow$

    $A = 15$. Let $x = 1 \Rightarrow -12 = C(1)(2) \Rightarrow C = -6$. Let $x = -1 \Rightarrow -10 = B(-1)(-2) \Rightarrow B = -5$.

    The expression can be written $\dfrac{15}{x} + \dfrac{-5}{x+1} + \dfrac{-6}{x-1}$.

13. By long division $\dfrac{x^2}{x^2 + 2x + 1} = 1 + \dfrac{-2x - 1}{(x + 1)^2}$.

Multiply $\dfrac{-2x - 1}{(x + 1)^2} = \dfrac{A}{x + 1} + \dfrac{B}{(x + 1)^2}$ by $(x + 1)^2 \Rightarrow -2x - 1 = A(x + 1) + B$.

Let $x = -1 \Rightarrow 1 = B$. Let $x = 0$ with $B = 1 \Rightarrow -1 = A + 1 \Rightarrow A = -2$.

The expression can be written $1 + \dfrac{-2}{x + 1} + \dfrac{1}{(x + 1)^2}$.

15. By long division $\dfrac{2x^5 + 3x^4 - 3x^3 - 2x^2 + x}{2x^2 + 5x + 2} = x^3 - x^2 + \dfrac{x}{2x^2 + 5x + 2} = x^3 - x^2 + \dfrac{x}{(2x + 1)(x + 2)}$.

Multiply $\dfrac{x}{(2x + 1)(x + 2)} = \dfrac{A}{2x + 1} + \dfrac{B}{x + 2}$ by $(2x + 1)(x + 2) \Rightarrow x = A(x + 2) + B(2x + 1)$.

Let $x = -\dfrac{1}{2} \Rightarrow -\dfrac{1}{2} = A\left(\dfrac{3}{2}\right) \Rightarrow A = -\dfrac{1}{3}$. Let $x = -2 \Rightarrow -2 = B(-3) \Rightarrow B = \dfrac{2}{3}$.

The expression can be written $x^3 - x^2 + \dfrac{-1}{3(2x + 1)} + \dfrac{2}{3(x + 2)}$.

17. By long division $\dfrac{x^3 + 4}{9x^3 - 4x} = \dfrac{1}{9} + \dfrac{\frac{4}{9}x + 4}{9x^3 - 4x} = \dfrac{1}{9} + \dfrac{\frac{4}{9}x + 4}{x(3x + 2)(3x - 2)}$.

Multiply $\dfrac{\frac{4}{9}x + 4}{x(3x + 2)(3x - 2)} = \dfrac{A}{x} + \dfrac{B}{3x + 2} + \dfrac{C}{3x - 2}$ by $x(3x + 2)(3x - 2) \Rightarrow$

$\dfrac{4}{9}x + 4 = A(3x + 2)(3x - 2) + B(x)(3x - 2) + C(x)(3x + 2)$. Let $x = 0 \Rightarrow 4 = A(-4) \Rightarrow A = -1$.

Let $x = -\dfrac{2}{3} \Rightarrow -\dfrac{8}{27} + 4 = B\left(-\dfrac{2}{3}\right)(-4) \Rightarrow \dfrac{100}{27} = \dfrac{8}{3}B \Rightarrow B = \dfrac{25}{18}$.

Let $x = \dfrac{2}{3} \Rightarrow \dfrac{8}{27} + 4 = C\left(\dfrac{2}{3}\right)(4) \Rightarrow \dfrac{116}{27} = \dfrac{8}{3}C \Rightarrow C = \dfrac{29}{18}$.

The expression can be written $\dfrac{1}{9} + \dfrac{-1}{x} + \dfrac{25}{18(3x + 2)} + \dfrac{29}{18(3x - 2)}$.

19. Multiply $\dfrac{-3}{x^2(x^2 + 5)} = \dfrac{A}{x} + \dfrac{B}{x^2} + \dfrac{Cx + D}{x^2 + 5}$ by $x^2(x^2 + 5) \Rightarrow$

$-3 = A(x)(x^2 + 5) + B(x^2 + 5) + (Cx + D)(x^2) \Rightarrow -3 = Ax^3 + 5Ax + Bx^2 + 5B + Cx^3 + Dx^2$.

Equate coefficients. For x^3: $0 = A + C$.

For x^2: $0 = B + D$. For x: $0 = 5A \Rightarrow A = 0$. For the constants: $-3 = 5B \Rightarrow B = -\dfrac{3}{5}$.

Substitute $A = 0$ in the first equation. $C = 0$. Substitute $B = -\dfrac{3}{5}$ in the second equation. $D = \dfrac{3}{5}$.

The expression can be written $\dfrac{-3}{5x^2} + \dfrac{3}{5(x^2 + 5)}$.

21. Multiply $\dfrac{3x - 2}{(x + 4)(3x^2 + 1)} = \dfrac{A}{x + 4} + \dfrac{Bx + C}{3x^2 + 1}$ by $(x + 4)(3x^2 + 1) \Rightarrow$

$3x - 2 = A(3x^2 + 1) + (Bx + C)(x + 4) \Rightarrow 3x - 2 = 3Ax^2 + A + Bx^2 + 4Bx + Cx + 4C.$

Let $x = -4 \Rightarrow -14 = 49A \Rightarrow A = -\dfrac{2}{7}.$ Equate coefficients.

For x^2: $0 = 3A + B \Rightarrow 0 = -\dfrac{6}{7} + B \Rightarrow B = \dfrac{6}{7}.$ For x: $3 = 4B + C \Rightarrow 3 = \dfrac{24}{7} + C \Rightarrow C = -\dfrac{3}{7}.$

The expression can be written $\dfrac{-2}{7(x + 4)} + \dfrac{6x - 3}{7(3x^2 + 1)}.$

23. Multiply $\dfrac{1}{x(2x + 1)(3x^2 + 4)} = \dfrac{A}{x} + \dfrac{B}{2x + 1} + \dfrac{Cx + D}{3x^2 + 4}$ by $x(2x + 1)(3x^2 + 4) \Rightarrow$

$1 = A(2x + 1)(3x^2 + 4) + B(x)(3x^2 + 4) + (Cx + D)(x)(2x + 1).$

Let $x = 0 \Rightarrow 1 = A(1)(4) \Rightarrow A = \dfrac{1}{4}.$ Let $x = -\dfrac{1}{2} \Rightarrow 1 = B\left(-\dfrac{1}{2}\right)\left(\dfrac{19}{4}\right) \Rightarrow B = -\dfrac{8}{19}.$

Multiply the right side out. $1 = A(6x^3 + 3x^2 + 8x + 4) + 3Bx^3 + 4Bx + 2Cx^3 + Cx^2 + 2Dx^2 + Dx \Rightarrow$

$1 = 6Ax^3 + 3Ax^2 + 8Ax + 4A + 3Bx^3 + 4Bx + 2Cx^3 + Cx^2 + 2Dx^2 + Dx.$ Equate coefficients.

For x^3: $0 = 6A + 3B + 2C \Rightarrow 0 = 6\left(\dfrac{1}{4}\right) + 3\left(-\dfrac{8}{19}\right) + 2C \Rightarrow 0 = \dfrac{9}{38} + 2C \Rightarrow C = -\dfrac{9}{76}.$

For x^2: $0 = 3A + C + 2D \Rightarrow 0 = \dfrac{3}{4} - \dfrac{9}{76} + 2D \Rightarrow 0 = \dfrac{48}{76} + 2D \Rightarrow D = -\dfrac{24}{76}.$

The expression can be written $\dfrac{1}{4x} + \dfrac{-8}{19(2x + 1)} + \dfrac{-9x - 24}{76(3x^2 + 4)}.$

25. Multiply $\dfrac{3x - 1}{x(2x^2 + 1)^2} = \dfrac{A}{x} + \dfrac{Bx + C}{2x^2 + 1} + \dfrac{Dx + E}{(2x^2 + 1)^2}$ by $x(2x^2 + 1)^2 \Rightarrow$

$3x - 1 = A(2x^2 + 1)^2 + (Bx + C)(x)(2x^2 + 1) + (Dx + E)(x).$

Let $x = 0 \Rightarrow -1 = A(1) \Rightarrow A = -1.$ Multiply the right side out.

$3x - 1 = A(4x^4 + 4x^2 + 1) + 2Bx^4 + Bx^2 + Cx + 2Cx^3 + Dx^2 + Ex \Rightarrow$

$3x - 1 = 4Ax^4 + 4Ax^2 + A + 2Bx^4 + Bx^2 + Cx + 2Cx^3 + Dx^2 + Ex.$

Equate coefficients. For x^4: $0 = 4A + 2B \Rightarrow 0 = -4 + 2B \Rightarrow B = 2.$ For x^3: $0 = 2C \Rightarrow C = 0.$

For x^2: $0 = 4A + B + D \Rightarrow 0 = -4 + 2 + D \Rightarrow D = 2.$ For x: $3 = C + E \Rightarrow 3 = 0 + E \Rightarrow$

$E = 3.$ The expression can be written $\dfrac{-1}{x} + \dfrac{2x}{2x^2 + 1} + \dfrac{2x + 3}{(2x^2 + 1)^2}.$

27. Multiply $\dfrac{-x^4 - 8x^2 + 3x - 10}{(x+2)(x^2+4)^2} = \dfrac{A}{x+2} + \dfrac{Bx+C}{x^2+4} + \dfrac{Dx+E}{(x^2+4)^2}$ by $(x+2)(x^2+4)^2 \Rightarrow$

$-x^4 - 8x^2 + 3x - 10 = A(x^2+4)^2 + (Bx+C)(x+2)(x^2+4) + (Dx+E)(x+2).$

Let $x = -2 \Rightarrow -64 = A(64) \Rightarrow A = -1.$ Multiply the right side out.

$-x^4 - 8x^2 + 3x - 10 = Ax^4 + 8Ax^2 + 16A + Bx^4 + 2Bx^3 + 4Bx^2 + 8Bx + Cx^3 + 2Cx^2 + .$

$4Cx + 8C + Dx^2 + 2Dx + Ex + 2E.$

Equate coefficients.

For x^4: $-1 = A + B \Rightarrow -1 = -1 + B \Rightarrow B = 0.$ For x^3: $0 = 2B + C \Rightarrow 0 = 0 + C \Rightarrow C = 0.$

For x^2: $-8 = 8A + 4B + 2C + D \Rightarrow -8 = -8 + 0 + 0 + D \Rightarrow D = 0.$

For x: $3 = 8B + 4C + 2D + E \Rightarrow 3 = 0 + 0 + 0 + E \Rightarrow E = 3.$

The expression can be written $\dfrac{-1}{x+2} + \dfrac{3}{(x^2+4)^2}.$

29. By long division $\dfrac{5x^5 + 10x^4 - 15x^3 + 4x^2 + 13x - 9}{x^3 + 2x^2 - 3x} = 5x^2 + \dfrac{4x^2 + 13x - 9}{x^3 + 2x^2 - 3x} = 5x^2 + \dfrac{4x^2 + 13x - 9}{x(x+3)(x-1)}.$

Multiply $\dfrac{4x^2 + 13x - 9}{x(x+3)(x-1)} = \dfrac{A}{x} + \dfrac{B}{x+3} + \dfrac{C}{x-1}$ by $x(x+3)(x-1) \Rightarrow$

$4x^2 + 13x - 9 = A(x+3)(x-1) + B(x)(x-1) + C(x)(x+3)$ Let $x = 0 \Rightarrow -9 = A(-3) \Rightarrow$

$A = 3.$ Let $x = -3 \Rightarrow -12 = B(-3)(-4) \Rightarrow B = -1.$ Let $x = 1 \Rightarrow 8 = C(4) \Rightarrow C = 2.$

The expression can be written $5x^2 + \dfrac{3}{x} + \dfrac{-1}{x+3} + \dfrac{2}{x-1}.$

Appendix D: Percent Change and Exponential Functions

1.   (a)   Let R = 35. Then $r = \dfrac{35}{100} = 0.35$.

      (b)   Let $R = -0.07$. Then $r = -\dfrac{0.07}{100} = -0.0007$. A negative percentage generally corresponds to a quantity decreasing rather than increasing.

      (c)   Let R = 721. Then $r = \dfrac{721}{100} = 7.21$

      (d)   Let $R = \dfrac{3}{10}$. Then $r = \dfrac{\frac{3}{10}}{100} = \dfrac{3}{1000} = 0.003$

3.   (a)   Let $R = -5.5$. Then $r = -\dfrac{5.5}{100} = -0.055$.

      (b)   Let $R = -1.54$. Then $r = -\dfrac{1.54}{100} = -0.0154$.

      (c)   Let $R = 120$. Then $r = \dfrac{120}{100} = 1.2$.

      (d)   Let $R = \dfrac{3}{20}$. Then $r = \dfrac{\frac{3}{20}}{100} = \dfrac{3}{2000} = 0.0015$.

5.   (a)   Let $r = 0.37$. Then $R = 100(0.37) = 37\%$

      (b)   Let $r = -0.095$. Then $R = 100(-0.095) = -9.5\%$

      (c)   Let $r = 1.9$. Then $R = 100(1.9) = 190\%$

      (d)   Let $r = \dfrac{7}{20} = 0.35$ Then $R = 100(0.35) = 35\%$

7.   (a)   Let $r = -0.121$. Then $R = 100(-0.121) = -12.1\%$

      (b)   Let $r = 1.4$. Then $R = 100(1.4) = 140\%$

      (c)   Let $r = 3.2$. Then $R = 100(3.2) = 320\%$

      (d)   Let $r = -\dfrac{1}{4} = -0.25$ Then $R = 100(-0.25) = -25\%$

9.   (a) $\dfrac{B-A}{A} \times 100 \Rightarrow \dfrac{1000-500}{500} \times 100 \Rightarrow \dfrac{500}{500} \times 100 \Rightarrow 1 \times 100 \Rightarrow 100\%$

      (b) $\dfrac{A-B}{B} \times 100 \Rightarrow \dfrac{500-1000}{1000} \times 100 \Rightarrow -\dfrac{500}{1000} \times 100 \Rightarrow -\dfrac{1}{2} \times 100 \Rightarrow -50\%$

11.   (a) $\dfrac{B-A}{A} \times 100 \Rightarrow \dfrac{1.3-1.27}{1.27} \times 100 \Rightarrow \dfrac{0.03}{1.27} \times 100 \Rightarrow 0.0236 \times 100 \Rightarrow 2.36\%$

      (b) $\dfrac{A-B}{B} \times 100 \Rightarrow \dfrac{1.27-1.30}{1.30} \times 100 \Rightarrow \dfrac{-0.03}{1.30} \times 100 \Rightarrow -0.0231 \times 100 \Rightarrow -2.31\%$

13.   (a) $\dfrac{B-A}{A} \times 100 \Rightarrow \dfrac{65-45}{45} \times 100 \Rightarrow \dfrac{20}{45} \times 100 \Rightarrow 0.4444 \times 100 \Rightarrow 44.44\%$

      (b) $\dfrac{A-B}{B} \times 100 \Rightarrow \dfrac{45-65}{65} \times 100 \Rightarrow \dfrac{-20}{65} \times 100 \Rightarrow -0.3077 \times 100 \Rightarrow -30.77\%$

15. (a)   Let $A = 1500$ and $r = 1.2$. The increase is $rA = 1.2 \cdot 1500 = \$1800$.

(b)   The final value of the account is $A + rA = 1500 + 1800 = \$3300$

(c)   The account increased in value by a factor of $1 + r = 1 + 1.2 = 2.2$.

17. (a)   Let $A = 4000$ and $r = -0.55$. The decrease is $rA = -0.55 \cdot 4000 = -\$2200$.

(b)   The final value of the account is $A + rA = 4000 + (-2200) = \$1800$

(c)   The account decreased in value by a factor of $1 + r = 1 + (-0.55) = 0.45$.

19. (a)   Let $A = 7500$ and $r = -0.60$. The decrease is $rA = -0.60 \cdot 7500 = -\$4500$.

(b)   The final value of the account is $A + rA = 7500 + (-4500) = \$3000$

(c)   The account decreased in value by a factor of $1 + r = 1 + (-0.60) = 0.4$.

21.   The initial value is $C = 9500$, the rate of decrease is $r = -0.35$, and the decay factor is $a = 1 + (-0.35) = 0.65$. The sample of insects contains $f(x) = 9500 \cdot 0.65^x$ insects after x weeks.

23.   The initial value is $C = 2500$, the rate of increase is $r = 0.05$, and the growth factor is $a = 1 + (0.05) = 1.05$. The sample of fish contains $f(x) = 2500 \cdot 1.05^x$ fish after x months.

25.   The initial value is $C = 1000$, the rate of decrease is $r = -0.065$, and the decay factor is $a = 1 + (-0.065) = 0.935$. The mutual fund contains $f(x) = 1000 \cdot 0.935^x$ dollars after x years.

27.   For $f(x) = 8 \cdot 1.12^x$ the initial value is $C = 8$ and the growth factor is $a = 1.12$. Because $a = 1 + r$, it follows that $r = a - 1 \Rightarrow r = 1.12 - 1 = 0.12$ or 12%.

29.   For $f(x) = 1.5 \cdot 0.35^x$ the initial value is $C = 1.5$ and the decay factor is $a = 0.35$. Because $a = 1 + r$, it follows that $r = a - 1 \Rightarrow r = 0.35 - 1 = -0.65$ or -65%.

31.   For $f(x) = 0.55^x$ the initial value is $C = 1$ and the decay factor is $a = 0.55$. Because $a = 1 + r$, it follows that $r = a - 1 \Rightarrow r = 0.55 - 1 = -0.45$ or -45%.

33.   For $f(x) = 7 \cdot e^x$ the initial value is $C = 7$ and the growth factor is $a = e$. Because $a = 1 + r$, it follows that $r = e - 1 \Rightarrow r \approx 1.718$ or 171.8%.

35.   For $f(x) = 6 \cdot 3^{-x}$ the initial value is $C = 6$ and the decay factor is $a = 3^{-1} = \dfrac{1}{3}$. Because $a = 1 + r$, it follows that $r = \dfrac{1}{3} - 1 \Rightarrow r = -\dfrac{2}{3}$ or -66.7%.

37.   The population of the city is 35,000 and is increasing by a factor of 9.8% every 2 years. Thus $A_0 = 35,000, b = 1.098, k = 2$, and $f(t) = 35,000 \cdot 1.098^{t/2}$. After 8 years the population is $f(5) = 35,000 \cdot 1.098^{5/2} \approx 44,215$.

39.   The sample of bacteria is 1000 and is increasing by a factor of 3 every 7 hours. Thus $A_0 = 1000, b = 3, k = 7$, and $f(t) = 1000 \cdot 3^{t/7}$. After 11 hours the bacteria population is $f(11) = 1000 \cdot 3^{11/7} \approx 5620$.

41.   The intensity of light is I_0 and is decreasing by a factor of $\dfrac{1}{3}$ every 2 millimeters. Thus $A_0 = I_0, b = 1 - \dfrac{1}{3} = \dfrac{2}{3}, k = 2$, and $f(t) = I_0 \left(\dfrac{2}{3}\right)^{t/2}$. After 4.3 millimeters the intensity is $f(4.3) = I_0 \left(\dfrac{2}{3}\right)^{4.3/2} \approx 0.418 I_0$.

43.   The initial amount is 5000 and is increasing by a factor of 4 every 35 years. Thus
$A_0 = 5000, b = 4, k = 35$, and $f(t) = 5000 \cdot 4^{t/35}$. After 8 years the amount is
$f(8) = 5000 \cdot 4^{8/35} = \$6864.10$.

45. $T = \dfrac{70}{7} = 10$ years, $A = 2000e^{0.07(10)} = \$4027.51$

47. $T = \dfrac{70}{20} = 3.5$ years, $A = 500e^{0.2(3.5)} = \$1006.88$

49. $T = \dfrac{70}{25} = 2.8$ years, $A = 1500e^{0.25(2.8)} = \$3020.63$

51. $R = \dfrac{70}{40} = 1.75\%$

53. $R = \dfrac{70}{35} = 2\%$

55. $R = \dfrac{70}{70} = 1\%$

57.   The sample of bacteria is increasing by a factor of 6% every 8 hours. Thus
$A_0 = 1, b = 1.06, k = 8$, and $f(t) = 1 \cdot 1.06^{t/8}$. After 3 hours the percent increase is
$f(3) = 1.06^{3/8} \approx 1.02209$. Thus, the increase is about 2.21%.

59.   The number of cell phone subscribers increases by 25% in one year. Thus,
$A_0 = 1, b = 1.25, k = 1$, and $f(t) = 1 \cdot 1.25^{t}$. $f(1) = 1 \cdot 1.25^{1} = 1.25$ and we know that after one year the percent of cell phone users has increased to 125%.
Now after the second year the number of cell phone subscribers has decreased 20%. Thus,
$A_0 = 1.25, b = 0.8, k = 1$, and $f(t) = 1.25 \cdot 0.8^{t}$. $f(1) = 1.25 \cdot 0.8^{1} = 1$. Therefore, the number of subscribers at the beginning of the first year and the end of the second year is the same.

61.   The initial amount is 9.81 and is decreasing by a factor of 9% in one year. Thus
$A_0 = 9.81, b = 1 - 0.09 = 0.91, k = 1$, and $f(t) = 9.81 \cdot 0.91^{t}$. After 3 years the new wage is
$f(3) = 9.81 \cdot 0.91^{3} = \$7.39$.

63.   The initial amount of the element is 100% and has decreased to 40% after 2 years. Thus
$A_0 = 1, b = 0.4, k = 2$, and $f(t) = 1 \cdot 0.4^{t/2}$. After 8 years the remaining percentage is
$f(8) = 1 \cdot 0.4^{8/2} = 0.0256 = 2.56\%$.

Appendix E: Rotation of Axes

1. $4x^2 + 3y^2 + 2xy - 5x = 8 \Rightarrow B^2 - 4AC = 2^2 - 4(4)(3) = 4 - 48 < 0$. Circle, ellipse, or a point.

3. $2x^2 + 3xy - 4y^2 = 0 \Rightarrow B^2 - 4AC = 3^2 - 4(2)(-4) = 9 + 32 > 0$. Hyperbola or 2 intersecting lines.

5. $4x^2 + 4xy + y^2 + 15 = 0 \Rightarrow B^2 - 4AC = 4^2 - 4(4)(1) = 16 - 16 = 0$. Parabola, one line, or 2 parallel lines.

7. $2x^2 + \sqrt{3}xy + y^2 + x = 5 \Rightarrow \cot 2\theta = \dfrac{A - C}{B} = \dfrac{2 - 1}{\sqrt{3}} \Rightarrow \cot 2\theta = \dfrac{1}{\sqrt{3}} \Rightarrow 2\theta = 60° \Rightarrow \theta = 30°$

9. $3x^2 + \sqrt{3}xy + 4y^2 + 2x - 3y = 12 \Rightarrow \cot 2\theta = \dfrac{A - C}{B} = \dfrac{3 - 4}{\sqrt{3}} \Rightarrow \cot 2\theta = -\dfrac{1}{\sqrt{3}} \Rightarrow$

   $2\theta = 120° \Rightarrow \theta = 60°$

11. $x^2 - 4xy + 5y^2 = 18 \Rightarrow \cot 2\theta = \dfrac{A - C}{B} = \dfrac{1 - 5}{-4} = 1 \Rightarrow \cot 2\theta = 1 \Rightarrow 2\theta = 45° \Rightarrow \theta = 22.5°$

13. $x^2 - xy + y^2 = 6$ [1]; $\theta = 45°$; $x = x'\cos\theta - y'\sin\theta = \dfrac{\sqrt{2}}{2}x' - \dfrac{\sqrt{2}}{2}y'$ [2];

   $y = x'\sin\theta + y'\cos\theta = \dfrac{\sqrt{2}}{2}x' + \dfrac{\sqrt{2}}{2}y'$ [3]. Substitute [2] and [3] in [1].

   $$\left(\dfrac{\sqrt{2}}{2}x' - \dfrac{\sqrt{2}}{2}y'\right)^2 - \left(\dfrac{\sqrt{2}}{2}x' - \dfrac{\sqrt{2}}{2}y'\right)\left(\dfrac{\sqrt{2}}{2}x' + \dfrac{\sqrt{2}}{2}y'\right) + \left(\dfrac{\sqrt{2}}{2}x' + \dfrac{\sqrt{2}}{2}y'\right)^2 = 6 \Rightarrow$$

   $$\dfrac{1}{2}x'^2 - x'y' + \dfrac{1}{2}y'^2 - \dfrac{1}{2}x'^2 + \dfrac{1}{2}y'^2 + \dfrac{1}{2}x'^2 + x'y' + \dfrac{1}{2}y'^2 = 6 \Rightarrow \dfrac{1}{2}x'^2 + \dfrac{3}{2}y'^2 = 6 \Rightarrow$$

   $$\dfrac{x'^2}{12} + \dfrac{y'^2}{4} = 1.$$ See Figure 13.

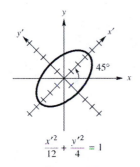

$\dfrac{x'^2}{12} + \dfrac{y'^2}{4} = 1$

Figure 13

15. $8x^2 - 4xy + 5y^2 = 36$ [1]; $\sin\theta = \dfrac{2}{\sqrt{5}}, y = 2, r = \sqrt{5}, x = \sqrt{5-4} = 1 \Rightarrow \cos\theta = \dfrac{1}{\sqrt{5}};$

$x = x'\cos\theta - y'\sin\theta = \dfrac{1}{\sqrt{5}}x' - \dfrac{2}{\sqrt{5}}y'$ [2]; $y = x'\sin\theta + y'\cos\theta = \dfrac{2}{\sqrt{5}}x' + \dfrac{1}{\sqrt{5}}y'$ [3].

Substitute [2] and [3] in [1].

$$8\left(\dfrac{1}{\sqrt{5}}x' - \dfrac{2}{\sqrt{5}}y'\right)^2 - 4\left(\dfrac{1}{\sqrt{5}}x' - \dfrac{2}{\sqrt{5}}y'\right)\left(\dfrac{2}{\sqrt{5}}x' + \dfrac{1}{\sqrt{5}}y'\right) + 5\left(\dfrac{2}{\sqrt{5}}x' + \dfrac{1}{\sqrt{5}}y'\right)^2 = 36 \Rightarrow$$

$$8\left(\dfrac{1}{5}x'^2 - \dfrac{4}{5}x'y' + \dfrac{4}{5}y'^2\right) - 4\left(\dfrac{2}{5}x'^2 - \dfrac{3}{5}x'y' - \dfrac{2}{5}y'^2\right) + 5\left(\dfrac{4}{5}x'^2 + \dfrac{4}{5}x'y' + \dfrac{1}{5}y'^2\right) = 36 \Rightarrow$$

$$\dfrac{8}{5}x'^2 - \dfrac{32}{5}x'y' + \dfrac{32}{5}y'^2 - \dfrac{8}{5}x'^2 + \dfrac{12}{5}x'y' + \dfrac{8}{5}y'^2 + 4x'^2 + 4x'y' + y'^2 = 36 \Rightarrow$$

$$4x'^2 + 9y'^2 = 36 \Rightarrow \dfrac{x'^2}{9} + \dfrac{y'^2}{4} = 1. \text{ See Figure 15.}$$

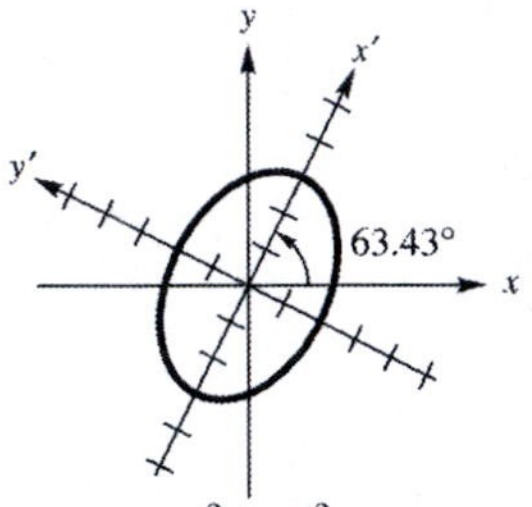

$$\dfrac{x'^2}{9} + \dfrac{y'^2}{4} = 1$$

Figure 15

17. $3x^2 - 2xy + 3y^2 = 8$ [1]; $\cot 2\theta = \dfrac{A - C}{B} = \dfrac{3 - 3}{-2} = 0 \Rightarrow 2\theta = 90° \Rightarrow \theta = 45°;$

$x = x'\cos\theta - y'\sin\theta = \dfrac{\sqrt{2}}{2}x' - \dfrac{\sqrt{2}}{2}y'$ [2]; $y = x'\sin\theta + y'\cos\theta = \dfrac{\sqrt{2}}{2}x' + \dfrac{\sqrt{2}}{2}y'$ [3].

Substitute [2] and [3] in [1].

$$3\left(\dfrac{\sqrt{2}}{2}x' - \dfrac{\sqrt{2}}{2}y'\right)^2 - 2\left(\dfrac{\sqrt{2}}{2}x' - \dfrac{\sqrt{2}}{2}y'\right)\left(\dfrac{\sqrt{2}}{2}x' + \dfrac{\sqrt{2}}{2}y'\right) + 3\left(\dfrac{\sqrt{2}}{2}x' + \dfrac{\sqrt{2}}{2}y'\right)^2 = 8 \Rightarrow$$

$$3\left(\dfrac{1}{2}x'^2 - x'y' + \dfrac{1}{2}y'^2\right) - 2\left(\dfrac{1}{2}x'^2 - \dfrac{1}{2}y'^2\right) + 3\left(\dfrac{1}{2}x'^2 + x'y' + \dfrac{1}{2}y'^2\right) = 8 \Rightarrow$$

$$\dfrac{3}{2}x'^2 - 3x'y' + \dfrac{3}{2}y'^2 - x'^2 + y'^2 + \dfrac{3}{2}x'^2 + 3x'y' + \dfrac{3}{2}y'^2 = 8 \Rightarrow 2x'^2 + 4y'^2 = 8 \Rightarrow \dfrac{x'^2}{4} + \dfrac{y'^2}{2} = 1$$

See Figure 17.

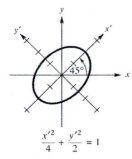

$$\frac{x'^2}{4} + \frac{y'^2}{2} = 1$$

Figure 17

19. $x^2 - 4xy + y^2 = -5$ [1]; $\cot 2\theta = \dfrac{A - C}{B} = \dfrac{1 - 1}{-4} = 0 \Rightarrow 2\theta = 90° \Rightarrow \theta = 45°$;

$x = x' \cos\theta - y' \sin\theta = \dfrac{\sqrt{2}}{2}x' - \dfrac{\sqrt{2}}{2}y'$ [2]; $y = x' \sin\theta + y' \cos\theta = \dfrac{\sqrt{2}}{2}x' + \dfrac{\sqrt{2}}{2}y'$ [3].

Substitute [2] and [3] in [1].

$$\left(\frac{\sqrt{2}}{2}x' - \frac{\sqrt{2}}{2}y'\right)^2 - 4\left(\frac{\sqrt{2}}{2}x' - \frac{\sqrt{2}}{2}y'\right)\left(\frac{\sqrt{2}}{2}x' + \frac{\sqrt{2}}{2}y'\right) + \left(\frac{\sqrt{2}}{2}x' + \frac{\sqrt{2}}{2}y'\right)^2 = -5 \Rightarrow$$

$$\frac{1}{2}x'^2 - x'y' + \frac{1}{2}y'^2 - 4\left(\frac{1}{2}x'^2 - \frac{1}{2}y'^2\right) + \frac{1}{2}x'^2 + x'y' + \frac{1}{2}y'^2 = -5 \Rightarrow$$

$$\frac{1}{2}x'^2 - x'y' + \frac{1}{2}y'^2 - 2x'^2 + 2y'^2 + \frac{1}{2}x'^2 + x'y' + \frac{1}{2}y'^2 = -5 \Rightarrow -x'^2 + 3y'^2 = -5 \Rightarrow$$

$$\frac{x'^2}{5} - \frac{3y'^2}{5} = 1. \text{ See Figure 19.}$$

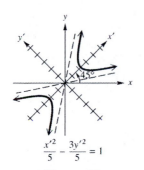

$$\frac{x'^2}{5} - \frac{3y'^2}{5} = 1$$

Figure 19

21. $7x^2 + 6\sqrt{3}\,xy + 13y^2 = 64$ [1]; $\cot 2\theta = \dfrac{A-C}{B} = \dfrac{7-13}{6\sqrt{3}} = \dfrac{-6}{6\sqrt{3}} = -\dfrac{1}{\sqrt{3}} \Rightarrow 2\theta = 120° \Rightarrow \theta = 60°$;

$x = x'\cos\theta - y'\sin\theta = \dfrac{1}{2}x' - \dfrac{\sqrt{3}}{2}y'$ [2]; $y = x'\sin\theta + y'\cos\theta = \dfrac{\sqrt{3}}{2}x' + \dfrac{1}{2}y'$ [3].

Substitute [2] and [3] in [1].

$7\left(\dfrac{1}{2}x' - \dfrac{\sqrt{3}}{2}y'\right)^2 + 6\sqrt{3}\left(\dfrac{1}{2}x' - \dfrac{\sqrt{3}}{2}y'\right)\left(\dfrac{\sqrt{3}}{2}x' + \dfrac{1}{2}y'\right) + 13\left(\dfrac{\sqrt{3}}{2}x' + \dfrac{1}{2}y'\right)^2 = 64 \Rightarrow$

$7\left(\dfrac{1}{4}x'^2 - \dfrac{\sqrt{3}}{2}x'y' + \dfrac{3}{4}y'^2\right) + 6\sqrt{3}\left(\dfrac{\sqrt{3}}{4}x'^2 - \dfrac{1}{2}x'y' - \dfrac{\sqrt{3}}{4}y'^2\right) +$

$13\left(\dfrac{3}{4}x'^2 + \dfrac{\sqrt{3}}{2}x'y' + \dfrac{1}{4}y'^2\right) = 64 \Rightarrow$

$\dfrac{7}{4}x'^2 - \dfrac{7\sqrt{3}}{2}x'y' + \dfrac{21}{4}y'^2 + \dfrac{18}{4}x'^2 - \dfrac{6\sqrt{3}}{2}x'y' - \dfrac{18}{4}y'^2 + \dfrac{39}{4}x'^2 + \dfrac{13\sqrt{3}}{2}x'y' + \dfrac{13}{4}y'^2 = 64 \Rightarrow$

$16x'^2 + 4y'^2 = 64 \Rightarrow \dfrac{x'^2}{4} + \dfrac{y'^2}{16} = 1$. See Figure 21.

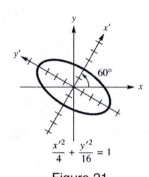

$\dfrac{x'^2}{4} + \dfrac{y'^2}{16} = 1$

Figure 21

23. $3x^2 - 2\sqrt{3}\,xy + y^2 - 2x - 2\sqrt{3}y = 0$ [1]; $\cot 2\theta = \dfrac{A-C}{B} = \dfrac{3-1}{-2\sqrt{3}} = -\dfrac{1}{\sqrt{3}} \Rightarrow$

$2\theta = 120° \Rightarrow \theta = 60°$;

$x = x'\cos\theta - y'\sin\theta = \dfrac{1}{2}x' - \dfrac{\sqrt{3}}{2}y'$ [2]; $y = x'\sin\theta + y'\cos\theta = \dfrac{\sqrt{3}}{2}x' + \dfrac{1}{2}y'$ [3].

Substitute [2] and [3] in [1]. $3\left(\dfrac{1}{2}x' - \dfrac{\sqrt{3}}{2}y'\right)^2 - 2\sqrt{3}\left(\dfrac{1}{2}x' - \dfrac{\sqrt{3}}{2}y'\right)\left(\dfrac{\sqrt{3}}{2}x' + \dfrac{1}{2}y'\right) +$

$\left(\dfrac{\sqrt{3}}{2}x' + \dfrac{1}{2}y'\right)^2 - 2\left(\dfrac{1}{2}x' - \dfrac{\sqrt{3}}{2}y'\right) - 2\sqrt{3}\left(\dfrac{\sqrt{3}}{2}x' + \dfrac{1}{2}y'\right) = 0 \Rightarrow 3\left(\dfrac{1}{4}x'^2 - \dfrac{\sqrt{3}}{2}x'y' + \dfrac{3}{4}y'^2\right) -$

$2\sqrt{3}\left(\dfrac{\sqrt{3}}{4}x'^2 - \dfrac{1}{2}x'y' - \dfrac{\sqrt{3}}{4}y'^2\right) + \left(\dfrac{3}{4}x'^2 + \dfrac{\sqrt{3}}{2}x'y' + \dfrac{1}{4}y'^2\right) - x' + \sqrt{3}y' - 3x' - \sqrt{3}y' = 0 \Rightarrow$

$\dfrac{3}{4}x'^2 - \dfrac{3\sqrt{3}}{2}x'y' + \dfrac{9}{4}y'^2 - \dfrac{3}{2}x'^2 + \sqrt{3}x'y' + \dfrac{3}{2}y'^2 + \dfrac{3}{4}x'^2 + \dfrac{\sqrt{3}}{2}x'y' + \dfrac{1}{4}y'^2 - 4x' = 0 \Rightarrow$

$4y'^2 - 4x' = 0 \Rightarrow 4y'^2 = 4x' \Rightarrow y'^2 = x'$. See Figure 23.

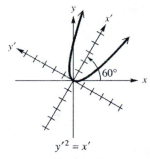

Figure 23

25. $x^2 + 3xy + y^2 - 5\sqrt{2}y = 15$ [1]; $\cot 2\theta = \dfrac{A - C}{B} = \dfrac{1 - 1}{3} = 0 \Rightarrow 2\theta = 90° \Rightarrow \theta = 45°$;

$x = x' \cos\theta - y' \sin\theta = \dfrac{\sqrt{2}}{2}x' - \dfrac{\sqrt{2}}{2}y'$ [2]; $y = x' \sin\theta + y' \cos\theta = \dfrac{\sqrt{2}}{2}x' + \dfrac{\sqrt{2}}{2}y'$ [3].

Substitute [2] and [3] in [1]. $\left(\dfrac{\sqrt{2}}{2}x' - \dfrac{\sqrt{2}}{2}y'\right)^2 + 3\left(\dfrac{\sqrt{2}}{2}x' - \dfrac{\sqrt{2}}{2}y'\right)\left(\dfrac{\sqrt{2}}{2}x' + \dfrac{\sqrt{2}}{2}y'\right) +$

$\left(\dfrac{\sqrt{2}}{2}x' + \dfrac{\sqrt{2}}{2}y'\right)^2 - 5\sqrt{2}\left(\dfrac{1}{2}x' - \dfrac{\sqrt{2}}{2}y'\right) = 15 \Rightarrow$

$\dfrac{1}{2}x'^2 - x'y' + \dfrac{1}{2}y'^2 + 3\left(\dfrac{1}{2}x'^2 - \dfrac{1}{2}y'^2\right) + \dfrac{1}{2}x'^2 + x'y' + \dfrac{1}{2}y'^2 - 5x' - 5y' = 15 \Rightarrow$

$\dfrac{1}{2}x'^2 - x'y' + \dfrac{1}{2}y'^2 + \dfrac{3}{2}x'^2 - \dfrac{3}{2}y'^2 + \dfrac{1}{2}x'^2 + x'y' + \dfrac{1}{2}y'^2 - 5x' - 5y' = 15 \Rightarrow$

$\dfrac{5}{2}x'^2 - \dfrac{1}{2}y'^2 - 5x' - 5y' = 15 \Rightarrow 5x'^2 - 10x' - y'^2 - 10y' = 30 \Rightarrow$

$5(x'^2 - 2x' + 1) - (y'^2 + 10y' + 25) = 30 + 5 - 25 \Rightarrow 5(x' - 1)^2 - (y' + 5)^2 = 10 \Rightarrow$

$\dfrac{(x' - 1)^2}{2} - \dfrac{(y' + 5)^2}{10} = 1$. See Figure 25.

The graph of the equation is a hyperbola with its center at $(1, -5)$. By translating the axes of the $x'y'$-system

down 5 units and right 1 unit, we get an $x''y''$-coordinate system, in which the hyperbola is centered at the

origin. Thus $\dfrac{x''^2}{2} - \dfrac{y''^2}{10} = 1$.

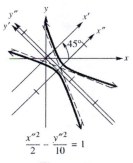

Figure 25

27. $4x^2 + 4xy + y^2 - 24x + 38y - 19 = 0$ [1]; $\cot 2\theta = \dfrac{A - C}{B} = \dfrac{4 - 1}{4} = \dfrac{3}{4} \Rightarrow$

$2\theta \approx 53.15° \Rightarrow \theta \approx 26.57°$; For 2θ; $x = 3, y = 4, r = \sqrt{3^2 + 4^2} = 5 \Rightarrow \cos 2\theta = \dfrac{3}{5}$.

$\sin \theta = \sqrt{\dfrac{1 - \cos 2\theta}{2}} = \sqrt{\dfrac{1 - \frac{3}{5}}{2}} = \sqrt{\dfrac{2}{10}} = \dfrac{\sqrt{5}}{5}$; $\cos \theta = \sqrt{\dfrac{1 + \cos 2\theta}{2}} = \sqrt{\dfrac{1 + \frac{3}{5}}{2}} = \sqrt{\dfrac{8}{10}} = \dfrac{2\sqrt{5}}{5}$;

$x = x' \cos \theta - y' \sin \theta = \dfrac{2\sqrt{5}}{5}x' - \dfrac{\sqrt{5}}{5}y'$ [2]; $y = x' \sin \theta + y' \cos \theta = \dfrac{\sqrt{5}}{5}x' + \dfrac{2\sqrt{5}}{5}y'$ [3].

Substitute [2] and [3] in [1]. $4\left(\dfrac{2\sqrt{5}}{5}x' - \dfrac{\sqrt{5}}{5}y'\right)^2 + 4\left(\dfrac{2\sqrt{5}}{5}x' - \dfrac{\sqrt{5}}{5}y'\right)\left(\dfrac{\sqrt{5}}{5}x' + \dfrac{2\sqrt{5}}{5}y'\right) -$

$\left(\dfrac{\sqrt{5}}{5}x' + \dfrac{2\sqrt{5}}{5}y'\right)^2 - 24\left(\dfrac{2\sqrt{5}}{5}x' - \dfrac{\sqrt{5}}{5}y'\right) + 38\left(\dfrac{\sqrt{5}}{5}x' + \dfrac{2\sqrt{5}}{5}y'\right) = 19 \Rightarrow$

$4\left(\dfrac{4}{5}x'^2 - \dfrac{4}{5}x'y' + \dfrac{1}{5}y'^2\right) + 4\left(\dfrac{2}{5}x'^2 + \dfrac{3}{5}x'y' - \dfrac{2}{5}y'^2\right) + \left(\dfrac{1}{5}x'^2 + \dfrac{4}{5}x'y' + \dfrac{4}{5}y'^2\right) -$

$\dfrac{48\sqrt{5}}{5}x' + \dfrac{25\sqrt{5}}{5}y' + \dfrac{38\sqrt{5}}{5}x' + \dfrac{76\sqrt{5}}{5}y' = 19 \Rightarrow$

$\dfrac{16}{5}x'^2 - \dfrac{16}{5}x'y' + \dfrac{4}{5}y'^2 + \dfrac{8}{5}x'^2 + \dfrac{12}{5}x'y' - \dfrac{8}{5}y'^2 + \dfrac{1}{5}x'^2 + \dfrac{4}{5}x'y' + \dfrac{4}{5}y'^2 - \dfrac{48\sqrt{5}}{5}x' + \dfrac{24\sqrt{5}}{5}y' +$

$\dfrac{38\sqrt{5}}{5}x' + \dfrac{76\sqrt{5}}{5}y' = 19 \Rightarrow 5x'^2 - 2\sqrt{5}x' + 20\sqrt{5}y' = 19 \Rightarrow$

$5\left(x'^2 - \dfrac{2\sqrt{5}}{5}x' + \dfrac{1}{5}\right) + 20\sqrt{5}y' = 19 + 1 \Rightarrow 5\left(x' - \dfrac{\sqrt{5}}{5}\right)^2 = 20 - 20\sqrt{5}y \Rightarrow$

$\left(x' - \dfrac{\sqrt{5}}{5}\right)^2 = 4 - 4\sqrt{5}y' \Rightarrow \left(x' - \dfrac{\sqrt{5}}{5}\right)^2 = -4\sqrt{5}\left(y' - \dfrac{\sqrt{5}}{5}\right)$. See Figure 27.

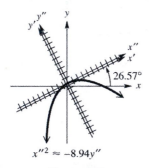

$x''^2 \approx -8.94y''$

Figure 27

29. $16x^2 + 24xy + 9y^2 - 130x + 90y = 0$ [1]; $\cot 2\theta = \dfrac{A - C}{B} = \dfrac{16 - 9}{24} = \dfrac{7}{24} \Rightarrow$

$2\theta \approx 73.74° \Rightarrow \theta \approx 36.87°$; For 2θ; $x = 7, y = 24, r = \sqrt{7^2 + 24^2} = 25 \Rightarrow \cos 2\theta = \dfrac{7}{25}.$

$\sin \theta = \sqrt{\dfrac{1 - \cos 2\theta}{2}} = \sqrt{\dfrac{1 - \frac{7}{25}}{2}} = \sqrt{\dfrac{18}{50}} = \dfrac{3}{5}$; $\cos \theta = \sqrt{\dfrac{1 + \cos 2\theta}{2}} = \sqrt{\dfrac{1 + \frac{7}{25}}{2}} = \sqrt{\dfrac{32}{50}} = \dfrac{4}{5}$;

$x = x' \cos \theta - y' \sin \theta = \dfrac{4}{5}x' - \dfrac{3}{5}y'$ [2]; $y = x' \sin \theta + y' \cos \theta = \dfrac{3}{5}x' + \dfrac{4}{5}y'$ [3].

Substitute [2] and [3] in [1]. $16\left(\dfrac{4}{5}x' - \dfrac{3}{5}y'\right)^2 + 24\left(\dfrac{4}{5}x' - \dfrac{3}{5}y'\right)\left(\dfrac{3}{5}x' + \dfrac{4}{5}y'\right) +$

$9\left(\dfrac{3}{5}x' + \dfrac{4}{5}y'\right)^2 - 130\left(\dfrac{4}{5}x' - \dfrac{3}{5}y'\right) + 90\left(\dfrac{3}{5}x' + \dfrac{4}{5}y'\right) = 0 \Rightarrow$

$16\left(\dfrac{16}{25}x'^2 - \dfrac{24}{25}x'y' + \dfrac{9}{25}y'^2\right) + 24\left(\dfrac{12}{25}x'^2 + \dfrac{7}{25}x'y' - \dfrac{12}{25}y'^2\right) + 9\left(\dfrac{9}{25}x'^2 + \dfrac{24}{25}x'y' + \dfrac{16}{25}y'^2\right) -$

$104x' + 78y' + 54x' + 72y' = 0 \Rightarrow \dfrac{256}{25}x'^2 - \dfrac{384}{25}x'y' + \dfrac{144}{25}y'^2 + \dfrac{288}{25}x'^2 + \dfrac{168}{25}x'y' - \dfrac{288}{25}y'^2 +$

$\dfrac{81}{25}x'^2 + \dfrac{216}{25}x'y' + \dfrac{144}{25}y'^2 - 50x' + 150y' = 0 \Rightarrow 25x'^2 - 50x' + 150y' = 0 \Rightarrow$

$25(x'^2 - 2x' + 1) = -150y' + 25 \Rightarrow 25(x' - 1)^2 = -150\left(y' - \dfrac{1}{6}\right) \Rightarrow (x' - 1)^2 = -6\left(y' - \dfrac{1}{6}\right).$

See Figure 29. The graph of the equation is a parabola with its vertex at $\left(1, \dfrac{1}{6}\right)$. By translating the axes of the

$x'y'$-system up $\dfrac{1}{6}$ units and right 1 unit, we get an $x''y''$-coordinate system, in which the parabola is centered at

the origin. Thus $x''^2 = -6y''$.

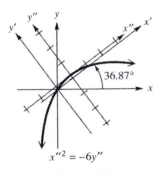

$x''^2 = -6y''$

Figure 29